Lecture Notes in Computer Science 13970

The series Lecture Notes in Computer Science (LNCS), including its subseries Lecture Notes in Artificial Intelligence (LNAI) and Lecture Notes in Bioinformatics (LNBI), has established itself as a medium for the publication of new developments in computer science and information technology research, teaching, and education.

LNCS enjoys close cooperation with the computer science R & D community, the series counts many renowned academics among its volume editors and paper authors, and collaborates with prestigious societies. Its mission is to serve this international community by providing an invaluable service, mainly focused on the publication of conference and workshop proceedings and postproceedings. LNCS commenced publication in 1973.

Michael Emmerich · André Deutz · Hao Wang ·
Anna V. Kononova · Boris Naujoks · Ke Li ·
Kaisa Miettinen · Iryna Yevseyeva
Editors

Evolutionary Multi-Criterion Optimization

12th International Conference, EMO 2023
Leiden, The Netherlands, March 20–24, 2023
Proceedings

Springer

Editors
Michael Emmerich
Leiden University
Leiden, The Netherlands

André Deutz
Leiden University
Leiden, The Netherlands

Hao Wang
Leiden University
Leiden, The Netherlands

Anna V. Kononova
Leiden University
Leiden, The Netherlands

Boris Naujoks
TH Köln
Gummersbach, Germany

Ke Li
University of Exeter
Exeter, UK

Kaisa Miettinen
University of Jyvaskyla
Jyvaskyla, Finland

Iryna Yevseyeva
De Montfort University
Leicester, UK

ISSN 0302-9743 ISSN 1611-3349 (electronic)
Lecture Notes in Computer Science
ISBN 978-3-031-27249-3 ISBN 978-3-031-27250-9 (eBook)
https://doi.org/10.1007/978-3-031-27250-9

This Springer imprint is published by the registered company Springer Nature Switzerland AG
The registered company address is: Gewerbestrasse 11, 6330 Cham, Switzerland

Preface

Welcome to the Proceedings of the 12th Conference on Evolutionary Multi-Criterion Optimization (EMO), held in Leiden, The Netherlands, March 20–24, 2023 in hybrid format.

Why hold EMO conferences? This question was discussed at EMO 2007 by its founders. The doubts regarding the viability of the conferences were fortunately cast away as the importance, need, and ubiquity of multi-criterion optimization keeps growing each year at a tremendous pace, impacting other areas and being influenced itself by them.

For millennia optimization (improving things) has played a crucial role for humans. In more recent times, EMO (and optimization in general) has become important in science in areas such as physics, biology, economics, social sciences, medical sciences, and mathematics. For instance, Snell's law was discovered by Willebrord Snellius through experimentation, only later it was realized that it can be derived from Fermat's principle of least time, stating that light always chooses the path that is traveled in the least time. As such, the laws of nature can often be perceived as a process of optimization and optimal decision-making. Secondly, another use of optimization is the following. Many insights can be gained by looking at extremal objects (for instance, given 2n points in the plane no three of which lie on a line, n of them blue and n of them red, it is always possible to create n line segments by using the given points such that the endpoints have different colors and no two segments intersect – this can be understood by looking at the appropriate extremal object). Thirdly, methodologies and techniques developed in the EMO community have been empowering many practical scenarios: from finding the best taxation system, the best returns on investments while avoiding too high risks, discovering potent drug candidates with few side effects, to designing engineering structures that optimally balance the energy consumption and the environmental impact (e.g., minimizing the CO_2 or CH_4 emission).

In the EMO conferences, we focus mainly on the evolutionary approaches to solving multi-criterion optimization and decision-making problems since the applicability of analytical/deterministic methods is often limited. For the scenarios where both categories of approaches are applicable, the hybridizations of analytical and evolutionary algorithms have appeared over the years, combining the strengths of both categories. Such hybridizations were also covered in the EMO conferences. In recent years, the EMO community has been bridged with the Multi-Criterion Decision-Making (MCDM) community, which focuses more on the decision-making aspects of the same problem. According to the EMO tradition, also in this year's event, many works are dedicated to designing and studying algorithms, ranging from novel algorithmic operators to the theoretical analysis of existing ones. Notably, there are some contributions that connect EMO with Machine Learning/Artificial Intelligence, which draws more and more research interests nowadays. Also, appropriate attention – also as a tutorial – is paid to benchmarking and empirical performance assessment, for instance, new benchmarking problem sets. Furthermore, some submissions address real-world problems using EMO methodologies, which nicely complete the scope of the conference.

The following five distinguished keynote speakers agreed to share their expertise at EMO 2023: 1) Yaochu Jin, Alexander von Humboldt Professor at Bielefeld University, Germany, and Surrey Distinguished Chair, Professor of Computational Intelligence, University of Surrey, UK. Professor Jin's contributions are in the cross-fertilization of AI and Multi-Criterion Optimization, among others. 2) Heike Trautmann, Professor of Data Science: Statistics and Optimization, both at the Department of Information Systems, the University of Münster, Germany, and the University of Twente, The Netherlands. Her research mainly focuses on Data Science, Automated Algorithm Selection and Configuration, Exploratory Landscape Analysis, (Multiobjective) Evolutionary Optimization, and Data Stream Mining. 3) Frank Neumann, Professor and leader of the Optimisation and Logistics Group and an Honorary Professorial Fellow at the University of Melbourne, Australia. Professor Neumann's work focuses on theoretical aspects of combinatorial and multi-objective optimization as well as high-impact applications in the areas of cybersecurity, renewable energy, logistics, and mining. 4) Kalyanmoy Deb is the Koenig Endowed Chair Professor at the Department of Electrical and Computer Engineering at Michigan State University (MSU), East Lansing, USA. Professor Deb's main research interests are in evolutionary optimization algorithms and their application in optimization and machine learning. He is largely known for his seminal research in Evolutionary Multi-Criterion Optimization. 5) Aneta Neumann, who is a researcher at the University of Adelaide, Australia, School of Computer and Mathematical Sciences, Faculty of Sciences, Engineering and Technology, is known for work on applications of optimization, analysis of stochastic optimization, and diversity optimization.

Ten esteemed scholars prepared five very instructive tutorials, continuing the recent tradition of having tutorials at EMO: Dimo Brockhoff, Tea Tusar "Benchmarking Multi-objective Optimizers 2.0"; Amiram Moshaiov "Evolutionary Multi-Concept Optimization"; Erella Eisenstadt-Matalon, Amiram Moshaiov, Kalyan Deb "Multi-Objective Games"; Kalyan Deb, Dhish Saxena, Erik Goodman "Machine Learning Assisted Evolutionary Multi-Objective Optimization"; Christian Grimme, Lennart Schaepermeier, Pascal Kerschke "Continuous Multimodal Multi-Objective Optimization".

Sixty-five papers were submitted to EMO 2023, of which forty-four were accepted. The acceptance rate was 67.5%, and each contribution was peer-reviewed (single-blind) by at least two experts in the field. Papers that were sent to the MCDM track, chaired by Kaisa Miettinen and Iryna Yeveyeva, were reviewed by a dedicated panel of reviewers. Mimicking EMO algorithms, which seek diversity among the candidate solutions, we also achieved great geographic, topic-wise diversity. The accepted papers distribute across five continents: Australia 4, Austria 2, Brazil 2, China 19, Finland 5, France 6, Germany 13, India 15, Italy 1, Japan 5, Mexico 2, Poland 1, Portugal 2, Slovenia 3, The Netherlands 13, United Kingdom 18, and the United States 5. The following topics were represented: Algorithm Design and Engineering; Machine Learning and Multi-criterion Optimization; Benchmarking and Performance Assessment; Indicator Design and Complexity Analysis; Applications in Real World Domains; and Multi-Criteria Decision Making and Interactive Algorithms.

EMO conferences have been held on four continents so far: EMO 2001 in Zürich, Switzerland (LNCS 1993), EMO 2003 in Faro, Portugal (LNCS 2632), EMO 2005 in Guanajuato, Mexico (LNCS 3410), EMO 2007 in Matsushima, Japan (LNCS 4403), EMO 2009 in Nantes, France (LNCS 5467), EMO 2011 in Ouro Preto, Brazil (LNCS 6576), EMO 2013 in Sheffield, UK (LNCS 7811), EMO 2015 in Guimarães, Portugal (LNCS 9019), EMO 2017 in Münster, Germany (LNCS 10173), EMO 2019 in East Lansing, USA (LNCS 11411), EMO 2021 in Shenzhen, China (LNCS 12654), and EMO 2023 in Leiden, Netherlands (LNCS 13970).

EMO 2023 gave us the opportunity to celebrate a special birthday, the 60th birthday of Kalyanmoy Deb, who is the creator of the biannual EMO conference series. Professor Deb is well known for his major, pioneering, and fundamental contributions to the field of Multi-Criterion Optimization. Also, Kalyan is a charismatic force in advancing research and promoting researchers in the EMO community.

It goes without saying that a conference consists of the work of authors, reviewers, keynote speakers, tutorial presenters, and all organizers, and the publishing company (that is, Springer Nature): to all these contributors, a big, heartfelt thank you!

Finally, we would like to thank Springer Nature for financing the Best Paper Award and for publishing the proceedings, moreover the Leiden Institute of Advanced Computer Science (LIACS), and Leiden University for financial support.

January 2023

Michael Emmerich
André Deutz
Hao Wang
Anna V. Kononova
Boris Naujoks
Ke Li
Kaisa Miettinen
Iryna Yevseyeva

Organization

General Chairs

Michael Emmerich	Leiden University, The Netherlands
André Deutz	Leiden University, The Netherlands
Hao Wang	Leiden University, The Netherlands

Program Committee Chairs

Hao Wang	Leiden University, The Netherlands
Ke Li	University of Exeter, UK

MCDM Track Chairs

Kaisa Miettinen	University of Jyvaskyla, Finland
Iryna Yevseyeva	De Montfort University, UK

Proceedings Chairs

Anna V. Kononova	Leiden University, The Netherlands
Boris Naujoks	Cologne University of Applied Sciences, Germany

Tutorial Chair

Ofer M. Shir	Tel-Hai College, Israel

Industrial Liaison Chair

Thomas Bäck	Leiden University, The Netherlands

Financial Chair

Hestia Tamboer	Leiden University, The Netherlands

Online Conference Chair

Roy de Winter	Leiden University, The Netherlands

Local Organizing Committee

Diederick Vermetten	Leiden University, The Netherlands
Ksenia Pereverdieva	Leiden University, The Netherlands
Roy de Winter	Leiden University, The Netherlands
Jacob de Nobel	Leiden University, The Netherlands
Marcel Tichelaar	Leiden University, The Netherlands
Ilse Driessen	Leiden University, The Netherlands

Steering Committee

David W. Corne	Heriot-Watt University, UK
Kalyanmoy Deb	Michigan State University, USA
Michael Emmerich	Leiden University, The Netherlands
Carlos M. Fonseca	University of Coimbra, Portugal
Hisao Ishibuchi	Southern University of Science and Technology, China
Robin Purshouse	University of Sheffield, UK
Joshua D. Knowles	University of Birmingham, UK
Kaisa Miettinen	University of Jyvaskyla, Finland
J. David Schaffer	Binghamton University, USA
Eckart Zitzler	PH Bern, Switzerland

Keynote Speakers

Kalyanmoy Deb	Michigan State University, USA
Yaochu Jin	Bielefeld University, Germany
Aneta Neumann	University of Adelaide, Australia
Frank Neumann	University of Adelaide, Australia
Heike Trautmann	University of Münster, Germany

Program Committee

Saurabh Kumar Agarwal	Malaviya National Institute of Technology Jaipur, India
Richard Allmendinger	University of Manchester, UK

Kirill Antonov	Leiden University, The Netherlands
Slim Bechikh	University of Carthage, Tunisia
Jürgen Branke	Warwick Business School, UK
Dimo Brockhoff	École Polytechnique, France
Xinye Cai	Nanjing University of Aeronautics and Astronautics, China
Ran Cheng	Southern University of Science and Technology, China
Sung-Bae Cho	Yonsei University, Republic of Korea
Tinkle Chugh	University of Exeter, UK
Carlos A. Coello Coello	CINVESTAV-IPN, Mexico
Lino Costa	Universidade do Minho, Portugal
Roy de Winter	Leiden University, The Netherlands
Alexandre Delbem	University of Sao Paulo, Brazil
Bilel Derbel	University of Lille, France
Tome Eftimov	Jožef Stefan Institute, Slovenia
Saber Elsayed	University of New South Wales Canberra, Australia
Jonathan Fieldsend	University of Exeter, UK
Guangtao Fu	University of Exeter, UK
Wenyin Gong	China University of Geosciences, China
Jin-Kao Hao	Université d'Angers, France
Cheng He	Huazhong University of Science and Technology, China
Carlos Henggeler Antunes	University of Coimbra, Portugal
Qi Huang	Leiden University, The Netherlands
Hisao Ishibuchi	Southern University of Science and Technology, China
Shouyong Jiang	University of Aberdeen, UK
Pascal Kerschke	Dresden University of Technology, Germany
Sandeep Kulkarni	Michigan State University, USA
Ke Li	University of Exeter, UK
Hao Li	Xidian University, China
Fan Li	Huazhong University of Science and Technology, China
Arnaud Liefhooghe	University of Lille, France
Manuel López-Ibáñez	University of Manchester, UK
Yi Mei	Victoria University of Wellington, New Zealand
Olaf Mersmann	Cologne University of Applied Sciences, Germany
Efrén Mezura-Montes	University of Veracruz, Mexico
Sanaz Mostaghim	Otto von Guericke Universität Magdeburg, Germany

Boris Naujoks	Cologne University of Applied Sciences, Germany
Antonio J. Nebro	University of Málaga, Spain
Frank Neumann	University of Adelaide, Australia
Bach Nguyen	Victoria University of Wellington, New Zealand
Lie Meng Pang	Southern University of Science and Technology, China
Luís Paquete	University of Coimbra, Portugal
Ksenia Pervedieva	Leiden University, The Netherlands
Pasqualina Potena	RISE Research Institutes of Sweden AB, Sweden
Proteek Roy	Michigan State University, USA
Günter Rudolph	TU Dortmund, Germany
Sergio Santander-Jiménez	University of Extremadura, Spain
Marc Sevaux	Université de Bretagne-Sud, France
Ke Shang	Southern University of Science and Technology, China
Pradyumn Shukla	Karlsruhe Institute of Technology, Germany
Jana Siebert	University Palacky Olomouc, Czech Republic
Johannes Ulrich Siebert	University of Bayreuth, Germany
Hemant Singh	University of New South Wales, Australia
Chaoli Sun	Taiyuan University of Science and Technology, China
Ricardo Takahashi	Universidade Federal de Minas Gerais, Brazil
Heike Trautmann	Westfälische Wilhems-Universität Münster, Germany
Tea Tušar	Jožef Stefan Institute, Slovenia
Bas van Stein	Leiden University, The Netherlands
Miguel A. Vega-Rodríguez	University of Extremadura, Spain
Diederick Vermetten	Leiden University, The Netherlands
Handing Wang	Xidian University, China
Thomas Weise	University of Science and Technology of China, China
Ka-Chun Wong	University of Toronto, Canada
Bai Yan	Southern University of Science and Technology, China
Furong Ye	Leiden University, The Netherlands
Xingyi Zhang	Anhui University, China
Almin Zhou	East China Normal University, China

Program Committee MCDM Track

Carlos Henggeler Antunes	University of Coimbra, Portugal
Vitor Basto Fernandes	University Institute Lisbon, Portugal

Contents

Benchmarking and Performance Assessment

Indicator Design and Complexity Analysis

Applications in Real World Domains

Multi-criteria Decision Making and Interactive Algorithms

Algorithm Design and Engineering

Visual Exploration of the Effect of Constraint Handling in Multiobjective Optimization

Tea Tušar[1,2(✉)], Aljoša Vodopija[1,2], and Bogdan Filipič[1,2]

[1] Jožef Stefan Institute, Ljubljana, Slovenia
[2] Jožef Stefan International Postgraduate School, Ljubljana, Slovenia
{tea.tusar,aljosa.vodopija,bogdan.filipic}@ijs.si

Abstract. Constraint handling in multiobjective optimization is more complex than in single-objective optimization, where the values of the objective and constraints are easier to combine. To gain insight into the characteristics of constraint handling techniques (CHTs) for multiobjective optimization, we explore their effect independently from search methods. We regard CHTs as transformations that alter the problem landscape and visualize these modified landscapes. This helps us predict potential strengths and weaknesses for search methods. We then use a simple local search technique to test our predictions. Results of the experiments with six CHTs applied on 12 test problems show specific properties of the studied CHTs that can help us devise better CHTs in the future, as well as find suitable search methods for them.

Keywords: Constrained multiobjective optimization · Constraint handling technique · Problem landscape · Visualization

1 Introduction

Constraint handling in multiobjective optimization requires taking into account multiple (conflicting) objectives as well as constraints (often represented by the overall constraint violation). As such, it is more demanding than constraint handling in single-objective optimization, where the values of the sole objective and the overall constraint violation can be combined more naturally. Possibly for this reason, many constraint handling techniques (CHTs) in multiobjective optimization are closely intertwined with the search method [4,9,17,19], which makes it is hard to understand how much, when and why a particular CHT is more efficient than some other.

For example, as Ma and Wang show in [18], the efficiency of constrained multiobjective optimization algorithms heavily depends on the type of the problem. However, their study does not decouple CHTs from the optimization methods, meaning that its findings are tied to the frameworks of NSGA-II [5] and MOEA/D [27] that encompass the examined CHTs. Similarly holds for the work

M. Emmerich et al. (Eds.): EMO 2023, LNCS 13970, pp. 3–16, 2023.
https://doi.org/10.1007/978-3-031-27250-9_1

by Alsouly et al. [1] and our previous work [25], which connect problem landscape features with algorithm performance, but the considered algorithms differ in multiple mechanisms, not just in the CHT.

A notable exception in this regard is the study by Fukumoto and Oyama [11], which proposes a generic framework for incorporating CHTs into multiobjective optimization algorithms. It views a CHT separately from the search method and introduces a way to combine the two that covers dominance-based (e.g., NSGA-II [5]), decomposition-based (e.g., MOEA/D [27]), and indicator-based (e.g., IBEA [29]) multiobjective optimization algorithms. The experiments are then performed on different combinations of search methods and CHTs.

In this work, we explore the effect of CHTs independently from search methods, that is, as independently as possible. The goal is to enhance the understanding of their workings and provide intuition that can help guide the improvement of existing CHTs as well as find suitable search methods for particular CHTs. To this end, we regard CHTs as transformations that alter the problem landscape. We compute the CHT-based ranking of solutions from a grid approximation of the problem landscape to visualize it for various constrained multiobjective optimization problems (CMOPs). In this way, we are able to gain insight into the problem as 'seen' by an algorithm that uses a particular CHT. The CHT-based problem landscapes help us predict potential advantages and disadvantages for search methods. We then use a simple deterministic grid-traversing local search to test our predictions. The CMOPs used in this study are a combination of eight well known test CMOPs and four new, relatively simple problems with known properties that can help understand the characteristics of CHTs.

2 Background

2.1 Constrained Multiobjective Optimization Problems

We formulate a CMOP as follows:

$$\begin{aligned} \text{minimize} \quad & f(x) = (f_1(x), \ldots, f_m(x)) \\ \text{subject to} \quad & g_i(x) \le 0, \quad i = 1, \ldots, p, \end{aligned} \tag{1}$$

where $x = (x_1, \ldots, x_n) \in S$ is a search vector from the search space S, $f_i : S \to \mathbb{R}$ are objective functions and $g_i : S \to \mathbb{R}$ are inequality constraint functions. We do not explicitly include equality constraints as they can be formulated as inequality constraints with the help of a user-defined tolerance value.

The overall constraint violation of solution x is computed with

$$v(x) = \sum_{i=1}^{p} v_i(x), \tag{2}$$

where $v_i(x) = \max(0, g_i(x))$ is the constraint violation for constraint $g_i(x)$. Given that in this work we do not consider the constraints separately, we will be

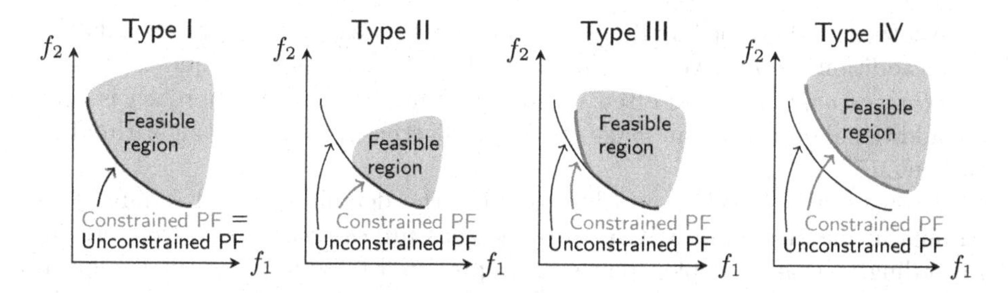

Fig. 1. The four types of CMOPs (adapted from [18]). The Pareto fronts (PFs) of the unconstrained/constrained problems are shown with thin black/thick orange lines. (Color figure online)

using the shorter term *constraint violation* instead of overall constraint violation to refer to $v(x)$ in the rest of this paper.

A solution x is *feasible* when it satisfies all constraints, that is, when $v(x) = 0$. The set of all feasible solutions is called the *feasible region*. A solution $x \in S$ *dominates* another solution $y \in S$ when $f_i(x) \leq f_i(y)$ for all $i = 1, \ldots, m$ and $f_j(x) < f_j(y)$ for at least one $j = 1, \ldots, m$. Additionally, a feasible solution $x^* \in S$ is *Pareto optimal* if there are no feasible solutions $x \in S$ that dominate x^*. All nondominated feasible solutions represent the *Pareto set*, and its image in the objective space is called the *Pareto front*.

When constraints are added to an otherwise unconstrained multiobjective optimization problem, this can affect the size and position of its Pareto set and front. The constraints that influence the Pareto set and front are called *active constraints*, while the remaining ones are termed *inactive constraints*. The degree of this change is the basis for the classification of CMOPs into types as proposed by Ma and Wang [18]. Figure 1 shows the four types, which range from no change to the Pareto front (Type I), to a reduced Pareto front (Type II), a partially displaced Pareto front (Type III), and finally an entirely different Pareto front (Type IV).

2.2 Constraint Handling Techniques

Our study comprises six methods for handling constraints in multiobjective optimization. In the following, we describe these CHTs and their known strengths and weaknesses.

One possible way of handling constraints (or rather, not handling them) is to simply ignore them and solve the problem as if it was an unconstrained one. We refer to this technique as *constraint violation ignored*. While such a strategy cannot be expected to yield good results on problems with active constraints and is therefore mostly omitted from comparison studies, it can be rather powerful for solving CMOPs where the constraints do not severely affect the optima, that is, problems of Type I (and, to some degree, Type II) [11].

Another method for handling constraints is to treat the constraint violation as an additional objective to be minimized[1]. We call this technique *constraint violation as objective.* One often mentioned drawback of this approach is that the additional objective can make the multiobjective optimization algorithm less efficient [17].

A very popular technique (due to being the default way of handling constraints in the algorithm NSGA-II) is the *constrained-domination principle* [5]. According to this principle, solution x is preferred to solution y if: (i) solution x is feasible and solutions y is infeasible, (ii) both solutions are feasible and x dominates y, or (iii) both solutions are infeasible and x has a lower constraint violation than y. The method is known to work rather well, except on problems with multimodal constraint functions [28].

The multiobjective version of the *epsilon-constraint method* [22] could be viewed as a relaxed variant of the constrained-domination principle, where solutions with the constraint violation lower than a predefined $\varepsilon \geq 0$ threshold are treated as feasible. More formally, the epsilon-constraint method prefers solution x to solution y when: (i) solution x dominates solution y and both have a small constraint violation ($v(x) \leq \varepsilon$ and $v(y) \leq \varepsilon$) or the same constraint violation, or (ii) solution x has a lower constraint violation than solution y. The optimization methods using the epsilon-constraint CHT usually gradually lower the value of ε during the algorithm run [2]. Choosing the appropriate starting value for ε as well as the mechanism to update it is nontrivial and problem-dependent.

Contrary to the methods that keep the objectives separate from the constraints, the *penalty function* transforms the objective values of infeasible solutions x to $f_i'(x)$ by either using the constraint violation (when there are no feasible solutions in the current population) or some penalty value that depends on the value of the objective, the constraint violation and the proportion r of feasible individuals in the current population [26]:

$$f_i'(x) = \begin{cases} v(x), & \text{if } r = 0 \\ (1-r)v(x) + rf_i(x) + \sqrt{f_i(x)^2 + v(x)^2}, & \text{if } r > 0 \end{cases}. \tag{3}$$

Suitably setting/adjusting the penalty value is recognized as a difficult task [17].

Finally, we also consider *stochastic ranking*, where the comparison of feasible solutions is done based on the dominance relation, while the infeasible solutions are compared either w.r.t. the constraint violation or the dominance relation—the decision between the two is done randomly [13].

3 Methodology

3.1 Test Problems

In order to explore the effect of CHTs, we need to select some test CMOPs. Because we aim to understand and visualize their landscapes, we choose problems with only two variables and two objectives. Ideally, the problems should

[1] The alternative variant, where each separate constraint violation is regarded as a new objective, is not considered in this work.

have various properties and be of different types [18]. We select eight problems from the existing well-known CMOP suites (C-DTLZ [15], DAS-CMOP [7], DC-DTLZ [16] and MW [18]) as well as create four new ones, CBB1–4, where CBB stands for Constrained BBOB [14] Biobjective (problem).

All four CBB problems were created by adding constraints to the first instance of the 2-D `bbob-biobj` problem F_1 (the double sphere problem) [3]. This is one of the easiest biobjective problems to solve as the Pareto set and front are linear and the problem landscape is unimodal (but not separable). However, when adding constraints to such a problem, it can become more difficult to solve while at the same time still easy to understand and interpret, which is why we created the CBB problems and added them to our test problem set.

The constraint function used in CBB1 is linear. It intersects the Pareto set of F_1 in such a way that the Pareto set of CBB1 consists of two connected linear parts. The constraint functions in the case of CBB2 and CBB3 are created by slightly shifting a single Gaussian peak function [12] with the same mean but a different covariance matrix, yielding in one case a problem of Type III (the Pareto set of CBB2 is formed by two linear parts of the original problem and one spherical that connects them) and in the other case a problem of Type IV (the entire Pareto set of F_1 is infeasible, the Pareto set of CBB3 consists of three disconnected spherical regions). Finally, CBB4 uses the inverted Gaussian peak function with three peaks as the constraint function (because the function is inverted, the peaks now form the feasible region). Again, the entire Pareto set of the original problem is infeasible, which yields a Type IV problem, whose Pareto set consists of two disconnected spherical regions. The exact definitions of constraints for problems CBB1–4 are provided in the supplementary material [24].

Thus we have 12 test problems in total, three of each type: Type I: DAS-CMOP3, DAS-CMOP5, MW14, Type II: C2-DTLZ2, DAS-CMOP1, DC1-DTLZ1, Type III: CBB1, CBB2, MW3, and Type IV: CBB3, CBB4 and MW11.

3.2 CMOP Landscape Visualization

First, we wish to visualize the problem landscapes of our 12 test problems (see Fig. 2). We can do so by approximating the search space with a grid of points. In this study, we always use a grid of 301×301 points[2]. We handle separately the feasible and infeasible regions of each problem. The feasible regions are visualized using the dominance rank ratio [3,10], which computes for each point on the grid the number of other grid points that dominate it and then visualizes them as a ratio of all grid points—using blue hues in the logarithmic scale to emphasize smaller values. The darker the color, the closer a point is to the Pareto set. Points with a domination rank of zero are Pareto optimal and visualized in black. The points in the infeasible regions are colored in red hues according

[2] Note that using a grid approximation inevitably results in some artifacts. For example, a linear Pareto set is in reality a line, but because of the approximation, some points adjacent to this line also result as nondominated, yielding a 'thick line'. The coarser the grid, the larger the artifacts.

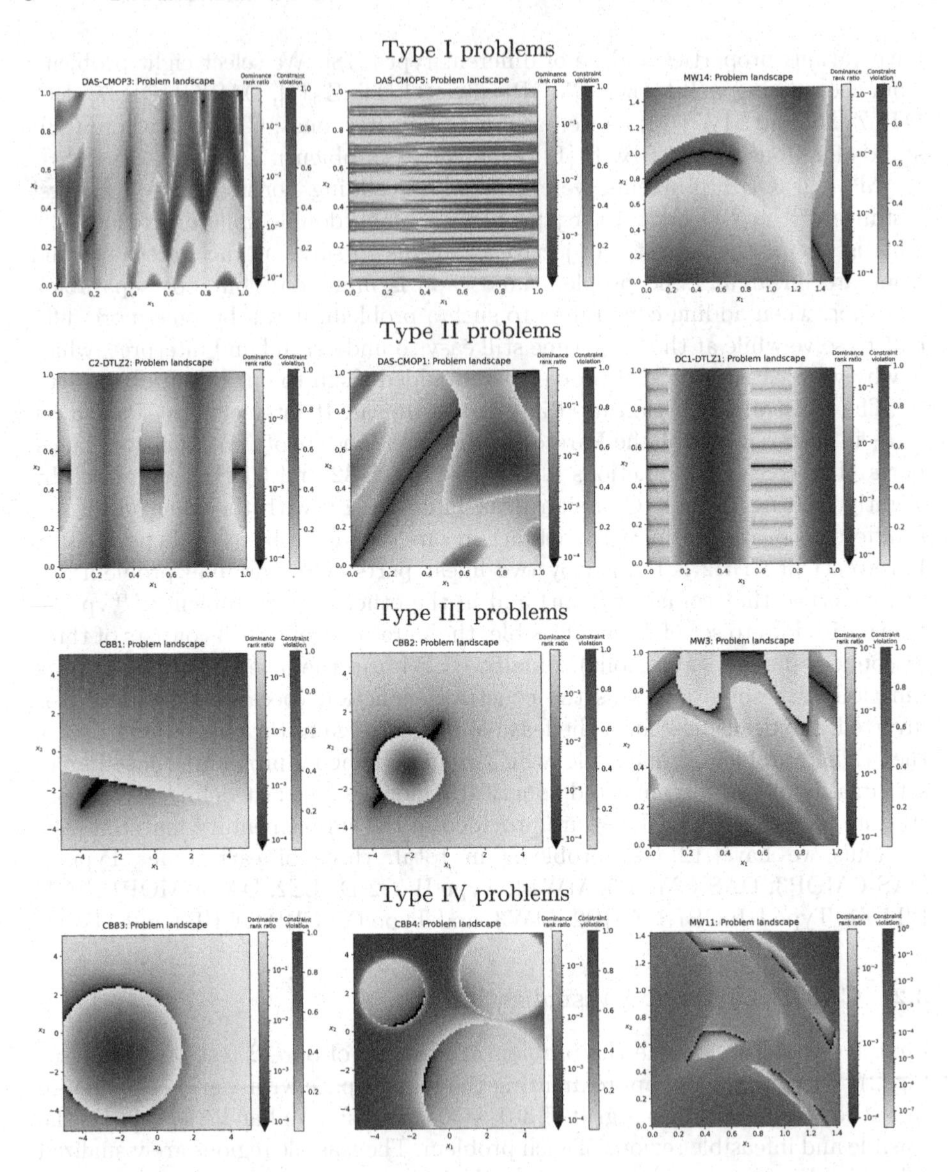

Fig. 2. Problem landscape plots for all 12 CMOPs used in this study. Each three problems of the same type are placed in the same row (from Type I at the top to Type IV at the bottom). Blue hues show the dominance rank ratio [3,10] in the feasible regions with black denoting the Pareto set. Red hues show the constraint violation in the infeasible regions. (Color figure online)

to their constraint violation values. Here, darker colors signify larger constraint violations. As we can see from Fig. 2, the chosen problems form a diverse selection of landscapes with various properties and Pareto set shapes.

Next, we want to see what becomes of the problem landscape when viewed from the perspective of a particular CHT. To this end, we compute for each CHT the rank that the CHT would assign to each point on the grid (compared to other points). Then, we visualize the ratio of this rank in blue hues (similarly as for the feasible problem regions of the original problem). Again, black is used to denote the grid points with the lowest rank—the optimal points according to the CHT. In this way, we gain a CHT-based problem landscape that assigns a single value to each grid point. We will show the visualizations of these landscapes in Sect. 4.2.

Note that if the Pareto set according to the CHT does not contain the entire Pareto set of the original problem, we can expect that an optimization algorithm using this CHT will have issues with convergence to the Pareto set.

3.3 Local Search

While already the visualization of a CHT-based landscape and comparison to the original problem landscape gives a good idea of some of the issues that a search method would encounter if it was used to find the optimum of such a problem, we wish to quantify these effects. Since any mechanism of a search method affects the behavior and interpretation of its results, we resort to a very simple, deterministic procedure—local search with a Moore neighborhood (each inner grid point in 2-D has eight neighbors).

Given a starting point on the grid, the local search iteratively moves to the best neighboring grid point that is not worse that the current point until a stopping criterion has been reached. The stopping criteria are: (i) the current point is optimal in the CHT-based landscape, (ii) the current point is better than all neighbors (it is a local optimum) (iii) all neighboring points have already been visited (to avoid cycling). In order to assure that this procedure is deterministic, the neighbors are always inspected in the same order (the north neighbor first then the rest in clockwise order) and an earlier neighbor always takes precedence over a later one when the ranks are tied among neighbors.

We can compute several quantities from a local search path on a CHT-based problem landscape. First, we can check (and visualize) if the final point of the path is Pareto optimal in the original landscape. If so, the path is denoted as successful (shown in orange) and the final point is visualized with a star. Otherwise, the path is deemed unsuccessful (shown in red) and the final point is denoted as a cross. In addition, simulating an optimization algorithm that chooses the best solution from its entire archive, we also record how many of the points on the path are Pareto optimal in the original landscape and how many are feasible. Of course, we also measure the path length (the number of points on the path).

4 Experiments

4.1 Experimental Setup

In our experiments, we apply the six CHTs from Sect. 2.2 to the 12 test problems from Sect. 3.1. We normalize both objectives and constraint violations to [0, 1] before computing the CHT-based landscapes. There are no parameters to be set for the first three CHTs: constraint violation ignored, constraint violation as objective, and constraint-domination principle. We set the ε of the epsilon-constraint method to the 5th percentile of the constraint violation value of all infeasible grid points to mimic the initial parameter setting from [6]. Note that we do not vary the ε, therefore the epsilon-constraint method landscape in our study should be regarded as the landscape seen by the search method at the beginning of the optimization. Given that we do not use a population-based algorithm, we set the proportion r to the proportion of feasible points on the grid for the penalty function [26]. Finally, we use the recommended setting of 0.45 for the probability of comparing infeasible solutions according to the objective values in stochastic ranking [20].

We repeat local search 100 times, starting from 100 equally-spaced points on the grid for each combination of CHT and test problem. While the CHT-based landscapes are static for all steps of the local search for the first five CHTs, we use ten different stochastic ranking landscapes (in a loop) to mimic its stochastic behavior (each local search step uses one of the landscapes in turn).

4.2 Results and Discussion

CHT-Based Landscapes. We first inspect the CHT-based landscapes of the Type II problem CBB1, which is the easiest to understand (see the blue-hued landscapes in Fig. 3 and ignore the orange and red lines for now). When the constraint violation is ignored, the landscape obviously matches that of the original problem F_1, for which the Pareto set is linear. As approximately 2/3 of the apparent (as perceived by the CHT) Pareto set lie in the infeasible region, any search method that would ignore the constraint violation would spend a lot of effort in the infeasible region, making it inefficient.

When the constraint violation is treated as an objective, something interesting happens. The Pareto set of this CHT-based landscape contains not only the original Pareto set, but also a large region of otherwise infeasible solutions, which are nevertheless nondominated in the resulting 3-D objective space. While this does not happen on our problems of Type I and II, it appears on all six problems of Type III and IV. This would likely mislead a search method to regard a part of the infeasible region as optimal, which means that any optimization algorithm that uses this CHT needs to additionally check for feasibility of the apparent optimal solutions in order to be efficient.

Next, the landscapes of the constraint-domination principle, the epsilon-constraint method and the penalty function look very similar. However, note that the 'line' that we see in the landscape of the epsilon-constraint method

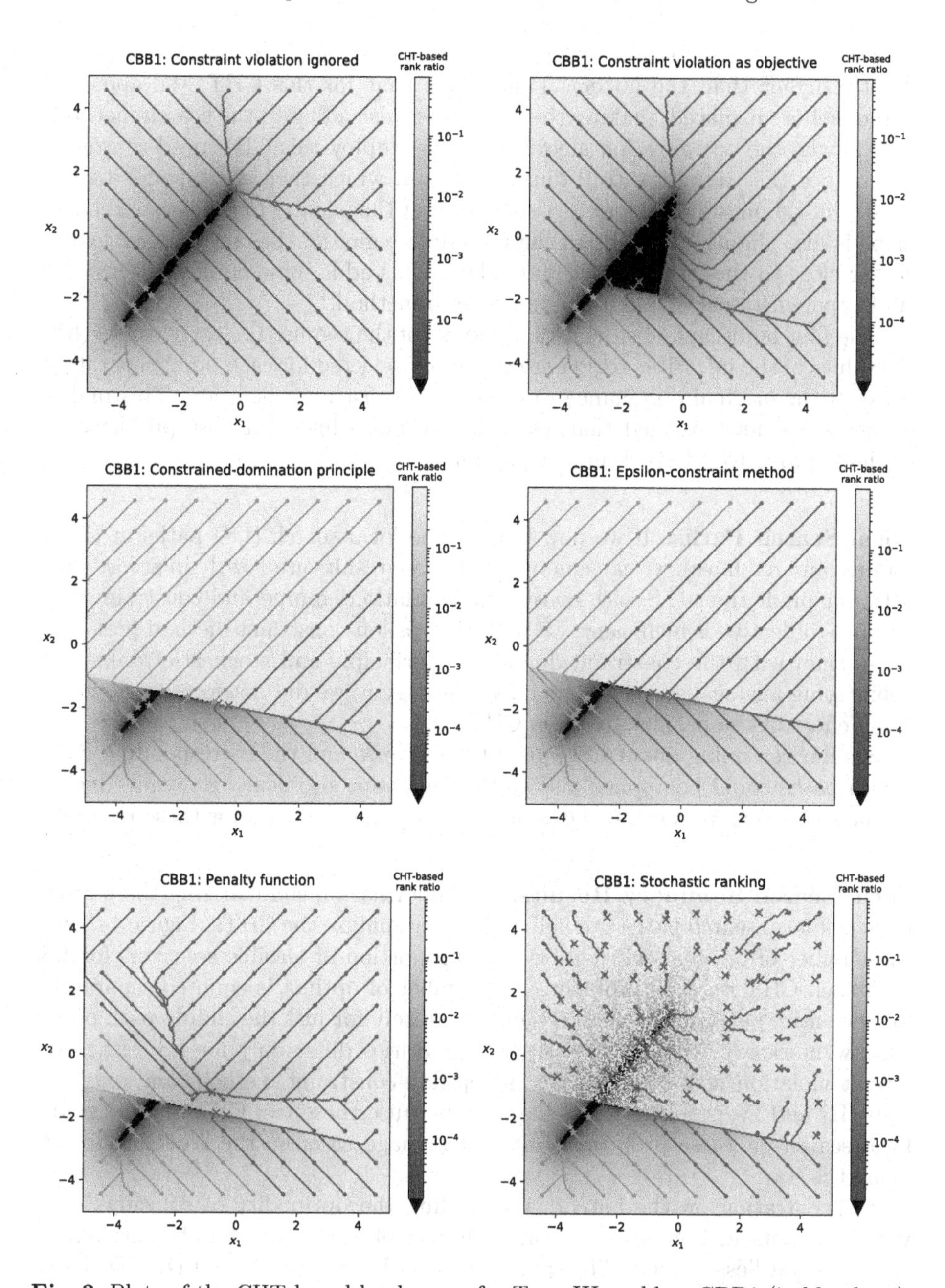

Fig. 3. Plots of the CHT-based landscapes for Type III problem CBB1 (in blue hues) for the six considered CHTs. Black denotes the Pareto set of these landscapes. Orange and red lines show the paths of local optimization starting in 100 different points shown with dots. If the path ends in a point that is optimal in the original problem landscape, the line is orange and it ends with a star, otherwise the line is red and it ends with a cross. (Color figure online)

does not match the feasible space boundary from Fig. 2 (the former is placed slightly higher than the latter). This means that for this CHT, the apparent Pareto set is misplaced, making the landscape misleading to the search method. This is why the optimization algorithms that employ this CHT need to gradually reduce the value of ε to 0 during the run, which then corresponds to the constraint-domination principle. Also, note that the penalty function-based landscape is also slightly different as the infeasible region of the original problem is darker close to the feasible space boundary. This adds some nonlinearity to the landscape with unclear influence on a search method.

The first landscape of stochastic ranking (of the ten used) clearly shows that the values of the infeasible region are randomly selected for each point separately between the original dominance rank and the constraint violation. This makes its landscape more rugged than the original one, which can pose problems to methods prone to get stuck in local optima.

Local Search Paths. If we now look at the local search (LS) paths in Fig. 3 (orange and red lines), we can confirm that these results are mostly in accordance with our predictions (LS with constrained violation ignored and constraint violation as objective is inefficient, LS with the epsilon-constraint method performs worse than with the constraint-domination principle and stochastic ranking is debilitating for local search in the infeasible region; we did not foresee the damaging effect of the penalty function CHT).

Similar reasoning about CHT-based landscapes and the corresponding local search paths could be applied also to the remaining problems. However, due to the lack of space we refer to the supplementary material [24] for these results.

Local Search Summary Results. The information summarizing the performance of local search paths can help us further analyze the CHTs. Figure 4 shows the number of optimal solutions vs. the proportion of feasible solutions for LS with each CHT on each problem. The number of optima is counted separately for the entire path (filled markers) and separately for just the ending path point (hollow markers). Note that these two quantities differ only for LS with constraint violation ignored and with the epsilon-constraint method, and only for Type III and IV problems. This happens because these two CHTs fail to guide local search on these problems, but still manage to cross the true Pareto set along the way.

Concentrating on the outcomes regarding the optimality of solutions (the y axis of plots in Fig. 4) we can immediately observe that the absolute worst results (regardless of the CHT) are achieved on DAS-CMOP5 and DC1-DTLZ1, which are multimodal and thus detrimental to local search. These two problems therefore do not help our analysis. Disregarding them, we can see the trend that the number of optimal solutions diminishes with increasing problem type, which could be expected. The relatively poor performance of local search with all CHTs except of the constrained-domination principle on the most basic problem

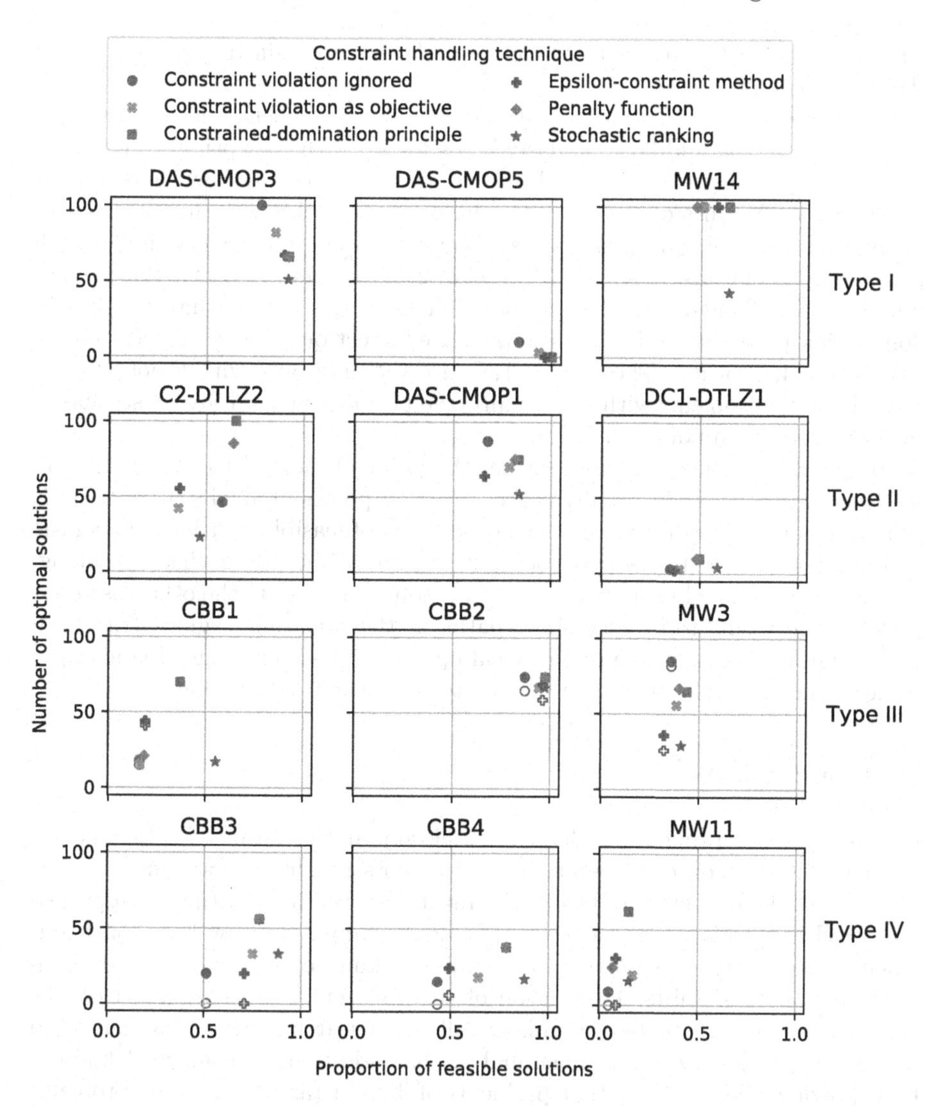

Fig. 4. The number of optimal solutions (y axis) vs. the proportion of feasible solutions (x axis) for local search with each CHT on each problem. Filled markers denote the number of all optimal solutions on the path, while the hollow markers show the number of final optimal solutions (the two differ only for constraint violation ignored and the epsilon-constraint method on problems of Type III and IV). (Color figure online)

CBB1 is quite disappointing. It shows that the constrained-domination principle is hard to beat and the other CHTs still have room for improvement.

We can further see that ignoring constraint violations is a very good strategy for solving problems of Type I, which confirms the results from [11]. Not so surprisingly, it is also one of the best CHTs for some Type II and III problems (those for which the intersection between the unconstrained and constrained Pareto set is large).

The performance of LS with constraint violation as objective never stands out (it is always in the middle). Similarly holds for LS with the epsilon-constrained method on problems of Type I, II and III (for Type IV, it shows a very poor performance). We also observe that the performances of LS with the constrained-domination principle and with the penalty function are mostly very similar with just a few exceptions. There (on C2-DTLZ2, CBB1 and MW11), the penalty function-based landscape is visibly different from the one by the constrained-domination principle, which has the undesired effect of guiding the local search away from the optima. These two CHTs are the only ones with a potential to solve Type IV problems with LS. Finally, the performance of LS with stochastic ranking is solidly among the worst.

If we look at the same results from the point of view of feasibility, we can see that, due to the 100 equally-spaced starting points of local search, there is generally not a large difference in the proportion of feasible solutions among the different CHTs. One (not so obvious) outlier here is LS with stochastic ranking, whose relatively good proportion of feasible solutions despite the otherwise poor performance stems from very short paths in the infeasible regions (the local search quickly becomes trapped in local optima of this very rugged landscape) rather than the CHT guiding the search towards the feasible region.

5 Conclusions

In this paper we proposed to look more closely at the various CHTs used for solving CMOPs in order to gain insight into their strenghts and weaknesses. This can help us devise better CHTs in the future, as well as find (more) appropriate search methods for particular CHTs. For example, we saw that constraint violation as objective requires additionally checking for feasibility, the epsilon-constraint method shifts the location of the apparent Pareto set and that the rugged landscape of stochastic ranking calls for a search method that can avoid being stuck in local optima. Our analysis has additionally confirmed findings from previous work [11,23] that problems of Type I (as well as some problems of Type II and III) are not helpful for benchmarking optimization algorithms on CMOPs as simply ignoring the constrains performs equally well.

As this work was limited to 2-D search and objective spaces we will consider generalizing our methodology to higher dimensions in the future. We would also like to similarly visualize the effects of dynamically changing CHTs and put more focus on local Pareto sets (possibly by using visualizations from [21] or [8]).

Acknowledgment. The authors acknowledge financial support from the Slovenian Research Agency (project no. N2-0254, young researcher program and research core funding no. P2-0209).

References

1. Alsouly, H., Kirley, M., Muñoz, M.A.: An instance space analysis of constrained multi-objective optimization problems (2022). https://arxiv.org/abs/2203.00868
2. Asafuddoula, M., Ray, T., Sarker, R.A., Alam, K.: An adaptive constraint handling approach embedded MOEA/D. In: IEEE Congress on Evolutionary Computation, CEC'12, pp. 1–8. IEEE (2012). https://doi.org/10.1109/CEC.2012.6252868
3. Brockhoff, D., Auger, A., Hansen, N., Tušar, T.: Using well-understood single-objective functions in multiobjective black-box optimization test suites. Evol. Comput. **30**(2), 165–193 (2022). https://doi.org/10.1162/evco_a_00298
4. Coello Coello, C.A.: Constraint-handling techniques used with evolutionary algorithms. In: Genetic and Evolutionary Computation Conference, GECCO'22, Companion Material, pp. 1310–1333. ACM (2022). https://doi.org/10.1145/3520304.3533640
5. Deb, K., Agrawal, S., Pratap, A., Meyarivan, T.: A fast and elitist multiobjective genetic algorithm: NSGA-II. IEEE Trans. Evol. Comput. **6**(2), 182–197 (2002). https://doi.org/10.1109/4235.996017
6. Fan, Z., Li, W., Cai, X., Huang, H., Fang, Y., You, Y., Mo, J., Wei, C., Goodman, E.: An improved epsilon constraint-handling method in MOEA/D for CMOPs with large infeasible regions. Soft. Comput. **23**(23), 12491–12510 (2019). https://doi.org/10.1007/s00500-019-03794-x
7. Fan, Z., et al.: Difficulty adjustable and scalable constrained multiobjective test problem toolkit. Evol. Comput. **28**(3), 339–378 (2019). https://doi.org/10.1162/evco_a_00259
8. Fieldsend, J.E., Chugh, T., Allmendinger, R., Miettinen, K.: A visualizable test problem generator for many-objective optimization. IEEE Trans. Evol. Comput. **26**(1), 1–11 (2022). https://doi.org/10.1109/TEVC.2021.3084119
9. Filipič, B., Vodopija, A.: Constraint handling in multiobjective optimization: tutorial. In: Presented at the IEEE World Congress on Computational Intelligence, WCCI'22. https://dis.ijs.si/filipic/wcci2022tutorial/
10. Fonseca, C.M.: Multiobjective genetic algorithms with application to control engineering problems. Ph.D. thesis, University of Sheffield (1995)
11. Fukumoto, H., Oyama, A.: A generic framework for incorporating constraint handling techniques into multi-objective evolutionary algorithms. In: Sim, K., Kaufmann, P. (eds.) EvoApplications 2018. LNCS, vol. 10784, pp. 634–649. Springer, Cham (2018). https://doi.org/10.1007/978-3-319-77538-8_43
12. Gallagher, M., Yuan, B.: A general-purpose tunable landscape generator. IEEE Trans. Evol. Comput. **10**(5), 590–603 (2006). https://doi.org/10.1109/TEVC.2005.863628
13. Geng, H., Zhang, M., Huang, L., Wang, X.: Infeasible elitists and stochastic ranking selection in constrained evolutionary multi-objective optimization. In: Wang, T.-D., Li, X., Chen, S.-H., Wang, X., Abbass, H., Iba, H., Chen, G.-L., Yao, X. (eds.) SEAL 2006. LNCS, vol. 4247, pp. 336–344. Springer, Heidelberg (2006). https://doi.org/10.1007/11903697_43
14. Hansen, N., Auger, A., Ros, R., Finck, S., Pošík, P.: Comparing results of 31 algorithms from the black-box optimization benchmarking BBOB-2009. In: Genetic and Evolutionary Computation Conference, GECCO'10, Companion Material, pp. 1689–1696. ACM (2010). https://doi.org/10.1145/1830761.1830790
15. Jain, H., Deb, K.: An evolutionary many-objective optimization algorithm using reference-point based nondominated sorting approach, Part II: handling constraints

and extending to an adaptive approach. IEEE Trans. Evol. Comput. **18**(4), 602–622 (2014). https://doi.org/10.1109/TEVC.2013.2281534
16. Li, K., Chen, R., Fu, G., Yao, X.: Two-archive evolutionary algorithm for constrained multiobjective optimization. IEEE Trans. Evol. Comput. **23**(2), 303–315 (2019). https://doi.org/10.1109/TEVC.2018.2855411
17. Liang, J., et al.: A survey on evolutionary constrained multi-objective optimization. In: IEEE Transactions on Evolutionary Computation (2022). https://doi.org/10.1109/TEVC.2022.3155533
18. Ma, Z., Wang, Y.: Evolutionary constrained multiobjective optimization: test suite construction and performance comparisons. IEEE Trans. Evol. Comput. **23**(6), 972–986 (2019). https://doi.org/10.1109/TEVC.2019.2896967
19. Rahimi, I., Gandomi, A.H., Chen, F., Mezura-Montes, E.: A review on constraint handling techniques for population-based algorithms: from single-objective to multi-objective optimization (2022). https://arxiv.org/abs/2206.13802
20. Runarsson, T., Yao, X.: Stochastic ranking for constrained evolutionary optimization. IEEE Trans. Evol. Comput. **4**(3), 284–294 (2000). https://doi.org/10.1109/4235.873238
21. Schäpermeier, L., Grimme, C., Kerschke, P.: One PLOT to show them all: visualization of efficient sets in multi-objective landscapes. In: Bäck, T., et al. (eds.) PPSN 2020. LNCS, vol. 12270, pp. 154–167. Springer, Cham (2020). https://doi.org/10.1007/978-3-030-58115-2_11
22. Takahama, T., Sakai, S.: Constrained optimization by the ϵ constrained differential evolution with gradient-based mutation and feasible elites. In: IEEE Congress on Evolutionary Computation, CEC'06, pp. 1–8. IEEE (2006). https://doi.org/10.1109/CEC.2006.1688283
23. Tanabe, R., Oyama, A.: A note on constrained multi-objective optimization benchmark problems. In: IEEE Congress on Evolutionary Computation, CEC'17, pp. 1127–1134. IEEE (2017). https://doi.org/10.1109/CEC.2017.7969433
24. Tušar, T., Vodopija, A., Filipič, B.: Visual exploration of the effect of constraint handling in multiobjective optimization: supplementary material (2022). https://doi.org/10.5281/zenodo.7440416
25. Vodopija, A., Tušar, T., Filipič, B.: Characterization of constrained continuous multiobjective optimization problems: a feature space perspective. Inf. Sci. **607**, 244–262 (2022). https://doi.org/10.1016/j.ins.2022.05.106
26. Woldesenbet, Y.G., Yen, G.G., Tessema, B.G.: Constraint handling in multiobjective evolutionary optimization. IEEE Trans. Evol. Comput. **13**(3), 514–525 (2009). https://doi.org/10.1109/TEVC.2008.2009032
27. Zhang, Q., Li, H.: MOEA/D: a multiobjective evolutionary algorithm based on decomposition. IEEE Trans. Evol. Comput. **11**(6), 712–731 (2007). https://doi.org/10.1109/TEVC.2007.892759
28. Zhu, Q., Zhang, Q., Lin, Q.: A constrained multiobjective evolutionary algorithm with detect-and-escape strategy. IEEE Trans. Evol. Comput. **24**(5), 938–947 (2020). https://doi.org/10.1109/TEVC.2020.2981949
29. Zitzler, E., Künzli, S.: Indicator-based selection in multiobjective search. In: Yao, X., et al. (eds.) PPSN 2004. LNCS, vol. 3242, pp. 832–842. Springer, Heidelberg (2004). https://doi.org/10.1007/978-3-540-30217-9_84

A Two-Stage Algorithm for Integer Multiobjective Simulation Optimization

Fei Liu[1(✉)] and Qingfu Zhang[1,2]

[1] Department of Computer Science, City University of Hong Kong, Hong Kong, China
fliu36-c@my.cityu.edu.hk, qingfu.zhang@cityu.edu.hk
[2] The City University of Hong Kong Shenzhen Research Institute, Shenzhen, China

Abstract. Multiobjective discrete optimization via simulation (MDOvS) has received considerable attention from both academics and industry due to its wide application. This paper proposes a two-stage fast convergent search algorithm for MDOvS. In its first stage, the multiobjective optimization problem under consideration is decomposed into several single-objective optimization subproblems, and a Pareto retrospective approximation method is used to generate an approximated optimal solution for each subproblem. In the second stage, from the solutions generated in the first stage, a multiobjective local stochastic search with a revised simulation allocation rule is used to explore the entire Pareto front. Our experimental studies show that the proposed method outperforms the state-of-the-art MO-COMPASS on a set of test instances with noisy evaluations and a bi-objective bus scheduling problem. Our proposed method is up to ten times faster than MO-COMPASS.

Keywords: Multiobjective optimization · Simulation optimization · Integer-ordered · Decomposition · Stochastic

1 Introduction

Multiobjective discrete optimization via simulation (MDOvS) involves optimizing several conflicting objectives in a discrete design space. The objective function evaluations are conducted by computer or physical simulation experiments. Each simulation on an objective at a candidate solution $\boldsymbol{x}$ can produce a noisy estimate value of the objective function at $\boldsymbol{x}$. To reduce the estimation variance, one can do multiple independent simulation experiments at $\boldsymbol{x}$ and use the average of the obtained noisy function values as its estimate value [10]. Many real-life applications can be modeled as MDOvS. Examples can be found in manufacturing [1], aviation [15], medical [3], and transportation [22].

MDOvS methods can be roughly classified into two categories according to their goals. The methods in one category assume that the number of alternatives (i.e., candidate solutions) is small. These methods evaluate all the alternatives and aim to identify the best solutions. The key design issue in these methods is

M. Emmerich et al. (Eds.): EMO 2023, LNCS 13970, pp. 17–28, 2023.
https://doi.org/10.1007/978-3-031-27250-9_2

how to allocate the simulation budget to these alternatives, i.e., to decide the number of simulations for each alternative for maximizing the probability that the true Pareto optimal solutions are correctly identified or other performance metrics [2,13,16].

The methods in the other category are for the problem with a large number of candidate solutions. It is impractical, if not impossible, to evaluate all the candidate solutions. These methods only select a small number of candidate solutions for evaluation. Thus, besides simulation budget allocation, they need efficient search strategies for identifying promising candidate solutions for evaluation. Global search and local search including gradient methods have been implemented as search strategies. For example, multiobjective partition-based random search [20] adopts a global random search method, and it can be readily hybridized with other multiobjective optimization methods. MO-COMPASS [12,14], a multiobjective version of COMPASS [9,21], a fast local search for MDOvS, mainly exploits neighborhoods of some promising solutions that have been identified during the search, and it uses a simple rule for simulation budget allocation. R-PERLE [4,5] uses a retrospective search strategy and a pseudo-gradient-based line search for fast searching promising solutions. It adopts an epsilon-constraint method to scale the multiobjective optimization problem into several single-objective problems. The epsilon constraints are hard to design.

This paper proposes a two-stage algorithm for solving integer-ordered MDOvS. In the first stage, we decompose the problem into several single-objective optimization problems and perform retrospective approximation on each sub-problem to find alternatives near the Pareto front. In the second stage, we use a revised multiobjective local stochastic search to explore the entire Pareto front. The main contribution is twofold: 1) A decomposition-based Pareto retrospective approximation is proposed. It generates approximated non-dominated solutions efficiently. 2) A two-stage framework is designed. The framework takes advantage of both the efficient Pareto retrospective approximation and the local stochastic search to balance optimization convergence and diversity.

The remainder of this paper is organized as follows. Section 2 gives the problem formulation, Sect. 3 introduces the proposed algorithm in detail, Sect. 4 presents the experimental studies and discussion, and the last section concludes the paper.

2 Problem Definition

Consider the following multiobjective simulation optimization problem with integer-ordered variables:

$$\begin{aligned} &\min\ (f_1(\boldsymbol{x}),\dots,f_m(\boldsymbol{x})),\\ &\text{s.t. } \boldsymbol{x}\in\Theta, \end{aligned} \tag{1}$$

where $f_j(\boldsymbol{x}) = E[F_j(\boldsymbol{x})]$ is the expected value of the j-th random objective function $F_j(\boldsymbol{x})$ at $\boldsymbol{x}$ for $j = 1, \ldots, m$. Independent samples (i.e., simulation values) of $F_j(\boldsymbol{x})$ can be obtained to estimate the value of $f_j(\boldsymbol{x})$ at each alternative $\boldsymbol{x}$. Θ is a finite integer-ordered design space, which is typically too large to be evaluated exhaustively. We assume that $F_j(\boldsymbol{x})$ on each alternative $\boldsymbol{x}$ follows a normal distribution $N(f_j(\boldsymbol{x}), \sigma^2)$ and the σ is the same for all $\boldsymbol{x}$ [10].

Pareto optimality is adopted in this paper to define optimal solutions. $\boldsymbol{x}_1$ is said to dominate $\boldsymbol{x}_2$, denoted as $\boldsymbol{x}_1 \prec \boldsymbol{x}_2$, if and only if $f_j(\boldsymbol{x}_1) \leq f_j(\boldsymbol{x}_2)$ for each $j \in \{1, \ldots, m\}$. and $f_j(\boldsymbol{x}_1) < f_j(\boldsymbol{x}_2)$ for at least one $j \in \{1, \ldots, m\}$. $\boldsymbol{x}^*$ is Pareto optimal if no $\boldsymbol{x} \in \Theta$ dominates $\boldsymbol{x}^*$. The set of all the Pareto optimal points is called the Pareto set (PS) and the set of their corresponding objective vectors is called the Pareto front (PF). The goal is to approximate the entire PF as close and as diverse as possible.

3 Algorithm

We propose a two-stage multiobjective simulation optimization method (TSMOSO). It consists of a Pareto retrospective approximation method for the first stage and a multiobjective local stochastic search for the second stage. Its basic idea is to quickly obtain some approximated Pareto optimal solutions in the first stage and then from them to explore the entire Pareto front in the second stage.

3.1 Pareto Retrospective Approximation Method for the First Stage

We proposed a decomposition-based Pareto retrospective approximation method (PRA) for the first stage. Its pseudocode is shown in Algorithm 1. The initial alternatives are generated by Latin Hypercube Sampling [8]. After initialization, the multiobjective optimization problem is decomposed into several sub-problems defined by K uniformly distributed weight vectors $\boldsymbol{\lambda}_1, \ldots, \boldsymbol{\lambda}_K$. The weight vectors are generated using the method proposed by Das et al. [6]. The decomposition function used in this paper is Tchebycheff aggregation [17]

$$f^{te}(\boldsymbol{x}|\boldsymbol{\lambda}) = \max_{1 \leq j \leq m} \{\lambda_j (f_j(\boldsymbol{x}) - z_j{}^*)\}, \tag{2}$$

where $\mathbf{z}^* = (z_1{}^*, \ldots, z_m{}^*)$ is a reference point. In our implementation, the reference point is set as the minimum objective value vector among all the evaluated alternatives.

To optimize each sub-problem, a retrospective approximation (RA) [11] approach is used, which is an improved version of sample average approximation (SAA) [18]. In SAA, the objective function value is replaced by its empirical mean and then a deterministic algorithm can be used. SAA uses a fixed number of replications to calculate the empirical mean. To reduce the number of replications and computational cost, RA adaptively increases the number of replications in each iteration.

RA starts at the current best alternative in the initial population P_0. It is an iterative method. In each iteration of RA, the optimization starts from the optimal solution in the last iteration. At the end of each iteration, the number of replications r is enlarged to be $r = r * t_2$, where t_2 is a scale factor. When r reaches a threshold rt, the RA search on the current sub-problem i stops. In each iteration of RA, a pseudo-gradient-based search is performed. It iteratively performs two procedures 1) direction updating and 2) line search to minimize $\bar{f}^{te}(\boldsymbol{x}|\boldsymbol{\lambda}_i)$.

Algorithm 1. Pareto Retrospective Approximation (PRA)

Input:
initial population P_0; weight vectors $\{\boldsymbol{\lambda}_1, \ldots, \boldsymbol{\lambda}_K\}$;
initial step size s_0; scale factor t_1; scale factor t_2;
initial replication number r_0; threshold of replication number rt.
Output: updated population P.

```
procedure PRA(P_0, s_0, t_1, t_2, {λ_1, ..., λ_K})
    P ← P_0, r ← r_0
    for i = 1 to K do
        x ← arg min_{x∈P} f̄^te(x|λ_i)
        while r ≤ rt do
            while not local optimal do
                s ← s_0
                D, P ← UPDATE DIRECTION(x, λ_i, r, P)
                x, P ← LINE SEARCH(x, λ_i, r, s, t_1, D, P)
            end while
            r ← r * t_2
        end while
    end for
end procedure
```

```
procedure UPDATE DIRECTION(x, λ_i, r, P)
    for j ← 1 to d do
        x_new ← x + e^j
        D_j ← f̄_r^te(x_new|λ_i) − f̄_r^te(x|λ_i)
        P ← P ∪ x_new
    end for
    D ← D/Norm(D)
end procedure
```

```
procedure LINE SEARCH(x, λ_i, r, s, t_1, D, P)
    x_old ← x
    x_new ← round(x_old + D * t_1)
    while f̄_r^te(x_new|λ_i) < f̄_r^te(x_old|λ_i) do
        P ← P ∪ x_new
        s ← s * t_1
        x_old ← x_new
        x_new ← round(x_old + D * s)
    end while
    x ← x_new
end procedure
```

In direction updating, search direction $\boldsymbol{D}$ on current alternative $\boldsymbol{x}$ is updated to be the negative pseudo gradient direction $\boldsymbol{D}$ on discrete space. Each dimension of $\boldsymbol{D}$ is calculated according to $D_j = -(\bar{f}^{te}(\boldsymbol{x} + \boldsymbol{e}^j|\boldsymbol{\lambda}_i) - \bar{f}^{te}(\boldsymbol{x}|\boldsymbol{\lambda}_i))$, where $j \in \{1, \ldots, m\}$, and $\boldsymbol{e}^j$ is a unit vector along j-th dimension. The direction is then normalized to be $\boldsymbol{D} = \boldsymbol{D}/Norm(\boldsymbol{D})$.

In line search, the new alternative $\boldsymbol{x}_{new}$ is calculated from the old alternative $\boldsymbol{x}_{old}$ along the direction $\boldsymbol{D}$ with a step size of s and rounded to a nearest feasible discrete position $\boldsymbol{x}_{new} = round(\boldsymbol{x}_{old} + s * \boldsymbol{D})$. If the fitness of the new alternative $\bar{f}^{te}(\boldsymbol{x}_{new}|\boldsymbol{\lambda}_i)$ is better than the old one $\bar{f}^{te}(\boldsymbol{x}_{old}|\boldsymbol{\lambda}_i)$, the step size is enlarged to be $s = s * t_1$ and the line search continues along current direction $\boldsymbol{D}$, otherwise, the line search procedure along current direction $\boldsymbol{D}$ stops. If the local optimal point is reached, the current iteration of RA stops.

In the RA of each sub-problem i, we record a sufficient replication number r_i. A r is called sufficient if the following pseudo-gradient-based search with a larger replication number doesn't change the local optimal. A small r_i generally reflects a low uncertainty level of the problem and vice versa. We calculate the average sufficient replication number $r_{ave} = \sum r_i/K$ and use it in the second stage to determine the simulation allocation budget.

3.2 Local Stochastic Search for the Second Stage

We adopt a multiobjective local stochastic search (LSS) to explore the entire Pareto front in the second stage. Algorithm 2 shows the pseudocode of LSS. Firstly, non-dominated sorting is performed on the tested alternative population P. The P is sorted into approximated non-dominated alternatives P_{nd} and dominated alternatives P_d according to their empirical means. Then, a promising area $\mathcal{P}r$ is generated around P_{nd} according to the euclidean distance in the design space. A detailed formulation of $\mathcal{P}r$ can be found in [14]. n_{new} new alternatives P_{new} are selected randomly from the promising area $\mathcal{P}r$ and the population P is updated. New simulations are allocated on the updated population P according to the simulation allocation rule (SAR). The algorithm stops when the total number of simulations N exceeds the budget B.

Algorithm 2. Local Stochastic Search (LSS)

Input:
simulation budget B; new population size n_{new};
evaluated alternatives P.
Output:
approximated non-dominated alternatives P_{nd}.

while $N < B$ **do**
  $P_{nd}, P_d \leftarrow NondSort(P)$
  $\mathcal{P}r \leftarrow PromisingArea(P_{nd}, P_d)$
  $P_{new} \leftarrow RandomSampling(\mathcal{P}r, n_{new})$
  $P \leftarrow P \cup P_{new}$
  $P, N \leftarrow SAR(P)$
end while
$P_{nd} \leftarrow NondSort(P)$

The main difference between the local stochastic search used in our method and that used in MO-COMPASS is on the SAR. MO-COMPASS [14] uses a simple two-level SAR. In i-th iteration, it first ensures $min\{1, log(i)\}$ samples are allocated to each alternative, and distributes $|P_{nd}| \times min\{1, log(i)\}$ on the predicted non-dominated alternatives P_d and the alternatives selected from the promising area P_{new} to emphasis these critical solutions.

We design a multi-level SAR. It has two differences compared with the original one [14]: 1) The number of samples allocated on the critical solutions is set to be $s_{ave} \times min\{1, log(i)\}$, where s_{ave} is the average sufficient number calculated in the first stage. The reason is that $|P_{nd}|$ cannot reflect the uncertainty of the problem while s_{ave} can. It can avoid using too many simulations on the problem with small uncertainty. 2) A multi-level allocation method is used instead of the two-level one. The first-level alternatives are the original critical alternatives $P^1 = P_d \cup P_{new}$, the second-level alternatives are the predicted non-dominated alternatives of the rest $P^2 = NondSort(P/P^1)$ and so on. The number of replications for the n-th-level alternatives P^n is set to be $s_{ave} \times min\{1, log(i)\}/n$. In this way, the simulation allocation is more smooth. It improves the robustness on problems with large uncertainty.

Compared with the well-known MO-COMPASS, which is a pure local stochastic search, the proposed method divides the optimization process into two stages. Stage one provides a few high-quality approximate Pareto optimal solutions and Stage two starts from these solutions and generates more solutions to approximate the whole Pareto front. This strategy makes our method more efficient when the Pareto optimal solutions are close to each other as in many applications. Stage one uses our designed decomposition-based retrospective approximation. Compared with the retrospective approximation based on epsilon constraints, e.g., R-PERLE [5], our approach is easy to implement and can be generalized to many-objective simulation problems. Stage two uses the multiobjective local stochastic search. Compared with MO-COMPASS, it adopts a new simulation allocation rule and can save the number of simulations.

4 Experimental Studies

We have compared the proposed TSMOSO with the current state-of-the-art MO-COMPASS. First, experimental studies are carried out on test instances with noisy evaluations. An ablation study is conducted to validate the contributions of different algorithmic components. Then, the effectiveness of TSMOSO is further demonstrated on a real-world biobjective bus scheduling problem.

4.1 Experiments on Test Instances

We used ZDT instances [7,14]. We modify the Pareto set of ZDT instances to be a nonlinear function of variables and scale the Pareto front into the region $[0, 1]$. The design space is discretized into $\Theta = \{0, \dots, L\}^d$, where L and d are the discretization level and the number of variables, respectively. In this paper, they are set to be $L = 10, d = 5$, which results in 161051 possible alternatives. Apparently, it is too expensive to perform an exhaustive evaluation of every alternative, which reflects the importance of the fast convergent search. Figure 1

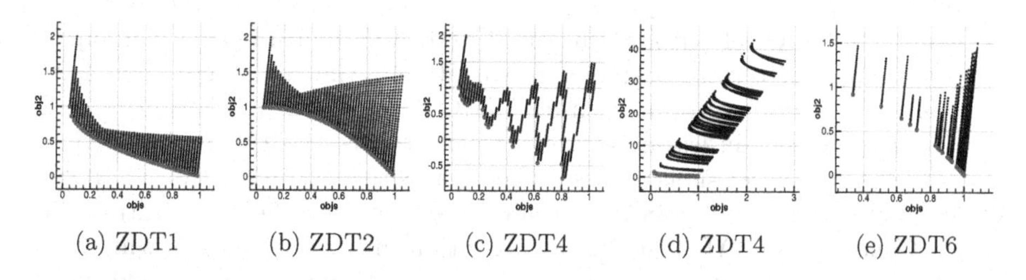

(a) ZDT1 (b) ZDT2 (c) ZDT4 (d) ZDT4 (e) ZDT6

Fig. 1. An illustration of the discrete objective spaces of the test instances, where the red sets are the Pareto fronts. (Color figure online)

illustrates the discrete objective spaces (in black) and the Pareto fronts (in red) of different test instances.

Each instance has four different noise levels $\sigma = \{0.01, 0.1, 0.2, 0.5\}$, which results in 20 test cases. The noise level is the standard error of normal distribution. In each case, we assume the noise levels on different objectives at different alternatives are the same. 30 independent runs with different initial samplings are carried out. Experiment settings for TSMOSO are: 1) initial population size $|P_0| = 20$, 2) the number of sub-problems $K = 3$, 3) the initial linear search step length and its scale factor are $s_0 = 1.2$ and $t_1 = 1.5$, respectively, 4) the initial number of replications and its scale factor are $r_0 = 4$ and $t_2 = 2$, respectively, and 5) the threshold of the number of replications $rt = 128$.

There are several additional algorithmic components in TSMOSO compared with MO-COMPASS. An ablation study is carried out to show the importance of different components. We identify the three most important ones: 1) the Pareto retrospective approximation, 2) the indicator s_{ave} (denoted as Savg), which decides the number of simulations in the second-stage optimization, and 3) the multi-level simulation allocation rule (denoted as mlSAR). By deleting them one by one from TSMOSO, we obtain the following three versions of TSMOSO:

- TSMOSO without mlSAR and Savg (V1)
- TSMOSO without mlSAR (V2)
- TSMOSO (V3)

Table 1 shows the simulation number (SN) and optimization efficiency score (ES) of different algorithms on ZDT test instances. SN is the number of simulations spent by different algorithms when they reach the same level (threshold) of HV. In this paper, we choose to use 99% HV_{PF} as the threshold, where HV_{PF} is the hypervolume of the Pareto front. ES is defined as the NS of the proposed TSMOSO divided by that of the compared method. The larger the ES, the better the performance. The last row is the four average ES values with respect to four different noise levels $\sigma = \{0.01, 0.1, 0.2, 0.5\}$.

The results indicate that: 1) The proposed algorithm surpasses MO-COMPASS in all the 40 test cases and the efficiency improvement is up to tenfold; 2) The number of simulations spent increases with the noise level; 3) The superiority of TSMOSO over MO-COMPASS is diminished as the noise level increases. TSMOSO V1 outperforms MO-COMPASS, which reflects the importance of using Pareto retrospective approximation in the first stage. There is an obvious performance improvement from TSMOSO V1 to V2. The benefit of using Savg is more obvious in less noisy cases, which reveals that the original SAR is too conservative on low-noise cases. When comparing the results of TSMOSO V2 and V3, interestingly, mlSAR doesn't always increase the optimization efficiency. Generally, mlSAR benefits the algorithm in high-noise cases while it is less effective in low-noise cases. The reason is that mlSAR spends additional simulations on middle-level alternatives. The additional simulations improve the robustness in high-noise cases but are useless in low-level cases, where a small number of simulations is already enough to identify optimal solutions.

Table 1. The simulaion number (SN) and efficiency score (ES) on ZDT instances using MO-COMPASS and different versions of TSMOSO

Instance	Noise level	MO-COMPASS		TSMOSO V1		TSMOSO V2		TSMOSO V3
		SN	ES	SN	ES	SN	ES	SN
zdt1	0.01	58531	0.13	13824	0.55	6383	1.20	7671
zdt1	0.05	98729	0.24	37153	0.64	21287	1.12	23838
zdt1	0.1	101857	0.36	51034	0.73	34640	1.07	37131
zdt1	0.2	125289	0.46	90503	0.63	68991	0.83	57394
zdt2	0.01	76674	0.10	18997	0.41	7057	1.11	7847
zdt2	0.05	109663	0.20	48433	0.46	19280	1.16	22294
zdt2	0.1	218195	0.17	49883	0.73	37037	0.98	36233
zdt2	0.2	185315	0.57	127389	0.82	108892	0.96	104782
zdt3	0.01	73217	0.10	13897	0.53	6670	1.11	7389
zdt3	0.05	101084	0.18	37163	0.49	22937	0.80	18331
zdt3	0.1	102096	0.32	38974	0.85	34982	0.94	33012
zdt3	0.2	126062	0.42	89348	0.59	48975	1.08	52803
zdt4	0.01	49666	0.10	19070	0.27	4723	1.09	5143
zdt4	0.05	46008	0.14	21529	0.29	5851	1.08	6349
zdt4	0.1	43265	0.25	17129	0.62	8675	1.23	10666
zdt4	0.2	53095	0.39	33226	0.63	23212	0.90	20907
zdt6	0.01	60851	0.14	27967	0.30	8230	1.01	8307
zdt6	0.05	62848	0.16	33293	0.30	16473	0.62	10145
zdt6	0.1	74629	0.19	65071	0.22	20592	0.70	14505
zdt6	0.2	175454	0.28	73125	0.67	54111	0.90	48635
Average ES		{0.11, 0.19, 0.26, 0.42}		{0.41, 0.44, 0.63, 0.67}		{1.10, 0.96, 0.99, 0.93}		

Figure 2 compares the converge curves with a noise level of 0.1 on ZDT instances. The colored curve is the average HV and the vertical error bar is the normal deviation. Apparently, TSMOSO convergences much faster than MO-COMPASS. In addition, the error bars reach zero at the end of optimization, which means the algorithms can converge to the Pareto front in different independent runs.

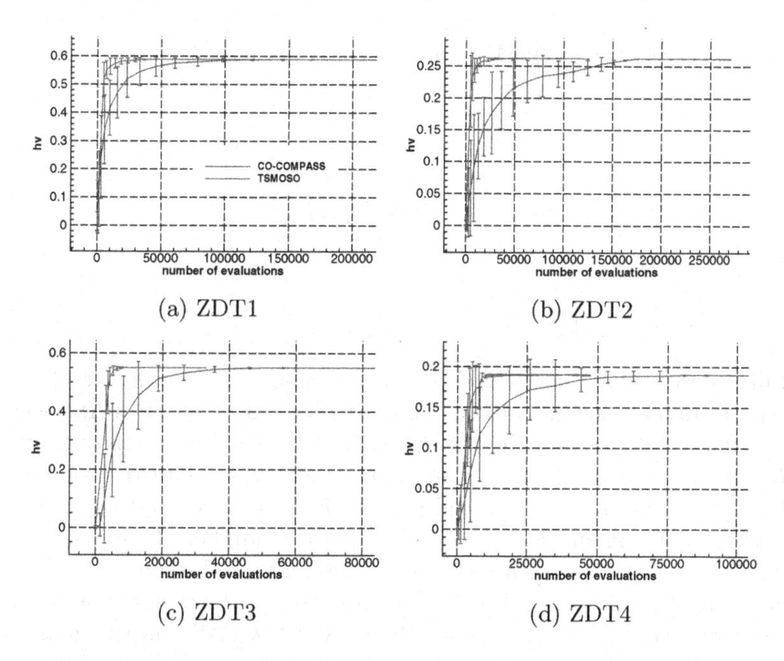

(a) ZDT1 (b) ZDT2

(c) ZDT3 (d) ZDT4

Fig. 2. Average convergence curves of TSMOSO (red) and MO-COMPASS (blue) on ZDT instances (Color figure online)

4.2 Biobjective Bus Scheduling

We demonstrate the algorithm on a biobjective integer bus scheduling problem [5,19]. The problem requires scheduling the arrival times at the bus station of a fleet of buses to minimize the expected operating cost of buses and the expected waiting time of passengers. We suppose the passengers arrive at the bus station according to a Poisson distribution $P(k, \lambda)$, where $\lambda = 10$ is the arrival rate per time unit. There are $\tau = 100$ time units per day. During a day, a fleet of buses $b \in \{1, 2, \ldots, q\}$ with infinite capacity is scheduled to visit the bus station. Therefore, the decision variables are the integer arrival times of buses at the station $\boldsymbol{x} = \{x_1, x_2, \ldots, x_q\}$. We assume there is a no-cost bus at time 0 and a pre-scheduled bus at time τ and a bus is defined as not used if it is scheduled at the same time as any other buses. The design space is selected to be $\Theta = \{0, 10, 20, \ldots, 100\}^q$, where $q = 9$, i.e., its a nine-bus scheduling problem. The two objectives are formulated as:

$$
\begin{aligned}
F_1(\boldsymbol{x}) &= \sum_{\ell=1}^{q+1} c_0 \mathbb{I}\left\{x_\ell - x_{\ell-1} > 0\right\} + \left(N_i\left(x_\ell\right) - N_i\left(x_{\ell-1}\right)\right)^\gamma, \\
F_2(\boldsymbol{x}) &= \sum_{j=1}^{N_i(\tau)} W_{ij},
\end{aligned}
\tag{3}
$$

where $N_i(x_l) - N_i(x_{l-1})$ denotes the number of passenger arrivals between bus $l-1$ and bus l on the i-th day, where $l = 1, 2, \ldots, q+1$. W_{ij} is the wait time of the j-th passenger for $j = 1, 2, \ldots, N_i(\tau)$ passengers on the i-th day.

At any feasible $\boldsymbol{x}$ the expected value of objectives are:

$$\begin{aligned} f_1(\boldsymbol{x}) &= \mathbb{E}\left[F_1(\boldsymbol{x})\right] \\ &= \sum_{\ell=1}^{q+1} c_0 \mathbb{I}\left\{x_\ell - x_{\ell-1} > 0\right\} + \mathbb{E}\left[\sqrt{N_i(x_\ell) - N_i(x_{\ell-1})}\right] \\ &\approx \sum_{\ell=1}^{q+1} c_0 \mathbb{I}\left\{x_\ell - x_{\ell-1} > 0\right\} + \sqrt{\lambda(x_\ell - x_{\ell-1})}, \\ f_2(\boldsymbol{x}) &= \mathbb{E}\left[F_2(\boldsymbol{x})\right] \\ &= \mathbb{E}\left[\sum_{j=1}^{N_i(\tau)} W_{ij}\right] = (\lambda/2)\sum_{\ell=1}^{q+1}(x_\ell - x_{\ell-1})^2 . \end{aligned} \quad (4)$$

A detailed description and analysis of the problem can be found in [5].

We perform 30 independent runs of TSMOSO and MO-COMPASS with different initial populations on the bus scheduling problem. The experimental settings are the same as that used on numerical instances. The two objectives are scaled to [0,1] according to the ideal and nadir points provided in [5]. Figure 3 (a) shows the convergence curves of HV versus the number of simulations in 30 independent runs. Although there is high randomness and some of the runs have not yet converged, we can still easily observe from the results that TSMOSO shows significantly faster convergence than MO-COMPASS on this multiobjective bus scheduling problem. Figure 3 (b) compares the approximated PFs of the two methods with a computational budget of $B = 1 \times 10^5$ simulations. Results show that TSMOSO outperforms MO-COMPASS in terms of both convergence and diversity.

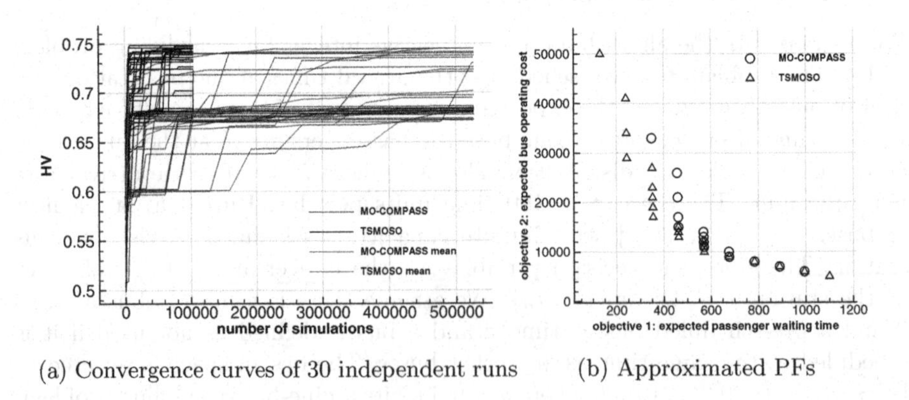

(a) Convergence curves of 30 independent runs (b) Approximated PFs

Fig. 3. Comparison of TSMOSO and MO-COMPASS on the biobjective bus scheduling problem

5 Conclusion

This paper has proposed a two-stage algorithm called TSMOSO for solving integer-ordered MDOvS. In the first stage, It uses a Pareto retrospective approximation to generate a set of approximated non-dominated solutions. In the second stage, it uses a local stochastic search with a revised multi-level simulation allocation rule to explore the entire PF. The proposed algorithm is compared with the state-of-the-art MO-COMPASS on noisy test instances and a biobjective bus scheduling problem. Results demonstrate its effectiveness and efficiency. In the future, we will investigate the combination of global search methods with TSMOSO and apply it to real-world problems.

Acknowledgements. This work was supported by the National Natural Science Foundation of China under grant number 62276223.

References

1. Andersson, M., Grimm, H., Persson, A., Ng, A.: A web-based simulation optimization system for industrial scheduling. In: 2007 Winter Simulation Conference, pp. 1844–1852. IEEE (2007)
2. Applegate, E.A., Feldman, G., Hunter, S.R., Pasupathy, R.: Multi-objective ranking and selection: optimal sampling laws and tractable approximations via score. J. Simul. **14**(1), 21–40 (2020)
3. Chen, T., Wang, C.: Multi-objective simulation optimization for medical capacity allocation in emergency department. J. Simul. **10**(1), 50–68 (2016)
4. Cooper, K., Hunter, S.R.: Pymoso: software for multiobjective simulation optimization with R-PERLE and R-MinRLE. INFORMS J. Comput. **32**(4), 1101–1108 (2020)
5. Cooper, K., Hunter, S.R., Nagaraj, K.: Biobjective simulation optimization on integer lattices using the epsilon-constraint method in a retrospective approximation framework. INFORMS J. Comput. **32**(4), 1080–1100 (2020)
6. Das, I., Dennis, J.E.: Normal-boundary intersection: a new method for generating the pareto surface in nonlinear multicriteria optimization problems. SIAM J. Optim. **8**(3), 631–657 (1998)
7. Deb, K.: Multi-objective genetic algorithms: problem difficulties and construction of test problems. Evol. Comput. **7**(3), 205–230 (1999)
8. Giunta, A., Wojtkiewicz, S., Eldred, M.: Overview of modern design of experiments methods for computational simulations. In: 41st Aerospace Sciences Meeting and Exhibit, p. 649 (2003)
9. Hong, L.J., Nelson, B.L.: Discrete optimization via simulation using compass. Oper. Res. **54**(1), 115–129 (2006)
10. Hunter, S.R., et al.: An introduction to multiobjective simulation optimization. ACM Trans. Model. Comput. Simul. (TOMACS) **29**(1), 1–36 (2019)
11. Kim, S., Pasupathy, R., Henderson, S.G.: A guide to sample average approximation. In: Fu, M.C. (ed.) Handbook of Simulation Optimization. ISORMS, vol. 216, pp. 207–243. Springer, New York (2015). https://doi.org/10.1007/978-1-4939-1384-8_8
12. Lee, L.H., Chew, E.P., Li, H.: Multi-objective compass for discrete optimization via simulation. In: Proceedings of the 2011 Winter Simulation Conference (WSC), pp. 4065–4074. IEEE (2011)
13. Lee, L.H., Chew, E.P., Teng, S., Goldsman, D.: Finding the non-dominated pareto set for multi-objective simulation models. IIE Trans. **42**(9), 656–674 (2010)
14. Li, H., Lee, L.H., Chew, E.P., Lendermann, P.: Mo-compass: a fast convergent search algorithm for multi-objective discrete optimization via simulation. IIE Trans. **47**(11), 1153–1169 (2015)

15. Li, H., Zhu, Y., Chen, Y., Pedrielli, G., Pujowidianto, N.A.: The object-oriented discrete event simulation modeling: a case study on aircraft spare part management. In: 2015 Winter Simulation Conference (WSC), pp. 3514–3525. IEEE (2015)
16. Li, J., Liu, W., Pedrielli, G., Lee, L.H., Chew, E.P.: Optimal computing budget allocation to select the nondominated systems-a large deviations perspective. IEEE Trans. Autom. Control **63**(9), 2913–2927 (2017)
17. Miettinen, K.: Nonlinear Multiobjective Optimization, vol. 12. Springer Science & Business Media, Berlin (2012)
18. Pasupathy, R., Ghosh, S.: Simulation optimization: a concise overview and implementation guide. Theor. Driven Influential Appl. pp. 122–150 (2013)
19. Wang, H., Pasupathy, R., Schmeiser, B.W.: Integer-ordered simulation optimization using R-SPLINE: retrospective search with piecewise-linear interpolation and neighborhood enumeration. ACM Trans. Model. Comput. Simul. (TOMACS) **23**(3), 1–24 (2013)
20. Weizhi, L.: On Solving multi-objective simulation optimization by optimal computing budget allocation and random search. Ph.D. thesis, National University of Singapore (Singapore) (2018)
21. Xu, J., Nelson, B.L., Hong, J.L.: Industrial strength compass: a comprehensive algorithm and software for optimization via simulation. ACM Trans. Model. Comput. Simul. (TOMACS) **20**(1), 1–29 (2010)
22. Zhou, C., Li, H., Lee, B.K., Qiu, Z.: A simulation-based vessel-truck coordination strategy for lighterage terminals. Transp. Res. Part C: Emerg. Technol. **95**, 149–164 (2018)

RegEMO: Sacrificing Pareto-Optimality for Regularity in Multi-objective Problem-Solving

Ritam Guha(✉) and Kalyanmoy Deb

Computational Optimization and Innovation (COIN) Laboratory,
Michigan State University, East Lansing, Michigan 48824, USA
{guharita,kdeb}@msu.edu

Abstract. Multi-objective optimization problems give rise to a set of Pareto-optimal (PO) solutions, each of which makes a certain trade-off among objectives. When multiple PO solutions are to be considered for different scenarios as platform-based solutions, a common structure in them, if available, is highly desired for easier understanding, standardization, and management purposes. In this paper, we propose a modified optimization methodology to avoid converging to theoretical PO solutions having no common structure and converging to a set of near-Pareto solutions having simplistic common principles with regularity where the common principles are extracted from the PO solutions in an automated fashion. After proposing the methodology, we first demonstrate its working principle on a number of constrained and unconstrained multi-objective test problems. Thereafter, we demonstrate the practical significance of the proposed approach to a number of popular engineering design problems. Searching for a set of solutions with regularity-based principles for different platforms is a practically important task. This paper should encourage more similar algorithmic developments in the near future.

Keywords: Regularity · Pareto-optimal solutions · Platform-based designs · Evolutionary multi-objective optimization · Decision making

1 Introduction

The general structure of any optimization problem involves minimizing or maximizing single or multiple objective functions, representing the key performance indicators (KPIs) of the problem, and satisfying a number of constraint functions, imposing certain relationships among variables for solutions to be meaningful. The first task is to mathematically formulate the resulting optimization problem and then apply a suitable optimization algorithm to find the optimal solution(s). Based on the number of objectives, the task can be categorized as a single-objective [1–3], multi-objective (2–3 objectives) [4–7], or, many-objective (>3 objectives) [8–10] optimization. In most real-world multi or many-objective problems, it is not possible to find a single solution that is the best in terms of all specified objectives. So, typically multi- and many-objective optimization algorithms attempt to find a set of Pareto-optimal (PO) solutions.

M. Emmerich et al. (Eds.): EMO 2023, LNCS 13970, pp. 29–42, 2023.
https://doi.org/10.1007/978-3-031-27250-9_3

Classical point-based optimization algorithms use a *generative* solution methodology in which the multi- or many-objective problem is scalarized to a parametric single-objective problem. PO solutions are then generated one by one using different parameter values. However, due to the basic operation with a population of solutions and their implicit parallelism property, evolutionary algorithms are increasingly being used to solve multi- and many-objective optimization problems. It has been argued that since every PO solution must satisfy certain optimality conditions [11,12], collectively they are expected to follow certain common properties involving decision variables, objectives, and constraint values [13], resulting from the satisfaction of the equilibrium optimality condition. The common principles extracted from a PO solution set can provide valuable information to the user, as they exhibit explicit knowledge about the properties of optimal solutions. A procedure of finding such common principles from Pareto-optimal solutions is termed as a task of *innovization* – deciphering innovative solution principles through optimization [13]. While "innovized" principles were observed to exist in many practical problems, not every problem may exhibit such common principles. Even if such principles exist, they can be quite complex for human users to comprehend and make use of.

In this study, we argue that in practical problems, users would be willing to sacrifice optimality in solutions with a certain type of *regularity*, particularly if true PO solutions do not possess any simplistic pattern involving variables, objective, and constraint values. In order not to deviate too much from the true PO set for regularity, we propose a bi-objective optimization task that attempts to find trade-off solutions that are not far from the true PO solutions but possess regularity in terms of common patterns of features within certain specified complexity. Besides providing an easier understanding of trade-off solutions, regularity-based solutions would also facilitate an easier maintenance and switching methodology from one trade-off solution to another in practice.

The rest of the paper is organized as follows: Sect. 2 provides the motivation behind the present study. A brief overview of the literature along a similar direction is provided in Sect. 3. Section 4 describes the proposed methodology in detail. The experimental outcome and corresponding discussion are presented in Sect. 5. Finally, Sect. 6 concludes the paper and provides additional direction for further research on this topic.

2 Motivation for Proposed Study

The goal of a multi-objective optimization process is to find a set of trade-off feasible PO solutions to achieve two main purposes. First, each PO solution is a high-quality candidate solution that in principle can be adopted in practice, and hence they, collectively, provide an idea of alternate solutions pertaining to a problem. Second, the trade-off information of PO solutions can be integrated with users' decision-making priorities to choose a single preferred solution for implementation. In certain scenarios, the knowledge of alternate PO solutions may be used to switch from one PO solution to another, if the circumstances demand. In other scenarios, different PO solutions can be suitable for different computing platforms or environments. Thus, if the final solution set from a multi-objective optimization task possesses certain regularity principles (with simplistic variable-objective-constraint relationships), switching or maintenance of PO solutions under different computing platforms or other scenarios can be easily achieved. Such a solution philosophy is akin to platform-based design principles [14–16] which accelerated standardized design solutions to be adopted during the late nineties due to easier maintenance and re-usability considerations.

Figure 1 illustrates the concept of a *regular-front*, in comparison to a PO front, introduced in this study. PO solutions may not have any regularity or the desired regularity in them, because no regularity requirement is usually enforced as an optimization goal. The figure illustrates that efficient solutions can be widely different from each other and may not possess any easy-to-comprehend common principles. Every PO solution can come from a unique combination of variables without much common pattern from one solution to another. This may require every PO solution to be interpreted differently with its own inventory, maintenance, and operating conditions. If such PO solutions are to be used in a platform-based application scenario in which a solution is needed for different platforms (having different compute powers or differently scaled applications), it is desired that solutions have certain common properties so that an easier inventory, maintenance or similar operating conditions can be adopted. The figure shows that solutions lying on a regular front can have common properties (circular cross-section), but cause a small worsening of performance metrics compared to PO solutions. We argue here that such regularity-based solutions will be more desired in practice than PO solutions, for achieving a better understanding and control of dealing with the solutions. The implementations of the concept of regularity might be different in different problem scenarios, but the high-level idea remains the same. For example, in numerical optimization problems, we may like to have a constant value for certain variables to all regular solutions or have a simplistic relationship, such as $x_1 \leq x_2$ among all regular solutions. In the case of neural architecture search, we may want to have certain common repeating blocks of connections (known as micro-architecure [17]) in all trade-off neural network architectures.

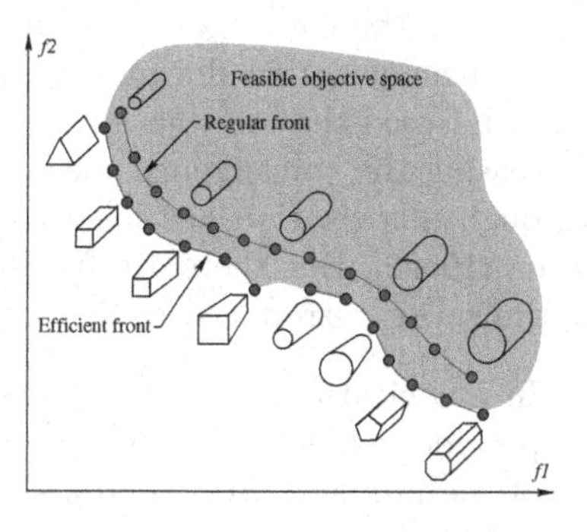

Fig. 1. A regular front contains solutions with common simplistic features but may be dominated by the PO front. It may be worth sacrificing original efficient solutions having no easily-comprehensible features for solutions with some regularity.

3 Past Studies

The concept of regularity in multi-objective optimization is novel and there is not much study yet in this direction. However, the concept is similar to the task of *innovization*, which was introduced by the second author in 2006 [13]. Innovization deals with finding solution properties which are common to PO solutions. After its introduction in [13], it has gained popularity over the years as a process to get useful information about different design problems. Using innovization, the authors were able to extract innovative design principles for three design problems: multiple-disk clutch brake design, spring design, and welded beam design problems. Since its inception, innovization has evolved and found innovative applications. In [18], the authors have utilized innovization as a way of improving the convergence speed through repair operations. The authors have applied innovization during optimization to discover interesting design principles and used the information to guide the search in a better direction thereby increasing the convergence speed. This idea of extracting the design principles through innovization has been used by multiple researchers in the subsequent years [19–22]. But there are fundamental distinctions between regularity-based optimization introduced in this paper and innovization task. For example,

- In innovization, common properties of PO solutions are sought, so they can provide vital knowledge about PO solutions to reach the original Pareto front of the problem. In regularity-based optimization, the goal is to find a set of trade-off solutions with certain simplistic properties of variables, objectives, and constraints. The resulting solutions need not be PO solutions but are expected to be close to the PO set in the objective space.
- Even though innovization attempts to extract important design information from the intermediate/final PO solutions, all the PO solutions may or may not follow the extracted information because it does not enforce all the solutions to follow the pattern. But, in regularity-based optimization, the goal is to find a set of regular solutions that exactly follow the obtained regularity principles.

Platform-based design studies [14–16] are close to the concept of regularity-based optimization, but the former do not usually use any optimization method to arrive at common properties among the platform of solutions.

4 Regularity-Based Optimization (RegEMO) Procedure

As discussed in Sect. 2, the goal of the proposed algorithm is to search for solutions having two properties: (i) they possess some simple regularity principles, and (ii) they are as close as possible to the PO solutions. The most intuitive starting point of the approach is to look for some common principles that are already existing in the majority of the PO solutions.

Let us illustrate the concept through a simple constrained two-variable, two-objective test problem (BNH). The PO solutions obtained by NSGA-II [23] are shown in Fig. 2b in blue points. Figure 2a shows the complexity and Pareto deviation of different regularity principles considered by the RegEMO process. The red stars in Fig. 2b are the regular points corresponding to the preferred regularity principle from Fig. 2a. This process is described in more detail in Sect. 5.1 By analyzing the solutions, we observe that for AB, $x_1 = x_2 \in [0, 3]$ and for

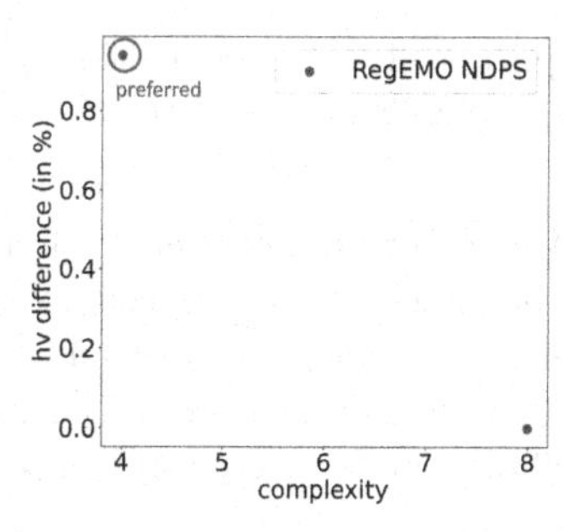

(a) Selection of the final regular front.

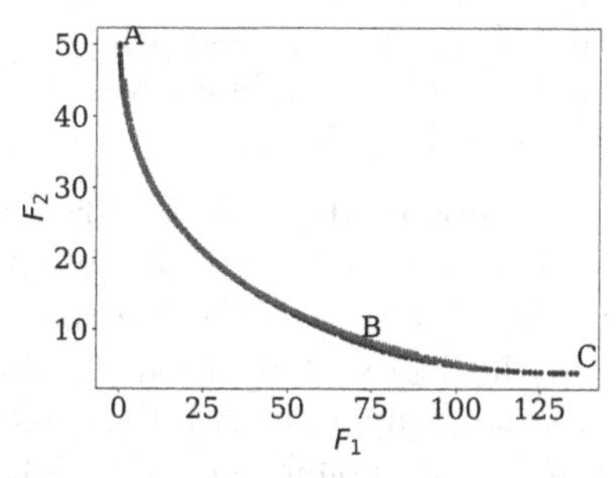

(b) PO and regular efficient fronts.

Fig. 2. Proposed RegEMO procedure is illustrated on BNH problem. (Color figure online)

BC, $x_2 = 3$ with $x_1 \in [3, 5]$. Such information on the properties of PO solutions is useful to practitioners and the concept of discovering such knowledge of PO solutions was termed the task of innovization [13]. If the above information is comprehensible to the users so that they can be used for inventory, management, or operation of the problem, there is no need for any further study and we shall call these PO solutions as regular solutions.

However, if the division of properties is somehow complex to comprehend or use, the user may be interested in finding a new set of solutions, not far from the PO solutions, but possessing a more simplistic relationship, maybe within the maximum desired complexity provided by the user. For example, our proposed regularity-based EMO (RegEMO) has found a new set of trade-off solutions (shown in red stars) that is close to the original PO set but has the following a single simplistic linear property for the entire set:

$$x_2 = 0.6x_1 + 0.49, \quad x_1 \in [0, 4.18]. \tag{1}$$

The above principle sacrifices the extreme parts of the PO front and makes a slight deviation from the original PO front in the lower left part of the PO front, but provides a simple linear relationship for users to have a better comprehension and use of the knowledge.

4.1 Steps of Proposed RegEMO Procedure

The RegEMO procedure consists of six steps, as described below.

Step 1: Discovery and Clustering of the Pareto Front: The first step of the regularity search process is to find a set of PO solutions using an evolutionary multi-objective optimization (EMO) or an evolutionary many-objective optimization (EMaO) procedure. Thereafter, a clustering operation is applied to cluster the Pareto front based on the design space representations of the PO solutions. In this study, we have pre-specified the number of clusters (n_c) to be found using the k-means [24] clustering approach, but later it can be replaced by other clustering processes which do not require such pre-specification. If the user is interested in finding a set of regularity principles common to the entire PO set, $n_c = 1$ can be set.

Step 2: Identification of Non-fixed Variables: For each cluster of solutions (say C_j), we try to identify variables that are not fixed to some specific values, rather they vary across the PO solution subset. One simple idea would be to measure the degree of variation of a variable (say i-th) in the subset ($x_i^{(k)}$, where $k \in C_j$) and compare it with the original specified search space using variable's lower and upper bounds (x_i^L and x_i^U):

$$\Delta_i = \frac{\max_{k \in C_j} x_i^{(k)} - \min_{k \in C_j} x_i^{(k)}}{x_i^U - x_i^L}. \tag{2}$$

By checking if Δ_i is within a pre-specified threshold (ζ), the variable can be declared as a fixed variable. However, there is a problem with this approach. A variable may converge to two or a few widely different values on the search space, producing a large value of the numerator of the above equation. Although the variable has settled to a few values, the above metric will not declare it as a fixed variable. To alleviate this, we propose a binning procedure. We divide the range ($x_i^U - x_i^L$) in a certain number of bins (n_{bins}). If a variable has representations in equal or more than 50% bins, we declare it a non-fixed variable, else it becomes a fixed variable.

Step 3: Regularity Search in Fixed Variables: The variables which are not identified as non-fixed variables ($\mathbf{x} \in \bar{\mathcal{F}}$) are termed as fixed variables ($\mathbf{x} \in \mathcal{F}$). Next, we attempt to look for any regularity (piece-wise or complete) among the fixed variables. The process starts by computing the average of fixed variable values ($x_{i,\text{avg}}$) in the population and arranging them in a non-decreasing order: $[\sim, S] = \texttt{ascend\_sort}(x_{i,\text{avg}})$. The set S contains the variable ID of the fixed variables in ascending order of the average variable value. Thereafter, we fit a regression function $r(s)$ (polynomial of degree η) through the average variable values as a function of sorted variable ID (s) representing the fixed variable x_{S_s}. If the regression fit does not produce a small error, we divide the sorted variables into smaller pieces and find a piece-wise regression fit within the desired maximum error (ϵ_f).

Step 4: Regularity Search in Non-Fixed Variables: Non-fixed variables do not have a convergence to any fixed value(s), hence finding regularity in non-fixed variables is more challenging. However, despite the variations in non-fixed variables, they can be related to each other in a specific way and follow certain simplistic relationships. Next, we attempt to decipher any such relationships among non-fixed variables. Several procedures are possible, but in our current implementation, we divide the non-fixed variables into three categories: (i) non-fixed dependent, (ii) non-fixed independent with which non-fixed dependent variables have a relationship, and (iii) orphan non-fixed variables with no apparent relationship with other non-fixed variables.

The following tasks are performed on non-fixed variables only. The first step is to identify these three different categories of non-fixed variables. For the identification, we have used Pearson correlation coefficient (PCC). Each non-fixed variable is assigned a correlation score obtained by summing up the PCC scores of that variable with every other variable. The variables having higher correlation scores are candidates for becoming non-fixed dependent variables as they are more related to the rest of the variables. So, we select the top K non-fixed variables having higher correlation scores to become non-fixed dependent variables. Each non-fixed dependent variable is then represented as a linear combination of the remaining non-fixed variables where the coefficients (multiples of κ) of the linear combination denote coefficients of linear regressor fitted for the non-fixed dependent variables with respect to the other non-fixed variables. The non-fixed variables having non-zero coefficients are termed as non-fixed independent variables and the ones having zero coefficients are termed as non-fixed orphan variables. A relationship is validated with an MSE bound ϵ_{nf}.

Note that the besides classifying the variables into four categories (fixed and non-fixed variables together), the above process also assigns values for the fixed variables and relationships among certain non-fixed variables.

Step 5: New Optimization Problem Formulation to Find Candidate Regular Solution Set: The above regularity relationships, although obtained from PO solutions, are on one hand simplistic (constant or linearly dependent on each other), but appear in an approximate manner with tolerances specified above. Since they capture a simple and approximate relationship (justifying regularity), users may be interested in knowing what *new* trade-off solutions would be most appropriate to satisfy the relationships so that they are not far from the actual PO front.

For this purpose, we formulate a new optimization problem by enforcing the obtained relationships. To determine the variables of the new optimization problem, first, all fixed variables are set to their observed fixed values and are not considered as variables for the new optimization problem. Second, non-fixed dependent variables are set by the obtained relationships (as constraints) as functions of non-fixed independent variables and are also not considered as variables of the new optimization problem. The non-fixed independent variables and orphan variables are chosen as variables of the new optimization problem and their variable bounds are adjusted to the lower and upper bounds of their variations in the PO set.

The objective function of the new optimization problem is identical to the original problem. Constraints of the original problems are also included. We employ an EMO/EMaO algorithm to again solve the new optimization problem. It is expected that the obtained regular solution set will be inferior to the original PO set, and therefore, their acceptability of them must be traded based on the gain in simplicity in obtained regularity principles. We execute the following final step for this purpose.

Step 6: Bi-objective Parametric Search and Choice of the Best Regular Solution Set: The above process of arriving at regularity principles involves a

Table 1. RegEMO parameters and allowable values for the bi-objective search.

RegEMO parameters	Description	Search space
Fixed var. regression degree (η)	Highest degree for polynomial fitting regressor	1, 2, 3
Non-fixed independent var. factor (κ)	Coefficients in multiples of	0.1, 0.3, 0.5
Non-fixed dependent equations (K)	Max. number of non-fixed dependent vars. allowed	1, 2
Threshold for Δ_i (ζ)	Threshold used for deciding if a variable is fixed	0.2, 0.5
Fixed var. MSE bound (ϵ_f)	Upper bound on MSE for regularity requirement	0.1, 0.3, 0.5
Non-fixed var. MSE bound (ϵ_{nf})	Upper bound on MSE for regularity requirement	0.1, 0.3, 0.5
Number of clusters in the PO set (n_c)	Number of clusters for dividing the PO set	1, 2, 3

number of parameters mentioned in Table 1. A change in any of these parameters ($\mathbf{p}$-vector) will produce a different set of regularity principles ($\mathbf{R}(\mathbf{p})$) having a different complexity ($\mathbf{C}(\mathbf{p})$) estimate and the new optimization will produce a different regular solution set ($\mathbf{Y}(\mathbf{p})$) with a different deviation ($\mathbf{d}(\mathbf{p})$) estimate from the original PO set. The first task will be to employ an unconstrained bi-objective search in which parameters ($\mathbf{p}$) are variables and two objectives ($\mathbf{d}$ and $\mathbf{C}$) are minimized. This will ensure that the final regular solution set $\mathbf{Y}^*$ is minimally away from the original PO set under the added constraints and also have a minimal complexity estimate.

We now define metrics for two objectives. The deviation from the PO set is simply defined as the percentage difference in hypervolume (HV) metrics of the original PO set and the obtained $\mathbf{Y}$-set, from the original PO set hypervolume: $\mathbf{d} = 1-\text{HV}(\mathbf{Y})/\text{HV}(\text{PO})$. However, the complexity metric objective is computed from the structure of fixed and non-fixed variable relationships. For an n-variable vector having n_f fixed, n_{ni} non-fixed independent, n_{nd} non-fixed dependent variables, we assign the following complexity metric value for each variable type: (i) fixed var.: $c_1 = 0.5$, (ii) non-fixed indep. var.: $c_2 = 6n - 11$, (iii) non-fixed dep. var.: $c_3 = 3\,n_{ni}$, and (iv) orphan var.: $c_4 = c_2(n-2)+4$. The above assignments are chosen by comparing different pairwise scenarios of relationships and enforcing an intuitive preference to the more desired choice for each scenario. The complexity of the regular solution set is then computed as follows:

$$\mathbf{C} = c_1 n_f + c_2 n_{ni} + c_3 n_{nd} + c_4(n - n_f - n_{ni} - n_{nd}). \tag{3}$$

Since the total number of parametric combinations is small (972), we execute an exhaustive bi-objective search and identify the non-dominated parametric solution (NDPS) set. Then, a pre-specified decision-making approach is used to choose the preferred NDPS. We select the knee point (having largest trade-off [25] in the neighborhood) of the NDPS set if it has an HV deviation $\mathbf{d}$ less than or equal to 2%, else we use the principle that leads to the least HV deviation. Every step of the proposed six-step procedure is illustrated in Fig. 4.

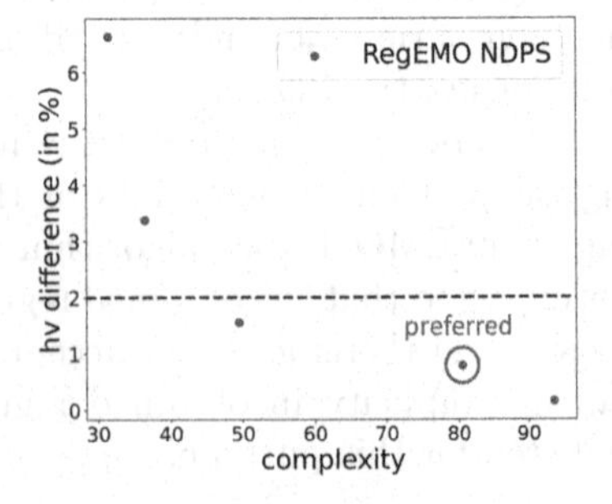

Fig. 3. Bi-objective front for decision-making on the welded-beam problem.

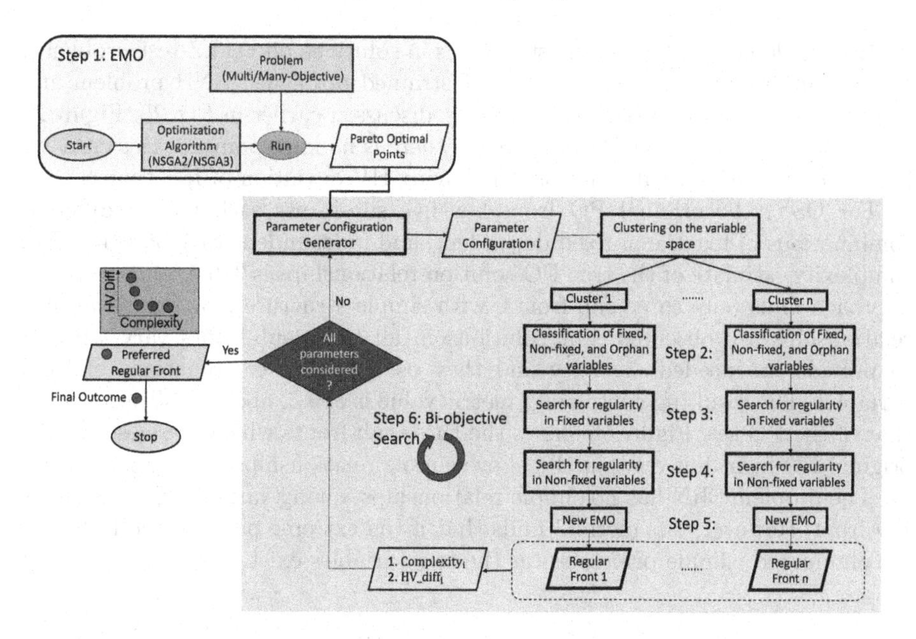

Fig. 4. Proposed steps for finding the regular front are depicted.

5 Results and Discussion

To visualize the effectiveness of the proposed approach, we apply it to a number of constrained and unconstrained test problems as well as a few engineering design problems. The regularity search uses NSGA-II for two-objective problems and NSGA-III for three-objective problems. To keep the computational cost on the same level, we have used 40, 000 function evaluations for all problems.

5.1 Test Problems

The final regularity principles obtained from the proposed six-step procedure are presented in Table 2. Three-objective DTLZ2, DTLZ5 and DTLZ7 problems have 10 variables each with certain known structures of PO solutions. Hence, they are ideal problems to test our procedure. The first two problems are constructed with x_3 to x_{10} taking a value of 0.5, while the first two variables change within their lower and upper bounds $[0, 1]$ uniformly to provide diversity in solutions. Table 2(a) shows that the best-selected solution $\mathbf{Y}$ by our proposed procedure has identified the above facts. Figure 5a shows that the original PO set and our finally selected regular set are almost identical, meaning that the original PO solutions already possess simple relationships and there is no need to find any further approximate solutions close to the PO set. In the case of DTLZ5, the first two objectives are correlated, hence only one variable causes the diversity in the entire PO front, while other variables including x_2 get fixed to 0.5. Here too, the obtained regular solution set is close to the original PO set. For DTLZ7 problem, variables x_3 to x_{10} are fixed to zero and there are two orphan variables that produce the entire NDPS set.

Having demonstrated simple structures in solutions on DTLZ test problems, now we apply our procedure to four constrained problems. BNH problem and its results from our approach were already discussed earlier in Fig. 2b. Figure 2a shows the bi-objective NDPS front. Since there is no knee point, in this case, we have selected the top-left point as it is having HV deviation of less than 2%.

For OSY, the original PO front has five sub-fronts each having different combinations of fixed, non-fixed dependent, and independent variables [12]. The complexity estimate of the true PO solution relationships is 379.5. Our proposed approach finds only three sub-fronts, with simple structures. No multi-variable relationships are observed, but solutions in all three sub-fronts vary with x_3 to provide the needed diversity and they use different combinations of fixed variables. The resulting complexity metric value is 319.5, about 16% better than that of the PO set. Figure 5b shows the three sub-fronts which are closer to the original PO front but contain the above simple relationships.

The problem SRN has no simple relationships among variables for the entire PO front. However, our method finds that if the extreme parts of the PO front are eliminated, simple principles for the two variables exist.

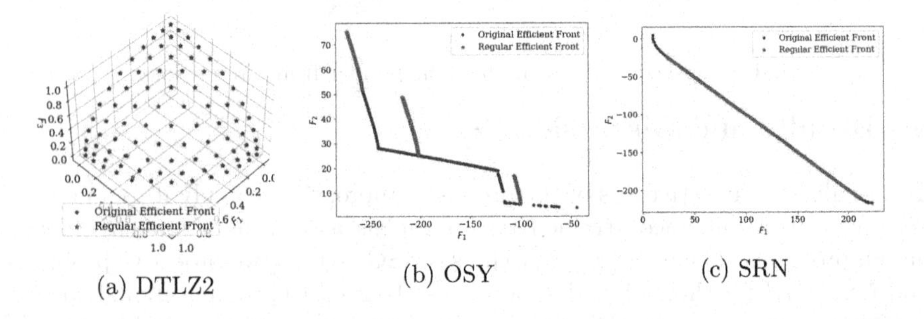

Fig. 5. Original and regular efficient sets for constrained and unconstrained test problems show minimal deviation and regularity principles depicted in Table 2.

5.2 Engineering Problems

Finally, we apply RegEMO procedure to three engineering problems. The final sets of regular solutions are presented in Fig. 6, while the regularity principles embedded in the solutions are shown in Table 3.

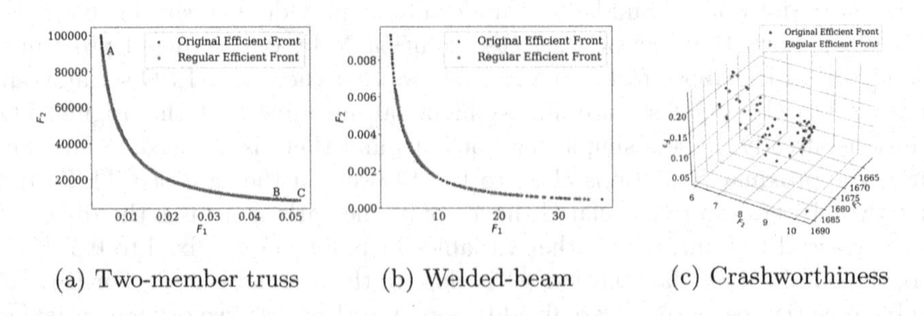

Fig. 6. Original and regular efficient sets for three engineering problems show minimal deviation with regularity principles presented in Table 3.

Table 2. Regularity principles extracted by RegEMO for constrained and unconstrained test problems.

Problem	Regularity principles
(a) DTLZ2	Fixed variables: $x_3 = 0.50$, $x_4 = 0.50$, $x_5 = 0.51$, $x_6 = 0.50$, $x_7 = 0.50$, $x_8 = 0.50$, $x_9 = 0.50$, $x_{10} = 0.50$ Orphan variables: $x_1 \in [0.00, 1.00]$, $x_2 \in [0.00, 1.00]$ Non-fixed independent variables: None Non-fixed dependent variables: None
(b) DTLZ5	Fixed variables: $x_2 : 0.47$, $x_3 : 0.50$, $x_4 : 0.50$, $x_5 : 0.50$, $x_6 : 0.51$, $x_7 : 0.50$, $x_8 : 0.50$, $x_9 : 0.50$, $x_{10} : 0.49$ Orphan variables: $x_1 \in [0.00, 1.00]$ Non-fixed independent variables: None Non-fixed dependent variables: None
(c) DTLZ7	Fixed variables: $x_3 = 0.00$, $x_4 = 0.00$, $x_5 = 0.00$, $x_6 = 0.00$ $x_7 = 0.00$, $x_8 = 0.00$, $x_9 = 0.00$, $x_{10} = 0.00$ Orphan variables: $x_1 \in [0.00, 0.88]$, $x_2 \in [0.00, 0.88]$ Non-fixed independent variables: None Non-fixed dependent variables: None
(d) BNH	Fixed variables: None Orphan variables: None Non-fixed independent variables: $x_1 \in [0.03, 4.18]$ Non-fixed dependent variables: $x_2 = (0.60 \times x_1) + 0.49$
(f) OSY	Regular Front 1 Fixed variables: Piece 1: $x_2 = 0.92$, $x_4 = 0.00$, $x_6 = 0.11$ Piece 2: $x_1 = 4.72$, $x_5 = 1.00$ Orphan variables: $x_3 \in [1.00, 4.96]$ Non-fixed independent variables: None Non-fixed dependent variables: None Regular Front 2 Fixed variables: $x_1 = 5.00$, $x_2 = 1.00$, $x_4 = 0.00$, $x_5 = 4.99$, $x_6 = 0.08$ Orphan variables: $x_3 \in [1.01, 4.92]$ Non-fixed independent variables: None Non-fixed dependent variables: None Regular Front 3 constant variables: Piece 1: $x_1 = 0.16$, $x_4 = 0.00$ Piece 2: $x_2 = 1.84$, $x_5 = 1.01$, $x_6 = 0.10$ Orphan variables: $x_3 \in [1.0, 3.54]$ Non-fixed independent variables: None Non-fixed dependent variables: None
(g) SRN	Fixed variables: $x_1 = -2.36$ Orphan variables: $x_2 \in [2.55, 14.81]$ Non-fixed independent variables: None Non-fixed dependent variables: None

Table 3. Regularity principles extracted by RegEMO for engineering problems.

(a) Two-Member Truss	(b) Welded Beam Design	(c) Crashworthiness
Regular Front 1 Fixed variables: None Orphan variables: None Non-fixed independent variables: $x_2 \in [0.001, 0.01]$ Non-fixed dependent variables: $x_1 = (0.5 \times x_2) + 0.0$ $x_3 = (9.9 \times x_2) + 1.99$	Regular Front 1 Fixed variables: $x_2 = 0.40$, $x_3 = 9.98$ Orphan variables: $x_1 \in [0.98, 1.10]$, $x_4 \in [1.76, 5.00]$ Non-fixed independent variables: None Non-fixed dependent variables: None	Regular Front 1 Fixed variables: $x_1 = 1.07$, $x_2 = 2.99$ Orphan variables: $x_3 \in [1.00, 3.00]$, $x_4 \in [1.00, 1.08]$, $x_5 \in [1.01, 2.97]$ Non-fixed independent variables: None Non-fixed dependent variables: None
Regular Front 2 Fixed variables: $x_1 = 0.004$, $x_2 = 0.01$ Orphan variables: $x_3 \in [2.57, 3.0]$ Non-fixed independent variables: None Non-fixed dependent variables: None	Regular Front 2 Fixed variables: $x_3 = 9.99$ Orphan variables: None Non-fixed independent variables: $x_4 \in [0.35, 1.33]$ Non-fixed dependent variables: $x_1 = (0.70 \times x_4) + 0.05$ $x_2 = (-1.20 \times x_4) + 1.99$	Regular Front 2 Fixed variables: $x_1 = 1.00$, $x_3 = 1.00$, $x_4 = 1.00$ Orphan variables: $x_2 \in [1.00, 1.52]$, $x_5 \in [1.03, 2.68]$ Non-fixed independent variables: None Non-fixed dependent variables: None

For the truss problem, the PO front has two parts with a transition point at $f_1 = 0.045$. For $f_1 \leq 0.045$, $x_3 = 2$ and $x_1 = 0.5x_2$ are two principles for the PO solutions reported in [13]. Our procedure is able to discover the same. While x_3 seems to be related to x_2, but in the original range, $x_3 \in [1, 3]$, the variation of x_3 in the regular set is [1.99, 2.09], which is almost acceptable as a good convergence near $x_3 = 2$. This front is represented by region A-B in Fig. 6a. On the right side of the transition point (marked by region B-C), RegEMO discovers that x_2 is fixed at 0.01 which is its upper bound. x_1 also gets fixed to 0.004, but x_3 becomes an orphan variable in the range [2.57, 3.00]. These findings are simple and similar to the ones observed before by analyzing PO solutions. For these reasons, the original PO front and our regular front are quite close. For the welded-beam design problem, the selected RegEMO solution comes from the knee point (Fig. 3). The entire PO set has $x_3 = 10$, which is found by our procedure. As shown in [13], x_2 stays constant at 0.40 on a part of the PO front but increases with f_2 at another part with simplistic regularity principles. For the crashworthiness problem, no innovization study has been made before to discover any apparent principles among variables. Clearly, the PO set has two distinct sub-fronts and our procedure discovers two simple relationships among variables. Although some parts of the original PO set are not covered by these simplistic principles, Fig. 6c shows that NDPS points cover major parts of both sub-fronts.

6 Conclusions

In this paper, we have questioned the practical validity of Pareto-optimal (PO) solutions which have no simplistic common properties among their variable values. We then argued that a trade-off solution set near the original PO set, possessing certain simple relationships of variables, may be desired in lieu of PO solutions. We have called them *regular* solutions and proposed here a six-step procedure to identify them. The proposed procedure starts with the PO solutions obtained by an EMO or EMaO algorithm and analyzes them to classify all variables into four types depending on their constancy and dependencies on each other. Thereafter, a parametric new optimization problem is formulated with a reduced variable space (restricted by the variable relationships obtained by the analysis steps). Finally, a bi-objective analysis of each parameter combination's effect on resulting non-dominated regular solution sets is performed with two conflicting criteria: minimizing hypervolume difference between the true PO front and the resulting regular front and minimizing the complexity of the resulting variable relationships. A preferred regular solution set is then chosen based on a trade-off between the two criteria.

The working of the proposed procedure has been demonstrated on a few test problems for which the original PO set was designed to have simplistic relationships among variables. Thereafter, the procedure has been applied to a number of constrained problems and engineering design problems to discover near PO solutions but possessing simplistic variable relationships.

The concept is practical and the proposed procedure is novel, but there can be other ways to achieve the same, which can be pursued next. The proposed procedure, being algorithmic, keeps the final choice of a preferred regular front to the decision-maker to make the approach further appealing to practitioners. We believe that sacrificing Pareto-optimality for regularity in solutions demonstrated in this paper should encourage more such studies in the coming years.

References

1. Singh, H.K., Isaacs, A., Nguyen, T.T., Ray, T., Yao, X.: Performance of infeasibility driven evolutionary algorithm (IDEA) on constrained dynamic single objective optimization problems. In: 2009 IEEE Congress on Evolutionary Computation, pp. 3127–3134. IEEE (2009)
2. Elsayed, S., Hamza, N., Sarker, R.: Testing united multi-operator evolutionary algorithms-II on single objective optimization problems. In: 2016 IEEE Congress on Evolutionary Computation (CEC), pp. 2966–2973. IEEE (2016)
3. Gwiazda, T.D.: Crossover for Single-Objective Numerical Optimization Problems, vol 1. Tomasz Gwiazda (2006)
4. Deb, K.: Multi-objective optimization. In: Burke, E., Kendall, G. (eds.) Search Methodologies, pp. 403–449. Springer, Boston (2014). https://doi.org/10.1007/978-1-4614-6940-7_15
5. Taboada, H.A., Espiritu, J.F., Coit, D.W.: MOMS-GA: a multi-objective multi-state genetic algorithm for system reliability optimization design problems. IEEE Trans. Reliab. **57**(1), 182–191 (2008)
6. Obayashi, S., Jeong, S., Chiba, K.: Multi-objective design exploration for aerodynamic configurations. In: 35th AIAA Fluid Dynamics Conference and Exhibit, p. 4666 (2005)
7. Yildiz, A.R., Solanki, K.N.: Multi-objective optimization of vehicle crashworthiness using a new particle swarm based approach. Int. J. Adv. Manufact. Technol. **59**(1), 367–376 (2012). https://doi.org/10.1007/s00170-011-3496-y
8. Zou, X., Chen, Y., Liu, M., Kang, L.: A new evolutionary algorithm for solving many-objective optimization problems. IEEE Trans. Syst. Man Cybern. Part B (Cybernetics) **38**(5), 1402–1412 (2008)
9. Mane, S., Narasinga, M.R.: Many-objective optimization: problems and evolutionary algorithms-a short review. Int. J. Appl. Eng. Res. **12**(20), 9774–9793 (2017)
10. Cui, Z., et al.: A pigeon-inspired optimization algorithm for many-objective optimization problems. Sci. China Inf. Sci. **62**(7), 1–3 (2019). https://doi.org/10.1007/s11432-018-9729-5
11. Miettinen, K.: Nonlinear Multiobjective Optimization (1999)
12. Deb, K.: Multi-Objective Optimization Using Evolutionary Algorithms. John Wiley & Sons Inc., USA (2001)
13. Deb, K., Srinivasan, A.: Innovization: innovating design principles through optimization. In: Proceedings of the 8th Annual Conference on Genetic and Evolutionary Computation, pp. 1629–1636 (2006)
14. Keutzer, K., Newton, A.R., Rabaey, J.M., Sangiovanni-Vincentelli, A.: System-level design: orthogonalization of concerns and platform-based design. IEEE Trans. Comput. Aided Des. Integr. Circ. Syst. **19**(12), 1523–1543 (2000)
15. Sangiovanni-Vincentelli, A.: Defining platform-based design. EEDesign (2002)

16. Sangiovanni-Vincentelli, A., Carloni, L., De Bernardinis, F., Sgroi, M.: Benefits and challenges for platform-based design. In: Proceedings of the 41st Annual Design Automation Conference, pp. 409–414 (2004)
17. Elsken, T., Metzen, J.H., Hutter, F.: Neural architecture search: a survey. J. Mach. Learn. Res. **20**(1), 1997–2017 (2019)
18. Gaur, A., Deb, K.: Effect of size and order of variables in rules for multi-objective repair-based innovization procedure. In: 2017 IEEE Congress on Evolutionary Computation (CEC), pp. 2177–2184. IEEE (2017)
19. Ghosh, A., Goodman, E., Deb, K., Averill, R., Diaz, A.: A large-scale bi-objective optimization of solid rocket motors using innovization. In: 2020 IEEE Congress on Evolutionary Computation (CEC), pp. 1–8. IEEE (2020)
20. Bandaru, S., Aslam, T., Ng, A.H., Deb, K.: Generalized higher-level automated innovization with application to inventory management. Eur. J. Oper. Res. **243**(2), 480–496 (2015)
21. Mkaouer, M.W., Kessentini, M., Bechikh, S., Deb, K., Ó Cinné, M.: Recommendation system for software refactoring using innovization and interactive dynamic optimization. In: Proceedings of the 29th ACM/IEEE International Conference on Automated Software Engineering, pp. 331–336 (2014)
22. Mittal, S., Saxena, D.K., Deb, K.: Learning-based multi-objective optimization through ANN-assisted online innovization. In: Proceedings of the 2020 Genetic and Evolutionary Computation Conference Companion, pp. 171–172 (2020)
23. Deb, K., Pratap, A., Agarwal, S., Meyarivan, T.A.M.T.: A fast and elitist multi-objective genetic algorithm: NSGA-II. IEEE Trans. Evol. Comput. **6**(2), 182–197 (2002)
24. Pham, D.T., Dimov, S.S., Nguyen, C.D.: Selection of k in k-means clustering. In: Proceedings of the Institution of Mechanical Engineers, Part C: Journal of Mechanical Engineering Science, vol. 219, no. 1, pp. 103–119 (2005)
25. Mittal, S., Kumar, D., Deb, S.K.: A unified automated innovization framework using threshold-based clustering. In: 2020 IEEE Congress on Evolutionary Computation (CEC), pp. 1–8. IEEE (2020)

Cooperative Coevolutionary NSGA-II with Linkage Measurement Minimization for Large-Scale Multi-objective Optimization

Rui Zhong[1(✉)] and Masaharu Munetomo[2]

[1] Graduate School of Information Science and Technology, Hokkaido University, Sapporo 060-0814, Japan
rui.zhong.u5@elms.hokudai.ac.jp

[2] Information Initiative Center, Hokkaido University, Sapporo 060-811, Japan
munetomo@iic.hokudai.ac.jp

Abstract. In this paper, we propose a novel decomposition method based on cooperative coevolution (CC) to deal with large-scale multi-objective optimization problems (LSMOPs) named Linkage Measurement Minimization (LMM), and after decomposition, NSGA-II is employed to optimize the subcomponents separately. CC is a mature and efficient framework for solving large-scale optimization problems (LSOPs), which decomposes LSOPs into multiple nonseparable subcomponents and solves them alternately based on a divide-and-conquer strategy. The essence of the successful implementation of the CC framework is the design of decomposition methods. However, in LSMOPs, variables in different objective functions may have different interactions, and the design of a proper decomposition method for LSMOPs is more difficult than for single objective optimization problems. Our proposed LMM can identify the relatively strong interactions and search the better decomposition iteratively. We evaluate our proposal on 21 benchmark functions of 500-D and 1000-D, and numerical experiments show that our proposal is quite competitive with the current popular decomposition methods.

Keywords: Cooperative coevolution (CC) · Linkage Measurement Minimization (LMM) · large-scale multi-objective optimization problems (LSMOPs) · NSGA-II

1 Introduction

The performance of canonical multi-objective evolutionary algorithms (MOEAs) degenerates rapidly when solving large-scale optimization problems (LSOPs). This is mainly due to the presence of the curse of dimensionality, which means that the search space of optimization problems increases exponentially as the number of variables increases. Thus, solving LSOPs faces a huge challenge. Many

M. Emmerich et al. (Eds.): EMO 2023, LNCS 13970, pp. 43–55, 2023.
https://doi.org/10.1007/978-3-031-27250-9_4

research efforts in computer science ranging from computational linear algebra [11] and machine learning [4] to numerical optimization [7] have been published to alleviate the curse of dimensionality.

Cooperative coevolution (CC) [14] is a flexible and efficient framework to deal with LSOPs. CC decomposes the LSOPs into several nonseparable subcomponents and EAs are applied to solve the subcomponents alternately. Since the solution found in a sub-problem cannot form a complete solution for evaluation, representative solutions of the other subcomponents are required, which compose the context vector [2]. The context vector is updated iteratively and acts as the context in which cooperation occurs.

Many published research state that the CC framework is sensitive to decomposition [12,13]. Thus, the key to the successful implementation of the CC framework is the design of the decomposition method. In this paper, we design a novel decomposition method named Linkage Measurement Minimization (LMM). We regard the decomposition problem as a combinatorial optimization problem and design an objective function based on Linkage Identification by the Nonlinearity Check on Real-Coded GA (LINC-R) to lead the direction of optimization.

The remainder of this paper is organized as follows, Sect. 2 covers preliminaries and a brief review of decomposition methods in LSMOPs. Section 3 introduces our proposal CC-NSGA-LMM in detail. Section 4 shows the experiments and analysis. Section 5 discusses future research. Finally, Sect. 6 concludes this paper.

2 Preliminaries and Related Works

We first introduce the preliminaries in this section, including the definition of LSMOPs, the separability of variables, and NSGA-II. Then, a brief review of decomposition methods in LSMOPs is involved.

2.1 Preliminaries

Large-Scale Multi-objective Problems. Without loss of generality, a Multi-objective Problem (MOP) can be mathematically defined as Eq. (1):

$$\begin{aligned} \min\ F(\mathrm{x}) = (f_1(\mathrm{x}), f_2(\mathrm{x}), ..., f_M(\mathrm{x})), \\ s.t.\ \mathrm{x} \in \Omega, \end{aligned} \tag{1}$$

where $f : \Omega \rightarrow \Lambda \subseteq \mathbb{R}^M$ consists of M objectives, Λ is the objective space, $\Omega \subseteq \mathbb{R}^D$ is the decision space, and $\mathrm{x} = (x_1, x_2, ..., x_D) \in \Omega$ is a solution consisting of D decision variables. The dominance relation between two solutions can be defined as $(\forall i \in 1, ..., M, f_i(\mathrm{x}) \leqslant f_i(\mathrm{y}) \wedge (\exists j \in 1, ..., M, f_i(\mathrm{x}) \leq f_i(\mathrm{y}))$. If this formula is satisfied, we can say that x dominates y. A Pareto optimal solution is a solution that is not dominated by any solution in Ω. In LSMOPs, a trial solution x) contains a large number of decision variables (e.g., $D \geq 1000$).

Separability of Variables. Separability of variables refers to the fact that the setting of one variable has an impact on the fitness landscapes of other variables. Given two variables x_i and x_j, if $\frac{\partial^2 f}{\partial x_i \partial x_j} = 0$, then x_i and x_j are separable. A partially separable function can be formulated as:

$$f(\mathrm{X}) = \sum_{i=1}^{m} f_i(\mathrm{X}_i) \tag{2}$$

m is the number of subcomponents. Notice that interaction is not rigorous equivalent the nonseparability. A simple example is $f(x) = (x_i + x_j)^2$, $x_i, x_j \in [0, 10]$. Although x_i and x_j interact, x_i and x_j are still separable in limited search space considering monotonicity. Thus, the ultimate goal is to identify the separability of all variables and to develop a decomposition method to assign the variables into subcomponents properly.

NSGA-II. Non-dominated Sorting Genetic Algorithm (NSGA-II) is proposed in paper [6]. With three special characteristics, the fast non-dominated sorting approach, the fast crowded distance estimation procedure, and the simple crowded comparison operator, NSGA-II becomes one of the most popular MOEAs. More details can ref to [6].

2.2 Decomposition Methods in LSMOPs

CC is first applied to deal with large-scale single-objective problems (LSSOPs) and extended to LSMOPs from 2013 [1]. However, LSMOPs often contain multiple conflicting objective functions, and the interactions between variables in different objective functions may be diverse, and both the convergence and diversity of the population should be considered when optimizing each group of decision variables. Hence, the implementation of a divide-and-conquer strategy on LSMOPs is much more difficult. Up to now, there are mainly three categories of decomposition techniques: random grouping, differential grouping, and variable analysis.

Random Grouping in LSMOPs. CCGDE3 [1] first adopts the random grouping and assigns the variables into several groups with equal sizes. Although the first work on LSMOPs is relatively naive and the random grouping seems not reliable, it obtained satisfactory performance on some LSMOPs with up to 5,000 decision variables in comparison to conventional MOEAs, and the mathematical analysis in DECC-G [17] proves that it is quite efficient for random grouping to capture the interactions without information about fitness landscape. Besides, many random grouping strategies designed for LSSOPs such as MLCC [18] are also introduced to LSMOPs and form a new algorithm MOEA/D-RDG [16].

Differential Grouping in LSMOPs. Although the random grouping technique has the advantage of environmental insensitiveness and easy implementation, it does not consider the interactions between variables at all, thus, the direction of optimization may be misled. To address this problem, the differential grouping is extended from LSSOPs to LSMOPs. When identifying the interaction between x_i and x_j, Eq. (3) defines the mechanism of differential grouping in LSSOPs.

$$\begin{aligned} &if\ |(f(s_{ij}) - f(s_i)) - (f(s_j) - f(s))| < \epsilon \\ &then\ x_i\ and\ x_j\ are\ separable \end{aligned} \tag{3}$$

s_i, s_j, and s_{ij} are perturbed on s with δ in respective dimension(s), ϵ is a allowable error. TS [15] applies the Eq. (3) to all objective functions, and separability between x_i and x_j is identified only Eq. (3) is satisfied in all objective functions.

Variable Analysis in LSMOPs. Both random grouping and differential grouping were originally proposed for LSSOPs, which concentrate on assigning the decision variables into subcomponents but ignore the population diversity in the objective space. MOEA/DVA [10] perturbs the random samples, if all of the perturbed solutions are nondominated with each other, the decision variable is regarded as a position variable, whereas if each perturbed solution is dominated by or dominates all of the others, the decision variable is regarded as a distance variable; otherwise, it is regarded as a mixed variable. The position variables influence the population diversity but do not change population convergence. Hence they need only to be slightly adjusted for maintaining the population diversity. On the contrary, the distance variables influence population convergence but do not impact population diversity, which deserves lots of computational costs to be deeply optimized for the best convergence. Numerical experiments show that MOEA/DVA significantly outperforms many other MOEAs on LSMOPs in benchmarks.

3 CC-NSGA-LMM

In this section, we will introduce the details of our proposal and the techniques. Here in Fig. 1, we demonstrate the flowchart of the main steps. There are two stages in CC-NSGA-LMM. In the decomposition stage, we regard the decomposition problem as a combinatorial optimization problem, and the objective function LMM is designed based on LINC-R. Elitist GA is applied to optimize this objective function. In the optimization stage, the subcomponents are optimized by NSGA-II.

Next, we will give a simple mathematical explanation of our designed linkage measurement function (LMF).

The original LINC-R is defined in Eq. (4):

$$\begin{aligned} &\exists s \in Pop: \\ &if\ |(f(s_{ij}) - f(s_i)) - (f(s_j) - f(s))| > \epsilon \\ &then\ x_i\ and\ x_j\ are\ nonseparable \end{aligned} \tag{4}$$

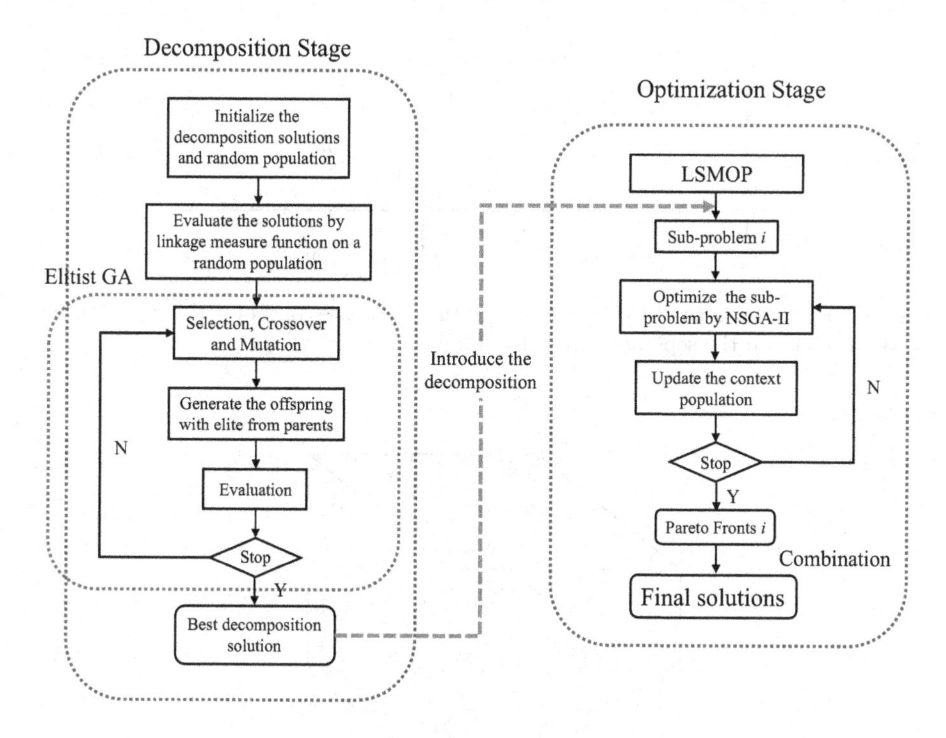

Fig. 1. The flowchart of CC-NSGA-LMM

And we notice that the original LINC-R can be understood as the additive form of vector. Thus, Eq. (4) can be written to Eq. (5)

$$\begin{aligned} &\exists s \in Pop: \\ if\ &|(f(s_{ij}) - f(s)) - ((f(s_i) - f(s)) + (f(s_j) - f(s)))| > \epsilon \\ &then\ x_i\ and\ x_j\ are\ nonseparable \end{aligned} \tag{5}$$

Figure 2 shows how original LINC-R and variant LINC-R work on separable variables. The original LINC-R compares the equivalence between fitness difference to identify the interaction while the variant LINC-R identify the interaction depending on the establishment of vector addition.

Then, we derive the variant LINC-R to 3-D or higher dimensions. In 3-D space, the schematic diagram is shown in Fig. 3. Similarly, we define the fitness difference in 3-D space in Eq. (6)

$$\begin{aligned} s &\in Pop: \\ \Delta f_i &= f(s_i) - f(s) \\ \Delta f_j &= f(s_j) - f(s) \\ \Delta f_k &= f(s_k) - f(s) \\ \Delta f_{ijk} &= f(s_{ijk}) - f(s) \end{aligned} \tag{6}$$

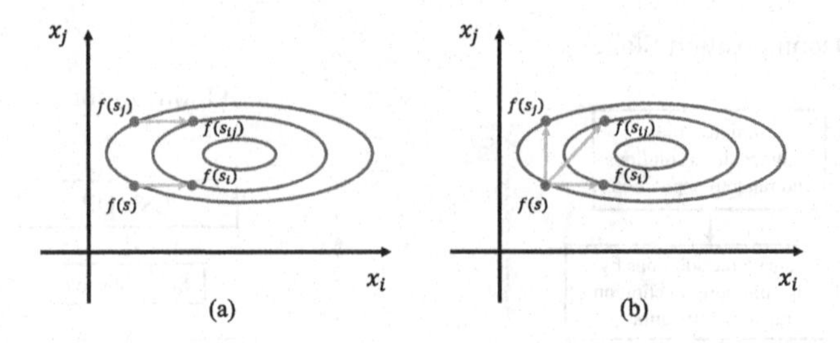

Fig. 2. (a) The original LINC-R works on the separable variables (b) The variant LINC-R works on the separable variables [19]

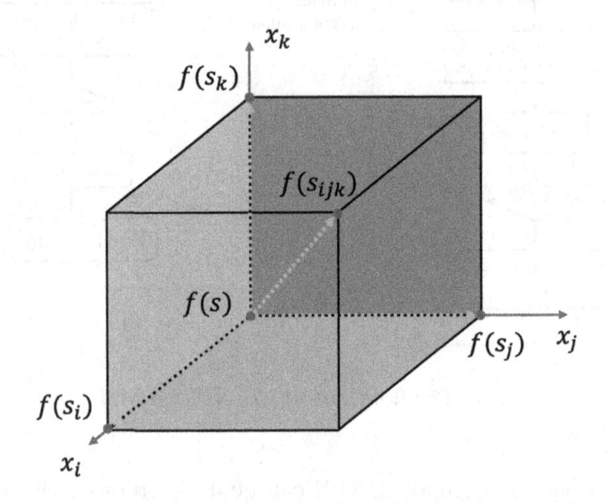

Fig. 3. The variant LINC-R works on 3-D space [19]

And Eq. (7) defines the variant LINC-R in 3-D space

$$\begin{aligned} &\exists s \in Pop: \\ if\ &|\Delta f_{ijk} - (\Delta f_i + \Delta f_j + \Delta f_k)| > 2\epsilon \\ &then\ interaction(s)\ exist\ in\ x_i, x_j, x_k \end{aligned} \tag{7}$$

Notice that the allowable error becomes 2ϵ. Therefore, we can reasonably infer the variant LINC-R in n-D space on Eq. (8).

$$\begin{aligned} &\exists s \in Pop: \\ if\ &|\Delta f_{1,2,...,n} - (\Delta f_1 + \Delta f_2 + ... + \Delta f_n)| > (n-1)\epsilon \\ &then\ interaction(s)\ exist\ in\ x_1, x_2, ..., x_n \end{aligned} \tag{8}$$

Notice that we only detect the interactions based on the finite individuals, which means when $|\Delta f_{1,2,...,n} - (\Delta f_1 + \Delta f_2 + ... + \Delta f_n)| < (n-1)\epsilon$ is satisfied at all individuals, then we consider this function is a fully separable function by default. Although this strategy is limited especially for trap functions, it is impossible to check the interactions on the whole fitness landscape. Thus, Eq. (9) is approximately correct in LSSOPs.

$$\begin{aligned} &\forall s \in Pop: \\ if\ |\Delta_{1,2,\dots,n} - (\Delta_1 + \Delta_2 + \dots + \Delta_n)| &< (n-1)\epsilon \\ then\ x_1, x_2, \dots, x_n\ &are\ separable \end{aligned} \tag{9}$$

However, when Eq. (9) is not satisfied, we only know that interactions exist in some variable pairs, but we cannot know the interactions exist in which pairs, so we can actively detect the interactions between variables. Taking the 3-D space as an example,

$$\begin{aligned} if\ \exists s \in Pop: |\Delta f_{ijk} - (\Delta f_i + \Delta f_j + \Delta f_k)| &> 2\epsilon \\ and\ \forall s \in Pop: |\Delta f_{ijk} - (\Delta f_{ij} + \Delta f_k)| &< \epsilon \\ then\ x_i, x_j\ are\ nonseparbale& \\ and\ x_k\ is\ separable\ from\ x_i, x_j& \end{aligned} \tag{10}$$

Therefore, Our target is to apply the heuristic algorithm to find the interactions between all variables as much as possible. According to the above explanation, in the n-D problem, LMF is formulated as Eq. (11)

$$\mathrm{LMF(s)} = \frac{|\Delta f_{1,2,\dots,n} - \sum_{i,j,\dots}^{m} \Delta f_{i,j,\dots}|}{m-1} \tag{11}$$

m is the number of subcomponents. Equation (11) calculates the detected linkage strength. We extend Eq. (11) to LSMOPs:

$$\mathrm{MOLMF(s)} = \sum_{s \in Pop} \sum_{j=1}^{M} w_j \frac{|\Delta f_{1,2,\dots,n} - \sum_{i,j,\dots}^{m} \Delta f_{i,j,..}|}{m-1}, \ \sum_{j=1}^{M} w_j = 1 \tag{12}$$

w_j is the weight of the jth objective function, and M is the number of objective functions in LSMOPs. We apply averaging weights in Eq. (12).

To find a suitable decomposition, we apply the Elitist Genetic Algorithm (EGA) [5] to optimize the Eq. (12). The elitist reservation strategy directly replicates the best individual without crossover, mutation, and selection to the next generation. This strategy can prevent the optimal individual from destroying the superior gene and chromosome structure during crossover and mutation.

4 Numerical Experiments

In this Section, we ran experiments to evaluate our proposal. In Sect. 4.1, we introduce the experiment settings, including benchmark functions, comparing methods, and parameters of algorithms. In Sect. 4.2, we provide the experiment results. Finally, we analyze our proposal in Sect. 4.3.

4.1 Experiment Settings

Benchmark Functions. We conduct our experiments on benchmarks of ZDT1-6, DTLZ1-7, UF1-2, and WFG1-5, 7 up to 500-D and 1000-D. We did not apply high-dimensional WFG6, WFG8, and WFG9 because these functions are not suitable for extending to high dimensions due to high computational cost. All benchmark functions are provided by geatpy [9] and pymoo [3].

Comparing Methods and Parameters. We combine our proposal in decomposition with NSGA-II (CC-NSGA-LMM) and compare it with Random Grouping (CC-NSGA-G) [1], Differential Grouping (CC-NSGA-DG) [15], and Monotonicity Detection (CC-NSGA-LIMD) [8] with 30 trial runs. Table 1 shows the parameters of our proposal in the grouping stage, and Table 2 shows the parameters of subcomponents optimization. The total FEs include the FEs consumed in problem decomposition and subcomponents optimization.

Table 1. The parameters of decomposition optimization

Parameter	Value
Optimizer	Elitist GA
Population size	20
Max iteration	20
Gene length	6 and 7

Table 2. The parameters of subcomponents optimization

Parameter	Value
Dimension	500-D and 1000-D
Total FEs	750,000 and 1,500,000
Optimizer	NSGA-II
Population size	50
Crossover rate	0.9
Mutation rate	0.2

4.2 Performance of CC-NSGA-LMM

In this section, the performance of CC-NSGA-LMM is studied. We randomly choose one trial run result in 30 trial runs and draw the Pareto Front (PF) graph within compared methods and reference sets. Due to space limitations, we select some representative PF graphs in Fig. 4. The mean of HV calculated in 30 trial runs is shown in Table 3. The best solution among CC-NSGA-G, CC-NSGA-DG, CC-NSGA-LIMD, and CC-NSGA-LMM these 4 methods is in bold in 500-D and 1000-D respectively to show the performance of our proposed decomposition method.

Table 3. The mean of HV among 4 methods in 30 trial runs

Func	CC-NSGA-G		CC-NSGA-DG		CC-NSGA-LIMD		CC-NSGA-LMM	
	500-D	1000-D	500-D	1000-D	500-D	1000-D	500-D	1000-D
ZDT1	0.27	0.22	0.60	0.57	0.32	0.21	**0.99**	**1.35**
ZDT2	0.74	0.40	1.04	0.90	0.79	0.55	**1.50**	**1.54**
ZDT3	0.46	0.22	0.75	0.44	0.52	0.23	**1.13**	**0.79**
ZDT4	1012.58	673.10	1961.57	1247.62	1085.54	630.41	**2833.41**	**4022.32**
ZDT5	**752.32**	**2749.27**	600.79	1791.75	750.34	2574.25	603.45	1786.29
ZDT6	**0.75**	**0.44**	0.42	0.14	0.66	0.37	0.46	0.24
DTLZ1	3.14e11	2.75e12	1.87e11	7.39e11	**3.27e11**	**3.07e12**	1.87e11	2.44e11
DTLZ2	**2309.57**	**23917.32**	36.70	6954.45	1720.39	23656.01	36.12	518.45
DTLZ3	2.48e12	3.57e13	1.58e13	1.29e14	2.41e12	3.75e13	**1.59e13**	**1.43e14**
DTLZ4	**4235.78**	**35457.28**	6.16	4.49	2798.33	30458.98	6.92	12.92
DTLZ5	1400.88	11725.02	12071.98	**18025.29**	1900.68	12555.10	**12091.59**	15468.05
DTLZ6	4.67e6	2.08e7	4.43e6	1.87e7	**5.56e6**	**2.23e7**	2.04e6	1.11e7
DTLZ7	0.34	0.35	1.27	1.22	0.42	0.39	**1.30**	**1.46**
UF1	0.20	0.20	1.82	1.35	0.56	0.33	**1.90**	**1.57**
UF2	0.10	0.08	0.20	0.14	0.15	0.10	**0.40**	**0.38**
WFG1	0.01	0.02	0.90	0.16	0.01	0.01	**1.09**	**0.80**
WFG2	3.11	0.50	2.48	0.58	**3.20**	0.65	2.84	**0.69**
WFG3	**3.47**	0.63	2.80	0.80	3.02	0.51	2.56	**0.88**
WFG4	0.73	1.23	2.67	2.31	0.88	0.65	**2.84**	**2.66**
WFG5	1.01	1.15	2.19	1.90	1.17	0.94	**2.66**	**2.40**
WFG7	1.86	0.55	2.00	1.36	1.60	0.94	**2.12**	**2.01**

4.3 Analysis

From Table 3, we can see our proposed CC-NSGA-LMM outperforms the compared three methods in the majority of benchmark functions. This is mainly due to the following aspects. (1). We minimize the LMF based on multiple samples. Although this is a necessary condition for the implementation of LINC-R and LIMD in low-dimensional space, the interactions are only possible to be identified around a sample in high-dimensional space due to the FEs limitation, such as DG. Therefore, our proposal is more robust for solving LSMOPs based on CC. (2). Due to the performance of EGA and limited computational resource allocation, the whole interactions cannot be detected. Thus, the individual containing

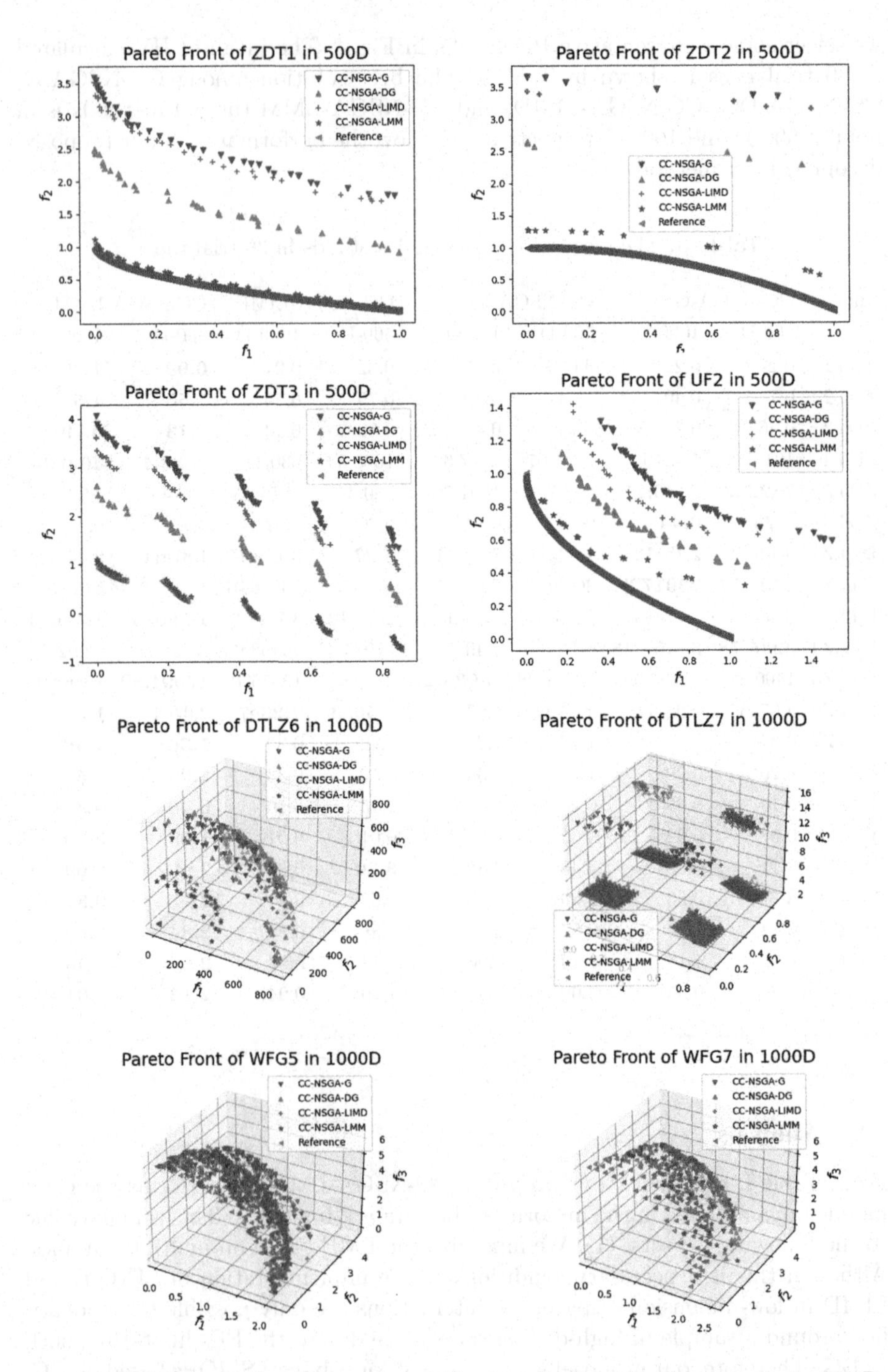

Fig. 4. The representative PF graphs within 4 methods and reference sets in 500-D and 1000-D.

the strong interactions has better fitness, and the genome structure has a higher probability to be preserved. Meanwhile, some weak interactions between variables are ignored. This process will increase the error in the optimization stage but accelerate the subcomponents optimization, especially under the limitation of FEs.

5 Discussion

The above analysis shows our proposal has broad prospects for solving LSMOPs, however, there are still many aspects for improvement. Here, we list a few open topics for potential and future research.

5.1 Self-adaptation of Weight in LMF

In this paper, we apply the averaging weights in LMF. Actually, the importance of objective functions is different. In future research, determining the weights by the information on the fitness landscape is a topic of our future research.

5.2 More Powerful MOEAs

The improvement of the performance of MOEAs is also one of the themes of our research. The design of a novel search scheme or local search strategy combined with MOEAs is an interesting topic.

5.3 The Scalability of CC-NSGA-LMM

The extension of our proposal to deal with very large-scale optimization problems (VLSOPs), LSOPs with constraints, and LSOPs in noisy environments are our future research topics.

6 Conclusion

In this paper, we propose a novel decomposition method for LSMOPs. we treat the decomposition problem as a combinatorial optimization problem and design a linkage measurement function to lead the optimization. Experiments show that our proposal is a promising study to solve LSMOPs.

Acknowledgement. This work was supported by JSPS KAKENHI Grant Number JP20K11967.

References

1. Antonio, L.M., Coello, C.A.C.: Use of cooperative coevolution for solving large scale multiobjective optimization problems. In: 2013 IEEE Congress on Evolutionary Computation, pp. 2758–2765. IEEE (2013)
2. van den Bergh, F., Engelbrecht, A.: A cooperative approach to particle swarm optimization. IEEE Trans. Evol. Comput. **8**(3), 225–239 (2004). https://doi.org/10.1109/TEVC.2004.826069
3. Blank, J., Deb, K.: Pymoo: multi-objective optimization in python. IEEE Access **8**, 89497–89509 (2020)
4. Bottou, L., Curtis, F.E., Nocedal, J.: Optimization methods for large-scale machine learning. SIAM Rev. **60**(2), 223–311 (2018). https://doi.org/10.1137/16M1080173
5. De Jong, K.A.: An Analysis of the Behavior of a Class of Genetic Adaptive Systems. University of Michigan, Ann Arbor (1975)
6. Deb, K., Pratap, A., Agarwal, S., Meyarivan, T.: A fast and elitist multiobjective genetic algorithm: NSGA-II. IEEE Trans. Evol. Comput. **6**(2), 182–197 (2002)
7. Gould, N., Orban, D., Toint, P.: Numerical methods for large-scale nonlinear optimization. Acta Numer. **14**, 299–361 (2005). https://doi.org/10.1017/S0962492904000248
8. Izumiya, K., Munetomo, M.: Multi-objective evolutionary optimization based on decomposition with linkage identification considering monotonicity. In: 2017 IEEE Congress on Evolutionary Computation (CEC), pp. 905–912 (2017). https://doi.org/10.1109/CEC.2017.7969405
9. Jazzbin, E.: Geatpy: the genetic and evolutionary algorithm toolbox with high performance in python (2020)
10. Ma, X., Liu, F., Qi, Y., Wang, X., Li, L., Jiao, L., Yin, M., Gong, M.: A multiobjective evolutionary algorithm based on decision variable analyses for multiobjective optimization problems with large-scale variables. IEEE Trans. Evol. Comput. **20**(2), 275–298 (2015)
11. Martinsson, P.G., Tropp, J.A.: Randomized numerical linear algebra: foundations and algorithms. Acta Numer. **29**, 403–572 (2020). https://doi.org/10.1017/S0962492920000021
12. Omidvar, M.N., Li, X., Yao, X.: A review of population-based metaheuristics for large-scale black-box global optimization: Part A. In: IEEE Transactions on Evolutionary Computation, pp. 1–1 (2021). https://doi.org/10.1109/TEVC.2021.3130838
13. Omidvar, M.N., Li, X., Yao, X.: A review of population-based metaheuristics for large-scale black-box global optimization: Part B. In: IEEE Transactions on Evolutionary Computation, pp. 1 (2021). https://doi.org/10.1109/TEVC.2021.3130835
14. Potter, M.A., De Jong, K.A.: A cooperative coevolutionary approach to function optimization. In: Davidor, Y., Schwefel, H.-P., Männer, R. (eds.) PPSN 1994. LNCS, vol. 866, pp. 249–257. Springer, Heidelberg (1994). https://doi.org/10.1007/3-540-58484-6_269
15. Sander, F., Zille, H., Mostaghim, S.: Transfer strategies from single- to multi-objective grouping mechanisms. In: Proceedings of the Genetic and Evolutionary Computation Conference, pp. 729–736. GECCO'18, Association for Computing Machinery (2018). https://doi.org/10.1145/3205455.3205491
16. Song, A., Yang, Q., Chen, W.N., Zhang, J.: A random-based dynamic grouping strategy for large scale multi-objective optimization. In: 2016 IEEE Congress on Evolutionary Computation (CEC), pp. 468–475. IEEE (2016)

17. Yang, Z., Tang, K., Yao, X.: Large scale evolutionary optimization using cooperative coevolution. Inf. Sci. **178**(15), 2985–2999 (2008)
18. Yang, Z., Tang, K., Yao, X.: Multilevel cooperative coevolution for large scale optimization. In: 2008 IEEE Congress on Evolutionary Computation (IEEE World Congress on Computational Intelligence), pp. 1663–1670 (2008). https://doi.org/10.1109/CEC.2008.4631014
19. Zhong, R., Munetomo, M.: Random population-based decomposition method by linkage identification with non-linearity minimization on graph. In: 2022-MPS-139, pp. 1–4 (2022)

Data-Driven Evolutionary Multi-objective Optimization Based on Multiple-Gradient Descent for Disconnected Pareto Fronts

Renzhi Chen[1] and Ke Li[2(✉)]

[1] National Innovation Institute of Defence Technology, Beijing, China
[2] Department of Computer Science, University of Exeter, Exeter EX4 4QF, UK
k.li@exeter.ac.uk

Abstract. Data-driven evolutionary multi-objective optimization (EMO) has been recognized as an effective approach for multi-objective optimization problems with expensive objective functions. The current research is mainly developed for problems with a 'regular' triangle-like Pareto-optimal front (PF), whereas the performance can significantly deteriorate when the PF consists of disconnected segments. Furthermore, the offspring reproduction in the current data-driven EMO does not fully leverage the latent information of the surrogate model. Bearing these considerations in mind, this paper proposes a data-driven EMO algorithm based on multiple-gradient descent. By leveraging the regularity information provided by the up-to-date surrogate model, it is able to progressively probe a set of well distributed candidate solutions with a convergence guarantee. In addition, its infill criterion recommends a batch of promising candidate solutions to conduct expensive objective function evaluations. Experiments on 33 benchmark test problem instances with disconnected PFs fully demonstrate the effectiveness of our proposed method against four selected peer algorithms.

Keywords: Data-driven optimization · Multiple-gradient descent · Evolutionary multi-objective optimization

1 Introduction

Many real-world scientific and engineering applications involve multiple conflicting objectives, a.k.a. multi-objective optimization problems (MOPs). For example, tuning a water distribution system to optimize its financial and operational costs [22], minimizing the energy consumption while maximizing locomotion speed in a complex robotic system [2]. In multi-objective optimization, there does not exist a solution that optimizes all conflicting objectives simultaneously. Instead, we are looking for a set of representative, with a promising convergence and diversity, trade-off solutions that compromise one objective for another.

This work was supported by UKRI Future Leaders Fellowship (MR/S017062/1, MR/X011135/1), NSFC (62076056), Royal Society (IES/R2/212077), EPSRC (2404317), and Amazon Research Award.

M. Emmerich et al. (Eds.): EMO 2023, LNCS 13970, pp. 56–70, 2023.
https://doi.org/10.1007/978-3-031-27250-9_5

Due to the population-based characteristics, evolutionary algorithms (EAs) have been widely recognized as an effective approach for MO [4]. However, one of the major criticisms of EAs is its daunting amount of function evaluations (FEs) required to obtain a set of reasonable solutions. This is unfortunately unacceptable in practice since FEs are either computationally or financially demanding, e.g., computational fluid dynamic simulations can take from minutes to hours to carry out a single FE [16]. To mitigate this issue, data-driven evolutionary optimization[1], guided by surrogate models of computationally expensive objective functions, have become a powerful approach for solving expensive optimization problems [17]. For example, some researchers considered various ways to build a surrogate model of the expensive objective functions, either collectively [1,3,11,21] or as a weighted aggregation [19,23,31,34]. According to the ways of surrogate modeling, bespoke model management strategies are developed to select promising candidate solution(s) for conducting expensive FEs. In particular, this can either be driven by the surrogate model directly [1,23,31] or an acquisition function inferred from the model uncertainty [3,11,19,21,34]. There are two gaps in the current literature that hinder the further uptake of data-driven evolutionary multi-objective optimization (EMO) in practice.

- Most, if not all, existing studies are mainly designed and validated on prevalent test problems (e.g., DTLZ1 to DTLZ4 [7] and WFG4 to WFG9 [13]) characterized as 'regular' triangle-like Pareto-optimal fronts (PFs). Unfortunately, this is unrealistic in the real-world optimization scenarios [14]. On the contrary, it is not uncommon that the PFs of real-world applications are featured as disconnected, incomplete, degenerated, and/or badly-scaled (partially due to the complex and nonlinear relationship between objectives), it is surprising that the research on handling MOPs with irregular PFs is lukewarm in the context of data-driven EMO, except for [11].
- In addition, the evolutionary operators for offspring reproduction are directly derived from the EA (e.g., crossover and mutation [12], differential evolution [30], particle swarm optimization [18]) or conventional mathematical programming (e.g., simplex [9], Nelder-Mead [25], and trust-region methods [27]). By this means, the regularity information of the underlying MOP embedded in the surrogate model(s) is unfortunately yet exploited. Note that such information can be beneficial to navigate a more effective exploration of the search space.

Bearing these considerations in mind, this paper proposes a data-driven evolutionary multi-objective optimization based on multiple-gradient descent [8] for expensive MOPs with disconnected PFs. Its basic idea is to leverage the gradient information of the surrogate models to explore promising candidate solutions. It consists of the following two distinctive components.

- `MGD-based evolutionary search`: As the main crux of our proposed algorithm, it generates a set of candidate solutions guided by the multiple-gradient descent of the surrogate model of each computationally expensive objective

[1] It is also known as surrogate-assisted EA interchangeably in the literature [15].

function. In a nutshell, these candidate solutions are first randomly sampled in the decision space. Then, they are gradually guided to interpolate well distributed potential solutions along the manifold of the surrogate PS.
- `Infill criterion`: It recommends a batch of promising candidate solutions obtained by the `MGD-based evolutionary search` step to carry out expensive FEs for the model management.

Our experiments on 33 benchmark test problem instances with disconnected PFs fully demonstrate the effectiveness and outstanding performance of our proposed $\texttt{D}^2\texttt{EMO/MGD}$ against four selected peer algorithms.

The rest of this paper is organized as follows. Section 2 gives some preliminary knowledge pertinent to this paper. The technical details of our proposed method are introduced in Sect. 3. The experimental setup is given in Sect. 4 and the results are presented and discussed in Sect. 5. Section 6 concludes this paper and sheds some lights on potential future directions.

2 Preliminaries

In this section, we give some basic definitions pertinent to this paper.

2.1 Basic Definitions in Multi-objective Optimization

The MOP considered in this paper is defined as:

$$\begin{array}{ll} \text{minimize} & \mathbf{F}(\mathbf{x}) = (f_1(\mathbf{x}), \cdots, f_m(\mathbf{x}))^T \\ \text{subject to} & \mathbf{x} \in \Omega \end{array}, \tag{1}$$

where $\mathbf{x} = (x_1, \cdots, x_n)^T$ is a decision vector and $\mathbf{F}(\mathbf{x})$ is an objective vector. $\Omega = [x_i^L, x_i^U]_{i=1}^n \subseteq \mathbb{R}^n$ defines the search space. $\mathbf{F} : \Omega \to \mathbb{R}^m$ defines the mapping from the search space Ω to the objective space $\mathbb{R}^m$.

Definition 1. *Given two solutions $\mathbf{x}^1, \mathbf{x}^2 \in \Omega$, $\mathbf{x}^1$ is said to Pareto dominate $\mathbf{x}^2$, denoted as $\mathbf{x}^1 \preceq \mathbf{x}^2$, if and only if $f_i(\mathbf{x}^1) \leq f_i(\mathbf{x}^2)$ for all $i \in \{1, \cdots, m\}$ and $\mathbf{F}(\mathbf{x}^1) \neq \mathbf{F}(\mathbf{x}^2)$.*

Definition 2. *A solution $\mathbf{x}^* \in \Omega$ is said to be Pareto-optimal if and only if $\nexists \mathbf{x}' \in \Omega$ such that $\mathbf{x}' \preceq \mathbf{x}^*$.*

Definition 3. *The set of all Pareto-optimal solutions is called the Pareto-optimal set (PS), i.e., $PS = \{\mathbf{x}^* | \nexists \mathbf{x}' \in \Omega \text{such that} \mathbf{x}' \preceq \mathbf{x}^*\}$ and their corresponding objective vectors form the Pareto-optimal front (PF), i.e., $PF = \{\mathbf{F}(\mathbf{x}^*) | \mathbf{x}^* \in \mathcal{PS}\}$.*

2.2 Gaussian Process Regression Model

In view of the continuously differentiable property, we consider the Gaussian process regression (GPR) [28] as the surrogate model of each expensive objective function. Given a set of training data $\mathcal{D} = \{(\mathbf{x}^i, f(\mathbf{x}^i)\}_{i=1}^N$, a GPR model aims to learn a latent function $g(\mathbf{x})$ by assuming $f(\mathbf{x}^i) = g(\mathbf{x}^i) + \epsilon$ where $\epsilon \sim \mathcal{N}(0, \sigma_n^2)$

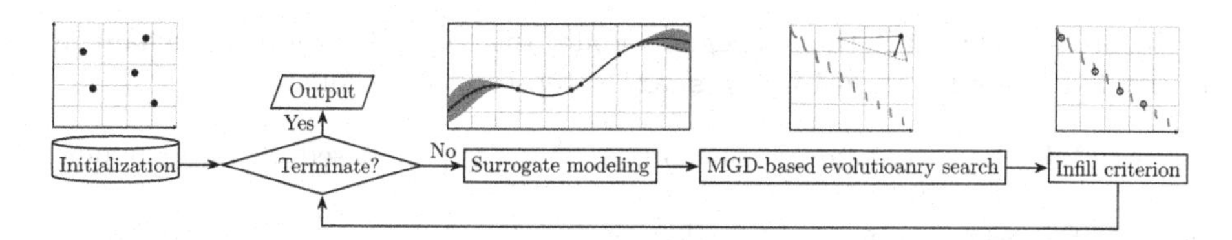

Fig. 1. The flow chart of the proposed D^{2}EMO/MGD.

is an independently and identically distributed Gaussian noise. For each testing input vector $\mathbf{z}^* \in \Omega$, the mean and variance of the target $f(\mathbf{z}^*)$ are predicted as:

$$\overline{g}(\mathbf{z}^*) = m(\mathbf{z}^*) + \mathbf{k}^{*T}(K + \sigma_n^2 I)^{-1}(\mathbf{f} - \mathbf{m}(X)), \quad (2)$$

$$\mathbb{V}[g(\mathbf{z}^*)] = k(\mathbf{z}^*, \mathbf{z}^*) - \mathbf{k}^{*T}(K + \sigma_n^2 I)^{-1}\mathbf{k}^*, \quad (3)$$

where $X = (\mathbf{x}^1, \cdots, \mathbf{x}^N)^T$ and $\mathbf{f} = (f(\mathbf{x}^1), \cdots, f(\mathbf{x}^N))^T$. $\mathbf{m}(X)$ is the mean vector of X, $\mathbf{k}^*$ is the covariance vector between X and $\mathbf{z}^*$, and K is the covariance matrix of X. In this paper, we use the radial basis function as the covariance function to measure the similarity between a pair of two solutions $\mathbf{x}$ and $\mathbf{x}' \in \Omega$:

$$k(\mathbf{x}, \mathbf{x}') = \gamma \exp(-\frac{\|\mathbf{x} - \mathbf{x}'\|^2}{\ell}), \quad (4)$$

where $\|\cdot\|$ is the Euclidean norm and γ and length scale ℓ are two hyperparameters. The predicted mean $\overline{g}(\mathbf{z}^*)$ is directly used as the prediction of $f(\mathbf{z}^*)$, and the predicted variance $\mathbb{V}[g(\mathbf{x}^*)]$ quantifies the uncertainty. In practice, the hyperparameters associated with the mean and covariance functions are learned by maximizing the log marginal likelihood function as recommended in [28]. For the sake of simplicity, here we assume that the mean function is a constant 0 and the inputs are noiseless.

3 Proposed Method

In this section, we plan to delineate the implementation of our proposed data-driven evolutionary multi-objective optimization based on multiple-gradient descent (dubbed D^{2}EMO/MGD). As the flowchart shown in Fig. 1, D^{2}EMO/MGD starts with an `initialization` step based on an experimental design method such as Latin hypercube sampling [29]. Note that these initial samples will be evaluated based on the computationally expensive objective functions. During the main loop, the `surrogate modeling` step builds a surrogate model by using the GPR for each expensive objective function based on the data collected so far. The other two steps will be delineated in the following paragraphs.

3.1 MGD-Based Evolutionary Search

This step aims to search for a set of promising candidate solutions $\mathcal{P} = \{\hat{\mathbf{x}}^i\}_{i=1}^{\tilde{N}}$, which are assumed to be an appropriate approximation to the PF, based on the surrogate model built in the `surrogate modeling` step. The working mechanism of this `MGD-based evolutionary search` step is given as follows.

Step 1: Initialize a candidate solution set $\mathcal{P} = \{\hat{\mathbf{x}}^i\}_{i=1}^{\tilde{N}}$ based on Latin hypercube sampling upon Ω.

Step 2: For each solution $\hat{\mathbf{x}}^i \in \mathcal{P}$, do

Step 2.1: Calculate the gradient of the predicted mean of each objective function $\nabla\overline{g}_j(\hat{\mathbf{x}}^i)$ where $j \in \{1, \cdots, m\}$.

Step 2.2: Find a nonnegative unit vector $\mathbf{w}^* = (w_1^*, \cdots, w_m^*)^\top$ that satisfies:

$$\mathbf{w}^* = \underset{\mathbf{w}}{\text{argmin}} \left\| \sum_{j=1}^{m} w_j \nabla\overline{g}_j(\hat{\mathbf{x}}^i) \right\|, \tag{5}$$

where $\mathbf{w} = (w_1, \cdots, w_m)^\top$, $\sum_{i=1}^{m} w_i = 1$ and $w_i \geq 0$, $i \in \{1, \cdots, m\}$.

Step 2.3: Obtain a directional vector $\mathbf{u}^*$ as:

$$\mathbf{u}^* = \begin{cases} \underset{1\leq j\leq m}{\text{argmax}} \|\nabla\overline{g}_j(\hat{\mathbf{x}}^i)\|, & \text{if} \sum_{j=1}^{m} w_j^* \nabla\overline{g}_j(\hat{\mathbf{x}}^i) = 0 \\ \underset{1\leq j\leq m}{\text{argmin}} \|\nabla\overline{g}_j(\hat{\mathbf{x}}^i)\|, & \text{if} \exists i, j \in \{1, \cdots, m\}, \langle \nabla\overline{g}_j(\hat{\mathbf{x}}^i), \nabla\overline{g}_j(\hat{\mathbf{x}}^i) \rangle > \delta \\ \sum_{j=1}^{m} w_j^* \nabla\overline{g}_j(\hat{\mathbf{x}}^i), & \text{otherwise} \end{cases} \tag{6}$$

where $\langle *, * \rangle$ measures the acute angle between two vectors, and $\delta = \min \left\{ \|\nabla\overline{g}_k(\hat{\mathbf{x}}^i)\| \right\}_{k=1}^{m}$.

Step 2.4: Amend the updated solution to $\mathcal{P} \leftarrow \mathcal{P} \bigcup \{\hat{\mathbf{x}}^i + \eta\mathbf{u}^*\}$

Step 3: Remove the dominated solutions in $\mathcal{P}$ according to their predicted objective functions.

Step 4: If the stopping criterion is met, then stop and output $\mathcal{P}$. Otherwise, go to Step 2.

Remark 1. As discussed in [8], the multiple-gradient descent (MGD) is a natural extension of the single-objective gradient to finding a PF. In a nutshell, its basic idea is to iteratively update a solution $\mathbf{x}$ along a 'specified' direction so that all objective functions can thus be improved. Different from the linear weighted aggregation, the MGD works for non-convex PF. Therefore, we can expect a satisfactory diversity in case the initial population is well distributed. Note that since the objective functions are assumed to be as a black box a priori, the MGD is not directly applicable in our context.

Remark 2. In this paper, since the computationally expensive objective functions are modeled by GPR, which is continuously differentiable, we can derive the gradient of the predicted mean function w.r.t. a solution $\mathbf{x}$ as:

$$\frac{\partial\overline{g}(\mathbf{x})}{\partial\mathbf{x}} = \frac{\partial\mathbf{k}^*}{\partial\mathbf{x}} K^{-1}\mathbf{f}, \tag{7}$$

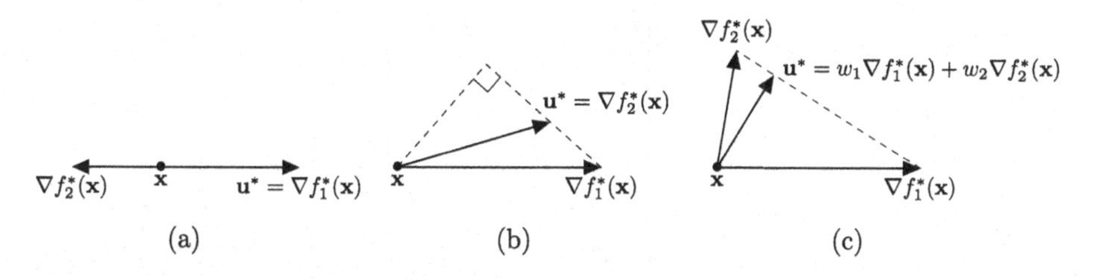

Fig. 2. Illustrative examples of the calculation of $\mathbf{u}^*$ in Eq. (6).

where the first-order derivative of $\mathbf{k}^*$, i.e., the covariance vector between $\mathcal{P}$ and $\mathbf{x}$, is calculated as:

$$\frac{\partial \mathbf{k}^*}{\partial \mathbf{x}} = -\frac{\partial \|\mathbf{x} - \mathbf{x}'\|}{\partial \mathbf{x}} \frac{\overline{g}(\mathbf{x})}{\ell}. \tag{8}$$

Remark 3. The optimization problem in (5) is essentially equivalent to finding a minimum-norm point in the convex hull. When $m = 2$, we have the closed form solution as:

$$w_1^* = \frac{\left(\nabla \overline{g}_2(\hat{\mathbf{x}}^i) - \nabla \overline{g}_1(\hat{\mathbf{x}}^i)\right)^\top \nabla \overline{g}_2(\hat{\mathbf{x}}^i)}{\left\|\nabla \overline{g}_2(\hat{\mathbf{x}}^i) - \nabla \overline{g}_1(\hat{\mathbf{x}}^i)\right\|^2}, w_2^* = 1 - w_1^*. \tag{9}$$

Remark 4. Figure 2 gives an illustrative example for each of the three conditions given in equation (6) when $m = 2$. More specifically, when the gradients of two objective functions are in opposite directions as shown in Fig. 2(a), $\mathbf{u}^*$ is chosen as the one with a larger Euclidean norm. If the gradients are too close to each other as shown in Fig. 2(b), $\mathbf{u}^*$ is chosen as the one with a smaller Euclidean norm. On the contrary, $\mathbf{u}^*$ is set as the weighted aggregation of two gradients as shown in Fig. 2(c). In particular, the weights are obtained from Step 2.2.

Remark 5. According to the Karush-Kuhn-Tucker (KKT) conditions [20], we have $\forall \mathbf{x}^* \in \mathcal{PS}$, $\exists \boldsymbol{\alpha} = (\alpha_1, \cdots, \alpha_m)^T$, where $\alpha_i \geq 0$, $i \in \{1, \cdots, m\}$ and $\sum_{i=1}^m \alpha_i = 1$, such that $\sum_{i=1}^m \alpha_i \nabla f_i(\mathbf{x}^*) = 0$. In this case, we come up with Corollary 1, the proof of which can be found in the supplemental document of this paper[2].

Corollary 1. *Considering the m objective functions defined in (1), $\forall \mathbf{x} \in \mathcal{PS}$, $\exists \mathbf{w}^*$ that satisfies (5) and $\mathbf{u}^* = \sum_{j=1}^m w_j^* \nabla f_j(\mathbf{x}) = 0$, we can obtain a new solution $\mathbf{x}' = \mathbf{x} + \eta \mathbf{u}^*$ such that $\mathbf{x}'$ is still on the PS.*

Remark 6. According to Corollary 1, the `MGD-based evolutionary search` step can be understood as pushing a solution towards the PS first before implementing a random walk along the PS as an illustrative example shown in Fig. 3.

Remark 7. In Step 2.4, $\eta \in (0, 1]$ is a random scaling factor along the direction vector $\mathbf{u}^*$. The stopping criterion in Step 4 is the number of iterations (here it is set as 100 in our experiments) of this `MGD-based evolutionary search` step.

[2] The supplemental document can be found from https://tinyurl.com/2s3takpd.

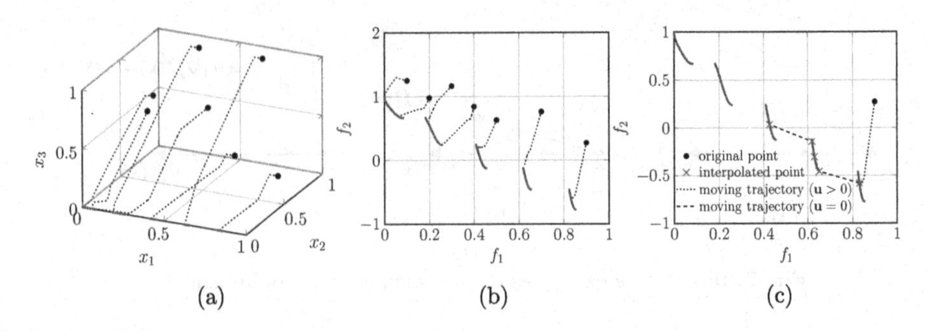

Fig. 3. Illustrative examples of MGD w.r.t. different solutions moving towards different (a) PS segments and (b) PF segments, and (c) the working mechanism **MGD-based evolutionary search** step.

3.2 Infill Criterion

This step aims to pick up $\xi \geq 1$ promising solutions from $\mathcal{P}$ and evaluate them by using the computationally expensive objective functions. These newly evaluated solutions are then used to update the training dataset for the next iteration. In a nutshell, there are two main differences w.r.t. many existing works on data-driven EA [17] and Bayesian optimization [10]. First, even though we use the GPR as the surrogate model, our infill criterion does not rely on an uncertainty quantification measure, a.k.a. acquisition function. Second, instead of recommending one solution for the computationally expensive function evaluation in a sequential manner, the `infill criterion` step of D^{2}EMO/MGD proposes to select a batch of samples at a time. Under a limited computational budget, we can expect to reduce the number of iterations of the main loop in Fig. 1 by ξ times. In addition, since many physical experiments can be carried out in parallel given the availability of more than one infrastructure (e.g., the training and validation of machine learning models are usually distributed into multiple cores or GPUs for hyper-parameter optimization in automated machine learning), such batched recommendation provides an actionable way for parallelization. Therefore, we can anticipate the practical importance to save the computational overhead. In this paper, we propose a simple infill criterion based on the individual Hypervolume [36] contribution (IHV). Specifically, the IHV of each candidate solution $\mathbf{x} \in \mathcal{P}$ is calculated as:

$$\mathtt{IHV}(\mathbf{x}) = \mathtt{HV}(\mathcal{P}) - \mathtt{HV}(\mathcal{P} \setminus \{\mathbf{x}\}), \tag{10}$$

where $\mathtt{HV}(\mathcal{P})$ evaluates the Hypervolume of $\mathcal{P}$. Then, the top ξ solutions in $\mathcal{P}$ with the largest IHV are picked up for the expensive function evaluations.

4 Experimental Setup

This section introduces our experimental setup including the benchmark test problems, the peer algorithms along with their parameter settings, and the performance metrics and statistical tests.

4.1 Benchmark Test Problems

In our empirical study, we consider benchmark test problems with disconnect PF segments to constitute our benchmark suite, including ZDT3 [35], DTLZ7 [7] and WFG2 [13] along with their variants dubbed ZDT3⋆, DTLZ7⋆ and WFG2⋆. Their mathematical definitions and characteristics can be found in the supplemental document. For each benchmark test problem, we set the number of objectives as $m = 2$ and the number of variables as $n \in \{3, 5, 8\}$ respectively in our empirical study.

4.2 Peer Algorithms and Parameter Settings

To validate the competitiveness of our proposed algorithm, we compare its performance with `ParEGO` [19], `MOEA/D-EGO` [34], `K-RVEA` [3], and `HSMEA` [11] widely used in the literature. We do not intend to delineate their working mechanisms here while interested readers are referred to their original papers for details. The parameter settings are listed as follows.

- Number of function evaluations (FEs): The initial sample size is set to $11 \times n - 1$ for all algorithms and the maximum number of FEs is capped as 250.
- Reproduction operators: The parameters associated with the simulated binary crossover [5] and polynomial mutation [6] are set as $p_c = 1.0$, $\eta_c = 20$, $p_m = 1/n$, $\eta_m = 20$.
- Kriging models: As for the algorithms that use Kriging for surrogate modeling, the corresponding hyperparameters of the MATLAB Toolbox DACE [26] is set to be within the range $[10^{-5}, 10^5]$.
- Batch size ξ: It is set as $\xi = 10$ for our proposed algorithms and $\xi = 5$ is set in `MOEA/D-EGO`.
- Number of repeated runs: Each algorithm is independently run on each test problem for 31 times with different random seeds.

4.3 Performance Metric and Statistical Tests

To have a quantitative evaluation of the performance of different algorithms, we use the widely used HV as the performance metric. To have a statistical interpretation of the significance of comparison results, we use the following three statistical measures in our empirical study.

- Wilcoxon signed-rank test [33]: This is a widely used non-parametric statistical test to conduct a pairwise comparison. In our experiments, we set the significance level as $p = 0.05$.
- A_{12} effect size [32]: This is an effect size measure that evaluates the probability of one algorithm is better than another. Specifically, given a pair of peer algorithms, $A_{12} = 0.5$ means they are *equivalent*. $A_{12} > 0.5$ denotes that one is better for more than 50% of the times. $0.56 \leq A_{12} < 0.64$ indicates a *small* effect size while $0.64 \leq A_{12} < 0.71$ and $A_{12} \geq 0.71$ mean a *medium* and a *large* effect size, respectively.

- Scott-knott test [24]: This is used to rank the performance of different peer algorithms over 31 runs on each test problem. In a nutshell, it uses a statistical test and effect size to divide the performance of peer algorithms into several clusters. In particular, the performance of peer algorithms within the same cluster are statistically equivalent. The smaller the rank is, the better performance of the algorithm achieves.

5 Experimental Results

In this section, our empirical study aims to investigate: 1) the performance of our proposed D^{2}EMO/MGD compared against the selected peer algorithms; and 2) the effectiveness of the `MGD-based evolutionary search` and the `infill criterion` steps of D^{2}EMO/MGD.

5.1 Performance Comparisons with the Peer Algorithms

The statistical comparison results of the Wilcoxon signed-rank test of the HV values between our proposed D^{2}EMO/MGD against the other peer algorithms are given in Table 1. From these results, it is clear to see that the HV values obtained by D^{2}EMO/MGD are statistically significantly better than the other four peer algorithms in all comparisons. As the selected results of the population distribution obtained by different algorithms shown in Fig. 4, it is clear to see that the solutions obtained by D^{2}EMO/MGD not only have a good convergence on the PF, but also have a descent distribution on all disconnected PF segments. In contrast, the other peer algorithms either struggle to converge to the PF or hardly approximate all segments. It is interesting to note that the performance of HSMEA and K-RVEA are acceptable on ZDT3 and DTLZ7$*$, which have a relatively small number of disconnected segments, whereas their performance deteriorate significantly when the number of disconnected segments becomes large.

In addition to the pairwise comparisons, we apply the Scott-knott test to classify their performance into different groups to facilitate a better ranking among these algorithms. Due to the large number of comparisons, it will be messy if we list all ranking results ($11 \times 3 = 33$ in total). Instead, we pull all the Scott-knott test results together and show their distribution and variance as the bar charts with error bar in Fig. 5(a). From this results, we can see that our proposed D^{2}EMO/MGD is the best algorithm in all comparisons, which confirm the observations from Table 1. In addition, to have a better understanding of the performance difference of D^{2}EMO/MGD w.r.t. the other peer algorithms, we investigate the comparison results of A_{12} effect size. From the bar charts shown in Fig. 5(b), it is clear to see that the better results achieved by D^{2}EMO/MGD is consistently classified as statistically large.

Table 1. Comparison results of D^{2}EMO/MGD with the peer algorithms

	n	D^{2}EMO/MGD	HSMEA	K-RVEA	MOEA/D-EGO	ParEGO
ZDT3	3	**1.3199(3.34E-3)**	1.2914(5.99E-2)†	1.1814(1.77E-1)†	1.1831(9.85E-2)†	1.1549(4.84E-2)†
	5	**1.3271(1.52E-3)**	1.2888(7.93E-2)†	1.1842(7.44E-2)†	1.0566(1.58E-1)†	1.0364(1.57E-1)†
	8	**1.3260(4.29E-3)**	1.1901(1.53E-1)†	1.2394(1.16E-1)†	1.0050(1.14E-1)†	0.7974(1.09E-1)†
ZDT31	3	**1.3813(2.64E-2)**	1.3183(2.06E-1)†	1.0556(2.54E-1)†	1.0983(2.84E-1)†	1.0839(9.35E-2)†
	5	**1.3706(2.74E-2)**	1.3064(6.99E-2)†	1.1107(1.72E-1)†	1.0351(2.13E-1)†	1.0068(8.73E-2)†
	8	**1.3655(7.06E-2)**	1.1851(1.14E-1)†	1.1098(1.65E-1)†	0.9112(1.21E-1)†	0.8659(2.16E-1)†
ZDT32	3	**1.2818(2.18E-2)**	1.2309(9.73E-2)†	1.1852(7.96E-2)†	1.1265(4.74E-2)†	1.1170(1.52E-2)†
	5	**1.2610(8.73E-2)**	1.1801(9.62E-2)†	1.1359(6.86E-2)†	1.0960(6.78E-2)†	1.0801(1.07E-1)†
	8	**1.2860(2.12E-2)**	1.1604(9.41E-2)†	1.1550(8.02E-2)†	1.0355(1.04E-1)†	1.0175(1.15E-1)†
ZDT33	3	**0.9089(6.31E-4)**	0.8643(5.41E-2)†	0.8683(7.07E-2)†	0.8578(3.22E-2)†	0.8249(2.46E-2)†
	5	**0.9075(1.04E-3)**	0.8508(3.32E-2)†	0.8845(6.85E-2)†	0.8148(1.83E-2)†	0.7487(5.64E-2)†
	8	**0.9036(7.43E-2)**	0.8130(9.20E-2)†	0.8630(6.48E-2)†	0.7911(3.25E-2)†	0.7126(4.45E-2)†
WFG2	3	**5.9298(3.99E-2)**	5.7944(2.10E-1)†	5.7083(1.48E-1)†	4.9327(3.51E-1)†	4.6709(2.84E-1)†
	5	**5.9799(3.14E-2)**	5.8221(1.11E-1)†	5.4672(1.40E-1)†	4.6521(4.99E-1)†	4.5911(6.76E-1)†
	8	**5.7457(6.63E-2)**	5.1535(3.36E-1)†	5.1191(3.18E-1)†	4.0520(4.96E-1)†	3.8378(4.09E-1)†
WFG21	3	**6.2155(4.87E-2)**	5.5538(4.16E-1)†	5.5407(3.49E-1)†	4.9540(3.22E-1)†	4.8377(2.97E-1)†
	5	**6.2630(2.37E-2)**	5.0867(5.17E-1)†	5.5867(9.45E-2)†	4.7417(4.15E-1)†	4.6174(3.08E-1)†
	8	**6.0309(5.79E-2)**	4.5420(3.14E-1)†	5.2068(3.06E-1)†	4.0720(3.10E-1)†	3.9503(4.48E-1)†
WFG22	3	**2.8400(6.29E-2)**	2.8154(1.42E-1)†	2.5958(1.55E-1)†	2.0903(1.65E-1)†	2.0581(1.68E-1)†
	5	**2.8638(6.03E-2)**	2.6386(1.21E-1)†	2.3946(6.85E-2)†	1.8671(1.47E-1)†	1.9506(1.78E-1)†
	8	**2.6378(1.14E-1)**	2.0694(2.54E-1)†	1.8419(1.50E-1)†	1.3691(5.16E-1)†	1.3248(2.54E-1)†
WFG23	3	**5.7030(3.45E-1)**	5.6479(4.02E-1)†	5.4574(2.62E-1)†	4.9859(2.54E-1)†	5.0298(4.56E-1)†
	5	**5.9445(9.55E-2)**	5.3802(3.19E-1)†	5.4525(1.42E-1)†	4.9615(2.60E-1)†	4.7642(5.69E-1)†
	8	**5.6221(9.38E-2)**	4.7431(2.40E-1)†	4.8954(2.38E-1)†	4.2413(3.71E-1)†	4.0819(2.20E-1)†
DTLZ7	3	**1.3234(7.52E-3)**	1.3195(1.84E-2)†	1.2407(1.18E-2)†	1.2550(2.56E-2)†	1.2076(6.05E-2)†
	5	**1.3330(1.29E-3)**	1.3138(1.73E-2)†	1.2371(2.14E-2)†	1.2131(1.20E-1)†	1.1303(5.55E-2)†
	8	**1.3297(2.99E-3)**	1.3123(1.09E-2)†	1.2263(2.52E-2)†	1.1591(1.13E-1)†	1.0667(1.21E-1)†
DTLZ71	3	**1.3996(2.64E-3)**	1.3959(1.02E-2)†	1.3289(1.79E-2)†	1.2916(7.39E-1)†	1.2735(2.84E-2)†
	5	**1.4040(2.96E-3)**	1.3936(9.43E-3)†	1.3190(1.26E-2)†	1.2746(3.82E-1)†	1.0609(1.67E-1)†
	8	**1.4034(4.42E-3)**	1.3762(1.07E-1)†	1.3146(1.50E-2)†	1.0508(1.41E-1)†	0.9420(2.51E-1)†
DTLZ72	3	**1.4013(9.67E-3)**	1.4013(1.69E-2)†	1.3292(2.58E-2)†	1.3006(1.44E-1)†	1.2863(4.18E-2)†
	5	**1.4075(2.62E-3)**	1.3926(1.00E-2)†	1.3313(1.72E-2)†	1.1908(1.14E-1)†	1.2150(6.59E-2)†
	8	**1.4052(2.34E-3)**	1.3875(1.20E-1)†	1.3170(1.53E-2)†	1.0411(8.06E-2)†	0.9540(2.36E-1)†

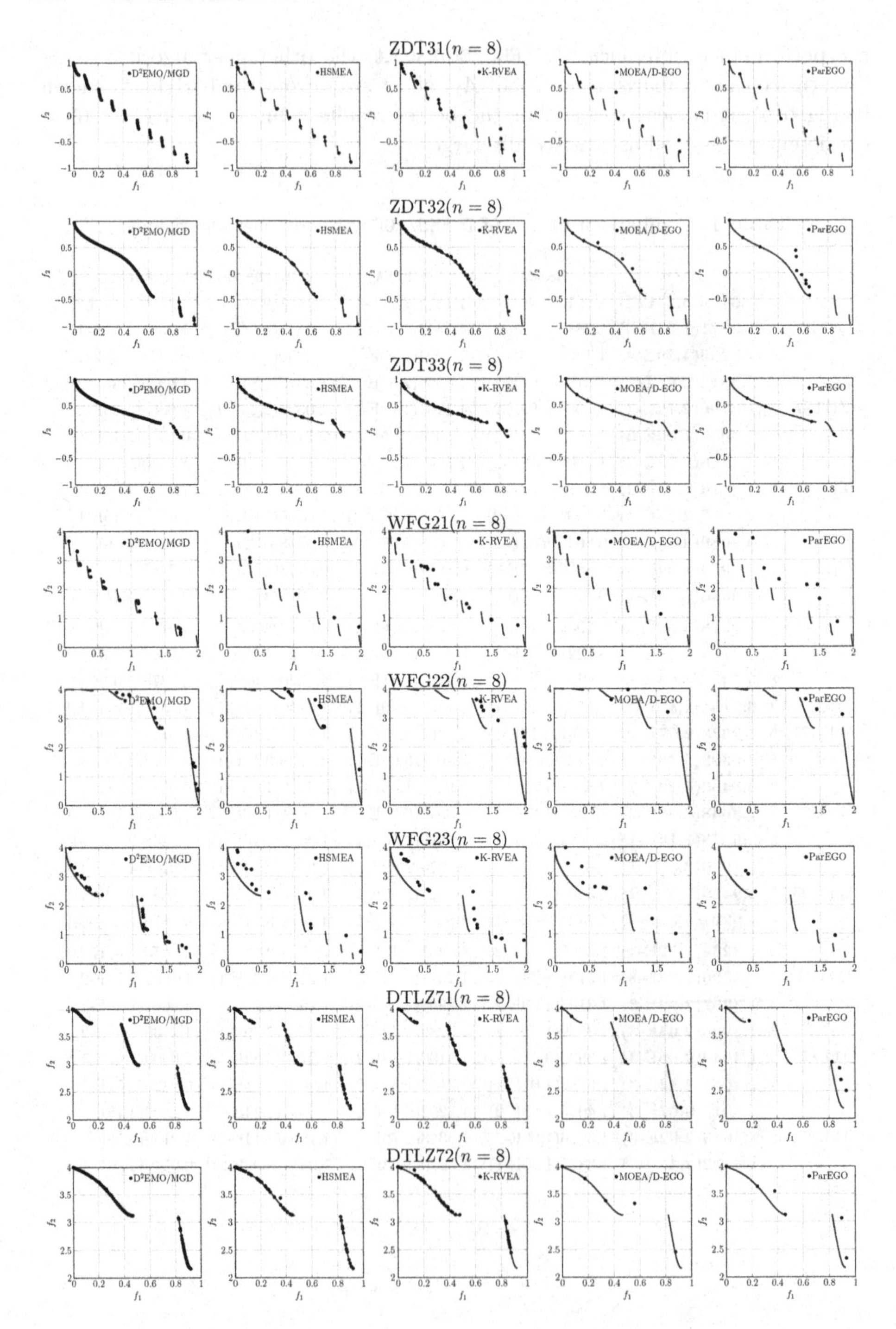

Fig. 4. Non-dominated solutions obtained by different algorithms (with the medium HV values) on ZDT3∗, WFG2∗, and DTLZ7∗ ($n = 8$), respectively.

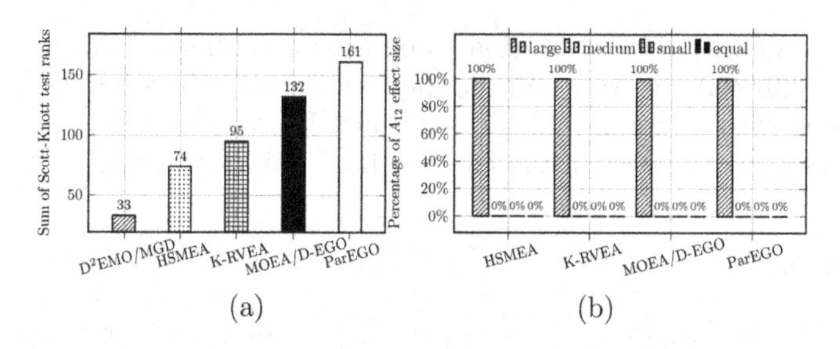

Fig. 5. Statistical test results: (a) sum of the Scott-knott test results on all comparisons; (b) percentage of the *equal*, *large*, *medium* and *small* A_{12} effect size, respectively, when comparing D^{2}EMO/MGD against the other four peer algorithms.

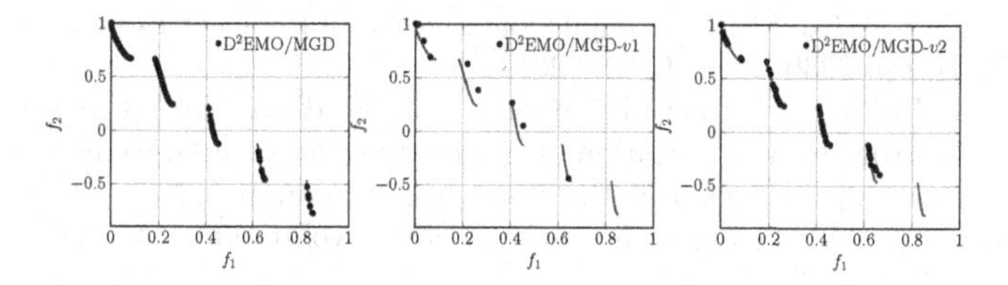

Fig. 6. Non-dominated solutions (with the medium HV value) found by different variants of D^{2}EMO/MGD on ZDT3 with $n = 8$.

5.2 Ablation Study

In this subsection, we empirically investigate the effectiveness of two key algorithmic components of D^{2}EMO/MGD. More specifically, we first compare the performance of D^{2}EMO/MGD w.r.t. the variant D^{2}EMO/MGD-$v1$ without using the MGD-based evolutionary search step. Instead, it applies the widely used simulated binary crossover [5] as the alternative operator for offspring reproduction. From the selected results shown in Fig. 6, we can see that D^{2}EMO/MGD-$v1$ can only find a very limited number of solutions on the PF with a poor diversity. Thereafter, we compare the performance of D^{2}EMO/MGD w.r.t. the variant D^{2}EMO/MGD-$v2$ without using the infill criterion introduced in Sect. 3.2. In particular, it uses a random selection to recommend the candidates for expensive function evaluations. From the selected results shown in Fig. 6, we can see that D^{2}EMO/MGD-$v2$ can only find some of the disconnected PF segments.

6 Conclusion

Most, if not all, existing data-driven EMO algorithms, directly apply evolutionary operators for offspring reproduction, while the regularity information embedded in the surrogate models has been unfortunately ignored. Bearing this consideration in mind, this paper, for the first time, investigates the use of MGD to leverage the latent information of the surrogate models to accelerate the convergence of the evolutionary population towards the PF. Due to the extra diversity provided by the exploration along the approximated PS manifold, our proposed D^{2}EMO/MGD has shown strong performance on the selected benchmark test problems with disconnected PF segments. In addition, the ablation study also confirms the usefulness of the batched infill criterion guided by the IHV.

References

1. Akhtar, T., Shoemaker, C.A.: Multi-objective optimization of computationally expensive multi-modal functions with RBF surrogates and multi-rule selection. J. Glob. Optim. **64**(1), 17–32 (2016). https://doi.org/10.1007/s10898-015-0270-y
2. Ariizumi, R., Tesch, M., Choset, H., Matsuno, F.: Expensive multiobjective optimization for robotics with consideration of heteroscedastic noise. In: Proceedings of the 2014 IEEE/RSJ International Conference on Intelligent Robots and Systems (RSJ 2014), pp. 2230–2235. IEEE (2014)
3. Chugh, T., Jin, Y., Miettinen, K., Hakanen, J., Sindhya, K.: A surrogate-assisted reference vector guided evolutionary algorithm for computationally expensive many-objective optimization. IEEE Trans. Evol. Comput. **22**(1), 129–142 (2018)
4. Deb, K.: Multi-Objective Optimization Using Evolutionary Algorithms. Wiley, New York (2001)
5. Deb, K., Agrawal, R.B.: Simulated binary crossover for continuous search space. Complex Syst. **9**(2), 115–148 (1994)
6. Deb, K., Goyal, M.: A combined genetic adaptive search (GeneAS) for engineering design. Comput. Sci. Inf. **26**, 30–45 (1996)
7. Deb, K., Thiele, L., Laumanns, M., Zitzler, E.: Scalable test problems for evolutionary multiobjective optimization. In: Abraham, A., Jain, L., Goldberg, R. (eds.) Evolutionary Multiobjective Optimization. Advanced Information and Knowledge Processing, pp 105–145. Springer, London (2005). https://doi.org/10.1007/1-84628-137-7_6
8. Désidéri, J.-A.: Multiple-gradient descent algorithm for pareto-front identification. In: Fitzgibbon, W., Kuznetsov, Y.A., Neittaanmäki, P., Pironneau, O. (eds.) Modeling, Simulation and Optimization for Science and Technology. CMAS, vol. 34, pp. 41–58. Springer, Dordrecht (2014). https://doi.org/10.1007/978-94-017-9054-3_3
9. Dickson, J.C., Frederick, F.P.: A decision rule for improved efficiency in solving linear programming problems with the simplex algorithm. Commun. ACM **3**(9), 509–512 (1960)
10. Frazier, P.I.: A tutorial on Bayesian optimization. arXiv preprint arXiv:1807.02811 (2018)
11. Habib, A., Singh, H.K., Chugh, T., Ray, T., Miettinen, K.: A multiple surrogate assisted decomposition-based evolutionary algorithm for expensive multi/many-objective optimization. IEEE Trans. Evol. Comput. **23**(6), 1000–1014 (2019)
12. Holland, J.H.: Genetic algorithms and the optimal allocation of trials. SIAM J. Comput. **2**(2), 88–105 (1973)
13. Huband, S., Hingston, P., Barone, L., While, R.L.: A review of multiobjective test problems and a scalable test problem toolkit. IEEE Trans. Evol. Comput. **10**(5), 477–506 (2006)
14. Ishibuchi, H., He, L., Shang, K.: Regular Pareto front shape is not realistic. In: Proceedings of the 2019 IEEE Congress on Evolutionary Computation (CEC 2019), pp. 2034–2041. IEEE (2019)

15. Jin, Y.: A comprehensive survey of fitness approximation in evolutionary computation. Soft. Comput. **9**(1), 3–12 (2005)
16. Jin, Y., Sendhoff, B.: A systems approach to evolutionary multiobjective structural optimization and beyond. IEEE Comp. Int. Mag. **4**(3), 62–76 (2009)
17. Jin, Y., Wang, H., Chugh, T., Guo, D., Miettinen, K.: Data-driven evolutionary optimization: an overview and case studies. IEEE Trans. Evol. Comput. **23**(3), 442–458 (2019)
18. Kennedy, J., Eberhart, R.: Particle swarm optimization. In: Proceedings of International Conference on Neural Networks (ICNN 1995), Perth, WA, Australia, 27 November–1 December 1995, pp. 1942–1948. IEEE (1995)
19. Knowles, J.D.: ParEGO: a hybrid algorithm with on-line landscape approximation for expensive multiobjective optimization problems. IEEE Trans. Evol. Comput. **10**(1), 50–66 (2006)
20. Kuhn, H.W., Tucker, A.W.: Nonlinear programming. In: Proceedings of the 2nd Berkeley Symposium on Mathematical Statistics and Probability, pp. 481–492. University of California Press, Berkeley, CA (1951)
21. Loshchilov, I., Schoenauer, M., Sebag, M.: Dominance-based Pareto-surrogate for multi-objective optimization. In: Deb, K., et al. (eds.) SEAL 2010. LNCS, vol. 6457, pp. 230–239. Springer, Heidelberg (2010). https://doi.org/10.1007/978-3-642-17298-4_24
22. Marques, J., Cunha, M.C., Savic, D.A.: Multi-objective optimization of water distribution systems based on a real options approach. Environ. Model. Softw. **63**, 1–13 (2015)
23. Martínez, S.Z., Coello, C.A.C.: MOEA/D assisted by RBF networks for expensive multi-objective optimization problems. In: Proceedings of the 2013 Genetic and Evolutionary Computation Conference (GECCO 2013), pp. 1405–1412. ACM (2013)
24. Mittas, N., Angelis, L.: Ranking and clustering software cost estimation models through a multiple comparisons algorithm. IEEE Trans. Softw. Eng. **39**(4), 537–551 (2013)
25. Nelder, J.A., Mead, R.: A simplex method for function minimization. Comput. J. **7**(4), 308–313 (1965)
26. Nielsen, H.B., Lophaven, S.N., Søndergaard, J.: DACE – A MATLAB Kriging toolbox (2002)
27. Powell, M.J.D.: On the global convergence of trust region algorithms for unconstrained minimization. Math. Program. **29**(3), 297–303 (1984)
28. Rasmussen, C.E., Williams, C.K.I.: Gaussian processes for machine learning. In: Adaptive Computation and Machine Learning, MIT Press (2006)
29. Santner, T.J., Williams, B.J., Notz, W.I.: The Design and Analysis of Computer Experiments. SSS, Springer, New York (2018). https://doi.org/10.1007/978-1-4939-8847-1
30. Storn, R., Price, K.V.: Differential evolution - a simple and efficient heuristic for global optimization over continuous spaces. J. Glob. Optim. **11**(4), 341–359 (1997)
31. Sun, G., Li, G., Gong, Z., He, G., Li, Q.: Radial basis functional model for multi-objective sheet metal forming optimization. Eng. Optim. **43**(12), 1351–1366 (2011)
32. Vargha, A., Delaney, H.D.: A critique and improvement of the CL common language effect size statistics of McGraw and Wong. J. Educ. Behav. Stat. **25**(2), 101–132 (2000)
33. Wilcoxon, F.: Individual comparisons by ranking methods (1945)

34. Zhang, Q., Liu, W., Tsang, E.P.K., Virginas, B.: Expensive multiobjective optimization by MOEA/D with Gaussian process model. IEEE Trans. Evol. Comput. **14**(3), 456–474 (2010)
35. Zitzler, E., Deb, K., Thiele, L.: Comparison of multiobjective evolutionary algorithms: empirical results. Evol. Comput. **8**(2), 173–195 (2000)
36. Zitzler, E., Thiele, L.: Multiobjective evolutionary algorithms: a comparative case study and the strength Pareto approach. IEEE Trans. Evol. Comput. **3**(4), 257–271 (1999)

Eliminating Non-dominated Sorting from NSGA-III

Balija Santoshkumar[1(✉)], Kalyanmoy Deb[1], and Lei Chen[2]

[1] Computational Optimization and Innovation (COIN) Laboratory, Michigan State University, East Lansing, MI 48864, USA
{balijasa,kdeb}@egr.msu.edu
[2] Guangdong University of Technology, Guangzhou, China
https://www.coin-lab.org

Abstract. The series of non-dominated sorting based genetic algorithms (NSGA-series) has clearly shown their niche in solving multi- and many-objective optimization problems since mid-nineties. Of them, NSGA-III was designed to solve problems having three or more objectives efficiently. It is well established that with an increase in number of objectives, an increasingly large proportion of a random population stays non-dominated, thereby making only a few population members to remain dominated. Thus, in many-objective optimization problems, the need for a non-dominated sorting (NDS) procedure is questionable, except in early generations. In support of this argument, it can also be noted that most other popular evolutionary multi- and many-objective optimization algorithms do not use the NDS procedure. In this paper, we investigate the effect of NDS procedure on the performance of NSGA-III. From simulation results on two to 10-objective problems, it is observed that an elimination of the NDS procedure from NSGA-III must accompany a penalty boundary intersection (PBI) type niching method to indirectly emphasize best non-dominated solutions. Elimination of the NDS procedure from NSGA-III will open up a number of avenues for NSGA-III to be modified for different scenarios, such as, for parallel implementations, surrogate-assisted applications, and others, more easily.

Keywords: Non-dominated sorting · Multi-objective optimization · Evolutionary computation · NSGA-III

1 Introduction

The first non-dominated sorting based genetic algorithm (NSGA) was proposed in 1995 [15]. It ushered in a new era of computational optimization methods for handling two-objective problems along with a few other contemporary algorithms [6,8,9]. In 2002, an elitist and parameter-less version of NSGA, called NSGA-II, was proposed to solve primarily two and three-objective problems [4]. Thereafter, in 2014, a reference vector based extension, called NSGA-III, was proposed to handle three and more objective problems. They all have one operation in common: non-dominated sorting (NDS) of the population based on

M. Emmerich et al. (Eds.): EMO 2023, LNCS 13970, pp. 71–85, 2023.
https://doi.org/10.1007/978-3-031-27250-9_6

the partial ordering of their objective vectors. The NSGA-series of procedures require that every population member to be classified into a different NDS level. To achieve the sorting procedure within a population of solutions, pairwise comparison of solutions are made with their objective vectors to identify the set of population members which are not dominated by any other population member. This set of non-dominated solutions belong to the first NDS-level. To obtain the second NDS-level members, the first NDS-level members are discounted from the population and another round of pair-wise domination check is performed. The members which do not get dominated by any remaining population members belong to the second NDS-level. This process is continued until all population members are classified into a distinct NDS level. The NSGA series of procedures were based on these sorted classes of population members and emphasized a lower NDS level to be infinitely more important than the next higher NDS level. All NSGA operations were applied by keeping the hierarchy of NDS level of population members. Thus, NDS is intricately linked to the core of NSGA series of algorithms.

It has also been established that when many-objective optimization problems (with more than three conflicting objectives) are to solved, a randomly created population contains increasingly more NDS-level one solutions. For a 10-objective problem, the number of non-dominated (ND) solutions in a random population of size 100 is about 95 [6], The argument can be extended to state that number of NDS level-two members will be significantly small compared to NDS level-one members, and so on. Thus, the effectiveness of executing the NDS procedure for many-objective problems can be questionable. Whether a solution belongs to second or third level of non-domination may not matter on the overall progress of the search algorithm, as there are not many population members exist in the dominated levels altogether. Moreover, two dominated solutions of different levels may stay close in the objective space, hence a classification of one solution to a relatively lower class and the other to a higher class may not produce any significant difference in the performance of the search algorithm.

Thus, it is worth an investigation to eliminate NDS from the NSGA-III procedure and classify the entire population into two classes: non-dominated and dominated classes. If the performance stays similar to the original NSGA-III procedure, the modified search procedure can be beneficial in a number of scenarios. First, the NDS sorting procedure takes $O(MN^2)$ [4] (where M is the number of objectives and N is the population size), which is more complex than identifying the NDS level one members ($O(MN(\log N)^{M-2})$) [13]. Second, when an EMO or EMaO algorithm is to be implemented with surrogates for handling computationally expensive problems, the NDS procedure may have to be performed on a population evaluated with a mix of high-fidelity and surrogate-assisted evaluations. In such a scenario, a classification of every population member to a precise NDS level may be an overkill, particularly since the objective values are noisy.

Having made the argument against a full effectiveness of NDS in a search algorithm, the next important question pertains to the dependencies of NSGA-III's other operators on the NDS procedure. In this paper, we eliminate the NDS procedure from NSGA-III and study the changes that must be introduced in other NSGA-III operators to bring the modified NSGA-III's performance at least at par to that of the original NSGA-III procedure.

In the rest of the paper, we present the modified NSGA-III procedure (without NDS procedure) in Sect. 2. Results on two to 10-objective problems of the modified NSGA-III procedure is presented and compared against the original NSGA-III and other EMaO algorithms in Sect. 3. Conclusions are drawn in Sect. 4.

2 Proposed Algorithm: N$\tilde{\text{S}}$GA-III(NSGA-III\NDS)

The basic framework of the proposed algorithm is similar to NSGA-III [7] with significant changes in the way (i) domination check is executed, (ii) the survival selection operator is modified, and (iii) a few other minor changes are adopted resulting from the change in domination check procedure.

Instead of using Das-Dennis method of creating reference vectors, we use Riesz s-Energy based method proposed in [2,3]. This allows any population size (N) to be used for any objective dimension. Like in NSGA-III [7], first, we initialize the population P_t of size N. We generate an offspring population Q_t of size N with the standard genetic operators and without care of each population member's association to any reference vector. However, the mating selection operator requires the domination status of a population member, which we describe in the next paragraph. The combined population (parent and offspring) is R_t of size $2N$. The survival selection operator then chooses N solutions from R_t and save to P_{t+1}. Iterations proceed until a termination criterion is satisfied. The overall algorithm is provided in Algorithm 1.

2.1 Classification of Pop. Members

We classify all the population members (R_t) into three hierarchical classes: Class 1: non-dominated and feasible solutions; Class 2: feasible and dominated solutions, and Class 3: infeasible solutions. Notice that all feasible solutions from the second ND front are combined into Class 2. Figure 1 illustrates the classification process. We use a hierarchical classification process for both mating and survival selection operators. A solution belonging to a lower class is better. Thus, for binary tournament selection in the mating selection operator above classification helps to select the better candidate.

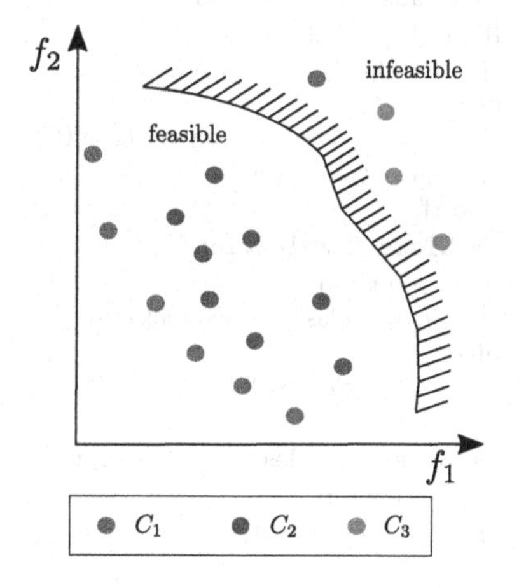

Fig. 1. Classification of population members into three hierarchical classes.

2.2 Association of Population Members

Population members are normalized using the same procedure as in NSGA-III. Thereafter, each population member is associated with a particular reference

---

Algorithm 1: Generation t in $\tilde{\text{NSGA}}$-III=NSGA-III\NDS

```
   Data: Predefined set of reference directions W_r and parent population P_t
   Result: P_{t+1}
1  S_t = ∅, no of selections remaining (n_r) = N
   /* offspring population generation                                        */
2  Q_t = Recombination + Mutation (P_t)
3  R_t = P_t ∪ Q_t
4  I=find-non-dominated (R_t)
   /* finding ideal point and nadir point for normalization of
      objective space                                                       */
5  ideal, nadir=hyperplane-boundary-estimation(R_t, I)
   /* Association each s in R_t with a reference direction                  */
6  [π(s), d_2(s)]=associate-to-niches(R_t, W_r, ideal, nadir)
   /* π(s)=closest reference direction, d_2(s)= perpendicular distance
      between π(s) and s                                                    */
7  d(s) =pbi-decomposition(R_t, W_r, θ = 5, ideal, nadir)
   /* d(s) = d_1 + θd_2 distance between π(s) and s                         */
8  Set attribute to each population member non-dominated ND = 1 and dominated
   ND = 0
9  Set attributes each population member R_t(CV, ND, d(s))
   /* classifying population members                                        */
10 Class 1 (C_1) : R_t(CV = 0 ∩ ND = 1)
11 Class 2 (C_2) : R_t(CV = 0 ∩ ND = 0)
12 Class 3 (C_3) : R_t(CV > 0)
   /* Class 1 selection                                                     */
13 if (|C_1| ≤ N) then
14 |   S_t = S_t ∪ C_1; n_r = N − |S_t|
15 else
16 |   S_t, n_r= class-survivor-selection(C_1, S_t, n_r, W_r, π(s), d(s));
17 |   P_{t+1} = S_t, break
18 end if
   /* Class 2 selection                                                     */
19 if n_r > 0 then
20 |   S_t, n_r= class-survivor-selection(C_2, S_t, n_r, W_r, π(s), d(s));
21 else
22 |   P_{t+1} = S_t, break
23 end if
   /* Class 3 selection for constrained problems                            */
24 if n_r > 0 then
25 |   S_t= tournament-selection(C_3);
26 else
27 |   P_{t+1} = S_t, break
28 end if
```

---

vector based on the d_2 distance metric, which is the perpendicular distance from the population member's normalized objective vector to the reference line. This association principle is followed for all classes of population members.

2.3 Class-wise Mating Selection

In mating selection, two population members are picked at random and a winner must be selected to act as a parent for mating. A lower class member is always the winner. This allows a feasible solution to be better than infeasible solution and a non-dominated feasible solution to be better than a feasible dominated solution. But if both picked members belong to Class 3, the one with smaller overall normalized constrained violation (CV) value [6] wins. For highly constrained problems, it is likely that most population members are infeasible. In this case, preferring a smaller constraint violated solution provides a good signal to the EMO to gradually progress toward the PO front. When both solutions belong to either Class 1 or Class 2, one is randomly chosen as a parent. This disallows any competition between solutions from different classes. Since the number of reference vectors (desired number of final PO solutions) is identical to the population size (N), mating selection, involving exactly N members, must not encourage any competition among N population members.

2.4 Reference Vector Based Niching in Survival Selection

In the survival selection, there are $2N$ population members and exactly N better diverse members must be chosen as the next generation's starting population. Competitions between associated feasible solutions of a reference vector may be allowed here. For this purpose, we follow a similar niching procedure as in NSGA-III and choose a single member for each reference vector.

For this purpose, all associated members of a reference vector are considered and instead of the d_2 metric, following procedure is used. Associated members are considered in a hierarchical manner based on their class (from Class 1 to Class 3 in the order). Among all the members of the best class, the best solution is chosen as follows. If the best class is 3 (meaning all infeasible associated members), the member with smallest CV is chosen. If the best class is 1 or 2, all associated Class 1 or Class 2, as the case may be, members are compared with the PBI distance metric: $d(\mathbf{s}) = d_1 + \theta d_2$, where d_1 is the distance along reference direction to origin (equivalent to the ideal point), θ (5 used here) is a parameter, and d_2 is the perpendicular distance (see Fig. 2). We choose the one having smaller $d(\mathbf{s})$ as the winner. After performing the above process for all

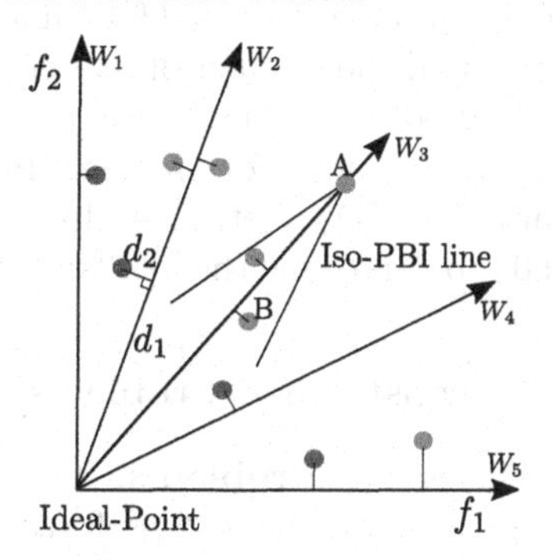

Fig. 2. Association of population members to reference directions (**W**) and their distance metrics.

N reference vectors (note that some may not have any associated population member), the process is repeated to pick the next round of solutions. This is repeated until N members are selected. The reason for choosing the PBI metric for selection can be given from Fig. 2. Since all dominated classes of solutions are clubbed together, the association principle can allow a distant member lying close to the reference vector may be judged to be better based on the d_2 metric. But the use if PBI metric makes a combined weighted distance of d_1 and d_2 and thus, even being close to the reference vector, a near ideal-point member may be judged better. When NDS was conducted in NSGA-III, member A was never allowed to compare against member B for closeness to the reference vector (as B dominates A and they will be in different non-dominated levels), hence PBI metric was not needed. But without the NDS, a distance metric with combined d_1 and d_2 is must to compare two associated population members for the same reference vector. Hence, we replace NSGA-III's d_2 vector with the PBI metric for choosing the winner.

3 Experimental Results

In this section, we present the simulation results of the proposed method N$\tilde{\text{S}}$GA-III with NSGA-III [7,12] and MOEA/D [17] on ZDT [18], BNH [1], OSY [14], SRN [15], TNK [16], DTLZ test suite [5] and WFG test suite [10] with objectives ranging from 2 to 10. To support the use of the PBI metric in the survival selection operator, we also replace it with the d_2 metric (call it N$\tilde{\text{S}}$GA-III-d_2) and compare with N$\tilde{\text{S}}$GA-III.

For each problem, we run all the algorithms 31 times with different initial population members. The population size for the 2 and 3-objective problems is set to 100, for the 5 and 8-objective-problems to 200, and for the 10-objectives problem to 300. Each run is executed for a maximum of 100,000 solution evaluations (SEs). We have used the number of reference directions the same as the population size. We have used IGD+ [11] as the performance metric, as it measures both convergence and diversity. For all algorithm, the final generation members are used to compute the IGD+ metric. We have used Wilcoxon signed-rank test with at most $p = 0.05$ to determine the best and statistically similar methods.

3.1 Unconstrained Problems

Two-Objective Problems: The performance metrics for ZDT problems are given in Table 1. The best performing method is marked in bold, and the other methods which are statistically similar to the best method are in marked in italics. The representative objective vectors for some representative ZDT problems are presented in Fig. 3. It can be observed that N$\tilde{\text{S}}$GA-III-d_2 performs the best in three out of the six problems, despite not executing NDS. Interestingly, removal of NDS operation from NSGA-III is found to be more effective than performing the NDS operation, but the use of NDS is not found too detrimental. The bottom-right figure shows that although in initial generations N$\tilde{\text{S}}$GA-III-d_2

converges slowly because of the d_2 metric selection but after a certain number of generations (around 150) it performs better.

Table 1. ZDT problems IGD+ performance metrics for 31 runs. The best performing method is in bold and the other methods which are statistically similar to the best method are in italics with Wilcoxon signed-rank test having $p = 0.05$. NS̃GA-III (5-th and 6-th column) is NSGA-III without the sorting process.

Problem	M	MOEA/D-TCH	NSGA-III	NS̃GA-III-d_2	NS̃GA-III
ZDT1	2	3.9720e−3	4.2770e−3	**2.8750e−3**	3.2900e−3
ZDT2	2	**2.6940e−3**	4.2370e−3	3.0290e−3	3.7760e−3
ZDT3	2	3.2490e−3	3.1050e−3	**2.1120e−3**	*1.9180e−3*
ZDT4	2	7.4550e−3	4.0090e−3	**3.0570e−3**	3.9250e−3
ZDT5	2	8.7824e−2	**8.3809e−2**	9.1272e−2	*9.6077e−2*
ZDT6	2	2.7570e−3	**2.3190e−3**	2.3440e−3	2.3570e−3
Best/similar/total →		1/0/6	2/0/6	**3**/0/6	0/2/6

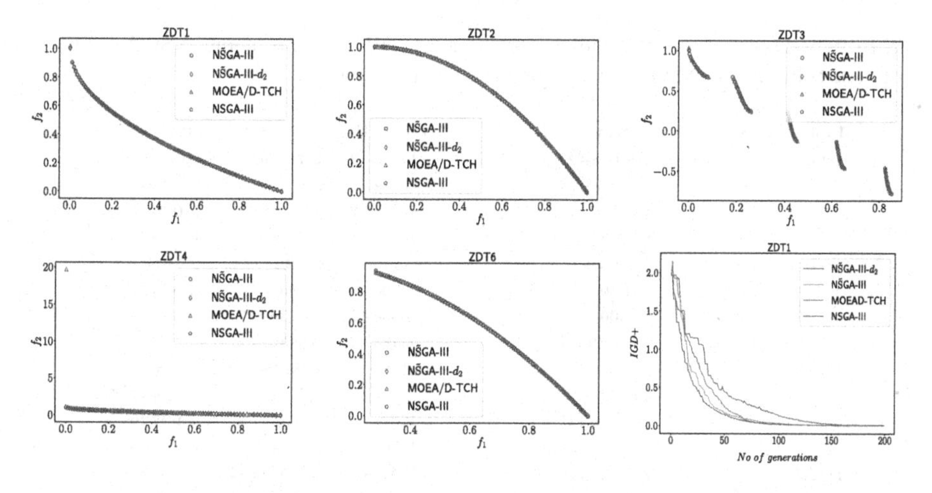

Fig. 3. Obtained solutions by all methods on some ZDT problems.

Three and Many-Objective Problems: The number of variables for DTLZ test suite is chosen with $k = 10$ where $k = (n - M + 1)$ and for WFG we have chosen $n = 30$ with $k = 2(M - 1)$ and $k + l = n \geq M$. The performance metrics for DTLZ and WFG problems are presented in Tables 2 and 3, respectively. Representative solutions on some DTLZ and WFG problems are shown in Figs. 4 and 5, respectively.

It is reported (and consistent with the literature) that MOEA/D performs better than NSGA-III on DTLZ problems mainly due to the similarly-scaled objective values for all objectives. MOEA/D reports all ND solutions from the final generation as an outcome, while NSGA-III reports a single best population member for each active reference line from the final generation. While the number of reported solutions from both these algorithms are more or less identical for

DTLZ2 type of problems, MOEA/D will report more ND solutions for DTLZ5 (degenerate PO front dimension or constrained problems) than NSGA-III. It is important to highlight that NSGA-III's final population may have more ND solutions than its reported number of solutions, but one solution per reference vector is reported to provide a widely distributed set of ND solutions. The IGD+ values shown for DTLZ5 and DTLZ6 (degenerate problems) in Table 2 with brackets are the IGD+ values computed by taking only PBI-metric associated solutions for each active reference vectors instead of taking all final ND members. Comparative IGD+ values with N$\tilde{S}$GA-III are now observed.

It is also clear from Table 3 that MOEA/D does not work well on WFG problems, mainly due to non-uniform scaling of objectives in these problems. N$\tilde{S}$GA-III works better than all other methods on WFG problems (26 best of 36 problems) and the combined DTLZ and WFG problems (36 best of 64 problems), followed by N$\tilde{S}$GA-III-d_2 method (14 best of 64 problems). Thus, it is interesting to conclude from the two tables that NDS was not a very important operation for NSGA-III for solving three and many-objective problems. While in two-objective problems N$\tilde{S}$GA-III-d_2 works better, for three and many-objective problems, N$\tilde{S}$GA-III works much better with the PBI metric, rather than d_2 metric.

Table 2. IGD+ performance metrics for 31 runs on DTLZ problems. N$\tilde{S}$GA-III (5-th and 6-th columns) is NSGA-III without the sorting process.

Problem	M	MOEA/D-PBI	NSGA-III	N$\tilde{S}$GA-III-d_2	N$\tilde{S}$GA-III
DTLZ1	3	3.4964e−2	3.3075e−2	**2.7376e−2**	3.0468e−2
	5	1.3433e−2	2.4264e−2	2.3384e−2	**9.8980e−3**
	8	3.5368e−2	6.8785e−2	4.4221e−2	**2.0986e−2**
	10	8.0652e−2	1.0557e−1	7.2443e−2	**5.3484e−2**
DTLZ2	3	2.6866e−2	2.6114e−2	**2.5251e−2**	2.5400e−2
	5	4.8030e−3	4.2260e−3	*2.8530e−3*	**2.5850e−3**
	8	9.5400e−3	1.0530e−2	1.1239e−2	**7.5900e−3**
	10	**1.1309e−2**	3.0248e−2	2.9952e−2	1.7584e−2
DTLZ3	3	4.1858e−2	3.5138e−2	**2.9578e−2**	3.1881e−2
	5	2.6342e−2	6.3278e−2	6.0382e−2	**1.9904e−2**
	8	6.5717e−1	1.4607e−1	9.8674e−2	**3.9972e−2**
	10	*7.6384e−1*	2.4370e−1	2.2173e−1	**8.2418e−2**
DTLZ4	3	2.3108e−1	2.5742e−2	**2.5200e−2**	2.5293e−2
	5	1.2806e−1	7.1360e−3	**2.4350e−3**	4.7470e−3
	8	1.6967e−1	1.2310e−2	**8.4650e−3**	1.4287e−2
	10	1.9387e−1	*2.2136e−2*	**2.2127e−2**	2.4918e−2
DTLZ5	3	**1.0821e−2** (1.9361e−2)	1.8919e−2	8.3213e−2	7.3286e−2
	5	**5.4780e−3** (4.3105e−2)	1.7890e−1	1.7765e−1	5.6062e−2
	8	**1.4013e−2** (3.6413e−1)	3.7251e−1	3.2480e−1	9.8550e−2
	10	**2.0389e−2** (3.7156e−1)	3.9726e−1	3.0289e−1	8.0261e−2
DTLZ6	3	**1.0750e−2** (1.8531e−2)	1.9141e−2	2.4254e−2	1.9901e−2
	5	**6.8450e−3** (3.2681e−2)	2.3767e+0	2.2249e+0	8.4766e−2
	8	**1.3850e−2** (3.6379e−1)	4.1552e+0	4.0039e+0	5.5816e−1
	10	**2.0448e−2** (3.7158e−1)	5.7369e+0	5.5902e+0	1.8406e+0
DTLZ7	3	5.9116e−2	**3.3175e−2**	3.5356e−2	3.5393e−2
	5	1.3871e−1	1.1134e−1	1.1303e−1	**1.0425e−1**
	8	1.2302e+0	1.8461e−1	1.8533e−1	**1.7900e−1**
	10	1.6129e+0	*1.8760e−1*	**1.8816e−1**	*1.8847e−1*
Best/similar/total →		9/1/28	1/2/28	8/1/28	**10/1/28**

With more objectives, the proportion of non-dominated members in a finite population becomes more [6]. It also likely that at later generations, most reference vectors will have a single associated member, particularly for problems having every reference vector leading to a distinct PO solution. In such cases, the use of d_2 or PBI metric may not matter much. However, early on, this may

Table 3. IGD+ performance metrics for 31 runs on WFG problems. N$\tilde{\text{S}}$GA-III (5-th and 6-th columns) is NSGA-III without the sorting process.

Problem	M	MOEA/D-PBI	NSGA-III	N$\tilde{\text{S}}$GA-III-d_2	N$\tilde{\text{S}}$GA-III
WFG1	3	4.3446e−1	4.1909e−1	3.7651e−1	**3.6318e−1**
	5	4.5475e−1	5.1397e−1	4.8717e−1	**4.7428e−1**
	8	**4.5695e−1**	5.4412e−1	5.0332e−1	4.8024e−1
	10	4.8536e−1	4.8279e−1	4.3790e−1	**4.1891e−1**
WFG2	3	7.7834e−2	2.7083e−2	1.6556e−2	**1.3154e−2**
	5	1.3579e−1	4.4112e−2	4.4695e−2	**3.0848e−2**
	8	1.5377e−1	4.4840e−2	4.5205e−2	**3.2744e−2**
	10	1.6877e−1	*2.8194e−2*	**2.6693e−2**	2.9477e−2
WFG3	3	1.6463e−1	4.7558e−2	6.1073e−2	**4.4320e−2**
	5	2.3672e+0	3.5530e−1	2.5779e−1	**2.0911e−1**
	8	6.5851e+1	1.9058e+0	4.9633e−1	**3.7363e−1**
	10	5.8560e+2	2.8201e+1	4.3064e+0	**3.5071e+0**
WFG4	3	5.9218e−2	4.5232e−2	3.0611e−2	**2.9883e−2**
	5	2.8312e−1	9.9648e−2	**8.8945e−2**	*8.9776e−2*
	8	7.1960e−1	1.5102e−1	*1.4690e−1*	**1.4659e−1**
	10	7.4449e−1	3.4704e−1	*3.1872e−1*	**3.1720e−1**
WFG5	3	6.2305e−2	5.6573e−2	*4.4542e−2*	**4.4398e−2**
	5	1.8382e−1	8.3703e−2	**7.5884e−2**	*7.6385e−2*
	8	8.3541e−1	1.7123e−1	*1.6990e−1*	**1.6945e−1**
	10	1.9304e+0	1.8594e−1	*1.8349e−1*	**1.8225e−1**
WFG6	3	6.5468e−2	5.4180e−2	*4.0747e−2*	**3.9925e−2**
	5	2.7583e−1	7.6170e−2	*6.5350e−2*	**6.4881e−2**
	8	6.5052e−1	1.2266e−1	**1.1600e−1**	*1.1656e−1*
	10	7.7136e−1	1.5115e−1	*1.4879e−1*	**1.4749e−1**
WFG7	3	6.9492e−2	3.8074e−2	2.7420e−2	**2.7256e−2**
	5	2.2374e−1	5.6599e−2	**4.8126e−2**	*4.8975e−2*
	8	6.3725e−1	1.1505e−1	*1.1071e−1*	**1.1030e−1**
	10	7.5967e−1	1.5021e−1	1.4735e−1	**1.4209e−1**
WFG8	3	8.8971e−2	7.3944e−2	5.9074e−2	**5.8133e−2**
	5	2.4557e−1	1.2901e−1	*1.2277e−1*	**1.2269e−1**
	8	8.6388e−1	**2.0446e−1**	2.1285e−1	*2.0935e−1*
	10	1.0092e+0	**2.1857e−1**	2.5549e−1	2.5674e−1
WFG9	3	9.1279e−2	6.4961e−2	*4.9177e−2*	**4.6800e−2**
	5	1.8287e−1	1.1333e−1	*1.1344e−1*	**1.0975e−1**
	8	5.2175e−1	**1.5261e−1**	1.6783e−1	*1.5377e−1*
	10	7.2420e−1	*1.9624e−1*	**1.9613e−1**	*1.9981e−1*
Best/similar /total →		1/0/36	3/2/36	6/12/36	**26**/7/36
DTLZ + WFG →		10/1/64	4/4/64	14/13/64	**36**/8/64

not be the case and the difference between d_2 and PBI metric may show up. If NS̃GA-III and NS̃GA-III-d_2 IGD+ metric values are compared for the DTLZ5 problem having a few active reference vectors leading to a PO solution, many associated population members are expected for each of the active reference vectors. The performance of NS̃GA-III is better than NS̃GA-III-d_2. To support this argument, we plot the variation of IGD+ value versus generations in Fig. 6 for five-objective DTLZ2 and DTLZ5 problems. It can observed that while the performance of all three NSGA-III methods are more or less the same (with a slight edge for NS̃GA-III), for DTLZ5, NS̃GA-III performs the best.

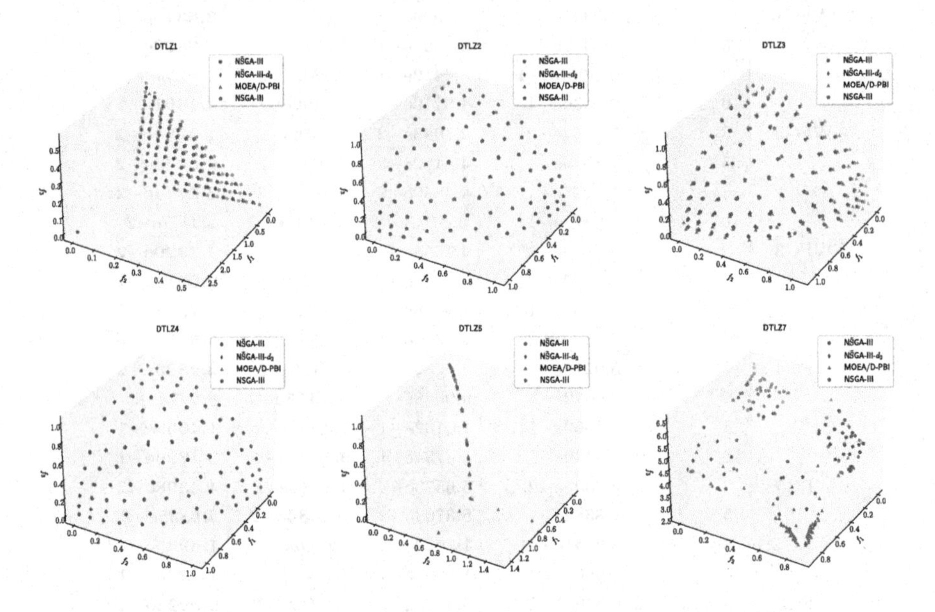

Fig. 4. Obtained solutions by all methods on some DTLZ problems.

3.2 Constrained Problems

Next, we apply all three NSGA-III methods to constrained problems. Since MOEA/D is not usually used for constrained problems, we ignore it here. Our NS̃GA-III method includes constraint violation as Class 3 solutions and are well-equipped to solve constrained problems.

Two-Objective Problems: First, we consider two-objective test problems: BNH, OSY, SRN and TNK [6]. Results are presented in Table 4. It is clear that NS̃GA-III-d_2 performs the best for the two-objective problems, as in the case of unconstrained two-objective problems, shown in Table 1.

Table 4. IGD+ performance metrics for 31 runs on two-objective constrained test problems. N$\tilde{\text{S}}$GA-III (4-th and 5-th columns) is NSGA-III without the sorting process.

Problem	M	NSGA-III	N$\tilde{\text{S}}$GA-III-d_2	N$\tilde{\text{S}}$GA-III
BNH	2	**2.8870e−3**	2.9970e−3	5.4130e−3
OSY	2	2.4169e−2	**4.7080e−3**	9.3770e−3
SRN	2	*2.9650e−3*	3.1560e−3	**2.9390e−3**
TNK	2	4.3210e−3	**3.5450e−3**	5.1930e−3
Best/similar /total →		1/1/4	**2**/0/4	1/0/4

Three and Many-Objective Problems: Table 5 presents the results on three and many-objective (5, 8 and 10-obj.) constrained optimization problems. It is clear that both NSGA-III versions without NDS operation works better than the original NSGA-III, with a slight edge for N$\tilde{\text{S}}$GA-III (10 best out of 20 problems), followed by N$\tilde{\text{S}}$GA-III-d_2 (8 best out of 20 problems).

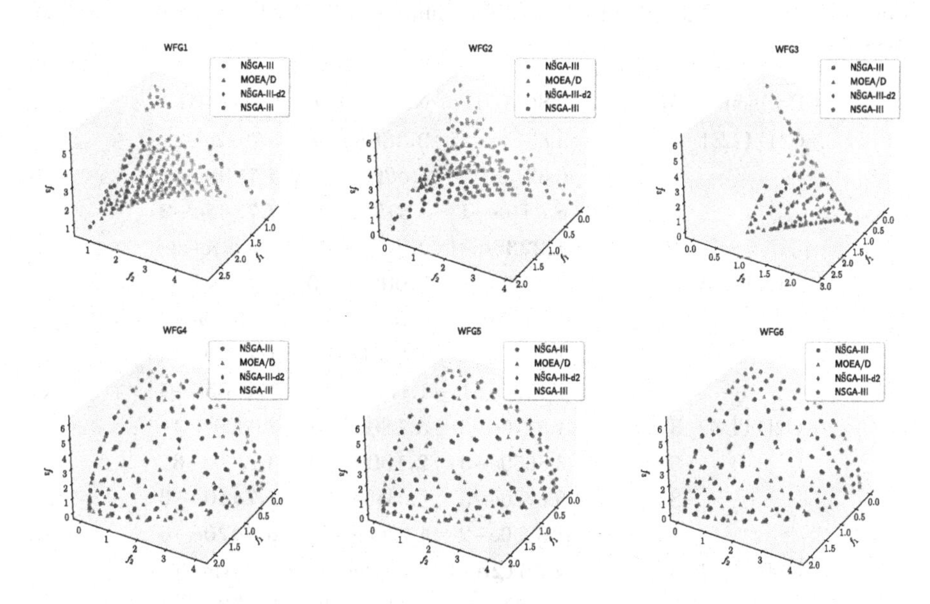

Fig. 5. Obtained solutions by all methods on some WFG problems.

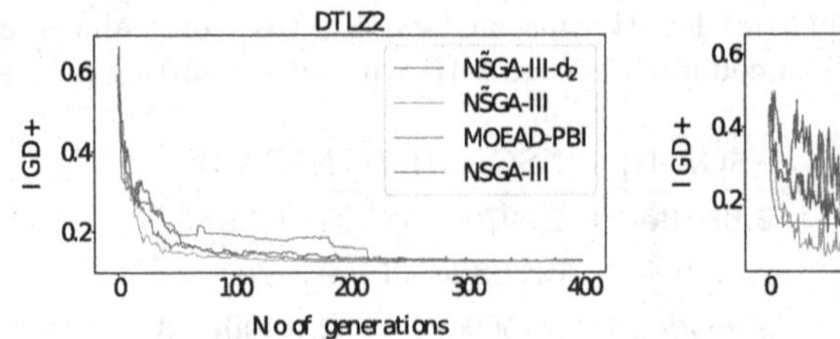

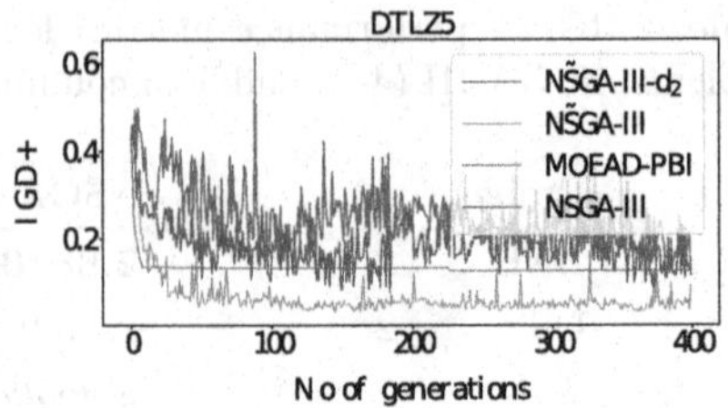

Fig. 6. Convergence history of algorithms on DTLZ2-5obj and DTLZ5-5obj problems. N$\tilde{S}$GA-III produces better IGD+ values.

Figure 7 shows the performance of all NSGA-III versions on the C2DTLZ2 problem with 10-objectives, showing similar distributions, but convergence by N$\tilde{S}$GA-III is slightly better (see also Table 5).

Table 5. IGD+ performance metrics for 31 runs on three and many-objective constrained problems. N$\tilde{S}$GA-III (4-th and 5-th columns) is NSGA-III without the sorting process.

Problem	M	NSGA-III	N$\tilde{S}$GA-III-d_2	N$\tilde{S}$GA-III
C1DTLZ1	3	3.8573e−2	**2.3388e−2**	3.7922e−2
	5	6.3775e−2	*5.5396e−2*	**4.7830e−2**
	8	*8.0516e−2*	*8.3178e−2*	**7.7365e−2**
	10	**1.2233e−1**	*1.2774e−1*	*1.3327e−1*
C1DTLZ3	3	8.0129e+0	**8.0080e+0**	8.0109e+0
	5	1.1587e+1	1.1583e+1	**1.1570e+1**
	8	1.1677e+1	1.1673e+1	**1.1620e+1**
	10	1.1726e+1	1.1721e+1	**1.1665e+1**
C2DTLZ2	3	1.8170e−3	**3.1800e−4**	8.0800e−4
	5	3.8650e−3	**2.1900e−3**	3.1260e−3
	8	5.6720e−3	5.0310e−3	**1.7410e−3**
	10	1.3350e−2	1.2366e−2	**5.4420e−3**
C3DTLZ1	3	**4.5512e−1**	*5.8319e−1*	5.8754e−1
	5	5.3615e−1	5.3443e−1	**5.2738e−1**
	8	5.8136e−1	5.7790e−1	**5.4731e−1**
	10	5.6870e−1	5.5841e−1	**4.7341e−1**
C3DTLZ4	3	9.7930e−3	**3.3400e−3**	9.3730e−3
	5	*1.9092e−2*	**1.7264e−2**	2.0938e−2
	8	*2.5991e−2*	**2.5469e−2**	2.8631e−2
	10	*3.4809e−2*	**3.4504e−2**	*3.7774e−2*
Best/similar/total →		2/4/20	8/4/20	**10**/2/20

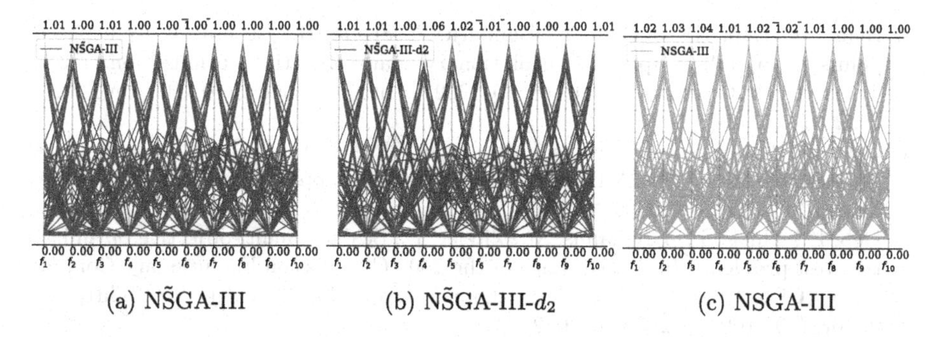

(a) NS̃GA-III (b) NS̃GA-III-d_2 (c) NSGA-III

Fig. 7. Obtained solutions using PCP plot for CDTLZ2 problem.

4 Conclusions

This paper has questioned the use of non-dominated sorting (NDS) operation in NSGA-III method for solving two to 10-objective problems. Since population members are not divided into different non-dominated levels for performing mating and survival selection operators, the choice of an appropriate solution within the associated members for a reference direction becomes important. We have investigated two approaches: (i) NS̃GA-III-d_2, which uses the original orthogonal distance metric d_2 and (ii) NS̃GA-III, which uses the well-known PBI metric. Based on 87 different problems, following two conclusions can be made:

- The NDS operation is not absolutely necessary and for many-objective problems, NSGA-III without NDS performs better than the original NSGA-III.
- For two objective problems, NS̃GA-III with the d_2-metric has a slow progress in the beginning, but can catch up with the performance of NS̃GA-III or the original NSGA-III with enough generations.

These observations are important for making NSGA-III more computationally efficient. Since NDS operation is not essential, domination check can be completed with a smaller computational time. Moreover, since convergence rate is faster for NS̃GA-III, it can be used with more effectiveness to build better surrogate-assisted NSGA-III methods with a limited number of solution evaluations. We plan to pursue some of these extensions in the near future.

Acknowledgments. Authors acknowledge the financial support from Koenig Endowed Chair funding from the Department of Electrical and Computer Engineering at Michigan State University, East Lansing, USA.

References

1. Binh, T.T., Korn, U.: MOBES: a multiobjective evolution strategy for constrained optimization problems. In: Proceedings of the Third International Conference on Genetic Algorithms (MENDEL 1997), pp. 176–182 (1997)
2. Blank, J., Deb, K.: Pymoo: multi-objective optimization in Python. IEEE Access **8**, 89497–89509 (2020). https://doi.org/10.1109/access.2020.2990567

3. Blank, J., Deb, K., Dhebar, Y., Bandaru, S., Seada, H.: Generating well-spaced points on a unit simplex for evolutionary many-objective optimization. IEEE Trans. Evol. Comput. **25**(1), 48–60 (2021). https://doi.org/10.1109/tevc.2020.2992387
4. Deb, K., Pratap, A., Agarwal, S., Meyarivan, T.: A fast and elitist multiobjective genetic algorithm: NSGA-II. IEEE Trans. Evol. Comput. **6**(2), 182–197 (2002). https://doi.org/10.1109/4235.996017
5. Deb, K., Thiele, L., Laumanns, M., Zitzler, E.: Scalable multi-objective optimization test problems. In: Proceedings of the 2002 Congress on Evolutionary Computation (CEC 2002) (Cat. No.02TH8600), vol. 1, pp. 825–830. IEEE (2002). https://doi.org/10.1109/cec.2002.1007032
6. Deb, K.: Multi-Objective Optimization Using Evolutionary Algorithms. Wiley, London (2001)
7. Deb, K., Jain, H.: An evolutionary many-objective optimization algorithm using reference-point-based nondominated sorting approach, Part I: solving problems with box constraints. IEEE Trans. Evol. Comput. **18**(4), 577–601 (2014). https://doi.org/10.1109/tevc.2013.2281535
8. Fonseca, C.M., Fleming, P.J.: Genetic algorithms for multiobjective optimization: formulation, discussion, and generalization. In: Proceedings of the Fifth International Conference on Genetic Algorithms, pp. 416–423. Morgan Kaufmann, San Mateo, CA (1993)
9. Horn, J., Nafpliotis, N., Goldberg, D.E.: A niched Pareto genetic algorithm for multiobjective optimization. In: Proceedings of the First IEEE Conference on Evolutionary Computation. IEEE World Congress on Computational Intelligence, pp. 82–87. IEEE (1994). https://doi.org/10.1109/icec.1994.350037
10. Huband, S., Barone, L., While, L., Hingston, P.: A scalable multi-objective test problem toolkit. In: Coello Coello, C.A., Hernández Aguirre, A., Zitzler, E. (eds.) EMO 2005. LNCS, vol. 3410, pp. 280–295. Springer, Heidelberg (2005). https://doi.org/10.1007/978-3-540-31880-4_20
11. Ishibuchi, H., Masuda, H., Tanigaki, Y., Nojima, Y.: Modified distance calculation in generational distance and inverted generational distance. In: Gaspar-Cunha, A., Henggeler Antunes, C., Coello, C.C. (eds.) EMO 2015. LNCS, vol. 9019, pp. 110–125. Springer, Cham (2015). https://doi.org/10.1007/978-3-319-15892-1_8
12. Jain, H., Deb, K.: An evolutionary many-objective optimization algorithm using reference-point based nondominated sorting approach, Part II: handling constraints and extending to an adaptive approach. IEEE Trans. Evol. Comput. **18**(4), 602–622 (2014). https://doi.org/10.1109/tevc.2013.2281534
13. Kung, H.T., Luccio, F., Preparata, F.P.: On finding the maxima of a set of vectors. J. ACM **22**(4), 469–476 (1975). https://doi.org/10.1145/321906.321910
14. Osyczka, A., Kundu, S.: A new method to solve generalized multicriteria optimization problems using the simple genetic algorithm. Struct. Optim. **10**(2), 94–99 (1995). https://doi.org/10.1007/BF01743536
15. Srinivas, N., Deb, K.: Muiltiobjective optimization using nondominated sorting in genetic algorithms. Evol. Comput. **2**(3), 221–248 (1994). https://doi.org/10.1162/evco.1994.2.3.221
16. Tanaka, M., Watanabe, H., Furukawa, Y., Tanino, T.: GA-based decision support system for multicriteria optimization. In: 1995 IEEE International Conference on Systems, Man and Cybernetics. Intelligent Systems for the 21st Century, vol. 2, pp. 1556–1561. IEEE (1995)

17. Zhang, Q., Li, H.: MOEA/d: a multiobjective evolutionary algorithm based on decomposition. IEEE Trans. Evol. Comput. **11**(6), 712–731 (2007). https://doi.org/10.1109/tevc.2007.892759
18. Zitzler, E., Deb, K., Thiele, L.: Comparison of multiobjective evolutionary algorithms: empirical results. Evol. Comput. **8**(2), 173–195 (2000). https://doi.org/10.1162/106365600568202

Scalability of Multi-objective Evolutionary Algorithms for Solving Real-World Complex Optimization Problems

António Gaspar-Cunha[1](✉), Paulo Costa[1], Francisco Monaco[2], and Alexandre Delbem[2]

[1] IPC-Institute for Polymers and Composites, University of Minho, Guimarães, Portugal
agc@dep.uminho.pt

[2] Institute of Mathematics and Computer Science, University of São Paulo, São Paulo, Brazil

Abstract. The use Multi-Objective Evolutionary Algorithms (MOEAs) to solve real-world multi-objective optimization problems often finds a problem designated by the curse of dimensionality. This is mainly because the progression of the algorithm along successive generations is based on non-dominance relations that practically do not exist when the number of objectives is high. Also, the existence of many objectives makes the choice of a solution to the problem under study very difficult. Several methods have been proposed in the literature to reduce the number of objectives to use during the optimization process. In the present work, a methodology to reduce the number of objectives is proposed. This method is based on DAMICORE (DAta MIning of COde REpositories), a machine-learning algorithm proposed by the authors. A theoretical comparison with a similar machine learning approach is made, pointing out some advantages of using the proposed algorithm using a benchmark problem designated by DTLZ5. Also, a real problem is used to show the effectiveness of the methodology.

Keywords: Objectives reduction · Data mining · MOEAs · Many objectives

1 Introduction

Real-world optimization problems have very often many objectives that must be satisfied simultaneously. Multi-Objective Evolutionary Algorithms (MOEAs) showed to be very efficient in solving this type of problems. However, the use of MOEAs to tackle these problems suffers from a difficulty designated by the curse of dimensionality, in which the increase of the number of objectives makes the progression towards the Pareto-Optimal Frontier (POF) very challenging [1–4]. Also, the existence of many objectives makes the visualization of the POF almost impossible, which puts difficult to help the decision maker in selecting a solution and, more importantly, to explain his decision. Associated with this, often the decision space is also of large size, which, additionally, worsens the difficulties for the MOEAs [5]. Therefore, a scalability problem arises, not only due to the high number of decision variables and/or objectives but also due to the

M. Emmerich et al. (Eds.): EMO 2023, LNCS 13970, pp. 86–100, 2023.
https://doi.org/10.1007/978-3-031-27250-9_7

complex interrelations existent between the decision variables, between the objectives, and between the decision variables and the objectives.

To deal with the curse of dimensionally due to the high number of objectives several methodologies to reduce the number of objectives during the optimization have been proposed in the literature. This is a challenging problem in which a balance must be made between the requirement of preserving the original (with all objectives) problem characteristics and the possibility of obtaining acceptable solutions when the number of objectives used during the optimization is reduced.

Another important question is to know if the individual objectives are relevant during the entire optimization process (along the generations), i.e., it is important to know if it is possible to start the optimization using some objectives and during the successive generations change the objectives used. This is, as the solutions approach the POF the complex structure between the decision variables and the objectives can change.

The objective of this work is to study the applicability of MOEAs to complex real problems based on objectives reduction using a data mining technique, and not to compare the performance of different methodologies. Thus, using the data mining methodology proposed the aim is to be able to use the existing complex relations inside the problem to help the optimization process. However, must be clear that the aim is not to obtain a reduced objective set which represents the original set without errors, but the reduced set will be a very good approximation to the originally defined set of objectives with a reduced error. This will be very useful for real problems in which it is not possible to reduce the number of objectives due to the strong interaction between these same objectives, but that makes it possible to facilitate the search and help the DM to understand the considered process. In fact, real problems often have some characteristics that are not present in benchmark problems. In such real problems, there is a complexity associated with the relations between the decision variables and the objectives that are not present in benchmark test problems, where those relations are placed on purpose [5].

This text is organized as follows: in Sect. 2 a state-of-the-art related to objectives reduction will be made, in Sect. 3 the data mining technique used will be presented and discussed, in Sect. 4 the real-world problem to study is presented, in Sect. 5 the results are presented and discussed, and in Sect. 6 the conclusions and some suggestion for further work will be stated.

2 Related Work with Objectives Reduction

Brockoff and Zitzler [1, 2] proposed two different algorithms for objectives reduction based on the definition of two types of problems, the first, to obtain the minimum objective subset that produces a certain error, and the second, to obtain an objective subset of a predefined size with the minimum possible error. Deb and Saxena [3, 4] based on principal component analysis proposed a method for reducing the number of objectives by maintaining the objectives that can explain most of the variance in the objective space, but without explaining clearly how the objective reduction alters the dominance structure. Jaimes et al. [6] proposed two algorithms to address the two problems identified by Brockoff and Zitzler above based on a feature selection method. The algorithms were validated by comparing the results with the algorithms of Brockoff and Zitzler. Saxena et al. [7] presented a framework for using both linear and nonlinear objective reduction algorithms. The authors proposed to develop a general framework, taking into account the possibility of the data being noisy, to reduce the number of algorithms parameters and to propose an error measure. The algorithms were applied to a broad range of problems and the results were compared with others in the literature. Sinha et al. [8] proposed a methodology to reduce the objectives before presenting the solutions to the decision maker that iteratively chose the best solutions. The methodology was applied to solve some real world problems. Finally, Duro et al. [9] proposed a machine learning methodology with the aim of learning the preference-structure different objectives present in the problem in order to obtain the smallest set of objectives that can originate the same POF, the smallest objective set corresponding to a minimum pre-defined error and the objective sets of a certain size that originates a minimum error. The characteristics of this methodology were compared with the one proposed in the present paper at the end of the next section.

3 Approach Based on Data Mining

3.1 Data Mining Methodology Adopted - FS-OPA

Main concepts. First introduced by Sanches et al. [10], DAMICORE, borrowed from Theory, Complex Networks, and Phylogenetic Inference, aiming at revealing the hidden hierarchical relationship of objects from an unstructured (raw) dataset. It uses three principal steps: S1) given a metric of similarity, build a distance matrix comparing every two objects; S2) convert the matrix into a phylogenetic tree by connecting close objects according to hierarchical levels of similarity; S3) apply a community detection process to group close subtrees into clusters. Figure 1 shows a set of objects xi, the elements d_{ij} of the distance matrix correspond to a measure of dissimilarities between objects x_i and x_j, according to some given metric. The matrix is broken down into a tree, where the distance between any two objects (leaves) corresponds to the sum of the lengths of the branches connecting them. Finally, the third step groups the objects that are strongly connected (according to the tree topology) into a community, generating a set of distinguishable similarity clusters. The original DAMICORE method selects three specific algorithms for the Steps S1, S2, and S3 (Fig. 1), respectively, Normalized Compression Distance (NCD) [12, 13] (since it works with any data type and mixed types); Neighbor Joining [14] (NJ) (widely employed in bioinformatics), and Fast Newman (FN) (that constructs a graph partition by a greedy algorithm that uses a bottom-up strategy to maximize the graph modularity function [15]).

The pipeline with NCD, NJ, possesses relevant properties. NCD makes DAMICORE a data-type agnostic method, in the sense that it works with any kind of object (texts, images, audio, etc.) and mixed data types. Moreover, DAMICORE can be used without any data pre-processing, such as filtering, outlier detection, feature extraction, parameter setup and knowledge of the problem domain. DAMICORE requires no parameters setup to run (although some execution options may improve its performance). DAMICORE has been successfully employed in a variety of fields, for example, software-hardware co-design [16–18], compiler optimization [19, 20], student profiling in e-learning environments [21, 22], identification of phytopathology from sensor data [18], systematic literature review, identification of cross-cut concerns [23], electrical distribution systems [24], and novel methods in bioinformatics [25].

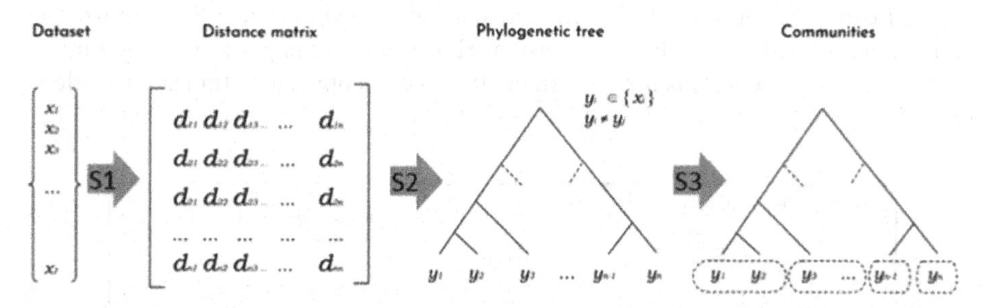

Fig. 1. The tree-steps pipeline called DAMICORE (reproduced with permission from [11]).

Feature Sensitivity Analysis. A Feature Sensitivity (FS) analysis aims at salienting the principal features of a problem, taking into account common real-world challenges (such as the quality of data acquired and the database consistency and representativeness), its feature interactions, and their contribution to a target or objective. Such a scope differs from those where the standard feature selection algorithms have succeeded. In other words, an FS strategy is expected to benefit the learning of a problem from scratch. Such learning can induce a model for optimization algorithms (such as Estimation of Distribution Algorithms). We use phylogram-based models since they can work with small datasets and there is an optimization approach prepared to use such models: Optimization based on Phylogram Analysis (OPA).

Figure 2 shows a diagram synthesizing OPA with the use of the FS analysis by it; such a combination is called FS-OPA. The two principal FS steps are: *A*) Salienting Samples (SS) according to a criterion; and *B*) applying DAMICORE to construct a phylogram-based model. SS ranks the samples according to each of the *M* criteria (or non-dominated fronts), producing the sets of selected samples (Fig. 3), denoted BC1 (the samples in the best quantile according to Criterion 1), BC2, ..., BCM. DAMICORE constructs a phylogram (a rough model) from BCi, $i = 1, \ldots, M$, generating *M* models (BC1-based model, ..., BCM-based model). Then, a consensus strategy produces a unified

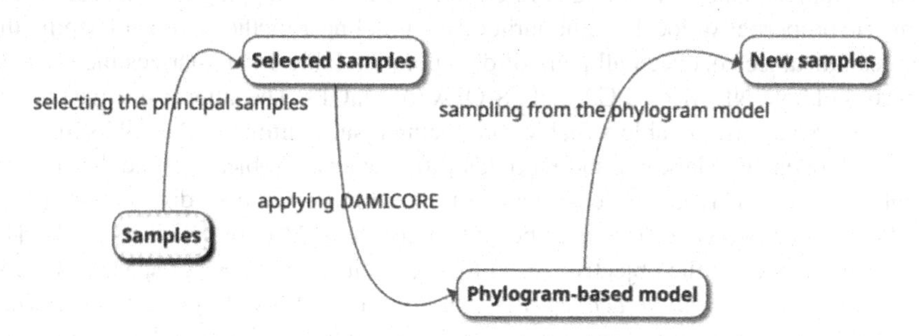

Fig. 2. Diagram of the optimization based on phylogram analysis - OPA.

phylogram-based model. An OPA cycle completes when the unified model generates new samples.

This paper instantiates the procedures from Steps *A* and *B* for the modelling of polymer extrusion (Sect. 4). The learned model is expected to benefit the optimization problem associated with polymer extrusion. However, the sampling from the unified model (the last OPA step) is not performed, thus, not a complete optimization cycle is run.

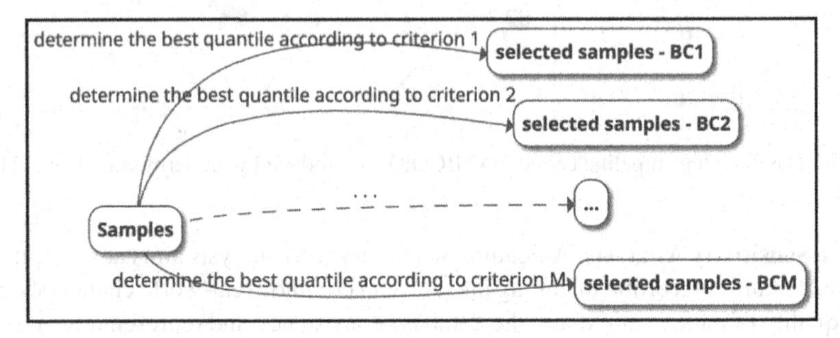

Fig. 3. SS procedure that obtains the selected samples shown in Fig. 2.

Soares et al. [24] and Martins et al. [18] show some experimental results and proofs related to the performance of OPA for challenging combinatorial mono- and multi-objective optimization problems. The main mechanisms of FS-OPA, relevant to the scope of a data-driven design of an extruder concentrated on the DAMICORE method, are introduced in Sect. 3.2.

3.2 Comparison of FS-OPA with NL-MVU-PCA for MaOPs Data-Driven Structural Learning

NL-MVU-PCA (Non Linear – Maximum Variance Unfolding - Principal Component Analysis) is the main method investigated for MaOPs by [9]. The non-linear (NL) approach performs the optimization of the Kernel (Gram) matrix values by minimizing the Maximum Variance Unfolding (MVU) to find the best mapping (that preserves the geometric properties of local neighbourhoods; while linear methods aim at keeping the Euclidean distances between all pairs of data points). Table 1 summarizes the relevant properties of both NL-MVU-PCA and FS-OPA for MaOPs. The latter investigates three types of associations: variable-variable (producing results similar to the Gibbs measure for Ising Models, or Markov Random fields [26]), objective-objective (the dissimilarities when found can favour the construction of (non-dominated) front distributions [25]), and the variable-objective (that may benefit inference as Markov Blankets [27]). The former only works on the objective space for the exclusive purpose of space reduction to determine *the essential objective set* [9]. Moreover, the FS-OPA preserves the original variable space, which favours non-expert human interpretability (relevant for some classes of real-world problems); it also has a relatively low-time complexity and has shown beneficial results when applied to small datasets [10, 16–23].

The Feature Sensitivity (FS) analysis of FS-OPA aims at finding the variable and/or objective interactions that benefit inferences through structural (graph-based) and probabilistic modelling. Probabilistic results are fundamental due to the odds of bias in observed data or small-data sampling. Interpretability is fundamental for some classes of real-world problems, mainly involving decisions by experts from an application domain. Variable-variable and variable-objective interactions also benefit practitioners' comprehension of founds (The Why), increasing their confidence in a decision.

Table 1. NL-MVU-PCA and DAMICORE for multidimensional data-driven structural learning applied to MaOPs.

Category	Types		NL-MVU-PCA	FS-OPA
Analyses	Objective-objective		X	X
	Variable-variable			X
	Variable-objective			X
	Objective space reduction		X	
	Sensitivity			X
Priors	Kernel function choice		X	Not necessary
	Parameter optimization		X	Not necessary
Interpretability - The Why	Non-expert practitioner			X
Scalability	Time-complexity	Usual cases	$O(M^3 q^3)$*	$O(l^3)$**
		The worst case	$O(M^6)$	$O(n\, l^2 + l^2)$
	Sample-size support		Empirical	Theoretical and empirical

* M is the number of objectives and q is the number of clusters.
** l is the number of variables and objectives, and n is the number of data resamples.

The time complexity for usual cases and the worst-case estimates the overhead of the two procedures for the multidimensional data-driven structural learning when applied to MaOPs. The number of clusters in NL-MVU-PCA relates to the number of constraints to maintain the local isometry (Mq; in the worst case, $q = M-1$ and M^2) [9]. FS-OPA with usual resampling is $O(l^3)$ since $n \leq l$ (as in leave-one-out resampling) [24]. Moreover, $l = M$ for a space analysis only uses objectives. Thus, the time complexities of FS-OPA and NL-MVU-PCA have a ratio of $(n + M)/M^4$ $(1/q^3)$ of running time for $l = M$ in the worst case (in the usual case).

Another relevant factor for reliable results is the minimal amount of samples required. Usually, the sample size for PCA is empirically obtained. FS-OPA has a theoretical model to determine the minimal amount of samples to reach reliable results [24], which can also be empirically corroborated.

Figure 4 illustrates an FS-OPA output for DTLZ5 (2,10), also used by Duro et al. [9] for explaining the capacity of the method of them to find redundant objectives (objectives f1, f2, f3, f4, f5, f6, f7, f8, and f9 are linearly correlated in DTLZ5(2,10)). A random population of size 31 with samples normalized and Euclidian distance was used to obtain a distance matrix. SS procedure in Fig. 3 was not applied. The output of Fig. 4 shows variables and objectives arranged into a phylogram with leaf nodes (the objects under analysis) composing clusters (similarly to the end of the pipeline in Fig. 1) - they are identified by the same colour.

Objective functions f1, …, f9 are partitioned into three neighbour clusters ({f1, f2}, {f3, f4, f5, f6} and {f7, f8, f9}) in the phylogram structure; while f10 is together with the leaf nodes corresponding to variables. The phylogram structure aggregates f1, …, and f9 into the same subtree, while f10 is isolated from the other objectives in the complementary subtree. The unique node with the label "100" (another type of result from a tree consensus) splits the phylogram into those two subtrees. Such a label ("100") means that the leaf nodes f10 and x1,.., x10, and f10 were in the same subtree (with the remaining leaf nodes in the complementary subtree) in 100% of all the constructed phylograms, independently on each subtree topology in a phylogram. Such an interpretation suggests a hypothesis: f10 is weakly correlated to the other objectives, which are significantly associated with themselves. Thus, f10 and one of the other objectives could compose an *essential objective set*; this result is consistent with the DTLZ5(2,10) problem structure. The FS-OPA also produces other outputs (useful for human comprehension of some classes of real-world problems), which are explored in Sections related to the extrusion problem.

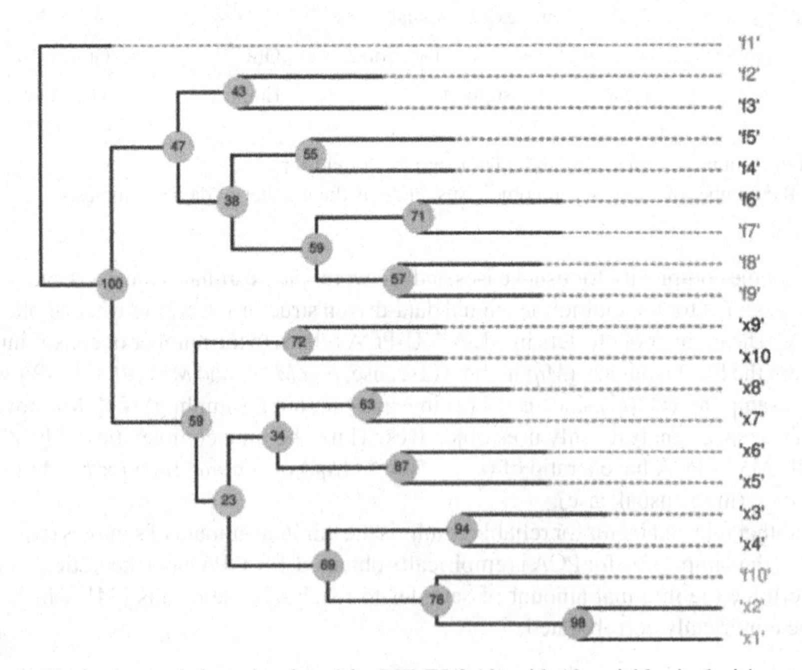

Fig. 4. Phylogram and clustering found for DTLZ5(2,10) with 10 variables in decision space.

4 Case Study

The methodology proposed will be applied to a real-world problem in polymer extrusion in which the complex thermomechanical environment involved is described in detail in Gaspar-Cunha and Covas [28, 29] encompassing the possibility of using inside the same extruder alternately two different types of screws, a conventional and a Maillefer barrier screw. The complexity of the thermomechanical phenomena taking place, including the flow of solid pellets and melted polymer and the coexistence of both, is described in detail in reference [28], while in reference [29] the numerical model used was assessed experimentally. Table 2 shows the geometrical parameters involved in the description of both types of screws. Considering that only one screw can be used in the machine, an additional decision variable was added, case, to activate the decision variables correspondent to one of the types of screws, i.e., when case ranges in the interval [0.0,0.5] the decision variables of the conventional screw are used while when case ranges in the interval]0.5,1.0] the other screw is considered. Thus, the total number of decision variables is 15.

Table 2. Geometrical parameters of both conventional and Maillefer barrier screws.

Screw type	Decision variables								
Conventional screw	Case	L1	L2	H1	H3	P	e		
Maillefer barrier screw		L1_	L2_	H1_	H3_	P_	e_	Hf	wf
Range of variation	[0,1]	[100,400]	[170.400]	[18, 22]	[22, 26]	[25,35]	[3, 4]	[0.1,0.6]	[3, 4]

In the extrusion system, the performance of the process depends on the polymer properties, machine operating conditions and geometry. In the present example, a Low Density Polyethylene (LDPE) is used and the operating conditions were fixed and include: screw rotational speed (N = 40 rpm) and barrel temperature profile in three zones (Tb1 = 140 °C, Tb2 = 150 °C and Tb3 = 160 °C). The geometrical parameters are the decision variables defined randomly by the optimization methodology in the range identified in Table 3. The performance of the machine was quantified using six objectives, two to maximize (machine output, Q (kg/hr) and degree of mixing, WATS) and four to minimize (length of screw required to melt the polymer, L (m), melt temperature at the exit, T (°C), mechanical power consumption required to rotate the screw, Power (W), and viscous dissipation quantified as the ratio between the melt temperature and the fixed barrel temperature, TTb). Due to a lack of space, more details about this process and all the decision variables and objectives involved the reader is referred to references [28, 29].

The optimizations were made using the SMS- EMOA algorithms. For comparison, and due to the stochastic nature, 11 independent runs with different random numbers were done. For all test runs the population size and the maximum number of generations were set to 50 and 100, respectively.

5 Results and Discussion

The following strategy for analyzing the potentiality of DAMICORE in reducing the number of objectives required for optimization will be pursued:

1. Perform 11 optimization runs using 6 objectives;
2. Apply DAMICORE to the initial population of one of the previous runs;
3. Obtain the phylogram and the distance tables;
4. Perform an analysis of the phylogram and tables and select the minimum number of objectives;
5. Perform 11 optimization runs using the minimum number of objectives;
6. Taking into account the distance between the objectives check if some other objectives can be added;
7. Perform 11 optimization runs using the number of objectives defined in 6;
8. Compare the optimization results using the hypervolume metric.

The application of DAMICORE to the random initial population of one of the optimization runs (steps 2 and 3 above) produces the results presented in Fig. 5 and in Tables 2 and 3. Table 3 shows the distances between the decision variables and the objectives obtained from the phylogram of Fig. 5. The average distance can quantify the degree of influence of each decision variable in the objectives.

Figure 5 shows the complex relations existent between the decision variables and the objectives, but also between the different objectives. The objectives, identified by black boxes, are assembled in different clusters and in pairs, having the minimum distance between themselves, specifically (Q, L), (Power, WATS) and (T, TTb). Simultaneously, it is possible to see that some decision variables are grouped together with these pairs of objectives, respectively (Q, L, L1, L2), (Power, WATS, L1_, L2_) and (T, TTb, Hf).

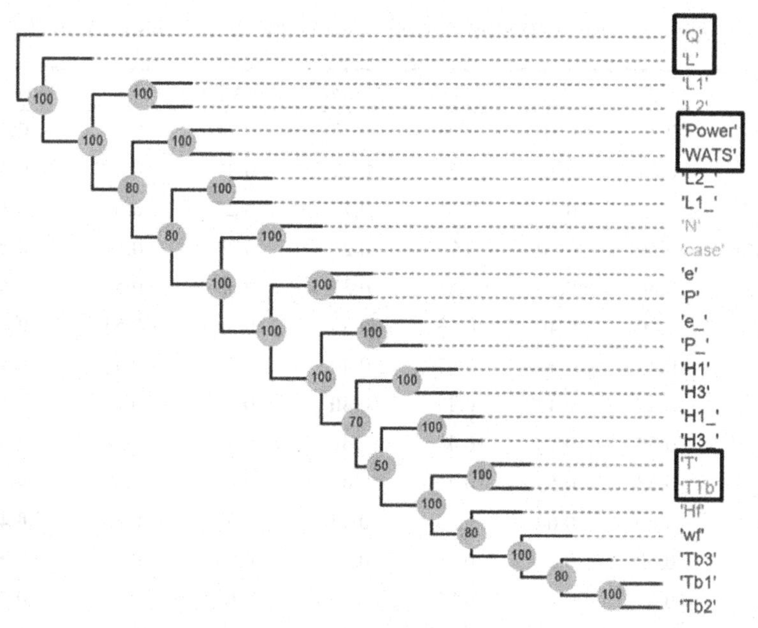

Fig. 5. Phylogram and clustering found for the extrusion problem – initial optimization of the optimization run (the cluster are identified by different colours).

Taking into account the knowledge about the process those relations are not easy to explain. In fact, if only the conventional screw is considered, Q and L depend mainly on the value of H3. Therefore, the application of DAMICORE constitutes, in this apparently simple example, a methodology that is able to capture some other information when the Maillefer barrier screw is also considered. This is, the technique applied is able to capture the indirect influence of the H3 on Q and L through the value of L1 and L2 because these variables (and also L1_ and L2_ for Power and WATS) determine the length of the barrier (since it finishes after L1 + L2) and as consequence, the location of the screw were H3 acts.

Table 3. Distances between the decision variables and the objectives for the phylogram of Fig. 5.

	'Q'	'L'	'T'	'Power'	'WATS'	'TTb'	Average
'L2'	0.20	0.20	0.73	0.27	0.27	0.73	**0.40**
'L2_'	0.33	0.33	0.60	0.27	0.27	0.60	**0.40**
'L1_'	0.33	0.33	0.60	0.27	0.27	0.60	**0.40**
'L1'	0.20	0.20	0.73	0.27	0.27	0.73	**0.40**
'case'	0.40	0.40	0.53	0.33	0.33	0.53	**0.42**
'N'	0.40	0.40	0.53	0.33	0.33	0.53	**0.42**
'e'	0.47	0.47	0.47	0.40	0.40	0.47	**0.44**
'P'	0.47	0.47	0.47	0.40	0.40	0.47	**0.44**
'e_'	0.53	0.53	0.40	0.47	0.47	0.4	**0.46**
'P_'	0.53	0.53	0.40	0.47	0.47	0.4	**0.46**
'H3_'	0.60	0.60	0.33	0.53	0.53	0.33	**0.48**
'H1_'	0.60	0.60	0.33	0.53	0.53	0.33	**0.48**
'H3'	0.67	0.67	0.27	0.60	0.60	0.27	**0.51**
'H1'	0.67	0.67	0.27	0.60	0.60	0.27	**0.51**
'Hf'	0.73	0.73	0.20	0.67	0.67	0.20	**0.53**
'wf'	0.80	0.8	0.27	0.73	0.73	0.27	**0.60**
'Tb3'	0.87	0.87	0.33	0.80	0.80	0.33	**0.66**
'Tb1'	0.93	0.93	0.40	0.87	0.87	0.40	**0.73**
'Tb2'	1.00	1.00	0.47	0.93	0.93	0.47	**0.80**
Average	**0.5647**	**0.5647**	**0.4384**	**0.5126**	**0.5126**	**0.4384**	

Some other relations can be explained, for example, the cluster (T, TTb, Hf) exists because the melt temperature (T) value is controlled by the viscous dissipation generated in the gap with size Hf.

A conclusion from the simultaneous analysis of Fig. 5 and Table 4 is that, instead of using the 6 objectives, can be possible to use during the optimization only one objective of each pair (steps 4 and 5 above), for example, Q, WATS and T, given that the objectives belonging to the same cluster are strongly related.

Finally, an analysis of Table 4 (steps 6 and 7 above) shows that the minimum distance between objectives is 0.325 for Power and WATS. This clearly means that these two objectives (and also Q and L, since their distance is also low, 0.345) have a strong connection with all the others, i.e., they aggregate more information of all objectives and, as a consequence, of the decision variables. The idea is to include one of these objectives in the optimization process. In the end, instead of having 6 objectives, the optimization can be performed with three of four objectives.

Therefore, the following cases were selected to perform 11 optimization runs using SMS-EMO to compare the corresponding performance using the hypervolume metric:

Case 1- 6 objectives, Q, L, Power, WATS, T and TTb;
Case 2- 4 objectives: Q, L, WATS and T;
Case 3- 4 objectives: Q, Power, WATS and T;
Case 4: 3 objectives: Q, WATS and T.

The idea is to conclude what information is lost when the number of objectives is reduced taking into account the methodology proposed here.

Table 4. Distances between the objectives for the phylogram of Fig. 5.

	'Q'	'L'	'T'	'Power'	'WATS'	'TTb'	Average
'Q'	0.00	0.07	0.73	0.27	0.27	0.73	**0.345**
'L'	0.07	0.00	0.73	0.27	0.27	0.73	**0.345**
'T'	0.73	0.73	0.00	0.67	0.67	0.07	**0.478**
'Power'	0.27	0.27	0.67	0.00	0.07	0.67	**0.325**
'WATS'	0.27	0.27	0.67	0.07	0.00	0.67	**0.325**
'TTb'	0.73	0.73	0.07	0.67	0.67	0.00	**0.478**

6 Conclusions

A methodology based on machine learning was presented and applied in the reduction of the number of objectives for multiobjective optimization using MOEAs. This approach has some important characteristics that are an important improvement concerning similar state-of-the-art methodologies, namely: it allows analysis of the relations variable-variable and variable-objectives (and not only objective-objective), does not need kernel function choice and parameters optimization, allows to interpret of the results to help the decision maker, its time complexity is low and supports theoretical and empirical sample-size.

The application of the methodology to the well-known DTZL5(2,10) benchmark problem showed its potential to reduce the number of objectives by capturing the complex relations between the different objectives, with an additional possibility that is to capture the relations objectives-variables.

The resolution of a difficult real-world problem using the approach proposed has proven that almost automatically it is possible to reduce the number of objectives by losing only less than ten per cent of the Pareto-optimal frontier obtained. Additionally, the intervention of the decision maker during the process, e.g., when selecting the objectives to be considered in the optimization, can be very useful because the person interested is able to see how the process works and to interpret the results obtained. Finally, an

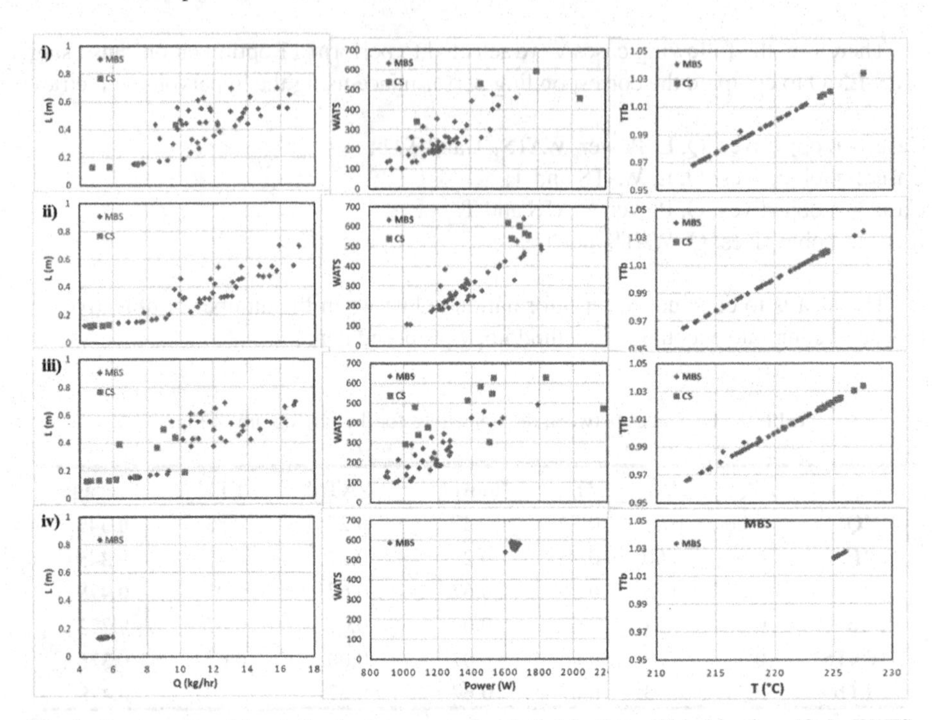

Fig. 6. Pareto-optimal fronts for the 4 cases studies: i) 6 objectives; ii) 4 objectives (Q, L, WATS, T), iii) 4 objectives (Q, Power, WATS, T); iv) 3 objectives (Q, WATS, T).

important characteristic of the method proposed is the capacity to explain the results obtained.

As a future work, the methodology will be compared with other approaches in the literature (Fig. 6 and Table 5).

Table 5. Hypervolume for the 4 cases studied and relative loss percentage.

6 Objectives	4 Objectives (Q, L, WATS, T)	4 Objectives (Q, Power, WATS, T)	3 Objectives (Q, WATS, T)
0.24636	0.22754	0.22215	0.02117
– —	7.6%	9.8%	91.4%

References

1. Brockhoff, D., Zitzler, E.: Are all objectives necessary? on dimensionality reduction in evolutionary multiobjective optimization. In: Runarsson, T.P., Beyer, H.-G., Burke, E., Merelo-Guervós, J.J., Whitley, L.D., Yao, X. (eds.) PPSN 2006. LNCS, vol. 4193, pp. 533–542. Springer, Heidelberg (2006). https://doi.org/10.1007/11844297_54
2. Brockhoff, D., Zitzler, E.: Objective reduction in evolutionary multiobjective optimization: theory and applications. In: Evolutionary Computation **17**(2), 135–166 (2009). https://doi.org/10.1162/evco.2009.17.2.135

3. Deb, K., Saxena, D.K.: Searching for pareto-optimal solutions through dimensionality reduction for certain large-dimensional multi-objective optimization problems. In: 2006 IEEE Congress on Evolutionary Computation (CEC'2006), pp. 3353–3360, IEEE, Vancouver, BC, Canada (2006)
4. Saxena, D.K., Deb, K.: Non-linear dimensionality reduction procedures for certain large-dimensional multi-objective optimization problems: employing correntropy and a novel maximum variance unfolding. In: Obayashi, S., Deb, K., Poloni, C., Hiroyasu, T., Murata, T. (eds.) EMO 2007. LNCS, vol. 4403, pp. 772–787. Springer, Heidelberg (2007). https://doi.org/10.1007/978-3-540-70928-2_58
5. Hong, W.-J., Yang, P., Tang, K.: Evolutionary computation for large-scale multi-objective optimization: a decade of progresses. Int. J. Autom. Comput. **18**(2), 155–169 (2020). https://doi.org/10.1007/s11633-020-1253-0
6. López J., Coello C., Chakraborty, D.: Objective reduction using a feature selection technique. In: Proceedings of the 10th Annual Conference on Genetic and Evolutionary Computation – GECCO'08 (2008). https://doi.org/10.1145/1389095.1389228
7. Saxena, D.K., Duro, J.A., Tiwari, A., Deb, K., Zhang, Q.: Objective reduction in many-objective optimization: linear and nonlinear algorithms. In: IEEE Transactions on Evolutionary Computation **17**(1), 77–99 (2013). https://doi.org/10.1109/tevc.2012.2185847
8. Sinha, A., Saxena, D.K., Deb, K., Tiwari, A.: Using objective reduction and interactive procedure to handle many-objective optimization problems. In: Applied Soft Computing **13**(1), 415–427 (2013). https://doi.org/10.1016/j.asoc.2012.08.030
9. Duro, J.A., Saxena, K.D., Deb, K., Zhang, Q.: Machine learning based decision support for many-objective optimization problems. In: Neurocomputing **146**, 30–47 (2014). https://doi.org/10.1016/j.neucom.2014.06.076
10. Sanches, A., Cardoso, J.M., Delbem, A.C.: Identifying merge-beneficial software kernels for hardware implementation. In: Reconfigurable Computing and FPGAs (ReConFig), International Conference on, IEEE, pp. 74–79 (2011)
11. Gaspar-Cunha, A., Monaco, F., Sikora, J., Delbem, A.: Artificial intelligence in single screw polymer extrusion: learning from computational data. In: Engineering Applications of Artificial Intelligence, **116**, 105397 (2022). https://doi.org/10.1016/j.engappai.2022.105397
12. Li, M., Vitányi, P.: An Introduction to Kolmogorov Complexity and its Applications. In: Springer Science & Business Media (2013). https://doi.org/10.1007/978-0-387-49820-1
13. Lui, L.T., Terrazas, G., Zenil, H., Alexander, C., Krasnogor, N.: Complexity measurement based on information theory and kolmogorov complexity. In: Artificial Life **21**(2), 205224 (2015)
14. Newman, M.E.: Modularity and community structure in networks. In: Proceedings of the National Academy of Sciences **103**(23), 8577–8582 (2006)
15. Newman, M.E.: Fast algorithm for detecting community structure in networks. In: Physical Review **69**(6), 066133 (2004)
16. Silva, B.D.A., Cuminato, L.A., Delbem, A.C.B., Diniz, P.C., Bonato, V.: Application-oriented cache memorybconfiguration for energy efficiency in multi-cores. In: IET Computers Digital Techniques **9**(1), 73–81 (2015)
17. Silva, B.A., Delbem, A.C.B., Deniz, P.C., Bonato, V.: Runtime mapping and scheduling for energy efficiency in heterogeneous multi-core systems. In: International Conference on Reconfigurable Computing and FPGAs, pp. 1–6, Mayan Riviera (2015)
18. Martins, L.G.A., Nobre, R., Delbem, A.C.B., Marques, E., Cardoso, J.M.P.: A clustering-based approach for exploring sequences of compiler optimizations. In: 2014 IEEE Congress on Evolutionary Computation (CEC), pp. 2436–2443 (2014) Beijing. https://doi.org/10.1109/CEC.2014.6900634

19. Martins, L.G., Nobre, R., Delbem, A.C., Marques, E., Cardoso, J. M.: Exploration of compiler optimization sequences using clustering-based selection. In: The 2014 SIGPLAN/SIGBED conference on Languages, compilers and tools for embedded systems, pp. 63, Edinburgh (2014)
20. Moro, L.F.S., Lopes, A.M.Z., Delbem, A.C.B., Isotani, S.: Os desafios para minerar dados educacionais de forma rápida e intuitiva: o caso da damicore e a caracterização de alunos em ambientes de elearning. In: Workshop de desafios da computação aplicada à educação XXXIII Congresso da Sociedade Brasileira de Computação, pp. 1–10, Maceio (2013)
21. Moro, L.F., Rodriguez, C.L., Andrade, F.R.H., Delbem, A.C.B., Isotani, S.: Caracterização de alunos em ambientes de ensino online, in: Workshop de Mineração de Dados em Ambientes Virtuais do Ensino/Aprendizagem, Anais do Congresso Brasileiro de Informática na Educação, pp. 1–10, Dourados (2014)
22. Ferreira, E.J., Melo, V.V., Delbem, A.C.B.: Algoritmos de estimação de distribuição em mineração de dados: Diagnóstico do greening in citrus. In: II Escola Luso-Brasileira de Computação Evolutiva, p. 1, Guimarães, Portugal (2010)
23. Mansour, M.R., Alberto, L.F.C., Ramos, R.A., Delbem, A.C.: Identifying groups of preventive controls for a set of critical contingencies in the context of voltage stability. In: Circuits and Systems (ISCAS), IEEE International Symposium on, IEEE, pp. 453–456 (2013)
24. Soares, A., Râbelo, R., Delbem, A.: Optimization based on phylogram analysis. In: Expert Systems with Applications **78**, pp. 32–50, ISSN 0957–4174 (2017). https://doi.org/10.1016/j.eswa.2017.02.012
25. Fonseca, C.M., Guerreiro, A.P., López-Ibáñez, M., Paquete, L.: On the computation of the empirical attainment function. In: Takahashi, R.H.C., Deb, K., Wanner, E.F., Greco, S. (eds.) EMO 2011. LNCS, vol. 6576, pp. 106–120. Springer, Heidelberg (2011). https://doi.org/10.1007/978-3-642-19893-9_8
26. Goutsias, J.K.: Mutually compatible Gibbs random fields. In: IEEE Transactions on Information Theory **35**(6), 1233–1249 (1989)
27. Pearl, J., Geiger, D., Verma, T.: Conditional independence and its representations. In: Kybernetika **25**(7), 33–44 (1989)
28. Gaspar-Cunha, A., Covas, J.A.: The plasticating sequence in barrier extrusion screws part i: modeling. Polym. Eng. Sci. **54**(8), 1791–1803 (2014)
29. Gaspar-Cunha, A., Covas, J.A.: The plasticating sequence in barrier extrusion screws part ii: experimental assessment. Polym.-Plast. Technol. Eng. **53**(14), 1456–1466 (2014)

Machine Learning and Multi-criterion Optimization

Multi-objective Learning Using HV Maximization

Timo M. Deist[1(✉)], Monika Grewal[1], Frank J.W.M. Dankers[2], Tanja Alderliesten[2], and Peter A.N. Bosman[1,3]

[1] Centrum Wiskunde and Informatica, Amsterdam, The Netherlands
timo.deist@cwi.nl
[2] Leiden University Medical Center, Leiden, The Netherlands
[3] Delft University of Technology, Delft, The Netherlands

Abstract. Real-world problems are often multi-objective, with decision-makers unable to specify a priori which trade-off between the conflicting objectives is preferable. Intuitively, building machine learning solutions in such cases would entail providing multiple predictions that span and uniformly cover the Pareto front of all optimal trade-off solutions. We propose a novel approach for multi-objective training of neural networks to approximate the Pareto front during inference. In our approach, we train the neural networks multi-objectively using a dynamic loss function, wherein each network's losses (corresponding to multiple objectives) are weighted by their hypervolume maximizing gradients. Experiments on different multi-objective problems show that our approach returns well-spread outputs across different trade-offs on the approximated Pareto front without requiring the trade-off vectors to be specified a priori. Further, results of comparisons with the state-of-the-art approaches highlight the added value of our proposed approach, especially in cases where the Pareto front is asymmetric.

Keywords: Multi-objective optimization · Neural networks · Pareto front · Hypervolume · Multi-objective learning

1 Introduction

Multi-objective (MO) optimization refers to finding Pareto optimal solutions for multiple, often conflicting, objectives. In MO optimization, a solution is Pareto optimal if none of the objectives can be improved without a simultaneous detriment in performance on at least one of the other objectives [35]. MO optimization is used for MO decision-making in many real-world applications [32] e.g., e-commerce recommendation [21], treatment plan optimization [25,27], and aerospace engineering [29]. In this paper, we focus on learning-based MO decision-making i.e., MO training of machine learning (ML) models so that MO decision-making is possible during inference. Specifically, we focus on training neural networks to generate a finite number of Pareto optimal solutions for each

T. M. Deist and M. Grewal—contributed equally.

M. Emmerich et al. (Eds.): EMO 2023, LNCS 13970, pp. 103–117, 2023.
https://doi.org/10.1007/978-3-031-27250-9_8

sample[1], so that they together provide a discrete approximation of the Pareto front[2].

The most straightforward approach for MO optimization is linear scalarization, i.e., optimizing a linear combination of different objectives according to scalarization weights. The scalarization weights are based on the desired trade-off between multiple objectives which is often referred to as 'user preference'. A major issue with linear scalarization is that user preferences cannot always be straightforwardly translated to linear scalarization weights. Recently proposed approaches have tackled this issue and find solutions on the average Pareto front for conflicting objectives according to a pre-specified user preference vector [20,23]. However, in many real-world problems, the user preference vector cannot be known a priori and decision-making is only possible *a posteriori*, i.e., after multiple solutions are generated that are (near) Pareto optimal for a specific sample[3]. For example, in neural style transfer [11] where photos are manipulated to imitate an art style from a selected painting, the user preference between the amount of semantic information (the photo's content) and artistic style can only be decided by looking at multiple different resultant images on the Pareto front (Fig. 5). Moreover, defining multiple trade-offs, typically by defining multiple scalarizations, to evenly cover the Pareto front is far from trivial, e.g., if the Pareto front is asymmetric. Here, we define asymmetry in Pareto fronts as asymmetry in the distribution of Pareto optimal solutions in the objective space on either side of the 45°-line, the line which represents the trade-off of equal marginal benefit along all objectives (see Pareto fronts in Fig. 1). We demonstrate and discuss this further in Sect. 4. To enable a posteriori decision-making per sample, multiple solutions spanning the Pareto front need to be generated without requiring the user preference vectors beforehand.

Despite many developments in the direction of MO training of neural networks with pre-specified user preferences, research on MO learning allowing for a posteriori decision-making is still scarce. Here, we present a novel method to multi-objectively train a set of neural networks to this end, leveraging the concept of hypervolume. Although we present our approach for training neural networks, the proposed formulation can be used for a wide range of ML models.

The hypervolume (HV) – the objective space dominated by a given set of solutions – is a popular metric to compare the quality of different sets of solutions approximating the Pareto front. It has its origins in the field of evolutionary algorithms [39], which are commonly accepted to be state of the art for multi-objective optimization. Theoretically, if the HV is maximal for a set of solutions, these solutions are on the Pareto front [9]. Additionally, HV not only encodes the proximity of a set of solutions to the Pareto front but also their diversity, which means that HV maximization provides a straightforward way for finding diverse solutions on the Pareto front. Therefore, we use hypervolume maximization for

[1] Note that, during inference, only *near* Pareto optimal solutions can be generated due to the generalization gap between training and inference.

[2] The Pareto front is the set of all Pareto optimal solutions in objective space.

[3] For more information on a posteriori decision-making, please refer to [14].

MO training of neural networks. We train the set of neural networks with a dynamically weighted combination of loss functions corresponding to multiple objectives, wherein the weight of each loss is based on the HV-maximizing gradients. In summary, our paper has the following main contributions:

- An MO approach for training neural networks
  - using gradient-based HV maximization
  - predicting Pareto optimal and diverse solutions on the Pareto front per sample without requiring specification of user preferences
  - enabling learning-based a posteriori decision-making.
- Experiments to demonstrate the added value of the proposed approach, specifically in asymmetric Pareto fronts.

2 Related Work

MO optimization has been used in machine learning for hyperparameter tuning of machine learning models [2,18], multi-objective classification of imbalanced data [33], and discovering the complete Pareto set starting from a single Pareto optimal solution [22]. [15] used MO optimization for finding configurations of deep neural networks for conflicting objectives. [13] proposed optimizing the weights of an autoencoder multi-objectively for finding the Pareto front of sparsity and reconstruction error. [24] used the Tchebycheff procedure for multi-objective optimization of a single neural network with multiple heads for multi-task text classification. Although we do not focus on these directions, our proposed approach can be used in similar applications.

MO training of a set of neural networks such that their predictions approximate the Pareto front of multiple objectives is closely related to the work presented in this paper. Similar to our work, [20,23] describe approaches with dynamic loss formulations to train multiple networks such that the predictions from these multiple networks together approximate the Pareto front. However, in these approaches, the trade-offs between conflicting objectives are required to be known in advance whereas our proposed approach does not require knowing the set of trade-offs beforehand. Other approaches [19,28] involve training a "hypernetwork" to predict the weights of another neural network based on a user-specified trade-off. Recently, it has been proposed to condition a neural network for an input user preference vector to allow for predicting multiple points near the Pareto front during inference [31]. While these approaches can approximate the Pareto front by iteratively predicting neural network weights or outputs based on multiple user preference vectors, the process of sampling the user preference vectors may still be intensive for an unknown Pareto front shape. Another approach proposes growing dense Pareto fronts from sparse Pareto optimal solutions [22], for which our approach can provide baseline solutions.

Gradient-based HV maximization is a key component of our approach. [26] have described gradient-based HV maximization for single networks and formulated a dynamic loss function treating each sample's error as a separate loss. [1]

applied this concept for training in generative adversarial networks. HV maximization is also applied in reinforcement learning [34,38]. While these approaches use HV maximizing gradients to optimize the weights of a single neural network, our proposed approach formulates a dynamic loss based on HV maximizing gradients for a *set* of neural networks. Different from our approach, other concurrent approaches for HV maximization are based on transformation to $(m-1)D$ (where m is the number of objectives) integrals by use of polar coordinates [7], random scalarization [12], and a q-Expected hypervolume improvement function [3].

3 Approach

MO learning of a network parameterized by a vector θ can be formulated as minimizing a vector of n losses $\mathcal{L}(\theta, s_k) = [L_1(\theta, s_k), \ldots, L_n(\theta, s_k)]$ for a given set of samples $S = \{s_1, \ldots, s_k, \ldots, s_{|S|}\}$. These loss functions form the loss space, wherein the subspace attainable by a sample's losses is bounded by its Pareto front. To learn multiple networks with loss vectors on each sample's Pareto front, we replace θ by a set of parameters $\Theta = \{\theta_1, \ldots, \theta_p\}$, where each parameter vector θ_i represents a network. The corresponding set of loss vectors is $\{\mathcal{L}(\theta_1, s_k), \ldots, \mathcal{L}(\theta_p, s_k)\}$ and is represented by a stacked loss vector $\mathfrak{L}(\Theta, s_k) = [\mathcal{L}(\theta_1, s_k), \ldots, \mathcal{L}(\theta_p, s_k)]$. **Our goal is to learn a set of p networks such that loss vectors in $\mathfrak{L}(\Theta, s_k)$ corresponding to the networks' predictions for sample s_k lie on and span the Pareto front of loss functions for sample** s_k. In other words, each network's loss vector is Pareto optimal and lies in a distinct subsection of the Pareto front for each sample. To achieve this goal, we train networks so that the loss subspace Pareto dominated by the networks' predictions (i.e., the HV) is maximal.

The HV of a loss vector $\mathcal{L}(\theta_i, s_k)$ for a sample s_k is the volume of the subspace $D_r(\mathcal{L}(\theta_i, s_k))$ in loss space dominated by $\mathcal{L}(\theta_i, s_k)$. This is illustrated in Fig. 1a. To keep this volume finite, the HV is computed with respect to a reference point r which bounds the space to the region of interest[4]. Subsequently, the HV of multiple loss vectors $\mathfrak{L}(\Theta, s_k)$ is the HV of the union of dominated subspaces $D_r(\mathcal{L}(\theta_i, s_k)), \forall i \in \{1, 2, ..., p\}$. The MO learning problem to maximize the mean HV of all $|S|$ samples is as follows:

$$\text{maximize} \frac{1}{|S|} \sum_{k=1}^{|S|} \text{HV}\left(\mathfrak{L}(\Theta, s_k)\right) \tag{1}$$

The update direction of gradient ascent for parameter vector θ_i of network i is:

$$\frac{\partial \frac{1}{|S|} \sum_{k=1}^{|S|} \text{HV}(\mathfrak{L}(\Theta, s_k))}{\partial \theta_i} \tag{2}$$

[4] The reference point is generally set to large coordinates in loss space to ensure that it is always dominated by all loss vectors.

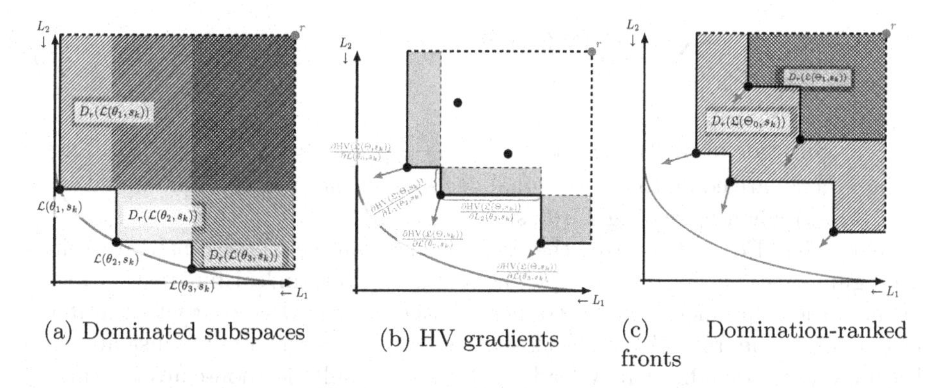

(a) Dominated subspaces (b) HV gradients (c) Domination-ranked fronts

Fig. 1. **(a)** Three Pareto optimal loss vectors $\mathcal{L}(\theta_i, s)$ on the Pareto front (green) with dominated subspaces $D_r(\mathcal{L}(\theta_i, s_k))$ with respect to reference point r. The union of dominated subspaces is the dominated hypervolume (HV) of $\mathfrak{L}(\Theta, s_k)$. **(b)** Gray markings illustrate the computation of the HV gradients $\frac{\partial \text{HV}(\mathfrak{L}(\Theta,s))}{\partial \mathcal{L}(\theta_i,s)}$ (gray arrows) in the three non-dominated solutions. **(c)** The same five solutions grouped into two domination-ranked fronts Θ_0 and Θ_1 with corresponding HV, equal to their dominated subspaces $D_r(\mathcal{L}(\theta_i, s_k))$, and HV gradients. (Color figure online)

By exploiting the chain rule decomposition of HV gradients as described in [8], the update direction in Eq. (2) for parameter vector θ_i of network i can be written as follows:

$$\frac{1}{|S|} \sum_{k=1}^{|S|} \frac{\partial \text{HV}\left(\mathfrak{L}(\Theta, s_k)\right)}{\partial \mathcal{L}(\theta_i, s_k)} \cdot \frac{\partial \mathcal{L}(\theta_i, s_k)}{\partial \theta_i} \quad \forall i \in \{1, \ldots, p\} \tag{3}$$

The dot product of $\frac{\partial \text{HV}(\mathfrak{L}(\Theta,s_k))}{\partial \mathcal{L}(\theta_i,s_k)}$ (the HV gradients with respect to loss vector $\mathcal{L}(\theta_i, s_k)$) in loss space, and $\frac{\partial \mathcal{L}(\theta_i,s_k)}{\partial \theta_i}$ (the matrix of loss vector gradients in the network i's parameters θ_i) in parameter space, can be decomposed to

$$\frac{1}{|S|} \sum_{k=1}^{|S|} \sum_{j=1}^{n} \frac{\partial \text{HV}\left(\mathfrak{L}(\Theta, s_k)\right)}{\partial L_j(\theta_i, s_k)} \frac{\partial L_j(\theta_i, s_k)}{\partial \theta_i} \quad \forall i \in \{1, \ldots, p\} \tag{4}$$

where $\frac{\partial \text{HV}(\mathfrak{L}(\Theta,s_k))}{\partial L_j(\theta_i,s_k)}$ is the scalar HV gradient in the single loss function $L_j(\theta_i, s_k)$, and $\frac{\partial L_j(\theta_i,s_k)}{\partial \theta_i}$ are the gradients used in gradient descent for single-objective training of network i for loss $L_j(\theta_i, s_k)$. Based on Eq. (4), one can observe that mean HV maximization of loss vectors from a set of p networks for $|S|$ samples can be achieved by weighting their gradient descent directions for loss functions $L_j(\theta_i, s_k)$ with their corresponding HV gradients $\frac{\partial \text{HV}(\mathfrak{L}(\Theta,s_k))}{\partial L_j(\theta_i,s_k)}$ for all i, j. In other terms, the MO learning of a set of p networks can be achieved by minimizing[5] the following dynamic loss function for each network i:

[5] Minimizing the dynamic loss function maximizes the HV because the reference point r is in the positive quadrant ("to the right and above 0").

$$\frac{1}{|S|}\sum_{k=1}^{|S|}\sum_{j=1}^{n}\frac{\partial \mathrm{HV}\left(\mathfrak{L}(\Theta, s_k)\right)}{\partial L_j(\theta_i, s_k)} L_j(\theta_i, s_k) \quad \forall i \in \{1, \dots, p\} \tag{5}$$

The computation of the HV gradients $\frac{\partial \mathrm{HV}(\mathfrak{L}(\Theta,s_k))}{\partial L_j(\theta_i,s_k)}$ is illustrated in Fig. 1b. These HV gradients are equal to the marginal decrease in the subspace dominated only by $\mathcal{L}(\theta_i, s_k)$ when increasing $L_j(\theta_i, s_k)$.

Note that Eq. 5 maximizes the HV for each sample's losses instead of first averaging losses on the set of samples as commonly done in learning tasks. Consequently, the neural networks are trained on each sample's Pareto front separately, instead of on the front of averages losses. In [5], we experimentally illustrate that learning an average front may lead to undesired results for non-convex fronts.

3.1 HV Maximization of Domination-Ranked Fronts

A relevant caveat of gradient-based HV maximization is that HV gradients $\frac{\partial \mathrm{HV}(\mathfrak{L}(\Theta,s_k))}{\partial L_j(\theta_i,s_k)}$ in strongly dominated solutions are zero (because no movement in any direction will affect the HV, Fig. 1b) and in weakly dominated solutions are undefined [8]. To resolve this issue, we follow [37]'s approach, which avoids the problem of dominated solutions by sorting all loss vectors into separate fronts Θ_l of mutually non-dominated loss vectors and optimizing each front separately (Fig. 1c). l is the domination rank, and $q(i)$ is the mapping of network i to domination rank l. By maximizing the HV of each front, trailing fronts with domination rank > 0 eventually merge with the non-dominated front Θ_0 and a single front is maximized by determining optimal locations for each loss vector on the Pareto front.

Furthermore, we normalize the HV gradients $\frac{\partial \mathrm{HV}\left(\mathfrak{L}(\Theta_{q(i)},s_k)\right)}{\partial \mathcal{L}(\theta_i,s_k)}$ as in [6] such that their length in loss space is 1. The dynamic loss function with domination-ranking of fronts and HV gradient normalization is:

$$\frac{1}{|S|}\sum_{k=1}^{|S|}\sum_{j=1}^{n}\frac{1}{w_i}\frac{\partial \mathrm{HV}\left(\mathfrak{L}(\Theta_{q(i)}, s_k)\right)}{\partial L_j(\theta_i, s_k)} L_j(\theta_i, s_k) \quad \forall i \in \{1, \dots, p\} \tag{6}$$

where $w_i = \left\| \frac{\partial \mathrm{HV}\left(\mathfrak{L}(\Theta_{q(i)},s_k)\right)}{\partial \mathcal{L}(\theta_i,s_k)} \right\|$.

3.2 Implementation

We implemented the HV maximization of losses from multiple networks, as defined in Eq. (6), in Python[6]. The neural networks were implemented using the PyTorch framework [30]. We used [10]'s HV computation reimplemented by Simon Wessing, available from [36]. The HV gradients $\frac{\partial \mathrm{HV}\left(\mathfrak{L}(\Theta_{q(i)},s_k)\right)}{\partial L_j(\theta_i,s_k)}$ were

[6] Code is available at https://github.com/timodeist/multi_objective_learning.

computed following the algorithm by [8]. Networks with identical losses were assigned the same HV gradients. For non-dominated networks with one or more identical losses (which can occur in training with three or more losses), the left- and right-sided limits of the HV function derivatives are not the same [8], and they were set to zero. Non-dominated sorting was implemented based on [4].

3.3 A Toy Example

Consider an example of MO regression with two conflicting objectives: given a sample $x_k \in S$, from input variable $X \in [0, 2\pi]$, predict the corresponding output z_k that matches y_k^1 from target variable Y_1 and y_k^2 from target variable Y_2, simultaneously. The relation between X, Y_1, and Y_2 is as follows:

$$Y_1 = \cos(X), \quad Y_2 = \sin(X)$$

The corresponding mean square error formulations for loss functions are $L_j = \frac{1}{|S|}\sum_{k=1}^{|S|}(y_k^j - z_k)^2; j \in \{1, 2\}$. We generated 200 samples of input and target variables for training and validation each. We trained five neural networks for 20000 iterations each with two fully connected linear layers of 100 neurons followed by ReLU nonlinearities. Figure 2a shows the HV over training iterations for the set of networks, which stabilizes visibly. Figure 2b shows predictions (y-axis) for validation samples evenly sampled from $[0, 2\pi]$ (x-axis). These predictions by five neural networks constitute Pareto front approximations for each sampled x_k, and correspond to precise predictions for $\cos(X)$ and $\sin(X)$, and trade-offs between both target functions. A network may generate predictions with changing trade-offs for different samples, as demonstrated Networks 2–5 in Fig. 2b for $x \in [\frac{3/2}{\pi}, 2\pi]$. Figure 2c shows the predictions for the highlighted samples in Fig. 2b in loss space, wherein they seem to be evenly distributed on the approximated Pareto front. It becomes clear from Figs. 2b & 2c that each x_k has a differently sized Pareto front which the networks are able to predict. Figure 2c also demonstrates an a posteriori decision-making scenario. Upon visualizing the different Pareto fronts per sample, a user might decide to select predictions corresponding to different trade-offs for different samples.

4 Experiments

We performed experiments with two MO problems: MO regression with differently shaped Pareto fronts and neural style transfer.[7] We compared the performance of our approach with **linear scalarization** and two state-of-the-art approaches:

[7] Additional experiments are provided in [5]: multi-observer medical image segmentation, MO regression with three losses, multi-style transfer, and a counter-example for initial loss normalization.

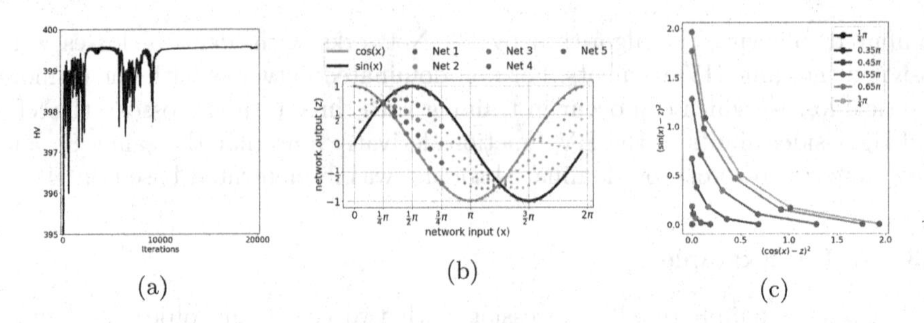

Fig. 2. MO regression on two losses. (a) HV values for a set of networks over training iterations. (b) Network outputs for $X \in [0, 2\pi]$. (c) Generated Pareto front predictions for a selection of six samples from $[\frac{1}{4}\pi, \frac{3}{4}\pi]$ in loss space.

Pareto MTL [20] and **EPO** [23]. Pareto MTL and EPO try to find Pareto optimal solutions on the average Pareto front for a given trade-off vector using dynamic loss functions. For a consistent comparison, we used the trade-offs used in the original experiments of EPO for Pareto MTL, EPO, and as fixed weights in linear scalarization.

Experiments were run on systems using Intel(R) Xeon(R) Silver 4110 CPU @ 2.10 GHz with NVIDIA GeForce RTX 2080Ti, or Intel(R) Core(R) i5-3570K @ 3.40 Ghz with NVIDIA GeForce GTX 1060 6 GB. For training, the Adam optimizer [17] was used. The learning rate and β_1 of Adam were tuned for each approach separately based on the maximal HV of validation loss vectors.

4.1 MO Regression

We considered three cases for the MO regression toy problem described in Sect. 3.3 each demonstrating a different Pareto front shape: the symmetric case with two MSE losses as in Fig. 2, and two asymmetric cases each with MSE as one loss and L1-norm or MSE scaled by $\frac{1}{100}$ as the second loss. The reference point for our proposed approach was set to $(20, 20)$.

Figure 3 shows Pareto front approximations for all three cases. Figures 3a & 3c show that fixed linear scalarizations and EPO produce networks generating well-distributed outputs with low losses that predict a sample's symmetric Pareto front for two conflicting MSE losses. The positions on the front approximated by linear scalarization seem to be far from the pre-specified trade-offs (gray lines). This is expected because, by definition of linear scalarization, the solutions should lie on the approximated Pareto front where the tangent is perpendicular to the search direction specified by the trade-offs. For Pareto MTL, networks are clustered closer to the center of the approximated Pareto front.

Optimizing MSE and L1-Norm (Figs. 3e–3h) results in an asymmetric Pareto front approximation. The predictions by our HV maximization-based approach remain well distributed across the fronts. EPO also still provides a decent spread albeit less uniform across samples whereas linear scalarization and Pareto MTL tend to both or mostly the lower extrema, respectively.

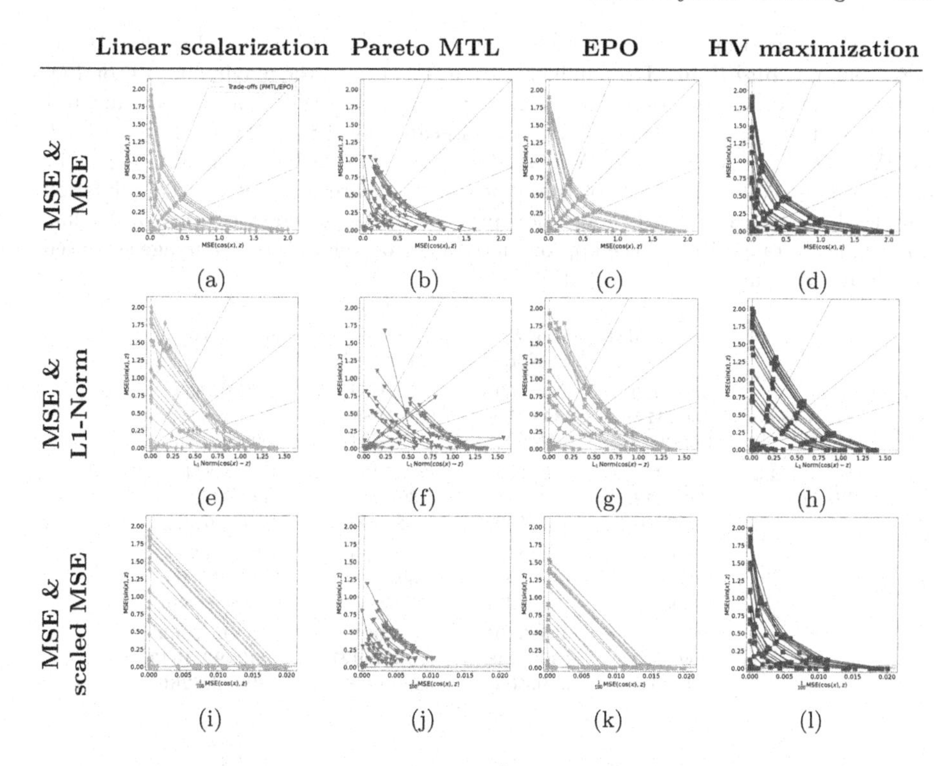

Fig. 3. Pareto front approximations on a random subset of validation samples by sets of five neural networks trained using four approaches. Three different pairs of loss functions are used: (a–d) MSE and MSE, (e–h) MSE and L1-Norm, and (i–l) MSE and scaled MSE.

The difficulty of manually pre-specifying the trade-offs without knowledge of the Pareto front becomes more evident when optimizing losses with highly different scales (Figs. 3i–3l). The pre-specified trade-offs do not evenly cover the Pareto fronts. Consequently, the networks trained by EPO do not cover the Pareto front evenly despite following the pre-specified trade-offs. Further, the networks optimized by Pareto MTL cover only the upper part of the fronts. Networks trained with fixed linear scalarizations tend towards both extrema. On the other hand, our approach trains networks that follow well-distributed trade-offs on the Pareto front. Normalizing losses from differing scales as in Figs. 3i–3l might not sufficiently improve methods based on pre-specified trade-offs (Pareto MTL, EPO) or fixed linear scalarizations [5].

The mean HV over 200 validation samples is computed for all approaches and Table 1 displays the median and inter-quartile ranges (IQR) over 25 runs. The magnitude of the HV is largely determined by the position of the reference point. For $r = (20, 20)$ the maximal HV equals 400 minus the area bounded by the utopian point $(0, 0)$ and a sample's Pareto front. Even poor approximations of a sample's Pareto front can yield a HV ≥ 390. For these reasons, HVs in Table 1 appear large and minuscule differences between HVs are relevant. As

Table 1. Comparison of HV across different approaches. The maximal median HV in each column is **highlighted**. Small increases in HV close to the maximum (10^6 or 400) matter: see Sect. 4.1. A statistically significant one-sided Wilcoxon signed rank test with correction for multiple comparison is indicated by: LS vs HV max. (*), PMTL vs HV max. (†), and EPO vs HV max. (‡). **Columns 1–3:** Median (inter-quartile range) values of the mean HV of the approximated Pareto fronts for 200 validation samples from 25 runs of MO regression problem are reported. **Column 4:** Median (inter-quartile range) HV of the approximated Pareto fronts for the 25 image sets used in neural style transfer are reported.

	MSE & MSE	MSE & L1-Norm	MSE & scaled MSE	Style & content
Linear scalarization (LS)	**399.5929*** **(399.5776, 399.6018)**	399.2909 (399.2738, 399.3045)	399.9859 (399.9857, 399.9864)	999990.7699 (999988.6580, 999992.5850)
Pareto MTL (PMTL)	397.1356 (396.3212, 397.6288)	392.2956 (392.0377, 393.4942)	398.3159 (397.4799, 398.6699)	997723.8748 (997583.5152, 998155.6837)
EPO	399.5135 (399.5051, 399.5348)	399.0884 (398.998, 399.1743)	399.9885 (399.9883, 399.9889)	999988.4297 (999984.4808, 999989.8338)
HV maximization	399.5823† ‡ (399.5619, 399.6005)	**399.3795*** † ‡ **(399.3481, 399.4039)**	**399.9954*** † ‡ **(399.9927, 399.9957)**	**999999.7069** **(999999.4543, 999999.8266)*** † ‡

expected, our approach finds higher HV values for the case of asymmetric front shapes (Table 1 columns 2 and 3, and Figs. 3e–3l). In case of the symmetric front shape (Fig. 3a), since the pre-specified trade-offs appear to span the Pareto front shape well, linear scalarization's training based on fixed loss weights is more efficient than training on a dynamic loss with varying weights as used by HV maximization. This increased efficiency of training using fixed weights that are suitable for symmetric MSE losses presumably results in a slightly higher HV for linear scalarization (Table 1 column 1).

4.2 Neural Style Transfer

We further considered the MO optimization problem of neural style transfer as defined in [11] (we reused and adjusted Pytorch's neural style transfer implementation [16]), where pixels of an image are optimized to minimize content loss (semantic similarity with a target image) and style loss (artistic similarity with a style image) simultaneously. We performed experiments with 25 image pairs (image sources as in [5]), obtained by combining 5 content and 10 style images to generate 6 solutions on the Pareto front. The reference point in our approach was chosen as (100, 10000) based on preliminary runs.

Figure 4 shows the obtained Pareto front estimates for 25 image sets by each approach. Linear scalarization (a) and EPO (c) determine solutions close to or on the chosen user preferences which, however, do not diversely cover the range of possible trade-offs. Pareto MTL (b) achieves sets of clustered and partly

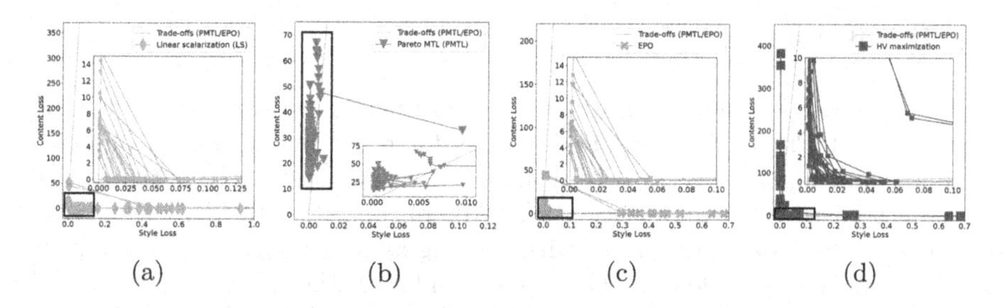

Fig. 4. Pareto front estimates in loss space by different approaches for neural style transfer using four approaches: (a) Linear scalarization (b) Pareto MTL, (c) EPO, and (d) HV maximization. Sections within the black frames are magnified.

dominated solutions, which do not cover trade-offs with low content loss. On the other hand, HV maximization (d) returns Pareto front estimates that broadly cover diverse trade-offs between style and content loss across different image sets without having to specify user preferences. This is also reflected in the significantly larger median HVs reported in Table 1.

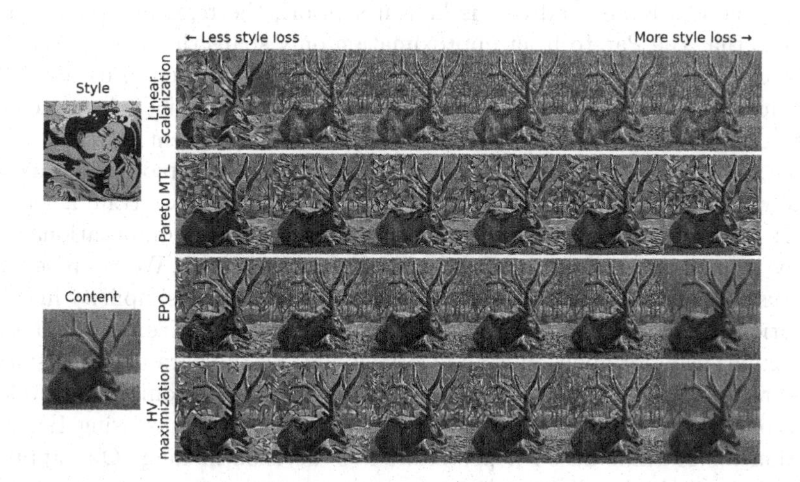

Fig. 5. Neural style transfer example by all four approaches for one image set.

Figure 5 shows the images generated by each approach for one of the image sets. This case was manually selected for its aesthetic appeal.[8] The images seen here match observations from Fig. 4, e.g., Pareto MTL's images show little diversity in style and content, many images by linear scalarization of EPO have too little style match ('uninteresting' images), and images by HV maximization show most interesting diversity.

[8] Generated images for all 25 image sets are available at https://github.com/timodeist/multi_objective_learning.

5 Discussion

We have proposed an approach to train a set of neural networks such that they jointly predict Pareto front approximations for each sample during inference, without requiring user-specified trade-offs. Our approach translates the concept of gradient-based HV maximization from MO optimization to MO learning. We provide experimental comparisons with existing approaches that require a priori specification of the trade-offs. The results highlight the advantage of our HV maximization approach, especially in MO problems that exhibit asymmetric Pareto front.

Our HV maximization based approach does not require specifying p trade-offs a priori (based on the number of predictions, p, required on the Pareto front), which essentially are $p(n-1)$ hyperparameters of the learning process for n losses. Choosing these trade-offs well requires knowledge of the Pareto front shapes, which is often not known a priori. HV maximization, however, introduces the n-dimensional reference point r and thus n additional hyperparameters. However, choosing a reference point such that the entire Pareto front gets approximated is not complex. It often suffices to use losses of randomly initialized networks rescaled by a factor ≥ 1 as the reference point. If only a specific section of the Pareto front is relevant and this is known a priori, the reference point can be chosen so that the Pareto front approximation only spans the chosen section.

HV-based training for sets of neural networks can, in theory, be applied to any number of networks, p, and loss functions, n. In practice, the time complexity of exact HV (exponential in n, [10]) and HV gradient (quadratic in p with $n \leq 4$, [8]) computations is limiting but may be overcome by algorithmic improvements using, e.g., HV approximations. Further, we train a separate network corresponding to each prediction. This increases computational load linearly if more predictions on the Pareto front are desired. We train separate networks instead of one multi-headed network for the sake of simplicity in experimentation and clarity when demonstrating our approach. It is expected that the HV maximization formulation would work similarly if the parameters of some of the neural network layers are shared, which would decrease computational load.

In conclusion, we describe MO training of neural networks using HV maximization for learning-based a posteriori MO decision-making. Our approach provided the desired well-spread Pareto front approximations on artificial MO regression problems. On the MO style transfer problem, our method yielded encouraging results that emphasize its usefulness for a posteriori decision-making.

Acknowledgements. We thank dr. Marco Virgolin (Chalmers University of Technology) for his valuable contributions. The research is funded by Open Technology Programme (15586) of the Dutch Research Council (NWO), Elekta, and Xomnia, and the public-private partnership allowance for top consortia for knowledge and innovation from the Ministry of Economic Affairs.

References

1. Albuquerque, I., Monteiro, J., Doan, T., Considine, B., Falk, T., Mitliagkas, I.: Multi-objective training of generative adversarial networks with multiple discriminators. arXiv preprint arXiv:1901.08680 (2019)
2. Avent, B., Gonzalez, J., Diethe, T., Paleyes, A., Balle, B.: Automatic discovery of privacy-utility Pareto fronts. Proc. Priv. Enh. Technol. **2020**(4), 5–23 (2020)
3. Daulton, S., Balandat, M., Bakshy, E.: Differentiable expected hypervolume improvement for parallel multi-objective Bayesian optimization. arXiv preprint arXiv:2006.05078 (2020)
4. Deb, K., Pratap, A., Agarwal, S., Meyarivan, T.: A fast and elitist multiobjective genetic algorithm: NSGA-II. IEEE Trans. Evol. Comput. **6**(2), 182–197 (2002)
5. Deist, T.M., Grewal, M., Dankers, F.J., Alderliesten, T., Bosman, P.A.: Multi-objective learning to predict Pareto fronts using hypervolume maximization. arXiv preprint arXiv:2102.04523 (2021)
6. Deist, T.M., Maree, S.C., Alderliesten, T., Bosman, P.A.N.: Multi-objective optimization by uncrowded hypervolume gradient ascent. In: Bäck, T., et al. (eds.) PPSN 2020. LNCS, vol. 12270, pp. 186–200. Springer, Cham (2020). https://doi.org/10.1007/978-3-030-58115-2_13
7. Deng, J., Zhang, Q.: Approximating hypervolume and hypervolume contributions using polar coordinate. IEEE Trans. Evol. Comput. **23**(5), 913–918 (2019)
8. Emmerich, M., Deutz, A.: Time complexity and zeros of the hypervolume indicator gradient field. In: Schuetze, O., et al. (eds.) EVOLVE - A Bridge between Probability, Set Oriented Numerics, and Evolutionary Computation III. Studies in Computational Intelligence, vol. 500, pp. 169–193. Springer, Heidelberg (2014). https://doi.org/10.1007/978-3-319-01460-9_8
9. Fleischer, M.: The measure of Pareto optima applications to multi-objective metaheuristics. In: Fonseca, C.M., Fleming, P.J., Zitzler, E., Thiele, L., Deb, K. (eds.) EMO 2003. LNCS, vol. 2632, pp. 519–533. Springer, Heidelberg (2003). https://doi.org/10.1007/3-540-36970-8_37
10. Fonseca, C.M., Paquete, L., López-Ibánez, M.: An improved dimension-sweep algorithm for the hypervolume indicator. In: 2006 IEEE International Conference on Evolutionary Computation, pp. 1157–1163. IEEE (2006)
11. Gatys, L.A., Ecker, A.S., Bethge, M.: Image style transfer using convolutional neural networks. In: Proceedings of the IEEE Conference on Computer Vision and Pattern Recognition, pp. 2414–2423 (2016)
12. Golovin, D., et al.: Random hypervolume scalarizations for provable multi-objective black box optimization. arXiv preprint arXiv:2006.04655 (2020)
13. Gong, M., Liu, J., Li, H., Cai, Q., Su, L.: A multiobjective sparse feature learning model for deep neural networks. IEEE Trans. Neural Netw. Learn. Syst. **26**(12), 3263–3277 (2015)
14. Hwang, C.L., Masud, A.S.M.: Multiple Objective Decision Making—Methods and Applications: A State-of-the-Art Survey, vol. 164. Springer, Heidelberg (2012). https://doi.org/10.1007/978-3-642-45511-7
15. Iqbal, M.S., Su, J., Kotthoff, L., Jamshidi, P.: FlexiBO: cost-aware multi-objective optimization of deep neural networks. arXiv preprint arXiv:2001.06588 (2020)
16. Jacq, A.: Neural style transfer using Pytorch (2017). https://pytorch.org/tutorials/advanced/neural_style_tutorial.html
17. Kingma, D.P., Ba, J.: Adam: a method for stochastic optimization. arXiv preprint arXiv:1412.6980 (2014)

18. Koch, P., Wagner, T., Emmerich, M.T., Bäck, T., Konen, W.: Efficient multi-criteria optimization on noisy machine learning problems. Appl. Soft Comput. **29**, 357–370 (2015)
19. Lin, X., Yang, Z., Zhang, Q., Kwong, S.: Controllable Pareto multi-task learning. arXiv preprint arXiv:2010.06313 (2020)
20. Lin, X., Zhen, H.L., Li, Z., Zhang, Q.F., Kwong, S.: Pareto multi-task learning. Adv. Neural. Inf. Process. Syst. **32**, 12060–12070 (2019)
21. Lin, X., et al.: A Pareto-efficient algorithm for multiple objective optimization in e-commerce recommendation. In: Proceedings of the 13th ACM Conference on Recommender Systems, pp. 20–28 (2019)
22. Ma, P., Du, T., Matusik, W.: Efficient continuous Pareto exploration in multi-task learning. In: International Conference on Machine Learning, pp. 6522–6531. PMLR (2020)
23. Mahapatra, D., Rajan, V.: Multi-task learning with user preferences: gradient descent with controlled ascent in Pareto optimization. In: International Conference on Machine Learning, pp. 6597–6607. PMLR (2020)
24. Mao, Y., Yun, S., Liu, W., Du, B.: Tchebycheff procedure for multi-task text classification. In: Proceedings of the 58th Annual Meeting of the Association for Computational Linguistics, pp. 4217–4226 (2020)
25. Maree, S.C., et al.: Evaluation of bi-objective treatment planning for high-dose-rate prostate brachytherapy—a retrospective observer study. Brachytherapy **18**(3), 396–403 (2019)
26. Miranda, C.S., Von Zuben, F.J.: Single-solution hypervolume maximization and its use for improving generalization of neural networks. arXiv preprint arXiv:1602.01164 (2016)
27. Müller, B., Shih, H., Efstathiou, J., Bortfeld, T., Craft, D.: Multicriteria plan optimization in the hands of physicians: a pilot study in prostate cancer and brain tumors. Radiat. Oncol. **12**(1), 1–11 (2017)
28. Navon, A., Shamsian, A., Chechik, G., Fetaya, E.: Learning the Pareto front with hypernetworks. arXiv preprint arXiv:2010.04104 (2020)
29. Oyama, A., Liou, M.S.: Multiobjective optimization of rocket engine pumps using evolutionary algorithm. J. Propul. Power **18**(3), 528–535 (2002)
30. Paszke, A., et al.: Automatic differentiation in PyTorch. Adv. Neural Inf. Process. Syst. (2017). https://github.com/pytorch/pytorch
31. Ruchte, M., Grabocka, J.: Efficient multi-objective optimization for deep learning. arXiv preprint arXiv:2103.13392 (2021)
32. Stewart, T., et al.: Real-world applications of multiobjective optimization. In: Branke, J., Deb, K., Miettinen, K., Słowiński, R. (eds.) Multiobjective Optimization. LNCS, vol. 5252, pp. 285–327. Springer, Heidelberg (2008). https://doi.org/10.1007/978-3-540-88908-3_11
33. Tari, S., Hoos, H., Jacques, J., Kessaci, M.-E., Jourdan, L.: Automatic configuration of a multi-objective local search for imbalanced classification. In: Bäck, T., et al. (eds.) PPSN 2020. LNCS, vol. 12269, pp. 65–77. Springer, Cham (2020). https://doi.org/10.1007/978-3-030-58112-1_5
34. Van Moffaert, K., Nowé, A.: Multi-objective reinforcement learning using sets of Pareto dominating policies. J. Mach. Learn. Res. **15**(1), 3483–3512 (2014)
35. Van Veldhuizen, D.A., Lamont, G.B.: Multiobjective evolutionary algorithms: analyzing the state-of-the-art. Evol. Comput. **8**(2), 125–147 (2000)
36. Wang, H., Deutz, A., Bäck, T., Emmerich, M.: Code repository: Hypervolume indicator gradient ascent multi-objective optimization. https://github.com/wangronin/HIGA-MO

37. Wang, H., Deutz, A., Bäck, T., Emmerich, M.: Hypervolume indicator gradient ascent multi-objective optimization. In: Trautmann, H., et al. (eds.) EMO 2017. LNCS, vol. 10173, pp. 654–669. Springer, Cham (2017). https://doi.org/10.1007/978-3-319-54157-0_44
38. Xu, J., Tian, Y., Ma, P., Rus, D., Sueda, S., Matusik, W.: Prediction-guided multi-objective reinforcement learning for continuous robot control. In: International Conference on Machine Learning, pp. 10607–10616. PMLR (2020)
39. Zitzler, E., Thiele, L.: Multiobjective evolutionary algorithms: a comparative case study and the strength pareto approach. IEEE Trans. Evol. Comput. **3**(4), 257–271 (1999)

Sparse Adversarial Attack via Bi-objective Optimization

Phoenix Williams(✉), Ke Li, and Geyong Min

Department of Computer Science, University of Exeter, Exeter EX4 4QF, UK
{pw384,k.li,g.min}@exeter.ac.uk

Abstract. Neural classifiers have achieved near human level performances when applied to several real-world tasks. Despite their successes, recent works have demonstrated their vulnerability to adversarial attacks. In particular, image classifiers have shown to be vulnerable to fine-tuned noise that perturb a small number of pixels, known as sparse attacks. To generate such perturbations current works either prioritise query efficiency by allowing the size of the perturbation to be unbounded or the minimization of its size by allowing a large number of pixels to be perturbed. Addressing the drawbacks of both approaches we propose a method of conducting query efficient sparse adversarial attacks that minimizes the number of perturbed pixels by formulating the attack as a constrained bi-objective optimization problem. Within the single objective unbounded query-efficient scenario our method is able to outperform state-of-the-art sparse attack algorithms in terms of success rate and query efficiency. When also minimizing the number of perturbed pixels in the bi-objective setting, the proposed method is able to generate adversarial perturbations that impact a fewer number of pixels than its state-of-the-art competitors.

Keywords: Adversarial attack · Multi-objective optimization · Evolutionary algorithms

1 Introduction

Deep neural networks (DNNs) have achieved state-of-the-art performance in various tasks and have been applied successfully to several real-world problems [5,7,34]. In particular, for image classification tasks they have been able to achieve near human-level accuracy [18,24,25,27,33,44,45,55]. Despite their success, works in the literature [19,30,38,48] have demonstrated that adding small optimized perturbations to correctly classified images can cause trained DNNs to misclassify. Such images are commonly referred to as adversarial images. Furthermore, specific adversarial images have demonstrated the ability to cause DNNs to misclassify to a particular class [1,40] and have shown to exist within the physical world [30]. Due to these vulnerabilities many concerns have been raised about their application to security-critical tasks [1]. It has been stated that a key

M. Emmerich et al. (Eds.): EMO 2023, LNCS 13970, pp. 118–133, 2023.
https://doi.org/10.1007/978-3-031-27250-9_9

area of addressing these concerns is the development of attack algorithms that generate strong adversarial images [3]. Consequently, there is a growing interest in developing new attack algorithms that generate such images. By formulating the attack as an optimization problem, many works apply existing and novel optimization methods to solving this particular task. Many works assume full access to an attacked model [9,10,19,30,35,42,48] i.e. white-box attacks, however attention must also be given to the black-box setting where the attacker can only access the outputs of the attacked neural network classifier.

Current black-box attack algorithms can be classified by the constraints they apply to the perturbation. Many works assume no limit to the number of pixels a perturbation can modify and therefore constrain its size by its l_∞ or l_2 norm. Such works include those that use surrogate models [13,22,38], where the surrogate is assumed to be highly similar to the targeted model. However, this strong assumption questions the validity of such attacks in a real-world scenario. Other approaches do not make such an assumption and adjust the perturbation solely based on the outputs of the targeted classification model. Some works estimate the gradient of the targeted model's loss function using finite-differences [4,11,26,52,53] and make use of gradient-based optimization algorithms. Meta-heuristics such as evolutionary algorithms [1,36,39] or random search [2] have also shown to have success when attacking DNNs.

Conversely, sparse-attacks [14,15,37,41,47,54] constrain the perturbation by its l_0 norm which limits the number of pixels it can modify. Prioritising query-efficiency [14,47], attacks in the literature constrain the l_0 norm of the perturbation to a small percentage of the total number of pixels and allow the size of the modifications to be unbounded. Despite the argument that these unbounded modifications do not alter the semantic content of the original image, they are easily detectable to the human eye. Alternatively, works have aimed to address this drawback by proposing methods of generating sparse perturbations with constrained size, namely sparse and imperceptible attacks [15,49,54]. By constraining the size of the perturbation by either its l_∞ or l_2 norm, sparse and imperceptible attacks aim to minimize the number of pixels that are modified. Despite their efforts, results have shown attack algorithms using this approach still perturb a large number of pixels and require a large number of model queries.

From the works in the current literature, it is clear that there is a trade-off between the efficiency of the sparse-attack and the imperceptibility of the perturbation. To address this trade-off we formulate the sparse attack as a bi-objective optimization problem such that the generated perturbation causes the desired misclassification whilst also minimally impacting the original image. To handle this bi-objective problem we adapt the NSGA-II algorithm of Deb et al. [16] to the adversarial attack setting.

Contributions. Motivated by works in the literature, we propose a bi-objective approach that aims to generate adversarial perturbations whose impact on the original image is minimal. To address our formulated task we adapt the NSGA-II [16] algorithm by proposing novel crossover and mutation operators as well as an updated domination definition to reflect our priorities of the task. From our experiments, we show that the proposed method outperforms current state-of-the-art sparse-attack algorithms and can generate adversarial images with less perturbed pixels than its competitors.

The rest of this paper is organized as follows. Section 2 provides a summary of related works and details the attack setting we address. Section 3 outlines the proposed attack method and gives details of its implementation. Description of the experimental setup is given in Sect. 4 with the results of our conducted experiments shown in Sect. 5. We conclude this paper in Sect. 6 with a discussion of the proposed method and directions for future work.

2 Preliminaries

2.1 Adversarial Attack

Let $f : \mathcal{X} \subseteq [0,1]^{h \times w \times 3} \rightarrow \mathbb{R}^K$ be a classifier that takes an image $\mathbf{x} \in \mathcal{X}$ with height h and width w, and assigns it a class $y = \underset{r=1,\cdots,K}{\arg\max}\ f_r(\mathbf{x})$ where f_r is the probability of input $\mathbf{x}$ being of class r. An untargeted attack aims to find a perturbation $\delta \in \mathbb{R}^{h \times w \times 3}$ such that

$$\underset{r=1,\cdots,K}{\arg\max}\ f_r(\mathbf{x}+\delta) \neq y, \quad g(\delta) \leq \epsilon \tag{1}$$

where y is the correctly predicted label of the input $\mathbf{x}$, g is the constraint function of δ and K is the total number of classes. To ensure the adversarial image is close to its original counterpart, works in literature constrain δ by its l_p norm. Hence the discovery of a δ that satisfies Eq. (1) can be described by the optimization of

$$\underset{\delta \in \mathbb{R}^{h \times w \times 3}}{\arg\min}\ L(f(\mathbf{x}+\delta), y), \quad ||\delta||_p \leq \epsilon \tag{2}$$

where the minimization of L leads to the desired misclassification. Setting $\mathbf{x}_{adv} = \mathbf{x} + \delta$, we consider the untargeted attack scenario and define our loss function L as the following,

$$L_u(\mathbf{x}_{adv}) = \log f_y(\mathbf{x}_{adv}) - \log f_q(\mathbf{x}_{adv}) \tag{3}$$

where $f_q(\cdot) = \underset{r \neq y}{\arg\max}\ f_r(\mathbf{x}_{adv})$ and $r = 1 \cdots, K$.

2.2 Related Works

l_2 and l_∞ constrained adversarial attacks aim to find adversarial images $\mathbf{x}_{adv}$ that satisfy Eq. (2) where $||\mathbf{x}_{adv} - \mathbf{x}||_2 \leq \epsilon$ and $||\mathbf{x}_{adv} - \mathbf{x}||_\infty \leq \epsilon$ respectively. Within the white-box setting [9,10,15,19,30,35,42,48] it is assumed that the attacker has access to the architecture, weights and gradients of f. Using back-propagation for gradient computation the attacker formulates the attack as an optimization problem and solves using gradient-based methods. A more restricted scenario is the black-box setting where only the outputted class probabilities of the attacked classifier is accessible. Most works addressing the black-box setting aim to discover a perturbation value for each pixel by flattening δ to form a high-dimensional vector and generates $\mathbf{x}_{adv} = \mathbf{x} + \delta$ by reshaping δ to match $\mathbf{x}$. Some works in the literature estimate the gradient of Eq. (2) using finite-differences [4,11,26,52,53], these estimations are then applied within gradient-based optimization algorithms. In particular, the ZOO [11] algorithm proposed by Chen et. al makes use of the Adam [28] optimization algorithm with estimated gradients to generate adversarial images. A core issue with these approaches is the number of model queries required to estimate the gradient of the current perturbation δ within the high-dimensional image space. To overcome the computational cost many works reduce the dimension of the search space through the use of bi-linear interpolation [1,11,26,52] or auto-encoders [52]. Despite embeddings reducing the dimension of the search space, they have been shown to warp the search space such that it may not contain the optimum solution [31]. Alternatively, works have attempted to alleviate the computational cost by using natural evolutionary search approach's. Such works [26,52] sample points from zero-mean Gaussian distributions with unit-variances and have achieved competitive performances. Meta-heuristics such as evolutionary approaches [1,21,36,39] and random search [2] have also shown to be efficient mechanisms for conducting adversarial attacks and have shown to outperform gradient estimation algorithms in the literature. In particular, the GenAttack [1] algorithm evolves a population of adversarial images using genetic operators and applies adaptive parameter scaling with bi-linear interpolation to handle the high dimension of the search space. Despite their success, the performance of meta-heuristics have shown to suffer in high dimensional search spaces. Li et al. [32] propose a novel method for handling the high dimension of the image space by assuming an adversarial image can be constructed by a weighted sum of neighbouring images with differential evolution [46] and CMA-ES [23] being applied by the authors to optimize their values. Attacks making use of Bayesian optimization have also been proposed [40,43] and have shown good performance in the low-query availability setting. Other approaches make use of substitute models [13,22,38] trained on similar data sets to the attacked model. By conducting white-box attacks on the substitute model a series of adversarial images are generated which are used to attack the originally attacked model. In a real-world scenario, the availability of a similar model is unlikely within the black-box setting. Other addressed scenarios include the limited information scenario where only a subset of outputs are accessible to the attacker [26]. This setting is a generalization of the decision-based attack [6,8,12,20] where only the classified class probability is returned.

The number of works addressing the l_0 constrained attack is far fewer than those constrained by the l_2 and l_∞ norms, such sparse attacks aim to generate adversarial images by perturbing at most k pixels. Many existing l_0 bounded attacks prioritise the minimization of the perturbation size over query efficiency [15,37,41,50,54] and so require many model queries whilst also allowing

k to be large. Alternatively, other works prioritise query efficiency and set k to be small whilst allowing the size of the perturbation to be unbounded [14,47]. The work of Su et. al [47] represents a solution as a $k \times 5$ dimensional vector that consists of the x, y position of each pixel and its perturbed r, g, b value. Applying differential evolution [46], the values are adjusted to solve Eq. (2). Su et al. were able to generate adversarial images even under the extreme condition of setting $k = 1$. Croce et. al [14] proposed a random search algorithm that outperformed all compared black-box and white-box attack algorithms by iteratively sampled a changing distribution of pixel positions and RGB values. For comparison, Croce et al. [14] modified the PGD$_0$ algorithm [15] to the black-box setting by applying the gradient estimation mechanism proposed by Ilyas et. al [26]. The vast majority of works addressing sparse and imperceptible attacks [15,37,41,49,54] initially allow all pixels to be perturbed and aim to reduce their l_0 norm during the optimization process. Notable works include the white-box attack algorithm of Zhu et al. [54] who apply an evolutionary-inspired homotopy algorithm with a gradient based optimizer to minimize the l_0 norm of the perturbation whilst constraining its l_∞ norm. Addressing the black-box setting, the work of Tian et al. [51] proposes a dual-population co-evolutionary algorithm where one population is evolved to find successful adversarial examples and the other to minimize the l_0 and l_2 norm of the perturbation, named DCEA-ISA.

Considered Adversarial Attack Scenario. As outlined in Sect. 2.1, works addressing the l_0 constrained attack consider one of two approaches, perturbation size minimization [15,37,41,54] or query efficiency [14,47]. In a real world scenario the ability to conduct several thousand model queries within the black-box setting is questionable, however allowing the perturbation size to be unbounded results in them being easily visible to the human-eye. Therefore we propose a method of conducting query efficient attacks that generate images with a minimal number of perturbed pixels. To achieve this, we formulate the adversarial attack as the following problem:

$$\begin{aligned} &\text{minimize} \quad \mathbf{F}(\delta) = (L_u(\mathbf{x}_{adv}), ||\mathbf{x}_{adv} - \mathbf{x}||_2)^T \\ &\text{subject to} \, ||\delta||_0 \leq k \end{aligned} \tag{4}$$

where $\mathbf{x}_{adv} = \mathbf{x} + \delta$ is the adversarial image and $\mathbf{F}(\cdot)$ is the objective vector.

3 Proposed Method

The flowchart of our proposed algorithm for sparse adversarial attack is shown in Fig. 1. To attack an image classification model f we first generate an initial population of N solutions by randomly sampling a set of k pixel positions and corresponding RGB values. We constrain the RGB value of the perturbation to the set $\{-1, 1, 0\} \in \mathbb{R}$ where the probability of sampling 0 is described by Pr_0. During the main loop of the algorithm, a population of offsprings Q_t is generated using crossover and mutation operations. Combining the current population P_t

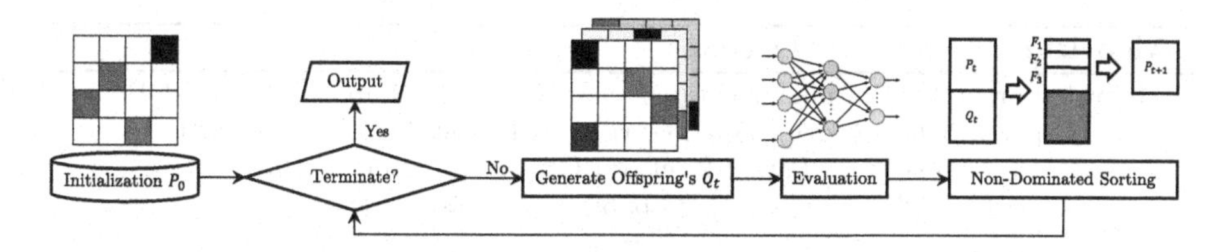

Fig. 1. Flowchart of the NSGA-II algorithm for sparse and imperceptible adversarial attack

and the offspring population Q_t, the next population P_{t+1} is constructed through non-dominated sorting. For the rest of this section, we give explicit detail about the implemented offspring generation and solution evaluation operators. The non-dominated sorting method follows that of Deb et al. with a modified domination operator defined in Sect. 3.2. We conclude this section by outlining how the minimization of the perturbations l_2 norms leads to the minimzation of its l_0 when we constrain its values to the set $\{-1, 1, 0\}$.

3.1 Solution Evaluation

Given a solution $\mathbf{s}_i$ that contains of a set of k pixel positions $\{p_0, p_1, \cdots, p_k\}$ and corresponding perturbations $\{\mathbf{rgb}_0, \mathbf{rgb}_1, \cdots, \mathbf{rgb}_k\} \in \mathbb{R}^{k\times 3}$, we iteratively adjust each pixel p_i of the target image $\mathbf{x}$ by adding $\mathbf{rgb}_i$, i.e. $\mathbf{x}_{p_i} + \mathbf{rgb}_i$. Finally, we project $\mathbf{x}_{adv}$ by applying the l_∞ clipping mechanism

$$Proj(\mathbf{x})_\infty = \begin{cases} 1 & \mathbf{x} > 1 \\ 0 & \mathbf{x} < 0 \\ \mathbf{x} & otherwise, \end{cases} \tag{5}$$

once $\mathbf{x}_{adv}$ is constructed we evaluate using Eq. (4).

3.2 Offspring Population Generation

Given a population P_t of solutions, we generate a population of offspring solutions by applying crossover and mutation operators onto a set of parents selected by tournament selection.

Tournament Selection: We make use of binary tournament selection for selecting parents to crossover. Specifically, we generate two non-overlapping solution sets and select a non-dominated solution from each set as parents. Whereas the original NSGA-II algorithm gives equal weight to each objective in their domination definition [16], our primary goal is to generate adversarial images and secondly to minimize its l_2 norm. To reflect our priorities, we define our domination criterion as follows;

Algorithm 1: Crossover Operator

Input: Parent 1 set of pixel locations $\{p_{10}, p_{11} \cdots p_{1k}\}$, Parent 1 set of RGB perturbation $\{\mathbf{rgb}_{10}, \mathbf{rgb}_{11} \cdots \mathbf{rgb}_{1k}\}$, Parent 2 set of pixel locations $\{p_{20}, p_{21} \cdots p_{2k}\}$, Parent 2 set of RGB perturbation $\{\mathbf{rgb}_{20}, \mathbf{rgb}_{21} \cdots \mathbf{rgb}_{2k}\}$, Probability of crossover P_c

```
1  for i = 0 ··· k do
2      if rand() < P_c then
3          pixel_temp ← p_1i
4          rgb_temp ← rgb_1i
5          if p_2i not in {p_10, p_11 ··· p_1k} then
6              pixel_temp ← p_2i
7              rgb_temp ← rgb_2i
8          if p_1i not in {p_20, p_21 ··· p_2k} then
9              p_2i ← p_1i
10             rgb_2i ← rgb_1i
11         p_1i ← pixel_temp
12         rgb_1i ← rgb_temp
```

Definition 1 (Domination). *Given two solutions* $\mathbf{s}^1, \mathbf{s}^2$ *that generate images* $\mathbf{x}^1_{adv}$ *and* $\mathbf{x}^2_{adv}$, $\mathbf{s}^1$ *is said to dominate* $\mathbf{s}^2$ *if one of the following conditions is satisfied:*

- $\mathbf{x}^1_{adv}$ *satisfies Eq. (1) and* $\mathbf{x}^2_{adv}$ *does not*
- *Both* $\mathbf{x}^1_{adv}$ *and* $\mathbf{x}^2_{adv}$ *satisfy equation (1) and* $||\mathbf{x}^1_{adv} - \mathbf{x}||_2 < ||\mathbf{x}^2_{adv} - \mathbf{x}||_2$
- *Both* $\mathbf{x}^1_{adv}$ *and* $\mathbf{x}^2_{adv}$ *do not satisfy Eq. (1) and* $L_u(\mathbf{x}^1_{adv}) < L_u(\mathbf{x}^2_{adv})$.

If neither solution dominates the other we select the solution with the greater crowding distance as proposed by Deb et al. [16]. Once two parents are chosen, we generate two offspring solutions by applying crossover and mutation operations. We note that the use of this domination mechanism reduces Eq. (4) to the single-objective problem of l_2 norm minimization under the criterion that $\mathbf{x}^i_{adv}$ satisfies Eq. (2).

Crossover: Given two parent solutions, we generate two offspring solutions by exchanging a subset of each solution's pixel locations and corresponding RGB perturbations. For each perturbed pixel in the two solutions, the crossover operator exchanges the respective pixel position and RGB perturbation between the two solutions with a probability of P_c. To avoid duplicate pixels within a single solution, a pixel exchange is only conducted if the solution does not already perturb the inserted pixel. We provide the pseudo-code of the operator in Algorithm 1. Once completed, the mutation operator is applied to both offspring solutions.

Algorithm 2: Mutation Operator

Input: Solution set of pixel locations $\{p_0, p_1 \cdots p_k\}$, Solution set of RGB perturbations $\{\mathbf{rgb}_0, \mathbf{rgb}_1 \cdots \mathbf{rgb}_k\}$, Zero sampling probability Pr_0, Probability of mutation P_m, Set of all pixel positions U

1. $M \leftarrow \{0 \cdots k\}$
2. $d \leftarrow P_m \times k$
3. $A \leftarrow \mathcal{U}(M)^{k-d}$ // uniformly sample $k - d$–pixels to remain unchanged
4. $p_{new} \leftarrow p_A$ // store unchanged pixel positions to new set
5. $\mathbf{rgb}_{new} \leftarrow \mathbf{rgb}_A$ // store unchanged RGB perturbations to new set
6. $B \leftarrow \mathcal{U}(M \backslash \{p_0, p_1 \cdots p_k\})^d$ // uniformly randomly d–new pixel positions
7. $Pr_1 \leftarrow \frac{1-Pr_0}{2}$
8. **for** $i = 0 \cdots d$ **do**
9. $\mathbf{rgb}_B \leftarrow \{\}$
10.   **for** $j = 0 \cdots 3$ **do**
11. $v \leftarrow rand()$
12.   **if** $v < Pr_0$ **then**
13. $\mathbf{rgb}_{Bj} \leftarrow 0$
14.   **else if** $v < Pr_0 + Pr_1$ **then**
15. $\mathbf{rgb}_{Bj} \leftarrow 1$
16.   **else**
17. $\mathbf{rgb}_{Bj} \leftarrow -1$
18. $\mathbf{rgb}_{new} \leftarrow \mathbf{rgb}_{new} \cup \{\mathbf{rgb}_B\}$
19. $p_{new} \leftarrow p_{new} \cup B$
20. **return** p_{new}, $\mathbf{rgb}_{new}$

Mutation Operator. To mutate a solution with pixel positions and RGB perturbations, we first determine the severity of the mutation by setting the number of changed pixel positions to $d = P_m \times k$, where P_m is the probability of mutation and k is the l_0 constraint. Once determined, the operator randomly selects $k-d$ pixel positions to be copied into the final offspring in addition to their corresponding RGB perturbations. To fill the remaining d pixel positions, the operator samples from the set $B \leftarrow \mathcal{U}(M \backslash \{p_0, p_1 \cdots p_k\})^d$ of all other possible pixel positions that can be sampled. For each sampled pixel position from the set B, we sample an RGB perturbation from the set $\{-1, 1, 0\}$ where Pr_0 is the probability of sampling 0. By sampling 0 we reduce the l_2 distance between the adversarial and targeted image. Hence, the introduction of Pr_0 gives us increased control when defining the priority of the attack.

3.3 l_2 Minimization Leads to l_0 Minimization

To minimize the l_2 distance between the adversarial and target image, the perturbation values $\mathbf{rgb}_i$ require to be close to 0. As the value of each perturbation is constrained to the set $\{-1, 1, 0\}$, the algorithm aims to generate a solution where its $\mathbf{rgb}$ values are mostly 0 whilst still satisfying Eq. (2). As a perturbation of all zeros removes its impact on a pixel, the minimization of the perturbations l_2 norms leads to the minimization of its l_0 norm.

4 Experimental Setup

This section introduces the experimental setting for validating the effectiveness of our proposed adversarial attack method against state-of-the-art algorithms.

4.1 Attack Setting

We attack two models trained on the Cifar-10 [29] and ImageNet [17] data-sets. The Cifar-10 data-set contains 60,000 images of size (32, 32, 3) from 10 categories, with 50,000 of the images coming from the training set and 10,000 images from the test set. ImageNet is a far larger data-set that contains 1,000 classes with each class containing 1300 images which are resized to (224, 224, 3). For the Cifar-10 data set, we attack two commonly used networks in the literature, the All Convolutional Network (AllConvNet) [45] and the Network in Network (NiN) [33] which achieve test accuracies of 87% and 85% respectively. For the ImageNet data set, we attack the Efficient Convolutional Neural Network (MobileNet) [25] and the Deep Residual Network [24] with 50 layers (ResNet50). The MobileNet achieves a top-1 accuracy of 70% and a top-5 accuracy of 89% with ResNet50 achieving 76% top-1 accuracy and a 93% top-5 accuracy.

Fig. 2. Adversarial images generated by the proposed and Sparse-rs methods by attacking the ResNet50 ImageNet model.

We allocate a maximum budget of 1000 queries when attacking Cifar-10 models and 5000 queries for ImageNet models due to their larger images. For each Cifar-10 model, we randomly select 500 correctly classified test-set images to attack and 500 correctly classified validation-set images for ImageNet models.

4.2 Performance Metric

In our experiments we use an algorithm's attack success rate (ASR) to assess its performance. Given a set of A_{total} attacked images with A_s being successfully attacked, we define the ASR value as

$$ASR = \frac{A_s}{A_{total}}. \tag{6}$$

We additionally compare algorithms using the average number of queries and l_0, l_2 distances of the adversarial images in A_{total}. In the scenario where the proposed method returns a set of adversarial solutions, we make use of the solution with the smallest l_2 norm that satisfies Eq. (1) within the performance metric.

4.3 Parameter Settings

The proposed algorithm contains several parameters that are set before attacking a model, unless otherwise stated we maintain the following setup;

- N: We set the population size to $N = 5$ for all runs of the proposed algorithm.
- P_c: The probability of exchanging a pixel between two parents is set to $P_c = 0.5$.
- P_m: The probability of replacing each pixel with a randomly sampled pixel outside the current solution is set to $P_m = 0.3$.
- Pr_0: The probability of sampling a 0 perturbation value for each RGB channel is set to $Pr_0 = 0.3$.

5 Experimental Results

We evaluate the performance of the proposed method when applied to the query efficiency and minimization of perturbed pixels in the subsequent sections.

Query Efficient Setting. To apply the proposed algorithm to the single objective query efficient setting we ignore the l_2 distance objective function and set the probability of sampling zero $Pr_0 = 0$. We set the l_0 constraint $k = 24$ when attacking Cifar-10 models and $k = 150$ when attacking ImageNet models which is 2.34% and 0.299% of the image's total number of pixels, respectively.

Table 1. ASR and Queries of attacking Cifar-10 (top) and ImageNet (bottom) models with $k = 24$ and $k = 150$ respectively.

Method	AllConvNet			NiN		
	ASR	l_0	l_2	ASR	l_0	l_2
QSA-NSGA-II (ours)	100%	**22.41**	2.98	100%	**16.96**	2.03
Homotopy-Attack	100%	52.67	0.421	100%	43.25	0.382
DCEA-ISA	100%	92.73	**0.319**	100%	78.82	**0.278**

Method	MobileNet			ResNet50		
	ASR	l_0	l_2	ASR	l_0	l_2
QSA-NSGA-II (ours)	100%	**53.38**	19.87	100%	**57.49**	20.77
Homotopy-Attack	100%	2773.83	2.134	100%	2981.22	2.662
DCEA-ISA	100%	174.23	**0.46**	100%	226.02	**0.57**

Competitors. As outlined by Croce et al. [14], many existing l_0-attacks do not aim at query efficiency and alternatively aim to minimize the size of the perturbations. Hence, we compare our proposed method to the Sparse-rs algorithm of Croce et al. [14] that demonstrated superior performance over its competition and is considered state-of-the-art. We additionally compare with the one-pixel attack method of Su et al. [47] with a population size $N = 50$ and the black-box version of the PGD_0 [14,15].

Results. From the results in Table 1 we see the proposed method achieves better or equal success rate compared to its competitors. Where an improved success rate is not possible the proposed algorithm on average requires fewer model queries to generate adversarial images apart from the One Pixel Attack algorithm when attacking ImageNet models. Despite the better efficiency, the One Pixel Attack algorithm struggles to successfully attack both ImageNet models which lead us to believe that its low number of queries is only due to its ability to succeed when attacking easily attacked images. We compare adversarial images generated by the proposed and Sparse-rs algorithm in Fig. 2.

Perturbation Minimization Setting. Incorporating the l_2 distance between the adversarial and original image into our objective, we aim to generate adversarial images that have a minimal number of perturbed pixels. Following the query efficient scenario, we set the maximum number of perturbed pixels $k = 24$ and $k = 150$ when attacking Cifar-10 and ImageNet models, respectively. We keep the budget constant with that described in Sect. 5 when attacking each model.

Fig. 3. Comparison of adversarial images generated with and without l_2 minimization.

Competitors. As previously outlined the majority of works aiming to minmize the perturbation size allow a large number of pixels to be initial perturbed. In this work we compare the proposed method with the DCEA-ISA and Homotopy algorithms of Tian et al. [50] and Zhu et al. [54] respectively. We allow each algorithm to use the full query budget of 1000 and 5000 model queries for Cifar-10 and ImageNet models, respectively.

Results. The convergence plots shown in Fig. 4 demonstrate the ability of the proposed method to minimize the l_2 distance between the adversarial and original image once an adversarial perturbation has been generated. When attacking Cifar-10 trained models the l_2 distance plateau's before the budget is exhausted. In contrast, when attacking the ImageNet models we see that the l_2 distance is still decreasing as the budget is exhausted, this leads us to believe that the proposed

Table 2. ASR, and average l_0 and l_2 norms of adversarial images generated by attacking Cifar-10 (top) and ImageNet (bottom) models.

Method	AllConvNet		NiN	
	ASR	Queries	ASR	Queries
Sparse-rs	**100%**	44.33	**100%**	20.65
PGD_0	84.2%	182.72	95.2%	94.24
One Pixel Attack	43.34%	57.51	68.64%	53.92
QSA-NSGA-II (ours)	**100%**	**25.36**	**100%**	**13.85**

Method	MobileNet		ResNet50	
	ASR	Queries	ASR	Queries
Sparse-rs	98.64	520.34	98.33	459.34
PGD_0	85.89%	1163.67	82.1%	1045.89
One Pixel Attack	24.64%	**54.01**	26.11%	**57.41**
QSA-NSGA-II (ours)	**99.81**	358.18	**99.43**	319.06

method would be able to further minimize the distance given a larger budget. We show an example of an adversarial image generated with and without l_2 minimization in Fig. 3.

In comparison to state-of-the-art attack algorithms shown in Table 2 we see that all algorithms were able to successfully attack every image on all models, however the proposed method was able to generate adversarial images by perturbing fewer pixels. Compared with peer algorithms, the adversarial images generated by the proposed algorithm have greater l_2 distances from the original. A possible reason for this comes as a result of the difference in priorities between the algorithms. Whereas the proposed algorithm constrains the number of initially adjusted pixels the compared algorithms constrain the size of the perturbation added to each pixel whilst allowing a large number of pixels to be perturbed which results in a greater number of pixels being adjusted.

6 Conclusion and Future Directions

We propose an evolutionary method of conducting sparse adversarial attacks that minimize number of perturbed pixels by considering the task as a bi-objective problem. By conducting a series of attacks on four models, we show the proposed method's ability to quickly locate adversarial images and minimize their l_2 distance from the original. Despite its success, there are two areas of research we feel are prosperous directions for future work.

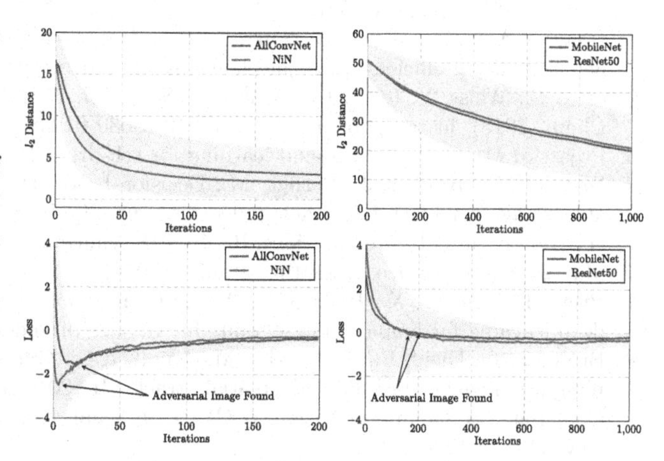

Fig. 4. Plots showing the average and standard deviation of the minimum loss (bottom) and l_2 distance (top) during the attacking process on Cifar-10 (right) and ImageNet (left) models.

1. l_∞ Minimization: The current version of the algorithm minimizes the l_2 distance by sampling an increased number of zeros for each channel of a perturbation until a pixel perturbation vanishes. By jointly reducing the l_∞ norm of the perturbation the l_2 distance of a solution would be minimized further.
2. Multi-Task Optimization: A natural conclusion from our experiments is that it is difficult to generate adversarial images with minimal l_0 and l_2 norms. By considering the minimization of each norm as an individual task, applying a multi-task optimization algorithm to jointly solve both problems could bring additional benefits by sharing information between them.

Acknowledgements. This work was supported by UKRI Future Leaders Fellowship (MR/S017062/1), EPSRC (2404317), NSFC (62076056), Royal Society (IES/R2/212077) and Amazon Research Award.

References

1. Alzantot, M., Sharma, Y., Chakraborty, S., Srivastava, M.B.: Genattack: practical black-box attacks with gradient-free optimization. CoRR (2018)
2. Andriushchenko, M., Croce, F., Flammarion, N., Hein, M.: Square attack: a query-efficient black-box adversarial attack via random search. In: Vedaldi, A., Bischof, H., Brox, T., Frahm, J.-M. (eds.) ECCV 2020. LNCS, vol. 12368, pp. 484–501. Springer, Cham (2020). https://doi.org/10.1007/978-3-030-58592-1_29
3. Bai, T., Luo, J., Zhao, J., Wen, B., Wang, Q.: Recent advances in adversarial training for adversarial robustness. In: Proceedings of the Thirtieth International Joint Conference on Artificial Intelligence, IJCAI 2021, Virtual Event/Montreal, Canada, 19–27 August 2021
4. Bhagoji, A.N., He, W., Li, B., Song, D.: Practical black-box attacks on deep neural networks using efficient query mechanisms. In: Ferrari, V., Hebert, M., Sminchisescu, C., Weiss, Y. (eds.) ECCV 2018. LNCS, vol. 11216, pp. 158–174. Springer, Cham (2018). https://doi.org/10.1007/978-3-030-01258-8_10
5. Bojarski, M., et al.: End to end learning for self-driving cars. CoRR (2016)
6. Brendel, W., Rauber, J., Bethge, M.: Decision-based adversarial attacks: reliable attacks against black-box machine learning models. In: 6th International Conference on Learning Representations, ICLR 2018, Vancouver, BC, Canada, April 30–3 May 2018, Conference Track Proceedings (2018)
7. Brigo, D., Huang, X., Pallavicini, A., de Ocariz Borde, H.S.: Interpretability in deep learning for finance: a case study for the heston model. CoRR (2021)
8. Brunner, T., Diehl, F., Truong-Le, M., Knoll, A.C.: Guessing smart: biased sampling for efficient black-box adversarial attacks. In: 2019 IEEE/CVF International Conference on Computer Vision, ICCV 2019, Seoul, Korea (South), October 27–2 November 2019
9. Carlini, N., Wagner, D.A.: Towards evaluating the robustness of neural networks. In: 2017 IEEE Symposium on Security and Privacy, SP 2017, San Jose, CA, USA, 22–26 May 2017. IEEE Computer Society (2017)
10. Chen, P., Sharma, Y., Zhang, H., Yi, J., Hsieh, C.: EAD: elastic-net attacks to deep neural networks via adversarial examples. In: Proceedings of the Thirty-Second AAAI Conference on Artificial Intelligence, (AAAI-18), the 30th innovative Applications of Artificial Intelligence (IAAI-18), and the 8th AAAI Symposium on Educational Advances in Artificial Intelligence (EAAI-18), New Orleans, Louisiana, USA, 2–7 February 2018
11. Chen, P., Zhang, H., Sharma, Y., Yi, J., Hsieh, C.: ZOO: zeroth order optimization based black-box attacks to deep neural networks without training substitute models. In: Proceedings of the 10th ACM Workshop on Artificial Intelligence and Security, AISec@CCS 2017, Dallas, TX, USA, 3 November 2017
12. Cheng, M., Le, T., Chen, P., Zhang, H., Yi, J., Hsieh, C.: Query-efficient hard-label black-box attack: an optimization-based approach. In: 7th International Conference on Learning Representations, ICLR 2019, New Orleans, LA, USA, 6–9 May 2019

13. Cheng, S., Dong, Y., Pang, T., Su, H., Zhu, J.: Improving black-box adversarial attacks with a transfer-based prior. In: Advances in Neural Information Processing Systems 32: Annual Conference on Neural Information Processing Systems 2019, NeurIPS 2019, 8–14 December 2019, Vancouver, BC, Canada (2019)
14. Croce, F., Andriushchenko, M., Singh, N.D., Flammarion, N., Hein, M.: Sparse-RS: a versatile framework for query-efficient sparse black-box adversarial attacks. In: Thirty-Sixth AAAI Conference on Artificial Intelligence, AAAI 2022. AAAI Press (2022)
15. Croce, F., Hein, M.: Sparse and imperceivable adversarial attacks. In: 2019 IEEE/CVF International Conference on Computer Vision, ICCV 2019, Seoul, Korea (South), October 27–2 November 2019. IEEE (2019)
16. Deb, K., Agrawal, S., Pratap, A., Meyarivan, T.: A fast and elitist multiobjective genetic algorithm: NSGA-II. IEEE Trans. Evol. Comput. (2002)
17. Deng, J., Dong, W., Socher, R., Li, L.J., Li, K., Fei-Fei, L.: Imagenet: a large-scale hierarchical image database. In: 2009 IEEE Conference on Computer Vision and Pattern Recognition, pp. 248–255 (2009)
18. Fang, X., Bai, H., Guo, Z., Shen, B., Hoi, S.C.H., Xu, Z.: DART: domain-adversarial residual-transfer networks for unsupervised cross-domain image classification. Neural Netw. (2020)
19. Goodfellow, I.J., Shlens, J., Szegedy, C.: Explaining and harnessing adversarial examples. In: 3rd International Conference on Learning Representations, ICLR 2015, San Diego, CA, USA, 7–9 May 2015, Conference Track Proceedings (2015)
20. Guo, C., Frank, J.S., Weinberger, K.Q.: Low frequency adversarial perturbation. In: Proceedings of the Thirty-Fifth Conference on Uncertainty in Artificial Intelligence, UAI 2019, Tel Aviv, Israel, July 22–25, 2019. Proceedings of Machine Learning Research (2019)
21. Guo, C., Gardner, J.R., You, Y., Wilson, A.G., Weinberger, K.Q.: Simple black-box adversarial attacks. In: Proceedings of the 36th International Conference on Machine Learning, ICML 2019, 9–15 June 2019, Long Beach, California, USA. Proceedings of Machine Learning Research, PMLR (2019)
22. Guo, Y., Yan, Z., Zhang, C.: Subspace attack: exploiting promising subspaces for query-efficient black-box attacks. In: Advances in Neural Information Processing Systems 32: Annual Conference on Neural Information Processing Systems 2019, NeurIPS 2019, 8–14 December 2019, Vancouver, BC, Canada (2019)
23. Hansen, N., Müller, S.D., Koumoutsakos, P.: Reducing the time complexity of the derandomized evolution strategy with covariance matrix adaptation (CMA-ES). Evol. Comput. (2003)
24. He, K., Zhang, X., Ren, S., Sun, J.: Identity mappings in deep residual networks. In: Leibe, B., Matas, J., Sebe, N., Welling, M. (eds.) ECCV 2016. LNCS, vol. 9908, pp. 630–645. Springer, Cham (2016). https://doi.org/10.1007/978-3-319-46493-0_38
25. Howard, A.G., et al.: MobileNets: efficient convolutional neural networks for mobile vision applications. CoRR (2017)
26. Ilyas, A., Engstrom, L., Athalye, A., Lin, J.: Black-box adversarial attacks with limited queries and information. In: Proceedings of the 35th International Conference on Machine Learning, ICML 2018, Stockholmsmässan, Stockholm, Sweden, 10–15 July 2018. Proceedings of Machine Learning Research, PMLR (2018)
27. Junior, F.E.F., Yen, G.G.: Particle swarm optimization of deep neural networks architectures for image classification. Swarm Evol. Comput. (2019)
28. Kingma, D.P., Ba, J.: Adam: a method for stochastic optimization. In: 3rd International Conference on Learning Representations, ICLR 2015, San Diego, CA, USA, 7–9 May 2015, Conference Track Proceedings (2015)

29. Krizhevsky, A., Hinton, G., et al.: Learning multiple layers of features from tiny images (2009)
30. Kurakin, A., Goodfellow, I.J., Bengio, S.: Adversarial examples in the physical world. CoRR (2016)
31. Letham, B., Calandra, R., Rai, A., Bakshy, E.: Re-examining linear embeddings for high-dimensional bayesian optimization. In: Advances in Neural Information Processing Systems 33: Annual Conference on Neural Information Processing Systems 2020, NeurIPS 2020, 6–12 December 2020, virtual (2020)
32. Li, C., Wang, H., Zhang, J., Yao, W., Jiang, T.: An approximated gradient sign method using differential evolution for black-box adversarial attack. IEEE Trans. Evolut. Comput. 1 (2022). https://doi.org/10.1109/TEVC.2022.3151373
33. Lin, M., Chen, Q., Yan, S.: Network in network. In: 2nd International Conference on Learning Representations, ICLR 2014, Banff, AB, Canada, 14–16 April 2014, Conference Track Proceedings (2014)
34. Liu, Y., Ling, J., Liu, Z., Shen, J., Gao, C.: Finger vein secure biometric template generation based on deep learning. Soft Comput. (2018)
35. Madry, A., Makelov, A., Schmidt, L., Tsipras, D., Vladu, A.: Towards deep learning models resistant to adversarial attacks. CoRR (2017)
36. Meunier, L., Atif, J., Teytaud, O.: Yet another but more efficient black-box adversarial attack: tiling and evolution strategies. CoRR (2019)
37. Narodytska, N., Kasiviswanathan, S.P.: Simple black-box adversarial attacks on deep neural networks. In: 2017 IEEE Conference on Computer Vision and Pattern Recognition Workshops, CVPR Workshops 2017, Honolulu, HI, USA, 21–26 July 2017. IEEE Computer Society (2017)
38. Papernot, N., McDaniel, P.D., Goodfellow, I.J.: Transferability in machine learning: from phenomena to black-box attacks using adversarial samples. CoRR (2016)
39. Qiu, H., Custode, L.L., Iacca, G.: Black-box adversarial attacks using evolution strategies. In: GECCO 2021: Genetic and Evolutionary Computation Conference, Companion Volume, Lille, France, 10–14 July 2021
40. Ru, B., Cobb, A.D., Blaas, A., Gal, Y.: Bayesopt adversarial attack. In: 8th International Conference on Learning Representations, ICLR 2020, Addis Ababa, Ethiopia, 26–30 April 2020
41. Schott, L., Rauber, J., Bethge, M., Brendel, W.: Towards the first adversarially robust neural network model on MNIST. In: 7th International Conference on Learning Representations, ICLR 2019, New Orleans, LA, USA, 6–9 May 2019. OpenReview.net (2019)
42. Sharma, Y., Chen, P.: Attacking the madry defense model with l_1-based adversarial examples. CoRR abs/1710.10733 (2017)
43. Shukla, S.N., Sahu, A.K., Willmott, D., Kolter, J.Z.: Black-box adversarial attacks with bayesian optimization. CoRR (2019)
44. Simonyan, K., Zisserman, A.: Very deep convolutional networks for large-scale image recognition. In: 3rd International Conference on Learning Representations, ICLR 2015, San Diego, CA, USA, 7–9 May 2015, Conference Track Proceedings (2015)
45. Springenberg, J.T., Dosovitskiy, A., Brox, T., Riedmiller, M.A.: Striving for simplicity: the all convolutional net. In: 3rd International Conference on Learning Representations, ICLR 2015, San Diego, CA, USA, 7–9 May 2015, Workshop Track Proceedings (2015)
46. Storn, R., Price, K.V.: Differential evolution - a simple and efficient heuristic for global optimization over continuous spaces (1997)

47. Su, J., Vargas, D.V., Sakurai, K.: One pixel attack for fooling deep neural networks. IEEE Trans. Evol. Comput. (2019)
48. Szegedy, C., et al.: Intriguing properties of neural networks. In: 2nd International Conference on Learning Representations, ICLR 2014, Banff, AB, Canada, 14–16 April 2014, Conference Track Proceedings (2014)
49. Tian, Y., Pan, J., Yang, S., Zhang, X., He, S., Jin, Y.: Imperceptible and sparse adversarial attacks via a dual-population based constrained evolutionary algorithm. IEEE Trans. Artif. Intell. (2022)
50. Tian, Y., Pan, J., Yang, S., Zhang, X., He, S., Jin, Y.: Imperceptible and sparse adversarial attacks via a dual-population based constrained evolutionary algorithm. IEEE Trans. Artif. Intell. (2022). https://doi.org/10.1109/TAI.2022.3168038
51. Tian, Y., Ha, D.: Modern evolution strategies for creativity: fitting concrete images and abstract concepts. In: Martins, T., Rodríguez-Fernández, N., Rebelo, S.M. (eds.) EvoMUSART 2022. LNCS, vol. 13221, pp. 275–291. Springer, Cham (2022). https://doi.org/10.1007/978-3-031-03789-4_18
52. Tu, C., et al.: Autozoom: autoencoder-based zeroth order optimization method for attacking black-box neural networks. In: The Thirty-Third AAAI Conference on Artificial Intelligence, AAAI 2019, The Thirty-First Innovative Applications of Artificial Intelligence Conference, IAAI 2019, The Ninth AAAI Symposium on Educational Advances in Artificial Intelligence, EAAI 2019, Honolulu, Hawaii, USA, January 27–1 February 2019 (2019)
53. Uesato, J., O'Donoghue, B., Kohli, P., van den Oord, A.: Adversarial risk and the dangers of evaluating against weak attacks. In: Proceedings of the 35th International Conference on Machine Learning, ICML 2018, Stockholmsmässan, Stockholm, Sweden, 10–15 July 2018. Proceedings of Machine Learning Research (2018)
54. Zhu, M., Chen, T., Wang, Z.: Sparse and imperceptible adversarial attack via a homotopy algorithm. In: Proceedings of the 38th International Conference on Machine Learning, ICML 2021, 18–24 July 2021, Virtual Event. Proceedings of Machine Learning Research, PMLR (2021)
55. Zoph, B., Vasudevan, V., Shlens, J., Le, Q.V.: Learning transferable architectures for scalable image recognition. In: 2018 IEEE Conference on Computer Vision and Pattern Recognition, CVPR 2018, Salt Lake City, UT, USA, 18–22 June 2018. Computer Vision Foundation/IEEE Computer Society (2018)

Investigating Innovized Progress Operators with Different Machine Learning Methods

Drishti Bhasin[1], Sajag Swami[1], Sarthak Sharma[1], Saumya Sah[1], Dhish Kumar Saxena[1](✉), and Kalyanmoy Deb[2]

[1] Indian Institute of Technology Roorkee, Roorkee, India
{drishti_b,sajag_s,sarthak_s,saumya_s,dhish.saxena}@me.iitr.ac.in
[2] BEACON Center, Michigan State University, East Lansing, MI, USA
kdeb@egr.msu.edu

Abstract. Recent studies have demonstrated that the performance of Reference vector (RV) based Evolutionary Multi- and Many-objective Optimization algorithms could be improved, through the intervention of Machine Learning (ML) methods. These studies have shown how *learning* efficient search directions from the intermittent generations' solutions, could be utilized to create pro-convergence and pro-diversity offspring, leading to better convergence and diversity, respectively. The entailing steps of data-set preparation, training of ML models, and utilization of these models, have been encapsulated as *Innovized Progress* operators, namely, IP2 (for convergence improvement) and IP3 (for diversity improvement). Evidently, the focus in these studies has been on proof-of-the-concept, and no exploratory analysis has been done to investigate, *if* and *how drastically* the operators' performance may be impacted, if their underlying ML methods (Random Forest for IP2, and kNN for IP3) are varied. This paper seeks to bridge this gap, through an exploratory analysis for both IP2 and IP3, based on eight different ML methods, tested against an exhaustive test suite comprising of seven multi-objective and 32 many-objective test instances. While the results broadly endorse the robustness of the existing IP2 and IP3 operators, they also reveal interesting tradeoffs across different ML methods, in terms of the Hypervolume (HV) metric and corresponding run-time. Notably, within the gambit of the considered test suite and different ML methods adopted, kNN emerges as a winner for both IP2 and IP3, based on conjunct consideration of HV metric and run-time.

Keywords: Multi-objective optimization · Many-objective optimization · Machine learning assisted optimization · *Innovized Progress*

1 Introduction

Reference vector (RV) based Evolutionary Multi- and Many-objective Optimization algorithms [2,6,8,11,17,18], hereafter denoted as RV-EMâOAs, are an

M. Emmerich et al. (Eds.): EMO 2023, LNCS 13970, pp. 134–146, 2023.
https://doi.org/10.1007/978-3-031-27250-9_10

important class of algorithms, that seek to find a set of well-distributed Pareto optimal solutions. RV-EMâOAs rely on the use of different dominance principles for convergence, within an RV-based framework which facilitates diversity [17].

Recent studies have shown that RV-EMâOAs are quite conducive for integration with Machine Learning (ML) methods, owing to the availability of multiple solution sets over successive generations, and also their tractability *along* and *across* the RVs. These studies leading to the proposition of *innovized progress* operators, namely, IP2 [13] and IP3 [14] have demonstrated as how the *convergence* and *diversity* capabilities of RV-EMâOAs, respectively, could be enhanced through ML intervention. Adopting a common template (Fig. 1), these operators have demonstrated that a judicious mapping of solutions in the objective (F) space, enables *learning* of efficient search directions in the variable (X) space, which could be utilized to create pro-convergence/diversity offspring. In that:

- *IP2 operator:* constructs a training-dataset by mapping *inter*-generational solutions *along* the RVs in the F space; trains a Random Forest (RF) [1] towards *learning* pro-convergence search directions in X space; and utilizes this learning for creation of pro-convergence offspring.
- *IP3 operator:* constructs a training-dataset by mapping *intra*-generational solutions *across* the RVs in the F space; trains a kNN [5] towards *learning* pro-diversity search directions in X space; and utilizes this learning for creation of pro-diversity offspring.

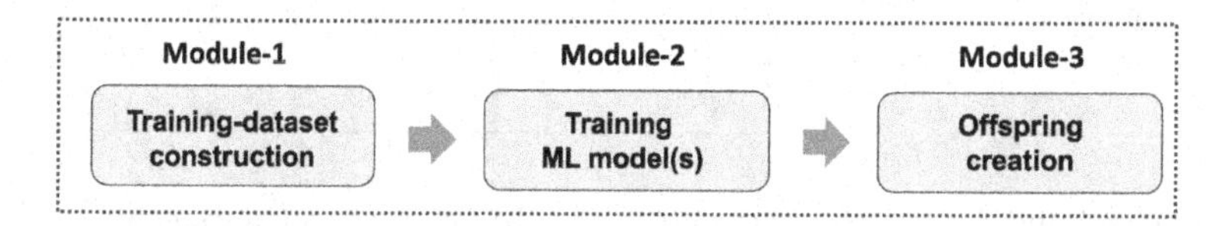

Fig. 1. The constitutive modules for the existing IP2 and IP3 operators

Notably, the focus in the above studies has been on proof-of-the-concept, and no exploratory analysis has been performed to analyze *if* and *how drastically* the performance of IP2 and IP3 operators may be impacted, if their underlying ML methods are changed. This paper recognizes this gap, and evaluates the performance of each of the IP2 and IP3 operators, based on seven alternative ML methods, against 39 multi- and many-objective test instances. The remaining paper is organized as follows. Section 2 shares more details about the existing IP2 and IP3 operators, and their integration with an RV-EMâOA. Then, Sect. 3 highlights the ML methods chosen to serve as alternatives to those being used in the existing IP2 and IP3 operators. The experimental settings are highlighted in Sect. 4, following which the experimental results are presented in Sect. 5. The paper concludes with Sect. 6.

2 Background

This section shares more facts on the IP2 and IP3 operators, in sequence, and how they integrate with an RV-EMâOA, leading to RV-EMâO-IP2 and RV-EMâO-IP3, respectively.

In any generation t of an RV-EMâO-IP2, the training-dataset is based on the mapping of (i) the target solutions: best-found solutions so far along different RVs, and (ii) the archive solutions: solutions from some earlier generations, along the respective RVs. Hence, the parent population P_t is used to update the target solutions. While the above mapping is depicted in Fig. 2a with respect to the F space, the mapping of the X vectors underlying the target and archive solutions constitutes the training-dataset, which is then used to train an ML (RF) model. First, all offspring are created from P_t using the natural variation operators, and then the trained ML model is used to advance 50% randomly picked offspring, creating 50% offspring towards better convergence. The advancement of a natural offspring using the trained ML model is symbolically depicted in Fig. 2b.

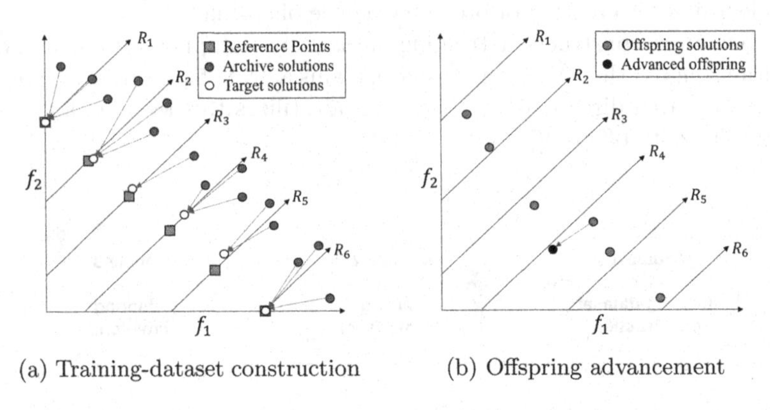

(a) Training-dataset construction (b) Offspring advancement

Fig. 2. Depicting the training-dataset construction and offspring advancement for the IP2 operator ($M = 2$).

It must be noted that unlike the case of RV-EMâO-IP2 (where a single ML model per generation sufficed), a generation t of RV-EMâO-IP3 requires as many ML models as the number of objectives (M). This was considered necessary towards a generic approach that could help address both aspects of diversity, namely, the *spread* of solutions, and their uniform *distribution*. The need for M ML models, in turn necessitate M training-datasets. As justified in [14], each dataset is designed to eventually empower any solution to undergo improvement in a distinct objective. For an illustration in a three-objective scenario ($M = 3$), Fig. 3a shows the projections of a handful of solutions in P_t, onto the unit simplex on which the RVs are sampled (rationale in [14]). Consider S_1 as the input solution whose contributory mappings to the training-dataset are to be identified. Hence, w.r.t. f_1, S_1 is mapped onto S_3 - that neighbouring solution (bounded by the dotted circles around S_1) which offers maximum improvement in f_1.

Similarly, S_1 gets mapped onto S_2 and S_4, w.r.t. f_2 and f_3, respectively. This process, when repeated for each solution, ideally leads to $N \times M$ mappings, of which M datasets are constituted (each, sized $N \times 1$), and correspondingly, M ML (kNN) models are trained. These models are used to advance 50% of the solutions in P_t, which are judiciously picked so that the resulting offspring equally contribute to a better spread and distribution. This could be appreciated through Fig. 3b. In that, the better spread could be achieved by advancing the boundary solution S_B, beyond the current unit-simplex. This can easily be realized by treating S_B as the input solution and subjecting it to the second ML model, so it gets advanced in the direction of improvement in f_2. Similarly, an improvement in distribution could be achieved by advancing S_G onto the empty RV $\mathcal{R}_G$ (which has no solution associated with it). This can be realized by treating S_G as the input solution and subjecting it to the third ML model, so it gets advanced in the direction of improvement in f_3.

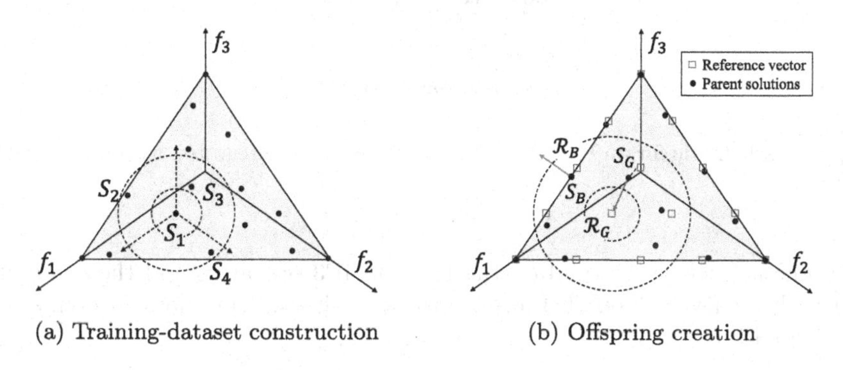

(a) Training-dataset construction

(b) Offspring creation

Fig. 3. Depicting the training-dataset construction and offspring creation for the IP3 operator ($M = 3$). The two dotted circles in (a) and (b) depict the neighbourhood for S_1 and $\mathcal{R}_G$, respectively.

Given the above, the flow of an RV-EMâOA generation, when integrated with the IP2/IP3 operator, is schematically depicted in Fig. 4. In that:

- first 100% (N) offspring are created using natural variation operators, denoted by Q^{V} (in boxes (a) and (b) under 'Offspring Population').
- if the underlying RV-EMâOA is integrated with the IP2 operator, it creates 50% pro-convergence offspring Q^{IP2} (in box (c)), by advancing 50% of randomly picked Q^{V}. The resulting offspring population is the collective set of solutions in boxes (b) and (c).
- if the underlying RV-EMâOA is integrated with the IP3 operator, it creates 50% pro-diversity offspring Q^{IP3} (in box (d)), by advancing 50% of the judiciously picked parent solutions. The resulting offspring population is the collective set of solutions in boxes (a) and (d).
- the resulting offspring population (sized N) is combined with the parent population, leading to the combined population, sized $2N$.

- finally, the survival selection of the underlying RV-EMâOA is executed, leading to the survived population (sized N), which serves as the parent population for the next generation.

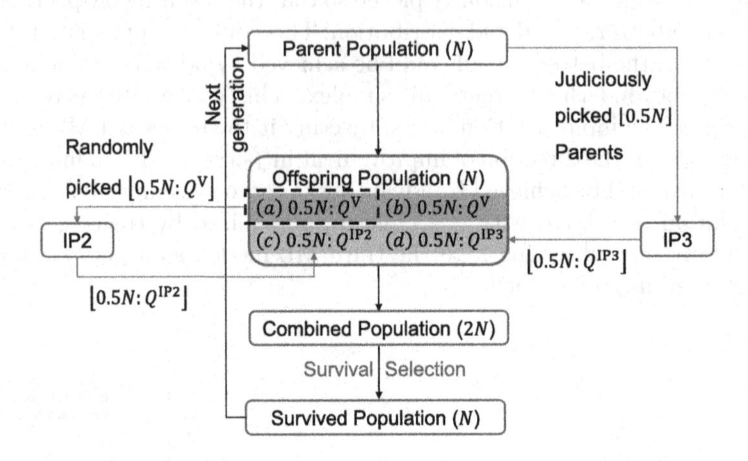

Fig. 4. A schematic for an RV-EMâOA generation, when integrated with the IP2/IP3 operator.

While a brief background on the IP2 and IP3 operators and their integration with an RV-EMâOA has been presented above, some notable points are highlighted below.

- the IP2 and IP3 operators are not invoked in every generation of an RV-EMâOA run. Instead, their invocation along intermittent generations is determined through a frequency parameter (determined on-the-fly).
- whenever IP2 is invoked, the creation of 50% pro-convergence offspring along the intermittent generations, eventually leads to improved convergence [13].
- whenever IP3 is invoked, the creation of 50% pro-diversity offspring along the intermittent generations, eventually leads to improved diversity [14].
- the hallmark of the ML-based IP2 and IP3 operators is that an RV-EMâOA integrated with either of these operators, *does not* necessitate any additional solution evaluations, compared to the base RV-EMâOA (completely relying on offspring produced by natural variation operators).

3 Alternative ML Methods for IP2 and IP3 Operators

This paper considers eight different ML methods, covering *Linear and Non-linear Modeling*, *Boosting*, and *Trees*, to investigate their suitability for the IP2 and IP3 operators. The chosen methods are highlighted below:

- from the family of linear methods - standard Linear Regression, Ridge Regression and Elastic Net Regression are included. The Linear Regression based on the least squares approach is used as the base model, but it can also be fitted in other ways, such as by minimizing the *lack of fit* through a penalized version of the least squares cost function as in Ridge Regression (L2-norm penalty) and Lasso (L1-norm penalty). Since Lasso is known to suffer where there are correlated features [10], it has not been considered. Hence, Ridge Regression, and Elastic Net Regression [15] based on a linear combination of the L2-norm and L1-norm penalties, are included.
- from the family of trees - Extra Trees Regressor and Random Forests are included, owing to their ability to efficiently handle complex, non-linear, high-dimensional data, with a lower tendency for *overfitting* [9]. Notably, Random Forests aggregate the results from many decision trees, each generated from a bootstrap sample of the data. Here, at each node, one feature is selected to split on, from a random subset of all features. While in the case of Random Forests, the optimum split is chosen, Extra Trees do it randomly.
- from the family of boosting algorithms - XGBoost is included. It is an implementation of gradient boosting that is scalable and optimized for execution speed and model performance. This is an ensemble technique, where each new model added sequentially learns from the previous models' errors [3].
- other non-linear methods such as k-Nearest Neighbours Regression and Support Vector Regression, are included. The former method identifies the k nearest inputs of the test instance in the original training dataset and returns the average of their respective targets [5]. The latter method seeks to find the hyperplane with the maximum number of points that lies within a threshold distance from the boundary line (unlike other Regression models that try to minimize the error between the real and predicted value).

For ease of reference in the subsequent sections (including Tables and Figures), the chosen ML methods have been abbreviated, as follows: Linear Regression (LR), Ridge Regression (Ridge), Elastic Net Regression (ENet), Extra Trees Regressor (ExTree), Random Forests (RF), XGBoost (XGB), k-Nearest Neighbours (kNN), and Support Vector Regression (SVR).

4 Experimental Settings

This section presents the experimental settings for the used - test suite; base RV-EMâOA; and performance indicator.

4.1 Test Suite and the Base RV-EMâOA

The considered test suite covers problems with a wide range of characteristics, including, bias, multi-modality, variable-linkages, and different PF shapes (convex, concave, mixed, inverted and disconnected). It includes the following:

- Multi-objective problems: these include: (a) $\tilde{\text{Z}}$DT [12] that are variants of ZDT [19] with modified $g(X)$-functions, leading to the PO solutions at $x_k = 0.5$, for $k = 2, \ldots, 30$, and (b) L1/L2 [12] with variables $n_{\text{var}} = 10$.

- Many-objective problems: these include: (a) DTLZ [7] with distance variables $k = 20$ and $M = 5$, 10, and (b) MaF [4] with $k = 20$ and $M = 5$, 10.

In this paper, NSGA-III has been used as the base RV-EMâOA, for which the SBX crossover ($p_c = 0.9$ and $\eta_c = 20$) and polynomial mutation ($p_m = 1/n_{\text{var}}$ and $\eta_m = 20$) are used. NSGA-III when integrated with the IP2 and IP3 operators, is referred to as NSGA-III-IP2 and NSGA-III-IP3, respectively.

4.2 Performance Indicator

The performance of NSGA-III-IP2 is to be compared amidst its eight variants corresponding to eight different ML methods (identified in Sect. 3). Since RF is the base method for the existing IP2 operator, NSGA-III-IP2-RF can be distinguished from the other NSGA-III-IP2-ML variants. Similarly NSGA-III-IP3-*k*NN can be distinguished from the remaining NSGA-III-IP3-ML variants. Since in this paper, NSGA-III is the sole RV-EMâOA, its name can be avoided for brevity, and the task in this paper translates to comparing the performance of: (a) IP2-RF with all other IP2-ML, and (b) IP3-*k*NN with all other IP3-ML, for which the Hypervolume (HV) measures are used. In that:

- the used Reference Points are given by $R_{1\times M} = [1 + \frac{1}{p}, \ldots, 1 + \frac{1}{p}]$, where p is the number of gaps set for the RV generating Das-Dennis method [14].
- first IP2-RF is run for 21 different seeds, and termination of each such run at say t_{TM} generations is governed by the stabilization tracking algorithm [16] with parameter settings, given by $\psi_{\text{TM}} \equiv \{3, 50\}$ [14]. Then the average of all 21 t_{TM} generations are computed, say $\hat{t}_{\text{TM}}$. Subsequently, each IP2-ML is run for 21 seeds until termination at $\hat{t}_{\text{TM}}$. Finally, the 21 HV measures available for IP2-RF are subjected to Wilcoxon ranksum test (for statistical significance) with a p-value of 0.05, against the 21 HV measures available for *each* IP2-ML. This test only infers if the difference between IP2-RF and any IP2-ML is statistically insignificant (denoted by =). If not, then their respective median values are directly compared, and the better/worse performing method is denoted by a +/− sign (as in Tables 1 and 2). Similar procedure is adopted for comparison of IP3-*k*NN with each IP3-ML.

5 Experimental Results

This section presents the comparative performance of: (a) seven IP2-ML variants versus IP2-RF (Table 1), and (b) seven IP3-ML variants versus IP3-*k*NN (Table 2), across multi- and many-objectives test instances. The dominant trend in both Tables 1 and 2, is that *even if the ML methods underlying the IP2 and IP3 operators are changed, the performance largely remains statistically equivalent (with a few exceptions, discussed later).* This indicates that *the original propositions of IP2 and IP3 are reasonably robust, and their performance may neither drastically deteriorate nor improve with a change in the underlying ML*

Table 1. HV-based comparison of IP2 with eight different ML methods, where RF is the base for pairwise comparisons. The symbols "+", "=" or "–" indicate whether the corresponding ML method is statistically better, equivalent, or worse, than RF. Here, $\hat{t}_{\mathrm{TM}}$ denotes the average of the termination generations for 21 runs of the base RF.

Problem	M	$\hat{t}_{\mathrm{TM}}$	kNN	ExTree	SVR	ENet	Ridge	LR	XGB	**RF**
L1	2	1058	0.520038+	0.510975=	0.490729=	0.487142–	0.492145=	0.488027=	0.505394=	0.506494
L2	2	1114	0.673404+	0.673393+	0.641642–	0.639171–	0.642941–	0.642229–	0.649486–	0.658637
Z̃DT1	2	1187	0.681860=	0.681859=	0.681860=	0.681860=	0.681860=	0.681860=	0.681860=	0.681860
Z̃DT2	2	1289	0.348794=	0.348794=	0.348794=	0.348794=	0.348794=	0.348794=	0.348794=	0.348794
Z̃DT3	2	1013	1.068370=	1.068457=	1.068402=	1.068368=	1.068490=	1.068559=	1.068454=	1.068502
Z̃DT4	2	1770	0.681860=	0.681860=	0.681860=	0.681860=	0.681860=	0.681859=	0.681859=	0.681860
Z̃DT6	2	1779	0.326889+	0.321376=	0.327059+	0.317623=	0.319643=	0.323048=	0.317781=	0.320871
Total (+/=/–)			3/4/0	1/6/0	1/5/1	0/5/2	0/6/1	0/6/1	0/6/1	of 7 probs.
DTLZ1	5	1206	2.485633=	2.486744=	2.486437=	2.486802=	2.486921=	2.486951=	2.486755=	2.479388
	10	2076	17.757704=	17.757704=	17.757704=	17.757707=	17.757711+	17.757708=	17.75771+	17.757703
DTLZ2	5	903	2.172478=	2.172430=	2.172397=	2.172605=	2.172648=	2.17262=	2.172606=	2.172439
	10	797	17.667911=	17.668915=	17.668607=	17.668416=	17.670702+	17.670357+	17.670246+	17.668409
DTLZ3	5	1196	2.111525=	2.110038=	2.086116=	2.103797=	2.102734=	2.109105=	2.083618=	0.975718
	10	1914	17.665819=	17.663740=	17.661898–	17.659894–	17.667285=	17.666322=	17.667097=	17.667027
DTLZ4	5	901	2.173324=	2.173321=	2.173259=	2.173090=	2.173194=	2.173259=	2.173285=	2.173402
	10	873	17.679661–	17.679996=	17.679292–	17.677214–	17.679034–	17.678961–	17.679912=	17.679992
MaF1	5	852	0.060541=	0.060107=	0.059505=	0.059808=	0.060784=	0.059723=	0.060170=	0.061855
	10	938	0.001951=	0.001928=	0.001968=	0.001973=	0.002021+	0.002039=	0.001882–	0.001984
MaF2	5	362	1.002621=	0.998740=	1.002171=	0.999238=	1.003607=	1.001332=	1.000969=	1.000973
	10	492	7.823111=	7.836465=	7.814538=	7.829009=	7.850499+	7.810496=	7.837603=	7.808070
MaF3	5	2509	2.488320=	2.488320=	2.488320=	2.488320=	2.488320=	2.488320=	2.488320=	2.488320
	10	1912	17.757727+	17.757727+	17.757727+	17.757727+	17.757727+	17.757727+	17.757727+	0
MaF4	5	646	2.475006=	2.474046=	2.483075+	2.472548=	2.466242=	2.477218=	2.480569=	2.462381
	10	1106	17.481657=	17.487127=	17.487090=	17.488453=	17.483530=	17.484489=	17.476629–	17.482483
MaF5	5	1117	2.487940=	2.487937=	2.487936=	2.487937–	2.487939=	2.487938=	2.487937=	2.487940
	10	1180	17.757727=	17.757727=	17.757727=	17.757727=	17.757727=	17.757727=	17.757727=	17.757727
MaF7	5	1095	0.968628=	0.970901=	0.969541=	0.969176=	0.969783=	0.972155=	0.970013=	0.971900
	10	746	7.634033=	7.626016=	7.620365=	7.579781=	7.637727=	7.631401=	7.619364=	7.611451
MaF8	5	1361	0.000504=	0.000456=	0.000491=	0.000495=	0.000475=	0.000506=	0.000481=	0.000509
	10	1041	0.000087=	0.000087=	0.000082=	0.000082=	0.000081=	0.000080=	0.000084=	0.000085
MaF9	5	2996	0.025761=	0.027336=	0.026762=	0.027222=	0.026980=	0.026736=	0.028677=	0.027412
	10	689	0.000303=	0.000282=	0.000317=	0.000324=	0.000322=	0.000319=	0.000332+	0.000296
MaF10	5	909	0.933306=	0.901095=	0.901880=	0.896384=	0.889783=	0.926323=	0.891255=	0.930609
	10	753	6.298122=	6.102056=	6.318167=	6.310587=	6.247213=	6.255246=	6.269511=	6.225195
MaF11	5	1006	2.452302+	2.442642=	2.449517+	2.437614=	2.421567–	2.420182–	2.448075+	2.439245
	10	1066	17.482878=	17.484652=	17.642898+	17.590201+	17.552494+	17.534068=	17.512623=	17.418106
MaF12	5	539	1.790848=	1.791929=	1.790169=	1.836252=	1.808378=	1.817218=	1.793968=	1.785985
	10	478	13.241563=	13.288255=	13.235779=	13.19385=	13.303345=	13.313089=	13.346762=	13.366043
MaF13	5	871	0.218946=	0.218760=	0.225984=	0.233687+	0.216365=	0.223561=	0.221912=	0.220197
	10	921	0.164408=	0.162449=	0.161884=	0.155932=	0.095460–	0.121185–	0.151681=	0.189759
Total (+/=/–)			2/29/1	1/31/0	4/26/2	3/26/3	6/23/3	2/27/3	5/25/2	of 32 probs.

method (Module 2, Fig. 1). Hence, future attempts to strengthen the Innovized Progress operators may need to focus on improving the other modules, namely, training-dataset generation and offspring creation.

For further insights into the results reported in Tables 1 and 2, the notion of *Hypervolume factor* is proposed for each ML method in conjunction with the multi-objective and many-objective test suite. For instance, in Table 1, intersection of Column 3 and shaded Row 9 with the $+/=/-$ entries as $3/4/0$, suggests that for the multi-objective test suite comprising of seven problem instances,

Table 2. HV-based comparison of IP3 with eight different ML methods, where kNN is the base for pairwise comparisons. The symbols "+", "=" or "–" indicate whether the corresponding ML method is statistically better, equivalent, or worse, than kNN. Here, $\hat{t}_{\text{TM}}$ denotes the average of the termination generations for 21 runs of the base kNN.

Problem	M	$\hat{t}_{\text{TM}}$	RF	ExTree	SVR	ENet	Ridge	LR	XGB	**kNN**
L1	2	1036	0.529465–	0.535453=	0.524300–	0.541379=	0.581979+	0.531304=	0.507512–	0.557609
L2	2	1157	0.661312–	0.658540–	0.659964–	0.659488–	0.661387–	0.664115+	0.660775–	0.662832
Z̃DT1	2	1200	0.681860=	0.681860=	0.681860=	0.681860=	0.681860=	0.681861=	0.681859=	0.681860
Z̃DT2	2	1263	0.348794=	0.348794=	0.348794=	0.348794=	0.348794=	0.348794=	0.348794=	0.348794
Z̃DT3	2	1003	1.068499=	1.068395=	1.068444=	1.068469=	1.068446=	1.068464=	1.068577+	1.068348
Z̃DT4	2	1755	0.681860=	0.681860=	0.681860=	0.681860=	0.681860=	0.681860=	0.681860=	0.681860
Z̃DT6	2	1812	0.332970=	0.330529=	0.327835–	0.328804=	0.336573=	0.328397–	0.336123=	0.333947
Total (+/=/–) ⟶			0/5/2	0/6/1	0/4/3	0/6/1	1/5/1	1/5/1	1/4/2	of 7 probs.
DTLZ1	5	1402	2.486749=	2.486892=	2.486839=	2.486717=	2.486795=	2.486833=	2.486964=	2.486885
	10	2023	17.757703=	17.757702=	17.757693=	17.757704=	17.757690=	17.757684=	17.757683=	17.757696
DTLZ2	5	866	2.171945=	2.172083+	2.171872=	2.172031+	2.172102+	2.172082+	2.172134+	2.171782
	10	757	17.664176=	17.664258=	17.664655=	17.664885=	17.664505=	17.664448=	17.664356=	17.664525
DTLZ3	5	1198	2.083931=	2.079724=	2.049543–	2.057957–	2.033053–	2.055357–	2.099911=	2.09651
	10	1851	17.659759=	17.660024=	17.658199–	17.658355=	17.656276–	17.659139=	17.663084=	17.660877
DTLZ4	5	915	2.173070–	2.173172=	2.173212=	2.173206=	2.173125=	2.173185=	2.173211=	2.173235
	10	800	17.676593–	17.67686=	17.67699=	17.676891=	17.676909=	17.676767=	17.677003=	17.676834
MaF1	5	853	0.059759=	0.059340=	0.059473=	0.059903=	0.059967=	0.059876=	0.05956=	0.060380
	10	1059	0.002074+	0.002155+	0.001959=	0.001957=	0.002004=	0.002149+	0.002159+	0.002000
MaF2	5	361	0.999336=	0.997776=	0.999999=	1.000740+	1.002444+	0.998881=	0.999796=	0.995748
	10	484	7.877194=	7.865008=	7.850738=	7.883851+	7.827113=	7.833383=	7.827931=	7.830099
MaF3	5	2416	2.488320=	2.488320=	2.488320=	2.488320=	2.488320=	2.488320=	2.488320=	2.488320
	10	3658	17.757727=	17.757727=	17.757727=	17.757727=	17.757727=	17.757727=	17.757727=	17.757727
MaF4	5	649	2.480958=	2.469060=	2.482613=	2.481608=	2.475904=	2.479125=	2.482338=	2.477745
	10	1094	17.529054=	17.534092=	17.541146=	17.497915–	17.565631=	17.535149=	17.542442=	17.557395
MaF5	5	1113	2.487942=	2.487944=	2.487942=	2.487946=	2.487946=	2.487943=	2.487947=	2.487944
	10	1197	17.757727=	17.757727=	17.757727=	17.757727=	17.757727=	17.757727=	17.757727=	17.757727
MaF7	5	1074	0.968845=	0.972045=	0.967674=	0.969739=	0.968382=	0.968788=	0.969065=	0.969353
	10	734	7.598982=	7.578182=	7.590612=	7.585287=	7.615941=	7.594704=	7.592734=	7.584186
MaF8	5	1361	0.000551=	0.000477=	0.000426–	0.000450–	0.000410-	0.000470=	0.000449–	0.000495
	10	1046	0.000065=	0.000067=	0.000076+	0.000077+	0.000074+	0.000069=	0.000076+	0.000068
MaF9	5	1820	0.028026=	0.028181=	0.025498–	0.026383–	0.027629=	0.028588=	0.026612–	0.028363
	10	725	0.000065=	0.000080=	0.000099=	0.000056=	0.000035=	0.000061=	0.000072=	0.000080
MaF10	5	1069	0.964888=	0.957266=	0.967990+	0.965795=	0.968337+	0.959917=	0.964241+	0.961056
	10	783	6.060485=	5.912519=	5.959022=	6.296865+	5.875247=	5.879183=	6.002479=	5.908347
MaF11	5	960	2.448849=	2.449593=	2.450035=	2.449182=	2.449345=	2.449839=	2.450457=	2.446909
	10	1231	17.561243=	17.571667=	17.567886=	17.551676=	17.550847=	17.558191=	17.549476=	17.557456
MaF12	5	546	1.782436=	1.784627=	1.785285=	1.782553=	1.783997=	1.78398=	1.785365=	1.784192
	10	481	13.118814=	13.118862=	13.130149=	13.165706=	13.122503=	13.077448=	13.243644+	13.041259
MaF13	5	911	0.246188=	0.246621=	0.237909=	0.242349=	0.236794=	0.238086=	0.241078=	0.236727
	10	904	0.237637=	0.236821=	0.230929=	0.192100=	0.225026=	0.229397=	0.235592=	0.226286
Total (+/=/–) ⟶			1/29/2	2/30/0	2/26/4	5/23/4	4/25/3	2/29/1	5/25/2	of 32 probs.

kNN performed statistically better in three, equivalent in four, and worse in none. In general, for both Tables 1 and 2, if the performance $(+/=/-)$ of any alternative ML method vis-à-vis the base method, for a multi or many-objective problem suite, is given by $B/E/W$, then it is proposed that the *Hypervolume factor* be defined as below.

$$Hypervolume\ factor = \frac{B - W}{B + E + W}. \tag{1}$$

Table 3. Run-time comparison for IP2 and IP3 operators, with eight different ML methods each. Here, RF and kNN are the base methods for the pairwise comparisons within IP2 and IP3, respectively. The entries in the format H/L indicates the number of times an ML method has a Higher/Lower run-time compared to the base method.

		RF	ExTree	SVR	ENet	Ridge	LR	XGB	kNN
IP2	Multi-objective	–	1/6	0/7	1/6	0/7	1/6	5/2	1/6
	Many-objective		6/26	5/27	4/28	4/28	8/24	14/18	8/24
IP3	Multi-objective	5/2	4/3	6/1	3/4	5/2	5/2	7/0	–
	Many-objective	27/5	12/20	17/15	9/23	14/18	10/22	29/3	

Hence, for any alternative ML method and a specific problem suite: a positive/negative *Hypervolume factor* would represent the percentage instances, where it is *relatively* better/worse than the base method.

Furthermore, the notion of *Hypervolume factor* is extended to computation of *Run-time factor*. As a precursor, it may be noted that the run-time for only the seed underlying the median HV run for each ML method was recorded[1]. Their relative summary formatted as H/L in Table 3, indicates the number of times an ML method has a Higher/Lower run-time compared to the base method. Given this, it is proposed that the *Run-time factor* be defined as below.

$$\textit{Run-time factor} = \frac{H - L}{H + L}. \tag{2}$$

Hence, for any alternative ML method and a specific problem suite: a positive/negative *Run-time factor* would represent the percentage instances, where it is *relatively* worse/better than the base method in terms of run-time.

In the wake of the above, the performance of different ML methods at the level of multi- and many-objective problem-suite is captured in Fig. 5. Understandably, RF being the base method for IP2, occupies the origin in Figs. 5a and 5b. Similarly, kNN being the base method for IP3, occupies the origin in Figs. 5c and 5d. Importantly, if any alternative ML method was to dominate the base method in case of IP2 or IP3, in terms of both HV and run-time, then it should occupy the fourth quadrant. *However, such occurrences are quite rare, as highlighted below*:

- For the IP2 operator applied to multi-objective suite: kNN and ExTree outperform the base RF, and seem to offer reasonably better HV in reasonably lower run-time.
- For the IP2 operator applied to many-objective suite: XGB, kNN, ExTree, SVR, and Ridge seem to offer only a marginally better HV than the base RF, but in reducing order of run-time.

[1] *For this paper, the 21 seed runs were executed in parallel to save the overall run-time, given which the exact run-time for each seed was not traceable. Hence, for run-time estimate, only the seed corresponding to the median hypervolume was executed again.*

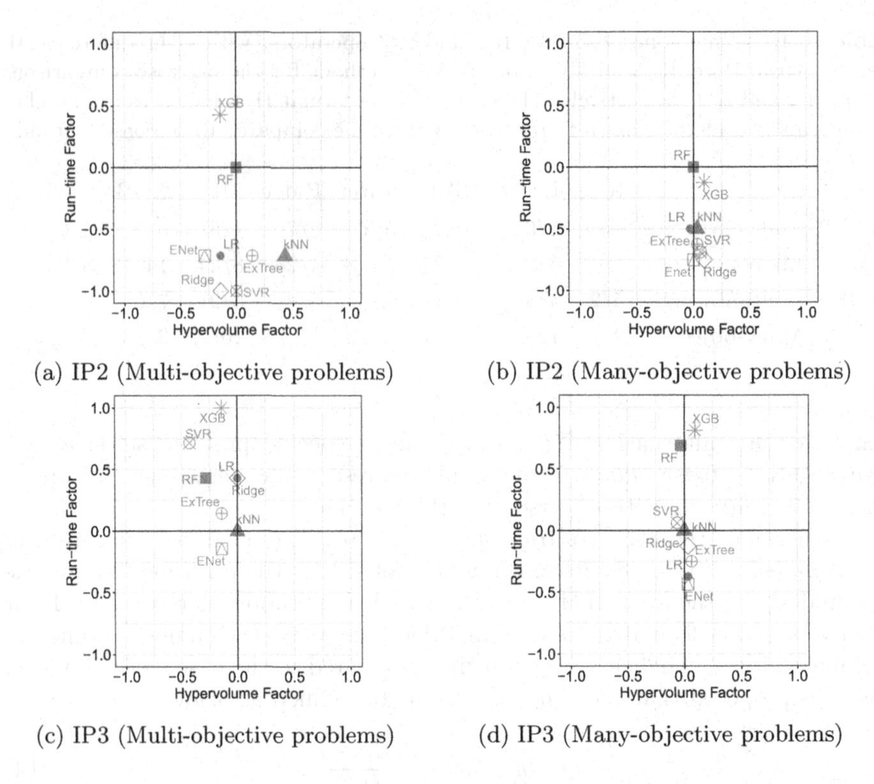

(a) IP2 (Multi-objective problems)

(b) IP2 (Many-objective problems)

(c) IP3 (Multi-objective problems)

(d) IP3 (Many-objective problems)

Fig. 5. Plots comparing HV factor and Run-time factor across the multi-objective and many-objective problem suite for IP2 and IP3 operators.

- For the IP3 operator applied to multi-objective suite: the base kNN seems to be the best choice, as it outperforms all other alternative ML methods.
- For the IP3 operator applied to many-objective suite: Ridge, ExTree, LR and ENet offer insignificantly better HV, but in reducing order of run-time.

Overall, if one winner has to be picked across all scenarios, then kNN can be inferred as the one.

The presented results also endorse the known characteristics of some the considered ML methods, as highlighted below. It can be observed that the considered linear ML models, namely LR, Ridge, and ENet have a comparable performance, particularly in terms of the HV measures. Within the family of trees, RF has a higher run-time than ExTree. This is consistent with the expectation, since RF chooses the optimal split, while ExTree relies on random split, saving some time. Overall, the boosting algorithm, namely, XGB is seen to have a higher run-time compared to the other ML methods. This could be attributed to the fact that during each split finding process, XGB iterates over all entries in the input data, making the process slower.

6 Conclusion

This paper sought to analyze, as to how sensitive the existing *Innovized Progress* operators (IP2 for convergence-improvement and IP3 for diversity-improvement) are to the choice of the underlying ML method. In that, the key concern was to investigate if by changing the underlying ML methods, the performance of these operators may drastically deteriorate, or if it could be significantly improved. Exhaustive experiments based on eight ML methods, tested against seven multi-objective and 32 many-objective test instances, suggest that the performance of IP2 could be marginally improved by replacing its base ML method, namely, Random forests, with kNN. However, on the whole, both the existing IP2 and IP3 operators are reasonably robust, and not too sensitive to the choice of underlying ML method. Besides the above inference, the results in this paper also endorsed some of the known characteristics of the considered ML methods. Finally, if one ML method is to be recommended for use, within the gambit of the existing IP2 and IP3 operators; considered test suite; and chosen ML methods, then kNN could be considered as the best performing method. Overall, this paper presents a systematic methodology to investigate the tradeoff associated with different ML methods in terms of their potential for performance enhancement of Evolutionary Multi- and Many-objective Optimization algorithms vis-à-vis the associated computational cost. It is hoped that this shall pave way for similar investigations in other existing ML-based enhancements, such as surrogate-modeling methods.

Acknowledgement. The authors wish to acknowledge Government of India, for supporting this research through an Indo-US SPARC project (code: P66). The authors also wish to thank Sukrit Mittal for his support through this work.

References

1. Breiman, L.: Random forests. Mach. Learn. **45**(1), 5–32 (2001). https://doi.org/10.1023/A:1010933404324
2. Chen, L., Liu, H.L., Tan, K.C., Cheung, Y.M., Wang, Y.: Evolutionary many-objective algorithm using decomposition-based dominance relationship. IEEE Trans. Cybern. **49**(12), 4129–4139 (2019). https://doi.org/10.1109/TCYB.2018.2859171
3. Chen, T., Guestrin, C.: XGBoost: a scalable tree boosting system, pp. 785–794 (2016). https://doi.org/10.1145/2939672.2939785
4. Cheng, R., et al.: A benchmark test suite for evolutionary many-objective optimization. Complex Intell. Syst. **3**(1), 67–81 (2017). https://doi.org/10.1007/s40747-017-0039-7
5. Cover, T.: Estimation by the nearest neighbor rule. IEEE Trans. Inf. Theory **14**(1), 50–55 (1968). https://doi.org/10.1109/TIT.1968.1054098
6. Deb, K., Jain, H.: An evolutionary many-objective optimization algorithm using reference-point-based nondominated sorting approach, part I: solving problems with box constraints. IEEE Trans. Evol. Comput. **18**(4), 577–601 (2014). https://doi.org/10.1109/TEVC.2013.2281535

7. Deb, K., Thiele, L., Laumanns, M., Zitzler, E.: Scalable test problems for evolutionary multiobjective optimization. In: Abraham, A., Jain, L., Goldberg, R. (eds.) Evolutionary Multiobjective Optimization: Theoretical Advances and Applications, pp. 105–145. Springer, London (2005). https://doi.org/10.1007/1-84628-137-7_6
8. Elarbi, M., Bechikh, S., Gupta, A., Ben Said, L., Ong, Y.: A new decomposition-based NSGA-II for many-objective optimization. IEEE Trans. Syst. Man Cybern.: Syst. **48**(7), 1191–1210 (2018). https://doi.org/10.1109/TSMC.2017.2654301
9. Goldstein, B.A., Polley, E.C., Briggs, F.B.S.: Random forests for genetic association studies. Stat. Appl. Genet. Mol. Biol. **10**(1) (2011). https://doi.org/10.2202/1544-6115.1691
10. Hussain, J.N.: High dimensional data challenges in estimating multiple linear regression. In: Journal of Physics: Conference Series, vol. 1591, no. 1, p. 012035 (2020). https://doi.org/10.1088/1742-6596/1591/1/012035
11. Liu, J., Wang, Y., Wei, S., Wu, X., Tong, W.: A parameterless penalty rule-based fitness estimation for decomposition-based many-objective optimization evolutionary algorithm. IEEE Access **7**, 81701–81716 (2019). https://doi.org/10.1109/ACCESS.2019.2920698
12. Mittal, S., Saxena, D.K., Deb, K., Goodman, E.D.: A learning-based *Innovized* progress operator for faster convergence in evolutionary multi-objective optimization. ACM Trans. Evol. Learn. Optim. **2**(1) (2021). https://doi.org/10.1145/3474059
13. Mittal, S., Saxena, D.K., Deb, K., Goodman, E.D.: Enhanced *Innovized* progress operator for evolutionary multi- and many-objective optimization. IEEE Trans. Evol. Comput. **26**(5), 961–975 (2022). https://doi.org/10.1109/TEVC.2021.3131952
14. Mittal, S., Saxena, D.K., Deb, K., Goodman, E.D.: A unified *Innovized* progress operator for performance enhancement in evolutionary multi- and many-objective optimization. Technical report. 2022006, Computational Optimization and Innovation Laboratory, Michigan State University, East Lansing, MI-48824, USA (2022). https://coin-lab.org/content/publications.html
15. Ogutu, J., Schulz-Streeck, T., Piepho, H.P.: Genomic selection using regularized linear regression models: ridge regression, lasso, elastic net and their extensions. BMC Proc. **6**(2), S10 (2012). https://doi.org/10.1186/1753-6561-6-S2-S10
16. Saxena, D.K., Kapoor, S.: On timing the nadir-point estimation and/or termination of reference-based multi- and many-objective evolutionary algorithms. In: Deb, K., et al. (eds.) EMO 2019. LNCS, vol. 11411, pp. 191–202. Springer, Cham (2019). https://doi.org/10.1007/978-3-030-12598-1_16
17. Saxena, D.K., Mittal, S., Kapoor, S., Deb, K.: A localized high-fidelity-dominance based many-objective evolutionary algorithm. IEEE Trans. Evol. Comput. 1 (2022). https://doi.org/10.1109/TEVC.2022.3188064
18. Yuan, Y., Xu, H., Wang, B., Yao, X.: A new dominance relation-based evolutionary algorithm for many-objective optimization. IEEE Trans. Evol. Comput. **20**(1), 16–37 (2016). https://doi.org/10.1109/TEVC.2015.2420112
19. Zitzler, E., Deb, K., Thiele, L.: Comparison of multiobjective evolutionary algorithms: empirical results. Evol. Comput. **8**(2), 173–195 (2000). https://doi.org/10.1162/106365600568202

End-to-End Pareto Set Prediction with Graph Neural Networks for Multi-objective Facility Location

Shiqing Liu[1], Xueming Yan[1,2(✉)], and Yaochu Jin[1(✉)]

[1] Faculty of Technology, Bielefeld University, 33619 Bielefeld, Germany
yaochu.jin@uni-bielefeld.de
[2] School of Information Science and Technology, Guangdong University of Foreign Studies, Guangzhou 510000, China
yanxm@gdufs.edu.cn

Abstract. The facility location problems (FLPs) are a typical class of NP-hard combinatorial optimization problems, which are widely seen in the supply chain and logistics. Many mathematical and heuristic algorithms have been developed for optimizing the FLP. In addition to the transportation cost, there are usually multiple conflicting objectives in realistic applications. It is therefore desirable to design algorithms that approximate a set of Pareto solutions efficiently without enormous search cost. In this paper, we consider the multi-objective facility location problem (MO-FLP) that simultaneously minimizes the overall cost and maximizes the system reliability. We develop a learning-based approach to predicting the distribution probability of the entire Pareto set for a given problem. To this end, the MO-FLP is modeled as a bipartite graph optimization problem and two graph neural networks are constructed to learn the implicit graph representation on nodes and edges. The network outputs are then converted into the probability distribution of the Pareto set, from which a set of non-dominated solutions can be sampled non-autoregressively. Experimental results on MO-FLP instances of different scales show that the proposed approach achieves a comparable performance to a widely used multi-objective evolutionary algorithm in terms of the solution quality while significantly reducing the computational cost for search.

Keywords: Combinatorial optimization · Multi-objective optimization · Graph neural network

1 Introduction

Multi-objective combinatorial optimization (MOCO) has received considerable attention over the past few decades due to its wide applications in the real-world. In MOCO, there are multiple conflicting objectives, and it is often non-trivial to optimize them simultaneously [1]. The multi-objective facility location problem (MO-FLP) is a typical NP-hard MOCO problem [23]. It aims to determine

M. Emmerich et al. (Eds.): EMO 2023, LNCS 13970, pp. 147–161, 2023.
https://doi.org/10.1007/978-3-031-27250-9_11

an optimal set of facility locations that can satisfy all the customer demands within certain constraints, while minimizing the total costs and maximizing the system reliability. Decisions made in facility location have a long-term impact on numerous operational and logistical strategies and are critical to both private and public firms [14].

A lot of work has been devoted to developing mathematical methods or handcrafted heuristic algorithms for solving MOCO problems. An intuitive approach is to reduce a multi-objective problem to a single-objective problem by calculating the weighted sum of multiple objectives. However, assigning a suitable weight to each objective introduces an additional hyperparameter optimization problem. Evolutionary algorithms (EAs) have been successful in approximating the Pareto set of MOCOs by maintaining and updating a set of solutions [6,7,34]. However, EAs and other population-based methods often require a large number of function evaluations during the search process, incurring prohibitive computing overhead when the objective functions are expensive to evaluate [17]. Moreover, it is difficult to reuse the knowledge about the optimal sets of solutions for other instances of the same problem that have already been solved.

Most existing work considers MOCO as constrained mixed-integer linear programming, overlooking the highly structured nature of the combinatorial optimisation problems. For example in the facility location problem (FLP), the locations of all facilities and customers can be represented by a set of nodes separately, and the transport overhead is the weight of the edge connecting two nodes from different sets. Generally, permutation-based COPs can be formulated as sequential decision-making tasks on graphs [18], and matching-based COPs can be considered as node and edge classification or prediction tasks on graphs. Therefore, machine learning methods can be used to extract high-dimensional characteristics of the graph-based problems and learn optimal policies to solve COPs instead of relying on handcrafted heuristics [1,31]. Graph neural networks (GNNs) can exploit the message passing scheme to learn the structural information of nodes and edges efficiently according to the graph topology. Consequently, GNNs are well-suited for tackling the MOCO problems [4,11,16]. However, most existing methods focus on solving permutation-based problems and only consider one single objective, neglecting the study of more commonly seen matching-based multi-objective COPs [9].

In this paper, we propose a learning-based approach leveraging graph convolutional networks (GCNs) to approximate the Pareto set distribution of the multi-objective facility location problem. The overall framework is shown in Fig. 1. The problem is formulated as a bipartite graph with edge connections between two independent sets of nodes. The model consists of two different residual gated GCNs for node classification and edge prediction tasks, respectively. The model takes bipartite graphs as the input, and transforms the original node and edge features into high-dimensional embeddings. Several residual gated graph convolutional layers are used to learn the structural information from the graph topology and update the embeddings iteratively. The output of the first GCN is a prediction of the probability of each factory being selected in the Pareto optimal solutions. The output of the second GCN is a probabilistic model in the form of an adjacency matrix, denoting the probability of each customer being

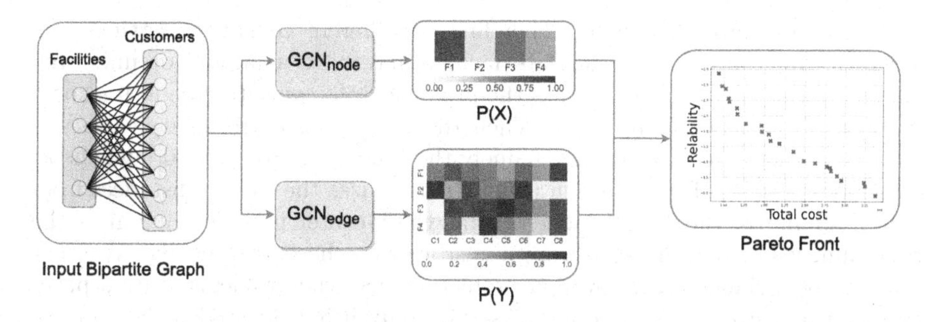

Fig. 1. An overview of the proposed approach. The MO-FLP instance is converted into a bipartite graph as the input to the GCNs. The two GCN models perform node classification and edge prediction and output the probability models, which are co-sampled to generate a set of non-dominated solutions in a one-shot manner.

assigned to each selected factory. The output probability models can be sampled directly to predict a set of Pareto solutions in a one-shot manner. The two networks are trained coordinately by supervised learning. The training data is a large set of MO-FLP instances with various approximated Pareto optimal solutions generated by a multi-objective evolutionary algorithm, e.g., the fast elitism non-dominated sorting genetic algorithm (NSGA-II) [7]. The main contributions of this paper include:

1. We formulate the MO-FLP as a bipartite graph optimization task and develop a novel learning-based combinatorial optimization method to directly approximate the Pareto set of new instances of the same problem without extra search.
2. We propose an end-to-end probabilistic prediction model based on two GCNs for node and edge predictions, respectively, and train the model with a supervised learning using data generated by a multi-objective evolutionary algorithm.
3. We demonstrate the efficiency of our proposed method for solving MO-FLP instances with different scales. Our experimental results show that the learning-based approach can approximate a set of Pareto optimal solutions without additional search, significantly reducing the computational cost compared to population-based algorithms.

2 Related Work

2.1 Facility Location Problem

FLPs are a typical class of NP-hard combinatorial problems in operations research [23]. FLPs consider choosing an optimal set of facilities among all the potential sites and determine an allocation scheme for all customers, under the constraints that all customer demands must be satisfied by the constructed facilities. A common objective of FLPs is to minimize the total costs, which consist of the transportation cost and the fixed cost.

FLPs have several variants depending on different constraint settings and objective functions. Each candidate facility may have a limited or unlimited maximum capacity, which classifies the problems into capacitated and uncapacitated facility location problems. When the number of established facilities is fixed to k, there are two variants, namely the k-median problem [30] and the k-center problem [5]. The k-median problem minimizes the sum of distances from each customer to the closest facility, while the k-center problem minimizes the maximum value of a distance from a customer to the closest facility. Another category of variants is the covering problem, where the problems share a property that a customer can receive the service only if it is located with a certain distance from the nearest facility [10]. The set covering problem aims to find a set of facilities with the minimum number that can satisfy all customers' demands. The maximum covering problem intends to find a set of facilities with a fixed number to maximize the total demands it covers. From an objective perspective, FLPs and its variants can be divided into single- and multi-objective problems. In addition to the overall costs, multi-objective facility location problems may also include other practical objectives such as the system reliability in logistics, which is quite desirable in real-world applications [9].

2.2 Graph Representation Learning

Graph-structured data is ubiquitous in daily life. Various kinds of data can be naturally expressed as graphs, such as social relationships, telecommunication networks, chemical molecules, and also combinatorial optimization tasks [33]. Generally, a graph is a collection of objects (nodes) along with a set of interactions (edges) between pairs of them [15]. With the development of machine learning techniques, graph representation learning has attracted increasing attention for in-depth analysis and effective utilization of graph data. Graph representation learning derives node and edge embeddings based on the graph topology for a variety of downstream tasks in machine learning, such as node classification [32], edge prediction [21], and graph clustering. The traditional graph representation methods neither use the node features nor share parameters in the encoder, and are not able to generalize to unseen nodes after training. To alleviate these limitations, graph neural networks are proposed to learn node embeddings in a more explainable way based on the topology and attributes of the input graph [33,36]. Early attempts made by Sperduti and Starita [29] dealt with arbitrary structured data as directed acyclic graphs with recursive neural networks. Gori [13] and Scarselli [28] generalized the recursive neural networks for other types of graph structures and introduced the concept of graph neural networks. With the compelling performance shown by convolutional neural networks in computer vision tasks, a lot of work has been devoted to the transfer of convolution operators to graph domain [35], which can be categorized into spectral-based methods [3,8,24] and spatial-based methods [12,25,26]. GNNs have been practically applied to various domains and achieved encouraging performance [11,16,18].

2.3 Machine Learning for Combinatorial Optimization on Graphs

NP-hard combinatorial optimization problems are non-trivial to solve, but the instances are relatively easy to generate. In many practical scenarios, decision-makers need to solve different instances of the same optimization task, where the instances share the same problem structure and only differ in data [1,4,20]. Traditional heuristic methods require extensive expert knowledge and a huge computational cost, and their solutions cannot be transferred to other instances. To address this limitation, recent years have seen a surge in research on machine learning approaches to combinatorial optimization to automate the solution of different instances of combinatorial optimization problems [4,31]. Combinatorial optimization problems often depict a collection of entities and their relations, which are graph-structured data in essence. Therefore, many GNN-based machine learning methods are proposed to solve combinatorial optimization problems [18–20]. Kool et al. developed a GNN model in an encoder-decoder architecture based on attention layers, and trained it using REINFORCE for solving routing problems [22]. In addition to solving COPs directly, machine learning techniques can also be used to provide valuable information to operation research algorithms [11]. Although a lot of effort has been devoted to developing ML methods for COPs, most work has focused on single-objective permutation-based problems, and little research on multi-objective matching-based problems has been reported.

3 Problem Formulation

This section begins with a formal definition of the multi-objective uncapacitated facility location problem, which is mathematically formulated as integer linear programming. Subsequently, we discuss how to measure the logistics system reliability in facility location problems.

Multi-objective Uncapacitated Facility Location. Consider a set of candidate facility locations and a set of demand points (customers) with fixed locations. Every customer has its own quantity of demand to be satisfied. Each potential facility has its own fixed cost for construction, and there are different transportation costs between facilities and customers associated with their distances. Each customer should be served by only one facility, while each facility can serve multiple customers simultaneously. The target is to identify the selected collection of facilities for construction and assign an allocation plan for all customers, in order to minimize the total costs and maximize the system reliability.

Mathematical Formulation. With the minimization of the total costs (including fixed costs and transportation costs) and the maximization of the system reliability as the two objectives, the multi-objective uncapacitated facility location problem can be defined as follows:

$$\min C_{\text{total}} = \sum_{i \in M} f_i X_i + \sum_{i \in M} \sum_{j \in N} q_j d_{ij} c_{ij} Y_{ij} \tag{1}$$

$$\max R_{\text{sys}} = \frac{\sum_{i\in M}\sum_{j\in N} q_j X_i \left[1 - F_{V_{ij}}\left(\frac{d_{ij}}{t_j}\right)\right]}{\sum_{j\in N} q_j} \tag{2}$$

$$\text{s.t.} \quad \sum_{i\in M} Y_{ij} = 1 \quad j \in N \tag{3}$$

$$X_i, Y_{ij} = \{0, 1\},\ Y_{ij} \le X_i \quad i \in M, j \in N \tag{4}$$

In the MO-FLP, assume there are m candidate facility locations denoted as a set $M = \{1, 2, \ldots, m\}$, and n customer points denoted as a set $N = \{1, 2, \ldots, n\}$. $f_i \in \mathbb{R}^+$ denotes the fixed cost of constructing the facility at candidate location $i\,(i \in M)$, and $q_j \in \mathbb{R}^+$ is the demand volume of customer $j\,(j \in N)$. $d_{ij} \in \mathbb{R}^+$ and $c_{ij} \in \mathbb{R}^+$ are the distance and the unit transportation cost between facility i and customer j respectively. V_{ij} denotes the speed for vehicles travelling from facility i to customer j, and $F_{V_{ij}}(\cdot)$ is a statistically regular velocity distribution. t_j is the delivery timescale required by customer j.

The decision variables are $X_i \in \{0, 1\}$, which denotes whether facility location j is selected ($X_i = 1$) or not ($X_i = 0$), and $Y_{ij} \in \{0, 1\}$, which denotes whether customer j is served by facility i ($Y_{ij} = 1$) or not ($Y_{ij} = 0$). The two objectives are the minimization of the total costs C_{total} and maximization of the system reliability R_{sys}.

Logistics System Reliability. Reliability is the probability that a system performs its intended function under the stated conditions [27]. Logistics system reliability is defined as the probability at which the system will successfully provide services to customers under certain conditions and within a specified time. System reliability is a common metric for assessing service levels in modern logistics. The service reliability R_{ij} between factory i and customer j is defined as:

$$R_{ij} = P\left(T_{ij} \le t_j\right) = P\left(\frac{d_{ij}}{V_{ij}} \le t_j\right) = P\left(V_{ij} \ge \frac{d_{ij}}{t_j}\right) = 1 - F_{V_{ij}}\left(\frac{d_{ij}}{t_j}\right) \tag{5}$$

where T_{ij} is the time cost for delivery from facility i to customer j, and $F_{V_{ij}}(\cdot)$ is a statistically regular velocity distribution function that usually follows the characteristics of a normal distribution. Based on this, the logistics system reliability of facilities serving multiple customers is calculated by:

$$R_{\text{sys}} = \frac{\sum_{i\in M}\sum_{j\in N} q_j R_{ij}}{\sum_{j\in N} q_j} = \frac{\sum_{i\in M}\sum_{j\in N} q_j \left[1 - F_{V_{ij}}\left(\frac{d_{ij}}{t_j}\right)\right]}{\sum_{j\in N} q_j} \tag{6}$$

4 Method

We first convert an MO-FLP instance to a bipartite graph based on its inherent structural properties, and then train a dual GCN-based model to directly output the probabilistic model of the Pareto optimal solutions for the given task. The proposed model consists of two graph convolutional networks GCN_{node} and GCN_{edge}. GCN_{node} learns high-dimensional representations of nodes and outputs a probabilistic prediction for each node via a simple multi-layer perceptron

(MLP) classifier. Meanwhile, GCN_{edge} learns high-dimensional edge representations and predicts the probability of each edge appearing in the Pareto optimal solutions in the form of an adjacency matrix. The entire model is trained in an end-to-end manner by minimizing the loss between predictions and ground-truth labels. During the test, the output probabilistic model is sampled and converted into a set of non-dominated solutions in a non-autoregressive manner, eliminating the requirement of further search when solving new instances.

4.1 Bipartite Optimization on MO-FLP

An instance of the MO-FLP is transformed into a bipartite graph $G = (U, V, E)$, whose vertices are divided into two independent sets U (including all candidate facilities) and V (including all customers), and these two parts are connected by a set of edges E. Within the graph, each facility in U contains the information of its fixed cost, while each customer in V contains its demand and delivery timescale information. The features of each edge in E contain the Euclidean distance, the transportation cost and the reliability between the facility and the customer it connects. The aim of converting the MO-FLP into a bipartite optimization is to derive high-dimensional embeddings in the latent space through graph representation learning, in order to predict optimal solutions by means of machine learning.

4.2 The Dual GCN-Based Model

Overall Framework. Note that a solution to an MO-FLP problem consists of two parts: $X = \{X_i \mid i \in M\}$ and $Y = \{Y_{ij} \mid i \in M, j \in N\}$. The decision variable X first determines a subset of facilities to be constructed from all candidate locations, then the decision variable Y identifies the allocation scheme between customers and the selected locations in X. According to the mathematical formulation in Sect. 3, the calculation of objective C_{total} in Eq. 1 requires both X and Y as the decision variables, while the second objective R_{sys} in Eq. 2 is only determined by X. Leveraging the structural properties of the MO-FLP problem discussed above, we propose to predict the two components X and Y by designing two GCN models, one for node prediction and the other for edge prediction.

As shown in Fig. 1, the proposed model consists of GCN_{node} and GCN_{edge}, which take the same bipartite graph as their input. More specifically, GCN_{node} loads node and edge information and computes H-dimensional representations for each node via iterative graph convolution operators. The last graph convolution layer is followed by a multi-layer perceptron (MLP) classifier, where the updated node embeddings are taken as its inputs to compute the probability of each node being selected in decision variable X. The output of the classifier is represented as a probabilistic model $P(X) \in \mathbb{R}^M$, where M is the number of all candidate facilities. Simultaneously, GCN_{edge} takes the same node and edge information as input attributes and derives H-dimensional representations for each edge. A following edge classifier is used to predict the probability of each edge occurring in the Pareto optimal solutions in the form of a heat-map over the adjacency matrix $P(Y) \in \mathbb{R}^{M \times N}$, where N is the number of all customers. The outputs of the two GCN models indicate the information of X and Y, respectively, which together constitute an approximation of the Pareto optimal solutions. The GCN architectures adopted in the proposed model consist of three building blocks: an embedding block, a graph convolution block and an MLP classifier.

Embedding Block. The inputs to the embedding block are a set of original node features $\mathbf{h_n} = \{\vec{u}_1, \vec{u}_2, \ldots, \vec{u}_M, \vec{v}_1, \vec{v}_2, \ldots, \vec{v}_N\}, \vec{u}_i \in \mathbb{R}^{F_u}, \vec{v}_j \in \mathbb{R}^{F_v}$ and edge features $\mathbf{h_e} = \{\vec{w}_{11}, \vec{w}_{12}, \ldots, \vec{w}_{MN}\}, \vec{w}_{ij} \in \mathbb{R}^{F_e}$. M and N are the numbers of facilities and customers, and F_u, F_v and F_e are the numbers of features for different nodes and edges. The outputs of the embedding block are node embeddings $\mathbf{n} = \{\vec{n}_1, \vec{n}_2, \ldots, \vec{n}_{M+N}\}, \vec{n}_i \in \mathbb{R}^H$ and edge embeddings $\mathbf{e} = \{\vec{e}_{11}, \vec{e}_{12}, \ldots, \vec{e}_{MN}\}, \vec{e}_{ij} \in \mathbb{R}^H$, where H is the dimension of the hidden space.

For node embeddings, each feature $a \in \mathbb{R}$ is first embedded in a d-dimensional vector $\vec{\alpha} \in \mathbb{R}^d$ by a learnable linear transformation to get adequate expressive power. Then all the feature vectors are concatenated together to get an embedding $\vec{n}_i$ for node i:

$$\vec{n}_i = \mathrm{concat}_{k=1}^{F_n}\left(\vec{\alpha}_i^k\right) \tag{7}$$

Similarly, the edge embedding $\vec{e}_{ij}$ for the edge between node i and node j is the concatenation of all the edge feature vectors:

$$\vec{e}_{ij} = \mathrm{concat}_{k=1}^{F_e}\left(\vec{\beta}_{ij}^k\right) \tag{8}$$

The selection of node and edge features as the input to the embedding layers depends on the problem's characteristics, which should have a significant impact on the objective function values. For the MO-FLP problem investigated in this work, there are several candidate node features of the bipartite graph served as input: the node category of the binary classification (i.e., whether a node belongs to the facility set or the customer set), the demand volume of a customer, the fixed cost of constructing a facility, the transportation costs and the reliability of all edges connected to a node. And the input edge features include the adjacency matrix of the bipartite graph, the transportation cost, and the reliability of an edge.

Graph Convolution Block. The message passing process mainly occurs in the graph convolution block by stacking several graph convolution layers sequentially. It leverages the structure and properties of the input graph in order to exchange information between neighbors and update node and edge embeddings without changing the connectivity. The graph convolution adopted in our model follows the framework of residual gated graph convolutional neural network [2], where additional edge features and residual gated operators are integrated to introduce heterogeneity in the message passing process.

In the graph convolution block, the inputs to the k-th layer are a set of node embeddings $\mathbf{n^k} = \left\{\vec{n}_1^k, \vec{n}_2^k, \ldots, \vec{n}_{M+N}^k\right\}$ and a set of edge embeddings $\mathbf{e^k} =$

$\{\vec{e}_{11}^{k}, \vec{e}_{12}^{k}, \ldots, \vec{e}_{MN}^{k}\}$ where $\vec{n}_i, \vec{e}_{ij} \in \mathbb{R}^H$. The k-th layer outputs an update set of both node and edge embeddings with the same dimension H.

Let $\vec{e}_{ij}^{k}$ denote the edge embedding between node i and node j at the k-th GCN layer. In the message passing of $\vec{e}_{ij}$ (the superscript k is omitted for simplicity), we first gather the associated node embeddings $\vec{n}_i$ and $\vec{n}_j$ as neighborhood information, and aggregate all the messages as $\vec{e}_{ij}'$. Then $\vec{e}_{ij}'$ is passed through a batch normalization layer BN and the rectified linear unit ReLU, to form the updated edge embedding $\vec{e}_{ij}^{k+1}$ together with the original input $\vec{e}_{ij}^{k}$:

$$\vec{e}_{ij}^{k+1} = \vec{e}_{ij}^{k} + \mathrm{ReLU}\left(\mathrm{BN}\left(\mathbf{U}\vec{e}_{ij}^{k} + \mathbf{V}\left(\vec{n}_i^{k} + \vec{n}_j^{k}\right)\right)\right) \tag{9}$$

where $\mathbf{U}, \mathbf{V} \in \mathbb{R}^{H \times H}$ are linear transformations. Suppose $\vec{n}_i^{k}$ denotes the node embedding of node i at the k-th layer. For updating $\vec{n}_i$, we first calculate the weight vector $\boldsymbol{\omega}_{ij}$ of each neighbor node j as:

$$\boldsymbol{\omega}_{ij} = \frac{\sigma\left(\vec{e}_{ij}\right)}{\sum_{j \in \mathcal{N}_i} \sigma\left(\vec{e}_{ij}\right) + \delta} \tag{10}$$

where $\mathcal{N}_i$ denotes all the first-order neighbors of node i. σ represents the sigmoid function, and $\delta > 0$ is a small value. Then we gather the neighbor embeddings $\vec{n}_j\,(j \in \mathcal{N}_i)$ and define the output of the k-th convolution layer as:

$$\vec{n}_i^{k+1} = \vec{n}_i^{k} + \mathrm{ReLU}\left(\mathrm{BN}\left(\mathbf{P}\vec{n}_i + \mathbf{Q}\sum_{j \in \mathcal{N}_i} \boldsymbol{\omega}_{ij}\vec{n}_j\right)\right) \tag{11}$$

where $\mathbf{P}, \mathbf{Q} \in \mathbb{R}^{H \times H}$ are linear transformations. The stack of graph convolution layers enables neighborhood messages to be progressively transferred within the graph. The dimensionality of the embeddings remains the same, however, the representation of each node and edge contains more local information in addition to its original features.

MLP Classifier. The updated representations are taken as inputs to an MLP for classification tasks. For node prediction in GCN_{node}, we consider $\vec{n}_i\,(i \in M)$ as the high-dimensional embedding of node i from the facility set M. For edge prediction in GCN_{edge}, we consider $\vec{e}_{ij}\,(i \in M, j \in N)$ as the embedding of edge between facility i and customer j. The probability $\hat{p}_i \in [0,1]$ of node i being selected as a constructed facility and the probability $\hat{p}_{ij} \in [0,1]$ of facility i serving customer j are predicted by:

$$\hat{p}_i = \mathrm{MLP}(\vec{n}_i),\ \hat{p}_{ij} = \mathrm{MLP}(\vec{e}_{ij}) \tag{12}$$

The weight parameters are trained in an end-to-end manner by minimizing the mean square error between the prediction $\hat{P}(X) = \{\hat{p}_i \mid i \in M\}$ and the ground-truth label $P(X) = \{p_i \mid i \in M\}$ via gradient descent methods.

Since each customer must be served by only one facility, we consider the edge prediction for each customer as a multi-class classification task and train the network parameters by minimizing the cross entropy loss between the prediction $\hat{P}(Y) = \{\hat{p}_{ij} \mid i \in M, j \in N\}$ and the ground-truth label $P(Y) = \{p_{ij} \mid i \in M, j \in N\}$, where $P(X)$ and $P(Y)$ are both derived from the approximated Pareto optimal solutions.

End-to-End Training. The dataset for training and testing the proposed model is generated by a multi-objective evolutionary algorithm. We generate MO-FLP instances of different scales (i.e., various numbers of facility and customer nodes) and approximate their Pareto fronts via the fast elitist non-dominated sorting genetic algorithm (NSGA-II) [7]. Then the probabilistic distributions $P(X)$ and $P(Y)$ for each instance are derived from a set of approximated Pareto optimal solutions, which serve as ground-truth labels for training and evaluating the proposed model.

5 Experiments

5.1 Dataset Generation and Hyperparameter Configurations

We consider MO-FLP problems with the following four different configurations: $M \times N$ are set to 20×50, 20×100, 50×100, and 50×200. We randomly generate 1000 instances for each problem scale and optimize them using NSGA-II until convergence to approximate the true Pareto fronts. Then the 1000 instances for each scale are divided into a training dataset, a validation dataset and a test dataset with 700, 200 and 100 pairs of instances and ground-truth labels, respectively. During each training epoch, the training data is split into mini-batches with a batch size $B = 20$ instances. The Adam optimizer is used to train the weights of the proposed model with an initial learning rate of $\gamma = 0.001$ and a maximum number of 300 epochs. Both GCN_{node} and GCN_{edge} consist of $l_{GCN} = 3$ graph convolutional layers and $l_{MLP} = 3$ classification layers. The dimension of the hidden space is set to $H = 128$ for node and edge embeddings. During the test, we sample 200 solutions from the output prediction for each instance and calculate the hypervolume (HV) and IGD value of the obtained non-dominated solution sets as the performance indicators.

5.2 Experimental Results

There are two variants of our proposed model adopted in the experiments, named *Dual_A* and *Dual_B* with different input features. *Dual_A* takes the node category, the customer demand and the fixed cost of each facility as the original node features, while *Dual_B* also considers the transportation costs and the service reliability of all the edges associated with the node. Both architectures share the same edge features as inputs. To investigate the model performance on MO-FLP with various scales, we compare it to NSGA-II with different numbers of function evaluations (MFEs). We set the number of independent runs to 20 for the compared algorithm, and calculate the mean and standard deviation of HV and IGD values as the performance indicators. The population size is set to 100 for all experiments.

Table 1 shows the performance of the proposed model compared to NSGA-II in different problem scales. We train the dual GCN-based models with different scales of the problem instances and evaluate them on test datasets. For each test case, 200 solutions are first sampled from the predicted probability distribution and evaluated by the objective functions to get a set of non-dominated solutions. Then we calculate the mean HV and IGD values. Finally, for HV and IGD values associated with each MFE configuration of NSGA-II, we count the percentage of the cases in which the proposed model performs better than NSGA-II out of the 100 test cases. The statistical results in Table 1 indicate that for an unseen instance, by only sampling 200 solutions from the model, the performance of the sampled solution set is already better than NSGA-II with more than 10000 function evaluations.

Table 1. The percentages of test instances where two variants of the proposed model perform better than NSGA-II with different MFEs in terms of the two indicators.

		MFEs	10000	20000	30000	40000	50000
20×50	*Dual_A*	HV	100%	98%	96%	75%	41%
		IGD	100%	90%	63%	31%	20%
	Dual_B	HV	100%	98%	88%	69%	29%
		IGD	100%	86%	55%	27%	8%
20×100	*Dual_A*	HV	100%	100%	94%	76%	51%
		IGD	100%	100%	86%	57%	45%
	Dual_B	HV	100%	100%	100%	100%	96%
		IGD	100%	100%	100%	98%	90%
50×100	*Dual_A*	HV	100%	100%	100%	73%	33%
		IGD	100%	94%	63%	29%	2%
	Dual_B	HV	100%	100%	96%	61%	27%
		IGD	100%	96%	53%	12%	0%
50×200	*Dual_A*	HV	100%	100%	100%	76%	14%
		IGD	100%	88%	43%	8%	0%
	Dual_B	HV	100%	100%	100%	100%	90%
		IGD	100%	100%	98%	86%	69%

Figure 2 depicts the differences between the HV values of the solution sets obtained by NSGA-II and the proposed model for different problem scales with different MFEs. A positive difference means that the proposed model performs better than NSGA-II. These results reveal that the proposed model outperforms NSGA-II when the MFEs is less than 40000 in most test cases for all scales. In some cases the model performance is even comparable to that of NSGA-II with 50000 MFEs.

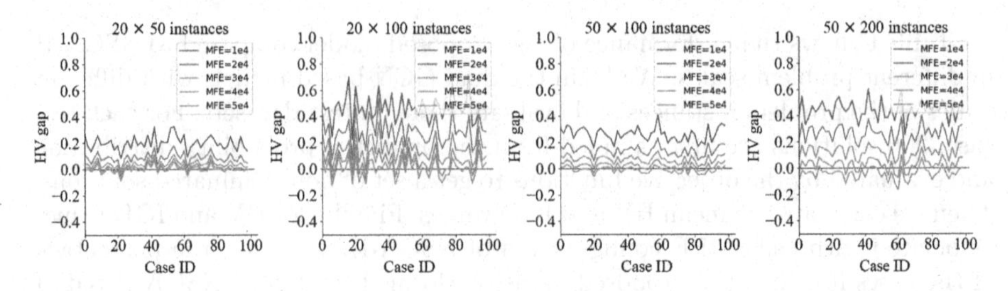

Fig. 2. The difference in HV values between the proposed model and NSGA-II with different MFEs.

5.3 Hyperparameter Sensitivity Analysis

We investigate the influence of different graph convolution layers and hidden dimensions on the two performance indicators, HV and IGD. We train the proposed model with different numbers of GCN layers on the 20×20 training dataset, and evaluate them on the test dataset with 100 unseen instances. The statistics of HV and IGD values are presented in the form of boxplots in Fig. 3(a). The results demonstrate that the increase in the number of GCN layers has a little impact on the model performance, and $l_{GCN} = 3$ achieves a slightly better performance. Similarly, we train the proposed model for different dimensions of the hidden space and plot the statistical results of the two indicators in Fig. 3(b). The model performance improves as the hidden dimension increases from 32 to 128. Note that a larger number of hidden layers and more GCN layers also lead to higher computational costs in the training process.

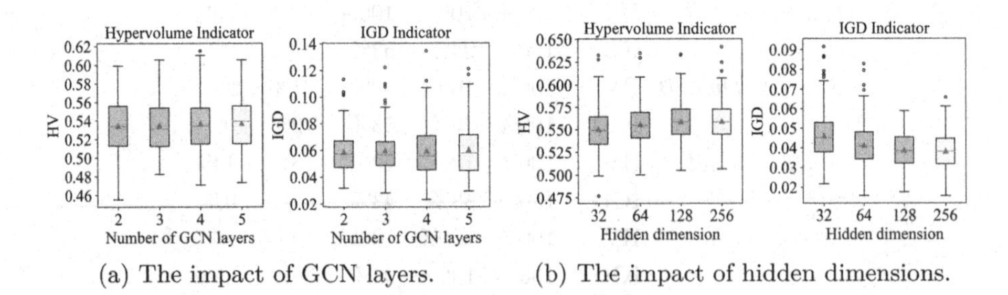

(a) The impact of GCN layers. (b) The impact of hidden dimensions.

Fig. 3. Sensitivity analysis. (a) The effect of different GCN layers. (b) The effect of different hidden dimensions.

6 Conclusion and Future Work

This paper proposes a learning-based approach to directly predicting a set of non-dominated solutions for multi-objective facility location. We convert the original combinatorial optimization problem into a bipartite graph, and train two GCN models for predicting Pareto optimal solutions for unseen instances by learning the distribution of Pareto optimal solutions in previously solved examples. Experimental results on different scales of MO-FLP instances demonstrate that by only sampling hundreds of solutions, the proposed dual GCN-based approach can achieve a performance comparable to NSGA-II using up to tens of thousands of function evaluations. Future work will focus on improving the model scalability and exploring the heterogeneity of the input graphs in order to generalize the proposed approach to more complex and realistic problems with conflicting objectives and multiple constraints.

Acknowledgment. This work was supported in part by the National Natural Science Foundation of China under Grant No. 62006053, in part by the Program of Science and Technology of Guangzhou under Grant No. 202102020878 and in part by a Ulucu PhD studentship. Y. Jin is funded by an Alexander von Humboldt Professorship for Artificial Intelligence endowed by the German Federal Ministry of Education and Research.

References

1. Bengio, Y., Lodi, A., Prouvost, A.: Machine learning for combinatorial optimization: a methodological tour d'horizon. Eur. J. Oper. Res. **290**(2), 405–421 (2021)
2. Bresson, X., Laurent, T.: Residual gated graph convnets. arXiv preprint arXiv:1711.07553 (2017)
3. Bruna, J., Zaremba, W., Szlam, A., LeCun, Y.: Spectral networks and locally connected networks on graphs. In: Proceedings of the International Conference on Learning Representations (ICLR) (2014)
4. Cappart, Q., Chételat, D., Khalil, E., Lodi, A., Morris, C., Veličković, P.: Combinatorial optimization and reasoning with graph neural networks. arXiv preprint arXiv:2102.09544 (2021)
5. Chakrabarty, D., Goyal, P., Krishnaswamy, R.: The non-uniform k-center problem. ACM Trans. Algorithms **16**(4), 1–19 (2020)
6. Cheng, R., Jin, Y., Olhofer, M., Sendhoff, B.: A reference vector guided evolutionary algorithm for many-objective optimization. IEEE Trans. Evol. Comput. **20**(5), 773–791 (2016)
7. Deb, K., Pratap, A., Agarwal, S., Meyarivan, T.: A fast and elitist multiobjective genetic algorithm: NSGA-II. IEEE Trans. Evol. Comput. **6**(2), 182–197 (2002)
8. Defferrard, M., Bresson, X., Vandergheynst, P.: Convolutional neural networks on graphs with fast localized spectral filtering. Adv. Neural Inf. Process. Syst. **29** (2016)
9. Farahani, R.Z., Fallah, S., Ruiz, R., Hosseini, S., Asgari, N.: OR models in urban service facility location: a critical review of applications and future developments. Eur. J. Oper. Res. **276**(1), 1–27 (2019)
10. García, S., Marín, A.: Covering location problems. In: Laporte, G., Nickel, S., da Gama, F.S. (eds.) Location Science, pp. 93–114. Springer, Cham (2015). https://doi.org/10.1007/978-3-319-13111-5_5
11. Gasse, M., Chételat, D., Ferroni, N., Charlin, L., Lodi, A.: Exact combinatorial optimization with graph convolutional neural networks. Adv. Neural Inf. Process. Syst. **32** (2019)
12. Gilmer, J., Schoenholz, S.S., Riley, P.F., Vinyals, O., Dahl, G.E.: Neural message passing for quantum chemistry. In: Proceedings of the International Conference on Machine Learning, pp. 1263–1272. PMLR (2017)

13. Gori, M., Monfardini, G., Scarselli, F.: A new model for learning in graph domains. In: Proceedings of the IEEE International Joint Conference on Neural Networks, vol. 2, pp. 729–734 (2005)
14. Hale, T.S., Moberg, C.R.: Location science research: a review. Ann. Oper. Res. **123**(1), 21–35 (2003)
15. Hamilton, W.L.: Graph representation learning. Synth. Lect. Artif. Intell. Mach. Learn. **14**(3), 1–159 (2020)
16. Hudson, B., Li, Q., Malencia, M., Prorok, A.: Graph neural network guided local search for the traveling salesperson problem. arXiv preprint arXiv:2110.05291 (2021)
17. Jin, Y.: A comprehensive survey of fitness approximation in evolutionary computation. Soft Comput. **9**(1), 3–12 (2005). https://doi.org/10.1007/s00500-003-0328-5
18. Joshi, C.K., Laurent, T., Bresson, X.: An efficient graph convolutional network technique for the travelling salesman problem. arXiv preprint arXiv:1906.01227 (2019)
19. Joshi, C.K., Laurent, T., Bresson, X.: On learning paradigms for the travelling salesman problem. arXiv preprint arXiv:1910.07210 (2019)
20. Khalil, E., Dai, H., Zhang, Y., Dilkina, B., Song, L.: Learning combinatorial optimization algorithms over graphs. Adv. Neural Inf. Process. Syst. **30** (2017)
21. Kim, J., Kim, T., Kim, S., Yoo, C.D.: Edge-labeling graph neural network for few-shot learning. In: Proceedings of the IEEE/CVF Conference on Computer Vision and Pattern Recognition, pp. 11–20 (2019)
22. Kool, W., Van Hoof, H., Welling, M.: Attention, learn to solve routing problems! In: Proceedings of the International Conference on Learning Representations (ICLR) (2019)
23. Laporte, G., Nickel, S., Saldanha-da-Gama, F.: Introduction to location science. In: Laporte, G., Nickel, S., Saldanha da Gama, F. (eds.) Location Science, pp. 1–21. Springer, Cham (2019). https://doi.org/10.1007/978-3-030-32177-2_1
24. Levie, R., Monti, F., Bresson, X., Bronstein, M.M.: CayleyNets: graph convolutional neural networks with complex rational spectral filters. IEEE Trans. Signal Process. **67**(1), 97–109 (2018)
25. Monti, F., Boscaini, D., Masci, J., Rodola, E., Svoboda, J., Bronstein, M.M.: Geometric deep learning on graphs and manifolds using mixture model CNNs. In: Proceedings of the IEEE Conference on Computer Vision and Pattern Recognition, pp. 5115–5124 (2017)
26. Niepert, M., Ahmed, M., Kutzkov, K.: Learning convolutional neural networks for graphs. In: Proceedings of the International Conference on Machine Learning, pp. 2014–2023. PMLR (2016)
27. Rausand, M., Hoyland, A.: System Reliability Theory: Models, Statistical Methods, and Applications, vol. 396. Wiley, New Jersey (2003)
28. Scarselli, F., Gori, M., Tsoi, A.C., Hagenbuchner, M., Monfardini, G.: The graph neural network model. IEEE Trans. Neural Netw. **20**(1), 61–80 (2008)
29. Sperduti, A., Starita, A.: Supervised neural networks for the classification of structures. IEEE Trans. Neural Netw. **8**(3), 714–735 (1997)
30. Vasilyev, I., Ushakov, A.V., Maltugueva, N., Sforza, A.: An effective heuristic for large-scale fault-tolerant k-median problem. Soft Comput. **23**(9), 2959–2967 (2018). https://doi.org/10.1007/s00500-018-3562-6
31. Vesselinova, N., Steinert, R., Perez-Ramirez, D.F., Boman, M.: Learning combinatorial optimization on graphs: a survey with applications to networking. IEEE Access **8**, 120388–120416 (2020)

32. Wang, Y., Wang, W., Liang, Y., Cai, Y., Liu, J., Hooi, B.: NodeAug: semi-supervised node classification with data augmentation. In: Proceedings of the 26th ACM SIGKDD International Conference on Knowledge Discovery & Data Mining, pp. 207–217 (2020)
33. Wu, Z., Pan, S., Chen, F., Long, G., Zhang, C., Philip, S.Y.: A comprehensive survey on graph neural networks. IEEE Trans. Neural Netw. Learn. Syst. **32**(1), 4–24 (2020)
34. Zhang, Q., Li, H.: MOEA/D: a multiobjective evolutionary algorithm based on decomposition. IEEE Trans. Evol. Comput. **11**(6), 712–731 (2007)
35. Zhang, S., Tong, H., Xu, J., Maciejewski, R.: Graph convolutional networks: a comprehensive review. Comput. Soc. Netw. **6**(1), 1–23 (2019). https://doi.org/10.1186/s40649-019-0069-y
36. Zhou, J., et al.: Graph neural networks: a review of methods and applications. AI Open **1**, 57–81 (2020)

Online Learning Hyper-Heuristics in Multi-Objective Evolutionary Algorithms

Julia Heise(✉) and Sanaz Mostaghim

Otto-von-Guericke University, Universitaetsplatz 2, Magdeburg 39106, Germany
{julia.heise,sanaz.mostaghim}@ovgu.de
https://www.ci.ovgu.de/

Abstract. Well-defined Hyper-Heuristics enhance the generalization of MOEAs and the blind usability on complex and even dynamic real-world application. Previous works already showed, that Hyper-Heuristics as selectors of crossover operators improve the performance of a single algorithm used on two opposing problem properties. In this paper, we present different selection mechanisms of Hyper-Heuristics, that are able to handle an expanded selection pool to cover more properties. We solve 20 benchmark problems with NSGA-II using those Hyper-Heuristics. By comparing the learning behaviour and the IGD trends of fixed crossover operator usages, we confirm that a combination of operators could outperform the best fixed operator. From the introduced Hyper-Heuristics in this paper, HHX-A made the best use of this advantage. It selects either all operators or a single operator alternately and learns fast which operators to prioritize to optimize the production. Due to periodic resets of the score, the Hyper-Heuristic is able to adapt fast to changes of the current state of the solving process. Although the pool is bigger and more diverse, we are able to show that HHX-A decides reasonably and fast. Therefore, it works well on a bigger set of problems with different properties.

Keywords: Multi-Objective Evolutionary Algorithm · Hyper-Heuristics · Selection mechanism · Crossover operator

1 Introduction

Hyper-Heuristics are high-level methods to construct heuristics. Whereas, heuristics are often used to find optimal solutions in a search space for a problem, a Hyper-Heuristic are designed to find an optimal heuristic to solve a specific problem [9]. In this paper, we intend to use Hyper-Heuristics as selectors of crossover operators in Multi-Objective Evolutionary Algorithms (MOEAs). For different problem properties, there are different optimal MOEAs [19]. To further generalize them, Hyper-Heuristics are designed and added into MOEAs to make them reliable on a wider variety of problems.

A well-known example of a problem that could benefit from the use of Hyper-Heuristic is job scheduling [19]. Although, it can be defined as one problem class

M. Emmerich et al. (Eds.): EMO 2023, LNCS 13970, pp. 162–175, 2023.
https://doi.org/10.1007/978-3-031-27250-9_12

because the solution spaces are alike, the properties and the complexity of these problems can differ and usually depend on the decision makers preferences. They can even be time-dependent, and their properties can change during the optimization process. In this case, it is nearly impossible to know which Meta-Heuristic would work best or even good enough. A Hyper-Heuristics could improve the situation by selecting the algorithm during the optimization process [10].

Hyper-heuristics have been applied to MOEAs in the past. In [24], Venske et al. propose a Hyper-Heuristic selection of crossover operators in MOEA/D for Many-Objective problems and could show that the proposed approach can improve the robustness and the quality of the algorithm. Similar to [15] by Pan et al. or [14] by Ono et al., they use a small pool of two to three crossover operators. Those are chosen in a way, that there is at least one good option for a specific problem class. Further approaches on applying Hyper-Heuristics are proposed by Pang et al. [16], who uses an offline learning to tune MOEA/D for specific problem properties, or Hong et al. [12], who uses a Hyper-Heuristic to generate polynomial mutation operators via training. An offline learner always needs training data sets, which should be at least similar to the actual problem. Both works showed, that otherwise the risk rises that the algorithm is outperformed by MOEAs that do not need training. Therefore, when using trained offline learners, more knowledge about the problem is again needed.

In this paper, we use an online learning Hyper-Heuristic to enhance the usability of the algorithm on unknown problems. We mainly focus on the selection mechanism and aim to expand the selection pool with more operators to cover several problem classes. We propose different variations of a learning mechanism applied to NSGA-II [5] to investigate their influence. We perform experiments on 20 test problems and compare our approach with NSGA-II using solely Simulated Binary Crossover (SBX) [6] and Uniform Crossover (UX) [21]. The results indicate that we could outperform the basic NSGA-II on most problems, although their properties are diverse.

This paper is organized as follows: first, we describe the general structure of our Hyper-Heuristics. In the second section, we propose our Hyper-Heuristics with the two basic selection methods. Afterwards, we analyse these algorithms in Sect. 4 and conclude that modification would be beneficial. Those modifications are presented in Sect. 5 and analysed in Sect. 6. We conclude our results and answer the questions about the generalization of MOEAs with Hyper-Heuristic in the last section.

2 Background

Hyper-Heuristics are generally defined as algorithms that optimize the selection of a heuristic or that constructs a new heuristic to solve an optimization problem. Therefore, they are part of a two-level framework. The high-level is the Hyper-Heuristic exploring the heuristic space H, and the low-level is a heuristic exploring the solution space S. The objective of a Hyper-Heuristic is to search in H for the optimal heuristic configuration $h*$, which generates the optimal solution $s*$ [17]. Considering, that in an optimization problem solutions are evaluated

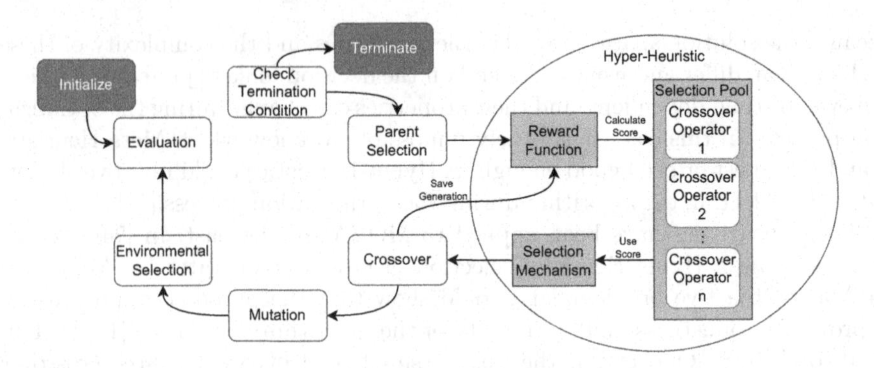

Fig. 1. Extended evolutionary algorithm using a Hyper-Heuristic instead of single crossover operation.

by function f, there has to be a mapping M so that $M : f(s) \rightarrow F(h)$, where F is the objective function for the Hyper-Heuristic. This leads to the formal definition:

$$F(h * | h* \rightarrow s*, h* \in H) \leftarrow f(s*, s* \in S) = \min\{f(s), s \in S\} \quad (1)$$

This general definition is applicable to multiple classes of Hyper-Heuristics. Burke et al. [2] classifies Hyper-Heuristics by dividing the feedback procession in online, offline and no learning and the heuristic search space into selection and generation. In this paper, we use exclusively online learning selection Hyper-Heuristics. These Hyper-Heuristic are applied to a MOEA to select the crossover operator. This use case is further investigated by Drake et al. [9]. They mention that in MOEAs, Hyper-Heuristics can select crossover operators so that they produce offspring optimized for the current problem. While this "Nature of how heuristics are grouped, chosen and applied" [9] differs for different Hyper-Heuristics, the basic structure remains the same. Figure 1, illustrates the main idea, which contains the application of this structure to a basic evolutionary algorithm (EA).

For each generation, the Hyper-Heuristic selects one or more crossover operators to produce the next generation. Thus, the heuristic space H is a set of crossover operators, the *selection pool*. The evaluation f of the solutions is done by the EA. To use this information for the learning process, we use a *reward function* that depends on the evaluation results. This function corresponds to the mapping M. The Hyper-Heuristic stores a cumulative score, which is updated by the reward function in every generation. This is utilized for the selection of a subset of operators from the selection pool for the current generation. This *selection mechanism* corresponds to the Hyper-Heuristics objective function F. Therefore, Hyper-Heuristics are mostly made up of those three exchangeable parts: *selection pool, reward function* and *selection mechanism*. In this paper, we focus on the selection mechanism and propose four different variants to evaluate its influence.

3 Selection Mechanisms: Single Selection and Distribution

In this section, we present two out of four developed Hyper-Heuristics. Both use the same reward function and the same selection pool. We use two different selection mechanisms: distribution and single selection. First, we introduce the reward function and the present the crossover operators in the selection pool, and afterwards the selection mechanisms are described.

For simplification, we use the expression *products of operator* to sum up the subset of offspring that were produced by a specific crossover operator. Furthermore, the set $\mathcal{X}$ describes the current generation and the set $\mathcal{Y}$ describes the offspring. $\mathcal{E}$ is the set of crossover operators, and $e \in \mathcal{E}$ is a specific operator. Products of e are therefore $\mathcal{Y}_e$ and becomes $\mathcal{X}_e$ after the environmental selection.

3.1 Reward Function

The reward function uses the survival rate of the offspring during the evolutionary cycle t. We consider the ratios between the latest offspring before $\mathcal{Y}_{e,t-1}$ and after the environmental selection $\mathcal{X}_{e,t}$, and the portion of products per operator in the current generation $\mathcal{X}_t$. The calculation is given in Eq. 2.

$$r_e = \frac{|\mathcal{X}_{e,t}|}{|\mathcal{Y}_{e,t-1}|} + \frac{|\mathcal{X}_{e,t}|}{|\mathcal{X}_t|} \tag{2}$$

Therefore, we use the survival of the fittest given by NSGA-II and no further evaluation is required for the learning process.

3.2 Selection Pool

The second component of the Hyper-Heuristic is the selection pool $\mathcal{E}$ populated with seven crossover operators. Simulated Binary Crossover (SBX) introduced by Deb and Agrawal in [6] and Uniform Crossover (UX) introduced by Syswerda and Gilbertin [21] are widely used and commonly known. We added Rotation-Based Simulated Binary Crossover (RSBX), which is derived from SBX with a rotational invariant property by Pan et al. [15]. We adapt two other known evolutionary operators, the Differential Evolution [20] and Covariance Matrix Adaption Evolutionary Strategy [11], to implement a Differential Evolution Crossover (DEX) and a Covariance Matrix Adaption Crossover (CMAX). Additionally, we derive from Simplex Crossover (SPX) introduced by Tsutsui et al. in [23] a simplified form using linear combinations of three parents, which we call LCX3. Lastly, we modified the Laplace Crossover (LX) presented by Deep et al. in [8] to get a new variation of a distribution-based crossover. With those operators, we cover a variety of self-adaptive behaviour as described by Beyer in [1], which is often added by a distribution-based crossover (SBX, RSBX, LX, LCX3), and different forms of centric production patterns, parent centric (SBX, RSBX, LX, UX, DEX) and mean centric (CMAX, LCX3), described by Deb et al. in [4].

Algorithm 1. Scoring Function.

```
▷ 𝒳_t: Set of Parent Individuals
▷ 𝒴_{t−1}: Set of Previous Individuals
▷ s⃗: current Score for each operator

function SCORING(𝒳_t, 𝒴_{t−1}, s⃗_{t−1})
    r⃗ ← RewardFunction(𝒳_t, 𝒴_{t−1})
    c⃗ ← Rank(r⃗)                    ▷ c: positions of elements in ascending sorted r
    s⃗ ← s⃗_{t−1} + (c⃗ − avg(c⃗))^3
    return s⃗
end function
```

Algorithm 2. Hyper-Heuristic Crossover Single Selector (HHX-S).

```
▷ e ∈ ℰ: Set of Evolutionary Operators
▷ 𝒳_t: Set of Parent Individuals
▷ 𝒴_{t−1}: Set of Previous Individuals
▷ s⃗_t: current Score for each e

function HYPERHEURISTICSINGLESELECTION(ℰ, 𝒳_t, 𝒴_{t−1}, s⃗_{t−1})
    s⃗_t ← SCORING(𝒳_t, 𝒴_{t−1}, s⃗_{t−1})
    p⃗ ← s⃗_t / sum(s⃗_t)                          ▷ score distribution
    e ← ROULETTEWHEEL(ℰ, p⃗)       ▷ common function for pobalistic selection
    𝒴_t ← e.CROSSOVER(𝒳_t)  ▷ execute crossover operation of selected operator
    𝒴_{t−1} ← 𝒴_t
    return [𝒴_t, 𝒴_{t−1}, s⃗_t]                  ▷ Save for next generation
end function
```

3.3 Selection Mechanism

The last component is the selection mechanism. We choose two different mechanisms to compare them and examine the impact of this part of the algorithm. Both variants update the score of each operator in the first step. The Scoring-Function described in Algorithm 1 utilizes the reward calculation given in Eq. 2 to measure the quality of the latest products of each crossover operator. Afterwards, the operators are ranked depending on the reward so that the score can be calculated by using a cubic function on the rank. The score is cumulative over the generations. We use a cubic function to ensure that the best performing operators receives a high score, and the bad performing operators get a decreased score. Therefore, the score enhances the online learning process.

After the update of the scores, the actual mechanism starts. We decided to implement a selection of a single operator per generation and a distribution of the whole generation to all operators. The selection is described in Algorithm 2 and is named HHX-S. It uses the well-known Roulette Wheel algorithm, to select the crossover depending on the current score.

The distribution is described in Algorithm 3 and is named HHX-D. In this case, the score is used to calculate the portions of the generation each operator

Algorithm 3. Hyper-Heuristic Crossover Distributor (HHX-D).

▷ $e \in \mathcal{E}$: Set of Evolutionary Operators
▷ $\mathcal{X}_t$: Set of Parent Individuals
▷ $\mathcal{Y}_{t-1}$: Set of Previous Individuals
▷ $\vec{s}_t$: current Score for each e

function HyperHeuristicDistribution($\mathcal{E}$, $\mathcal{X}_t$, $\mathcal{Y}_{t-1}$, $\vec{s}_{t-1}$)
  $\vec{s}_t \leftarrow$ Scoring($\mathcal{X}_t$, $\mathcal{Y}_{t-1}$, $\vec{s}_{t-1}$)
  $\vec{p} \leftarrow \frac{\vec{s}_t}{sum(\vec{s}_t)}$ ▷ score distribution
  for all e in $\mathcal{E}$ **do**
    $\mathcal{X}_{e,t} \leftarrow$ Distribute($\mathcal{X}$, p_e) ▷ subsets of parent generation for each operator
    $\mathcal{Y}_{e,t} \leftarrow$ e.Crossover($\mathcal{X}_{e,t}$) ▷ execute crossover operation with assigned subset
    $\mathcal{Y}_t \leftarrow \mathcal{Y}_t \bigcup \mathcal{Y}_{e,t}$ ▷ merge resulting subsets to offspring generation
  end for
  $\mathcal{Y}_{t-1} \leftarrow \mathcal{Y}_t$
  return [$\mathcal{Y}_t$, $\mathcal{Y}_{t-1}$, $\vec{s}_t$] ▷ Save for next generation
end function

receives. The decision, which individual is used with which operator, is random. The resulting child generation produced by each operator shall be the same size as the part of the parent generation they receive.

We compare both variants by implementing them in NSGA-II and solving 20 benchmark problems in the next section.

4 Evaluation and Experiments

In the experiments, we use 20 different benchmark problems. We compare both Hyper-Heuristics with NSGA-II [5] and different single Crossover operators. In prior experiments, we found out that the UX works best on most problems. Therefore, we use this and the classic version with SBX [6] for the analysis. We record their quality regarding the IGD [3] metric and additionally the selecting behaviour of the Hyper-Heuristics to evaluate their learning behaviour.

We selected DTLZ1-7 [7], RM1-4 [18], which are derived from ZDT1, ZDT2, ZDT6 [25] and DTLZ2, and WFG1-9 [13]. With these, we cover multiple variants of non-separability, modality and rotation on two and three objectives. We multiplied the number of decision variables by four to increase the complexity and emphasize the performance differences. We use the PlatEMO Framework [22]. It contains all the basic and many additional algorithms, benchmark problems and quality metrics.

During the evaluation, we aim to investigate the influence of the selection mechanism. We start with a comparison of the resulting IGD values, regard the over time development on example problems, and compare those afterwards to the selecting behaviour of both Hyper-Heuristics. We can figure out the advantages and disadvantages of each selection mechanism by analysing this behaviour.

Table 1. Inverted Generational Distance (IGD) of NSGA-II with HHX-D, HHX-S, UX and SBX as crossover operators on DTLZ, RM and WFG with increased number of dimensions.

Problem	M	D	HHX-D	HHX-S	UX	SBX
DTLZ1	3	28	3.9246e+1 (1.63e+1) ≈	1.5025e+2 (5.10e+1) −	2.8021e+1 (6.82e+0) +	4.1689e+1 (1.11e+1)
DTLZ2	3	48	1.1532e-1 (2.88e-2) −	8.6645e-2 (1.30e-2) +	8.4655e-2 (6.87e-3) +	1.0340e-1 (1.02e-2)
DTLZ3	3	48	4.4071e+2 (1.30e+2) ≈	1.0507e+3 (2.74e+2) −	2.6012e+2 (4.50e+1) +	3.8932e+2 (9.09e+1)
DTLZ4	3	48	1.1410e-1 (3.69e-2) ≈	1.0429e-1 (2.38e-2) +	9.1171e-2 (4.66e-1) +	1.2286e-1 (4.34e-1)
DTLZ5	3	48	3.2488e-2 (1.31e-2) +	2.0807e-2 (8.70e-3) +	2.9694e-2 (9.06e-3) +	3.8987e-2 (1.16e-2)
DTLZ6	3	48	5.9681e-3 (8.12e-1) +	8.1625e+0 (6.39e+0) +	2.7700e+1 (9.95e-1) −	1.7075e+1 (2.07e+0)
DTLZ7	3	88	9.0483e-1 (2.08e-1) ≈	4.0269e+0 (2.02e+0) −	6.3786e-1 (6.39e-2) +	8.4442e-1 (1.85e-1)
RM1	2	120	1.8299e-1 (5.90e-3) +	1.7186e-1 (1.06e-2) +	3.1063e-1 (5.89e-2) ≈	3.2392e-1 (6.06e-2)
RM2	2	120	2.9097e-1 (1.14e-2) +	2.8671e-1 (1.04e-2) +	5.0495e-1 (2.20e-2) +	5.1121e-1 (1.62e-2)
RM3	2	40	2.1076e+0 (2.04e-1) +	2.6314e+0 (3.79e-1) ≈	2.1805e+0 (3.55e-1) +	2.6871e+0 (3.79e-1)
⋮	⋮	⋮	⋮	⋮	⋮	⋮
RM4	3	48	5.1534e-1 (1.25e-1) +	3.5003e-1 (1.86e-1) +	5.4186e-1 (6.55e-2) ≈	5.4905e-1 (7.04e-2)
WFG1	3	48	1.2133e+0 (7.62e-2) +	1.5220e+0 (1.16e-1) −	1.5766e+0 (7.86e-2) −	1.2488e+0 (7.84e-2)
WFG2	3	48	2.7290e-1 (2.63e-2) +	2.8346e-1 (3.32e-2) ≈	2.4771e-1 (1.92e-2) +	2.9229e-1 (3.70e-2)
WFG3	3	48	3.0443e-1 (4.89e-2) +	2.8848e-1 (4.36e-2) +	2.5306e-1 (3.79e-2) +	3.2445e-1 (2.96e-2)
WFG4	3	48	3.1999e-1 (1.39e-2) ≈	3.2444e-1 (2.87e-2) ≈	2.6542e-1 (1.47e-2) +	3.1997e-1 (1.29e-2)
WFG5	3	48	2.8813e-1 (2.31e-2) +	3.7140e-1 (5.33e-2) −	3.0396e-1 (1.24e-2) +	3.4070e-1 (2.12e-2)
WFG6	3	48	3.6689e-1 (3.18e-2) +	3.7144e-1 (4.38e-2) ≈	3.2196e-1 (2.16e-2) +	3.7961e-1 (2.32e-2)
WFG7	3	48	3.7327e-1 (3.30e-2) +	3.4578e-1 (2.15e-2) +	4.1445e-1 (9.14e-2) ≈	4.3422e-1 (8.50e-2)
WFG8	3	48	4.4796e-1 (2.35e-2) −	4.3219e-1 (2.30e-2) −	3.3915e-1 (1.88e-2) +	3.8080e-1 (1.28e-2)
WFG9	3	48	3.1753e-1 (2.92e-2) +	3.0920e-1 (2.56e-2) +	3.8706e-1 (4.16e-2) ≈	4.0673e-1 (3.55e-2)
+/ − / ≈			13/2/5	10/6/4	14/2/4	

4.1 IGD Results of NSGA-II Using HHX-D, HHX-S, UX and SBX

In Table 1 our Hyper-Heuristics, HHX-S and HHX-D, implemented in NSGA-II are compared to NSGA-II with SBX and UX. We decided to use the rank sum test and highlight all algorithms that are considered as the best performing algorithms per benchmark. The data depicted in the cells are the median results of the IGD measurement over 31 runs. Furthermore, the rank sum test is used in a one-to-all comparison, which in our case compares the original NSGA-II to the other algorithms. The results are symbolized with markers to show, whether the algorithm performed significantly better than the original NSGA-II (+), significantly worse (−) or approximately equal (≈).

The data in Table 1 shows that the original NSGA-II is always dominated by at least one other algorithm. Prior experiments already showed that the NSGA-II with UX outperforms most other variations on most problems. A problem feature that is hard to handle for the UX is the rotation, thus, problems RM1-4. As RM3 is also multi-modal, most rotation invariant crossover operators still have difficulties on this problem, and UX performs similarly well due to its advantages on problems of this kind. Both variations with Hyper-Heuristics can compete with UX, but often lose in a direct comparison. Nevertheless, they outperform the original NSGA-II on most problems. On rotated problems, they also outperform the UX which is a hint, that they picked rotational invariant operators in those cases.

Assuming that the UX operator is the best performing operator in the pool, online learners would need to test all operators to learn this. Thus, extra effort is necessary, which we call *Learning Offset*. Due to that offset, it is very difficult to outperform the best operator in pool with our Hyper-Heuristics. In a setting, where the problem and its features are unknown, the user would also not know which crossover operator works best. Therefore, they would benefit from using the Hyper-Heuristic, that performs above the average. In Fig. 2 the IGD measurements of different variations of NSGA-II on two problems with different properties are visualized. On RM2 it is clear, that the UX and the SBX cannot

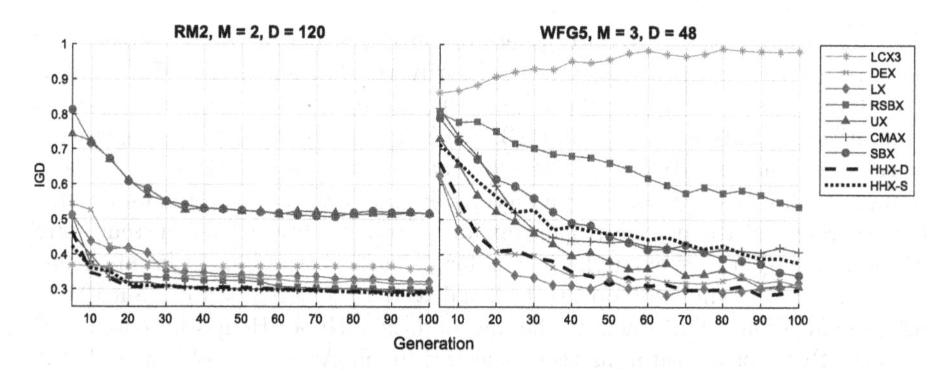

Fig. 2. IGD trends of the median runs of NSGA-II with single crossover operators and HHX-S and HHX-D on RM1 on the left and WFG8 on the right.

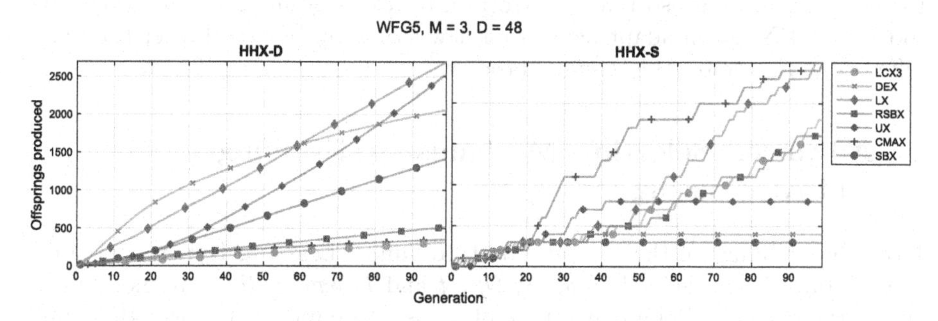

Fig. 3. Cumulative number of products of Crossover Operators selected by HHX-D on the left and HHX-S on the right on WFG5

achieve good results, whereas the rotational invariant operators perform similarly well. In this case, both Hyper-Heuristic had many suitable options to pick in the pool and thus the learning offset is minimal, and both algorithm outperform the other variations. On WFG5, most operators had difficulties producing sufficient offspring. In this case, it is not only hard to select a good operator, but also to learn this fast. The learning offset is therefore a lot bigger, especially for HHX-S as the learning is intrinsically slower than the learning of the HHX-D. This is also clear in the graph, as the HHX-S cannot achieve a sufficient result quality. HHX-D, on the other hand, outperformed the best operators on this problem.

4.2 Selection Behaviour of HHX-D and HHX-S

The question arises, how a selection of different operators can outperform the best operator within the selection pool. To answer this, we evaluate the behaviour of the Hyper-Heuristics. In Fig. 3 this behaviour is visualized by giving the cumulative number of products of the different crossover operators on WFG5. A difference in the selection is visible from the beginning on. Regarding the qualities of the single usages of the crossover operators in Fig. 2, it would be assumed that both Hyper-Heuristics would select primarily the LX and later DEX or UX. The HHX-D firstly prioritize the DEX, which changes to the LX after about 60 generations. Additionally, UX and SBX receive a big portion of the population in each generation. HHX-S on the other hand selects mostly the CMAX and changes to LX after about 50 generations. From those impressions, we assume that there is not the one best operator for a problem, but for the current state of the population. For example, if it is far or close to the Pareto Front. Thus, a combination of different operators could lead to a better performance than using a single one. This answers the question, how Hyper-Heuristics could outperform the best operator in their selection pool. Another problem with the current Hyper-Heuristic arises, on the other hand, which we call the *Learning Bias*. Especially the HHX-S could suffer from this, when it scores one operator too soon too high so that a correction or a change in preference would be too slow. HHX-D can adapt faster to a new situation, but the higher the score differences, the slower the learning process.

5 Advanced Selection Mechanisms: Evolving and Alternating

From the learnings of the first approach to Online Learning Hyper-Heuristics, new obstacles were defined: *Leaning Offset* and *Learning Bias*. We modify our Hyper-Heuristics and eliminate those obstacles, to improve the algorithms further. The learning offset is a big problem for the Hyper-Heuristic Selection, because of its bad explorative behaviour. Nevertheless, it performs very well on problems, where it can decide early on a good operator. Thus, the goal for a new Hyper-Heuristic would be, to have an improved exploration phase while keeping the exploiting of HHX-S. The first new Hyper-Heuristic uses an evolving approach and is therefore called HHX-E. It starts using the HHX-D to use its good exploration, then evolves to HHX-S to use a better exploitation and finally fully exploits the current best crossover operator. Assuming we know the number of the maximal function evaluations due to limited resources of the users, we can set the intervals in relation to this. With the results from previous tests, we decide to set the distribution portion to the first half and only use the last tenth for the exploitation of the current best crossover operator.

This approach might improve the Learning Offset, assuming the latest learning, that there is no best operator for a whole problem, but the best combination, the Learning Bias would not be defeated with this procedure. To fight this, a fourth approach on Hyper-Heuristic Online Learner is introduced. It uses HHX-D and HHX-S alternately and resets the score after each iteration. Again, we use the maximum number of function evaluations, to set the durations of each iteration. With the results from previous tests, we decide to use 10 iterations, with 70% using distribution and 30% using selection.

Table 2. Inverted Generational Distance (IGD) of NSGA-II with HHX-D, HHX-S, UX and SBX as crossover operators on DTLZ, RM and WFG with increased number of dimensions.

Problem	M	D	HHX-A	HHX-E	HHX-D	UX
DTLZ1	3	28	4.2961e+1 (2.14e+1) −	3.9691e+1 (1.56e+1) −	3.5948e+1 (1.51e+1) −	2.4737e+1 (9.57e+0)
DTLZ2	3	48	9.8555e-2 (1.48e-2) −	1.1057e-1 (2.39e-2) −	1.0440e-1 (1.81e-2) −	8.6873e-2 (6.46e-3)
DTLZ3	3	48	3.9664e+2 (1.93e+2) −	4.5935e+2 (1.11e+2) −	4.3394e+2 (1.50e+2) −	2.2892e+2 (5.21e+1)
DTLZ4	3	48	9.6731e-2 (1.34e-2) +	9.8610e-2 (2.27e-2) ≈	1.0498e-1 (2.98e-2) ≈	5.4353e-1 (4.58e-1)
DTLZ5	3	48	2.3905e-2 (7.92e-3) +	2.9282e-2 (1.04e-2) ≈	3.1162e-2 (7.59e-3) ≈	2.8337e-2 (1.18e-2)
DTLZ6	3	48	6.4028e-3 (9.52e-1) +	8.2228e-3 (9.82e-3) +	5.8280e-3 (9.30e-1) +	2.7478e+1 (8.34e-1)
DTLZ7	3	88	1.1940e+0 (2.33e-1) −	9.5052e-1 (3.62e-1) −	8.9030e-1 (2.75e-1) −	6.5082e-1 (1.09e-1)
RM1	2	120	1.7607e-1 (6.94e-3) +	1.8320e-1 (8.93e-3) +	1.8590e-1 (5.37e-3) +	2.9808e-1 (3.69e-2)
RM2	2	120	2.8288e-1 (7.22e-3) +	2.9107e-1 (6.81e-3) +	2.8974e-1 (1.12e-2) +	5.1133e-1 (1.80e-2)
RM3	2	40	2.1576e+0 (2.35e-1) ≈	2.2505e+0 (2.64e-1) ≈	2.2197e+0 (3.27e-1) ≈	2.1443e+0 (5.21e-1)
RM4	3	48	5.2202e-1 (9.81e-2) ≈	5.5466e-1 (1.20e-1) ≈	5.5549e-1 (7.95e-2) ≈	5.2676e-1 (5.38e-2)
WFG1	3	48	1.2513e+0 (9.47e-2) +	1.2236e+0 (1.46e-1) +	1.2408e+0 (6.78e-2) +	1.5849e+0 (7.04e-2)
WFG2	3	48	2.7654e-1 (2.15e-2) −	2.8238e-1 (2.05e-2) −	2.8108e-1 (2.64e-2) −	2.4958e-1 (2.75e-2)
WFG3	3	48	3.1925e-1 (2.43e-2) −	3.1558e-1 (4.59e-2) −	3.1363e-1 (3.88e-2) −	2.5565e-1 (3.47e-2)
WFG4	3	48	3.1918e-1 (1.43e-2) −	3.2702e-1 (2.10e-2) −	3.1987e-1 (1.96e-2) −	2.6904e-1 (1.62e-2)
WFG5	3	48	2.8768e-1 (1.86e-2) +	2.8894e-1 (3.00e-2) +	2.9196e-1 (1.71e-2) +	3.0737e-1 (7.77e-3)
WFG6	3	48	3.4394e-1 (3.49e-2) −	3.6968e-1 (4.57e-2) −	3.7308e-1 (2.73e-2) −	3.1948e-1 (1.81e-2)
WFG7	3	48	3.5342e-1 (3.56e-2) +	3.7009e-1 (3.68e-2) +	3.6724e-1 (4.73e-2) +	4.0745e-1 (9.64e-2)
WFG8	3	48	4.4250e-1 (1.22e-2) −	4.5588e-1 (1.79e-2) −	4.4599e-1 (1.87e-2) −	3.3756e-1 (1.72e-2)
WFG9	3	48	3.1322e-1 (1.72e-2) +	3.2711e-1 (2.96e-2) +	3.1805e-1 (2.56e-2) +	3.8912e-1 (3.97e-2)
+/ − / ≈			9/9/2	7/9/4	7/9/4	

6 Comparison of All Presented Algorithms

To examine the improvements of the named obstacles, we again use pairwise comparisons of IGD results on different benchmark problems. This time, we compare HHX-A, HHX-E, HHX-D and UX. We exclude HHX-S as it is mostly outperformed by HHX-D, and we still have the comparison to UX, to determine whether the new Hyper-Heuristics can compete with it on its best problems.

The results are shown in Table 2. HHX-A-NSGA-II performed better on nine different problems and worse on nine different problems than UX-NSGA-II. In comparison with HHX-D-NSGA-II, an improvement is noticeable on DTLZ3 and DTLZ4 and also on RM1 and RM2, where HHX-A-NSGA-II was the best performing algorithm between those four. HHX-E-NSGA-II, on the other hand, only differs a little from HHX-D-NSGA-II so that we can say it performs nearly equally well. To sum this up, HHX-A-NSGA-II is the best performing variant of all four Hyper-Heuristics. According to the IGD results, HHX-A performs nearly equally good to the UX operator. UX also has weaknesses on different problems, on which HHX-A finds the better option. To give an outlook and the possibilities of Hyper-Heuristics, we also examine the learning behaviour of all four Hyper-Heuristics. In Fig. 4 an exemplary development of the distribution of the scores on WFG5 are visualized.

According to the graphs given in Fig. 2 the use of DEX, LX, and UX led to the best results. It is remarkable, that HHX-D reaches a fixed distribution within the first third. HHX-S, on the other hand, changes its score distribution dramatically after about 40 generations. It weighted the LCX3 and the RSBX more than the UX operator on that point, although both did not perform well in the single usages. The development of the scores of HHX-E is similar to HHX-D. The most considerable changes happen in the first third. After the 50th generation, the selection mechanism changes to selection and the score behaves similar to HHX-S. Since the distribution mechanism weighted the SBX operator highly and the selection mechanism is very sensitive to biases, the SBX operator is still the most rewarded operator in the 90th generation and is solely selected for the last

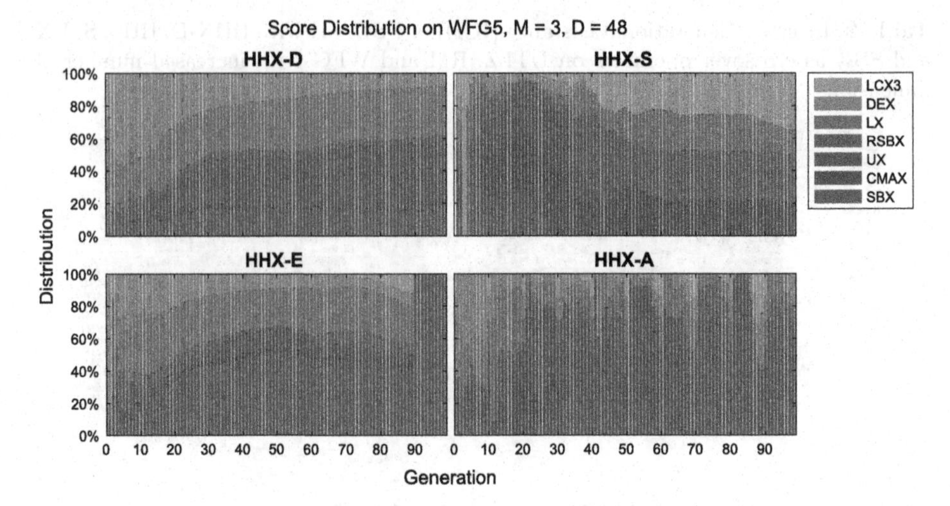

Fig. 4. Distribution of the Scores for each Crossover Operator during solving of WFG5 with NSGA-II using Hyper-Heuristics.

evolution of HHX-E. The HHX-A resets the scores in every 10th generation. Different phases of the problem are here more visible than in the other graphs. In the first iteration, DEX, LX and SBX achieve the highest rewards. In the second iteration the focus lies more on LX, UX and CMAX. In the following iterations, the tendency in choosing UX with some spikes on SBX and LXC3 is recognizable. This matches the trends illustrated in Fig. 2.

Considering the IGD measurements, HHX-A, HHX-D and HHX-E all performed similar well on WFG5. All of them have a distribution trend that matches the IGD trend in Fig. 2. Considering their functionality, HHX-E embraces the learning bias, which is an advantage on this problem because most operators perform well on it. It also resolves the problem with the learning offset, as it finds a good distribution as soon as HHX-D. HHX-A fulfils the expectations, that there is no bias, as it chooses in each phase the best option. The learning offset is also minimized, as the HHX-D gathers the necessary information very fast. Regarding, that there are still problems, where HHX-A does not perform as well as expected, the iterations might be too small. On some problems, the distribution might need more time gathering the information, so a better balancing of iterations and their lengths in this approach could still improve the performance. Nevertheless, HHX-A-NSGA-II offers a very well performing algorithm, that could be used on a variety of problems without knowing their properties.

7 Conclusion

In this paper, we proposed four different forms of Hyper-Heuristics as selectors of crossover operators in NSGA-II. We use two different selection mechanisms:

the selection of one operator per generation and the distribution of the generation to all operators. With the first, we build the HHX-S algorithm and with the latter, we build the HHX-D algorithm. In a first experimental evaluation, we concluded, that the distribution has a well explorative behaviour and the selection has a well exploiting behaviour. We named two different problems: The *Learning Bias* and the *Learning Offset*, which could both be minimized by using a combination of both algorithm. Therefore, we used an evolving approach with three stages (distribution, selection, exploitation) named HHX-E and an alternating approach with resets of the score in every 10th generation, named HHX-A. In a second experimental evaluation, we concluded that HHX-A is the most successful algorithm out of the presented four. It has a fast learning due to the distribution part, but it is not biased in different phases, so that new obstacles are faced without information that are not applicable any more.

We consider HHX-A as a successful Hyper-Heuristic, as it selects well performing crossover operators in every situation. As the *learning offset* is not resolved, it is still a difficulty to outperform the best crossover operator in the pool. Nevertheless, this method prevents the user from deciding about the operator by themselves and makes it easier to work with problems without any known properties.

Our future work is to look more into the other parts of the Hyper-Heuristic. The crossover operators in the pool could be further examined to select a better variety with fewer operators. The smaller the number of operators, the faster the Hyper-Heuristic can learn. Thus, it could be another method to minimize the *learning offset*.

References

1. Beyer, H.G., Deb, K.: On self-adaptive features in real-parameter evolutionary algorithms. IEEE Trans. Evol. Comput. **5**(3), 250–270 (2001). https://doi.org/10.1109/4235.930314
2. Burke, E.K., et al.: Hyper-heuristics: a survey of the state of the art. J. Oper. Res. Soc. **64**(12), 1695–1724 (2013). https://doi.org/10.1057/jors.2013.71
3. Coello Coello, C.A., Reyes Sierra, M.: A study of the parallelization of a coevolutionary multi-objective evolutionary algorithm. In: Monroy, R., Arroyo-Figueroa, G., Sucar, L.E., Sossa, H. (eds.) MICAI 2004. LNCS (LNAI), vol. 2972, pp. 688–697. Springer, Heidelberg (2004). https://doi.org/10.1007/978-3-540-24694-7_71
4. Deb, K., Joshi, D., Anand, A.: Real-coded evolutionary algorithms with parent-centric recombination. In: Proceedings of the 2002 Congress on Evolutionary Computation, CEC 2002 (Cat. No.02TH8600), vol. 1, pp. 61–66 (2002). https://doi.org/10.1109/CEC.2002.1006210
5. Deb, K., Pratap, A., Agarwal, S., Meyarivan, T.: A fast and elitist multiobjective genetic algorithm: NSGA-II. IEEE Trans. Evol. Comput. **6**(2), 182–197 (2002). https://doi.org/10.1109/4235.996017
6. Deb, K., Agrawal, R.B.: Simulated binary crossover for continuous search space. Complex Syst. **9**, 115–148 (1995)
7. Deb, K., Thiele, L., Laumanns, M., Zitzler, E.: Scalable test problems for evolutionary multiobjective optimization. In: Abraham, A., Jain, L., Goldberg, R. (eds.)

Evolutionary Multiobjective Optimization. Advanced Information and Knowledge Processing, pp. 105–145. Springer, London (2005). https://doi.org/10.1007/1-84628-137-7_6
8. Deep, K., Thakur, M.: A new crossover operator for real coded genetic algorithms. Appl. Math. Comput. **188**(1), 895–911 (2007). https://doi.org/10.1016/j.amc.2006.10.047, https://linkinghub.elsevier.com/retrieve/pii/S0096300306014287
9. Drake, J., Kheiri, A., Özcan, E., Burke, E.: Recent advances in selection hyper-heuristics. Eur. J. Oper. Res. **285**, 405–428 (2020). https://doi.org/10.1016/j.ejor.2019.07.073
10. Fritsche, G., Pozo, A.: A hyper-heuristic collaborative multi-objective evolutionary algorithm. In: 2018 7th Brazilian Conference on Intelligent Systems (BRACIS), pp. 354–359 (2018). https://doi.org/10.1109/BRACIS.2018.00068
11. Hansen, N., Ostermeier, A.: Completely derandomized self-adaptation in evolution strategies. Evol. Comput. **9**(2), 159–195 (2001). https://doi.org/10.1162/106365601750190398
12. Hong, L., Drake, J.H., Woodward, J.R., Özcan, E.: Automatically designing more general mutation operators of evolutionary programming for groups of function classes using a hyper-heuristic. In: Proceedings of the Genetic and Evolutionary Computation Conference 2016, pp. 725–732. ACM, Denver Colorado (2016). https://doi.org/10.1145/2908812.2908958, https://dl.acm.org/doi/10.1145/2908812.2908958
13. Huband, S., Barone, L., While, L., Hingston, P.: A scalable multi-objective test problem toolkit. In: Coello Coello, C.A., Hernández Aguirre, A., Zitzler, E. (eds.) EMO 2005. LNCS, vol. 3410, pp. 280–295. Springer, Heidelberg (2005). https://doi.org/10.1007/978-3-540-31880-4_20
14. Ono, I., Kita, H., Kobayashi, S.: A robust real-coded genetic algorithm using unimodal normal distribution crossover augmented by uniform crossover: effects of self-adaptation of crossover probabilities, p. 8
15. Pan, L., Xu, W., Li, L., He, C., Cheng, R.: Adaptive simulated binary crossover for rotated multi-objective optimization. Swarm Evol. Comput. **60**, 100759 (2021). https://doi.org/10.1016/j.swevo.2020.100759, https://www.sciencedirect.com/science/article/pii/S2210650220304120
16. Pang, L.M., Ishibuchi, H., Shang, K.: Using a genetic algorithm-based hyper-heuristic to tune MOEA/D for a set of various test problems. In: 2021 IEEE Congress on Evolutionary Computation (CEC), pp. 1486–1494 (2021). https://doi.org/10.1109/CEC45853.2021.9504748
17. Pillay, N., Qu, R.: Hyper-Heuristics: Theory and Applications. Natural Computing Series, Springer, Cham (2018). https://doi.org/10.1007/978-3-319-96514-7
18. Qingfu Zhang, Aimin Zhou, Yaochu Jin: RM-MEDA: a regularity model-based multiobjective estimation of distribution algorithm. IEEE Trans. Evol. Compu. **12**(1), 41–63 (2008). https://doi.org/10.1109/TEVC.2007.894202, http://ieeexplore.ieee.org/document/4358761/
19. Ross, P.: Hyper-heuristics. In: Burke, E.K., Kendall, G. (eds.) Search Methodologies: Introductory Tutorials in Optimization and Decision Support Techniques, pp. 529–556. Springer, Boston (2005). https://doi.org/10.1007/0-387-28356-0_17
20. Storn, R.: Differential evolution - a simple and efficient adaptive scheme for global optimization over continuous spaces, p. 16
21. Syswerda, G.: Uniform crossover in genetic algorithms (1989)
22. Tian, Y., Cheng, R., Zhang, X., Jin, Y.: PlatEMO: a MATLAB platform for evolutionary multi-objective optimization. IEEE Comput. Intell. Mag. **12**(4), 73–87 (2017)

23. Tsutsui, S., Yamamura, M., Higuchi, T.: Multi-parent recombination with simplex crossover in real coded genetic algorithms, p. 9
24. Venske, S.M., Almeida, C.P., Delgado, M.R.: Comparing selection hyper-heuristics for many-objective numerical optimization. In: 2021 IEEE Congress on Evolutionary Computation (CEC), pp. 1921–1928 (2021). https://doi.org/10.1109/CEC45853.2021.9504934
25. Zitzler, E., Deb, K., Thiele, L.: Comparison of multiobjective evolutionary algorithms: empirical results. Evol. Comput. **8**(2), 173–195 (2000). https://doi.org/10.1162/106365600568202, https://direct.mit.edu/evco/article/8/2/173-195/868

Surrogate-assisted Multi-objective Optimization via Genetic Programming Based Symbolic Regression

Kaifeng Yang(✉) and Michael Affenzeller

School of Informatics, Communications and Media University of Applied Sciences Upper Austria Softwarepark 11, 4232 Hagenberg, Austria
{Kaifeng.Yang,Michael.Affenzeller}@fh-hagenberg.at
https://heal.heuristiclab.com/

Abstract. Surrogate-assisted optimization algorithms are a commonly used technique to solve expensive-evaluation problems, in which a regression model is built to replace an expensive function. In some acquisition functions, the only requirement for a regression model is the predictions. However, some other acquisition functions also require a regression model to estimate the "uncertainty" of the prediction, instead of merely providing predictions. Unfortunately, very few statistical modeling techniques can achieve this, such as Kriging/Gaussian processes, and recently proposed genetic programming-based (GP-based) symbolic regression with Kriging (GP2). Another method is to use a bootstrapping technique in GP-based symbolic regression to estimate prediction and its corresponding uncertainty. This paper proposes to use GP-based symbolic regression and its variants to solve multi-objective optimization problems (MOPs), which are under the framework of a surrogate-assisted multi-objective optimization algorithm (SMOA). Kriging and random forest are also compared with GP-based symbolic regression and GP2. Experiment results demonstrate that the surrogate models using the GP2 strategy can improve SMOA's performance.

Keywords: Multi-objective optimization · Genetic programming · Symbolic regression · Surrogate model

1 Introduction

A common remedy to expensive optimization problems is to replace exact evaluations with approximations learned from past evaluations. Theoretically, any supervised learning techniques can be used for building up surrogate models, for instance, Gaussian processes or Kriging [16], random forest [7], supported vector machine [4], genetic programming based (GP-based) symbolic regression [9], to name a few. Among these techniques, the Kriging/Gaussian process is appealing both in academic studies and in applications due to its estimation of both prediction and the corresponding uncertainty as outputs. The utilization of these

M. Emmerich et al. (Eds.): EMO 2023, LNCS 13970, pp. 176–190, 2023.
https://doi.org/10.1007/978-3-031-27250-9_13

two outputs allows a balancing property of exploitation and exploration in an acquisition function (a.k.a infill criteria in some papers) during the optimization processes.

Compared with the black-box modeling technique of Kriging, the transparent-box modeling technique has been a hot topic recently due to its ability to interpret the relationship between inputs and predictions. Among many techniques of transparent-box modeling, GP-based symbolic regression (SR) attracts many researchers' attention, as it searches the space of all possible mathematical formulas for concise or closed-form mathematical expressions that best fit a dataset [1]. To quantify prediction uncertainty in GP-based symbolic regression, a simple technique is to utilize bootstrapping, which is widely used in the machine learning field. Bootstrapping is usually used in ensemble methods to reduce the bias of a (simple) predictor by using several predictors on different samples of a dataset and averaging their predictions. The set of predictions from each predictor can be used to estimate the properties of a predictor (such as prediction mean and its variance) by measuring those properties when sampling from an approximating distribution. For example, this technique is incorporated into a random forest (RF) in Sequential Model-Based Algorithm Configuration (SMAC) [12], where the variance is calculated using the predictions of the trees of the random forest. Following the same idea, the prediction variance of GP-based symbolic regression can also be estimated by the method of bootstrapping. Such similar works can be found in [2,8,10]. Another technique to quantify the uncertainty of the prediction's residual is achieved in [19], in which a so-called GP-based symbolic regression with Kriging (GP2) was proposed by summing a GP-based symbolic expression and one additional residual term to regulate the predictions of the symbolic expression, where the residual term follows a normal distribution and is estimated by Kriging.

The techniques mentioned above allow GP-based symbolic regression to employ acquisition functions that balance exploration and exploitation during the optimization process. Under the framework of surrogate-assisted multi-objective optimization algorithm (SMOA), this paper utilizes the most recently modeling technique GP2 [19] and GP-based symbolic regression incorporated with a bootstrapping technique as the surrogate models to solve multi-objective optimization problems (MOPs). These modeling techniques are compared with two state-of-the-art modeling techniques, Kriging, and random forest.

The main contribution of this paper is extending the surrogate models that allow an acquisition function's ability to balance exploration and exploitation in the field of SMOA. This paper is structured as follows: Sect. 2 describes the preliminaries of multi-objective optimization problems, and introduces the framework of surrogate-assisted multi-objective optimization algorithm; Sect. 3 explains three different techniques for surrogate models; Sect. 4 introduces the definition of upper confidence bound (UCB) [6,17] and its relating concepts; Sect. 5 shows the parameter settings and discusses the experiment results.

2 Surrogate-Assisted Multi-objective Optimization

This section introduces the preliminaries of the SMOA, including the MOP's definition in Sect. 2.1, and the framework of SMOA in Sect. 2.2.

2.1 Multi-objective Optimization Problem

A continuous MOP[1] is defined as minimizing multiple objective functions simultaneously and can be formulated as:

$$\arg\min_{\mathbf{x}} \quad \mathbf{f}(\mathbf{x}) := \big(f_1(\mathbf{x}), \cdots, f_m(\mathbf{x})\big), \qquad \mathbf{x} \in \mathcal{X} \subseteq \mathbb{R}^d \tag{1}$$

where m is the number of objective functions, f_i stands for the i-th objective functions $f_i : \mathcal{X} \to \mathbb{R}$, $i = 1, \ldots, m$, and $\mathcal{X}$ is a decision vector subset.

2.2 Framework of Surrogate-Assisted Multi-objective Optimization

The fundamental concept of SMOA is to firstly build a surrogate model $\mathcal{M}$ to reflect the relationship between a decision vector $\mathbf{x} = (x_1, \cdots, x_d)$ and its each corresponding objective value $y_i = f_i(\mathbf{x}), i \in \{1, \cdots, m\}$ for each objective. In SMOA, it is usually assumed that objective functions are mutually independent in an objective space[2].

SMOA starts with sampling an initial design of experiment (DoE) with a size of η (line 2 in Algorithm 1), $\mathrm{X} = \{\mathbf{x}^{(1)}, \mathbf{x}^{(2)}, \ldots, \mathbf{x}^{(\eta)}\} \subseteq \mathcal{X}$. By using the initial DoE, X and its corresponding objective values, $\mathbf{Y} = \{\mathbf{f}(\mathbf{x}^{(1)}), \mathbf{f}(\mathbf{x}^{(2)}), \cdots, \mathbf{f}(\mathbf{x}^{(\eta)})\} \subseteq \mathbb{R}^{m \times \eta}$ (line 3 in Algorithm 1), can be then utilized to construct surrogate models $\mathcal{M}_i$. Then, an SMOA starts with searching for a decision vector set $\mathbf{x}'$ in the search space $\mathcal{X}$ by maximizing the acquisition function $\mathscr{A}$ with parameters of γ and surrogate models $\boldsymbol{\mathcal{M}}$ (line 7 in Algorithm 1). Here, an acquisition function quantifies how good or bad a point $\mathbf{x}$ is in objective space. In this paper, $\mathscr{A}$ is the upper confidence bound (described in Sec. 4) due to its ability to balance exploration and exploitation, low computational complexity, and popularity in the deep learning fields.

By maximizing the acquisition function, a single-objective optimization algorithm is deployed to search for the optimal decision vector $\mathbf{x}^*$. In this paper, we use BI-population CMA-ES to search for the optimal $\mathbf{x}^*$ due to the favorable performance on BBOB function testbed [11]. The optimal decision vector $\mathbf{x}^*$ will then be evaluated by the 'true' objective functions $\mathbf{f}$. The surrogate models $\boldsymbol{\mathcal{M}}$ will be retrained by using the updated X and $\mathbf{Y}$. The main loop (as shown in Algorithm 1 from line 6 to line 12) will not stop until a stopping criterion is satisfied. Common stopping criteria include a number of iterations, convergence velocity, etc. In this paper, we specify the stopping criterion (T_c) as a number of function evaluations.

[1] Constraints are not considered in this paper.

[2] Recently, some work has considered the dependency in constructing the surrogate models, such as the so-called dependent Gaussian processes [3]. This paper does not consider a dependency between objectives for simplicity.

Algorithm 1: Surrogate-assisted Multi-objective Optimization Algorithm

```
1  SMOA(f, 𝒜, 𝒳, ℳ, γ, η, T_c)
   /* f: objective functions, 𝒜: acquisition function, 𝒳:
      search space, γ: parameters of 𝒜, ℳ: a surrogate model
      to train, T_c: maximum number of function evaluation     */
2  Generate the initial DoE: X = {x^(1), x^(2), ..., x^(η)} ⊂ 𝒳;
3  Evaluate Y ← {f(x^(1)), f(x^(2)), ..., f(x^(η))};
4  Train surrogate models ℳ_i on (X, Y_i), where i ∈ {1, ..., m};
5  g ← η;
6  while g < T_c do
7      x* ← arg max_x 𝒜(x; 𝓜, γ), where 𝓜 = {ℳ_1, ..., ℳ_m} and x ∈ 𝒳;
8      Y* ← f(x*);
9      X ← X ∪ {x*};
10     Y ← Y ∪ {Y*};
11     Re-train the surrogate models ℳ_i on (X, Y_i), where i ∈ {1, ..., m};
12     g ← g + 1
```

3 Surrogate Models

This section introduces a common surrogate model, Kriging in Sect. 3.1; a canonical GP-based symbolic regression in Sect. 3.2; and the GP-based symbolic regression with Kriging (GP2) in Sect. 3.3. Additionally, the method of quantifying prediction uncertainty is introduced in each subsection.

3.1 Kriging

Kriging is a statistical interpolation method and has been proven to be a popular surrogate model to approximate noise-free data in computer experiments. Considering a realization of y at n locations, expressed as the following vector $\mathbf{f}(\mathbf{X}) = (f(\mathbf{x}^{(1)}) \cdots, f(\mathbf{x}^{(n)}))$ and $\mathbf{X} = \{\mathbf{x}^{(1)}, \cdots, \mathbf{x}^{(n)}\} \subseteq \mathcal{X}$, Kriging assumes $\mathbf{f}(\mathbf{X})$ to be a realization of a random process f of the following form [5,13]:

$$f(\mathbf{x}) = \mu(\mathbf{x}) + \epsilon(\mathbf{x}), \tag{2}$$

where $\mu(\mathbf{x})$ is the estimated mean value and $\epsilon(\mathbf{x})$ is a realization of a Gaussian process with zero mean and variance σ^2.

The regression part $\mu(\mathbf{x})$ approximates the function $f(.)$ globally. The Gaussian process $\epsilon(\mathbf{x})$ takes local variations into account and estimates the uncertainty quantification. The correlation between the deviations at two decision vectors ($\mathbf{x}$ and $\mathbf{x}'$) is defined as:

$$Corr[\epsilon(\mathbf{x}), \epsilon(\mathbf{x}')] = R(\mathbf{x}, \mathbf{x}') = \prod_{i=1}^{m} R_i(x_i, x_i'),$$

where $R(.,.)$ is the correlation function that decreases with the distance between two points ($\mathbf{x}$ and $\mathbf{x}'$).

It is common practice to use a Gaussian correlation function (a.k.a., a squared exponential kernel):

$$R(\mathbf{x}, \mathbf{x}') = \prod_{i=1}^{m} \exp(-\theta_i (x_i - x_i')^2),$$

where $\theta_i \geq 0$ are parameters of the correlation model to reflect the variables' importance [21]. The covariance matrix can then be expressed through the correlation function: $Cov(\epsilon(\mathbf{x})) = \sigma^2 \mathbf{\Sigma}$, where $\mathbf{\Sigma}_{i,j} = R(\mathbf{x}_i, \mathbf{x}_j)$.

When $\mu(\mathbf{x})$ is assumed to be an unknown constant, the unbiased prediction is called ordinary Kriging (OK). In OK, the Kriging model determines the hyperparameters $\theta = (\theta_1, \theta_2, \cdots, \theta_n)$ by maximizing the likelihood over the observed dataset. The expression of the likelihood function is: $L = -\frac{n}{2}\ln(\sigma^2) - \frac{1}{2}\ln(|\mathbf{\Sigma}|)$.

Uncertainty Quantification: The maximum likelihood estimations of the mean $\hat{\mu}$ and the variance $\hat{\sigma}^2$ can be expressed by: $\hat{\mu} = \frac{\mathbf{1}_n^\top \mathbf{\Sigma}^{-1}\mathbf{y}}{\mathbf{1}_n^\top \mathbf{\Sigma}^{-1}\mathbf{1}_n}$ and $\hat{\sigma}^2 = \frac{1}{n}(\mathbf{y} - \mathbf{1}_n\hat{\mu})^\top \mathbf{\Sigma}^{-1}(\mathbf{y} - \mathbf{1}_n\hat{\mu})$. Then the prediction of the mean and the variance of an OK's prediction at a target point $\mathbf{x}^t$ can be derived as follows [13,20]:

$$\mu_{ok}(\mathbf{x}^t) = \hat{\mu} + \mathbf{c}^\top \mathbf{\Sigma}^{-1}(\mathbf{y} - \hat{\mu}\mathbf{1}_n), \tag{3}$$

$$\sigma^2_{ok}(\mathbf{x}^t) = \hat{\sigma}^2[1 - \mathbf{c}^\top \mathbf{\Sigma}^{-1}\mathbf{c} + \frac{1 - \mathbf{c}^\top \Sigma^\top \mathbf{c}}{\mathbf{1}_n^\top \Sigma^{-1}\mathbf{1}_n}], \tag{4}$$

where $\mathbf{c} = (Corr(y(x^t), y(x_1)), \cdots, Corr(y(x^t), y(x_n)))^\top$.

3.2 Genetic Programming Based Symbolic Regression

Genetic programming (GP) [15] is a typical evolutionary algorithm to solve optimization problems that can be formulated as: $\arg\min_{\mathbf{x}} f(\mathbf{x}), \mathbf{x} \in \mathcal{X}$, where $\mathbf{x} = (x_1, \cdots, x_n)$ represents a decision vector (also known as individual) in evolutionary algorithms (EAs). Similar to other EAs, GP evolves a population of solution candidates *Pop*(.) by following the principle of the survival of the fittest and utilizing biologically-inspired operators. Algorithm 2 is the pseudocode of GP, where *Variation* operator includes crossover/recombination and mutation operators.

The feature that distinguishes GP from other evolutionary algorithms is the variable-length representation for $\mathbf{x}$. The search space of GP-based symbolic regression problems is a union of a function space $\mathcal{F}$ and a terminal space $\mathcal{T} = \{\mathcal{S} \cup \mathbb{R}\}$, where $\mathcal{S}$ represents a variable symbol space (e.g., x_1) and $c \in \mathbb{R}$ represents a constant number. The function set $\mathcal{F}$ includes available symbolic functions (e.g., $\{+, -, \times, \div, \exp, \log\}$).

To distinguish the objective function and decision variables in MOPs, $f_{sr}(\cdot)$ and $\mathbf{t}$ are used to represent the objective function in GP-based symbolic regression and a GP individual for symbolic regression problems, respectively. The

Algorithm 2: Pseudocode of canonical genetic programming

Input: objective function $f(.)$, population size n_{pop}, crossover rate p_c, mutation rate p_m, maximum number of generation max_g ;
Output: optimal solution x^*

1 $g \longleftarrow 0$;
2 *Pop*(g) $\longleftarrow$ InitializePopulation (n_{pop}) ;
3 Evaluate(*Pop*(g), f);
4 **while** *termination criterion is not satisfied* **do**
5   *Pop*$'$(g) $\longleftarrow$ MatingSelection(*Pop*(g)) ;
6   *Pop*$''$(g) $\longleftarrow$ Variation(*Pop*$'$(g), p_c, p_m) ;
7   Evaluate(*Pop*$''$(g), f) ;
8   *Pop*($g+1$) $\longleftarrow$ EnviornmentalSelection(*Pop*$''$(g) $\cup$ *Pop*(g)) ;
9     $g \longleftarrow g+1$;
10 $x^* \longleftarrow \arg\min f(x)$ where $x \in$ *Pop*(g);

objective function $f_{sr}(\cdot)$ can be any performance metrics for supervised learning problems. For instance, *mean squared error* (MSE), the *Pearson correlation coefficient* (PCC), *Coefficient of determination* (R2), etc. In this paper, MSE, PCC, and R2 are used as the objective functions in GP-based symbolic regression, and they are defined as:
$\text{MSE} = \frac{1}{n}\sum_{i=1}^{n}(y_i - \hat{y}_i)^2$, $\text{PCC} = \frac{\sum_{i=1}^{n}(y_i-\bar{y})(\hat{y}_i-\bar{\hat{y}})}{\sqrt{\sum_{i=1}^{n}(y_i-\bar{y})^2}\sqrt{\sum_{i=1}^{n}(\hat{y}_i-\bar{\hat{y}})^2}}$, $\text{R2} = 1 - \frac{\sum_{i=1}^{n}(y_i-\hat{y}_i)^2}{\sum_{i=1}^{n}(y_i-\bar{y})^2}$,
where n is the number of samples, y_i and $\hat{y}_i$ represent the real response and predicted response of the i-th sample, $\bar{y} = \frac{1}{n}\sum_{i=1}^{n} y_i$, and $\bar{\hat{y}} = \frac{1}{n}\sum_{i=1}^{n} \hat{y}_i$.

Therefore, a GP-based symbolic regression problem in this paper can be well formulated as:

$$\begin{aligned}
&\arg\max_{\mathbf{t}} \ f_{sr}(\mathbf{t}), \quad t_i \in \{\mathcal{F} \cup \mathcal{T}\}|_{i=1,2,\cdots} && \text{if} \quad f_{sr} \in \{\text{PCC}, \text{R2}\} \ \text{or} \\
&\arg\min_{\mathbf{t}} \ f_{sr}(\mathbf{t}), \quad t_i \in \{\mathcal{F} \cup \mathcal{T}\}|_{i=1,2,\cdots} && \text{if} \quad f_{sr} = \text{MSE}.
\end{aligned}$$

Uncertainty Quantification: Similar to other bootstrapping techniques, each individual in the population of the GP-based symbolic regression is trained on a list of bootstrap samples. The variance error is estimated according to the error on the bootstrap samples. In this paper, the detailed procedure of bootstrapping for GP-based symbolic regression is as follows:

1. Resampling $n_b \leq n$ samples (to form a new dataset D') from original training dataset D of n samples;
2. Train the GP-based symbolic regression model ($\mathbf{t}$) based on dataset D';
3. Repeat (1) to (2) for n_{pop} times;

4. The bagging prediction and variance at a query vector $\mathbf{x}^t$ are then computed by the mean and variance over all n_{pop} predictions:

$$\mu_{sr}(\mathbf{x}^t) = \frac{1}{n_{pop}} \sum_{i=1}^{n_{pop}} \mathbf{t}_i(\mathbf{x}^t) \tag{5}$$

$$\sigma_{sr}^2(\mathbf{x}^t) = \frac{1}{n_{pop}} \sum_{i=1}^{n_{pop}} \left(\mathbf{t}_i(\mathbf{x}^t) - \mu_{sr}(\mathbf{x}^t)\right)^2 \tag{6}$$

3.3 GP-Based Symbolic Regression with Kriging

GP-based symbolic regression with Kriging (GP2) integrates GP-based symbolic regression and Kriging by summing a best-fitted expression ($symreg(\mathbf{x})$) on a training dataset and a residual term. The prediction of GP2 can be expressed as:

$$\mathrm{GP}^2_{symreg}(\mathbf{x}) = symreg(\mathbf{x}) + res(\mathbf{x}), \tag{7}$$

where $res(\mathbf{x})$ is the residual-regulating term that follows a normal distribution and is obtained by Kriging.

The pseudocode of the GP2 is shown in Algorithm 3. The algorithm starts from generating the training dataset D by computing the real responses $\mathbf{Y}$ of the decision set X. The best-fitted symbolic expression $symreg(\mathbf{x})$ on dataset D is generated by GP-based symbolic regression (at line 4 in Algorithm 3). The residuals of the best-fitted symbolic expression $symreg(\mathbf{x})$ on $\mathbf{X}$ can be computed by $\mathbf{Y} - \hat{\mathbf{Y}}$. $\mathbf{X}_{gp}$ (line 6 in Algorithm 3) represents the variables that affect the residuals of $symreg(\mathbf{x})$. The most intuitive way is to set $\mathbf{X}_{gp}$ as $\mathbf{X}$. In addition, it is also reasonable to set $\mathbf{X}_{gp}$ as $\hat{\mathbf{Y}}$ and $\{\mathbf{X} \cup \hat{\mathbf{Y}}\}$, due to the fact that residual of $symreg(\mathbf{x})$ highly depends on the the prediction $\hat{\mathbf{Y}}$. Then, Kriging models the relationship between the $\mathbf{X}_{gp}$ and the residuals of $symreg(\mathbf{x})$ in the dataset D (at line 7 in Algorithm 3). The residual of $symreg(\mathbf{x})$, noted as $res(\mathbf{x})$, also depends on $\mathbf{X}$, as $\mathbf{X}_{gp}$ depends on $\mathbf{X}$. The final surrogate model built by the GP2, noted as $\mathrm{GP}^2_{symreg}(\mathbf{x})$, is the sum of $symreg(\mathbf{x})$ and its corresponding residual distribution $res(\mathbf{x})$.

Uncertainty Quantification: The prediction of the symbolic expression of GP2 is a sum of symbolic expression and residual term that follows a normal distribution. The prediction and variance at a query vector $\mathbf{x}^t$ by using the GP2 strategy are:

$$\mu_{GP2}(\mathbf{x}^t) = symreg(\mathbf{x}^t) + \mu_{ok}(\mathbf{x}^t), \tag{8}$$

$$\sigma^2_{GP2}(\mathbf{x}^t) = s^2_{sr}(\mathbf{x}^t) + \sigma^2_{ok}(\mathbf{x}^t), \tag{9}$$

where $s^2_{sr}(\mathbf{x}^t)$ is the variance of the symbolic regression expression $symreg(\mathbf{x})$ at $\mathbf{x}^t$.

In this paper, we argued $s^2_{sr}(\mathbf{x}) = 0$ due to the fact that $symreg(\mathbf{x})$ is an explicit mathematical expression and there is no variance to $symreg(\mathbf{x})$ at an arbitrary point $\mathbf{x}$. Therefore, Eq. (9) can be simplified as $\sigma^2_{GP2}(\mathbf{x}^t) = \sigma^2_{ok}(\mathbf{x}^t)$.

Algorithm 3: GP2-based symbolic regression

Input: test problem $y(.)$, objective function $f(.)$, function set $\mathcal{F}$, terminal space $\mathcal{T}$;
Output: GP^2_{symreg}
1 $\mathbf{X} \longleftarrow \{\mathbf{x}^{(1)}, \cdots, \mathbf{x}^{(n)}\}$;
2 $\mathbf{Y} \longleftarrow \{y(\mathbf{x}^{(1)}), \cdots, y(\mathbf{x}^{(n)})\}$;
3 $D \longleftarrow \{\mathbf{X}, \mathbf{Y}\}$;
4 $symreg(\mathbf{x}) \longleftarrow \arg\max_{\mathbf{t}} f_{sr}(\mathbf{t}; D)$ or $\arg\min_{\mathbf{t}} f_{sr}(\mathbf{t}; D)$;
5 $\hat{\mathbf{Y}} \longleftarrow \{symreg(\mathbf{x}^{(1)}), \cdots, symreg(\mathbf{x}^{(n)})\}$;
6 $\mathbf{X}_{gp} \longleftarrow$ Choosing$(\mathbf{X}, \hat{\mathbf{Y}})$;
7 $res(\mathbf{x}) \longleftarrow$ Kriging$(\{\mathbf{X}_{gp}, \mathbf{Y} - \hat{\mathbf{Y}}\})$;
8 $\mathrm{GP}^2_{symreg}(\mathbf{x}) \longleftarrow symreg(\mathbf{x}) + res(\mathbf{x})$;

4 Acquisition Functions

The *Hypervolume Indicator* (HV), introduced in [22], is an essentially unary indicator for evaluating the quality of a Pareto-front approximation set in MOPs due to its *Pareto compliant* property[3] and no requirement of in-advance knowledge of the Pareto front. The maximization of HV leads to a high-qualified and diverse Pareto-front approximation set. The Hypervolume Indicator is defined as follows:

$$\mathrm{HV}(\mathcal{PF}, \mathbf{r}) := \lambda_m(\cup_{\mathbf{y} \in \mathcal{PF}}[\mathbf{y}, \mathbf{r}]), \tag{10}$$

where λ_m is the Lebesgue measure and the reference point $\mathbf{r}$ clips the space to ensure a finite volume.

The *Hypervolume improvement* (HVI) measures the improvement of a new solution in terms of HV and is defined as:

$$\mathrm{HVI}(\mathbf{y}; \mathcal{PF}, \mathbf{r}) := \mathrm{HV}(\mathcal{PF} \cup \{\mathbf{y}\}; \mathbf{r}) - \mathrm{HV}(\mathcal{PF}; \mathbf{r})$$

The *Upper confidence bound of the HVI* (UCB) [6,17] is an indicator based on the HVI and prediction uncertainty in a naïve way. Denoting $\boldsymbol{\mu}(\mathbf{x}) = (\mu_1(\mathbf{x}), \ldots, \mu_m(\mathbf{x}))$ as the prediction and $\boldsymbol{\sigma}(\mathbf{x}) = (\sigma_1(\mathbf{x}), \ldots, \sigma_m(\mathbf{x}))$ as the uncertainty of m independent surrogate models, the hypervolume improvement of upper confidence bound of a solution at $\mathbf{x}$, i.e., $\boldsymbol{\mu}(\mathbf{x}) - \omega\boldsymbol{\sigma}(\mathbf{x})$, $\omega \in \mathbb{R}_{\geq 0}$ specifies the confidence level:

$$\mathrm{UCB}(\mathbf{x}; \mathcal{PF}, \mathbf{r}, \omega) := \mathrm{HVI}(\boldsymbol{\mu}(\mathbf{x}) - \omega\boldsymbol{\sigma}(\mathbf{x}); \mathcal{PF}, \mathbf{r}). \tag{11}$$

The parameter ω controls the balancing weight between exploration and exploitation, and a positive ω rewards the high prediction uncertainty.

Example 1. Two examples of the landscape of HVI and UCB are shown in Fig. 1. The Pareto-front approximation set is denoted as $\mathcal{PF} = \{\mathbf{y}^{(1)} = (1, 3),\ \mathbf{y}^{(2)} =$

[3] $\forall A, B \subseteq \mathbb{R}^m : \mathrm{HV}(A, \mathbf{r}) > \mathrm{HV}(B, \mathbf{r}) \implies A \prec B$.

$(2,2), \mathbf{y}^{(3)} = (3,1)\}$, the reference point is $\mathbf{r} = (4,4)$, and the same uncertainty ($\boldsymbol{\sigma} = (1,1)$) is used for every point in the objective space, and ω is set as 1 for the UCB. The left and right figures show the landscape of HVI and UCB, respectively, in the objective space that ranges from -1 to 5 for both f_1 and f_2.

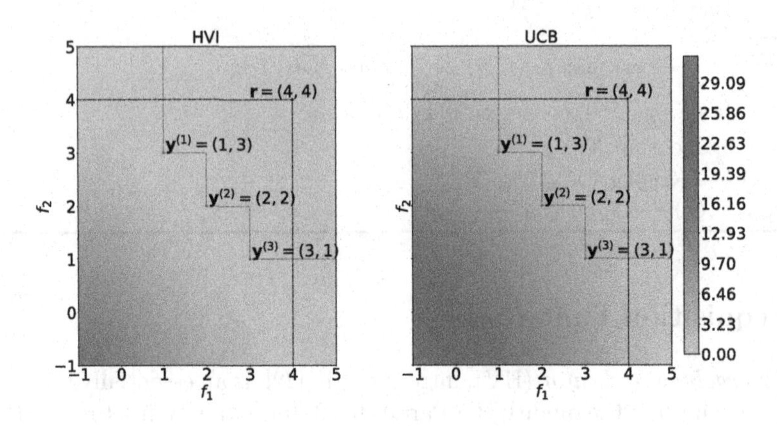

Fig. 1. Landscape of HVI and UCB.

5 Experiments

In this paper, the algorithms are performed on ZDT series problems [23] with the parameters of $d = 5$, $m = 2$, and a fixed reference point $\mathbf{r} = [15, 15]$. In this paper, four categories of surrogate models are compared, including Kriging (in Alg. I), five different GP2 variants by using different configurations (in Alg. II – VI), random forest (in Alg. VII), and GP-based symbolic regression (in Alg. VIII and IX). Each experiment consists of 15 independent runs.

5.1 Parameter Settings

The common parameters in all algorithms include DoE size $\eta = 30$ and the function evaluation budget $T_c = 200$. The stopping criteria of f_{sr} is 0.1, 0.99, 0.99 for MSE, PCC, and R2 metrics, respectively. The function evaluation of genetic programming in GP-based symbolic regression and GP2 is 4E3 [19] and 2E4 [18], respectively.

The parameter ω of UCB is set as $\sqrt{g/\log(g)}$ [14], where g is the number of function evaluation. The function space $\mathcal{F}$ for symbolic regression is $\{+, -, \times, \div, \sin, \cos, \text{pow}\}$. The acquisition function is the UCB and is optimized by CMA-ES in all algorithms. The hyper-parameters of CMA-ES follow: the number of restarts is 3, the number of initialized population size is 7, the maximum generation number is 2000, and the remaining parameters are set in default.

Table 1 shows other parameter settings in each algorithm, where n_{pop} represents the number of estimators in a random forest. The other parameters (e.g., p_m, p_c, etc.) are set as default values in GPLearn [18].

Table 1. Algorithm Parameter Configuration

	Alg. I	Alg. II	Alg. III	Alg. IV	Alg. V	Alg. VI	Alg. VII	Alg. VIII	Alg. IX
$\mathcal{M}$	Kriging	GP2	GP2	GP2	GP2	GP2	RF	GP-based SR	GP-based SR
n_{pop}	n.a	400	400	400	400	400	100	100	100
$\mathbf{X}_{gp}$	n.a	$\mathbf{X}$	$\mathbf{X}$	$\mathbf{X}$	$\{\mathbf{X} \cup \hat{\mathbf{Y}}\}$	$\mathbf{Y} - \hat{\mathbf{Y}}$	n.a	n.a	n.a
f_{sr}	n.a	MSE	PCC	R2	R2	PCC	n.a	MSE	R2

5.2 Empirical Results

In this section, we evaluate the Pareto-front approximation sets by using the HV indicator. Figure 2 shows the HV convergence curves of 15 independent runs on ZDT problems. To depict a slight difference in HV values, we discuss the average $\log(\Delta HV)$ convergence of HV relative to the reference HV value of 250 in this paper. Notice that the scale varies in different problems since all problems have different ranges of the Pareto fronts.

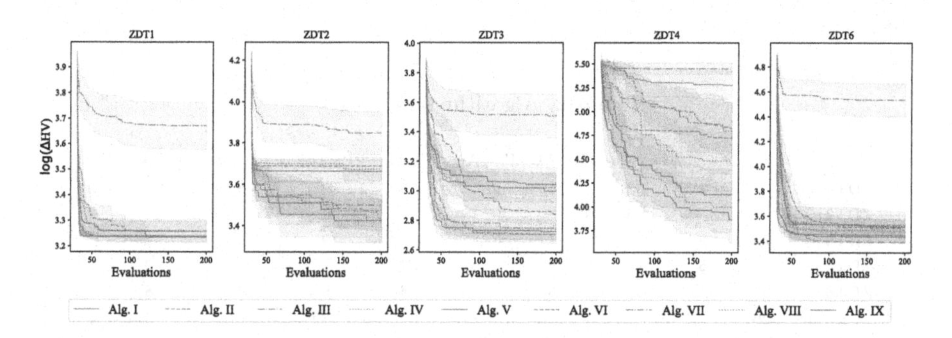

Fig. 2. Average log(ΔHV) convergence on ZDT1 – ZDT6. The shaded region represents the variance.

From Fig. 2, it is clear to see that Alg. VII of using the random forest as the surrogate model yields the worst experiment results w.r.t. mean HV and its corresponding standard deviation. Compared with Alg. I and other algorithms, the poor performance of Alg. VII is expected, because of the introduction of Kriging in other algorithms (except for Alg. VIII and IX), of which prediction is more accurate than that of the random forest when the training dataset has few samples. Compared with GP-based symbolic regression algorithms (Alg. VIII and IX), which use 'plain' GP, Alg. VII still can not outperform them, due to the efficient optimization mechanism in genetic programming.

On single-modal convex Pareto-front problem ZDT1, Fig. 2 shows that GP-based symbolic regression algorithms (Alg. VIII and IX) are only slightly better than random forest in Alg. VII, but are worse than the other algorithms of using Kriging. The performances of the algorithms (Alg. I – VI) are similar w.r.t. mean HV value over 15 runs. In addition, Alg. I, which merely uses Kriging as the surrogate model, converges the fastest at the early stage of optimization among all the test algorithms. Similarly, Alg. I also converges much faster than most of the test algorithms on the other two single-modal concave Pareto-front problems (ZDT2 and ZDT6). An explanation is the relatively simple landscape of these three test problems.

On discontinued Pareto-front problem ZDT3 and a multi-modal Pareto-front problem ZDT4, the algorithms' performance differs greatly. Firstly, the convergence of Alg. I is much slower than the other algorithms at the beginning stage of optimization. Since a relatively large[4] reference point is used in this paper, the HVI introduced by extreme solutions is definitely much larger than that of a knee point. Thus, we can conclude that the strategy of Kriging with UCB is difficult to search for extreme solutions at the early stage of optimization. Consequently, this will slow the convergence of Alg. I. Additionally, another intuitive explanation is that the landscape of the UCB based on Kriging models on ZDT3 and ZDT4 is not smooth enough to allow the optimizer CMA-ES to jump out of local optima. In other words, the UCB is hard to measure slight differences around the local optima based on Kriging's prediction because of the landscape's roughness.

Table 2. Algorithms ranking.

	Alg. I	Alg. II	Alg. III	Alg. IV	Alg. V	Alg. VI	Alg. VII	Alg. VIII	Alg. IX
ZDT1	6	2	3	5	1	4	9	8	7
ZDT2	2	4	5	3	1	8	9	7	6
ZDT3	7	2	4	1	3	5	9	6	8
ZDT4	8	6	5	2	1	7	9	4	3
ZDT6	3	7	1	5	2	4	9	6	8
Sum Rank	26	21	18	16	**8**	28	45	31	32

All algorithms' ranking (w.r.t. mean HV value over 15 runs) on each problem is shown in Table 2. From this table, it is easy to conclude that Alg. V, which utilizes GP2 method of $\mathbf{X}_{gp} = \{\mathbf{X} \cup \hat{\mathbf{Y}}\}$ and $f_{sr} = \text{R2}$, outperforms the other algorithms, and Alg. IV (GP2 method of $\mathbf{X}_{gp} = \{\mathbf{X}\}$ and $f_{sr} = \text{R2}$) is the second best algorithm. Since the search space of $\mathbf{x}$ of surrogate models in Alg. IV is the same search space of the objective function $\mathbf{f}(\mathbf{x})$ of MOPs, Alg. IV

[4] The ideal Pareto fronts on ZDT problems hold: $\max\left(f_1(\mathbf{x})\right) \leq 1, \max\left(f_2(\mathbf{x})\right) \leq 1$.

is more reasonable to interpret the residual regulation term $res(\mathbf{x})$ (in Eq. (7)) when it is compared with Alg. V. Additionally, there is no significant difference between the performance of Alg. IV and Alg. V on the test problems, see Table 3 for details. Considering these two aspects, it is recommended to use Alg. IV when the acquisition function is the UCB.

Table 3. The pairwise Wilcoxon's Rank-Sum test ($+/\approx/-$) matrix at a 0.05 significance level was performed among all the test algorithms, where algorithms in all the columns (except the first column) are pairwise compared with the algorithms in the first column.

	Alg. I	Alg. II	Alg. III	Alg. IV	Alg. V	Alg. VI	Alg. VII	Alg. VIII	Alg. IX
Alg. I	0/0/0	2/3/0	3/2/0	3/2/0	3/2/0	1/3/1	0/1/4	1/2/2	2/2/1
Alg. II	0/3/2	0/0/0	1/4/0	2/2/1	2/3/0	0/3/2	0/1/4	0/3/2	1/1/3
Alg. III	0/2/3	0/4/1	0/0/0	1/3/1	2/2/1	0/4/1	0/1/4	0/2/3	1/1/3
Alg. IV	0/2/3	1/2/2	1/3/1	0/0/0	0/5/0	0/3/2	0/0/5	0/4/1	0/3/2
Alg. V	0/2/3	0/3/2	1/2/2	0/5/0	0/0/0	0/1/4	0/0/5	0/1/4	0/2/3
Alg. VI	1/3/1	2/3/0	1/4/0	2/3/0	4/1/0	0/0/0	0/1/4	0/4/1	1/3/1
Alg. VII	4/1/0	4/1/0	4/1/0	5/0/0	5/0/0	4/1/0	0/0/0	4/1/0	5/0/0
Alg. VIII	2/2/1	2/3/0	3/2/0	1/4/0	4/1/0	1/4/0	0/1/4	0/0/0	1/3/1
Alg. IX	1/2/2	3/1/1	3/1/1	2/3/0	3/2/0	1/3/1	0/0/5	1/3/1	0/0/0
Sum of $+/\approx/-$	8/17/15	14/20/6	17/19/4	16/22/2	**23/16/1**	7/22/11	0/5/35	6/20/14	11/15/14

Table 3 shows the performance of pairwise Wilcoxon's Rank-Sum test ($+/\approx/-$) matrix among all test algorithms on the test problems. The sum of Wilcoxon's Rank-Sum test (sum of $+/\approx/-$) confirms that Alg. V performs best, as it significantly outperforms 23 pairwise instances between algorithms and problems. Besides, compared with Algo. I, most of the GP2 methods (Alg. II - Alg. V) perform at least similar to or significantly better than Alg. I.

6 Conclusion and Future Work

This paper utilizes two different techniques to quantify the prediction uncertainty of GP-based symbolic regression, namely, a bootstrapping technique in GP-based symbolic regression and using the Kriging to estimate the residual uncertainty in GP2. Two surrogate modeling techniques, GP-based symbolic regression, and GP2 of different configurations, are compared with Kriging and random forest on five state-of-the-art MOP benchmarks. The statistical results show the effectiveness of GP2 w.r.t. mean ΔHV convergence. Among five different variants of GP2, we recommend the usage of the GP2 by employing $\mathbf{X}$ or $\{\mathbf{X} \cup \hat{\mathbf{Y}}\}$ to predict residuals by setting R2 as the fitness function in genetic programming, because of their good statistical performance.

For future work, it is recommended to compare the performance of bootstrapping and Jackknife to estimate the variance of GP-based symbolic regression. It is also worthwhile to investigate the performance of the proposed methods on real-world applications.

Acknowledgments. This work is supported by the Austrian Science Fund (FWF – Der Wissenschaftsfonds) under the project (I 5315, 'ML Methods for Feature Identification Global Optimization).

References

1. Affenzeller, M., Winkler, S.M., Kronberger, G., Kommenda, M., Burlacu, B., Wagner, S.: Gaining deeper insights in symbolic regression. In: Riolo, R., Moore, J.H., Kotanchek, M. (eds.) Genetic Programming Theory and Practice XI. GEC, pp. 175–190. Springer, New York (2014). https://doi.org/10.1007/978-1-4939-0375-7_10
2. Agapitos, A., Brabazon, A., O'Neill, M.: Controlling overfitting in symbolic regression based on a bias/variance error decomposition. In: Coello, C.A.C., Cutello, V., Deb, K., Forrest, S., Nicosia, G., Pavone, M. (eds.) PPSN 2012. LNCS, vol. 7491, pp. 438–447. Springer, Heidelberg (2012). https://doi.org/10.1007/978-3-642-32937-1_44
3. Álvarez, M.A., Rosasco, L., Lawrence, N.D.: Kernels for Vector-Valued Functions. Review. Found. Trends Mach. Learn. **4**(3), 195–266 (2012). https://doi.org/10.1561/2200000036
4. Andres, E., Salcedo-Sanz, S., Monge, F., Pellido, A.: Metamodel-assisted aerodynamic design using evolutionary optimization. In: EUROGEN (2011)
5. Žilinskas, A., Mockus, J.: On one Bayesian method of search of the minimum. Avtomatica i Vicheslitel'naya Teknika **4**, 42–44 (1972)
6. Emmerich, M.T.M., Yang, K., Deutz, A.H.: Infill criteria for multiobjective Bayesian optimization. In: Bartz-Beielstein, T., Filipič, B., Korošec, P., Talbi, E.-G. (eds.) High-Performance Simulation-Based Optimization. SCI, vol. 833, pp. 3–16. Springer, Cham (2020). https://doi.org/10.1007/978-3-030-18764-4_1
7. Feurer, M., Klein, A., Eggensperger, K., Springenberg, J.T., Blum, M., Hutter, F.: Auto-sklearn: efficient and robust automated machine learning. In: Hutter, F., Kotthoff, L., Vanschoren, J. (eds.) Automated Machine Learning. TSSCML, pp. 113–134. Springer, Cham (2019). https://doi.org/10.1007/978-3-030-05318-5_6
8. Fitzgerald, J., Azad, R.M.A., Ryan, C.: A bootstrapping approach to reduce overfitting in genetic programming. In: Proceedings of the 15th Annual Conference Companion on Genetic and Evolutionary Computation, pp. 1113–1120. GECCO 2013 Companion, Association for Computing Machinery, New York, NY, USA (2013). https://doi.org/10.1145/2464576.2482690
9. Fleck, P., et al.: Box-type boom design using surrogate modeling: introducing an industrial optimization benchmark. In: Andrés-Pérez, E., González, L.M., Periaux, J., Gauger, N., Quagliarella, D., Giannakoglou, K. (eds.) Evolutionary and Deterministic Methods for Design Optimization and Control With Applications to Industrial and Societal Problems. CMAS, vol. 49, pp. 355–370. Springer, Cham (2019). https://doi.org/10.1007/978-3-319-89890-2_23
10. Folino, G., Pizzuti, C., Spezzano, G.: Ensemble techniques for parallel genetic programming based classifiers. In: Ryan, C., Soule, T., Keijzer, M., Tsang, E., Poli, R., Costa, E. (eds.) EuroGP 2003. LNCS, vol. 2610, pp. 59–69. Springer, Heidelberg (2003). https://doi.org/10.1007/3-540-36599-0_6
11. Hansen, N.: Benchmarking a bI-population CMA-ES on the BBOB-2009 function testbed. In: Proceedings of the 11th Annual Conference Companion on Genetic and Evolutionary Computation Conference: Late Breaking Papers. pp. 2389–2396.

GECCO 2009, Association for Computing Machinery, New York, NY, USA (2009). https://doi.org/10.1145/1570256.1570333
12. Hutter, F., Hoos, H.H., Leyton-Brown, K.: Sequential model-based optimization for general algorithm configuration. In: Coello, C.A.C. (ed.) LION 2011. LNCS, vol. 6683, pp. 507–523. Springer, Heidelberg (2011). https://doi.org/10.1007/978-3-642-25566-3_40
13. Jones, D.R., Schonlau, M., Welch, W.J.: Efficient global optimization of expensive black-box functions. J. Global Optim. **13**(4), 455–492 (1998)
14. Lukovic, M.K., Tian, Y., Matusik, W.: Diversity-guided multi-objective Bayesian optimization with batch evaluations. In: Larochelle, H., Ranzato, M., Hadsell, R., Balcan, M., Lin, H. (eds.) Advances in Neural Information Processing Systems, vol. 33, pp. 17708–17720. Curran Associates, Inc. (2020)
15. Koza, J.R.: Hierarchical genetic algorithms operating on populations of computer programs. In: Sridharan, N.S. (ed.) Proceedings of the Eleventh International Joint Conference on Artificial Intelligence IJCAI-89, vol. 1, pp. 768–774. Morgan Kaufmann, Detroit, MI, USA (20–25 Aug 1989). https://www.genetic-programming.com/jkpdf/ijcai1989.pdf
16. Li, N., Zhao, L., Bao, C., Gong, G., Song, X., Tian, C.: A real-time information integration framework for multidisciplinary coupling of complex aircrafts: an application of IIIE. J. Ind. Inf. Integr. **22**, 100203 (2021). https://doi.org/10.1016/j.jii.2021.100203
17. Srinivas, N., Krause, A., Kakade, S., Seeger, M.: Gaussian process optimization in the bandit setting: No regret and experimental design. In: Proceedings of the 27th International Conference on International Conference on Machine Learning, pp. 1015–1022. ICML2010, Omnipress, Madison, WI, USA (2010)
18. Stephens, T.: gplearn: Genetic programming in python (2019). https://gplearn.readthedocs.io/en/stable/
19. Yang, K., Affenzeller, M.: Quantifying uncertainties of residuals in symbolic regression via kriging. Procedia Comput. Sci. **200**, 954–961 (2022). 3rd International Conference on Industry 4.0 and Smart Manufacturing. https://doi.org/10.1016/j.procs.2022.01.293
20. Yang, K., Emmerich, M., Deutz, A., Bäck, T.: Efficient computation of expected hypervolume improvement using box decomposition algorithms. J. Global Optim. **75**(1), 3–34 (2019). https://doi.org/10.1007/s10898-019-00798-7
21. Yang, K., Emmerich, M., Deutz, A., Bäck, T.: Multi-objective Bayesian global optimization using expected hypervolume improvement gradient. Swarm Evol. Comput. **44**, 945–956 (2019). https://doi.org/10.1016/j.swevo.2018.10.007
22. Zitzler, E., Thiele, L.: Multiobjective evolutionary algorithms: a comparative case study and the strength pareto approach. IEEE Trans. Evol. Comput. **3**(4), 257–271 (1999)
23. Zitzler, E., Deb, K., Thiele, L.: Comparison of multiobjective evolutionary algorithms: empirical results. Evol. Comput. **8**(2), 173–195 (2000)

BY

Learning to Predict Pareto-Optimal Solutions from Pseudo-weights

Kalyanmoy Deb, Aryan Gondkar, and Suresh Anirudh(✉)

Computational Optimization and Innovation Laboratory, Michigan State University, East Lansing, MI 48824, USA
{kdeb,gondkara,suresha2}@msu.edu
https://www.coin-lab.org

Abstract. Evolutionary Multi-objective optimization (EMO) algorithms attempt to find a well-converged and well-diversified set close to true Pareto-optimal solutions. However, due to stochasticity involved in EMO algorithms, the uniformity in distribution of solutions cannot be guaranteed. Moreover, the follow-up decision-making activities may demand finding more solutions in specific regions on the Pareto-optimal front which may not be well-represented by the obtained EMO solutions. In this paper, we train machine learning algorithms to capture the relationship between pseudo-weight vectors, derived from location of EMO-obtained non-dominated objective vectors, and their respective decision variable vectors. The learnt system can then be utilized to predict the corresponding variable vector for any desired pseudo-weight vector. The proposed methodology is applied to a number of problem instances to demonstrate its working and usefulness in arriving at a desired distribution of near Pareto-optimal solutions. The methodology has the potential to be embedded within an EMO algorithm to produce a better distributed set of solutions, check the validity of apparent gaps in obtained fronts, and also to help find more non-dominated solutions at the preferred regions of the Pareto-optimal front for effective decision-making purposes.

Keywords: Deep neural networks · Multi-objective optimization · Pseudo-weights · Multi-criterion decision-making

1 Introduction

Evolutionary multi-objective optimization (EMO) algorithms are capable of finding multiple trade-off solutions near or on the Pareto-optimal (PO) front [2,5] for multi-objective optimization problems having more than one objective functions and multiple constraint functions. The algorithms have been extended to handle more than three-objective problems – known as many-objective optimization (MaO) problems [3,10,15]. However, EMO or EMaO algorithms are stochastic in nature and despite a great deal of effort in finding a well-distributed set of trade-off solutions, they often fail to find the desired distribution. Besides the

M. Emmerich et al. (Eds.): EMO 2023, LNCS 13970, pp. 191–204, 2023.
https://doi.org/10.1007/978-3-031-27250-9_14

stochasticity inherent in the algorithms, the complexity offered by a problem near the PO front is another reason for not arriving at a uniformly distributed set of trade-off solutions.

Moreover, the ensuing decision-making (DM) task in choosing a single preferred trade-off solution must focus on a specific region on the obtained trade-off solutions. It is not guaranteed that the EMO or EMaO-obtained trade-off solutions have enough solutions corresponding to the preferred region by the decision-makers. Often, the DM task cannot be pursued before the EMO or EMaO run is executed, as DMs may not be certain about the preference information without the knowledge of an optimized trade-off solution set.

Both the above scenarios demand that a computationally quick and simple procedure of creating a new trade-off solution easily at any part of the optimized front. This task is akin to various machine learning tasks in which input-output relationships are learnt from a set of training data and the learnt system is then used to predict output from a given and unseen input. In our task, the input-output training data are the EMO or EMaO-obtained trade-off solutions for which input is a unique indicator of a solution on the front and the output is the solution vector ($\mathbf{x}$). In this study, we use the pseudo-weight vector [5] as a unique indicator of a trade-off solution on the front. An advantage of using a pseudo-weight vector is that every component is normalized to lie in $[0, 1]$. After a machine learning system is learnt, any new pseudo-weight vector can be used as an input to find the respective $\mathbf{x}$-vector for which the corresponding $\mathbf{f}$-vector can be computed using the given objective functions.

In the remainder of this proof-of-principle study, we briefly outline a number of existing studies related to finding a uniformly distributed set of Pareto-optimal (PO) solutions, focusing on filling gaps in obtained fronts in Sect. 2. The proposed machine learning procedure for the post-optimality study is described in Sect. 3. Results on multi-objective and many-objective constrained and unconstrained test problems and engineering problems are presented in Sect. 4. Finally, Sect. 5 summarizes the findings of this study and proposes a number of viable extensions.

2 Existing Studies

Use of machine learning methods in EMO is well studied in recent literature. Machine learning methods are used for surrogate modeling ($\mathbf{x}$ to $\mathbf{f}$ mapping) [6], *innovization* ($\mathbf{x}$ to $\mathbf{x}$ mapping) [12], and as genetic operators (sampling in $\mathbf{x}$ space) [9].

Surrogate assisted optimization methods use machine learning methods as a computationally cheap alternative to expensive function evaluations. Exploitative optimization is performed using the surrogate function and high-fidelity evaluations are computed sparsely. Surrogate functions map from $\mathbf{x}$ to $\mathbf{f}$, thereby not helping our cause here. Inverse models [8] map $\mathbf{f}$ to $\mathbf{x}$, but require knowledge of true $\mathbf{f}$-vector to produce the requisite $\mathbf{x}$-vector, causing difficulties in DM efforts. Our proposed method maps $\mathbf{w}$ to $\mathbf{x}$ (where $\mathbf{w}$ are pseudo-weights [5], which represent place-holder information of a solution in the PO front), making it convenient for DM purposes.

Machine learning methods have been used to aid EMO algorithms in achieving better convergence. Mittal et al. [12] proposed a number of *innovization*-based progress operators that uses machine learning to learn meaningful search advances. These operators are based on intra-generational solutions that aid in diversity and convergence. He et al. [9] proposed a Generative Adversarial Network (GAN) driven optimization algorithm, where a GAN is trained on current solutions and is then used to sample off-spring. In our proposed method, the machine learning models are conditioned on pseudo-weights and hence lead to a more targeted approach for finding non-dominated solutions directly.

Gaps often exist on obtained PO front. It is important for DMs to confirm if the gap really exists or is a shortcoming of the chosen EMO algorithm. If a gap truly exists, it becomes interesting for DMs to know what causes a gap and what properties of $\mathbf{x}$-vector enables a gap on a PO front. Pellicer et al. [14] proposed a novel multi-step gap-finding method where converged solutions are clustered into gaps and a reference-direction based method [7] is used to validate lack of solutions in the gap. This method requires repeated runs of optimization algorithm, contrary to our proposed method where we do not need to optimize again to find new points in the apparent gaps.

3 Proposed Machine Learning Based EMO Procedure

An application of an EMO/EMaO algorithm can produce a set of H non-dominated (ND) solutions ($\mathbf{x}^{(k)} \in \mathbb{R}^n$, for $k = 1, 2, \ldots, H$). Each of these solutions can then be evaluated to compute the respective objective vectors $\mathbf{f}^{(k)} \in \mathbb{R}^M$, for $k = 1, 2, \ldots, H$. From the position of each objective vector on the entire ND set, the respective pseudo-weight vector $\mathbf{w}^{(k)} = \left(w_1^{(k)}, w_2^{(k)}, \ldots, w_M^{(k)}\right)$, ($\mathbf{w}^{(k)} \in \mathbb{R}^M$) can be computed as follows:

$$w_i^{(k)} = \frac{\left(f_i^{\max} - f_i^{(k)}\right) / \left((f_i^{\max} - f_i^{\min}\right)}{\sum_{j=1}^{M} \left(f_j^{\max} - f_j^{(k)}\right) / \left((f_j^{\max} - f_j^{\min}\right)}. \tag{1}$$

The pseudo-weight vector for a ND solution can be viewed as a representative identity on the ND set. For a two-objective problem the best f_1 solution corresponds to a pseudo-vector of $(1, 0)$, meaning that the solution is 100% importance for f_1 and no importance for f_2.

Clearly, Eq. 1 indicates that pseudo-weights are derived from the objective values of ND set. If the ND solutions are sorted according to objective vectors, the respective $\mathbf{w}$-vectors will also get sorted in a similar manner. The studies on "innovization" concept [4] have revealed Pareto-optimal (PO) solutions usually possess certain patterns or constancy with respect to certain variables. Thus, when the ND solutions are close enough to the true Pareto-optimal (PO) set, it is expected that for practical problems the respective $\mathbf{x}$-vectors would be somehow related to the pseudo-weight vectors. This then motivates us to capture the relationships between the derived $\mathbf{w}$-vectors and respective $\mathbf{x}$-vectors of the ND set using a machine learning method. Once the patterns, if any, are captured, a new ND solution ($\bar{\mathbf{x}}$-vector) is expected to be found by using a desired $\bar{\mathbf{w}}$-vector.

Figure 1 illustrates the training and testing procedure for developing a machine learning method. The procedure is presented in steps below:

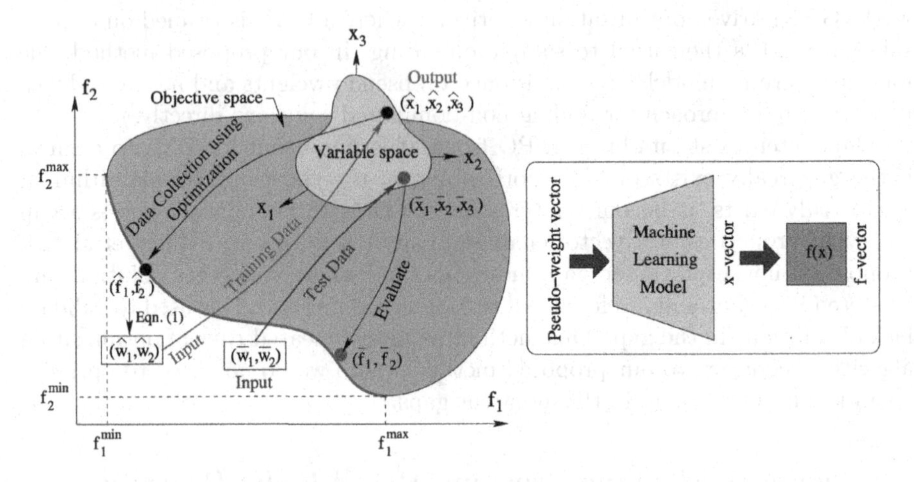

Fig. 1. Training data generation and test data to create new Pareto-optimal solution.

Step 1: From an EMO/EMaO-obtained ND set, calculate the pseudo-weight ($\mathbf{w}^{(k)}$)-vector for each solution (k) of the set using Eq. 1.
Step 2: Prepare training data ($\mathbf{w}^{(k)}, \mathbf{x}^{(k)}$) for $k = 1, 2, \ldots, H$, with $\mathbf{w}^{(k)}$ as input and $\mathbf{x}^{(k)}$ as output.
Step 3: Train a ML method using the training data.
Step 4: Use the trained ML model to find $\bar{\mathbf{x}}$ for any specific desired $\bar{\mathbf{w}}$ vector. Then, compute $\bar{\mathbf{f}}$ from $\bar{\mathbf{x}}$.

The resulting trained model can be used for various purposes.

Task 1: It can be used to check the validity of the obtained trained model. Pseudo-weights can be chosen at random from the entire PO set for testing purposes. Care is taken to not choose a training data as a test data.
Task 2: It can also be used to find new ND solutions in the apparent gaps in the EMO/EMaO-obtained ND set. This can be achieved by creating pseudo-weights in the gaps of the pseudo-weight space and then creating ND points by the trained model.
Task 3: It can be used to evaluate if an apparent gap in EMO/EMaO-obtained ND set is truly a gap. This can be determined by first creating pseudo-weights in the gaps and then finding resulting $\mathbf{x}$ using the trained model and then computing their $\mathbf{f}$ vectors to identify if they are dominated by the rest of the ND objective vectors.

Task 4: It can be used to quickly populate the ND set in a desired part of the ND set, mainly for decision-making or for a better visualization purposes. This can be achieved in two ways. First, creating suitable pseudo-weight vectors in the region of interest and then using the trained model to find new and hopefully non-dominated points.

Task 5: It can be used at intermediate generations of an EMO/EMaO run as an offspring creation mechanism. This can be achieved by first developing a trained model using the current ND points and then selecting pseudo-weights in less-dense areas of the pseudo-weight space and then using the model to create new solution.

In this study, we demonstrate the first four tasks here and leave the fifth task as an integral part of a new EMO algorithm requiring implementation and testing.

3.1 Training of Deep Neural Networks

Before we present results of our proposed method, we discuss machine learning methods considered in this study for developing the trained model. The problem here can be stated as a task of predicting an n-dimensional $\mathbf{x}$-vector from an M-dimensional pseudo-weight ($\mathbf{w}$) vector (where M is the number of objectives of the problem). Since usually $M \ll n$, the model development from a few input parameters to a large number of output parameters is a challenging task. We use two different machine learning methods for this purpose: (i) a deep neural network (DNN) approach and (ii) a Gaussian process regression (GPR) approach. In both DNN and GPR approaches, $\mathbf{x}$-vectors were normalized to zero mean and unit variance as required by the model. Pseudo-weight vectors are already within [0, 1] and are not normalized.

For every problem, the PO front is computed using NSGA-III [3], with a population size $N = 110M + 10$. These numbers are chosen with a trial-and-error study. Of these N PO points, $100M$ points are used as training points, and $10M$ points are taken as the test set. Finally, for the DNN, the remaining 10 points are used as a validation set, but discarded for GPR to maintain the similarity of the training data. Understanding the effect of size of dataset and other hyperparameters are left for future studies. Validation set was sampled uniformly from the training set for the purposes of model selection in the case of DNN. It is important to note that the proposed method is not conditional on the source of the training set. Hence, the PO front based training set can be replaced by the non-dominated set of an MOEA at the end of a particular generation.

Deep Neural Network (DNN) Approach: Multi-layer perceptrons (MLPs) with pseudo-weights as inputs and variables ($\mathbf{x}$) as outputs are implemented using the PyTorch [13] library and hyper-parameters are optimized using the Optuna software [1]. Owing to proof-of-concept nature of the study, DNNs with 1 to 6 hidden layers are used with ReLU activation and trained using Adam optimizer [11]. The complexity of the DNNs and granularity of hyper-parameters can be increased for handling more complex problems.

Gaussian Regression (GPR) Approach: An approach similar to surrogate modeling is used for training GPR models. Every output, x_i is modeled independently. A sundry of kernels, mean functions and other hyper-parameters are considered for a grid-search for finding the most suitable setting.

3.2 Handling Variable Bounds and Constraints

In an optimization problem, the variable vectors are usually bounded within lower and upper bounds. Since a DNN or a GPR approach does not usually restrict its output values automatically within any bound, the resulting output values for a test input data can be out of bounds, if a proper care is not taken. In this study, we normalize the output values as needed by the ML model. The output value from the system is then de-normalized and clipped to within their specified lower and upper bounds.

Constraint satisfaction is also a strict requirement in an optimization task. The resulting $\mathbf{x}$-vectors from the trained model may not guarantee to satisfy all constraints automatically. In this study, we simply discount infeasible $\mathbf{x}$ solutions, in case such a solution is created by the trained model; but a more sophisticated constrained handling method can be used during the training process. For example, constraint value of each constraint can be included as additional output to the DNN or GPR process. During testing, if any $\mathbf{w}$-vector (input) produces a positive constraint value (meaning a constraint violation), the solution is simply ignored. In this case, some training data with positive constraint violation, but non-dominated to the feasible ND set must be used to achieve a better training process. In this proof-of-principle study, we do not use any such sophisticated method.

4 Results

First, we present results on two-objective unconstrained and constrained test problems. Thereafter, we show results on three-objective problems, followed by a few many-objective problems. For each experiment, 11 runs are performed and their mean results are presented.

Tables 1, 2, 3 and 4 show Mean Absolute Errors (MAE) of test set on multi- and many-objective problems with different modeling approaches. The mean $\mathbf{x}$ is the average MAE scaled based on variable bounds between true $\mathbf{x}$-vectors (from the PO set) and model-obtained $\mathbf{x}$-vectors for the $10M$ test pseudo-weight vectors. Similarly, the mean $\mathbf{f}$ is the average MAE scaled based on range of objectives on the Pareto front ($\mathbf{f}^{\text{nadir}} - \mathbf{f}^{\text{ideal}}$) between the true $\mathbf{f}$-vectors and the model-obtained $\mathbf{x}$-vectors for the test pseudo-weight vectors. Their standard deviation values across the PO set are also presented in the table. 'Random' signifies that test pseudo-weight data are chosen at random on the entire PO front (Task 1). 'Continuous' and 'Edge' signify that a continuous region of a gap or a gap at one of the extreme parts of the PO set (Task 2) is used.

4.1 Two-Objective Problems

Table 1 presents prediction errors in **x** and **f**-vectors of test data by the DNN approach on two-objective unconstrained ZDT and constrained (BNH and OSY) problems. We can see that the error values are low for all the test problems.

Table 1. Performance of DNN approach on two-objective problems.

Problem	Test data	Mean **x**	Mean **f**	Std. dev. in **x**	Std. dev. in **f**
ZDT1	Continuous	6.281E−04	9.264E−03	1.051E−04	1.374E−03
	Edge	2.012E−03	1.864E−02	4.350E−04	4.332E−03
	Random	4.606E−04	6.918E−03	2.575E−04	3.689E−03
ZDT2	Continuous	7.387E−04	1.321E−02	1.242E−04	2.357E−03
	Edge	1.313E−03	3.986E−02	1.207E−04	3.680E−03
	Random	7.260E−04	1.387E−02	4.065E−04	6.786E−03
ZDT3	Continuous	1.026E−03	3.948E−02	2.030E−04	1.994E−02
	Random	4.619E−04	1.757E−02	2.591E−04	1.434E−02
BNH	Continuous	3.701E−03	2.760E−03	4.185E−03	5.378E−02
	Random	2.491E−03	1.407E−03	4.384E−03	7.347E−02
OSY	Continuous	2.172E−03	2.720E−02	3.685E−03	6.518E−01
	Random	7.833E−03	3.607E−02	3.323E−02	3.521E+00

Validating Task 1: Figure 2a shows that the randomly chosen pseudo-weight vectors produce **x**-vectors that make **f**-vectors fall on the PO front. The DNN-generated **f**-vectors (shown in red filled cirles) fall almost on top of the corresponding target **f**-vectors (shown with red open square). This is not an easy feat, as the DNN-learnt model produces **x**-vectors from supplied **w**-vectors and then the **x**-vectors are evaluated to compute **f**-vectors for plotting. This validates Task 1, proposed in Sect. 3 on ZDT1 problem. Note that the test data (red) are excluded from the training data (blue).

Next, we apply the GPR approach and results are tabulated in Table 2. Interestingly, much smaller error values are observed with GPR approach on the two-objective problems compared to the DNN approach. Figure 2b shows that GPR also can generate PO points very close to tagret **f**-vectors for the same set of pseudo-weights.

Validating Task 2: Next, we investigate if the proposed models can produce points on continuous gaps in the PO front produced by an EMO algorithm (Task 2). Outcome of DNN and GPR approaches are shown in the ZDT1 problem when the PO front has a gap on one of the extreme part of the PO front, as shown in Fig. 3. Error values are shown in Tables 1 and 2. Only blue points

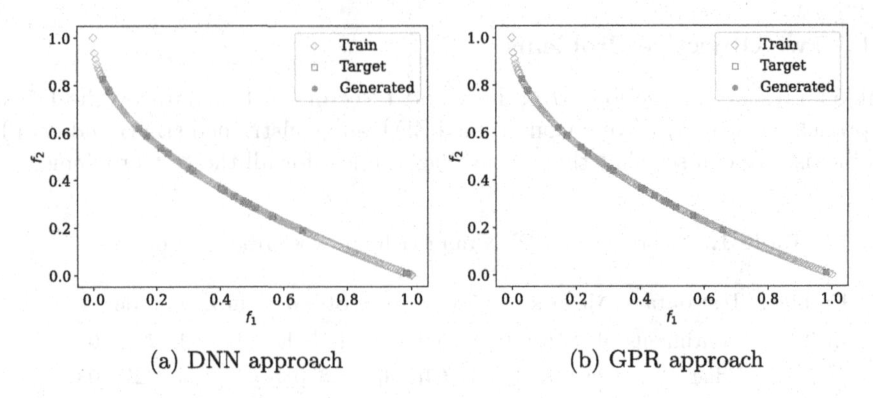

(a) DNN approach

(b) GPR approach

Fig. 2. DNN and GPR approaches on random test data for ZDT1 problem, demonstrating Task 1. (Color figure online)

Table 2. Performance of GPR approach on two-objective problems.

Problem	Test data	Mean **x**	Mean **f**	Std. dev. in **x**	Std. dev. in **f**
ZDT1	Continuous	2.625E−08	8.466E−08	5.694E−08	1.806E−07
	Edge	1.086E−06	2.447E−05	8.081E−07	1.820E−05
	Random	5.525E−09	3.462E−06	3.064E−09	7.511E−06
ZDT2	Continuous	1.083E−04	3.633E−03	3.903E−05	1.320E−03
	Edge	8.651E−04	3.811E−02	2.308E−04	1.023E−02
	Random	1.398E−05	4.200E−04	1.946E−05	6.093E−04
ZDT3	Continuous	3.182E−05	3.365E−03	9.706E−06	2.015E−03
	Random	3.659E−06	3.098E−04	4.985E−06	8.355E−04
BNH	Continuous	2.343E−03	1.274E−04	3.054E−03	2.224E−03
	Random	1.890E−03	2.073E−04	3.004E−03	9.670E−03
OSY	Continuous	1.471E−03	1.243E−03	4.562E−03	5.371E−02
	Random	1.491E−03	1.273E−03	1.146E−02	3.407E−01

are used to train the machine learning models, but red points are used to test. This requires ML models to learn how to extrapolate learnt relationships from training to test data and is always a harder task. Also, note that the models had to learn (i) how to create a suitable **x**-vector so that the resulting **f**-vector falls on the PO front, and (ii) make the **f**-vector fall at a place congruent to the chosen pseudo-weight vectors.

Figure 4 shows that GPR approach can also reproduce points at the middle part of the PO front on ZDT2 and ZDT3 problems. Here, too, training data did not include the points in the gap. Missing pseudo-weights at the gaps are estimated from the training data and are used to generate PO points in the gaps.

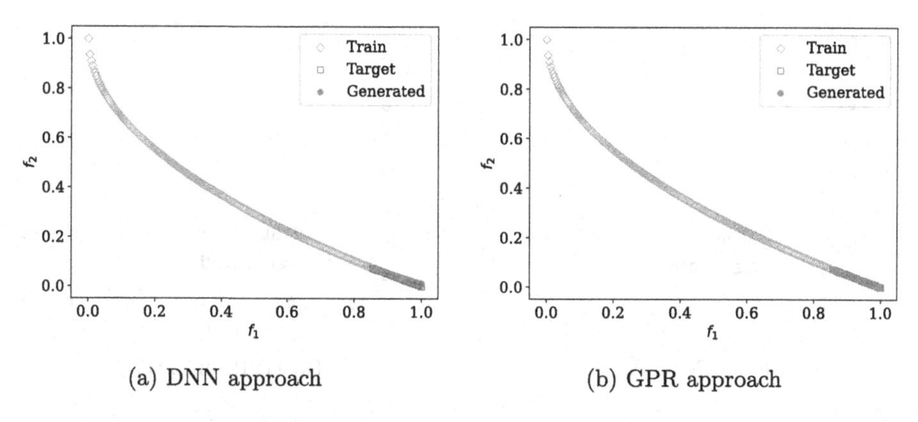

(a) DNN approach

(b) GPR approach

Fig. 3. DNN and GPR approaches in finding edge gap points on ZDT1 problem, demonstrating Task 2.

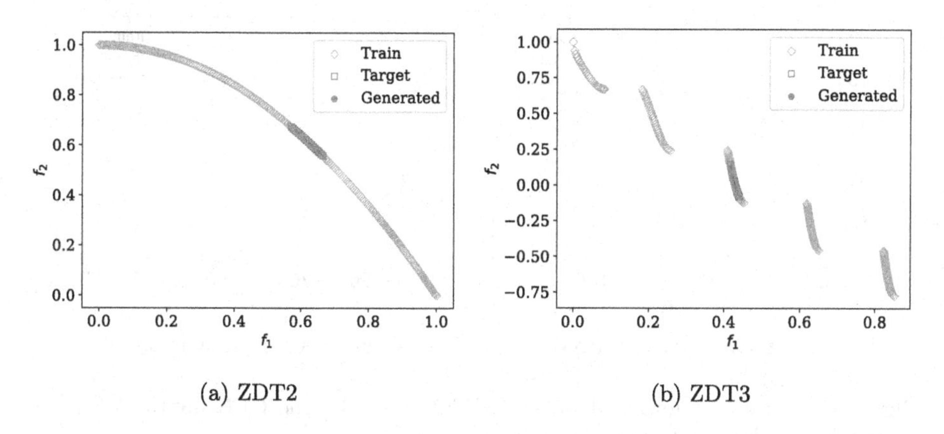

(a) ZDT2

(b) ZDT3

Fig. 4. GPR approach on continuous gap in middle of PO front for ZDT2 and ZDT3 problems, demonstrating Task 2.

Validating Task 3: In ZDT3 problem, there are natural gaps in the PO front. Thus, there will be gaps in the pseudo-weight space when they are computed from a set of EMO-obtained PO points. Next, we investigate what our trained DNN and GPR models would produce, if pseudo-weights in the natural gaps are chosen to find the respective **x**-vectors. Towards this task (Task 3), we create a uniform sample of 200 uniformly distributed set of pseudo-weight vectors and apply our trained models using EMO-obtained PO points to find respective **x**-vectors, compute their **f**-vectors, and then plot them in Fig. 5. It is clear that for pseudo-weights resulting a PO solution, our trained models are able to find them, but for pseudo-weights in natural gaps, trained models have produced dominated solutions, confirming that gaps observed in EMO solutions are true gaps and no PO solution exists there. Thus, our proposed approach can also be used to confirm reality of gaps in EMO-obtained PO fronts.

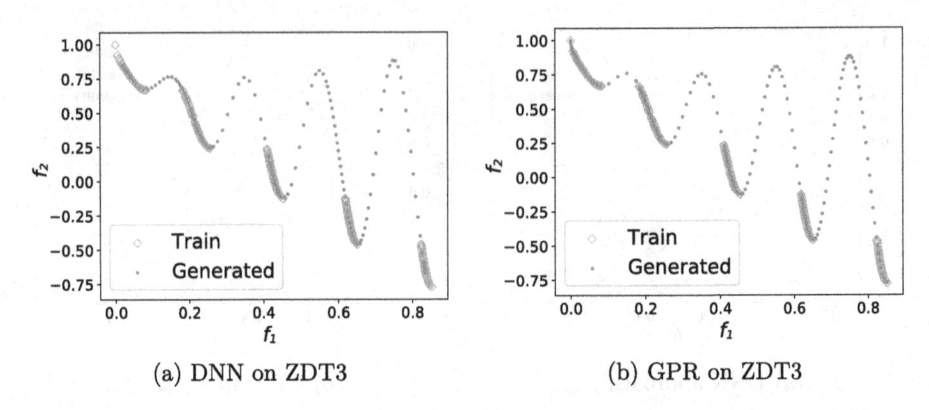

(a) DNN on ZDT3 (b) GPR on ZDT3

Fig. 5. Pseudo-weights on true gaps produce dominated solutions, but pseudo-weights on true PO front produce PO solutions, demonstrating Task 3.

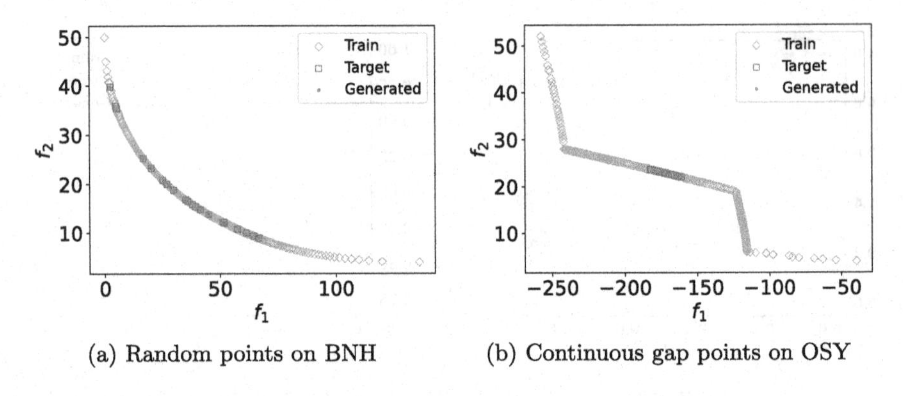

(a) Random points on BNH (b) Continuous gap points on OSY

Fig. 6. GPR results showing discovery of PO points by the GPR approach on two constrained problems, demonstrating Tasks 1 and 2.

Constrained Two-Objective Problems: Tables 1 and 2 show that the GPR approach is able to produce PO points from pseudo-weights with smaller error compared to DNN as well. To demonstrate, we present random and gap point discovery tasks (Tasks 1 and 2, respectively) for BNH and OSY problems in Fig. 6.

4.2 Three-Objective Problems

Owing to better results with GPR approach for two-objective problems, we use only GPR for three and many objective problems. Table 3 shows $\mathbf{x}$ and $\mathbf{f}$ errors along with their standard deviations.

Figure 7 shows random and continuous gap predictions for DTLZ2 problem. We see that the ML model is able to reasonably predict $\mathbf{x}$-vectors (hence, $\mathbf{f}$-vectors) for unseen pseudo-weights with low error.

Table 3. Performance of GPR approach on three-objective problems.

Problem	M	Gap type	Mean **x**	Mean **f**	Std. dev. in **x**	Std. dev. in **f**
DTLZ2	3	Continuous	6.981E−03	3.126E−02	2.449E−03	9.535E−03
		Edge	1.022E−02	5.968E−02	3.431E−03	1.962E−02
		Random	3.419E−03	1.153E−02	2.494E−03	7.168E−03
		Sparse	3.917E−03	2.238E−02	1.594E−03	8.891E−03
WFG2	3	Continuous	5.600E−02	6.514E−02	1.274E−01	1.703E−01
		Random	4.773E−02	4.922E−02	1.220E−01	1.732E−01
Carside	3	Random	1.171E−02	3.957E−03	9.080E−03	1.471E−02
Crashworthiness	3	Sparse	1.009E−02	7.785E−03	1.388E−02	3.903E−02

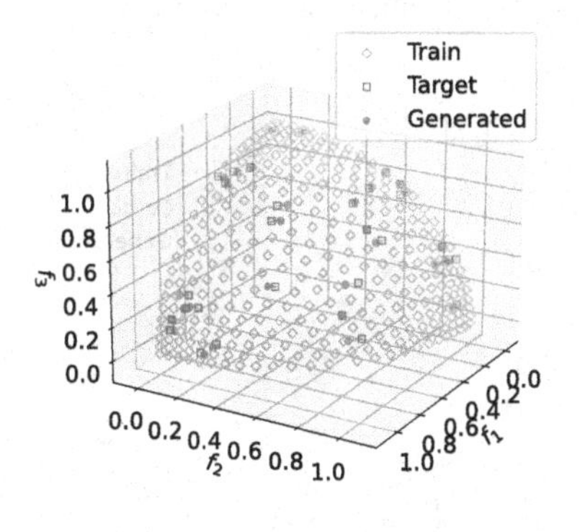

(a) Random points (Task 1)

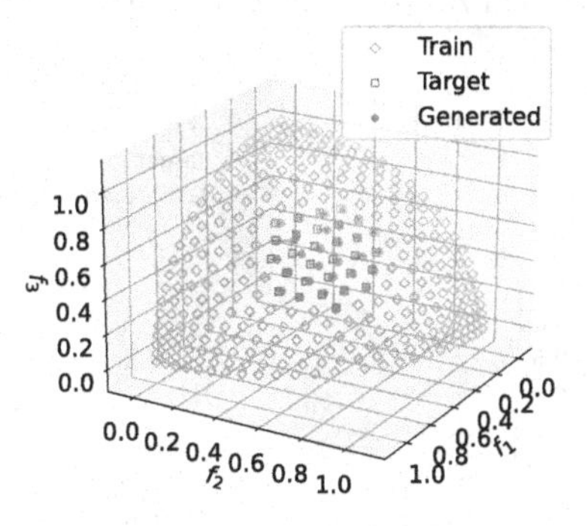

(b) Continuous gap points (Task 2)

Fig. 7. Random and continuous gap points by the GPR approach on three-objective DTLZ2 problem.

Constrained Three-Objective Problems: From Table 4, we see that prediction of PO solutions on constrained problems (carside impact and crashworthiness) can be achieved by the GPR approach with low errors in **x** and **f** space with low standard deviations.

Validating Task 4: Next, we consider a simulated case in which an EMO/EMaO produces a set of PO solutions which are not uniformly distributed. In this case as well, we can train a GPR model and supply pseudo-weights at the part with low-density of solutions and expect our model to predict PO points there. Figure 8 shows that the GPR approach can fill in additional points in such low-density regions for DTLZ2 and crashworthiness problems.

4.3 Many-Objective Optimization Problems

Finally, we apply our GPR approach to two five and 10-objective DTLZ2 and C2-DTLZ2 problems to demonstrate proof-of-principle results on many-objective optimization problems. Table 4 shows small error and standard deviation of error by the GPR approach.

Figure 9 shows the parallel coordinate plot (PCP) on these two problems with a few randomly chosen pseudo-weight vectors to demonstrate that the obtained **x**-vectors (and their **f**-vectors) produce near PO solutions.

Table 4. Performance of GPR approach on many-objective problems.

Problem	M	Test data	Mean **x**	Mean **f**	Std. dev in **x**	Std. dev in **f**
DTLZ2	5	Random	9.887E−03	1.382E−02	9.155E−03	8.668E−03
DTLZ2	10	Random	4.582E−02	1.418E−02	3.716E−02	8.500E−03
C2DTLZ2	5	Random	1.020E−02	1.323E−02	1.028E−02	7.448E−03
C2DTLZ2	10	Random	4.532E−02	1.418E−02	3.552E−02	9.490E−03

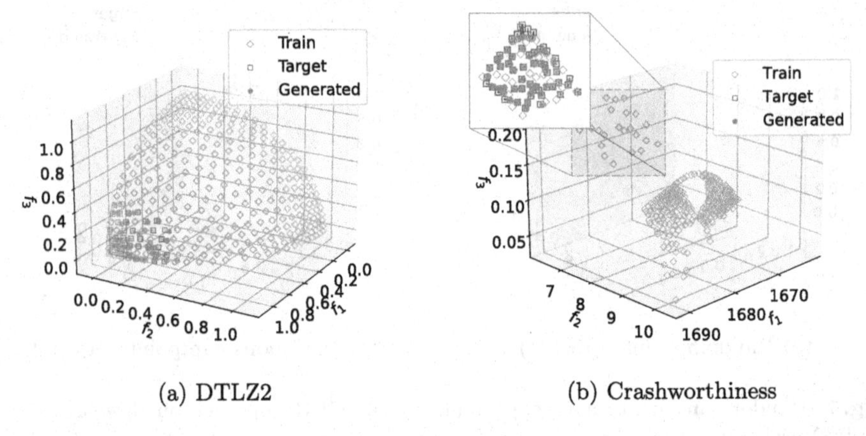

(a) DTLZ2 (b) Crashworthiness

Fig. 8. Additional solutions are supplied by the GPR approach for two three-objective problems at region of low-density of solutions, demonstrating Task 4. A few blue points were found by EMO on a part of PO front, but our ML approach has nicely replenished them. (Color figure online)

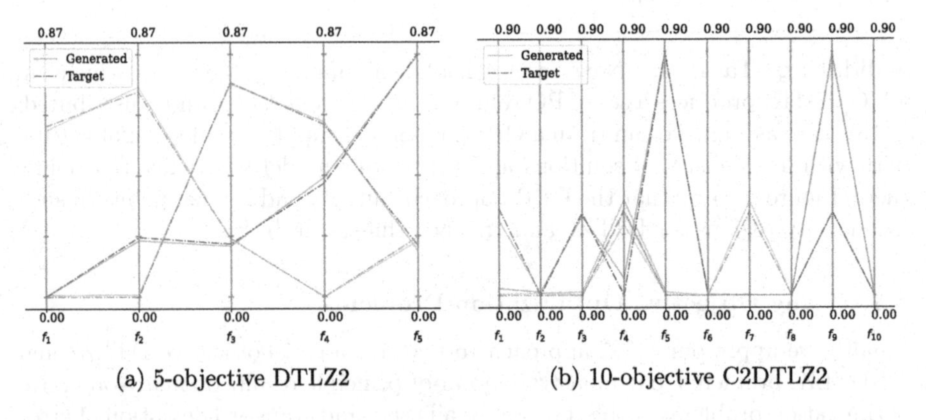

(a) 5-objective DTLZ2 (b) 10-objective C2DTLZ2

Fig. 9. PCP plots showing true PO and GPR-trained solutions are close for a few pseudo-weight vectors for 5-obj DTLZ2 and 10-obj. DTLZ2 problems.

5 Conclusions

We have shown that machine-learning (ML) models can be used to learn patterns between pseudo-weights and corresponding $\mathbf{x}$-vectors and generate new points on the Pareto front without doing an additional optimization. We have demonstrated that this method is scalable to many-objective test and real-world problems. This proof-of-concept study paves the way for using the proposed method as part of an EMO task for generating a better distributed ND fronts. Owing to the fact that the ML models are conditioned on pseudo-weights, the proposed method can be readily used for decision-making by the user without the need for further optimization.

The current study can be extended as a comparison with optimization based gap-filling methods, like reference direction based EMO algorithms. Applications to more complex and real-world problems will fully evaluate the potential of the proposed approach. Also, the modeling approaches can be improved to satisfy learning of constraints and variable bounds during the training process. For example, specific activation functions (such as, ReLU) can be used for the output layer of DNNs to restrict outputs to variable bounds. While constraint satisfaction was enforced in our results here, it would be a challenging task to include constraint satisfaction in the training process, since all PO solutions (training data) are expected to be feasible. Nevertheless, this proof-of-principle study opens up a unique use of machine learning methods in assisting multi-objective optimization.

References

1. Akiba, T., Sano, S., Yanase, T., Ohta, T., Koyama, M.: Optuna: a next-generation hyperparameter optimization framework. In: KDD 2019, pp. 2623–2631. Association for Computing Machinery, New York (2019). https://doi.org/10.1145/3292500.3330701
2. Coello, C.A.C., VanVeldhuizen, D.A., Lamont, G.: Evolutionary Algorithms for Solving Multi-objective Problems. Kluwer, Boston (2002)
3. Deb, K., Jain, H.: An evolutionary many-objective optimization algorithm using reference-point based non-dominated sorting approach, part I: solving problems with box constraints. IEEE Trans. Evol. Comput. **18**(4), 577–601 (2014)
4. Deb, K., Srinivasan, A.: Innovization: innovating design principles through optimization. In: Proceedings of the Genetic and Evolutionary Computation Conference (GECCO-2006), pp. 1629–1636. ACM, New York (2006)
5. Deb, K.: Multi-objective optimization using evolutionary algorithms (2001)
6. Deb, K., Roy, P.C., Hussein, R.: Surrogate modeling approaches for multiobjective optimization: methods, taxonomy, and results. Math. Comput. Appl. **26**(1) (2021). https://doi.org/10.3390/mca26010005, https://www.mdpi.com/2297-8747/26/1/5
7. Deb, K., Sundar, J.: Reference point based multi-objective optimization using evolutionary algorithms. In: Proceedings of the 8th Annual Conference on Genetic and Evolutionary Computation, pp. 635–642 (2006)

8. Farias, L.R.C., Araújo, A.F.R.: IM-MOEA/D: an inverse modeling multi-objective evolutionary algorithm based on decomposition. In: 2021 IEEE International Conference on Systems, Man, and Cybernetics (SMC), pp. 462–467 (2021). https://doi.org/10.1109/SMC52423.2021.9658650
9. He, C., Huang, S., Cheng, R., Tan, K.C., Jin, Y.: Evolutionary multiobjective optimization driven by generative adversarial networks (GANs). IEEE Trans. Cybern. **51**(6), 3129–3142 (2020)
10. Jain, H., Deb, K.: An evolutionary many-objective optimization algorithm using reference-point based non-dominated sorting approach, part II: handling constraints and extending to an adaptive approach. IEEE Trans. Evol. Comput. **18**(4), 602–622 (2014)
11. Kingma, D.P., Ba, J.: Adam: a method for stochastic optimization. In: Bengio, Y., LeCun, Y. (eds.) 3rd International Conference on Learning Representations, ICLR 2015, San Diego, CA, USA, 7–9 May 2015, Conference Track Proceedings (2015). http://arxiv.org/abs/1412.6980
12. Mittal, S., Saxena, D.K., Deb, K., Goodman, E.D.: Enhanced innovized progress operator for evolutionary multi- and many-objective optimization. IEEE Trans. Evol. Comput. **26**(5), 961–975 (2022). https://doi.org/10.1109/TEVC.2021.3131952
13. Paszke, A., et al.: PyTorch: an imperative style, high-performance deep learning library. In: Advances in Neural Information Processing Systems, vol. 32, pp. 8024–8035. Curran Associates, Inc. (2019)
14. Pellicer, P.V., Escudero, M.I., Alzueta, S.F., Deb, K.: Gap finding and validation in evolutionary multi-and many-objective optimization. In: Proceedings of the 2020 Genetic and Evolutionary Computation Conference, pp. 578–586 (2020)
15. Zhang, Q., Li, H.: MOEA/D: a multiobjective evolutionary algorithm based on decomposition. IEEE Trans. Evol. Comput. **11**(6), 712–731 (2007)

A Relation Surrogate Model for Expensive Multiobjective Continuous and Combinatorial Optimization

Hao Hao[1,2](✉) and Aimin Zhou[1,2]

[1] Shanghai Frontiers Science Center of Molecule Intelligent Syntheses, Shanghai, China
52194506007@stu.ecnu.edu.cn

[2] School of Computer Science and Technology, East China Normal University, Shanghai, China

Abstract. Currently, the research on expensive optimization problems mainly focuses on continuous problems and ignores combinatorial problems, which exist in many real-world applications. Since in surrogate model assisted evolution algorithms (SAEAs), the surrogate models from the community of machine learning are usually designed from continuous problems, and they are not suitable from combinatorial problems. For this reason, we propose a convolution relation model for both continuous and combinatorial problems. In the new relation model, a sample representation method of a relation map is proposed in the data preparation, and the convolution neural network is used to learn the relationships between pairs of candidate solutions. The new method is embedded into a basic multiobjective evolutionary algorithm and applied to a set of continuous and combinatorial problems. The experimental results suggest that the relation model with the same settings can solve continuous and combinatorial problems, and it has an advantage in terms of problem scalability.

Keywords: Combinatorial problems · Multiobjective problem · Expensive optimization · Relation model

1 Introduction

Expensive multiobjective optimization problems (EMOPs) exist widely in real-world applications [3,12]. As a kind of population-based optimization algorithm, evolution algorithms (EAs) can obtain a set of solutions that approximate the Pareto optimal front in a single run. Thus, EAs have been one of the most popular methods for solving general multiobjective optimization problems. However, general EAs are not suitable for EMOPs since EAs use a trial-and-error search strategy to generate a lot of candidate solutions for fitness evaluation, which is not affordable. For this reason, surrogate assisted evolutionary algorithms (SAEAs)

M. Emmerich et al. (Eds.): EMO 2023, LNCS 13970, pp. 205–217, 2023.
https://doi.org/10.1007/978-3-031-27250-9_15

have been proposed in the last decades. With the help of computationally efficient surrogate models, SAEAs can approximate the optimum of EMOPs with limited computational resources. SAEAs build a surrogate model based on the evaluated solutions and appropriate some newly generated solutions. In this way, only some promising candidate solutions will be evaluated by the real expensive functions, and the cost can thus be saved.

There is no doubt that surrogate models line in the center of SAEAs. Based on the type of surrogate model, SAEAs can be roughly classified into three categories, i.e., regression based SAEAs, classification based SAEAs, and relation based SAEAs. In the first category, regression models are used to approximate the objective value of variables. The Gaussian process (GP) [2], also known as the Kriging model, is the most widely used method. For example, the Kriging assisted RVEA (k-RVEA) [4] and the Kriging assigned two-archive evolution algorithm (KTA2) [15]. Other models such as random forest (RF) [17], radial basis function network (RBFN) [16] and neural network (NN) [11] are also used in regression based SAEAs. In the second category, the classification preselection based MOEA (CPS-MOEA) [20] trains the k-nearest neighbor model to distinguish good solutions from a set of trial solutions. In CSEA [14], a neural network is used to predict new candidate solutions belonging to a non-dominated or dominated set. In the third category, the relation model based method is a novel idea [8,9,19], using relationships between solutions to train the model rather than using the feature of a single point in the above methods. REMO [9] and θ-DEA-DP [19] show obvious advantage in solving on EOPs.

Most of the current research on SAEAs focus on continuous optimization problems. However, in real-world applications, many problems cannot be represented by continuous variables, such as the traveling salesman problem (TSP) [5] and the knapsack problem (KP) [23]. Furthermore, the evaluation cost may be expensive for solving these problems. As far as we know, there are very few works that focus on expensive multiobjective combinatorial problems. For this reason, this paper proposes a convolutional neural network based REMO algorithm called CREMO. The main contributions of this paper can be summarized as follows:

- A new perspective of relation pairs is proposed. The two solution vectors will be stitched together based on each dimension and form a relational feature map. The new feature map can represent richer features between each pair of solutions.
- Convolutional neural networks are used for the construction of relational models. On the one hand, features of the same dimension between solutions can be learned, and on the other hand, they can be learned for different input variables.
- A relation model based SAEAs is proposed for the continuous and combinatorial problem. To the best of our knowledge, this paper is the first to propose SAEAs with different problem solving ability.

The rest of this paper is summarized as follows. First, some preliminaries are presented in Sect. 2. The details of the proposed algorithm are illustrated in

Sect. 3. Then, Sect. 4 conducts experimental studies of continuous and combinatorial problems. Finally, Sect. 5 concludes this paper with some potential work in the future.

2 Preliminaries

2.1 Problem Definition

This paper considers expensive multiobjective continuous and combinatorial optimization problems, which can be defined as:

$$\begin{aligned} \min \quad & F(\mathbf{x}) = (f_1(\mathbf{x}), \cdots, f_M(\mathbf{x}))^T \\ \text{s.t.} \quad & x \in \textstyle\prod_{i=1}^{D} [a_i, b_i] \end{aligned} \tag{1}$$

where $\mathbf{x} = (x_1, \cdots, x_n) \in R^D$ is a decision variable vector; $a_i < b_i (i = 1, \cdots, D)$ are lower and upper boundary of the feasible region respectively in search space; $F : R^D \rightarrow R^M$ consists of M objective functions f_i $(i = 1, \cdots, M)$.

For combinatorial problems, the many-objective traveling salesman problem (MOTSP) [5] and the many-objective D-item knapsack problem (MOKP) [23] are taken count into the study. The MOTSP is according to the following mathematical program:

$$\begin{aligned} \min \quad & F(\mathbf{x}) = (f_1(\mathbf{x}), \cdots, f_M(\mathbf{x}))^T \\ \text{where} \quad & f_k(\rho) = \sum_{i=1}^{D-1} c^k_{\rho(i),\rho(i+1)} + c^k_{\rho(n),\rho(1)}, k = 1, 2, \ldots, M \end{aligned} \tag{2}$$

where D denotes the number of cities visited, the cost k for traveling from city i to city j is denoted by $c^k_{i,j}$, and ρ is the cyclic permutation of cities. A tour is defined by the cyclic permutation ρ of D cities. MOKP is defined as follows:

$$\begin{aligned} \max \quad & F(\mathbf{x}) = (f_1(\mathbf{x}), \cdots, f_M(\mathbf{x}))^T \\ \text{s.t.} \quad & \sum_{j=1}^{D} b_{ij} x_j \leq c_i, i = 1, 2, \ldots, M \\ & x_j \in \{0, 1\}, j = 1, 2, \ldots, n \\ \text{where} \quad & f_k(\mathbf{x}) = \sum_{j=1}^{n} a_{ij} x_j, i = 1, 2, \ldots, M \end{aligned} \tag{3}$$

where $\mathbf{x}$ is a D-dimensional binary vector, b_{ij} represents the weight of item j inside knapsack i, a_{ij} is the profit of item j inside knapsack i, and c_i is the capacity of knapsack i.

Due to the conflicting nature among the objectives in Eq. (1), (2), (3), usually no single solution can optimize all objectives at the same time. The tradeoff solutions, called Pareto optimal solutions, form a Pareto optimal set of a MOP.

2.2 Relation Learning

Relational learning is a novel class of methods in SAEAs [8–10,19], where relationships between solutions are used to train the model, which is different from the training process that uses a single solution in regression or classification based SAEAs. The basic idea of REMO [9] is to construct a relations model by three main steps, i.e., data preparation, model training, and model usage. In data preparation, the angle-based domination is used to split the current population $\mathcal{P}$ into two sub-populations according to their fitness values. Then the training set is denoted as $\mathcal{D} = \{(\langle \mathbf{x}_i, \mathbf{x}_j \rangle, l) | \mathbf{x}_i, \mathbf{x}_j \in \mathcal{P}\}$, where $\langle \mathbf{x}_i, \mathbf{x}_j \rangle$ is a feature vector combined by each two solutions and l is label of $\langle \mathbf{x}_i, \mathbf{x}_j \rangle$ that denotes the relation between $\mathbf{x}_i$ and $\mathbf{x}_j$. The l has three values, namely '−1', '0', and '+1', which means that the categories of $\mathbf{x}_i$ is worse than, same, and better than $\mathbf{x}_j$. Next, a variety of models are suitable for data fitting. Neural Network (NN), Random Forest (RF) and other models can learn the feature of relative good and bad for each solution in a relation pair. Finally, based on the 'voting-scoring' strategy, the model can appropriate the quality of the solution by counting some predicted results of relation pairs consisting of multiple evaluated solutions and one unevaluated solution.

3 Proposed Method

CREMO is given in Algorithm 1. It actually follows of REMO, which contains three main steps, which shall be introduced in detail.

- **Initialization (lines 1–3):** N_i initialization solutions are sampled from the search space. For a continuous problem, the Latin hypercube sampling method [13] is used to sample. The random initial method is used for combination problems.
- **Relation data preparation (lines 4–5):** The population is adaptively divided into two subpopulations $\mathcal{P}_n$ and $\mathcal{P}_d$. Then combine each two solutions and assign the label according to categories. The innovation is to combine the solutions according to the variable dimension instead of connecting them from end to end.
- **Relation model training (lines 6):** A convolutional network is used to fit the data $\mathcal{D}$.
- **Relation model usage (lines 8–12):** The model is integrated into a local evaluation evolutionary search without real evaluation. The newly generated trial solutions $\mathcal{Q}_t$ are prescreened by relation model $\mathcal{M}$ to form the next population $\mathcal{Q}$.
- **Environment selection (lines 13–15):** The promising solutions in Q' will be evaluated by real objective functions. Environment selection is executed to select N solutions from $\mathcal{P} \cup \mathcal{Q}'$ to be the new population $\mathcal{P}$ in the next generation.

Algorithm 1: Framework of CREMO

```
Input  : N (population size);
         N' (number of solutions for real evaluation);
         N_i (initial population size);
         w (maximum number of evaluations by with surrogate);
Output: P (Current population).
   // Initialization
 1 Initialize the population P = {x_1, x_2, ··· , x_{D_i}};
 2 Update fitness evaluation counter;
 3 while termeination condition is not satidfied do
       // Relation data preparation
 4     Split population P into two subpopulations P_n,P_d and get the current
         reference points P_ref;
 5     Construct relation data D form P_n and P_d;
       // Relation model training
 6     Training a relation model M by the data set D;
 7     Set Q = P, and t=0;
       // Relation model usage
 8     while t ≤ w do
 9         Generate trial solutions Q_t from {Q ∪ P_ref};
10         Estimate the value s(u) of u ∈ Q_t by Eq. (6) with P_n and P_d;
11         Select |P_ref| solutions from Q_t with the largest s(u) as the offspring
             population Q into next local iteration;
12     end
       // Re-evaluation and environment selection
13     Select N' solutions from Q with the largest s(u) as the promising solutions
         Q' for real evaluation;
14     Update fitness evaluation counter;
15     Select N solutions from {P ∪ Q'} as the next population P;
16 end
```

3.1 Data Preparation

The data construction method is as described in Sect. 2.2. Firstly, the $\mathcal{P}$ is divided into $\mathcal{P}_n$ and $\mathcal{P}_d$, where $\mathcal{P}_n$ is superior to $\mathcal{P}_d$. Then the solutions in $\mathcal{P}_n \cup \mathcal{P}_d$ come to combine to form the relationship pairs. Different from REMO, the feature vectors of two solutions are placed in parallel to form a relation feature map, as shown in Fig. 1, so that the same dimensional information can be incorporated into the model to improve the learning accuracy.

3.2 Model Training

The data set $\mathcal{D}$ has three classes, and we design a convolutional neural network to fit the data. The network structure is shown in Fig. 1. First, the input size of the network is $2 \times n$ for the relational feature map. Next is the Convolutional-Layer([2, 2], 100), Relu-Layer, Convolutional-Layer ([1, 2], 30), Relu-Layer,

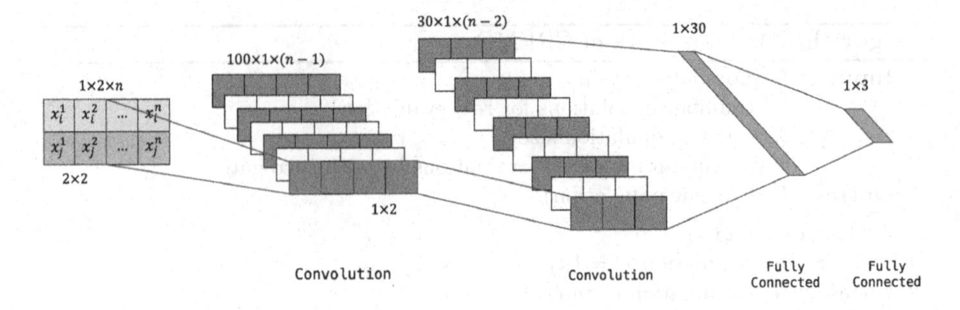

Fig. 1. Illustrations of network structure.

dropout-layer (0.2), FullyConnected-layer (30) and FullyConnected-layer (3). Finally, the network outputs the probability of label by a softmax function. The stochastic gradient descent with momentum (SGDM) with a 0.001 learning rate is used to optimize the cross-entropy loss function. The learning rate is set to 0.001, and the maximum epoch is set to 100. After the above model definition and training, we can obtain the triple classification model as described by Eq. 4:

$$[s_1, s_2, s_3] = \mathcal{M}([\mathbf{x}_i; \mathbf{x}_j]) \tag{4}$$

where s_1, s_2, s_3 denote probability that $\mathbf{x}_i \in \mathcal{P}_n, \mathbf{x}_j \in \mathcal{P}_d$, $\mathbf{x}_i$ and $\mathbf{x}_j$ are from the same subpopulation, or $\mathbf{x}_i \in P_d, \mathbf{x}_j \in P_n$ respectively. It also represents the probability that $\mathbf{x}_i$ is superior to, equal to, and inferior to $\mathbf{x}_j$.

3.3 Model Usage

This section introduce the 'voting-scoring' method [8]. For a newly generated solution $\mathbf{u}$, it combine all of the solution $\mathbf{x} \in \mathcal{P}$ to form relation maps $[\mathbf{u}; \mathbf{x}]$ and $[\mathbf{x}; \mathbf{u}]$. According to the categories of §, there are four types of combinations. We calculate the mean values of the four situations as follows:

$$\begin{aligned}
[\bar{s}_1^I, \bar{s}_2^I, \bar{s}_3^I] &= \underset{\mathbf{x} \in P_n}{mean}(\mathcal{M}([\mathbf{x}; \mathbf{u}])) \\
[\bar{s}_1^{II}, \bar{s}_2^{II}, \bar{s}_3^{II}] &= \underset{\mathbf{x} \in P_n}{mean}(\mathcal{M}([\mathbf{u}; \mathbf{x}])) \\
[\bar{s}_1^{III}, \bar{s}_2^{III}, \bar{s}_3^{III}] &= \underset{\mathbf{x} \in P_d}{mean}(\mathcal{M}([\mathbf{x}; \mathbf{u}])) \\
[\bar{s}_1^{IV}, \bar{s}_2^{IV}, \bar{s}_3^{IV}] &= \underset{\mathbf{x} \in P_d}{mean}(\mathcal{M}([\mathbf{u}; \mathbf{x}]))
\end{aligned} \tag{5}$$

Define a quality measurement of u as:

$$\begin{aligned}
s(\mathbf{u}) =& (\bar{s}_1^{II} + \bar{s}_1^{IV} + \bar{s}_2^{I} + \bar{s}_2^{II} + \bar{s}_3^{I} + \bar{s}_3^{III}) \\
&- (\bar{s}_1^{I} + \bar{s}_1^{III} + \bar{s}_2^{III} + \bar{s}_2^{IV} + \bar{s}_3^{II} + \bar{s}_3^{IV}).
\end{aligned} \tag{6}$$

$s(\mathbf{u})$ denotes the confidence that $\mathbf{u} \in \mathcal{P}_n$ via voting by the existing solutions in $\mathcal{P}$, and a higher value indicates a better quality of u.

4 Experimental Studies

This section studies the efficiency of the CREMO. In the following sections, Sect. 4.1 show the experimental settings. Section 4.2 compares the performances of CREMO with the other four at the state of art SAEAs on continuous problems. Section 4.3 studies the performances of CREMO on combinatorial problems, including MOTSP and MOKP, with various decision and objective sizes.

4.1 Experimental Settings

For fair comparisons, the seven compared algorithms and our proposed CREMO are all implemented in PlatEMO [18] using MATLAB. If not specified, the default parameters are used in the study.

- Continuous problems: CREMO and four SAEAs including CPS-MOEA [20], CSEA [14], K-RVEA [4] and KTA2 [15] are studied in DTLZ [7] and MaF [1] test suit. These four competitors represent the famous classification and regression based SAEAs. The decision space is set to $D = 30$, and the objective space is set to $M = 3, 6, 10$ for two test suits. The maximum fitness evaluation budget is set to 300. The population size N is set to 50, and if the algorithm uses the LHS method to get the initialization solution, the size of initial solutions N_i is set to 100.
- Combinatorial problems: It is hard to find an implementation of SAEAs for solving TSP and KP problems. Therefore, we chose three classical MOEAs, NSGA-II [6], MOEA/D [21] and IBEA [22], representing three classes of MOEAs. SAEAs, CPS-MOEA, and K-RVEA are chosen to represent regression and classification based methods, respectively. In order to make K-RVEA stable, we have used the *fitrgp* function in MATLAB instead of the original *dacefit* function. For CPS-MOEA, the reproduction operator is replaced by the GA operator to be consistent with other algorithms. For these two algorithms, we use * to indicate the modified versions. We set $D = 10, 30, 50, 100$ and $M = 2, 3, 6, 10$ for MOTSP and $D = 30, 50, 100, 250$, $M = 2, 3, 6$ for MOKP. The maximum fitness evaluation budget is set to 500. Other parameter settings of algorithms not mentioned above are consistent with the continuous problem.

The inverted generational distance (IGD) and hypervolume (HV) are used to evaluate the performance of each algorithm. Specifically, the IGD is used for continuous problems since an inadequate search will result in HV value of 0. Due to the lack of a real Pareto front, the HV indicator is used to measure the quality of solutions for combinatorial problems. The Wilcoxon rank-sum test is used to compare the experimental results, where '+', '−', and '≈' in the tables indicate that the value obtained by an algorithm is smaller than, greater than, or similar to that obtained by CREMO at a 95% significance level. The best mean IGD or HV values are highlighted by gray background for each row.

Table 1. The statistical results of the IGD metric values obtained by five algorithms with $D = 30$ on 36 test instances over 30 independent runs.

Problem	M	D	CPS-MOEA	CSEA	K-RVEA	KTA2	CREMO
DTLZ1	3	30	4.6991e+2≈	4.6662e+2≈	5.6634e+2 −	4.4598e+2 ≈	4.4958e+2
	6	30	4.6098e+2 −	3.0201e+2 ≈	4.0149e+2 −	3.5212e+2 −	3.2230e+2
	10	30	3.1911e+2 −	2.2270e+2 ≈	2.3553e+2 ≈	2.6764e+2 −	2.2392e+2
DTLZ2	3	30	1.0722e+0 −	8.8657e-1 ≈	1.5288e+0 −	6.8211e-1 +	8.2374e-1
	6	30	1.9690e+0 −	1.2255e+0 −	1.3797e+0 −	1.1949e+0 −	9.7444e-1
	10	30	1.9124e+0 −	1.2351e+0 −	1.2588e+0 −	1.2770e+0 −	1.0349e+0
DTLZ3	3	30	1.3606e+3 −	1.3466e+3 −	1.6175e+3 −	1.2840e+3 ≈	1.2645e+3
	6	30	1.4549e+3 −	1.0669e+3 +	1.3301e+3 −	1.3212e+3 −	1.1896e+3
	10	30	1.2557e+3 −	8.2882e+2 ≈	9.2955e+2 −	1.0737e+3 −	8.2485e+2
DTLZ4	3	30	1.4305e+0 −	9.2758e-1 −	1.6944e+0 −	9.4063e-1 −	8.1190e-1
	6	30	1.7952e+0 −	1.0980e+0 −	1.6091e+0 −	1.3505e+0 −	9.8425e-1
	10	30	1.8002e+0 −	1.1514e+0 ≈	1.6497e+0 −	1.3941e+0 −	1.1409e+0
DTLZ5	3	30	9.9558e-1 −	8.4869e-1 −	1.4307e+0 −	6.6019e-1 ≈	6.4730e-1
	6	30	1.5931e+0 −	8.8336e-1 −	1.0697e+0 −	1.0311e+0 −	7.5761e-1
	10	30	1.5062e+0 −	6.4316e-1 −	8.2529e-1 −	9.4336e-1 −	5.6524e-1
DTLZ6	3	30	1.6548e+1 +	2.2856e+1 −	1.9742e+1 +	1.7058e+1 +	2.1172e+1
	6	30	1.7377e+1 +	2.0581e+1 −	1.8992e+1 +	1.9677e+1 ≈	1.9871e+1
	10	30	1.4535e+1 +	1.6882e+1 −	1.5767e+1 ≈	1.6911e+1 −	1.5780e+1
DTLZ7	3	30	8.1860e+0 −	5.0128e+0 −	2.1749e-1 +	8.1524e-1 +	3.5656e+0
	6	30	1.5824e+1 −	1.7624e+1 −	9.1980e-1 +	3.2233e+0 +	1.1707e+1
	10	30	2.5347e+1 −	3.0997e+1 −	2.4127e+0 +	6.0435e+0 +	1.9620e+1
MaF1	3	30	1.0280e+0 −	8.1909e-1 −	1.2352e+0 −	3.7407e-1 +	6.9359e-1
	6	30	1.6757e+0 −	9.6509e-1 ≈	1.7256e+0 −	6.3262e-1 +	9.9831e-1
	10	30	1.7815e+0 −	1.0601e+0 ≈	1.5515e+0 −	5.8514e-1 +	1.0619e+0
MaF2	3	30	1.6538e-1 −	1.3049e-1 −	1.4473e-1 −	8.1919e-2 +	1.2050e-1
	6	30	2.2519e-1 ≈	2.3340e-1 −	2.2539e-1 ≈	1.4862e-1 +	2.2291e-1
	10	30	2.7730e-1 +	3.4928e-1 ≈	3.3911e-1 ≈	2.0389e-1 +	3.4507e-1
MaF3	3	30	2.8110e+6 +	7.4636e+6 −	4.5836e+6 ≈	1.8271e+7 −	5.3310e+6
	6	30	6.2377e+6 −	6.3649e+6 ≈	3.8457e+6 +	1.3348e+7 −	5.9000e+6
	10	30	2.2473e+9 ≈	5.1410e+6 −	3.2760e+6 ≈	9.6262e+6 −	3.7716e+6
MaF4	3	30	4.9818e+3 −	4.1749e+3 ≈	5.8379e+3 −	4.7260e+3 ≈	4.4221e+3
	6	30	4.3119e+4 −	3.1822e+4 −	4.1052e+4 −	3.9314e+4 −	2.8682e+4
	10	30	4.8882e+5 −	3.7518e+5 ≈	4.3565e+5 −	4.3507e+5 −	3.4962e+5
MaF5	3	30	6.0298e+0 −	3.6604e+0 ≈	5.5427e+0 −	3.6803e+0 ≈	3.4766e+0
	6	30	2.3000e+1 −	1.4155e+1 ≈	1.4925e+1 ≈	1.7504e+1 −	1.4386e+1
	10	30	2.2086e+2 −	1.4534e+2 ≈	1.4694e+2 ≈	1.5190e+2 ≈	1.5020e+2
+/ − / ≈			5/28/3	1/20/15	6/22/8	11/18/7	

4.2 Study on Continuous Problems

The statistical results of mean IGD values archived by five algorithms on DTLZ1-DTLZ7 and MaF1-MaF5 with $n = 30$ over 30 independent runs are summarized in Table 1. From the mean result, CREMO performs better than classification and regression based SAEAs. From the Wilcoxon rank-sum test result of all instances, we can see that CREMO obtains 28, 20, 22, 18 better, 5, 1, 6, 11 worse, and 3, 15, 8, 7 similar IGD values than CPS-MOEA, CSEA, K-RVEA, and KTA2 respectively.

Next, the running time versus problem size is analyzed. The KTA2, which has the best performance in the comparison algorithm according to Table 1, is selected as a comparator with CREMO. All algorithms are run on the same workstation (AMD 5800x, NVIDIA 3090Ti, and 32G Memory). We run each algorithm on the DTLZ test suit with different decision and objective space sizes under the same FEs. Next, the mean CPU time over DTLZ1-DTLZ7 on each size of the problem is recorded as the time in Fig. 2. As the problem scale expands, the running time of KTA2 grows rapidly. In contrast, the running time of CREMO grows slowly, especially on GPUs. The running time is not sensitive to the problem size. Experimental results show that CREMO is highly scalable in terms of problem size.

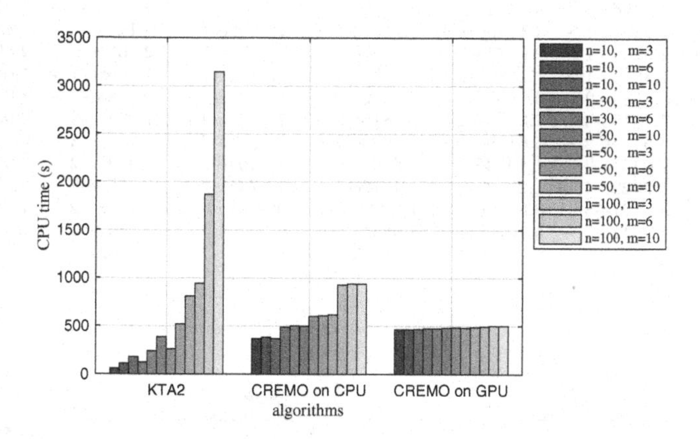

Fig. 2. Bar plot of mean running time of KTA2 and CREMO on DTLZ1-7 with different sizes under 300 fitness evaluations.

4.3 Study on Combinatorial Problems

This subsection compares CREMO with three representative MOEAs, i.e., NSGA-II, MOEA/D, and IBEA. And two variants of famous SAEAs, including CPS-MOEA* and K-RVEA*. The experimental results of the HV value provided by five comparison algorithms and CREMO on MOTSP are recorded in Table 2. It can be observed that CREMO obtains better HV values in most of the 16 different sizes of MOTSP with other comparison problems. Specifically, CREMO achieved the 15 best mean HV values. The same advantage also exists in MOKP, which achieved 15 best average HV indicators on 16 test problems. The runtime performance curves are shown in Fig. 3. The performance of CREMO shows the best convergence performance. In contrast, K-RVEA* based on the GP model performs poorly in boolean search space (MOKP) and many-objective ($m = 6$), and CPS-MOEA shows similar performance to model-free methods. The model does not provide adequate information in the search process. The non-dominated solutions form runs with the median HV values obtained by the compared algorithms on MOTSP and MOKP are shown in Fig. 4. For both minimization and maximization problems, the final solutions obtained by CREMO are closer to the ideal Pareto front (Fig. 4).

Table 2. The statistical results of the HV metric values obtained by six algorithms with on MOTSP test instances over 30 independent runs.

M	D	NSGA-II	MOEA/D	IBEA	CPS-MOEA*	KRVEA*	CREMO
2	10	5.7824e-1 ≈	5.5437e-1 −	5.7839e-1 ≈	5.6844e-1 ≈	5.7172e-1 ≈	5.7381e-1
2	30	5.4747e-1 −	5.3330e-1 −	5.4250e-1 −	5.2250e-1 −	5.3687e-1 −	5.7775e-1
2	50	4.5969e-1 −	4.5604e-1 −	4.6646e-1 −	4.5268e-1 −	4.7405e-1 −	5.0763e-1
2	100	4.1938e-1 −	4.1487e-1 −	4.2406e-1 −	4.1329e-1 −	4.3744e-1 −	4.5312e-1
3	10	4.8296e-1 ≈	4.6109e-1 −	4.8619e-1 ≈	4.8220e-1 ≈	4.8692e-1 ≈	4.9087e-1
3	30	3.7656e-1 −	3.7812e-1 −	3.7439e-1 −	3.5886e-1 −	3.8176e-1 −	4.0141e-1
3	50	3.0937e-1 −	3.1345e-1 −	3.1180e-1 −	2.9888e-1 −	3.1908e-1 −	3.4313e-1
3	100	2.6244e-1 −	2.6372e-1 −	2.6508e-1 −	2.5690e-1 −	2.6649e-1 −	2.8947e-1
6	10	1.3650e-1 −	1.2884e-1 −	1.5393e-1 −	1.3621e-1 −	1.4958e-1 −	1.6255e-1
6	30	8.6283e-2 −	9.0531e-2 −	9.7552e-2 −	8.5424e-2 −	9.5488e-2 −	1.0769e-1
6	50	7.1818e-2 −	7.3563e-2 −	7.6816e-2 −	7.1977e-2 −	7.7745e-2 −	9.0388e-2
6	100	5.5199e-2 −	5.6119e-2 −	5.8226e-2 −	5.5397e-2 −	5.7436e-2 −	6.4471e-2
10	10	2.0454e-2 −	1.8063e-2 −	2.5091e-2 −	2.0404e-2 −	2.4330e-2 −	3.0166e-2
10	30	1.2022e-2 −	1.2203e-2 −	1.4670e-2 −	1.1735e-2 −	1.5785e-2 −	2.0091e-2
10	50	9.5727e-3 −	8.8091e-3 −	1.0851e-2 −	9.5608e-3 −	1.1064e-2 −	1.4016e-2
10	100	6.7511e-3 −	5.7011e-3 −	7.0844e-3 −	6.7602e-3 −	7.1847e-3 −	8.8641e-3
+/ − / ≈		0/26/6	0/31/1	0/15/7	0/25/7	6/19/7	

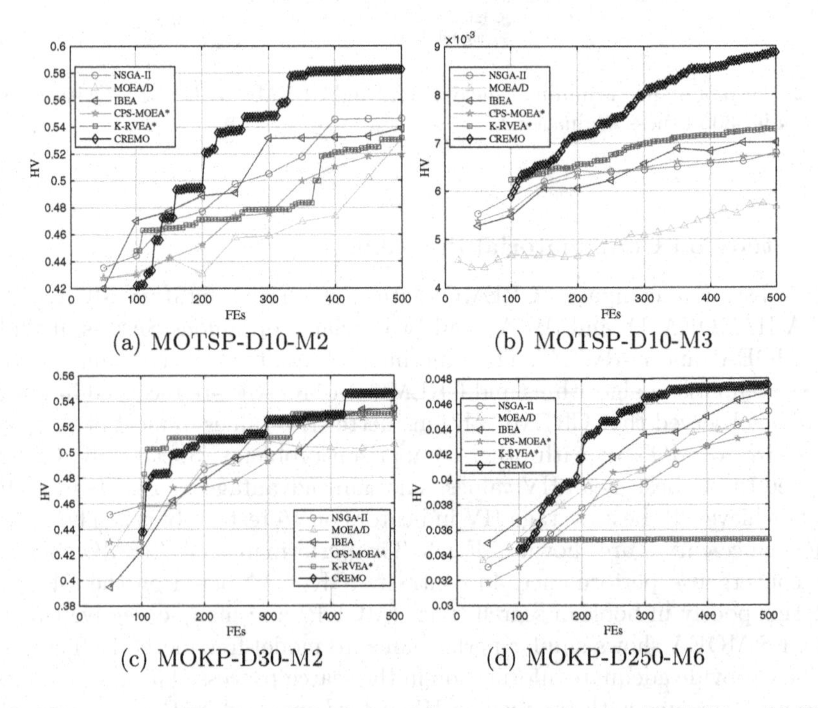

(a) MOTSP-D10-M2 (b) MOTSP-D10-M3

(c) MOKP-D30-M2 (d) MOKP-D250-M6

Fig. 3. The median HV values FEs obtained by six algorithms on MOTSP and MPKP over 30 independent runs.

Table 3. The statistical results of the HV metric values obtained by six algorithms with on MOKP test instances over 30 independent runs.

M	D	NSGA-II	MOEA/D	IBEA	CPS-MOEA*	K-RVEA*	CREMO
2	30	5.3307e-1 –	5.0650e-1 –	5.3389e-1 –	5.2695e-1 –	5.2428e-1 –	**5.4276e-1**
2	50	4.8839e-1 –	4.6364e-1 –	4.8505e-1 –	4.7352e-1 –	4.2985e-1 –	**4.9975e-1**
2	100	4.6206e-1 –	4.3285e-1 –	4.6345e-1 –	4.4451e-1 –	3.9610e-1 –	**4.7299e-1**
2	250	4.2974e-1 –	4.0487e-1 –	4.2990e-1 –	4.1092e-1 –	3.5585e-1 –	**4.4886e-1**
3	30	2.7820e-1 ≈	2.6751e-1 –	**2.8256e-1** ≈	2.7878e-1 ≈	2.4964e-1 –	2.7966e-1
3	50	3.0380e-1 ≈	2.9261e-1 –	3.0226e-1 –	2.9703e-1 –	2.6308e-1 –	**3.0758e-1**
3	100	2.6365e-1 –	2.6560e-1 –	2.6371e-1 –	2.5480e-1 –	2.1580e-1 –	**2.7799e-1**
3	250	2.5156e-1 –	2.5454e-1 –	2.5354e-1 –	2.3995e-1 –	2.0022e-1 –	**2.6612e-1**
6	30	5.5776e-2 ≈	4.5850e-2 –	5.6856e-2 ≈	5.5326e-2 –	4.7344e-2 –	**5.7801e-2**
6	50	5.3877e-2 –	4.8838e-2 –	5.4683e-2 –	5.1642e-2 –	4.3159e-2 –	**5.6987e-2**
6	100	5.5918e-2 –	5.4445e-2 –	5.8893e-2 –	5.5069e-2 –	5.0105e-2 –	**6.1912e-2**
6	250	4.5469e-2 –	4.4741e-2 –	4.6706e-2 ≈	4.3712e-2 –	3.5716e-2 –	**4.7343e-2**
+/ – / ≈		0/11/3	0/14/0	0/11/3	0/13/1	0/14/0	

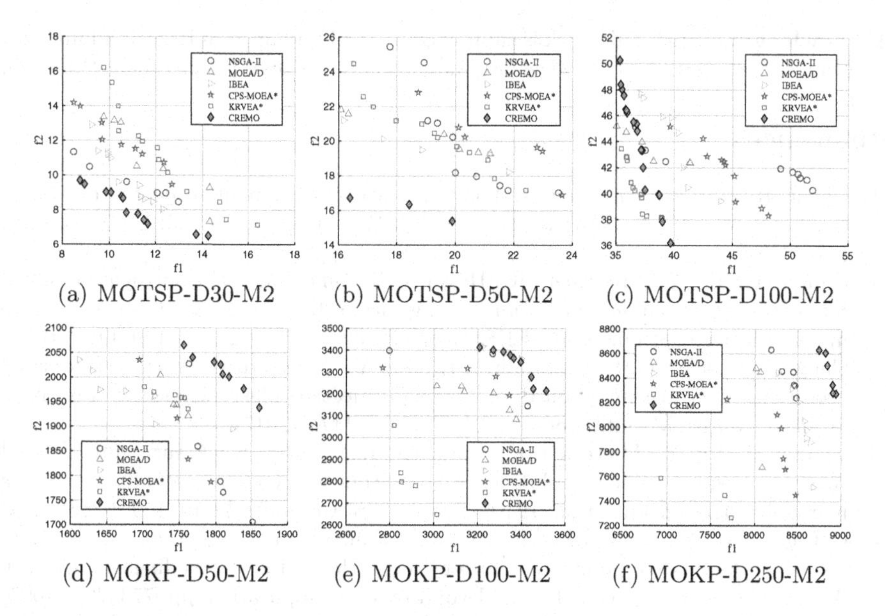

(a) MOTSP-D30-M2 (b) MOTSP-D50-M2 (c) MOTSP-D100-M2

(d) MOKP-D50-M2 (e) MOKP-D100-M2 (f) MOKP-D250-M2

Fig. 4. Nondominated solutions obtained by six algorithms in the run with median HV value for MOTPS and MOKP over 30 independent runs.

5 Conclusion

In this paper, we presented a convolution relation model assisted evolution optimization algorithm (CREMO), which can solve continuous and combinatorial problems with limited fitness evaluations. It is an extension of REMO on combinatorial optimization problems and still performs well on continuous problems. Specifically, in the data preparation stage, the REMO method divides the population into two sub-populations, and solutions are combined to construct relational data. We propose combining the features of two individuals according to the corresponding dimensions to form a relational feature map instead of concatenating from one end to another. This way, the model can capture solution features in the same dimension and improve prediction accuracy. Two special convolution kernels are used for feature extraction to learn relational features on different data classes, be they real, permutation, or boolean vectors.

The performance of CREMO on DTLZ and MaF test suits proves that CREMO performs well on continuous problems. Regarding problem scalability, CREMO works fast over regression model-assisted methods. Experimental studies are carried out on a set of problems from the MOTSP and MOKP test suites for combinatorial optimization problems. Results show the high efficiency of CREMO. Further, the proposed relation model can be embedded into other MOEAs dedicated to solving combinatorial optimization to further improve the algorithm's efficiency.

Acknowledgements. This work is supported by the Fundamental Research Funds for the Central Universities and the Defense Industrial Technology Development Program.

References

1. Cheng, R., et al.: A benchmark test suite for evolutionary many-objective optimization. Complex Intell. Syst. **3**(1), 67–81 (2017). https://doi.org/10.1007/s40747-017-0039-7
2. Chugh, T., Jin, Y., Miettinen, K., Hakanen, J., Sindhya, K.: A surrogate-assisted reference vector guided evolutionary algorithm for computationally expensive many-objective optimization. IEEE Trans. Evol. Comput. **22**(1), 129–142 (2018)
3. Chugh, T., Chakraborti, N., Sindhya, K., Jin, Y.: A data-driven surrogate-assisted evolutionary algorithm applied to a many-objective blast furnace optimization problem. Mater. Manuf. Process. **32**(10), 1172–1178 (2017)
4. Chugh, T., Jin, Y., Miettinen, K., Hakanen, J., Sindhya, K.: A surrogate-assisted reference vector guided evolutionary algorithm for computationally expensive many-objective optimization. IEEE Trans. Evol. Comput. **22**(1), 129–142 (2018)
5. Corne, D.W., Knowles, J.D.: Techniques for highly multiobjective optimisation: some nondominated points are better than others. In: Proceedings of the 9th Annual Conference on Genetic and Evolutionary Computation, pp. 773–780 (2007)
6. Deb, K., Pratap, A., Agarwal, S., Meyarivan, T.: A fast and elitist multiobjective genetic algorithm: NSGA-II. IEEE Trans. Evol. Comput. **6**(2), 182–197 (2002)
7. Deb, K., Thiele, L., Laumanns, M., Zitzler, E.: Scalable test problems for evolutionary multiobjective optimization. In: Abraham, A., Jain, L., Goldberg, R. (eds.) Evolutionary Multiobjective Optimization. Advanced Information and Knowledge Processing, pp. 105–145. Springer, London (2005). https://doi.org/10.1007/1-84628-137-7_6
8. Hao, H., Zhang, J., Lu, X., Zhou, A.: Binary relation learning and classifying for preselection in evolutionary algorithms. IEEE Trans. Evol. Comput. **24**(6), 1125–1139 (2020)
9. Hao, H., Zhou, A., Qian, H., Zhang, H.: Expensive multiobjective optimization by relation learning and prediction. IEEE Trans. Evol. Comput. **28**, 1157–1170 (2022)
10. Hao, H., Zhou, A., Zhang, H.: An approximated domination relationship based on binary classifiers for evolutionary multiobjective optimization. In: 2021 IEEE Congress on Evolutionary Computation (CEC), pp. 2427–2434. IEEE (2021)

11. Jin, Y., Sendhoff, B.: Reducing fitness evaluations using clustering techniques and neural network ensembles. In: Deb, K. (ed.) GECCO 2004. LNCS, vol. 3102, pp. 688–699. Springer, Heidelberg (2004). https://doi.org/10.1007/978-3-540-24854-5_71
12. Lu, Z., et al.: NSGA-net: neural architecture search using multi-objective genetic algorithm. In: Proceedings of the Genetic and Evolutionary Computation Conference (GECCO), pp. 419–427. ACM (2019)
13. Mckay, M.D., Beckman, R.J., Conover, W.J.: A comparison of three methods for selecting values of input variables in the analysis of output from a computer code. Technometrics **42**(1), 55–61 (2000)
14. Pan, L., He, C., Tian, Y., Wang, H., Zhang, X., Jin, Y.: A classification-based surrogate-assisted evolutionary algorithm for expensive many-objective optimization. IEEE Trans. Evol. Comput. **23**(1), 74–88 (2019)
15. Song, Z., Wang, H., He, C., Jin, Y.: A kriging-assisted two-archive evolutionary algorithm for expensive many-objective optimization. IEEE Trans. Evol. Comput. **25**, 1013–1027 (2021)
16. Sun, C., Jin, Y., Cheng, R., Ding, J., Zeng, J.: Surrogate-assisted cooperative swarm optimization of high-dimensional expensive problems. IEEE Trans. Evol. Comput. **21**(4), 644–660 (2017)
17. Sun, Y., Wang, H., Xue, B., Jin, Y., Yen, G.G., Zhang, M.: Surrogate-assisted evolutionary deep learning using an end-to-end random forest-based performance predictor. IEEE Trans. Evol. Comput. **24**(2), 350–364 (2020)
18. Tian, Y., Cheng, R., Zhang, X., Jin, Y.: PlatEMO: a MATLAB platform for evolutionary multi-objective optimization [educational forum]. IEEE Comput. Intell. Mag. **12**(4), 73–87 (2017)
19. Yuan, Y., Banzhaf, W.: Expensive multiobjective evolutionary optimization assisted by dominance prediction. IEEE Trans. Evol. Comput. **26**(1), 159–173 (2021)
20. Zhang, J., Zhou, A., Zhang, G.: A classification and pareto domination based multiobjective evolutionary algorithm. In: 2015 IEEE Congress on Evolutionary Computation (CEC), pp. 2883–2890. IEEE (2015)
21. Zhang, Q., Li, H.: MOEA/D: a multiobjective evolutionary algorithm based on decomposition. IEEE Trans. Evol. Comput. **11**(6), 712–731 (2007)
22. Zitzler, E., Künzli, S.: Indicator-based selection in multiobjective search. In: Yao, X., et al. (eds.) PPSN 2004. LNCS, vol. 3242, pp. 832–842. Springer, Heidelberg (2004). https://doi.org/10.1007/978-3-540-30217-9_84
23. Zitzler, E., Thiele, L.: Multiobjective evolutionary algorithms: a comparative case study and the strength pareto approach. IEEE Trans. Evol. Comput. **3**(4), 257–271 (1999)

Pareto Front Upconvert by Iterative Estimation Modeling and Solution Sampling

Tomoaki Takagi, Keiki Takadama, and Hiroyuki Sato(✉)

The University of Electro-Communications, 1-5-1 Chofugaoka, Chofu, Tokyo 182-8585, Japan
{tomtkg,h.sato}@uec.ac.jp, keiki@inf.uec.ac.jp

Abstract. For an efficient upconvert of the Pareto front resolution by utilizing a known candidate solution set, this paper proposed an algorithm that built the Pareto front and the Pareto set estimation models and repeated to sample a solution from them, evaluate it, and updated the estimation models with it. Conventional supervised multi-objective optimization algorithm (SMOA) built the Pareto front and the Pareto set estimation models with a known candidate solution set. SMOA sampled a set of well-distributed estimated points and evaluated them to upconvert the Pareto front resolution. However, depending on the distribution of the known candidate solutions, we could not expect the accuracy of the estimation models and the estimated points from them. The proposed method, the iterative SMOA (I-SMOA), gradually improved the accuracy of the estimation models through their iterative update with evaluated solutions. Experimental results on the DTLZ2 test problem showed that the proposed I-SMOA obtained solutions uniformly distributed more than the one by the conventional SMOA, and the proposed I-SMOA achieved higher robustness on the initially given candidate solutions.

Keywords: Pareto front estimation · Pareto set estimation · Solution aggregation · Response surface methodology · Pareto front upconvert

1 Introduction

Real-world optimization problems often involve multiple objective functions as conflicting concerns and require a high computational cost to evaluate each candidate solution. For these computationally expensive multi-objective optimization problems, a demand approximating the Pareto front with a limited number of objective function calls rises [1]. Also, it is not unusual that several candidate solutions with high optimality are already known before the optimization. For instance, they are products already developed and or released to the market. In this situation, new solution generation utilizing the known candidate solutions would be more efficient than a search from scratch, starting with randomized solutions involving inferior ones.

M. Emmerich et al. (Eds.): EMO 2023, LNCS 13970, pp. 218–230, 2023.
https://doi.org/10.1007/978-3-031-27250-9_16

In evolutionary optimization, the known candidate solutions can be included in the initial population as expert knowledge on the target problem, which is often called the directed initialization [2]. An alternative way to utilize the known candidate solutions was proposed as the supervised multi-objective optimization algorithm (SMOA) [3]. SMOA employs the response surface methodology such as Kriging [4], the radial basis neural network (RBNN) [5,6], etc., and builds the estimation models of the Pareto front and Pareto set with the known candidate solutions. SMOA samples a set of estimated objective vectors from the estimated Pareto front and evaluates their estimated variable vectors on the estimated Pareto set. As a result, the evaluated true objective vectors upconvert the Pareto front resolution efficiently. The upconvert quality of the Pareto front with this approach depends on the known candidate solution set, which is also called the training data. It is important to build accurate estimation models of the Pareto front and Pareto set as possible, even if the number of the known candidate solutions is small or they do not fully cover the Pareto front.

To improve the upconvert quality of the Pareto front resolution by utilizing the known candidate solutions, in this work, we propose a variant of SMOA, iterative SMOA (I-SMOA). The proposed I-SMOA builds the estimation models of the Pareto front and Pareto set as with the conventional SMOA. The proposed I-SMOA repeats to sample a single pair of estimated objective and variable vectors from the estimation models, evaluate it, and update the estimation models with it. The proposed I-SMOA continuously improves the quality of the estimation models through one-by-one solution evaluation and the update with it. We expect to improve the upconvert quality of the Pareto front and the robustness against various known candidate solutions. We verify the effects of the proposed I-SMOA on the DTLZ2 test problem [7] by comparing it to the conventional SMOA.

2 Multi-objective Optimization

For given variable space $\mathcal{X}$, variable vector $\boldsymbol{x} \in \mathcal{X}$, and objective functions f_i $(i = 1, 2, \ldots, m)$, a multi-objective optimization problem is defined as

$$\text{Minimize} \quad \boldsymbol{f}(\boldsymbol{x}) = (f_1(\boldsymbol{x}), f_2(\boldsymbol{x}), \ldots, f_m(\boldsymbol{x})). \tag{1}$$

For two solutions $\boldsymbol{x}, \boldsymbol{y} \in \mathcal{X}$, $\boldsymbol{x}$ dominates $\boldsymbol{y}$ ($\boldsymbol{x} \preceq \boldsymbol{y}$) iff $\forall i \in \{1, 2, \ldots, m\}$: $f_i(\boldsymbol{x}) \leq f_i(\boldsymbol{y})$ and $\exists i \in \{1, 2, \ldots, m\} : f_i(\boldsymbol{x}) < f_i(\boldsymbol{y})$. The goal of multi-objective optimization is to find the Pareto optimal solutions, the Pareto set, $\mathcal{P}_s = \{\boldsymbol{x} \in \mathcal{X} \mid \nexists \boldsymbol{y} \in \mathcal{X} : \boldsymbol{y} \preceq \boldsymbol{x}\}$, which are non-dominated solutions in the variable space $\mathcal{X}$. The Pareto front $\mathcal{P}_f = \{\boldsymbol{f}(\boldsymbol{x}) \mid \boldsymbol{x} \in \mathcal{P}_s\}$ is the objective vector set of Pareto set $\mathcal{P}_s$, which represents the optimal trade-off among objectives.

The larger number of well-distributed objective vectors of non-dominated solutions, the higher resolution of the Pareto front. However, we generally approximate the Pareto front with a limited number of objective vectors obtained by a search. Especially in problems with computationally expensive objective functions, we need to suppress the number of solutions to be evaluated.

3 Pareto Front and Pareto Set Estimation

3.1 Overview

Pareto front estimation estimates objective value change between known objective vectors in the objective space [8–11]. The estimated Pareto front is an upconverted high-resolution representation of the known objective vector set. The estimated Pareto front suggests objective value change even where objective vectors are unknown. It helps the decision-making and even the search [12–14]. Similarly, it has also been studied that the Pareto set estimation that suggests variable values change of the Pareto set in the variable space [15].

This work picks the L^1 unit hyperplane-based estimation [10] for both the Pareto front estimation and the Pareto set estimation. The L^1 unit hyperplane-based estimation is a solution aggregation methodology that converts a set of solutions into a model.

3.2 Pareto Front Estimation

We suppose to have a known candidate solution set $\mathcal{P}$, which are all evaluated and promising. Each in $\mathcal{P}$ is represented as $(\boldsymbol{f}, \boldsymbol{x})$, which is a pair of objective vector $\boldsymbol{f}$ and its corresponding variable vector $\boldsymbol{x}$, i.e., $\boldsymbol{f}$ is $\boldsymbol{f}(\boldsymbol{x})$. The known set can be represented like $\mathcal{P} = \{(\boldsymbol{f}^1, \boldsymbol{x}^1), (\boldsymbol{f}^2, \boldsymbol{x}^2), \dots\}$.

The L^1 unit hyperplane-based method [10] converts objective vector $\boldsymbol{f}$ of each known $(\boldsymbol{f}, \boldsymbol{x}) \in \mathcal{P}$ into L^1 norm $n = \sum_{i=1}^{m} f_i$ and L^1 unit vector $\boldsymbol{e} = \boldsymbol{f}/n$. The L^1 unit vector $\boldsymbol{e}$ represents the direction of objective vector $\boldsymbol{f}$ in the objective space. The L^1 norm n represents the distance to objective vector $\boldsymbol{f}$ in the direction $\boldsymbol{e}$. This method trains an estimation model, $\boldsymbol{f}$-model, by pairs of L^1 unit vector $\boldsymbol{e}$ as input and L^1 norm n as output in the known set $\mathcal{P}$. As the estimation model, this work employs Kriging [4] and RBNN [5,6]. The trained $\boldsymbol{f}$-model can output estimated norm $\hat{n}$ by inputting any L^1 unit vector $\boldsymbol{e}$, and estimated objective vector $\hat{\boldsymbol{f}}$ $(= \hat{n} \cdot \boldsymbol{e})$ can be obtained. That is, for any specified direction $\boldsymbol{e}$, the trained $\boldsymbol{f}$-model outputs the estimated distance $\hat{n}$ to promising objective vector $\hat{\boldsymbol{f}}$. We input a large set of uniformly distributed L^1 unit vectors $\mathcal{E} = \{\boldsymbol{e}^1, \boldsymbol{e}^2, \dots\}$ covering all over directions in the objective space to the trained $\boldsymbol{f}$-model and obtain their estimated L^1 norms $\{\hat{n^1}, \hat{n^2}, \dots\}$ and estimated objective vectors $\{\hat{\boldsymbol{f}^1}(= \hat{n^1} \cdot \boldsymbol{e}^1), \hat{\boldsymbol{f}^2}(= \hat{n}^2 \cdot \boldsymbol{e}^2), \dots\}$, which are the estimated Pareto front.

3.3 Pareto Set Estimation

We can estimate the Pareto set in the variable space similarly. For d dimensional variable space, we train d estimation models. For each variable element x_i $(i = 1, 2, \dots, d)$, we train an estimation model, x_i-model, by pairs of L^1 unit vector $\boldsymbol{e}$ as input and variable value x_i as output in the known set $\mathcal{P}$. The trained x_i-model can output estimated variable value $\hat{x_i}$ by inputting any L^1 unit vector $\boldsymbol{e}$. That is, the estimated variable vector $\hat{\boldsymbol{x}} = (\hat{x_1}, \hat{x_2}, \dots, \hat{x_d})$ can be obtained by inputting L^1 unit vector $\boldsymbol{e}$ to d kinds of x_i-models $(i = 1, 2, \dots, d)$. We input a

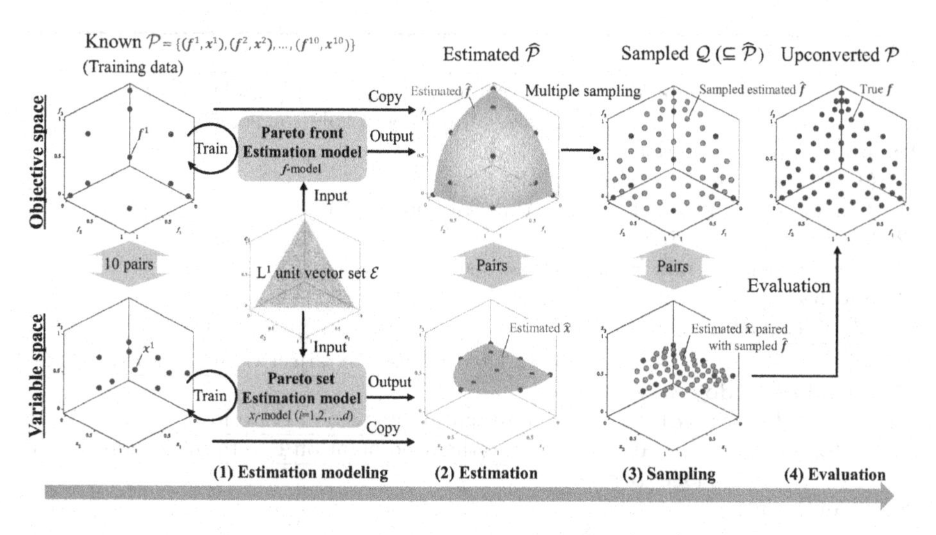

Fig. 1. Conventional SMOA [3] (Color figure online)

large set of uniformly distributed L^1 unit vectors $\mathcal{E} = \{e^1, e^2, \ldots\}$ covering all over directions in the objective space to the trained x_i-models ($i = 1, 2, \ldots, d$) and obtain the estimated variable vectors $\{\hat{x}^1, \hat{x}^2, \ldots\}$, which are the estimated Pareto set.

3.4 Estimated Set

For given known set $\mathcal{P}$ with L^1 unit vector set $\mathcal{E} = \{e^1, e^2, \ldots\}$, the Pareto front estimation outputs the estimated objective vectors $\{\hat{f}^1, \hat{f}^2, \ldots\}$ and the Pareto set estimation outputs the estimated variable vectors $\{\hat{x}^1, \hat{x}^2, \ldots\}$. As a result, the estimated set is represented as $\hat{\mathcal{P}} = \{(\hat{f}^1, \hat{x}^1), (\hat{f}^2, \hat{x}^2), \ldots\}$.

4 Conventional SMOA

The conventional SMOA samples and evaluates solutions by using the estimated Pareto front and the estimated Pareto set to upconvert the Pareto front resolution. Figure 1 shows a conceptual figure with $m = 3$ objectives and $d = 3$ variables, and Algorithm 1 shows a pseudo-code. The whole process is divided into (1) Estimation modeling, (2) Estimation, (3) Sampling, and (4) Evaluation.

The input of SMOA is a known set $\mathcal{P} = \{(f^1, x^1), (f^2, x^2), \ldots\}$. In (1) Estimation modeling process, we train the Pareto front estimation model, f-model, and the Pareto set estimation model, x_i-models ($i = 1, 2, \ldots, d$), with the known set $\mathcal{P}$ according to the procedure described in the previous section. Figure 1 shows a case with $|\mathcal{P}| = 10$ known candidate solutions with red points. In (2) Estimation process, we obtain the estimated set $\hat{\mathcal{P}} = \{(\hat{f}^1, \hat{x}^1), (\hat{f}^2, \hat{x}^2), \ldots\}$ by inputting a large set of L^1 unit vectors $\mathcal{E} = \{e^1, e^2, \ldots\}$ to the Pareto front

Algorithm 1 SMOA [3]

Require: Known candidate solutions (training data) $\mathcal{P} = \{(\boldsymbol{f}^1, \boldsymbol{x}^1), (\boldsymbol{f}^2, \boldsymbol{x}^2), \dots\}$, L^1 unit vector set $\mathcal{E} = \{\boldsymbol{e}^1, \boldsymbol{e}^2, \dots\}$, size of upconvert solution set N
Ensure: Upscaled solution set $\mathcal{P}$

(1) Estimation Modeling
1: $\boldsymbol{f}$-model $\leftarrow$ Train Pareto front estimation model ($\mathcal{P}$)
2: **for** $i \leftarrow 1, 2, \dots, d$ **do**
3: x_i-model $\leftarrow$ Train Pareto set estimation model ($\mathcal{P}$)
4: **end for**
(2) Estimation
5: $\hat{\mathcal{P}} \leftarrow \emptyset$ ▷ Estimated pair set of objective and variable vectors
6: **for all** $\boldsymbol{e} \in \mathcal{E}$ **do**
7: $\hat{\boldsymbol{f}} \leftarrow \boldsymbol{f}$-model ($\boldsymbol{e}$) ▷ An estimation point on Pareto front for direction $\boldsymbol{e}$
8: **for** $i \leftarrow 1, 2, \dots, d$ **do** ▷ An estimation point on Pareto set for direction $\boldsymbol{e}$
9: $\hat{x_i} \leftarrow x_i$-model ($\boldsymbol{e}$)
10: **end for**
11: $\hat{\mathcal{P}} \leftarrow \hat{\mathcal{P}} \cup \{(\hat{\boldsymbol{f}}, \hat{\boldsymbol{x}} = (\hat{x_1}, \hat{x_2}, \dots, \hat{x_d}))\}$
12: **end for**
(3) Sampling
13: $\mathcal{Q} \leftarrow \emptyset$ ▷ Sampled pair set, $\mathcal{Q} \subseteq \hat{\mathcal{P}}$
14: **while** $|\mathcal{Q}| \leq N$ **do** ▷ Sample N pairs
15: $(\hat{\boldsymbol{f}}, \hat{\boldsymbol{x}}) \leftarrow \mathop{\arg\max}\limits_{(\hat{\boldsymbol{f}}, \hat{\boldsymbol{x}}) \in \hat{\mathcal{P}}} \min\limits_{(\boldsymbol{f}, \boldsymbol{x}) \in \mathcal{P} \cup \mathcal{Q}} \sqrt{\sum_{j=1}^{m} (\hat{f}_j - f_j)^2}$
16: $\mathcal{Q} \leftarrow \mathcal{Q} \cup \{(\hat{\boldsymbol{f}}, \hat{\boldsymbol{x}})\}$, $\hat{\mathcal{P}} \leftarrow \hat{\mathcal{P}} \backslash \{(\hat{\boldsymbol{f}}, \hat{\boldsymbol{x}})\}$
17: **end while**
(4) Evaluation
18: **for all** $(\hat{\boldsymbol{f}}, \hat{\boldsymbol{x}}) \in \mathcal{Q}$ **do**
19: $\boldsymbol{f} \leftarrow$ Evaluate ($\hat{\boldsymbol{x}}$) ▷ Calls m objective functions on estimated $\hat{\boldsymbol{x}}$
20: $\mathcal{P} \leftarrow \mathcal{P} \cup \{(\boldsymbol{f}, \hat{\boldsymbol{x}})\}$
21: **end for**
22: **return** $\mathcal{P}$

estimation $\boldsymbol{f}$-model and the Pareto set estimation x_i-models ($i = 1, 2, \dots, d$). In Fig. 1, each estimated objective vector $\hat{\boldsymbol{f}}$ and estimated variable vector $\hat{\boldsymbol{x}}$ are respectively plotted as a small gray dot. In (3) Sampling process, we select N pairs of estimated objective and variable vectors as the sampled set $\mathcal{Q}$ from the estimated set $\hat{\mathcal{P}}$, i.e., $\mathcal{Q} \subseteq \hat{\mathcal{P}}$. This selection focuses only on the objective space and samples well-distributed N points as the sampled set $\mathcal{Q}$. For estimated objective vector $\hat{\boldsymbol{f}}$ of each estimated pair $(\hat{\boldsymbol{f}}, \hat{\boldsymbol{x}}) \in \hat{\mathcal{P}}$, we find the minimum distance to the known objective vectors in the known set $\mathcal{P}$ and the already sampled set $\mathcal{Q}$. We then take the estimated pair $(\hat{\boldsymbol{f}}, \hat{\boldsymbol{x}}) \in \hat{\mathcal{P}}$ with the maximum distance from the estimated set $\hat{\mathcal{P}}$ and add it to the sampled set $\mathcal{Q}$. We repeat the above N times. In (4) Evaluation process, we evaluate, call objective functions f_i ($i = 1, 2, \dots, m$), on estimated $\hat{\boldsymbol{x}}$ of each sampled pair $(\hat{\boldsymbol{f}}, \hat{\boldsymbol{x}}) \in \mathcal{Q}$ and add the obtained $(\boldsymbol{f}, \hat{\boldsymbol{x}})$ to the known set $\mathcal{P}$. In this way, the known set $\mathcal{P}$ is upconverted.

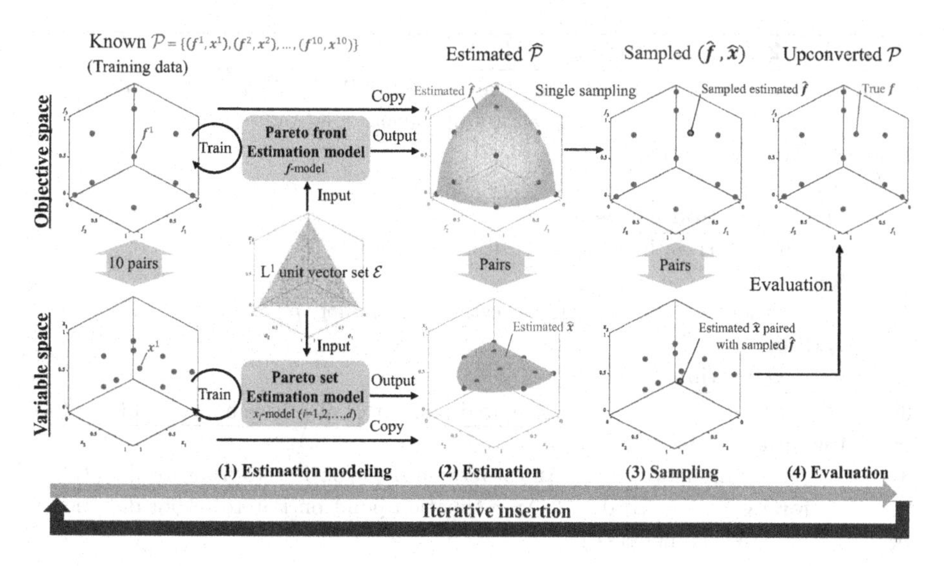

Fig. 2. Proposed iterative SMOA (I-SMOA) (Color figure online)

The conventional SMOA generates the estimation models once and samples N solutions all at once based on their estimation models. If the accuracy of the estimation models is low, the sampled points and the true evaluated points in the objective space are distanced. As a result, the approximation quality of the upconverted Pareto front cannot be expected.

5 Proposal: Iterative SMOA (I-SMOA)

In this work, we propose an iterative SMOA (I-SMOA) that repeats a single solution sampling, its evaluation, and estimation model update with the newly evaluated solution. The iteration gradually improves the accuracy of the estimation models and the sampling quality. Figure 2 shows a conceptual figure, and Algorithm 2 shows a pseudo-code. The proposed I-SMOA also involves (1) Estimation modeling, (2) Estimation, (3) Sampling, and (4) Evaluation processes. Differences from the conventional SMOA shown in Fig. 1 and Algorithm 1 are highlighted in blue. The differences are sampling size in (3) Sampling process and iterative executions of (1)–(4) processes.

In the proposed I-SMOA, the iterative loop in 1–17 lines of Algorithm 2 internally has (1)–(4) processes, which are respectively executed once in the conventional SMOA shown in Algorithm 1. The iteration is depicted as the blue arrow in the bottom of Fig. 2. In the proposed I-SMOA, (1) Estimation modeling and (2) Estimation processes are the same as the conventional SMOA. In (3) Sampling of the proposed I-SMOA, we select a single estimated pair $(\hat{\boldsymbol{f}}, \hat{\boldsymbol{x}}) \in \hat{\mathcal{P}}$ with the maximum distance to the nearest known set $\mathcal{P}$ in the objective space. In (4) Evaluation process, we evaluate the selected estimated pair $(\hat{\boldsymbol{f}}, \hat{\boldsymbol{x}})$ and add

Algorithm 2 Proposed Iterative SMOA

Require: Known candidate solutions (training data) $\mathcal{P} = \{(\boldsymbol{f}^1, \boldsymbol{x}^1), (\boldsymbol{f}^2, \boldsymbol{x}^2), \ldots\}$, L^1 unit vector set $\mathcal{E} = \{\boldsymbol{e}^1, \boldsymbol{e}^2, \ldots\}$, size of upconvert solution set N

Ensure: Upscaled solution set $\mathcal{P}$

1: **loop** N times

   (1) Estimation Modeling

2: $\boldsymbol{f}$-model $\leftarrow$ Train Pareto front estimation model ($\mathcal{P}$)

3: **for** $i \leftarrow 1, 2, \ldots, d$ **do**

4: x_i-model $\leftarrow$ Train Pareto set estimation model ($\mathcal{P}$)

5: **end for**

   (2) Estimation

6: $\hat{\mathcal{P}} \leftarrow \emptyset$ ▷ Estimated pair set of objective and variable vectors

7: **for all** $\boldsymbol{e} \in \mathcal{E}$ **do**

8: $\hat{\boldsymbol{f}} \leftarrow \boldsymbol{f}$-model ($\boldsymbol{e}$) ▷ An estimation point on Pareto front for direction $\boldsymbol{e}$

9: **for** $i \leftarrow 1, 2, \ldots, d$ **do** ▷ An estimation point on Pareto set for direction $\boldsymbol{e}$

10: $\hat{x_i} \leftarrow x_i$-model ($\boldsymbol{e}$)

11: **end for**

12: $\hat{\mathcal{P}} \leftarrow \hat{\mathcal{P}} \cup \{(\hat{\boldsymbol{f}}, \hat{\boldsymbol{x}} = (\hat{x_1}, \hat{x_2}, \ldots, \hat{x_d}))\}$

13: **end for**

   (3) Sampling

14: $(\hat{\boldsymbol{f}}, \hat{\boldsymbol{x}}) \leftarrow \underset{(\hat{\boldsymbol{f}},\hat{\boldsymbol{x}}) \in \hat{\mathcal{P}}}{\arg\max} \min_{(\boldsymbol{f},\boldsymbol{x}) \in \mathcal{P}} \sqrt{\sum_{j=1}^{m} (\hat{f}_j - f_j)^2}$ ▷ Sample a single pair

   (4) Evaluation

15: $\boldsymbol{f} \leftarrow$ Evaluate ($\hat{\boldsymbol{x}}$) ▷ Calls m objective functions on estimated $\hat{\boldsymbol{x}}$

16: $\mathcal{P} \leftarrow \mathcal{P} \cup \{(\boldsymbol{f}, \hat{\boldsymbol{x}})\}$

17: **end loop**

18: **return**$\mathcal{P}$

the obtained pair $(\boldsymbol{f}, \hat{\boldsymbol{x}})$ to the known set $\mathcal{P}$. We then go back to (1) Estimation modeling.

In this way, the proposed I-SMOA repeats to sample and evaluate a single solution while updating the estimation models with the evaluated result, the pair of the true objective vector and the examined variable vector. During the iteration, we expect the accuracy improvement of the estimation models and sampling quality minimizing the distance between the estimated objective vector $\hat{\boldsymbol{f}}$ and the true evaluated one $\boldsymbol{f}$. On the other hand, the iterative estimation modelings cost computationally. The proposed I-SMOA has suited to problems in that the computational cost of evaluations is much larger than the one of the iterative estimation modelings.

6 Experimental Settings

We used the DTLZ2 test problem [7] with $m = 3$ objectives and $d = 12$ variables. For the known candidate solutions $\mathcal{P}$, we first generated ten points on the unit plane by using two ways, which are the incremental lattice design (ILD) [16] and the uniform design using the Hammersley method (UDH) [17]. ILD and UDH provide different types of point distributions, respectively. We assumed two types of training data with them. We mapped the ten points to the Pareto front of DTLZ2 and employed the mapped ten points as the known objective vectors. Their variable vectors were calculated inversely, while using the DTLZ2 characteristic that all Pareto optimal solutions have $x_i = 0.5$ $(i = 3, 4, \ldots, 12)$.

We compare the conventional SMOA [3] and the proposed I-SMOA. As the response surface methodology, we employed Kriging [18] and RBNN [19] in this work. In both SMOAs, as the large L^1 unit vector set $\mathcal{E}$, we used uniformly distributed 26,335 L^1 unit vectors generated by ILD [16]. The upconvert size is set to $N = 150$, and the size of the output solution set becomes $|\mathcal{P}| + N = 10 + 150 = 160$. That is, 10 solutions are upconverted to 160 solutions with $N = 150$ function evaluations. Since the two algorithms are deterministic methods without random numbers, each algorithm ran once.

To evaluate the finally obtained solutions, we used *Hypervolume* (*HV*) [20] and *IGD* [21]. We utilized the PlatEMO implementation [22]. For *HV*, we first divided all objective vectors by 1.1 and then calculated *HV* of the divided ones and the reference point $(1.0, 1.0, 1.0)$. The higher *HV*, the better the Pareto front approximation. For *IGD*, we employed 9,870 reference points, which are mapped ones of points generated by the simplex-lattice design. For *IGD*, we employed 9,870 reference points, which are mapped ones of uniformly distributed points generated by the simplex-lattice design [23]. The lower *IGD*, the better the Pareto front approximation.

7 Experimental Results and Discussion

7.1 Obtained Upconverted Solution Set

Figure 3 shows finally obtained solutions when we input the known candidate solutions, the training data, generated by ILD. 10 red points are known candidate solutions. 150 gray points are generated solutions by the conventional SMOA and the proposed I-SMOA, respectively. The upper four figures are $m = 3$ dimensional objective spaces. The lower four figures are $x_1 - x_2$ variable spaces. The DTLZ2 problem used in this work also has variables x_i $(i = 3, 4, \ldots, 12)$. Since the optimal values of them are $x_i = 0.5$ $(i = 3, 4, \ldots, 12)$ due to the problem characteristic and obtained values are also the same, this figures only show the $x_1 - x_2$ variable space. Note that any point on the $x_1 - x_2$ variable space is the Pareto optimal in DTLZ2.

These results show that all four algorithms obtain solutions with high uniformity in the objective space. If we generate gray solutions in other ways as

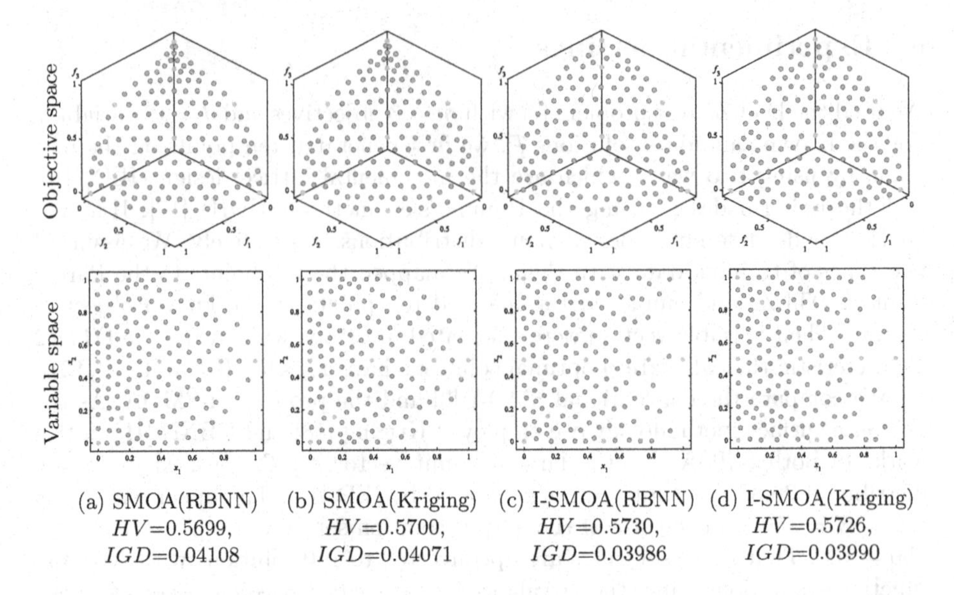

(a) SMOA(RBNN) HV=0.5699, IGD=0.04108 (b) SMOA(Kriging) HV=0.5700, IGD=0.04071 (c) I-SMOA(RBNN) HV=0.5730, IGD=0.03986 (d) I-SMOA(Kriging) HV=0.5726, IGD=0.03990

Fig. 3. ILD-based known solutions in red and upscaled solutions in gray (Color figure online)

mutations or perturbations based on random numbers for the upconvert, dominated solutions with $x_i \neq 0.5$ $(i = 3, 4, \ldots, 12)$ are generated, and solutions with low uniformity are generated [3]. On the other hand, we see SMOA and I-SMOA without the random factor obtain Pareto optimal solutions with high uniformity.

Next, we focus around $f_3 = 1$ in the objective space. In the case of the conventional SMOA, we see some upconverted solution circles in gray overlap, and the area is crowded with solutions. The absence of solutions in the upper right $(x_1, x_2) = (1, 1)$ and lower right $(x_1, x_2) = (1, 0)$ corners in the variable space is related to the solution distribution around $f_3 = 1$ in the objective space. On the other hand, in the case of the proposed I-SMOA, the distribution uniformity of solutions around $f_3 = 1$ in the objective space is higher than the one of the conventional SMOA. Also, in the variable space, the proposed I-SMOA obtained solutions near the upper and lower right corners. From the variable space, we see only two red points are known in the range $x_1 > 0.5$, and the two points commonly have $x_2 = 0.5$. The conventional SMOA trains the estimation models just once. The trained Pareto set estimation model could not estimate around the upper right $(x_1, x_2) = (1, 1)$ and lower right $(x_1, x_2) = (1, 0)$ corners in the variable space as the Pareto set since these areas are far from the known training data. On the other hand, the proposed I-SMOA could estimate near these areas as the Pareto set since the estimation models are iteratively updated.

Figure 4 shows finally obtained solutions when we input the known candidate solutions, the training data, generated by UDH. We see the two distributions of the known red point sets respectively by ILD in Fig. 3 and UDH in Fig. 4 are

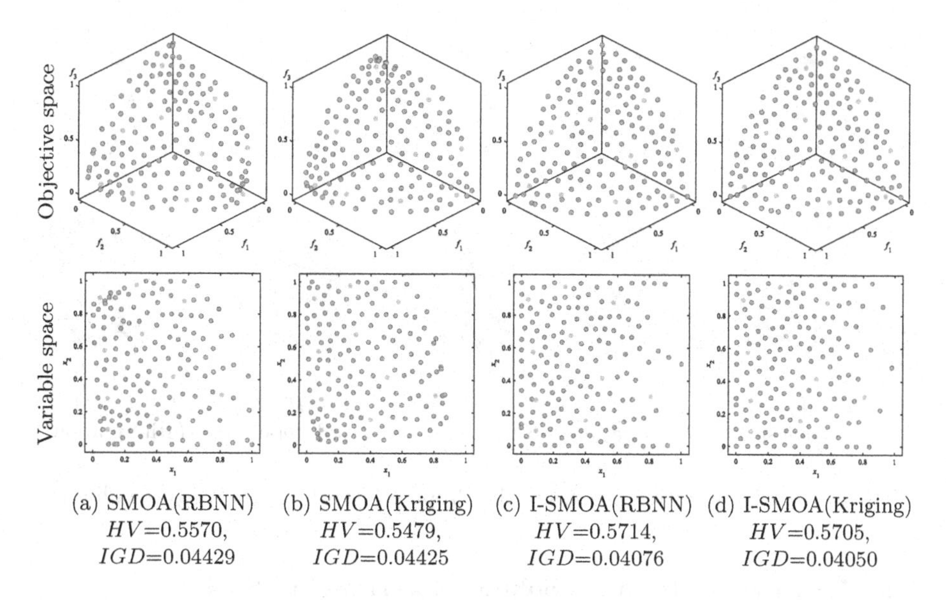

(a) SMOA(RBNN) HV=0.5570, IGD=0.04429 (b) SMOA(Kriging) HV=0.5479, IGD=0.04425 (c) I-SMOA(RBNN) HV=0.5714, IGD=0.04076 (d) I-SMOA(Kriging) HV=0.5705, IGD=0.04050

Fig. 4. UDH-based known solutions in red and upscaled solutions in gray (Color figure online)

different. In the case of ILD, 6 out of 10 points are on the edge of the Pareto front in the objective space. On the other hand, in the case of UDH, no points are on the edge of the Pareto front in the objective space.

Figure 4 show that the conventional SMOA faces difficulty in approximating the edge of the Pareto front. In other words, the conventional SMOA has difficulty generating solutions outside the known set in the objective space. On the other hand, the proposed I-SMOA is better than the conventional SMOA in the edge representation of the Pareto front. I-SMOA with RBNN provides some overlapped gray circles, but I-SMOA with Kriging provides non-overlapped points.

Figure 3 and Fig. 4 involve HV and IGD values in each caption. We see that results with UDH are worse than the ones with ILD in all four algorithms. That is, the Pareto front upconvert depends on the known solutions, the training data, and the UDH-based data is a more hard task than the ILD-based data. Also, we see the proposed I-SMOAs achieve better HV and IGD than the conventional SMOAs in both training data. The differences in metrics values between the two training data suggest that the proposed I-SMOA's dependency on the training data is lower than the conventional SMOA's one. That is, the proposed I-SMOA has robustness against the known training data compared with the conventional SMOA. Since the proposed I-SMOA iteratively updates the estimation models, the influence of the initial known training data can be decreased.

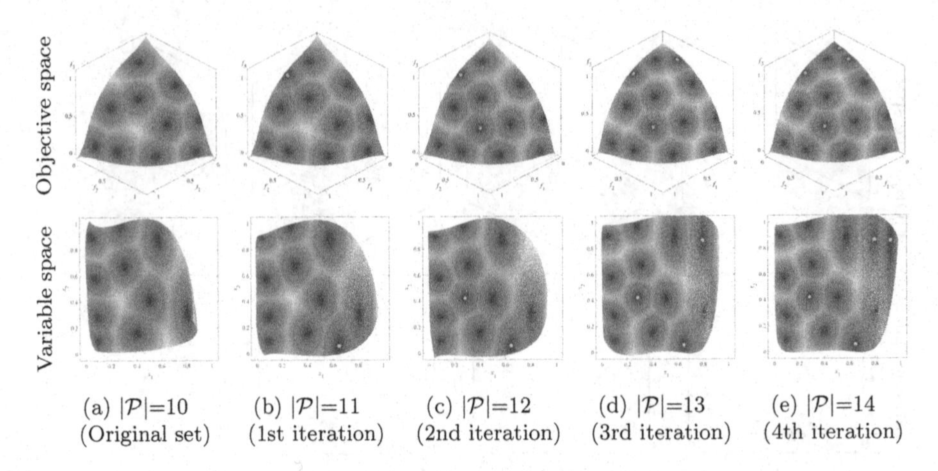

(a) $|\mathcal{P}|$=10 (Original set) (b) $|\mathcal{P}|$=11 (1st iteration) (c) $|\mathcal{P}|$=12 (2nd iteration) (d) $|\mathcal{P}|$=13 (3rd iteration) (e) $|\mathcal{P}|$=14 (4th iteration)

Fig. 5. Iterative insertion process of the Proposed I-SMOA (Color figure online)

7.2 Transition of Iterative Solution Insertion in I-SMOA

We focus on the proposed I-SMOA with Kriging shown in Fig. 4(d) and observe its algorithmic behavior. Figure 5 shows iterative solution insertion steps. Figure 5(a) only shows the known red solutions, Fig. 5(b) involves one gray solution inserted at the 1st iteration, Fig. 5(c) involves one more solution inserted at the 2nd iteration, and so on. In each figure, the estimated set is also plotted as very small dots with a color gradient. The color represents the minimum distance to the known set $\mathcal{P}$. The distance increases from blue to yellow. The most yellow estimated point in the objective space is sampled and evaluated in the proposed I-SMOA.

In Fig. 5(a) with $\mathcal{P} = 10$, two estimated sets of the conventional SMOA and the proposed I-SMOA are the same. The conventional SMOA and the proposed I-SMOA sample $N = 150$ estimated points. In the case of the conventional SMOA, the distance relation is changed based on the estimated values after every single point selection. On the other hand, the distance relation is changed based on the true values after every single point selection since the point is evaluated soon after the selection.

From Fig. 5(a) with $\mathcal{P} = 10$, we see the accuracy of the estimated Pareto front on edge is not enough in the objective space. It is desired to cover all $x_1 - x_2$ space in the variable space with colored small dots. However, the cover area with $\mathcal{P} = 10$ is not enough. Especially around the upper right $(x_1, x_2) = (1, 1)$ and the lower right $(x_1, x_2) = (1, 0)$ areas are far from the known points and recognized as the non Pareto set areas.

From the variable spaces shown in Fig. 5(a)–(d), we see the area covered by the colored estimated points is gradually expanded by adding new sampled and evaluated solutions. When the diversity of the known set is low, the expansion of the estimated area is needed to obtain widely distributed solutions.

8 Conclusions

To improve the upconvert quality of the Pareto front by utilizing the known candidate solutions, in this work, we proposed the iterative SMOA (I-SMOA), which iteratively repeated to sample a single promising solution, evaluate it, and update the estimation models with it. Experimental results using the DTLZ2 test problem with ten known candidate solutions, the training data, showed that the proposed I-SMOA achieved higher approximation performance of the Pareto front than the conventional SMOA. Also, the proposed I-SMOA obtained widely distributed solutions compared with the conventional SMOA in the objective space. The proposed I-SMOA also showed higher robustness against the initial known solutions, the training data, than the conventional SMOA.

In future works, we will verify the effectiveness of the proposed I-SMOA on problems with more than three objectives, a variety of Pareto front shapes, and complex variable relations. Also, we will compare the Pareto front approximation quality of the proposed I-SMOA with other efficient search methodologies, including the inversing model-based approach [24,25].

Acknowledgment. This work was supported by JSPS KAKENHI Grant Number 22H03660.

References

1. Chugh, T., Sindhya, K., Hakanen, J., Miettinen, K.: A survey on handling computationally expensive multiobjective optimization problems with evolutionary algorithms. Soft. Comput. **23**(9), 3137–3166 (2019)
2. Simon, D.: Evolutionary Optimization Algorithms Biologically-Inspired and Population-Based Approaches to Computer Intelligence, chap. 8.1 Initialization, p. 180. Wiley (2013)
3. Takagi, T., Takadama, K., Sato, H.: Supervised multi-objective optimization algorithm using estimation. In: 2022 IEEE Congress on Evolutionary Computation (CEC 2022), pp. 1–8 (2022)
4. Stein, M.L.: Interpolation of Spatial Data: Some Theory for Kriging. Springer, Heidelberg (2012)
5. Bishop, C.M., et al.: Neural Networks for Pattern Recognition. Oxford University Press, Oxford (1995)
6. Liu, J.: Radial Basis Function (RBF) Neural Network Control for Mechanical Systems: Design, Analysis and Matlab Simulation. Springer, Heidelberg (2013)
7. Deb, K., Thiele, L., Laumanns, M., Zitzler, E.: Scalable test problems for evolutionary multiobjective optimization. In: Abraham, A., Jain, L., Goldberg, R. (eds.) Evolutionary Multiobjective Optimization, pp. 105–145. Springer, London (2005). https://doi.org/10.1007/1-84628-137-7_6
8. Hartikainen, M., Miettinen, K., Wiecek, M.M.: PAINT: Pareto front interpolation for nonlinear multiobjective optimization. Comput. Optim. Appl. **52**(3), 845–867 (2012)
9. Kumar Singh, H., Shankar Bhattacharjee, K., Ray, T.: A projection-based approach for constructing piecewise linear Pareto front approximations. J. Mech. Des. **138**(9), 091404 (2016)

10. Takagi, T., Takadama, K., Sato, H.: Pareto front estimation using unit hyperplane. In: Ishibuchi, H., et al. (eds.) EMO 2021. LNCS, vol. 12654, pp. 126–138. Springer, Cham (2021). https://doi.org/10.1007/978-3-030-72062-9_11
11. Kobayashi, K., Hamada, N., Sannai, A., Tanaka, A., Bannai, K., Sugiyama, M.: Bézier simplex fitting: describing Pareto fronts of simplicial problems with small samples in multi-objective optimization. In: AAAI Conference on Artificial Intelligence, vol. 33, pp. 2304–2313 (2019)
12. Tian, Y., Si, L., Zhang, X., Tan, K.C., Jin, Y.: Local model-based Pareto front estimation for multiobjective optimization. IEEE Trans. Syst. Man Cybern. Syst. **53**, 623–634 (2022)
13. Zapotecas Martínez, S., Sosa Hernández, V.A., Aguirre, H., Tanaka, K., Coello Coello, C.A.: Using a family of curves to approximate the Pareto front of a multi-objective optimization problem. In: Bartz-Beielstein, T., Branke, J., Filipič, B., Smith, J. (eds.) PPSN 2014. LNCS, vol. 8672, pp. 682–691. Springer, Cham (2014). https://doi.org/10.1007/978-3-319-10762-2_67
14. Zhou, A., Zhang, Q., Jin, Y.: Approximating the set of Pareto-optimal solutions in both the decision and objective spaces by an estimation of distribution algorithm. IEEE Trans. Evol. Comput. **13**(5), 1167–1189 (2009)
15. Giagkiozis, I., Fleming, P.J.: Pareto front estimation for decision making. Evol. Comput. **22**(4), 651–678 (2014)
16. Takagi, T., Takadama, K., Sato, H.: Incremental lattice design of weight vector set. In: 2020 Genetic and Evolutionary Computation Conference (GECCO 2020), pp. 1486–1494 (2020)
17. Molinet Berenguer, J.A., Coello Coello, C.A.: Evolutionary many-objective optimization based on Kuhn-Munkres' algorithm. In: Gaspar-Cunha, A., Henggeler Antunes, C., Coello, C.C. (eds.) EMO 2015. LNCS, vol. 9019, pp. 3–17. Springer, Cham (2015). https://doi.org/10.1007/978-3-319-15892-1_1
18. Sacks, J., Welch, W.J., Mitchell, T.J., Wynn, H.P.: Design and analysis of computer experiments. Stat. Sci. **4**, 409–423 (1989)
19. Kim, P.: Matlab Deep Learning: With Machine Learning, Neural Networks and Artificial Intelligence, vol. 130, no. 21 (2017)
20. Zitzler, E., Thiele, L.: Multiobjective evolutionary algorithms: a comparative case study and the strength pareto approach. IEEE Trans. Evol. Comput. **3**(4), 257–271 (1999)
21. Coello, C.A.C., Cortés, N.C.: Solving multiobjective optimization problems using an artificial immune system. Genet. Program Evolvable Mach. **6**(2), 163–190 (2005)
22. Tian, Y., Cheng, R., Zhang, X., Jin, Y.: Platemo: a matlab platform for evolutionary multi-objective optimization [educational forum]. IEEE Comput. Intell. Mag. **12**(4), 73–87 (2017)
23. Das, I., Dennis, J.E.: Normal-boundary intersection: a new method for generating the pareto surface in nonlinear multicriteria optimization problems. SIAM J. Optim. **8**(3), 631–657 (1998)
24. Farias, L.R., Araújo, A.F.: IM-MOEA/D: an inverse modeling multi-objective evolutionary algorithm based on decomposition. In: 2021 IEEE International Conference on Systems, Man, and Cybernetics (SMC 2021), pp. 462–467. IEEE (2021)
25. Cheng, R., Jin, Y., Narukawa, K.: Adaptive reference vector generation for inverse model based evolutionary multiobjective optimization with degenerate and disconnected pareto fronts. In: Gaspar-Cunha, A., Henggeler Antunes, C., Coello, C.C. (eds.) EMO 2015. LNCS, vol. 9018, pp. 127–140. Springer, Cham (2015). https://doi.org/10.1007/978-3-319-15934-8_9

An Improved Fuzzy Classifier-Based Evolutionary Algorithm for Expensive Multiobjective Optimization Problems with Complicated Pareto Sets

Jinyuan Zhang, Linjun He, and Hisao Ishibuchi(✉)

Guangdong Provincial Key Laboratory of Brain-inspired Intelligent Computation, Department of Computer Science and Engineering, Southern University of Science and Technology, Shenzhen 518055, China
zhangjy@sustech.edu.cn , hisao@sustech.edu.cn

Abstract. Various surrogate-based multiobjective evolutionary algorithms (MOEAs) have been proposed to solve expensive multiobjective optimization problems (MOPs). However, these algorithms are usually examined on test suites with unrealistically simple Pareto sets (e.g., ZDT and DTLZ test suites). Real-world MOPs usually have complicated Pareto sets, such as a vehicle dynamic design problem and a power plant design optimization problem. Such MOPs are challenging to construct reliable surrogates for surrogate-based MOEAs. Constructed surrogates with low accuracy are likely to make incorrect predictions and even mislead the search direction. In this paper, we propose an improved fuzzy classifier-based MOEA by leveraging the accuracy information of the classifier. The proposed algorithm is compared with five state-of-the-art algorithms on two well-known test suites with complicated Pareto sets and four real-world problems. Experimental results demonstrate the effectiveness of the proposed algorithm in solving realistic MOPs with complicated Pareto sets when only a limited number of function evaluations are available.

Keywords: Expensive multiobjective optimization · Evolutionary algorithms · Fuzzy classifier · Surrogate models · Complicated Pareto set

1 Introduction

Engineering optimization problems usually have two or more conflicting objectives, known as multiobjective optimization problems (MOPs) [5,16] that need to be optimized simultaneously. A number of multiobjective evolutionary algorithms (MOEAs) have been proposed to solve MOPs [45]. Typically, MOEAs can be classified into three categories: dominance-based MOEAs [9,47], indicator-based MOEAs [2,33,46], and decomposition-based MOEAs [27,42].

M. Emmerich et al. (Eds.): EMO 2023, LNCS 13970, pp. 231–246, 2023.
https://doi.org/10.1007/978-3-031-27250-9_17

These MOEAs usually evaluate the quality of solutions based on the evaluated objective function values and require a large number of function evaluations (FEs) [23,33]. However, FEs are usually computationally expensive in engineering MOPs where the evaluation of a solution requires physical simulations that consume a large amount of time or resources [5]. The available number of FEs is usually limited for solving these expensive MOPs.

Several methods have been proposed for solving expensive MOPs. One of the most efficient methods is surrogate-based MOEAs [4,8,17]. Generally, surrogate-based MOEAs use computationally cheap surrogate models to replace the original objective functions or fitness functions to evaluate the quality of solutions. These surrogate-based MOEAs can be classified into two categories depending on the types of surrogate models: regression-based MOEAs [8,17,24,30] and classification-based MOEAs [26,29,41].

- Regression-based MOEAs use regression models to approximate the original objective functions or fitness functions of MOPs. The constructed models are used to evaluate the quality of solutions. Generally, the number of constructed models is the same as the number of objective functions, with one model for each objective function [3,21,43]. Therefore, the time consumption of model construction is high, and this consumption will increase with the increase in the number of objectives. Some algorithms have been proposed to reduce the number of constructed regression models [8,12].
- Classification-based MOEAs use classifiers to model the relation among solutions, e.g., the Pareto dominance relation among solutions. These classifiers are used to select promising solutions for subsequent optimization procedures. Since classification-based MOEAs usually build one classifier to model the relation among solutions, the number of constructed models is smaller than that in the regression-based methods.

However, these surrogate-based algorithms have usually been examined on test suites with simple Pareto sets. In Table 1, we summarize some typical surrogate-based MOEAs and the test suites used in their experimental studies. We can see that ZDT [7], DTLZ [11] and WFG [14] test suites are commonly used to examine the performance of surrogate-based MOEAs. However, the Pareto sets (PSs) of most of these test problems are linear and parallel to coordinate axes, which are simple and unrealistic [22,23]. Real-world MOPs, such as a vehicle dynamic design problem [19] and a power plant design optimization problem [13], usually have complicated PSs [10,15,23,28] due to the linkages between variables [10,28] and the nonlinear shape of PSs [23]. It is worth noting that real-world MOPs with complicated PSs are challenging to construct reliable surrogates for surrogate-based MOEAs. Constructed surrogates with low accuracy are likely to make incorrect predictions and even mislead the search direction. Although the accuracy of the surrogates can be measured during model construction, it is rarely used as an indicator to guide the search.

In this paper, we improve our previous work [39] and propose an improved fuzzy classifier-based multiobjective evolutionary algorithm (IFCS-MOEA) by leveraging the accuracy information of the classifier. A novel sorting mechanism

Table 1. Typical surrogate-based MOEAs and the test suites used in their experimental studies.

	Algorithm	Year	Test suites
Regression-based MOEAs	ParEGO [21]	2006	KNO1 [21], OKA [28], VLMOP [35], ZDT [7], DTLZ [11]
	MOEA/D-EGO [3]	2010	KNO1 [21], VLMOP2 [35], ZDT [7], LZ [23], DTLZ [11]
	K-RVEA [3]	2018	DTLZ [11], WFG [14]
	KTA2 [30]	2021	DTLZ [11], WFG [14]
	EDN-ARMOEA [12]	2022	DTLZ [11], WFG [14]
Classification-based MOEAs	CSEA [29]	2019	DTLZ [11], WFG [14]
	θ-DEA-DP [36]	2022	DTLZ [11], WFG [14]
	MCEA/D [31]	2022	DTLZ [11], WFG [14]

is proposed to consider the membership degree of each solution and the accuracy of the classifier simultaneously. The proposed algorithm is compared with five state-of-the-art surrogate-based algorithms on two well-known test suites with complicated PSs and four real-world optimization problems to show its superiority in dealing with realistic expensive MOPs.

The rest of this paper is organized as follows. Section 2 presents related work to this paper. Section 3 presents the proposed IFCS-MOEA framework in detail. Section 4 examines the effectiveness of the proposed framework and compares it with five state-of-the-art algorithms. Section 5 concludes this paper.

2 Related Work

2.1 Multiobjective Optimization Problems

Typically, an MOP can be expressed as follows:

$$\begin{aligned}&\text{Minimize } F(x) = (f_1(x), \cdots, f_M(x))^{\mathrm{T}},\\&\text{subject to } x \in \Omega \subset R^n,\end{aligned} \tag{1}$$

where x is an n-dimensional decision vector, Ω is the decision space, $F(x)$ is an M-dimensional objective vector, and $f_i(x)$, $i = 1, \ldots, M$ is the i-th objective function.

Since the objective functions in Eq. (1) are usually in conflict with each other, it is impossible to find a single optimal solution that can optimize all objective functions simultaneously. Therefore, Pareto optimal solutions are defined. Let u and v be two solutions to Eq. (1). u is said to dominate v, if $f_i(u) \leq f_i(v)$ for $i = 1, \ldots, M$ and $f_j(u) < f_j(v)$ for at least one $j \in \{1, \ldots, M\}$. Solution u is regarded as a Pareto optimal solution if there does not exist any solution that dominates u. The Pareto set (PS) is defined as the set of all Pareto optimal solutions. The Pareto front (PF) is defined as the image of the PS in the objective space.

2.2 Surrogate-Based MOEAs

Regression models are widely used in surrogate-based MOEAs to approximated the objective functions of MOPs. Knowles [21] proposed to use an efficient global optimization (EGO) algorithm [18] to solve expensive MOPs. The proposed algorithm constructed a Gaussian process model to mimic the landscape of MOPs. Zhang et al. [43] combined the EGO algorithm with MOEA/D to solve expensive MOPs. The proposed algorithm constructed a Gaussian model to mimic the landscape of each decomposed subproblem of an MOP. Chugh et al. [3] combined the Kriging model with a reference vector guided evolutionary algorithm to solve expensive MOPs. The proposed algorithm constructed each Kriging model to mimic each objective function of an MOP. Song et al. [30] combined the Kriging model with a two-archive evolutionary algorithm. The proposed method constructed each Kriging model to approximate each objective function of an MOP.

Generally, solutions in a population in MOEAs can be divided into two categories: non-dominated solutions and dominated solutions, based on the Pareto dominance relation among them. Therefore, classifiers can be built to mimic this relation among solutions and can be used to select promising solutions. Loshchilov et al. [26] combined a classifier with a regression model to predict the dominance relation between a new solution and the existing non-dominated solutions. Bandaru et al. [1] applied multi-class classifiers to mimic the dominance relation between each pair of solutions. Zhang et al. [40,41] employed classifiers to model the dominance relation among solutions and to pre-select promising offspring solutions. Lin et al. [25] used a classifier to pre-select promising offspring solutions, thereby reducing the required number of FEs of MOEA/D. Pan et al. [29] applied a classifier to predict the dominance relation between a new solution and the reference solutions. Zhang et al. [38,39] employed a fuzzy classifier to assist environmental selection of MOEAs. Class labels and membership degrees were used to select promising offspring solutions for function evaluations. Yuan et al. [36] proposed to use two feedforward neural network models for solving expensive MOPs. One model was used to predict the Pareto dominance relation between solutions, and another model was built to predict the θ-dominance relation among solutions. Sonoda et al. [31] proposed to use multiple classifiers for solving high-dimensional expensive MOPs. Each classifier was constructed for each subproblem in the MOEA/D-DE algorithm. Zhang et al. [37] proposed a dual fuzzy-classifier-based surrogate model. One fuzzy classifier was constructed to learn the Pareto dominance relation among solutions, and another fuzzy classifier was constructed to learn the crowdedness of solutions.

3 Our Proposed Algorithm

This section presents the details of our improved fuzzy classifier-based MOEA (IFCS-MOEA) framework. IFCS-MOEA is proposed by using an improved fuzzy classifier-based surrogate model (IFCS). The IFCS model is constructed for sorting unevaluated solutions. First, Sect. 3.1 presents the general framework of IFCS-MOEA. Then, Sect. 3.2 describes IFCS-based sorting strategy in detail.

Algorithm 1: Framework of IFCS-MOEA

```
1  Initialize the population P = {x^1, x^2, ..., x^N}, and evaluate the solutions in P;
2  Set Arc = P;
3  while termination condition is not satisfied do
4      Set A+ = Non-dominated_Selection(Arc) and A- = Arc\A+;
5      Construct a classifier [l, md+] = fuzzy_classifier_construction(x) by
       using A+ and A-;
6      Validate the accuracy of the classifier Accuracy = k-fold(Arc);
7      Set Q_p = ∅;
8      Mating_P = P;
9      while w < w_max do
10         Generate 2N offspring solutions Q = {y^1, ..., y^2N} by using Mating_P;
11         Sort the offspring solutions Q = IFCS_Sorting(Q, Accuracy);
12         Select the top N solutions Q_top from Q;
13         Q_p = Q_p ∪ Q_top;
14         Mating_P = Q_top;
15         w = w + 1;
16     end
17     Sort all solutions in Q_p by Q_p = IFCS_Sorting(Q_p, Accuracy);
18     Select the top η solutions Q_eval from Q_p and evaluate them;
19     Arc = Arc ∪ Q_eval;
20     P = Environmental_Selection(Arc, N);
21 end
```

3.1 Algorithm Framework

The framework of the proposed IFCS-MOEA is presented in Algorithm 1. It is composed of four main procedures as follows.

- Initialization: N solutions are initialized and evaluated in Line 1. All the evaluated solutions are collected in Arc in Line 2.
- Fuzzy classifier construction: All the solutions in the archive are used as training data to construct a fuzzy classifier. The Pareto dominance relation is used to define two classes of the training data in Line 4. The non-dominated solutions are positive, and the dominated solutions are negative. A fuzzy classifier is constructed in Line 5. This paper uses a Fuzzy-KNN classifier [20] to construct the IFCS model. The fuzzy-KNN uses fuzzy similarity to predict the class of each solution. When a fuzzy classifier is used to predict the quality of a new solution, the class label l of the new solution and the membership degree to each class are obtained. A membership degree indicates the degree to the class which a new solution belongs to. A new solution's membership degree is calculated based on its K nearest neighbor's membership degrees. In this paper, we use the classifier to deal with the two-class problem. The membership degree md_+ in Line 5 is only for the positive class while the

membership degree for the negative class is $1 - md_+$. $md_+ = 0.5$ is used as the classification boundary. If $md_+ \geq 0.5$, the solution is labeled as positive, otherwise it is negative. The k-fold cross-validation method is applied to validate the effectiveness of the classifier in Line 6. The accuracy of the classifier is obtained.

- Offspring generation: $2N$ offspring solutions are generated by using the mating population in Line 10. Next, the IFCS model is applied to sort the $2N$ offspring solutions in Line 11. The top N promising offspring solutions are selected and stored in Line 12. Then, these selected solutions are used as mating solutions to generate new offspring solutions. This offspring generation process is repeated w_{max} times.
- New population generation: The IFCS model is used to sort all selected $w_{max} \times N$ offspring solutions in Line 17. The top η solutions are selected and evaluated by the objective functions in Line 18. The archive is updated by using the newly evaluated solutions in Line 19. Finally, the environmental selection mechanism of an MOEA is used to select N solutions from Arc to form the new population for the next generation in Line 20.

3.2 IFCS-Based Sorting

After the fuzzy classifier is constructed, the k-fold cross-validation method is used to measure the reliability of the classifier. The mean accuracy ($Accuracy$) of the classifier is obtained after the validation. In our algorithm framework, we use $k = 10$ for experiments.

As mentioned in Sect. 3.1, for a solution, if its membership degree with respect to the positive class is $md_+ \geq 0.5$, the solution is classified as a positive solution by the classifier with small uncertainty. When the $0 \leq md_+ < 0.5$, the solution is classified as a negative solution with small uncertainty. When the md_+ value is close to 0.5, the classification result has a large uncertainty in the class prediction.

Based on the above considerations, we propose an IFCS-based sorting strategy to sort solutions based on the model accuracy and membership degrees. The details of the proposed IFCS-based sorting strategy are presented in Algorithm 2. The constructed fuzzy classifier is used to predict the label l and the membership degree md_+ (with respect to the positive class) of each solution in Q (Line 1). These solutions are ranked in different manners according to the accuracy of the classifier and the membership degree to the positive class.

Figure 1 plots the accuracy of the fuzzy classifier at each generation through the execution of IFCS-MOEA/D-DE on UF8 and LZ5 test problems with the median IGD values over 21 runs. The two figures show that the accuracy of the model is low at the beginning of optimization. The accuracy will increase along with the increase of generations. The reason is that since the size of the training data set is small in early generations, the model constructed by using these data cannot approach the true relation among solutions and is hard to make correct predictions. After several generations, the size of the training data set increases, the model can approach the true relation among solutions and the accuracy of the prediction increases. For this reason, we consider the following

Algorithm 2: Q = IFCS_Sorting(Q, *Accuracy*)

```
1  Predict the label and membership degree of each solution in Q by
   [l, md+] = fuzzy_classifier_prediction(y);
2  if Accuracy ≥ 70% then
3      Sort solutions in Q with respect to their membership degrees in descending
       order;
4  else if 30% ≤ Accuracy < 70% then
5      Qp = {y ∈ Q|l = 1};
6      Qn = {y ∈ Q|l ≠ 1};
7      Sort solutions in Qp with respect to their membership degrees in ascending
       order;
8      Sort solutions in Qn with respect to their membership degrees in descending
       order;
9      Q = Qp ∪ Qn, the solutions in Qp are ranked before the solutions in Qn;
10 else
11     Sort solutions in Q with respect to their membership degrees in ascending
       order;
12 end
```

three cases according to the model accuracy. We specify the threshold values as 30% and 70% since the accuracy of the model is usually larger than 70% in our experiments as shown in Fig. 1.

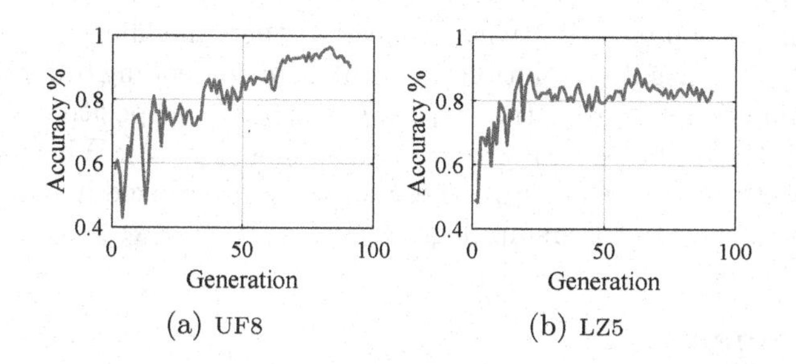

Fig. 1. The accuracy versus generation obtained by IFCS-MOEA/D-DE on UF8 and LZ5 with the median IGD values over 21 runs.

- *Accuracy* $\geq 70\%$: The unevaluated solutions are ranked in descending order with respect to md_+ values. This is because the model accuracy is high and we can trust the predictions of the classifier.
- $30\% \leq$ *Accuracy* $< 70\%$: First, positive solutions are ranked in ascending order with respect to md_+ values. This is because the model is more uncertain for the prediction of a solution with a smaller md_+ value than that with a larger md_+ value for the positive class. Evaluating uncertain solutions can improve the model accuracy (after evaluation, these solutions will be added to training data). Next, negative solutions are ranked in descending order with respect to md_+ values. This is because the model is more uncertain

for the prediction of the solution with a larger md_+ value than that with a smaller md_+ value for the negative class. Then, the positive solutions are ranked before the negative solutions.

- *Accuracy* < 30%: The unevaluated solutions are ranked in ascending order with respect to md_+ values. This is because the model accuracy is too small and we cannot trust the predictions of the classifier.

Table 2. Example of four unevaluated solutions.

	s_1	s_2	s_3	s_4
Label predicted by the classifier	0	1	1	0
Membership degree with respect to the positive class (md_+)	0.4	0.9	0.6	0.1

For example, suppose we have four solutions in Q as shown in Table 2. Each solution has a label and a membership degree with respect to the positive class predicted by the classifier. When the accuracy of the classifier is larger than or equal to 70%, these solutions are ranked in descending order with respect to their membership degrees (i.e., $s_2 > s_3 > s_1 > s_4$). When the accuracy of the classifier is larger than or equal to 30% and smaller than 70%, the positive solutions are ranked in ascending order with respect to their membership degrees (i.e., $s_3 > s_2$). Next, the negative solutions are ranked in descending order with respect to their membership degrees (i.e., $s_1 > s_4$). Then, the positive solutions are ranked before the negative solutions (i.e., $s_3 > s_2 > s_1 > s_4$). When the accuracy of the classifier is smaller than 30%, these solutions are ranked in ascending order with respect to their membership degrees (i.e., $s_4 > s_1 > s_3 > s_2$).

4 Experiments

This section examines the effectiveness of the proposed IFCS-MOEA framework. First, Sect. 4.1 presents the experimental settings. Second, Sect. 4.2 examines the effect of IFCS on MOEA/D-DE. Then, Sect. 4.3 compares the performance of IFCS-MOEA/D-DE with five state-of-the-art MOEAs on 19 test problems. Finally, Sect. 4.4 compares the performance of IFCS-MOEA/D-DE with five state-of-the-art MOEAs on four real-world application problems.

4.1 Experimental Settings

MOEA/D-DE [23] is integrated with the proposed framework for experiments, and the resulting algorithm is named IFCS-MOEA/D-DE. Five surrogate-based MOEAs, i.e., FCS-MOEA/D-DE [39], CPS-MOEA [41], CSEA [29], MOEA/D-EGO [43] and EDN-ARM-OEA [12] are used for comparison. UF1–10, LZ1–9 test problems [23,44] with complicated PSs are used for experiments. Among them,

UF1–7, LZ1–5, and LZ7–9 have 2 objectives, UF8–10, and LZ6 have 3 objectives. UF1–10, LZ1–5, and LZ9 are with 30 decision variables, and LZ6–8 are with 10 decision variables. The population size N is set to 45 for all compared algorithms. The maximum number of FEs is set as 500 since the problems are viewed as expensive MOPs [39]. For each test problem, each algorithm is executed 21 times independently. For IFCS-MOEA/D-DE, w_{max} is set to 30 and η is set to 5. For the other algorithms, we use the settings suggested in their papers. The IGD [6] metric is used to evaluate the performance of each algorithm. All algorithms are examined on PlatEMO [34] platform.

Table 3. The $mean_{std}$ IGD values of IFCS-MOEA/D-DE and MOEA/D-DE on UF1–10 and LZ1–9.

	IFCS-MOEA/D-DE	MOEA/D-DE
UF1	$8.87e{-}01_{1.03e-01}$	$1.04e{+}00_{1.48e-01}(-)$
UF2	$1.88e{-}01_{2.87e-02}$	$2.17e{-}01_{1.90e-02}(-)$
UF3	$6.04e{-}01_{2.57e-02}$	$6.51e{-}01_{4.11e-02}(-)$
UF4	$1.32e{-}01_{6.34e-03}$	$1.37e{-}01_{7.67e-03}(\sim)$
UF5	$4.20e{+}00_{3.67e-01}$	$4.49e{+}00_{3.60e-01}(-)$
UF6	$3.78e{+}00_{5.72e-01}$	$4.52e{+}00_{4.10e-01}(-)$
UF7	$9.50e{-}01_{1.11e-01}$	$1.11e{+}00_{1.21e-01}(-)$
UF8	$7.24e{-}01_{9.32e-02}$	$7.93e{-}01_{1.34e-01}(\sim)$
UF9	$7.57e{-}01_{8.23e-02}$	$8.99e{-}01_{9.73e-02}(-)$
UF10	$4.65e{+}00_{4.34e-01}$	$5.10e{+}00_{6.81e-01}(-)$
LZ1	$1.55e{-}01_{8.70e-03}$	$1.71e{-}01_{1.75e-02}(-)$
LZ2	$8.99e{-}01_{1.69e-01}$	$1.07e{+}00_{1.54e-01}(-)$
LZ3	$2.22e{-}01_{1.45e-02}$	$2.61e{-}01_{2.37e-02}(-)$
LZ4	$2.15e{-}01_{2.08e-02}$	$2.61e{-}01_{2.59e-02}(-)$
LZ5	$2.00e{-}01_{2.53e-02}$	$2.21e{-}01_{1.97e-02}(-)$
LZ6	$4.92e{-}01_{4.56e-02}$	$5.74e{-}01_{1.17e-01}(-)$
LZ7	$1.02e{+}00_{3.52e-01}$	$1.30e{+}00_{2.39e-01}(-)$
LZ8	$8.33e{-}01_{1.23e-01}$	$8.94e{-}01_{1.14e-01}(\sim)$
LZ9	$9.52e{-}01_{1.43e-01}$	$1.08e{+}00_{1.22e-01}(-)$
$+/-/\sim$		0/16/3

4.2 Effect of IFCS on MOEA/D-DE

This section examines the effectiveness of IFCS-MOEA framework on MOEA/D-DE. IFCS-MOEA is embedded with MOEA/D-DE (IFCS-MOEA/D-DE) and compared with the original MOEA/D-DE on UF1–10 and LZ1–9 test problems.

Table 3 shows the mean IGD values obtained by IFCS-MOEA/D-DE and MOEA/D-DE after 500 FEs on the 19 test problems. The Wilcoxon rank-sum test at the 5% significance level is used to evaluate the statistical difference between IFCS-MOEA/D-DE and MOEA/D-DE. In this table, "+, −, ∼" denote that the results obtained by MOEA/D-DE are better than, worse than, or similar to those obtained by IFCS-MOEA/D-DE, respectively. Table 3 shows that IFCS-MOEA/D-DE outperforms MOEA/D-DE on 16 test problems. On UF4, UF8, and LZ8, the two algorithms obtain similar results.

Figure 2 plots the mean IGD versus the number of FEs obtained by IFCS-MOEA/D-DE and MOEA/D-DE on the UF2, UF10, and LZ7 test problems. Figure 2 shows that IFCS-MOEA/D-DE converges faster and obtains better IGD values than MOEA/D-DE on these three test problems.

Based on the above results, we can conclude that IFCS-MOEA/D-DE is more efficient than MOEA/D-DE in solving these 19 MOPs with complicated PSs under a limited number of FEs.

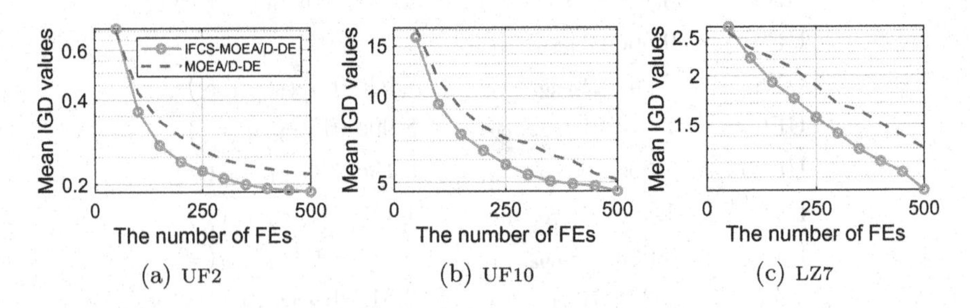

Fig. 2. The mean IGD values versus the number of FEs obtained by IFCS-MOEA/D-DE and MOEA/D-DE on UF2, UF10, and LZ7.

4.3 Performance Comparison with the State-of-the-art MOEAs

This section compares the performance of IFCS-MOEA/D-DE with five state-of-the-art surrogate-based MOEAs: FCS-MOEA/D-DE, CPS-MOEA, CSEA, MOEA/D-EGO, and EDN-ARMOEA. These algorithms are compared on the UF1–10 and LZ1–9 test problems.

Table 4 presents the mean IGD values obtained by the six algorithms after 500 FEs on the 19 test problems. The Wilcoxon rank-sum test is used for statistical test. The best result on each test problem is shaded. At the bottom of the table, we summarize the number of problems on which the performance of the compared algorithm is better than, worse than, and similar to that of

IFCS-MOEA/D-DE, respectively. In Table 4, IFCS-MOEA/D-DE outperforms FCS-MOEA/D-DE, CPS-MOEA, CSEA, MOEA/D-EGO, and EDN-ARMOEA on 9, 16, 11, 15, and 12 test problems, respectively. IFCS-MOEA/D-DE performs worse than FCS-MOEA/D-DE, CPS-MOEA, CSEA, MOEA/D-EGO, and EDN-ARMOEA on 6, 0, 6, 0, and 0 test problems, respectively.

Table 4. The *mean* IGD values of IFCS-MOEA/D-DE, FCS-MOEA/D-DE, CPS-MOEA, CSEA, MOEA/D-EGO, and EDN-ARMOEA on UF1–10 and LZ1–9.

	IFCS-MOEA/D-DE	FCS-MOEA/D-DE	CPS-MOEA	CSEA	MOEA/D-EGO	EDN-ARMOEA
UF1	8.87e-01$_{1.03e-01}$	5.73e-01$_{1.70e-01}$(+)	1.05e+00$_{1.56e-01}$(−)	5.98e-01$_{2.16e-01}$(+)	9.40e-01$_{1.68e-01}$(∼)	9.68e-01$_{1.69e-01}$(−)
UF2	1.88e-01$_{2.87e-02}$	3.04e-01$_{4.55e-02}$(−)	3.00e-01$_{3.73e-02}$(−)	3.26e-01$_{5.30e-02}$(−)	4.09e-01$_{5.39e-02}$(−)	4.12e-01$_{3.71e-02}$(−)
UF3	6.04e-01$_{2.57e-02}$	6.91e-01$_{5.89e-02}$(−)	7.30e-01$_{6.35e-02}$(−)	7.11e-01$_{7.26e-02}$(−)	7.63e-01$_{7.43e-02}$(−)	7.27e-01$_{5.08e-02}$(−)
UF4	1.32e-01$_{6.34e-03}$	1.70e-01$_{7.55e-03}$(−)	1.47e-01$_{6.80e-03}$(−)	1.59e-01$_{8.22e-03}$(−)	1.52e-01$_{8.43e-03}$(−)	1.72e-01$_{4.21e-03}$(−)
UF5	4.20e+00$_{3.67e-01}$	3.00e+00$_{4.19e-01}$(+)	4.49e+00$_{3.64e-01}$(−)	2.91e+00$_{5.98e-01}$(+)	4.90e+00$_{3.84e-01}$(−)	4.53e+00$_{3.80e-01}$(−)
UF6	3.78e+00$_{5.72e-01}$	2.48e+00$_{8.10e-01}$(+)	4.27e+00$_{6.09e-01}$(−)	1.79e+00$_{6.32e-01}$(+)	4.31e+00$_{8.82e-01}$(−)	4.03e+00$_{7.69e-01}$(∼)
UF7	9.50e-01$_{1.11e-01}$	6.03e-01$_{1.51e-01}$(+)	1.03e+00$_{1.75e-01}$(−)	4.21e-01$_{1.04e-01}$(+)	1.08e+00$_{1.88e-01}$(−)	1.06e+00$_{2.20e-01}$(∼)
UF8	7.24e-01$_{9.32e-02}$	1.35e+00$_{3.77e-01}$(−)	1.43e+00$_{2.03e-01}$(−)	1.40e+00$_{3.25e-01}$(−)	1.62e+00$_{3.32e-01}$(−)	1.93e+00$_{2.12e-01}$(−)
UF9	7.57e-01$_{8.23e-02}$	1.33e+00$_{2.04e-01}$(−)	1.38e+00$_{2.92e-01}$(−)	1.40e+00$_{2.99e-01}$(−)	1.87e+00$_{6.32e-01}$(−)	1.81e+00$_{2.13e-01}$(−)
UF10	4.65e+00$_{4.34e-01}$	7.13e+00$_{1.10e+00}$(−)	8.06e+00$_{1.20e+00}$(−)	7.93e+00$_{1.45e+00}$(−)	8.78e+00$_{1.40e+00}$(−)	9.72e+00$_{1.25e+00}$(−)
LZ1	1.55e-01$_{8.70e-03}$	1.59e-01$_{1.07e-02}$(∼)	1.52e-01$_{1.29e-02}$(∼)	1.62e-01$_{1.75e-02}$(∼)	1.75e-01$_{1.56e-02}$(−)	1.61e-01$_{1.41e-02}$(∼)
LZ2	8.99e-01$_{1.69e-01}$	5.54e-01$_{1.41e-01}$(+)	1.03e+00$_{1.31e-01}$(−)	4.92e-01$_{2.21e-01}$(+)	1.05e+00$_{1.90e-01}$(−)	9.98e-01$_{1.61e-01}$(∼)
LZ3	2.22e-01$_{1.45e-02}$	3.53e-01$_{4.84e-02}$(−)	3.49e-01$_{2.87e-02}$(−)	3.45e-01$_{6.22e-02}$(−)	4.46e-01$_{6.09e-02}$(−)	4.46e-01$_{3.44e-02}$(−)
LZ4	2.15e-01$_{2.08e-02}$	3.51e-01$_{5.51e-02}$(−)	3.40e-01$_{4.16e-02}$(−)	3.41e-01$_{4.85e-02}$(−)	4.24e-01$_{7.09e-02}$(−)	4.41e-01$_{4.02e-02}$(−)
LZ5	2.00e-01$_{2.53e-02}$	3.07e-01$_{4.22e-02}$(−)	3.10e-01$_{3.57e-02}$(−)	3.01e-01$_{4.34e-02}$(−)	4.20e-01$_{7.43e-02}$(−)	4.17e-01$_{4.52e-02}$(−)
LZ6	4.92e-01$_{4.56e-02}$	5.46e-01$_{1.50e-01}$(∼)	8.54e-01$_{2.24e-01}$(−)	6.32e-01$_{1.82e-01}$(−)	5.25e-01$_{8.00e-02}$(∼)	5.32e-01$_{1.07e-01}$(∼)
LZ7	1.02e+00$_{3.52e-01}$	8.12e-01$_{2.57e-01}$(∼)	1.52e+00$_{5.86e-01}$(−)	9.44e-01$_{2.66e-01}$(∼)	1.49e+00$_{5.47e-01}$(−)	1.50e+00$_{2.13e-01}$(−)
LZ8	8.33e-01$_{1.23e-01}$	7.54e-01$_{2.02e-01}$(∼)	8.82e-01$_{3.10e-01}$(∼)	9.71e-01$_{1.72e-01}$(−)	7.85e-01$_{2.73e-01}$(∼)	8.77e-01$_{8.86e-02}$(∼)
LZ9	9.52e-01$_{1.43e-01}$	5.78e-01$_{1.78e-01}$(+)	1.01e+00$_{1.37e-01}$(∼)	4.96e-01$_{1.57e-01}$(+)	9.46e-01$_{1.83e-01}$(∼)	9.50e-01$_{2.18e-01}$(∼)
+/−/∼		6/9/4	0/16/3	6/11/2	0/15/4	0/12/7

Figure 3 plots the non-dominated solutions obtained by IFCS-MOEA/D-DE, FCS-MOEA/D-DE, CPS-MOEA, CSEA, MOEA/D-EGO, and EDN-ARMOEA on UF2. For each algorithm, we choose a single run with the median IGD value over 21 runs. In this figure, the solutions obtained by each algorithm are shown by red circles and the PF is shown by a black curve. This figure shows that the solutions obtained by IFCS-MOEA/D-DE are closer to the PF than the solutions obtained by other five algorithms. The above results show that IFCS-MOEA/D-DE outperforms the five compared algorithms on most test problems. Therefore, we can conclude that IFCS-MOEA/D-DE is efficient in solving MOPs with complicated PSs.

4.4 Performance Comparison on Real-World Problems

This section compares the performance of IFCS-MOEA/D-DE and the five state-of-the-art MOEAs on four real-world MOPs [32]: reinforced concrete beam design problem (RCBD), pressure vessel design problem (PVD), coil compression spring design problem (CCSD), and gear train design problem (GTD). The first three MOPs have 2 objectives and the last one MOP have 3 objectives. Due to the page limit, readers can refer to [32] for the details of these real-world MOPs. In

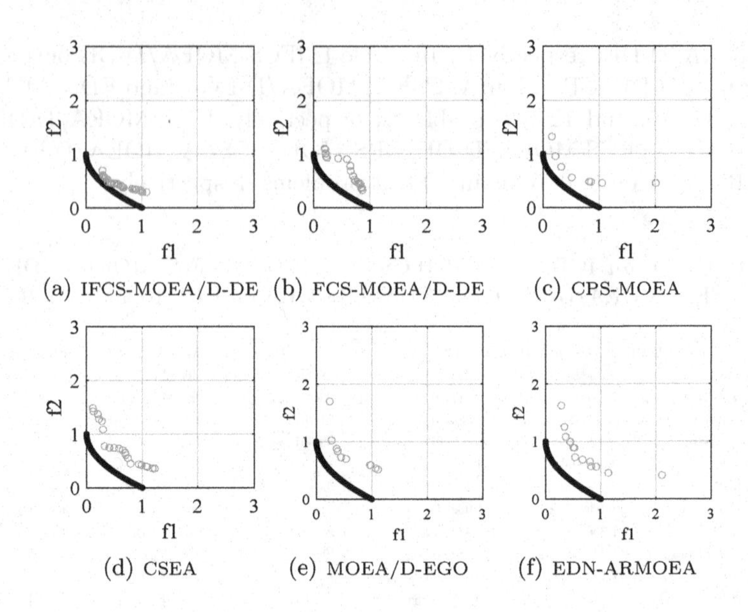

Fig. 3. The non-dominated solutions obtained by the five compared algorithms on UF2 with the median IGD value.

the experiments, the population size is $N = 45$. The maximal number of FEs is 500. Each algorithm executes 21 times on each test problem.

Table 5 shows the mean IGD values obtained by the five compared algorithms. The best result on each test problem is shaded. At the bottom of the table, we summarize the number of problems on which the performance of the compared algorithm is better than, worse than, and similar to that of IFCS-MOEA/D-DE, respectively. In Table 5, IFCS-MOEA/D-DE outperforms all the other algorithms on all test problems except for one case: there is no statistically significant difference between IFCS-MOEA/D-DE and EDN-ARMOEA on the GTD problem whereas the best average IGD value is obtained by IFCS-MOEA/D-DE. From the above results, we can conclude that the proposed IFCS-MOEA/D-DE algorithms outperforms the five state-of-the-art MOEAs in solving these real-world application problems under a limited number of FEs.

Table 5. The *mean* IGD values of IFCS-MOEA/D-DE, FCS-MOEA/D-DE, CPS-MOEA, CSEA, MOEA/D-EGO, and EDN-ARMOEA on four real-world problems.

	IFCS-MOEA/D-DE	FCS-MOEA/D-DE	CPS-MOEA	CSEA	MOEA/D-EGO	EDN-ARMOEA
RCBD	$1.26e\text{-}02_{7.25e-03}$	$3.09e\text{-}02_{2.07e-02}(-)$	$1.68e\text{-}02_{3.09e-03}(-)$	$2.80e\text{-}02_{7.70e-03}(-)$	$7.83e\text{-}02_{3.14e-02}(-)$	$1.89e\text{-}02_{6.17e-03}(-)$
PVD	$2.61e\text{-}02_{8.53e-03}$	$6.95e\text{-}02_{6.66e-02}(-)$	$1.27e\text{-}01_{1.34e-01}(-)$	$9.62e\text{-}02_{3.75e-02}(-)$	$1.16e\text{-}01_{5.48e-02}(-)$	$6.96e\text{-}02_{3.52e-02}(-)$
CCSD	$5.11e\text{-}03_{3.80e-03}$	$9.25e\text{-}02_{9.78e-02}(-)$	$3.74e\text{-}02_{3.60e-02}(-)$	$1.23e\text{-}01_{7.78e-02}(-)$	$1.78e\text{-}01_{1.17e-01}(-)$	$8.13e\text{-}02_{8.42e-02}(-)$
GTD	$5.15e\text{-}02_{1.28e-02}$	$1.23e\text{-}01_{1.15e-01}(-)$	$8.34e\text{-}02_{3.41e-02}(-)$	$1.56e\text{-}01_{6.79e-02}(-)$	$1.50e\text{-}01_{6.07e-02}(-)$	$9.17e\text{-}02_{7.25e-02}(\sim)$
$+/-/\sim$		0/4/0	0/4/0	0/4/0	0/4/0	0/3/1

5 Conclusion

This paper proposed an improved fuzzy classifier-based multiobjective evolutionary algorithm framework (IFCS-MOEA) for expensive MOPs. The IFCS-MOEA framework was developed based on an improved fuzzy classifier-based surrogate model. The IFCS model is used to sort unevaluated solutions based on the membership degrees and the model accuracy. Then, the promising offspring solutions are selected for function evaluations based on the sorting results. All the evaluated solutions are used for fuzzy classifier construction.

The proposed IFCS-MOEA framework was embedded with MOEA/D-DE for examination. The Fuzzy-KNN was used as the fuzzy classifier. The 10-fold cross-validation method was used to validate the quality of the classifier. IFCS-MOEA/D-DE was compared with the original MOEA/D-DE. The experimental results validated the effectiveness of IFCS in improving the performance of MOEA/D-DE on solving expensive MOPs under a limited number of FEs. Then, IFCS-MOEA/D-DE was compared with five state-of-the-art MOEAs on 19 test problems and four real-world application problems. The experimental results showed that IFCS-MOEA/D-DE outperformed the other five MOEAs in solving these problems under a limited number of FEs.

This paper validated the effectiveness of the IFCS-MOEA framework by embedding it with MOEA/D-DE. It is a future research topic to examine the effectiveness of IFCS-MOEA by embedding it with other MOEAs. It is also interesting to examine the performance of IFCS-MOEA/D-DE on other MOPs with complicated PSs. This paper used 30% and 70% as the accuracy threshold values in the proposed sorting strategy according to our preliminary results. It is necessary to further examine these threshold values on more test problems.

Acknowledgements. This work was supported by National Natural Science Foundation of China (Grant No. 62106099, 61876075), Guangdong Provincial Key Laboratory (Grant No. 2020B121201001), the Program for Guangdong Introducing Innovative and Enterpreneurial Teams (Grant No. 2017ZT07X386), The Stable Support Plan Program of Shenzhen Natural Science Fund (Grant No. 20200925174447003), Shenzhen Science and Technology Program (Grant No. KQTD2016112514355531).

References

1. Bandaru, S., Ng, A.H., Deb, K.: On the performance of classification algorithms for learning Pareto-dominance relations. In: Proceedings of 2014 IEEE Congress on Evolutionary Computation (CEC 2014), Beijing, China, pp. 1139–1146 (2014)
2. Beume, N., Naujoks, B., Emmerich, M.: SMS-EMOA: multiobjective selection based on dominated hypervolume. Eur. J. Oper. Res. **181**(3), 1653–1669 (2007)
3. Chugh, T., Jin, Y., Miettinen, K., Hakanen, J., Sindhya, K.: A surrogate-assisted reference vector guided evolutionary algorithm for computationally expensive many-objective optimization. IEEE Trans. Evol. Comput. **22**(1), 129–142 (2018)
4. Chugh, T., Sindhya, K., Hakanen, J., Miettinen, K.: A survey on handling computationally expensive multiobjective optimization problems with evolutionary algorithms. Soft. Comput. **23**(9), 3137–3166 (2019)

5. Chugh, T., Sindhya, K., Miettinen, K., Jin, Y., Kratky, T., Makkonen, P.: Surrogate-assisted evolutionary multiobjective shape optimization of an air intake ventilation system. In: Proceedings of the 2017 IEEE Congresson Evolutionary Computation (CEC 2017), Donostia, Spain, pp. 1541–1548 (2017)
6. Coello Coello, C.A., Reyes Sierra, M.: A study of the parallelization of a coevolutionary multi-objective evolutionary algorithm. In: Monroy, R., Arroyo-Figueroa, G., Sucar, L.E., Sossa, H. (eds.) MICAI 2004. LNCS (LNAI), vol. 2972, pp. 688–697. Springer, Heidelberg (2004). https://doi.org/10.1007/978-3-540-24694-7_71
7. Deb, K.: Multi-objective genetic algorithms: problem difficulties and construction of test problems. Evol. Comput. **7**(3), 205–230 (1999)
8. Deb, K., Hussein, R., Roy, P.C., Toscano-Pulido, G.: A taxonomy for metamodeling frameworks for evolutionary multiobjective optimization. IEEE Trans. Evol. Comput. **23**(1), 104–116 (2019)
9. Deb, K., Pratap, A., Agarwal, S., Meyarivan, T.: A fast and elitist multiobjective genetic algorithm: NSGA-II. IEEE Trans. Evol. Comput. **6**(2), 182–197 (2002)
10. Deb, K., Sinha, A., Kukkonen, S.: Multi-objective test problems, linkages, and evolutionary methodologies. In: Proceedings of the Genetic and Evolutionary Computation Conference (GECCO 2006), New York, NY, USA, pp. 1141–1148 (2006)
11. Deb, K., Thiele, L., Laumanns, M., Zitzler, E.: Scalable test problems for evolutionary multiobjective optimization. In: Abraham, A., Jain, L., Goldberg, R. (eds.) Evolutionary Multiobjective Optimization. Advanced Information and Knowledge Processing, pp. 105–145. Springer-Verlag, London, UK (2005). https://doi.org/10.1007/1-84628-137-7_6
12. Guo, D., Wang, X., Gao, K., Jin, Y., Ding, J., Chai, T.: Evolutionary optimization of high-dimensional multiobjective and many-objective expensive problems assisted by a dropout neural network. IEEE Trans. Syst. Man Cybern. Syst. **52**(4), 2084–2097 (2022)
13. Hillermeier, C.: Nonlinear Multiobjective Optimization: A Generalized Homotopy Approach. Birkhauser-Verlag, Basel (2000)
14. Huband, S., Barone, L., While, L., Hingston, P.: A scalable multi-objective test problem toolkit. In: Coello Coello, C.A., Hernández Aguirre, A., Zitzler, E. (eds.) EMO 2005. LNCS, vol. 3410, pp. 280–295. Springer, Heidelberg (2005). https://doi.org/10.1007/978-3-540-31880-4_20
15. Huband, S., Hingston, P., Barone, L., While, L.: A review of multiobjective test problems and a scalable test problem toolkit. IEEE Trans. Evol. Comput. **10**(5), 477–506 (2006)
16. Jia, L., Wang, Y., Fan, L.: Multiobjective bilevel optimization for production-distribution planning problems using hybrid genetic algorithm. Integr. Comput.-Aided Eng. **21**(1), 77–90 (2014)
17. Jin, Y.: Surrogate-assisted evolutionary computation: recent advances and future challenges. Swarm Evol. Comput. **1**(2), 61–70 (2011)
18. Jones, D.R., Schonlau, M., Welch, W.J.: Efficient global optimization of expensive black-box functions. J. Global Optim. **13**(4), 455–492 (1998)
19. Kasprzak, E.M., Lewis, K.E.: An approach to facilitate decision tradeoffs in pareto solution sets. J. Eng. Valuat. Cost Anal. **3**(1), 173–187 (2000)
20. Keller, J.M., Gray, M.R., Givens, J.A.: A fuzzy K-nearest neighbor algorithm. IEEE Trans. Syst. Man Cybern. **SMC-15**(4), 580–585 (1985)
21. Knowles, J.: ParEGO: a hybrid algorithm with on-line landscape approximation for expensive multiobjective optimization problems. IEEE Trans. Evol. Comput. **10**(1), 50–66 (2006)

22. Li, H., Zhang, Q.: A multiobjective differential evolution based on decomposition for multiobjective optimization with variable linkages. In: Runarsson, T.P., Beyer, H.-G., Burke, E., Merelo-Guervós, J.J., Whitley, L.D., Yao, X. (eds.) PPSN 2006. LNCS, vol. 4193, pp. 583–592. Springer, Heidelberg (2006). https://doi.org/10.1007/11844297_59
23. Li, H., Zhang, Q.: Multiobjective optimization problems with complicated Pareto sets, MOEA/D and NSGA-II. IEEE Trans. Evol. Comput. **13**(2), 284–302 (2009)
24. Lin, Q., Wu, X., Ma, L., Li, J., Gong, M., Coello, C.A.C.: An ensemble surrogate-based framework for expensive multiobjective evolutionary optimization. IEEE Trans. Evol. Comput. **26**(4), 631–645 (2022)
25. Lin, X., Zhang, Q., Kwong, S.: A decomposition based multiobjective evolutionary algorithm with classification. In: Proceedings of 2016 IEEE Congress on Evolutionary Computation (CEC 2016), Vancouver, Canada, pp. 3292–3299 (2016)
26. Loshchilov, I., Schoenauer, M., Sebag, M.: A mono surrogate for multiobjective optimization. In: Proceedings of Genetic and Evolutionary Computation Conference (GECCO 2010), Portland, USA, pp. 471–478 (2010)
27. Murata, T., Ishibuchi, H.: MOGA: multi-objective genetic algorithms. In: Proceedings of 1995 IEEE Congress on Evolutionary Computation (CEC 1995), Perth, Australia, pp. 289–294 (1995)
28. Okabe, T., Jin, Y., Olhofer, M., Sendhoff, B.: On test functions for evolutionary multi-objective optimization. In: Yao, X., et al. (eds.) PPSN 2004. LNCS, vol. 3242, pp. 792–802. Springer, Heidelberg (2004). https://doi.org/10.1007/978-3-540-30217-9_80
29. Pan, L., Cheng, H., Tian, Y., Wang, H., Zhang, X., Jin, Y.: A classification-based surrogate-assisted evolutionary algorithm for expensive many-objective optimization. IEEE Trans. Evol. Comput. **23**(1), 74–88 (2019)
30. Song, Z., Wang, H., He, C., Jin, Y.: A Kriging-assisted two-archive evolutionary algorithm for expensive many-objective optimization. IEEE Trans. Evol. Comput. **25**(6), 1013–1027 (2021)
31. Sonoda, T., Nakata, M.: Multiple classifiers-assisted evolutionary algorithm based on decomposition for high-dimensional multi-objective problems. IEEE Trans. Evol. Comput. **26**, 1581–1595 (2022)
32. Tanabe, R., Ishibuchi, H.: An easy-to-use real-world multi-objective optimization problem suite. Appl. Soft Comput. **89**, 106078 (2020)
33. Tian, Y., Cheng, R., Zhang, X., Cheng, F., Jin, Y.: An indicator-based multiobjective evolutionary algorithm with reference point adaptation for better versatility. IEEE Trans. Evol. Comput. **22**(4), 609–622 (2018)
34. Tian, Y., Cheng, R., Zhang, X., Jin, Y.: PlatEMO: a MATLAB platform for evolutionary multi-objective optimization [educational forum]. IEEE Comput. Intell. Mag. **12**(4), 73–87 (2017)
35. Veldhuizen, D.A.V., Lamont, G.B.: Multiobjective evolutionary algorithm test suites. In: Proceedings of the 1999 ACM Symposium on Applied Computing (SAC 1999), San Antonio, Texas, USA, pp. 351–357 (1999)
36. Yuan, Y., Banzhaf, W.: Expensive multi-objective evolutionary optimization assisted by dominance prediction. IEEE Trans. Evol. Comput. **26**(1), 159–173 (2022)
37. Zhang, J., He, L., Ishibuchi, H.: Dual fuzzy classifier-based evolutionary algorithm for expensive multiobjective optimization. IEEE Trans. Evol. Comput. (2022). (Early Access)

38. Zhang, J., Ishibuchi, H.: Multiobjective optimization with fuzzy classification-assisted environmental selection. In: Ishibuchi, H., et al. (eds.) EMO 2021. LNCS, vol. 12654, pp. 580–592. Springer, Cham (2021). https://doi.org/10.1007/978-3-030-72062-9_46
39. Zhang, J., Ishibuchi, H., Shang, K., He, L., Pang, L.M., Peng, Y.: Environmental selection using a fuzzy classifier for multiobjective evolutionary algorithms. In: Proceedings of the Genetic and Evolutionary Computation Conference (GECCO 2021), Lille, France, pp. 485–492 (2021)
40. Zhang, J., Zhou, A., Tang, K., Zhang, G.: Preselection via classification: a case study on evolutionary multiobjective optimization. Inf. Sci. **465**, 388–403 (2018)
41. Zhang, J., Zhou, A., Zhang, G.: A classification and Pareto domination based multiobjective evolutionary algorithm. In: Proceedings of 2015 IEEE Congress on Evolutionary Computation (CEC 2015), Sendai, Japan, pp. 2883–2890 (2015)
42. Zhang, Q., Li, H.: MOEA/D: a multiobjective evolutionary algorithm based on decomposition. IEEE Trans. Evol. Comput. **11**(6), 712–731 (2007)
43. Zhang, Q., Liu, W., Tsang, E., Virginas, B.: Expensive multiobjective optimization by MOEA/D with Gaussian process model. IEEE Trans. Evol. Comput. **14**(3), 456–474 (2009)
44. Zhang, Q., Zhou, A., Zhao, S., Suganthan, P.N., Liu, W., Tiwari, S.: Multiobjective optimization test instances for the CEC 2009 special session and competition. Technical report CES-487, The School of Computer Science and Electronic Engineering, University of Essex (2009)
45. Zhou, A., Qu, B.Y., Li, H., Zhao, S.Z., Suganthanb, P.N., Zhang, Q.: Multiobjective evolutionary algorithms: a survey of the state of the art. Swarm Evol. Comput. **1**(1), 32–49 (2011)
46. Zitzler, E., Künzli, S.: Indicator-based selection in multiobjective search. In: Yao, X., et al. (eds.) PPSN 2004. LNCS, vol. 3242, pp. 832–842. Springer, Heidelberg (2004). https://doi.org/10.1007/978-3-540-30217-9_84
47. Zitzler, E., Laumanns, M., Thiele, L.: SPEA2: improving the strength pareto evolutionary algorithm. Technical report 103, Computer Engineering and Networks Laboratory (TIK), Swiss Federal Institute of Technology (ETH) Zurich, Gloriastrasse 35, CH-8092 Zurich, Switzerland (2001)

Approximation of a Pareto Set Segment Using a Linear Model with Sharing Variables

Ping Guo[(✉)], Qingfu Zhang, and Xi Lin

Department of Computer Science, City University of Hong Kong, Kowloon, Hong Kong, China
{pingguo5-c,xi.lin}@my.cityu.edu.hk, qingfu.zhang@cityu.edu.hk

Abstract. In many real-world applications, the Pareto Set (PS) of a continuous multiobjective optimization problem can be a piecewise continuous manifold. A decision maker may want to find a solution set that approximates a small part of the PS and requires the solutions in this set share some similarities. This paper makes a first attempt to address this issue. We first develop a performance metric that considers both optimality and variable sharing. Then we design an algorithm for finding the model that minimizes the metric to meet the user's requirements. Experimental results illustrate that we can obtain a linear model that approximates the mapping from the preference vectors to solutions in a local area well.

1 Introduction

This paper considers the following continuous multiobjective optimization problem (MOP):

$$\begin{aligned} \text{minimize} \quad & F(x) = (f_1(x), \ldots, f_m(x)), \\ \text{subject to} \quad & x \in \Omega, \end{aligned} \tag{1}$$

where x is the decision variable, $\Omega \subseteq R^n$ is the decision space, $F : \Omega \rightarrow R^m$ contains m continuous objective functions $f_1(x), \ldots, f_m(x)$, and R^m is the objective space. Very often, the objectives in MOP (1) conflict with each other, and no single solution can optimize them simultaneously [9]. *Pareto optimality* is used to define the best trade-off candidate solutions. The set of all the Pareto optimal solutions is called the *Pareto Set* (PS). Its image in the objective space is called the *Pareto Front* (PF).

Aggregation is an important technique for solving MOPs [10]. Aggregation methods transform (1) into some single objective optimization problems. For a preference λ, an aggregation method aggregates all the f_i's into a scalar objective function, optimizes it and generates a Pareto optimal solution $x(\lambda)$ for the preference vector λ. Under some conditions, an aggregation method can find all the Pareto optimal solutions. In other words, the PS can be modeled by a function $x = x(\lambda)$. Moreover, under regularity conditions, it is piecewise continuous [11].

M. Emmerich et al. (Eds.): EMO 2023, LNCS 13970, pp. 247–259, 2023.
https://doi.org/10.1007/978-3-031-27250-9_18

Given a preference vector λ^0, a decision maker may be interested only in Pareto optimal solutions around $x(\lambda^0)$. It is reasonable to assume that $x(\lambda)$ is linear around a small neighborhood of λ^0. Let $x^1 = (x_1^1, \ldots, x_n^1)$,$x^2 = (x_1^2, \ldots, x_n^2) \in R^n$ be two candidate solutions, if $x_i^1 = x_i^2$, we say that x^1 and x^2 share variable x_i. In many real-life applications, when the preference changes, it is required to have an approximate Pareto optimal solution for the new preference with as many components the same as the current Pareto optimal solution. This requirement can be essential for reusing existing designs and reducing costs. In engineering design, shared components can support module design [2] and significantly reduce manufacturing costs. Deb et al. [2] advocate conducting data mining among the obtained Pareto optimal solutions to find useful patterns. To date, no research has been conducted on the integration of shared component requirements into the optimization process.

This paper makes a first attempt to address the issue of shared components. We model it as a problem to use a linear model to approximate a PS segment under the constraint of variable sharing. Much effort has been made to model the Pareto set using a math function [1,4,12]. However, all these existing works aim at approximating the actual Pareto set. Our approach considers the quality of solutions beyond Pareto optimality. We trade Pareto optimality for shared component requirements. Our major contributions can be summarized as follows:

- We study the optimality of a solution set under some shared component constraints instead of Pareto optimality.
- We incorporate the user's preference and the requirement on shared components to define a performance metric.
- We adopt the framework of MOEA/D to develop an algorithm for finding the model that optimizes the performance metric. This model can generate infinite solutions that satisfy the user's requirements.

The rest of the paper is organized as follows: We propose the original version of our performance metric and modification considering variable sharing in Sect. 2. Then we present the form of the linear model and the connection between variable sharing and the sparsity of the model in Sect. 3. We give out the framework of our algorithm and implementation details in Sect. 4. In Sect. 5, we conduct experiments to validate our algorithm. The last section summarizes the paper and list possible future work directions.

2 Performance Metric for Local Models

In this section, we introduce our performance metric that considers both optimality and variable sharing of solutions. We first define a preference vector distribution based on the user-provided preference vector. We then use the expected aggregation value of solutions output by a model as the first part of our metric. Finally, we implicitly define the second part of the metric with regards to variable sharing. Different implementations are possible for the second part. In the next section, we present our implementation using a linear model.

2.1 Local Approximation Metric

Consider MOP (1). Given:

- λ^0: a preference vector, from the $(m-1)$-D probability simplex;
- $N(\lambda^0)$: a neighborhood set of λ^0;
- $\mathcal{P}$: a probability distribution defined on $N(\lambda^0)$.

The neighborhood set can have different structures. In general, the neighborhood of λ defines a set of preference vectors that are close to λ in Euclidean space. The distribution $\mathcal{P}$ enables us to sample preference vectors from the neighborhood set. In this paper, we use a multi-variate normal distribution as the sampling distribution $\mathcal{P}$. This distribution puts more emphasis on the area that near the user's target solutions.

For any preference vector $\lambda \sim \mathcal{P}$, we can define a sub-problem using aggregation functions. In this paper, we use Chebyshev aggregation. Our metric can be generalized to other aggregation functions. The Chebyshev aggregation value of solution x with preference vector λ is by:

$$g(x, \lambda) = \max_i \lambda_i |f_i(x) - z_i|, \tag{2}$$

where z_i is the Utopian value for the i-th objective, and λ_i is the i-th component of the preference vector λ. The associated solution $x^*(\lambda)$ to the above sub-problem is as follows:

$$x^*(\lambda) = \arg\min_x g(x, \lambda). \tag{3}$$

We can denote the above mapping from the preference vector λ to the associated optimal solution as $x(\lambda)$. Now, we use a model $h_\theta(\lambda)$ parameterized by θ to predict the associated solution to the sub-problem defined with λ. Since the solution $x^*(\lambda)$ minimizes the aggregation value $g(x, \lambda)$, we can use the expected aggregation value of the model output to evaluate its optimality. For each preference vector $\lambda \sim \mathcal{P}$, the aggregation value computed using the model output is $g(h_\theta(\lambda), \lambda)$. Our metric $\mathcal{M}$ is as follows:

$$\mathcal{M}(h_\theta) = \mathbb{E}_{\lambda \sim \mathcal{P}}[g(h_\theta(\lambda), \lambda)]. \tag{4}$$

The above metric can be directly used to learn the Pareto set for different real-world applications, such as multi-task learning [7], neural multiobjective combinatorial optimization [6], and multiobjective Bayesian optimization [8]. In this work, we extend it to include the variable sharing constraint.

2.2 Shared Variable Metric

The above metric only considers the optimality of solutions under sub-problems defined with different preference vectors. We need to add extra terms to evaluate the model's performance according to special requirements from the user. In this paper, we consider variable sharing due to its importance in engineering design.

We denote the function that measures the degree of variable sharing of the model h_θ as $\mathcal{I}(h_\theta)$, called variable sharing degree (VSD). Without loss of generality, we assume a lower value of $\mathcal{I}$ indicates a higher degree of variable sharing. The explicit form of the VSD can be various, we will connect it with the sparsity of the model in the next section.

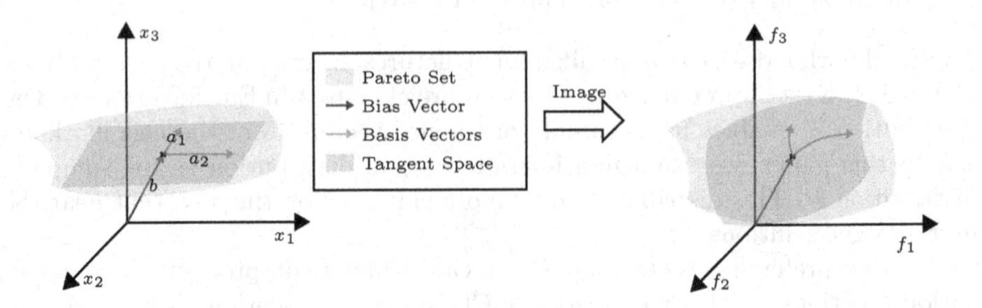

Fig. 1. Linear Model: Column vectors (e.g. a_1, a_2) in A can be regarded as basis vectors that span the subspace of a hyperplane. This hyperplane can be used to approximate the local PS segment since we expect it to be linear under mild conditions.

Now our goal is to find the model that minimizes both $\mathcal{M}$ and $\mathcal{I}$. We synthese these two metrics and reload the notion $\mathcal{M}$ to represent our final performance metric. The final version of the metric is as follows:

$$\mathcal{M}(h_\theta) = \mathbb{E}_{\lambda \sim \mathcal{P}}[g(h_\theta(\lambda), \lambda)] + \gamma \mathcal{I}(h_\theta), \tag{5}$$

where γ is the parameter that weighs the importance of the VSD. Larger values of γ lead to models that trade optimality for variable sharing. Our goal is to build a linear model that can approximate the local area of the PS and produce solutions with as many variables taking the same value across the solutions.

3 Linear Sparse Representation of the Local PS

In this section, we first give the analytical form of our linear model. Then, we discuss the sparsity of the parameters and use it as an implementation of the VSD.

3.1 Linear Model

Since the preference vectors are from a probability simplex, the sum of all the elements is equal to 1. So we only use the first $(m-1)$ elements of the preference vectors λ, denoted as $\lambda_{1:m-1}$ as the input. Our model designed using *first-order* approximation is as follows:

$$h_\theta(\lambda) = A(\lambda_{1:m-1} - \lambda^0_{1:m-1}) + b, \tag{6}$$

where $A \in R^{n\times(m-1)}$ and $b \in R^n$ are the parameters of the model. $\theta = (A, b)$ is still used to represent all the parameters of the model for brevity.

The interpretation of the column vectors of A can be a set of basis vectors of the tangent space at point x^0. The bias vector b is used to represent the associated solution x^0 with the preference vector w^0. We give an illustrative example of the linear model in Fig. 1.

The performance of the above model can be evaluated using metric (5). However, the form of VSD is not explicitly defined. In the following paragraphs, we give our implementation of VSD using sparsity of the model.

3.2 Variable Sharing and Sparsity of the Model

We notice that the linear approximation is actually a set of linear combinations of the column vectors in A and the bias vector b. Each non-zero row in A contributes to one dimension of output solutions the model. More "empty" rows in A lead to more shared variables in the solutions output by $h_\theta(\lambda)$. Therefore, we use the row sparsity of A as an implementation of the VSD. Specifically, we use the $(2, 1)$-norm of matrix A as the function to measure the degree of row sparsity. The $(2, 1)$-norm of matrix A is defined as:

$$\|A\|_{2,1} = \sum_{i=1}^{n} \|a_i\|_2, \tag{7}$$

where a_i is the i-th row of matrix A.

4 Method and Algorithm

4.1 An Alternative Problem

Using the linear model defined in the previous section, the optimization problem becomes:

$$\min_{A,b} \mathcal{M}(h_\theta) = \mathbb{E}_{\lambda\sim\mathcal{P}}[g(h_\theta(\lambda), \lambda)] + \gamma\|A\|_{2,1}. \tag{8}$$

If the explicit form of F is known, we can derive the form of g accordingly and apply a gradient descent algorithm to solve it. However, for real-world problems, the gradient information is often unavailable.

Our solution is to maintain a dynamic dataset of preference vector-solution pairs during our algorithm. We update this dataset in each iteration, assuming it converges to the local PS progressively. Then, the data from it can be regarded as noisy samples from the true PS. We replace the first term with the mean squared loss (MSE) between the samples and the output of the model.

Suppose we have a dataset $\{(\lambda^1, x^1), \ldots, (\lambda^N, x^N)\}$ at any iteration of our algorithm. We want to find a linear model to fit this dataset. More specifically, our goal is to minimize the following loss function:

$$\frac{1}{N}\sum_{i=1}^{N} \|x^i - (A(\lambda^i_{1:m-1} - \lambda^0_{1:m-1}) + b)\|^2 + \gamma\|A\|_{2,1}, \tag{9}$$

where λ^0 is the user-given preference vector.

The above problem is convex and easy to solve. In fact, we can minimize the above loss function by solving a series of simple regularized least square regression problems for each dimension of the data.

4.2 Algorithm Framework

Initialization of the Preference Set. In the previous section, we described how a dataset of preference vector-solution pairs is used to build the linear model. Here we illustrate the process of initializing the preference set using the user-given preference vector λ^0:

- Use a normal distribution $\mathcal{N}(0, \sigma^2 I)$ to sample noise vectors.
- Sample N noise vectors from $\mathcal{N}(0, \sigma^2 I)$, add λ^0 to them to generate N vectors.
- Project these vectors onto the probability simplex to normalize the disturbed vectors.

Through the above generation process, we obtain a preference set of N preference vectors. This process can also be viewed as sampling from the neighborhood of the preference vector on the simplex, as shown in Fig. 2.

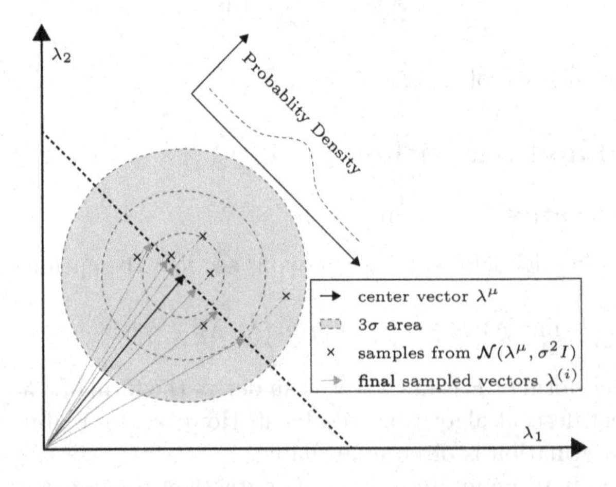

Fig. 2. Preference Vector Generation: This process is equivalent to sampling from $\mathcal{N}(\lambda^0, \sigma I)$ and projecting them on the simplex.

Main Algorithm. In general, we adopt the framework of MOEA/D [10] to design our algorithm. Details of our MOEA/D with local linear approximations, called MOEA/D-LLA (Local Linear Approximation), are shown in Algorithm 1. It takes the obtained preference vector set as its input. We first initialize a set of sub-problems using the preference set under Chebyshev aggregation and assign the value of reference point as done in MOEA/D. Then our algorithm maintains:

- a population X of size N, where the i-th individual is used to solve the sub-problem using λ^i,
- a set of decomposed value $\{g(x^i, \lambda^i)|1 \leq i \leq N\}$,
- and a reference point $z^* = (z_1^*, \ldots, z_m^*)^T$.

We generally push the population towards the PS by using genetic operators and our model to generate new solutions. We train our model by solving a regression task on the dataset of preference vector-solution pairs.

Algorithm 1: MOEA/D-LLA

Input: preference vector set $W = \{\lambda^1, \ldots, \lambda^N\}$
Parameters : regularization parameter γ, optimization step o
Output: matrix A, bias vector b, solution set X

1 Initialize matrix A, b, a population X
2 **while** *not terminated* **do**
   `// Optimization Step`
3   Using MOEA/D to optimize $g(x, \lambda^1), \ldots, g(x, \lambda^N)$ to obtain $(\lambda^1, x^1), \ldots, (\lambda^N, x^N)$;
   `// Regression Step`
4   **for** $i = 1 \rightarrow o$ **do**
5       $A^*, b^* = \arg\min_{A,b} \frac{1}{N} \sum_{i=1}^{N} \left\| x^i - (A(\lambda_{1:m-1}^i - \lambda_{1:m-1}^0) + b) \right\|_2^2 + \gamma \|A\|_{2,1}$
6   **end**
   `// Update Population Using Linear Model`
7   Generate new solutions using Algorithm 2 and use them to update population $(\lambda^1, x^1), \ldots, (\lambda^N, x^N)$.
8 **end**

Sampling New Solutions. In our algorithm, we use a hybrid strategy to generate new solutions. In Line 2, we use genetic operators to sample new solutions. In Line 6, we sample solutions from the linear model as in [12]. The sampling method is given in Algorithm 2.

Algorithm 2: Sampling New Solutions

---

1 Sample N noise vectors from $\mathcal{N}(0, \sigma_{noise}^2 I)$;
2 Add noise vectors on the preference set to obtain the noised set $\left\{\tilde{\lambda}^1, \ldots, \tilde{\lambda}^N\right\}$;
3 Generate N solutions using

$$x = A\tilde{\lambda}_{1:m-1} + b$$

---

5 Experimental Results

In this section, we study the optimality and the variable sharing aspect of our algorithm. Since standard test instances like ZDT [13] and DTLZ [3] naturally have shared variables in their Pareto set, we first design a problem with no shared variables and test our algorithm on it. Then we study the trade-off between optimality and variable sharing for this problem. Here, we use R-metric from [5] to incorporate the user-given preference vector into the evaluation of the solutions. The parameter δ for R-metric is set to be 6σ to target our preferred area.

5.1 Parameter Setting

All the experimental results are obtained from 10 independent runs of the algorithm. We add extra function evaluations in MOEA/D-DE as a compensation for not generating solutions from the model to ensure a fair comparison. The parameters of our algorithm are set as follows:

- The population size: It is 100 for all instances.
- The variance σ^2 of the sampling distribution $\mathcal{P}$: It is 0.02 for all instances.
- The variance σ^2_{noise} for sampling new solutions: It is 0.05 for all instances.
- The maximum number of generations: It is 300 for all instances.

5.2 Performance on None-shared Problem

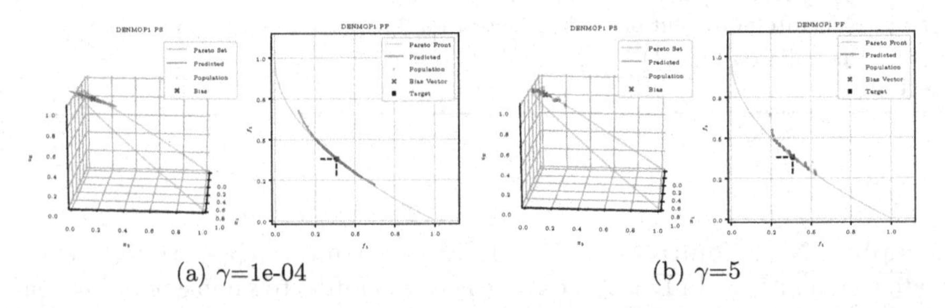

(a) γ=1e-04 (b) γ=5

Fig. 3. Linear Approximation for Local PS: An illustration of the population and the predictions of the model for a nonlinear Pareto set under different γ in both decision space and objective space.

To evaluate the effectiveness of our algorithm in finding solutions with shared variables, we create a new test instance, MOZDT1, by modifying ZDT1 by replacing $g(x)$. The form of our $g(x)$ is as follows:

$$\begin{aligned} l(x) &= ((1-2x_1)^2 - x_2)^2 + (x_3 + x_2 - 1)^2 \\ g(x) &= 1 + \frac{9}{n-3}\sum_{i=4}^{n} |x_i - x_1| + l(x) \end{aligned} \tag{10}$$

The Pareto set for this problem is defined as:

$$0 \leq x_1 \leq 1, x_2 = (1 - 2x_1)^2, x_3 = 1 - (1 - 2x_1)^2, \text{and } x_i = x_1 (i = 4, \ldots, n) \tag{11}$$

The Pareto optimal solutions in this set have no shared variables. Therefore, we cannot guarantee the optimality of the solutions if we desire more shared variables in them. As γ controls the importance of VSD in (5), we expect our model's output have more shared variables as γ increases.

We run our algorithms under different γ values and plot the results in Fig. 3. Additionally, we show the optimality and the degree of variable sharing in Fig. 4. In Fig. 4, we use R-IGD to evaluate the optimality of the solutions and the variances of decision variables to illustrate VSD.

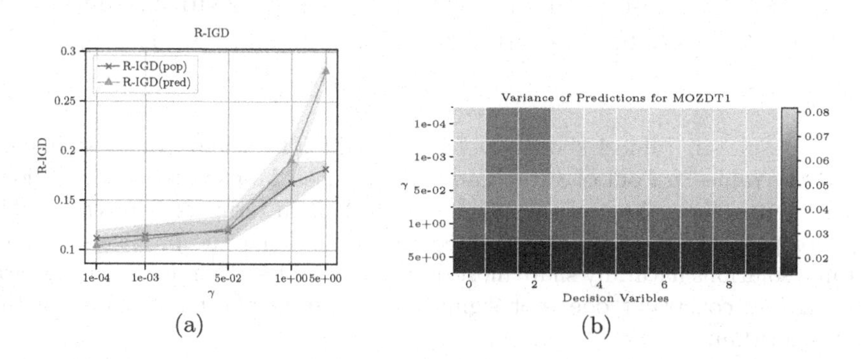

(a) (b)

Fig. 4. (a) The value of R-IGD calculated on the population from the last iteration and the predictions given by the model. The area within one standard deviation is shaded. (b) The variance of decision variables of the predictions. The model's predictions converge to a small area as γ increases.

Figure 4(a) shows that with a small gamma value, our algorithm is able to find a model that can generate solutions of high quality. However, we observe a significant deterioration of optimality when γ was increased to 5. Upon examining the predictions associated with $\gamma = 5$, we find that their variances reduce to almost zero. This indicates that most solutions output by the model are very similar to each other and can be considered the same solution. In future work, We will further investigate the impact of γ on different decision variables.

5.3 Performance on Standard Test Instances

The Pareto optimal solutions of standard test instances like ZDT [13] and DTLZ [3] have special structures in which the majority of their decision variables are shared. Therefore, if γ is set correctly, we expect VSD to act as a regularization term and help obtain a model that balances optimality and variable sharing.

We evaluate the quality of the solutions produced by MOEA/D-DE, our algorithm, and the model's predictions using the R-metric. The results are listed in Table 1.

Table 1. R-IGD and R-HV values obtained by MOEA/D-DE and LLA over 10 independent runs. The value of the population and the predictions are both evaluated.

Problem	MOEA/D-DE		LLA Pop		LLA Pred	
	R-IGD	R-HV	R-IGD	R-HV	R-IGD	R-HV
ZDT1	1.27e-01	5.56e-01	1.27e-01	5.56e-01	**7.76e-02**	**6.09e-01**
ZDT2	1.02e-01	2.89e-01	1.02e-01	2.89e-01	**6.60e-02**	**2.95e-01**
ZDT4	1.26e-01	5.55e-01	1.26e-01	5.55e-01	**7.78e-02**	**6.08e-01**
ZDT6	1.26e-01	3.37e-01	1.16e-01	3.42e-01	**8.27e-02**	**3.46e-01**
DTLZ1	1.90e-01	5.09e-01	1.92e-01	5.05e-01	**1.84e-01**	**6.36e-01**
DTLZ2	3.24e-01	1.90e-01	3.24e-01	1.90e-01	**2.82e-01**	**2.30e-01**
DTLZ3	3.23e-01	1.90e-01	3.25e-01	1.90e-01	**2.82e-01**	**2.29e-01**
DTLZ4	**3.97e-01**	**1.48e-01**	4.22e-01	1.35e-01	4.22e-01	1.42e-01

The results in Table 1 show that LLA's predictions achieve the best R-IGD and R-HV values in 7 out of 8 standard test instances. Moreover, the populations of LLA outperform those of the original MOEA/D in terms of R-metrics. This superior performance can be attributed to the fact that the problems' Pareto optimal solutions naturally share most decision variables. Therefore, adding variable sharing constraint does not significantly degrade the performance of the LLA algorithm.

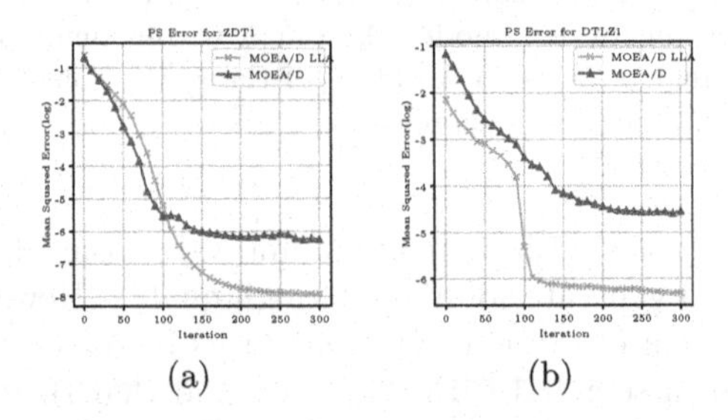

Fig. 5. Mean squared errors between the true optimal solutions and the model's predictions for ZDT1 and DTLZ1. The results are the average value of 10 independent runs and are plotted on logarithmic axis.

Approximation Error. To further evaluate the convergence of our algorithm on these instances, we plot the mean squared error between the output of our model and the true optimal solutions associated with the sub-problems defined by the preference vectors in Fig. 5. In these experiments, γ is set to be a small value (1e-03).

From Fig. 5, we can see that the residual error of MOEA/D-DE can not be further diminished by increasing the number of iterations. Our model's predictions improve quickly and exceed the performance of MOEA/D-DE after around 100 iterations. Furthermore, the quality of the solution for each sub-problem continues to improve with more function evaluations.

Variable Sharing. The weighting factor γ controls the trade-off between optimality and variable sharing. We plot the population from the last iteration and the hyper-plane that represents the output of the linear model obtained with different values of γ in Fig. 6.

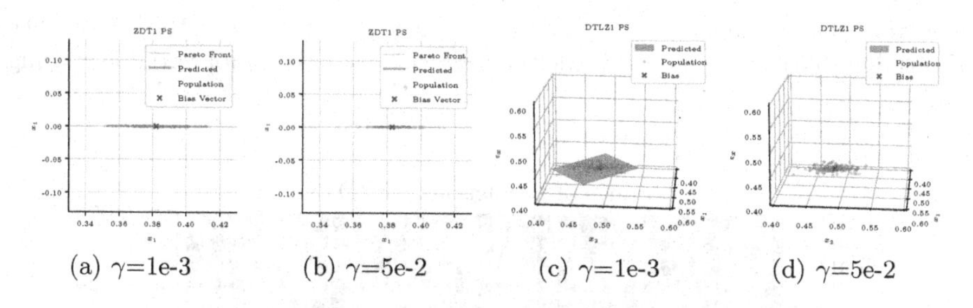

(a) γ=1e-3 (b) γ=5e-2 (c) γ=1e-3 (d) γ=5e-2

Fig. 6. The influence of γ on the variable sharing degree for ZDT1 and DTLZ1. The shape of the approximated PS shrinks as γ becomes bigger.

With larger γ, the variance of the decision variable x_2 becomes smaller in the output of the models. We see the same phenomenon for different test instances in Fig. 6(c) and Fig. 6(d). We can conclude now that our model has a higher degree of variable sharing when we increase γ. Moreover, these examples illustrate an enhancement of the influence of γ for PS in the higher-dimensional space.

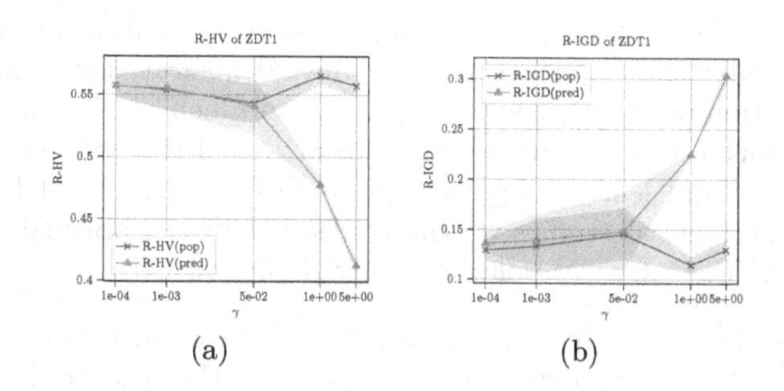

(a) (b)

Fig. 7. An illustration of the value of R-metric of the solutions set and the predictions of the linear model of our algorithm. The lines are the average value of 10 independent runs, while the shaded areas reflect the variance of the value.

We further investigate the influence of γ on optimality by plotting γ against the R-metric value of ZDT1 problem in Fig. 7. The R-metrics of the solution set fluctuate with the value of γ. However, we observe significant deterioration of performance of the linear model when we increase γ. Although ZDT1's Pareto optimal solutions naturally share the first decision variable, higher degree of variable sharing leads to worse optimality. We can conclude that the cost of sharing lies in the non-shared variables.

To better understand the imapct of γ on decision variables, we calculate the variance of each decision variable under different γ and show them in a heatmap in Fig. 8. Smaller variances indicate a higher degree of variable sharing. With a smaller γ, we can obtain a model that targets the correct shared decision variables without degrading the optimality of the solutions as shown in Fig. 8. However, when we increase γ to a large value (e.g. 5) to emphasize the importance of variable sharing, we indeed trade the performance (Fig. 7) for variable sharing (Fig. 8).

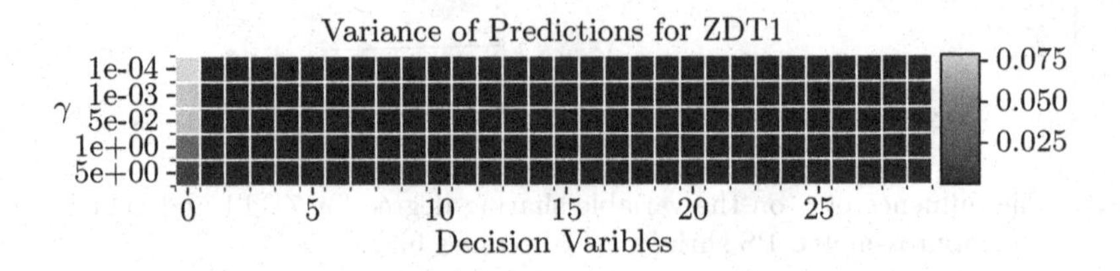

Fig. 8. Illustration of the variances of decision variables under different γ. For ZDT1, most of the decision variables are shared in the Pareto set.

6 Conclusion

In this paper, we have studied how to approximate a small part of the PS subject to the variable sharing constraint. We have defined its performance metric as the expectation of the aggregation value under Chebyshev aggregation. Our proposed algorithm can find the optimal linear model that minimizes this performance metric and learns a sparse representation of the local PS. We have conducted experimental studies on the trade-off between optimality and variable sharing. In the future, we plan to study the following:

- We will consider more test instances where no decision variables are shared in Pareto optimal solutions, further investigate the best trade-off between optimality and variable sharing in different problems settings.
- Instead of using regularized least square regression to learn the model parameters, we will explore more efficient and intelligent approaches such as deep learning and reinforcement learning.

Acknowledgments. This work is supported by the General Research Fund (GRF) grant (CityU 11215622) from the Research Grants Council of Hong Kong, China.

References

1. Deb, K., Bandaru, S., Greiner, D., Gaspar-Cunha, A., Tutum, C.C.: An integrated approach to automated innovization for discovering useful design principles: case studies from engineering. Appl. Soft Comput. **15**, 42–56 (2014)
2. Deb, K., Srinivasan, A.: Innovization: innovating design principles through optimization. In: Proceedings of the 8th Annual Conference on Genetic and Evolutionary Computation, pp. 1629–1636 (2006)
3. Deb, K., Thiele, L., Laumanns, M., Zitzler, E.: Scalable test problems for evolutionary multiobjective optimization. In: Abraham, A., Jain, L., Goldberg, R. (eds.) Evolutionary Multiobjective Optimization, pp. 105–145. Springer, London (2005). https://doi.org/10.1007/1-84628-137-7_6
4. Eichfelder, G.: An adaptive scalarization method in multiobjective optimization. SIAM J. Optim. **19**(4), 1694–1718 (2009)
5. Li, K., Deb, K., Yao, X.: R-metric: evaluating the performance of preference-based evolutionary multiobjective optimization using reference points. IEEE Trans. Evol. Comput. **22**(6), 821–835 (2017)
6. Lin, X., Yang, Z., Zhang, Q.: Pareto set learning for neural multi-objective combinatorial optimization. In: International Conference on Learning Representations (2022)
7. Lin, X., Yang, Z., Zhang, Q., Kwong, S.: Controllable pareto multi-task learning. arXiv preprint arXiv:2010.06313 (2020)
8. Lin, X., Yang, Z., Zhang, X., Zhang, Q.: Pareto set learning for expensive multi-objective optimization. In: Advances in Neural Information Processing Systems (2022)
9. Miettinen, K.: Nonlinear Multiobjective Optimization, vol. 12. Springer, New York (2012). https://doi.org/10.1007/978-1-4615-5563-6
10. Zhang, Q., Li, H.: MOEA/D: a multiobjective evolutionary algorithm based on decomposition. IEEE Trans. Evol. Comput. **11**(6), 712–731 (2007)
11. Zhang, Q., Zhou, A., Jin, Y.: RM-MEDA: a regularity model-based multiobjective estimation of distribution algorithm. IEEE Trans. Evol. Comput. **12**(1), 41–63 (2008)
12. Zhou, A., Zhao, H., Zhang, H., Zhang, G.: Pareto optimal set approximation by models: a linear case. In: Deb, K., et al. (eds.) EMO 2019. LNCS, vol. 11411, pp. 451–462. Springer, Cham (2019). https://doi.org/10.1007/978-3-030-12598-1_36
13. Zitzler, E., Deb, K., Thiele, L.: Comparison of multiobjective evolutionary algorithms: empirical results. Evol. Comput. **8**(2), 173–195 (2000)

Feature-Based Benchmarking of Distance-Based Multi/Many-objective Optimisation Problems: A Machine Learning Perspective

Arnaud Liefooghe[1](✉), Sébastien Verel[2], Tinkle Chugh[3], Jonathan Fieldsend[3], Richard Allmendinger[4], and Kaisa Miettinen[5]

[1] Univ. Lille, CNRS, Inria, Centrale Lille, UMR 9189 CRIStAL, 59000 Lille, France
arnaud.liefooghe@univ-lille.fr

[2] Univ. Littoral Côte d'Opale, LISIC, 62100 Calais, France
verel@univ-littoral.fr

[3] University of Exeter, Exeter EX4 4QD, UK
{t.chugh,j.e.fieldsend}@exeter.ac.uk

[4] The University of Manchester, Manchester M15 6PB, UK
richard.allmendinger@manchester.ac.uk

[5] University of Jyvaskyla, Faculty of Information Technology, FI-40014 University of Jyvaskyla, Finland
kaisa.miettinen@jyu.fi

Abstract. We consider the application of machine learning techniques to gain insights into the effect of problem features on algorithm performance, and to automate the task of algorithm selection for distance-based multi- and many-objective optimisation problems. This is the most extensive benchmark study of such problems to date. The problem features can be set directly by the problem generator, and include e.g. the number of variables, objectives, local fronts, and disconnected Pareto sets. Using 945 problem configurations (leading to 28 350 instances) of varying complexity, we find that the problem features and the available optimisation budget (i) affect the considered algorithms (NSGA-II, IBEA, MOEA/D, and random search) in different ways and that (ii) it is possible to recommend a relevant algorithm based on problem features.

Keywords: Multi/many-objective distance problems · Feature-based performance prediction · Automated algorithm selection

1 Introduction

Given a collection of problems and algorithms, the algorithm selection problem [24] is concerned with identifying an algorithm that is most suitable, in terms of some performance criteria, for a given problem at hand. An additional component is the availability of features characterising a problem. One can first extract problem features to generate a feature space, and then act on it as

M. Emmerich et al. (Eds.): EMO 2023, LNCS 13970, pp. 260–273, 2023.
https://doi.org/10.1007/978-3-031-27250-9_19

opposed to on the problem space. Significant research has been carried out on algorithm selection during the past two decades. Amongst others, tackling more efficiently a variety of continuous and mixed discrete/continuous [12,27], combinatorial [26] and multi-objective (continuous) optimisation problems [19], as well as supervised learning [22]. For algorithm selection, extracting problem features that drive algorithm performance is critical; doing this efficiently is the focus of fitness and exploratory landscape analysis [20,21]. Furthermore, understanding which and how problem features drive algorithm performance is valuable information when designing artificial problems of different complexity for benchmarking and tuning of algorithms. This information can then be used, e.g., to develop problem generators that can create test problems that meet user-defined problem characteristics. Such generators exist, for example, for many-objective distance-based optimisation [10] and cluster analysis [25].

Our focus is to advance the area of feature design and algorithm selection for multi- and many-objective optimisation problems. This is motivated by the prominence of problems in practice, combined with our limited understanding on suitable multi-objective problem features [12,20]. The most relevant work is given in [19], where features from landscape analysis (adapted from [17]) are applied on a benchmark set of 1 200 randomly-generated bi-objective interpolated continuous optimisation problems [30]. The study concluded that combining a classification model with a range of landscape features used as predictors can deliver a similar accuracy to predicting algorithm performance based on parameters used to generate the problems. It also investigated the relative importance of features for performance prediction and algorithm selection.

In this paper, we follow a similar approach to investigate the predictive power of parameters (problem features) used to generate distance-based multi/many-objective point problems (DBMOPPs) proposed in [10]. More precisely, we (i) generate 945 problems with different characteristics (as defined by 7 problem features), then (ii) test the correlation between the problem features and the performance of three popular multi-objective evolutionary algorithms and one baseline approach (random search), and finally assess the problem features as predictors for (iii) algorithm performance prediction and (iv) algorithm selection on machine learning regression and classification, respectively. This is the first study of DBMOPPs and the generator/problem features proposed in [10].

The paper is organised as follows. Section 2 introduces DBMOPPs, together with the generator and its parameters (problem features) used to control the generation of such problems. Section 3 gives the experimental setup and discusses algorithm performance. Section 4 presents an experimental analyses of the problem features for automated performance prediction and algorithm selection. Finally, Sect. 5 concludes the paper and discusses further research.

2 Distance-based Multi/Many-objective Problems

In multi-objective optimisation (with 2 or 3 competing objectives) and many-objective optimisation (with 4+ objectives), a *set* of solutions is typically sought that approximates the optimal trade-off combinations between the objectives, given any constraints on *decision vectors*. Formally, the tuple of a decision vector $\mathbf{x}$, and its evaluation under the objective vector $\mathbf{f}()$ define a *solution* $s = (\mathbf{x}, \mathbf{f}(\mathbf{x}))$. A solution s is said to dominate another s' if s performs better than s' on at least one objective, and no worse on all others. The maximal set of non-dominated decision vectors is known as the Pareto set, and its image in the objective space is known as the Pareto front.

A wide range of scalable test problem frameworks have been developed for multi- and many-objective optimisation, which are used (together with set quality indicators) to assess the performance of optimisers. These encapsulate a range of known problem characteristics (e.g. [3,6,11]). In addition, means for generating *instances* of problems have been created to prevent "tuning" of optimisers to particular suites of tests, to the detriment of performance on practical problems.

Frameworks for DBMOPPs have been developed over the last decade. They incorporate the range of features exhibited in other test suites (constraints, neutrality, multi-modality, dominance resistance regions, local fronts, etc.) and enable direct visualising of the search space in a plane. Initial work in this area includes [13,14], and [10] includes a summary of the features incorporated into this test problem design approach over the last 15 years. Arbitrarily many objectives can be defined. If $|\mathbf{x}| > 2$ the decision vector is projected into two dimensions via a pair of orthogonal vectors prior to function evaluation. We now describe the construction of DBMOPPs, and the generator we recently developed. We will then extensively investigate its characteristics.

Properties and Features of DBMOPPs. Point-based distance problems are parameterised by sets of attractor vectors, where the minimum distance to a member of the ith set, V_i, defines the ith objective value:

$$f_i(\mathbf{x}) = \min_{\mathbf{v} \in V_i} \texttt{dist}(\mathbf{v}, \mathbf{x}). \quad (1)$$

Further complexity can be added by imposing regions of constraint violation which locally adjust the distance function, and thereby can introduce discontinuities and neutrality to particular objectives, amongst other modifications.

In [10], we introduced a DBMOPP[1] instance generator, where problems can incorporate a range of properties, a subset of which are listed in Table 1 which we consider here. An example 3-objective problem with local fronts and dominance resistance regions is illustrated in Fig. 1, along with its local dominance landscape [9], PLOS-net [18] and PLON [8] network visualisations. In this work, we focus on box-constrained problems.

[1] Available in `Matlab` (https://github.com/fieldsend/DBMOPP_generator), and in `Python` (https://github.com/industrial-optimization-group/desdeo-problem/tree/master/desdeo_problem/testproblems/DBMOPP).

Table 1. Problem features considered from the DBMOPP generator [10].

Description	Name	Domain
Number of variables	`n_var`	$[\![2, 20]\!]$
Number of objectives	`n_obj`	$[\![2, 10]\!]$
Non-identical Pareto sets	`nonident_ps`	$\{\texttt{no}, \texttt{yes}\}$
Varying density	`var_density`	$\{\texttt{no}, \texttt{yes}\}$
Number of disconnected Pareto sets	`n_discon_ps`	$[\![0, 6]\!]$
Number of local fronts	`n_local_fronts`	$[\![0, 6]\!]$
Number of dominance resistance regions	`n_resist_regions`	$[\![0, 6]\!]$

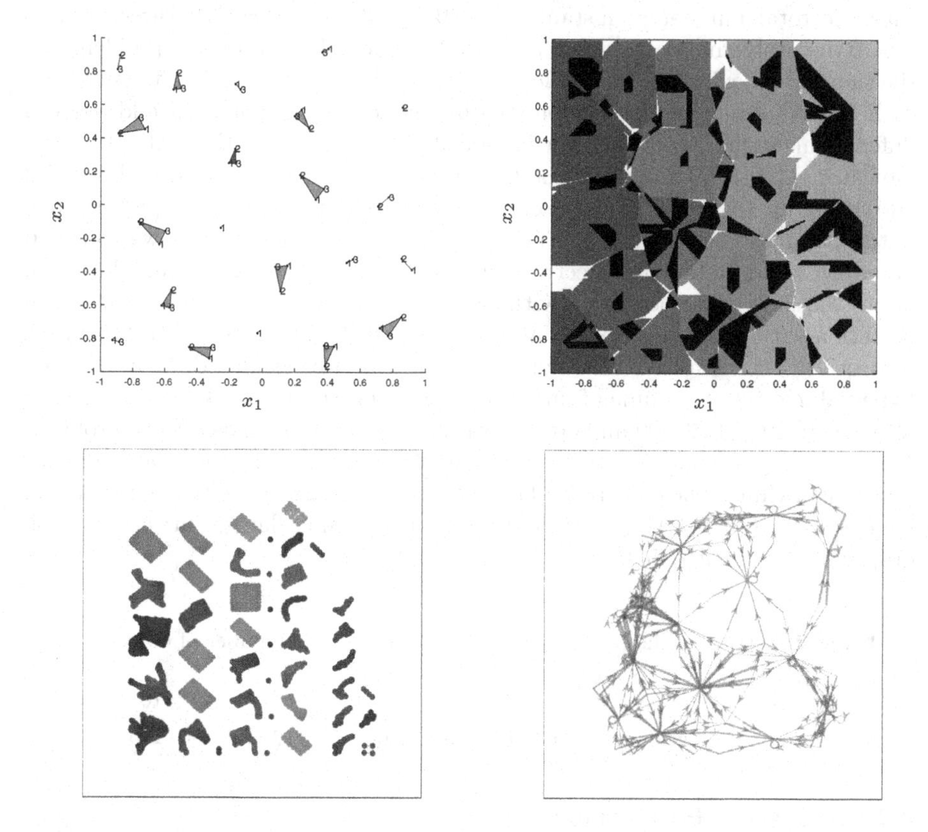

Fig. 1. *Top-left*: an example 3-objective problem with regions creating local fronts (green triangles), dominance resistance regions (points and lines) and Pareto set (red triangle). *Top-right*: its corresponding local dominance landscape – black regions are locally dominance neutral, shaded grey regions denote basins (for all basin members all neighbouring dominating moves lead to the same dominance neutral region), and white regions denote locations where immediate neighbours lead to different basins (saddles). *Bottom-left*: PLOS-net and *Bottom-right*: PLON network visualisations. (Color figure online)

3 Experimental Setup

This section describes our experimental setup covering the approach adopted to generate problems, algorithms and their settings, and performance metrics.

3.1 Dataset

We generated 945 problems with different parameter settings. A random latin hypercube sample [4] of size 1 000 was generated, among which 55 problems were discarded due to restrictions in the benchmark generator. For each parameter setting, 30 instances (folds) were independently created using the generator and thus the total number of instances was $945 \times 30 = 28\,350$. Features used to create the problems are provided in Table 1. All 30 folds for a given problem had the same complexity in terms of features. However, folds could be different from each other because of the randomness in the generator. For each fold, we ran different algorithms (one run per instance): NSGA-II [5], IBEA [32] with the ε indicator, MOEA/D [31] with the Chebyshev scalarising function and random search for up to 50 000 evaluations. For a fair comparison, we kept the same initial population for a fold and used the same population size. It was selected based on the number of objectives and is given in Table 2. We employed an out of the shelf implementation with simulated binary crossover and polynomial mutation with probability of 0.8 and $\frac{1}{\texttt{n\_var}}$ and distribution index of 20 and 20, respectively; $\kappa = 0.5$ for IBEA. For each algorithm, we calculated the normalised hypervolume (hypervolume of final solutions/hypervolume of the Pareto front[2]) after 5 000, 10 000, 30 000 and 50 000 evaluations for all 30 folds of each problem. Normalised hypervolume values were then averaged for each of the 945 problems and 4 algorithms. The code is available at: https://github.com/tichugh/Feature_Analysis_DBMOPP_EMO_2023, and the corresponding dataset at: https://doi.org/10.5281/zenodo.7155803.

Table 2. Setting of the population size according to the number of objectives.

Number of objectives	2	3	4	5	6	7	8	9	10
Population size	100	105	120	126	132	112	156	90	275

3.2 Algorithm Performance

We analysed results with `R` [23] using the `caret` [15], `rpart` [28] and `ggplot2` [29] packages. In Fig. 2 (left), we show the average normalised hypervolume (and the 95% confidence interval) for each algorithm with respect to the search budget over all considered 945 problems. Figure 2 (right) gives the proportion of problems where each algorithm obtained the best average performance, over the 30

[2] 1 000 members drawn from the Pareto front plus all non-dominated points found by the union of the algorithms' approximation sets for each instance. The reference point for hypervolume was $1.1 \times$ maximum of objective values on the Pareto front and estimated via Monte Carlo [7] with 50 000 samples for 4+ objectives.

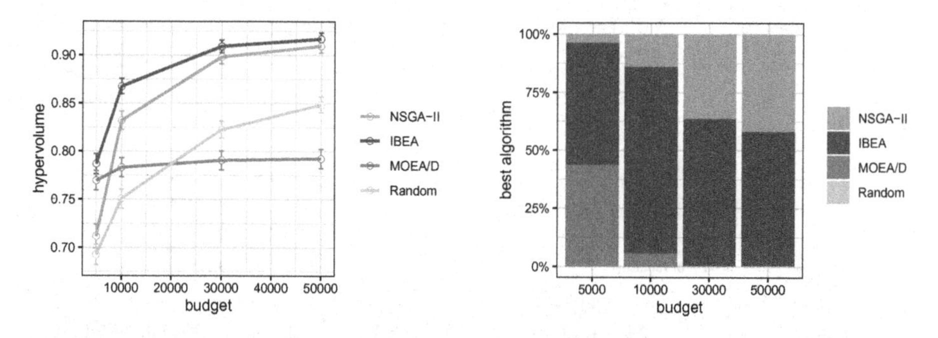

Fig. 2. Hypervolume (left) and proportion of problems where each algorithm obtains the best average hypervolume (right) with respect to the search budget.

folds, for the considered budgets. We observe that IBEA was consistently the best-performing algorithm, whatever the budget, and outperformed others in at least 50% of problems (for a budget of 5 000 evaluations), and at most 80% (10 000 evaluations). It was followed by NSGA-II (almost as good as IBEA for 50 000 evaluations). However, one should note that NSGA-II was not significantly better than random search for 5 000 evaluations. MOEA/D was efficient for the budget 5 000, but the increase of budget did not improve its performance as much as for the others. Indeed, the average hypervolume obtained by MOEA/D went from 0.77 for a budget of 5 000 to 0.79 for 50 000. A similar observation was reported in [10]. We conjecture that MOEA/D is significantly impacted by an increase in local Pareto fronts and multi-modality. Random search was dominated by other algorithms for the two lowest budgets. However, surprisingly, it surpassed MOEA/D for the highest budgets. In fact, random search was even the best for 1 problem for a budget of 30 000, and 2 problems for 50 000.

Interestingly, whatever the budget, there is no algorithm that outperforms the others for all problems. For the smallest budgets, IBEA and MOEA/D share the success almost equally on more than 95% of the 945 problems. For the largest budgets, NSGA-II and IBEA share the success on more than 99% of problems.

4 Experimental Study

This section uses a machine learning perspective to investigate the relationship of problem features and algorithm performance, the predictive power of features for performance prediction, and classification for feature-based algorithm selection.

4.1 Problem Features *vs* Algorithm Performance

We first investigate how problem features impact search performance. Figure 3 shows how the normalised hypervolume of algorithms is individually impacted by each of the 7 problem features. Due to space restrictions, we report only two budgets. In addition, Fig. 4 gives the Spearman's rank correlation coefficient between each problem feature and algorithm performance for all budgets. A larger hypervolume indicates a better performance and thus a positive correlation means that the problem feature has a favourable effect on algorithm performance.

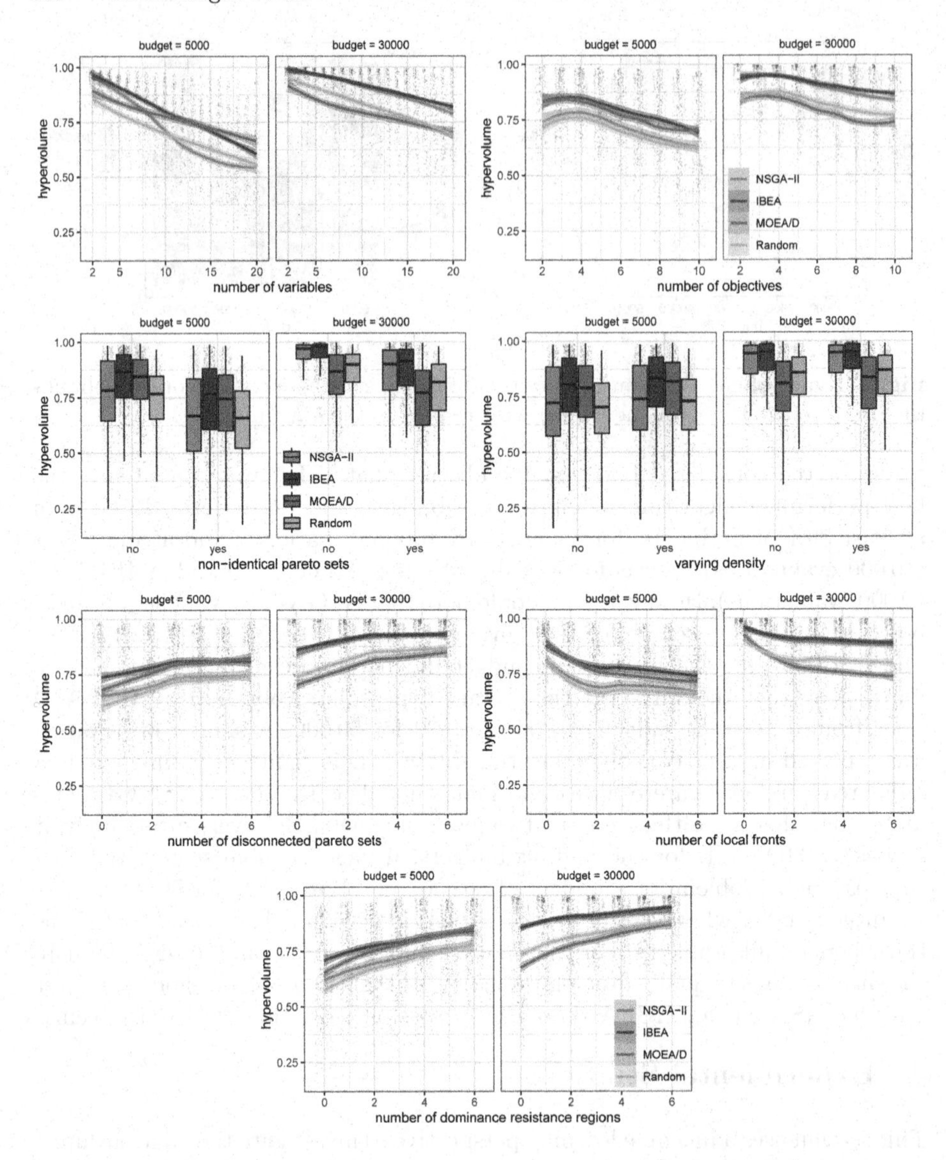

Fig. 3. Hypervolume vs. each problem feature and 5 000 and 30 000 evaluations.

For a given problem feature, the trend is similar for all algorithms and budgets (there is no feature with a positive effect for one algorithm and a negative effect for another algorithm). The same goes for the budgets. However, the strength of correlation is at times different. For instance, although a larger number of objectives (`n_obj`) means a worse performance for all algorithms and budgets, it is more impactful for NSGA-II than for other algorithms for large budgets.

The more variables (`n_var`), the worst the performance. However we see in Fig. 3 (top-left) that, for a budget of 5 000, NSGA-II is good for a small

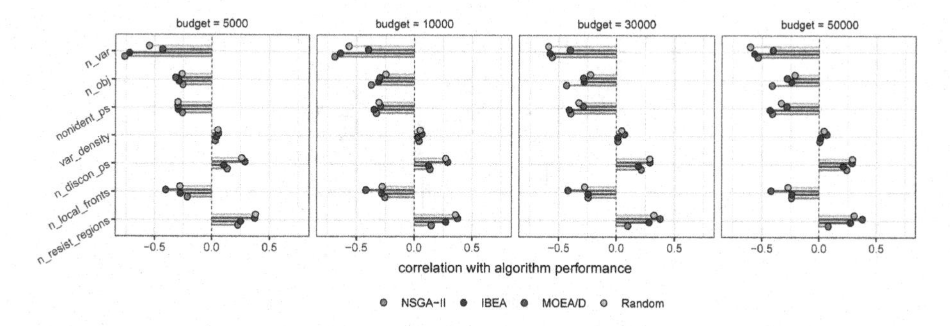

Fig. 4. Correlation between problem features and algorithm performance.

number of variables, and becomes worse than random search as the number of variables increases. Correspondingly, while the number of objectives and having non-identical Pareto sets (`nonident_ps`) negatively impact performance, the number of dominance resistance regions (`n_resist_regions`) has a positive effect. For a budget of 30 000, NSGA-II is as good as IBEA with few objectives and few dominance resistance regions, but IBEA gets better as both numbers increase. Besides, the fewer local fronts (`n_local_fronts`) and disconnected Pareto sets (`n_discon_ps`), the better the performance of all algorithms with all budgets. The varying density (`var_density`) has a minor impact on performance. Overall, problem features in the 945 problems often imply a significant difference between algorithms independently of other features.

4.2 Performance Prediction by Regression

The previous results concerned the *individual* effect of problem features on performance. We now investigate their *combined* effect by constructing a regression model for predicting the hypervolume reached by the algorithms under different budgets. We thus end up with 4 (algorithms) $\times$ 4 (budgets) $=$ 16 models aiming at predicting performance separately for each algorithm and budget, using the problem features as predictors. We consider random forest [1,16] with default parameters, a well-established state-of-the-art ensemble learning method that constructs multiple decision trees for regression. We start by evaluating the prediction accuracy of the trained models using 30 independent replicates of 10-fold cross-validation. We report the repeated cross-validated coefficient of determination (R^2) for each algorithm and budget in Fig. 5. We observe that the smallest R^2 obtained over all folds and repetitions is above 0.7, and the median R^2 is always above 0.85, whatever the algorithm and budget. This suggests that more than 85% of the variance in hypervolume values across all problems is explained by the model, and thus by problem features. The slight drop in the prediction accuracy for IBEA and NSGA-II as the budget increases is not significant and the R^2 values remain quite satisfactory. The prediction accuracy for random search is particularly high, regardless of the budget.

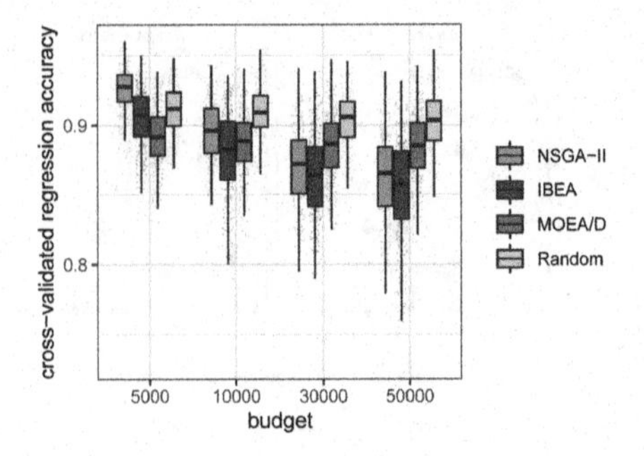

Fig. 5. Coefficient of determination (R^2) of regression models trained by algorithm and budget, calculated using 10-fold cross-validation with 30 repetitions.

For each regression model, we also compute the importance of predictors, commonly measured as the mean decrease of prediction accuracy with random forest [1,16]. The higher the value, the more important the predictor. Figure 6 shows the relative importance of problem features, scaled between 0 and 100. Overall, the most important problem features remain quite consistent with their correlation with algorithm performance reported in Sect. 4.1. The numbers of variables, objectives, dominance resistance regions and the presence of non-identical Pareto sets appear on top of the list, whereas the varying density only has a marginal contribution to the prediction accuracy. Interestingly, the presence of non-identical Pareto sets and the number of disconnected Pareto sets get more important as the budget increases. However, noticeable differences appear for MOEA/D, for which the numbers of local fronts and dominance resistance regions are highly important, regardless of the budget. They even surpass the number of variables as the most important features for larger budgets. Thus, it is interesting to note that even though the problem features coming from the problem generator have a similar effect on performance overall, their strength may be quite different depending on the considered algorithm and budget.

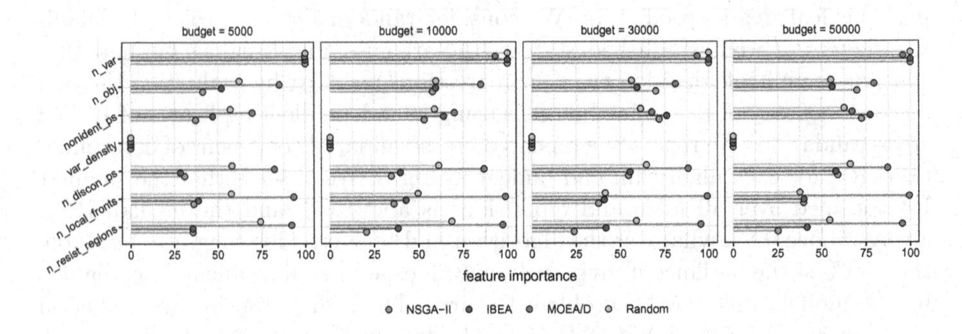

Fig. 6. Relative importance of problem features for regression models.

4.3 Algorithm Selection by Classification

Random Forest Classification. We now focus on feature-based algorithm selection to answer the following question: Given a problem, what is the recommended algorithm for solving it under a particular budget? The interest is no longer in which problem features have more impact on the performance of algorithms, but which features best distinguish algorithms from each other. We still apply random forest, this time for classification. We construct 4 classification models (one per budget) to predict the best algorithm using problem features as predictors. Their classification accuracy, based on 30 repetitions of 10-fold cross-validation, is reported in Fig. 7 (left). The lowest accuracy obtained over all folds and repetitions is 0.64, and the median accuracy is always above 0.75 for all budgets. This means that the classifier is able to predict the best algorithm in at least 75% of the cases, which is significantly better than a random classifier (with an accuracy of 25%), or a dummy classifier that would always select IBEA, the most frequent best algorithm for any budget, which outperformed other algorithms in 53%, 80%, 63%, and 58% of problems respectively, for budgets of 5 000, 10 000, 30 000, and 50 000 evaluations, as reported in Sect. 3.2. The accuracy of random forest is slightly higher for a budget of 10 000. We attribute this to the fact that IBEA outperforms other approaches more often under this budget, which makes the classification problem easier. We also report in Fig. 7 (right) the relative hypervolume deviation of the feature-based random forest classifier (`Auto`) from the virtual best algorithm, that is the ideal method, an oracle, that always selects the best algorithm. This measure is termed *regret*, as it indicates how far the obtained hypervolume is from an ideal classifier. It is compared against always selecting each one of the considered algorithms. Notice the log scale in the plot. The regret obtained by the feature-based classifier ranges from 0.0016 to 0.0044 and deviates from an ideal classifier by less than 0.5%. This is less than IBEA, the most frequent best algorithm for all budgets, by an order of magnitude. Compared to an ideal classifier, the relative performance of NSGA-II increases with the budget, while MOEA/D moves away from it.

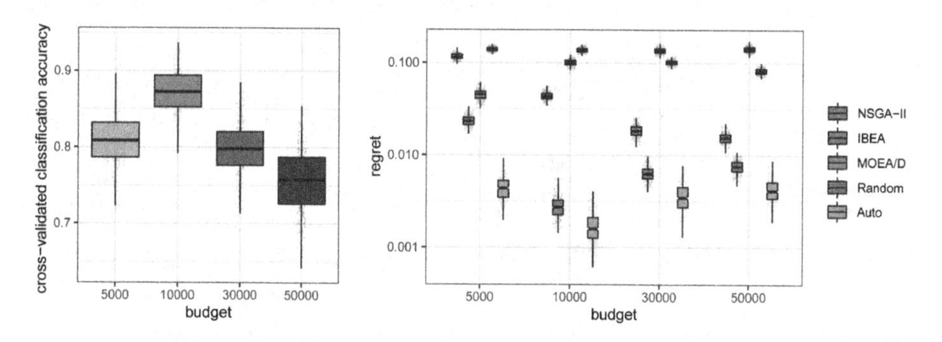

Fig. 7. Accuracy (left) and regret (right) of classification models trained by budget, calculated using 10-fold cross-validation with 30 repetitions.

In Fig. 8, we report the relative importance of problem features for the random forest classifiers. As suggested earlier, the importance of the number of variables decreases with the budget. By contrast, the importance of the numbers of objectives and of dominance resistance regions increases with the budget, and even exceeds the number of variables for larger budgets. The numbers of disconnected Pareto sets and local fronts are less important for algorithm selection than for performance regression. In fact, they have a low importance similar to the presence of non-identical Pareto sets and to the varying density for some budgets. Thus, although they have a significant effect on algorithm performance, the impact on all algorithms is the same. The feature analysis for classification shows that it is possible to recommend an algorithm based on problem features with a fairly high accuracy. Moreover, the features have a different impact on the choice of the algorithm depending on the budget.

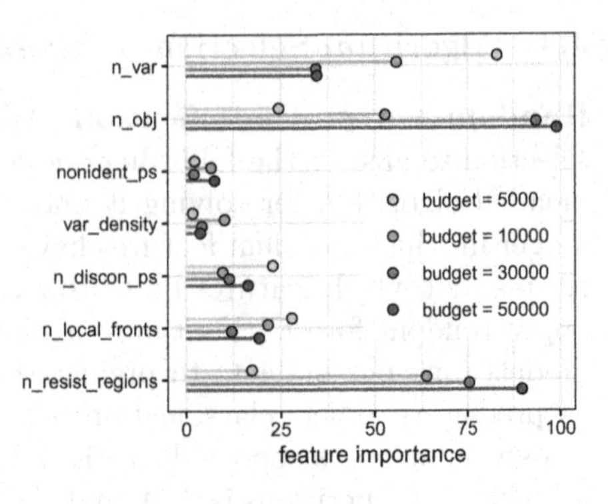

Fig. 8. Relative importance of problem features for classification models.

Decision Trees. We conclude with a basic classifier for algorithm selection based on a decision tree. Its construction follows the well-established CART algorithm [2,28], whose segmentation criterion is the Gini diversity index and which generates binary decision trees (i.e. a node has two children at most). In Fig. 9, we show decision trees for a budget of 5 000 (left) and 30 000 evaluations (right). Numbers below each node are the number of times NSGA-II, IBEA, MOEA/D and random search are each the best algorithm, respectively, followed by the proportion of problems covered by the node. There are only three values on the first rows for a budget of 5 000 since random search is never the best. Although the accuracy is slightly lower than that of a random forest (0.75

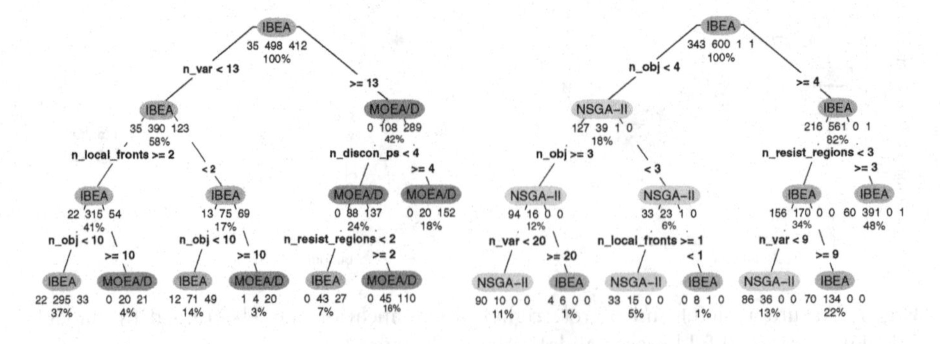

Fig. 9. Algorithm selection for budgets of 5 000 (left) and 30 000 evaluations (right).

and 0.79, respectively, for a budget of 5 000 and 30 000), we argue that this can provide a useful recommendation system for algorithm selection, with only three levels of decision in these examples.

The tree levels are consistent with the importance of features depicted by random forest, but their joint effect appears more explicitly. Under the smaller budget, the first decision is based on the number of variables. For example, with less than 13 variables, more than 2 local fronts, and less than 10 objectives, IBEA is the best. For a larger budget, and as expected from our previous comments, the feature that appears on top of the tree is the number of objectives. In this case, NSGA-II is the best with few objectives and few variables or few local fronts. Conversely, looking at the rightmost branch of the tree, IBEA is clearly the best with 4+ objectives and 3+ dominance resistance regions. This covers 48% of the problems. Such decision trees justify why an algorithm is recommended based on feature values. In addition, it points out the problem characteristics for which search mechanisms should be refined to improve performance.

5 Conclusions

We adopted a machine learning perspective to carry out the most extensive feature-based benchmark study of distance-based multi/many-objective optimisation problems to date. We generated 28 350 instances based on 945 problem configurations by varying the complexity controlled by 7 features. Random forests and decision trees were then used to understand correlations between the problem features and algorithm performance, predict algorithm performance, and automate the task of algorithm selection for a given problem and budget at hand. We find that, although the considered problem features affect the performance of algorithms in distinctive ways, when used as predictors in a random forest classifier we can predict the best algorithm with an accuracy of 75% or more. Thus, problem features can control the complexity of a problem, and lend themselves to selecting an algorithm when faced with a previously unseen problem. This is the first automated algorithm selection study for continuous problems with more than 2 objectives. We observed that the number of objectives is (i) negatively correlated with algorithm performance, (ii) one of the most important problem features for predicting algorithm performance, especially for larger budgets, and (iii) a key feature to making an accurate algorithm selection when faced with an unseen problem. Future work could investigate if considering additional problem and landscape features can help increase prediction accuracy further. Some expected algorithm behaviours with respect to specific problem features could also be corroborated based on a fine-grained analysis of the data produced in this work. At last, it would be worth studying the sensitivity of algorithm parameters, and their joint impact with problem features on performance.

Acknowledgements. This research is part of the thematic research area DEMO, Decision Analytics utilising Causal Models and Multiobjective Optimisation, jyu.fi/demo, at the University of Jyvaskyla.

References

1. Breiman, L.: Random forests. Mach. Learn. **45**(1), 5–32 (2001). https://doi.org/10.1023/a:1010933404324
2. Breiman, L., Friedman, J., Stone, C.J., Olshen, R.A.: Classification and Regression Trees. Taylor & Francis, Andover (1984)
3. Brockhoff, D., Tusar, T., Auger, A., Hansen, N.: Using well-understood single-objective functions in multiobjective black-box optimization test suites. arXiv CoRR (2016). https://doi.org/10.48550/arxiv.1604.00359
4. Carnell, R.: LHS: Latin hypercube samples (2022). r package version 1.1.5
5. Deb, K., Prarap, A., Agarwal, S., Meyarivan, T.: A fast and elitist multiobjective genetic algorithm: NSGA-II. IEEE Trans. Evol. Comput. **6**, 182–197 (2002)
6. Deb, K., Thiele, L., Laumanns, M., Zitzler, E.: Scalable test problems for evolutionary multiobjective optimization. In: Abraham, A., Jain, L., Goldberg, R. (eds.) Evolutionary Multiobjective Optimization: Theoretical Advances and Applications, pp. 105–145. Springer, London (2005). https://doi.org/10.1007/1-84628-137-7_6
7. Everson, R.M., Fieldsend, J.E., Singh, S.: Full elite sets for multi-objective optimisation. In: Parmee, I.C. (ed.) Adaptive Computing in Design and Manufacture V, pp. 343–354. Springer, London (2002). https://doi.org/10.1007/978-0-85729-345-9_29
8. Fieldsend, J.E., Alyahya, K.: Visualising the landscape of multi-objective problems using local optima networks. In: Proceedings of the Genetic and Evolutionary Computation Conference Companion, pp. 1421–1429 (2019)
9. Fieldsend, J.E., Chugh, T., Allmendinger, R., Miettinen, K.: A feature rich distance-based many-objective visualisable test problem generator. In: Proceedings of the Genetic and Evolutionary Computation Conference, pp. 541–549 (2019)
10. Fieldsend, J.E., Chugh, T., Allmendinger, R., Miettinen, K.: A visualizable test problem generator for many-objective optimization. IEEE Trans. Evol. Comput. **26**(1), 1–11 (2022)
11. Huband, S., Hingston, P., Barone, L., While, L.: A review of multiobjective test problems and a scalable test problem toolkit. IEEE Trans. Evol. Comput. **10**(5), 477–506 (2006)
12. Kerschke, P., Hoos, H.H., Neumann, F., Trautmann, H.: Automated algorithm selection: survey and perspectives. Evol. Comput. **27**(1), 3–45 (2019)
13. Köppen, M., Vicente-Garcia, R., Nickolay, B.: Fuzzy-pareto-dominance and its application in evolutionary multi-objective optimization. In: Coello Coello, C.A., Hernández Aguirre, A., Zitzler, E. (eds.) EMO 2005. LNCS, vol. 3410, pp. 399–412. Springer, Heidelberg (2005). https://doi.org/10.1007/978-3-540-31880-4_28
14. Köppen, M., Yoshida, K.: Substitute distance assignments in NSGA-II for handling many-objective optimization problems. In: Obayashi, S., Deb, K., Poloni, C., Hiroyasu, T., Murata, T. (eds.) EMO 2007. LNCS, vol. 4403, pp. 727–741. Springer, Heidelberg (2007). https://doi.org/10.1007/978-3-540-70928-2_55
15. Kuhn, M.: Building predictive models in R using the caret package. J. Stat. Softw. Art. **28**(5), 1–26 (2008)
16. Liaw, A., Wiener, M.: Classification and regression by randomforest. R News **2**(3), 18–22 (2002)
17. Liefooghe, A., Daolio, F., Verel, S., Derbel, B., Aguirre, H., Tanaka, K.: Landscape-aware performance prediction for evolutionary multiobjective optimization. IEEE Trans. Evol. Comput. **24**(6), 1063–1077 (2019)

18. Liefooghe, A., Derbel, B., Verel, S., López-Ibáñez, M., Aguirre, H., Tanaka, K.: On pareto local optimal solutions networks. In: Auger, A., Fonseca, C.M., Lourenço, N., Machado, P., Paquete, L., Whitley, D. (eds.) PPSN 2018. LNCS, vol. 11102, pp. 232–244. Springer, Cham (2018). https://doi.org/10.1007/978-3-319-99259-4_19
19. Liefooghe, A., Verel, S., Lacroix, B., Zăvoianu, A.C., McCall, J.: Landscape features and automated algorithm selection for multi-objective interpolated continuous optimisation problems. In: Proceedings of the Genetic and Evolutionary Computation Conference, pp. 421–429 (2021)
20. Malan, K.M.: A survey of advances in landscape analysis for optimisation. Algorithms **14**(2), 40 (2021)
21. Mersmann, O., Bischl, B., Trautmann, H., Preuss, M., Weihs, C., Rudolph, G.: Exploratory landscape analysis. In: Proceedings of the Genetic and Evolutionary Computation Conference, pp. 829–836 (2011)
22. Muñoz, M.A., Villanova, L., Baatar, D., Smith-Miles, K.: Instance spaces for machine learning classification. Mach. Learn. **107**(1), 109–147 (2018). https://doi.org/10.1007/s10994-017-5629-5
23. R Core Team: R: A language and environment for statistical computing. R Foundation for Statistical Computing, Vienna, Austria (2020)
24. Rice, J.R.: The algorithm selection problem. In: Advances in Computers, vol. 15, pp. 65–118. Elsevier (1976)
25. Shand, C., Allmendinger, R., Handl, J., Webb, A., Keane, J.: HAWKS: Evolving challenging benchmark sets for cluster analysis. IEEE Trans. Evol. Comput. **26**(6), 1206–1220 (2022)
26. Smith-Miles, K., Lopes, L.: Measuring instance difficulty for combinatorial optimization problems. Compute. Oper. Res. **39**(5), 875–889 (2012)
27. Smith-Miles, K.A.: Cross-disciplinary perspectives on meta-learning for algorithm selection. ACM Comput. Surv. **41**(1), 1–25 (2009)
28. Therneau, T., Atkinson, B.: rpart: Recursive partitioning and regression trees (2022). R package version 4.1.16
29. Wickham, H.: ggplot2: Elegant Graphics for Data Analysis. Springer, New York (2016)
30. Zăvoianu, A.-C., Lacroix, B., McCall, J.: Comparative run-time performance of evolutionary algorithms on multi-objective interpolated continuous optimisation problems. In: Bäck, T., et al. (eds.) PPSN 2020. LNCS, vol. 12269, pp. 287–300. Springer, Cham (2020). https://doi.org/10.1007/978-3-030-58112-1_20
31. Zhang, Q., Li, H.: MOEA/D: a multiobjective evolutionary algorithm based on decomposition. IEEE Trans. Evol. Comput. **11**, 712–731 (2007)
32. Zitzler, E., Künzli, S.: Indicator-based selection in multiobjective search. In: Yao, X., et al. (eds.) PPSN 2004. LNCS, vol. 3242, pp. 832–842. Springer, Heidelberg (2004). https://doi.org/10.1007/978-3-540-30217-9_84

Benchmarking and Performance Assessment

Partially Degenerate Multi-objective Test Problems

Lie Meng Pang, Yang Nan, and Hisao Ishibuchi(✉)

Guangdong Provincial Key Laboratory of Brain-Inspired Intelligent Computation, Department of Computer Science and Engineering, Southern University of Science and Technology, Shenzhen 518055, China

{panglm,hisao}@sustech.edu.cn, nany@mail.sustech.edu.cn

Abstract. Degenerate multi-objective test problems are included in test suites to evaluate EMO algorithms on a wide variety of test problems. However, it was pointed out in some studies that the frequently-used degenerate DTLZ5, DTLZ6 and WFG3 test problems do not have degenerate Pareto fronts. Their Pareto fronts are different from the originally intended degenerate shapes. Actually, they are partially degenerate test problems. Modified formulations of DTLZ5 and DTLZ6 were proposed to remove the non-degenerate parts of their Pareto fronts. However, the original formulations of DTLZ5, DTLZ6 and WFG3 continue to be used as degenerate test problems in many studies whereas they are not degenerate test problems. One issue in their use as degenerate test problems is that reference point sets for IGD calculation are sampled from the originally intended degenerate Pareto fronts whereas they are not the true Pareto fronts. Nevertheless, the original DTLZ5, DTLZ6 and WFG3 formulations are useful for performance evaluation of EMO algorithms since their Pareto front shapes are similar to some real-world problems and much more complicated than other test problems. That is, their use helps us to evaluate the performance of EMO algorithms on a wide variety of test problems including realistic and challenging test problems. In this paper, we clearly demonstrate the usefulness of the original DTLZ5, DTLZ6 and WFG3 formulations. Then, after pointing out the difficulty in their use in computational experiments, we explain how we can obtain reliable experimental results on those test problems.

Keywords: Evolutionary multi-objective optimization · Test problems · Degenerate Pareto fronts · Partially degenerate Pareto fronts · IGD indicator

1 Introduction

In the field of evolutionary multi-objective optimization (EMO), the performance of EMO algorithms is usually evaluated through computational experiments on benchmark test suites. Thus, it is highly desirable that a benchmark test suite consists of a wide variety of test problems with diverse characteristics including realistic test problems. In recent two decades, several benchmark test suites (e.g., ZDT [1], DTLZ [2], WFG [3], MaF [4], UF [5]) have been proposed to facilitate the growth of the EMO field.

M. Emmerich et al. (Eds.): EMO 2023, LNCS 13970, pp. 277–290, 2023.
https://doi.org/10.1007/978-3-031-27250-9_20

These test suites cover various problem characteristics. For example, these test suites include test problems with various fitness landscapes such as unimodal, multimodal, biased, and deceptive. They also include test problems with various Pareto front shapes such as linear, convex, concave, and disconnected. In some test suites, multi-objective test problems with degenerate Pareto fronts are included to increase the diversity of test suites. For example, in the DTLZ test suite [2], DTLZ5 and DTLZ6 were designed as degenerate test problems. In the WFG test suite [3], WFG3 was designed as a degenerate test problem.

An M-objective problem is generally considered degenerate if the dimension of its Pareto front is smaller than $(M - 1)$ [12], which can be a result of the existence of redundant objectives in its problem formulation [27]. Examples of degenerate Pareto fronts are illustrated in Fig. 1, in which the degenerate Pareto fronts of DTLZ5, DTLZ6 and WFG3 with three objectives are shown. These three test problems have been frequently used to demonstrate the ability of EMO algorithms to handle multi-objective problems with degenerate Pareto fronts (e.g., see [6–10]). If a problem contains both degenerate and non-degenerate parts of the Pareto front, it is referred to as a partially degenerate problem in this paper. In [27], it was demonstrated that partially redundant objectives can lead to a partially degenerate problem.

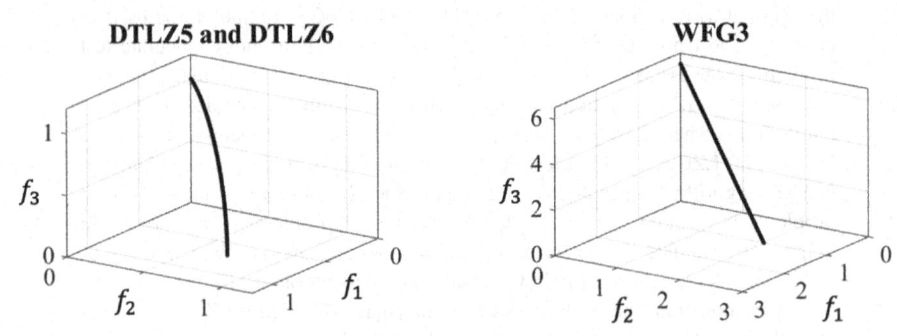

Fig. 1. The intended degenerate Pareto fronts for the three-objective DTLZ5, DTLZ6 and WFG3 test problems.

While the DTLZ5, DTLZ6 and WFG3 test problems have been frequently used to evaluate the performance of EMO algorithms on degenerate problems, it was pointed out in some studies that these three test problems are not degenerate test problems [3, 11, 12]. Their Pareto fronts are different from the originally intended shapes. Actually, they are partially degenerate test problems [12]. The true Pareto fronts for DTLZ5 and DTLZ6 have non-degenerate parts when they have more than three objectives [3, 11, 12]. WFG3 has a non-degenerate part of the Pareto front when it has three or more objectives [12]. In order to remove the non-degenerate parts, modified formulations of DTLZ5 and DTLZ6 were proposed in [11]. In [12], constraint conditions were derived to remove the non-degenerate part of the Pareto front of WFG3. Despite these efforts, the original formulations of DTLZ5, DTLZ6 and WFG3 are still used as degenerate test problems in many studies.

In this paper, we point out that the original formulations of DTLZ5, DTLZ6 and WFG3 with the partially degenerate Pareto fronts are good test problems. This is because their Pareto front shapes are similar to those of some real-world problems. This is also because their Pareto front shapes are much more complicated than those of the other DTLZ and WFG test problems. That is, the original formulations of DTLZ5, DTLZ6 and WFG3 are more realistic and challenging in performance evaluation of EMO algorithms than the other DTLZ and WFG test problems. One issue in their use is that the originally intended Pareto front of each test problem is often used to sample reference point sets for the inverted generational distance (IGD) [13] calculation. That is, the reference point sets and the test problems are not consistent. In other words, the original formulations of DTLZ5, DTLZ6 and WFG3 are not appropriately used for evaluating the performance of EMO algorithms. In this paper, we demonstrate the usefulness of the original DTLZ5, DTLZ6 and WFG3 test problems. We also provide suggestions on how to use them for performance evaluation of EMO algorithms.

The organization of this paper is as follows. Section 2 provides brief discussions on the Pareto fronts of DTLZ5, DTLZ6 and WFG3 with the original problem formulations. In Sect. 2, we also review the availability of these three test problems and their reference point sets for IGD calculation in frequently-used EMO experimental platforms: jMetal [22], PlatEMO [19] and pymoo [23]. Section 3 presents our experimental results for IGD-based performance evaluation. Section 4 concludes this paper.

2 DTLZ5, DTLZ6 and WFG3 Test Problems

2.1 Pareto Fronts of DTLZ5, DTLZ6 and WFG3

As shown in Fig. 1, the originally intended Pareto front shapes of the DTLZ5 and DTLZ6 test problems are one-dimensional curves independent of the number of objectives [2, 3, 12]. However, it was pointed out in [3, 11, 12] that the true Pareto fronts of DTLZ5 and DTLZ6 are not degenerate when the number of objectives is larger than three. The true Pareto front shapes of DTLZ5 and DTLZ6 are unknown for the case of four or more objectives. For WFG3, the originally intended Pareto front shape is a line as shown in Fig. 1. However, the true Pareto front of WFG3 includes the line part and other solutions [12], which gives rise to a flag-like shape in the three-objective case (see Fig. 2 (a)). In Fig. 2 (b), we show an approximated Pareto front of a real-world three-objective "reactive power optimization" problem called DDMOP5 in [24]. We can see that the two Pareto fronts in Fig. 2 have similar shapes. A similar partially degenerate flag-shaped Pareto front is also shown in [20] for a real-world three-objective "two-bar truss design" problem called RE3–3-1. For the case of four or more objectives, the true Pareto front shape of WFG3 is unknown.

To obtain clear pictures of the true Pareto front shapes of DTLZ5, DTLZ6 and WFG3 in a high-dimensional objective space, we use five EMO algorithms, i.e., MOEA/D with the PBI function [14], NSGA-III [15], θ-DEA [16], NSGA-II/SDR [17] and PREA [18] to approximate the Pareto front of each test problem in the five-objective space. These five algorithms are chosen based on the following considerations. MOEA/D and NSGA-III are frequently-used classic EMO algorithms. The other three are recently-proposed EMO algorithms which have shown promising performance on many-objective

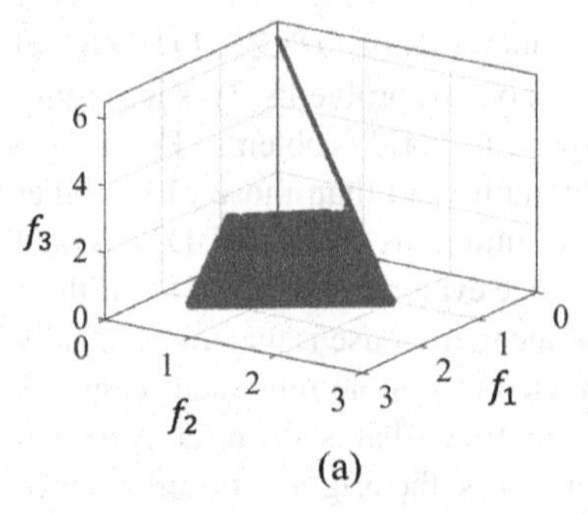

(a)

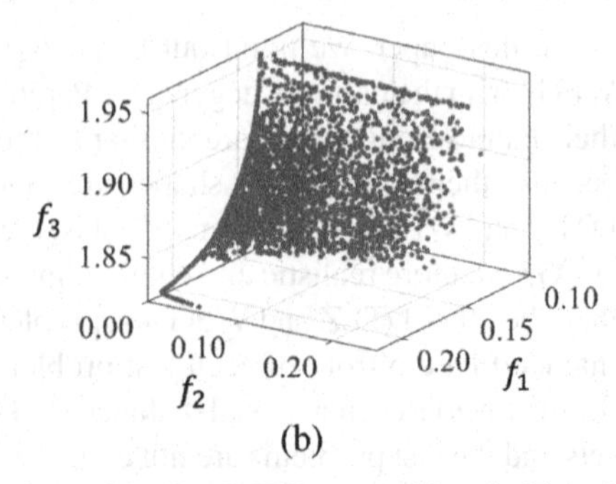

(b)

Fig. 2. The partially degenerate Pareto front of the three-objective WFG3 test problem in (a) and an approximated Pareto front of the real-world DDMOP5 problem [24] in (b).

problems. We use PlatEMO [19] for our experiments in this paper. The population size in each algorithm is specified as 210. Each algorithm is terminated after 1,000 generations. For other specifications in each algorithm, the default settings in PlatEMO are used. Each algorithm is executed 31 times on each test problem. To approximate the true Pareto front of each test problem, we use all non-dominated solutions among obtained solutions by 31 runs of the five algorithms (i.e., $31 \times 5 = 155$ runs in total).

Figure 3 shows an approximated Pareto front for the five-objective DTLZ5. Due to the paper length limitation, approximated Pareto fronts for the other two test problems are shown in the supplementary file (which is available from https://github.com/HisaoLabSUSTC/EMO2023). The approximated Pareto front in Fig. 2 (b) was created in the same manner as in Fig. 3 whereas the population size was 91 in Fig. 2 (b). As shown in Fig. 3 (and Figs. S1-S2 in the supplementary file), the approximated Pareto fronts of DTLZ5, DTLZ6 and WFG3 are highly irregular in the high-dimensional objective space. They are clearly different from the other test problems in the DTLZ and WFG test suites. Thus, their use increases the diversity of the test problems in these test suites.

A major challenge posed by DTLZ5, DTLZ6 and WFG3 is to find their entire Pareto fronts including the non-degenerate parts. Clearly different solution sets are often obtained by different EMO algorithms on these test problems even when almost the same results are obtained on other more standard test problems such as DTLZ1–4 and WFG4–9 with regular triangular Pareto fronts [21]. As an example, Fig. 4 shows the final population of a single run with the median IGD value among 31 runs of each algorithm on the five-objective DTLZ5 in the f_1–f_4 subspace (see Sect. 3 for IGD calculation). The f_1–f_4 subspace is shown here because it provides a clear demonstration of the search performance of the five EMO algorithms on the five-objective DTLZ5 problem with a partially degenerate Pareto front. The upper left subfigure shows the approximated Pareto front in the f_1–f_4 subspace, which is a copy from Fig. 3. In Fig. 4, clearly different solution sets are obtained from the five algorithms. No algorithms find a well-distributed solution set over the entire Pareto front. NSGA-III and PREA seem to find more diverse solutions on the non-degenerate part of the Pareto front than the other three algorithms. The main difference among the obtained solution sets in Fig. 4 is the diversity of solutions over the non-degenerate part (see also Fig. 7 for the ten-objective DTLZ5 and Fig. 8 for the

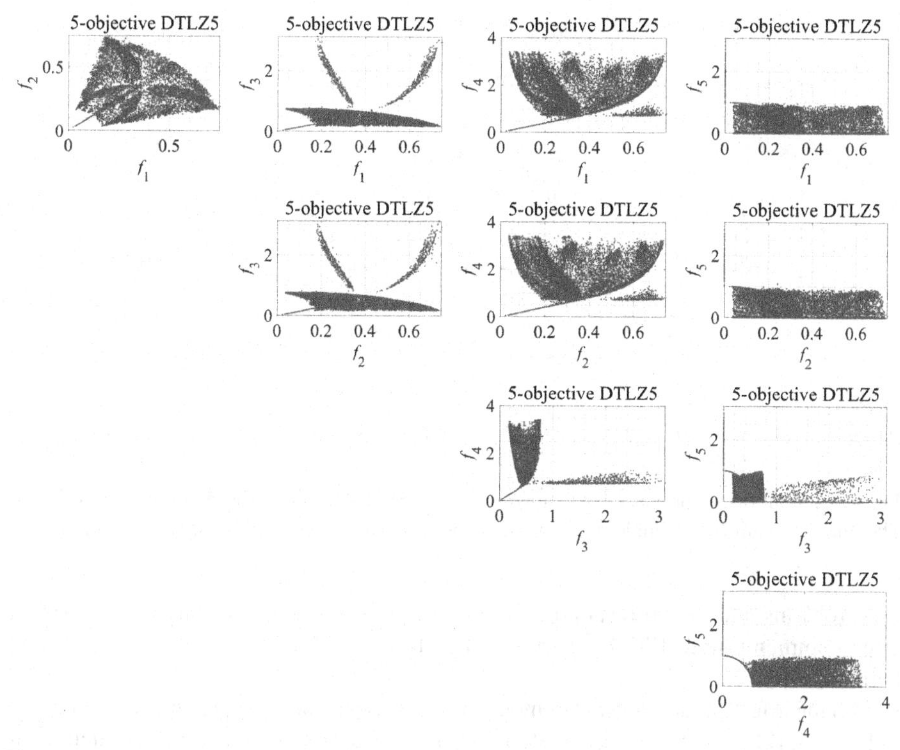

Fig. 3. An approximated Pareto front for the five-objective DTLZ5 test problem. Solutions are projected to the two-dimensional subspace.

ten-objective WFG3 in Sect. 3). Thus, the three partially degenerate test problems are useful for evaluating the diversification ability of EMO algorithms.

2.2 Availability of the Test Problems

In the previous subsection, we have discussed the usefulness of the original problem formulations of DTLZ5, DTLZ6 and WFG3 for performance evaluation of EMO algorithms. Whereas the three test problems are useful, one critical issue is that the originally intended degenerate Pareto fronts have been used to sample reference point sets for IGD calculation. For DTLZ5, DTLZ6 and WFG3 with the original problem formulations, IGD-based evaluation results are unreliable and misleading if reference point sets are sampled from the originally intended degenerate Pareto fronts. Under this reference point sampling mechanism, the calculated IGD values evaluate the approximation quality of the obtained solution sets only for the degenerate parts of the partially degenerate Pareto fronts.

In many EMO experimental platforms, the original formulations of DTLZ5, DTLZ6 and WFG3 are available. It is therefore necessary to check whether the reference point set for IGD calculation is sampled from the entire partially degenerate Pareto front of each test problem. In this subsection, we review the problem formulations of the three

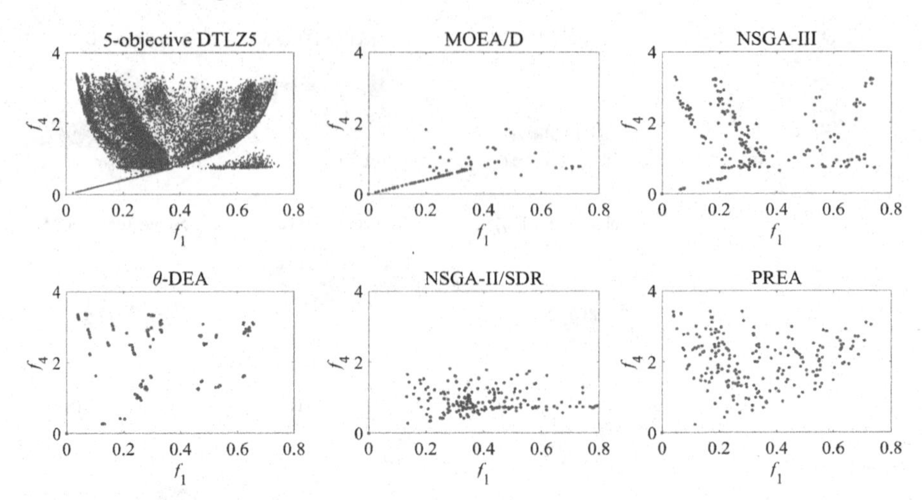

Fig. 4. A solution set obtained by a single run of each algorithm on the five-objective DTLZ5. The final population of a single run is projected to the two-dimensional subspace with f_1 and f_4.

test problems and the corresponding reference point sets for IGD calculation used in three commonly-used EMO experimental platforms: jMetal [22], PlatEMO [19] and pymoo [23].

All the jMetal, PlatEMO and pymoo platforms use the original problem formulations of DTLZ5, DTLZ6 and WFG3. Table 1 lists the reference point sets for IGD calculation for the three test problems in each platform. In jMetal and pymoo, the reference point sets for IGD calculation for DTLZ5, DTLZ6 and WFG3 are sampled from the originally intended degenerate Pareto fronts when the number of objectives (i.e., M) is three. This setting is appropriate for DTLZ5 and DTLZ6 since they have degenerate Pareto fronts when $M = 3$. However, this setting is not appropriate for WFG3 since its Pareto front is not degenerate when $M \geq 3$. For $M > 3$, the reference point sets for the three test problems are not provided in jMetal and pymoo. When reference point sets are not available in jMetal and pymoo, they can be constructed by combining the results of all runs of compared algorithms. This is a widely-used practice in the EMO field for unknown Pareto fronts (whereas this does not always lead to reliable comparison results [25]).

In PlatEMO, the provided reference point sets for IGD calculation for DTLZ5, DTLZ6 and WFG3 are sampled from the originally intended degenerate Pareto fronts regardless of the number of objectives. Thus, the reference points are not appropriate for DTLZ5 and DTLZ6 for $M > 3$ and WFG3 for $M \geq 3$. When the IGD indicator is used to evaluate the performance of EMO algorithms in PlatEMO for the three test problems, misleading results are likely to be obtained. It is therefore necessary to update the reference point sets for the three test problems in order to avoid creating unreliable IGD-based evaluation results. Moreover, it is important for users to be aware that the original problem formulations of DTLZ5, DTLZ6 and WFG3 are partially degenerate problems. When these three test problems are used for performance evaluation of EMO algorithms, we should always ensure that an appropriate reference point set for IGD calculation is used for each test problem.

Table 1. Reference point sets used for IGD calculation in jMetal, PlatEMO and pymoo.

Test problem	M	Reference point sets used for IGD calculation		
		jMetal [22]	PlatEMO [19]	pymoo [23]
DTLZ5, DTLZ6, WFG3	3	Sampled from the intended degenerate Pareto front	Sampled from the intended degenerate Pareto front	Sampled from the intended degenerate Pareto front
DTLZ5, DTLZ6, WFG3	>3	Not provided	Sampled from the intended degenerate Pareto front	Not provided

3 Performance Evaluation Results

In this section, we examine the performance of the five EMO algorithms (MOEA/D [14], NSGA-III [15], θ-DEA [16], NSGA-II/SDR [17] and PREA [18]) on DTLZ5, DTLZ6 and WFG3 with the original formulations. The population size for each algorithm is specified as 91 for three-objective problems, 210 for five-objective problems, and 275 for ten-objective problems. The termination condition of each algorithm is 1,000 generations. Each algorithm is executed 31 times on each test problem.

We report two types of IGD-based performance evaluation results for DTLZ5, DTLZ6 and WFG3. One is based on the reference point set sampled from the originally intended degenerate Pareto front (i.e., the reference point set provided in PlatEMO) for each test problem, and the other is based on the reference point set consisting of all non-dominated solutions among obtained solutions by 31 runs of the five algorithms. In the latter setting, the reference point set for each test problem is an approximation of the partially degenerate (i.e., true) Pareto front. In order to examine the reliability of the constructed reference point sets using the obtained solutions by the five algorithms, two different termination conditions are used to construct the reference point sets. One is 1,000 generations (which is the same as the termination condition for performance evaluation of the five algorithms), and the other is 10,000 generations. That is, we perform IGD-based comparison of the five algorithms using the three reference point sets for each test problem. Figures 5–6 show the three reference point sets for the ten-objective DTLZ5 and WFG3, respectively. The reference point sets provided in PlatEMO (i.e., the left figures in Figs. 5–6) are clearly different from the reference point sets obtained by the five algorithms with the two termination conditions (i.e., the center and right figures in Figs. 5–6). The difference in the reference point sets between the two termination conditions is small especially in Fig. 6 on the ten-objective WFG3. Reference point sets for the other many-objective test problems (i.e., the five-objective DTLZ5, DTLZ6 and WFG3, and the ten-objective DTLZ6) are shown in the supplementary file.

Experimental results using the three reference point sets are summarized in Tables 2, 3 and 4 using the ranking of the five algorithms ("1" is the best and "5" is the worst).

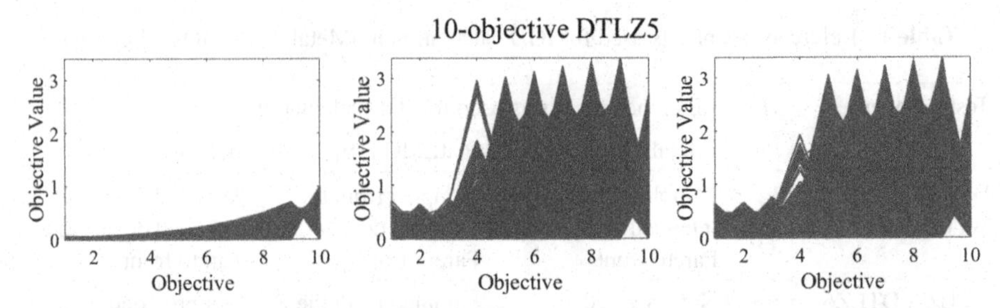

Fig. 5. Reference point sets used for IGD calculation for the ten-objective DTLZ5: (left) provided by PlatEMO, (middle) all non-dominated solutions obtained by the five algorithms with the termination condition of 1,000 generations, (right) all non-dominated solutions obtained by the five algorithms with the termination condition of 10,000 generations.

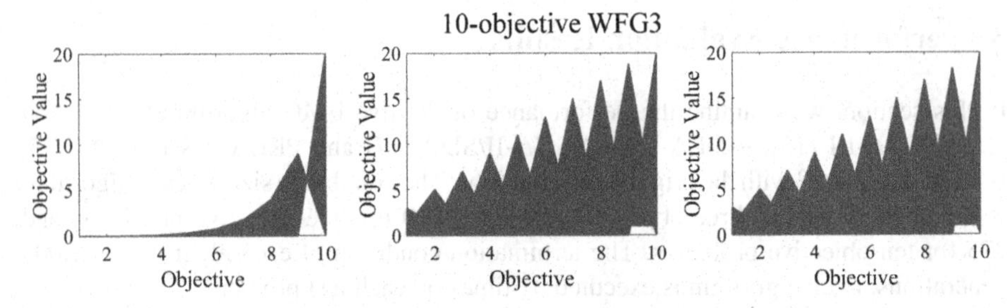

Fig. 6. Reference point sets used for IGD calculation for the ten-objective WFG3: (left) provided by PlatEMO, (middle) all non-dominated solutions obtained by the five algorithms with the termination condition of 1,000 generations, (right) all non-dominated solutions obtained by the five algorithms with the termination condition of 10,000 generations.

Table 2 is based on the reference point sets in PlatEMO. Tables 3–4 are based on the reference point sets obtained by the five algorithms (after 1,000 generations in Table 3 and 10,000 generations in Table 4). In each table, the best rank "1" is highlighted in bold. The average IGD value of each algorithm on each test problem is shown in Tables S1-S3 in the supplementary file for the three reference point sets.

In Table 2 with the PlatEMO reference point sets, NSGA-II/SDR has the best average rank over the three test problems (see the bottom line of Table 2). However, the difference in the average ranks among the five algorithms is small. A different algorithm has the best rank for a different test problem. For example, PREA has the best rank on the three-objective DTLZ5, DTLZ6 and WFG3 test problems whereas MOEA/D has the best rank on DTLZ5 and DTLZ6 with five and ten objectives.

In Tables 3–4, almost the same results are obtained. That is, Table 3 is almost the same as Table 4. For example, PREA always has the best rank for all test problems in these two tables. This is because similar reference point sets are obtained after 1,000 generations (in Table 3) and 10,000 generations (in Table 4) for each test problem as demonstrated in Figs. 5–6 (i.e., the center and right figures).

Table 2. The rank of each algorithm based on the average IGD value calculated using the reference point sets provided in PlatEMO.

Problem	M	MOEA/D	NSGA-III	θ-DEA	NSGA-II/SDR	PREA
DTLZ5	3	5	2	3	4	**1**
	5	**1**	3	5	2	4
	10	**1**	4	3	2	5
DTLZ6	3	3	2	4	5	**1**
	5	**1**	3	4	2	5
	10	**1**	5	2	3	4
WFG3	3	5	3	4	2	**1**
	5	5	3	4	**1**	2
	10	5	4	**1**	3	2
Average		3.00	3.22	3.33	**2.67**	2.78

Table 3. The rank of each algorithm based on the average IGD value calculated using the reference point set obtained by the five algorithms after 1,000 generations.

Problem	M	MOEA/D	NSGA-III	θ DEA	NSGA-II/SDR	PREA
DTLZ5	3	4	2	3	5	**1**
	5	5	2	3	4	**1**
	10	5	2	3	4	**1**
DTLZ6	3	3	2	4	5	**1**
	5	4	2	3	5	**1**
	10	5	3	2	4	**1**
WFG3	3	2	3	5	4	**1**
	5	2	3	4	5	**1**
	10	4	2	5	3	**1**
Average		3.78	2.33	3.56	4.33	**1.00**

One clear observation from Tables 2, 3 and 4 is that totally different results are obtained between Table 2 and Tables 3–4. For the five test problem instances shaded in Tables 3–4, the best algorithm on each test problem instance in Table 2 shows the worst performance in Table 3–4. Especially, on the ten-objective DTLZ5, the totally opposite rankings of the five algorithms are obtained between Table 2 (i.e., 1, 4, 3, 2, 5) and Tables 3–4 (i.e., 5, 2, 3, 4, 1). These different results are obtained since Table 2 is based on only the degenerated parts whereas Tables 3–4 are based on the entire Pareto fronts. For example, in Fig. 4, the degenerate part of the five-objective DTLZ5 is well covered by the solution set obtained by MOEA/D. Thus, MOEA/D is evaluated as the

Table 4. The rank of each algorithm based on the average IGD value calculated using the reference point set obtained by the five algorithms after 10,000 generations.

Problem	M	MOEA/D	NSGA-III	θ DEA	NSGA-II/SDR	PREA
DTLZ5	3	4	2	3	5	**1**
	5	5	2	3	4	**1**
	10	5	2	3	4	**1**
DTLZ6	3	3	2	4	5	**1**
	5	4	2	3	5	**1**
	10	5	3	2	4	**1**
WFG3	3	2	3	5	4	**1**
	5	2	3	4	5	**1**
	10	3	2	5	4	**1**
Average		3.67	2.33	3.56	4.44	**1.00**

best algorithm for the five-objective DTLZ5 in Table 2. However, the same solution set covers only a small region of the non-degenerate part in Fig. 4. Thus, MOEA/D is evaluated as the worst algorithm for the five-objective DTLZ5 in Tables 3–4.

To further examine the experimental results in Tables 2–4, the solution sets obtained by the five algorithms on the ten-objective DTLZ5 and WFG3 are shown as parallel coordinate plots in Figs. 7 and 8, respectively. For each algorithm on each test problem, a single run with the median IGD value among 31 runs is used in these figures. The reference point sets obtained after 10,000 generations in Table 4 are used for IGD calculation to choose a single run in Figs. 7 and 8 (and also in Fig. 4).

In Fig. 7, the solution set obtained by MOEA/D on the ten-objective DTLZ5 is similar to the PlatEMO reference point set in Fig. 5 (the left figure). Thus, MOEA/D is evaluated as the best algorithm on the ten-objective DTLZ5 in Table 2. However, the solution set obtained by MOEA/D is clearly different from the reference point sets obtained by the five algorithms after 1,000 and 10,000 generations in Fig. 5 (the center and right figures). Thus, MOEA/D is evaluated as the worst algorithm in Tables 3–4. Similar observations can be obtained for the solution sets of the other algorithms in Figs. 7–8 (e.g., the solution set by θ-DEA in Fig. 8 is similar to the PlatEMO reference point set in Fig. 6).

Our experimental results in Tables 2, 3 and 4 and Figs. 7–8 demonstrate that the reference point sets sampled from the originally intended degenerate Pareto front are not appropriate for DTLZ5, DTLZ6 and WFG3 with the original problem formulations (i.e., with the partially degenerate Pareto fronts). That is, IGD-based evaluation results on these test problems can be misleading when the reference point sets for IGD calculation are sampled from the originally intended degenerate Pareto front. Our suggestion is to use all non-dominated solutions among obtained solutions by all runs of all the examined algorithms as a reference point set for IGD calculation. Moreover, it is advisable to use an additional performance indicator (e.g., the hypervolume indicator) together with the IGD indicator for fair comparison of EMO algorithms. This is because performance comparison results based on a single indicator are not always reliable [26].

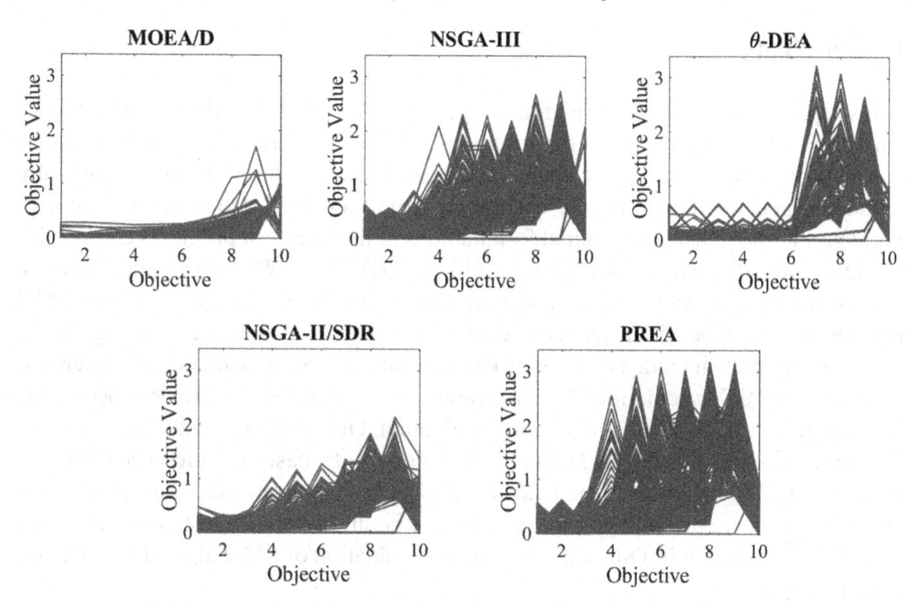

Fig. 7. The solution sets obtained by the five algorithms on the ten-objective DTLZ5 test problem. A single run with the median IGD value is selected from 31 runs of each algorithm.

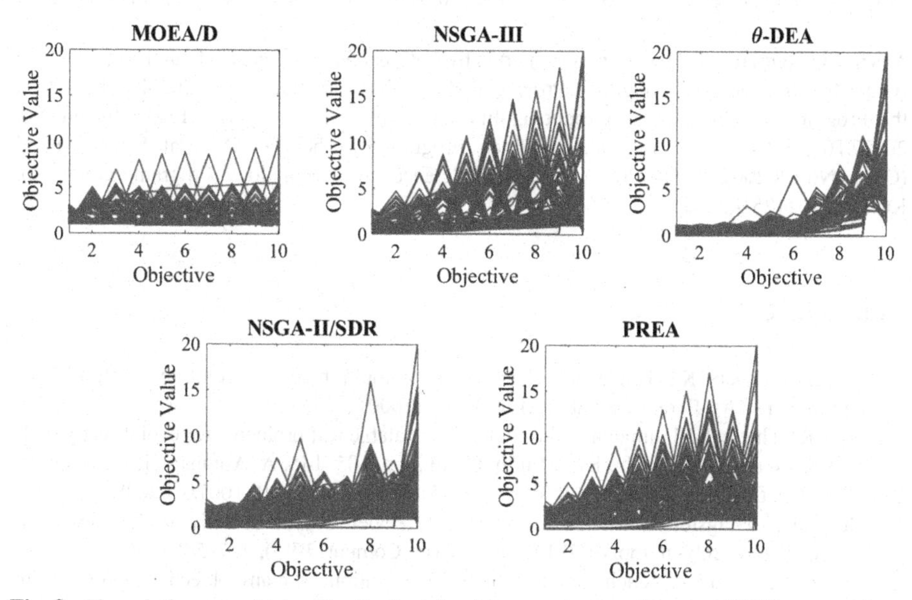

Fig. 8. The solution sets obtained by the five algorithms on the ten-objective WFG3 test problem. A single run with the median IGD value is selected from 31 runs of each algorithm.

4 Conclusions

In this paper, we showed that the partially degenerate Pareto fronts of the DTLZ5, DTLZ6 and WFG3 test problems with the original problem formulations are highly irregular in a high-dimensional objective space, which are clearly different from the originally intended degenerate Pareto fronts. Their Pareto fronts are similar to those of some real-world problems. Hence, the original formulations of the three test problems can be used to increase the diversity of test problems in the DTLZ and WFG test suites. That is, their original formulations are good test problems to evaluate the performance of EMO algorithms. One critical issue in their use for performance evaluation of EMO algorithms is that the originally intended degenerate Pareto fronts have been used to sample reference point sets for IGD calculation. That is, these three test problems have not been used appropriately in IGD-based performance evaluation. Our computational experiments in this paper demonstrated that IGD-based evaluation results based on reference point sets from the originally intended degenerate Pareto fronts are not reliable. Thus, it is always necessary to ensure that an appropriate reference point set for each test problem is used for IGD calculation in IGD-based performance evaluation of EMO algorithms on these test problems.

Since degenerate and partially degenerate problems are common in real-world applications [27], an interesting future research study would be to investigate the possibility of quantifying or measuring degeneracy through exploratory landscape analysis [28].

Acknowledgement. This work was supported by National Natural Science Foundation of China (Grant No. 61876075), Guangdong Provincial Key Laboratory (Grant No. 2020B121201001), the Program for Guangdong Introducing Innovative and Enterpreneurial Teams (Grant No. 2017ZT07X386), The Stable Support Plan Program of Shenzhen Natural Science Fund (Grant No. 20200925174447003), Shenzhen Science and Technology Program (Grant No. KQTD2016112514355531).

References

1. Zitzler, E., Deb, K., Thiele, L.: Comparison of multiobjective evolutionary algorithms: empirical results. Evol. Comput. **8**(2), 173–195 (2000)
2. Deb, K., Thiele, L., Laumanns, M., Zitzler, E.: Scalable test problems for evolutionary multiobjective optimization. In: Evol. Multi. Optim., pp. 105–145. A. Abraham, L. Jain, and R. Goldberg, Eds. London: Springer-Verlag (2005). https://doi.org/10.1007/1-84628-137-7_6
3. Huband, S., Hingston, P., Barone, L., While, L.: A review of multiobjective test problems and a scalable test problem toolkit. IEEE Trans. Evol. Comput. **10**(5), 477–506 (2006)
4. Cheng, R., et al.: A benchmark test suite for evolutionary many-objective optimization. Complex & Intelligent Syst. **3**(1), 67–81 (2017)
5. Zhang, Q., Zhou, A., Zhao, S.Z., Suganthan, P.N., Liu, W., Tiwari, S.: Multiobjective optimization test instances for the CEC-2009 special session and competition. Nanyang Technol. Univ., Singapore, Tech. Rep., pp. 1–30 (2008)
6. Liu, Y., Gong, D., Sun, J., Jin, Y.: A many-objective evolutionary algorithm using a one-by-one selection strategy. IEEE Trans. Cybn. **47**(9), 2689–2702 (2017)

7. Pan, L., Li, L., He, C., Tan, K.: A subregion division-based evolutionary algorithm with effective mating selection for many-objective optimization. IEEE Trans. Cybn. **50**(8), 3477–3490 (2020)
8. Gong, D., Liu, Y., Yen, G.G.: A meta-objective approach for many-objective evolutionary optimization. Evol. Comput. **28**(1), 1–25 (2020)
9. Hua, Y., Jin, Y., Hao, K., Cao, Y.: Generating multiple reference vectors for a class of many-objective optimization problems with degenerate Pareto fronts. Complex & Intelligent Syst. **6**(2), 275–285 (2020)
10. Zhang, K., Xu, Z., Xie, S., Yen, G.G.: Evolution strategy-based many-objective evolutionary algorithm through vector equilibrium. IEEE Trans. Cybn. **51**(11), 5455–5467 (2021)
11. Saxena, D.K., Duro, J.A., Tiwari, A., Deb, K., Zhang, Q.: Objective reduction in many-objective optimization: linear and nonlinear algorithms. IEEE Trans. Evol. Comput. **17**(1), 77–99 (2013)
12. Ishibuchi, H., Masuda, H., Nojima, Y.: Pareto fronts of many-objective degenerate test problems. IEEE Trans. Evol. Comput. **20**(5), 807–813 (2016)
13. Coello Coello, C.A., Reyes Sierra, M.: A study of the parallelization of a coevolutionary multi-objective evolutionary algorithm. In: Monroy, R., Arroyo-Figueroa, G., Sucar, L.E., Sossa, H. (eds.) MICAI 2004. LNCS (LNAI), vol. 2972, pp. 688–697. Springer, Heidelberg (2004). https://doi.org/10.1007/978-3-540-24694-7_71
14. Zhang, Q., Li, H.: MOEA/D: a multiobjective evolutionary algorithm based on decomposition. IEEE Trans. Evol. Comput. **11**(6), 712–731 (2007)
15. Deb, K., Jain, H.: An evolutionary many-objective optimization algorithm using reference-point-based non-dominated sorting approach, Part I: solving problems with box constraints. IEEE Trans. Evol. Comput. **18**(4), 577–601 (2014)
16. Yuan, Y., Xu, H., Wang, B., Yao, X.: A new dominance relation-based evolutionary algorithm for many-objective optimization. IEEE Trans. Evol. Comput. **20**(1), 16–37 (2016)
17. Tian, Y., Cheng, R., Zhang, X., Su, Y., Jin, Y.: A strengthened dominance relation considering convergence and diversity for evolutionary many-objective optimization. IEEE Trans. Evol. Comput. **23**(2), 331–345 (2019)
18. Yuan, J., Liu, H.-L., Gu, F., Zhang, Q., He, Z.: Investigating the properties of indicators and an evolutionary many-objective algorithm using promising regions. IEEE Trans. Evol. Comput. **25**(1), 75–86 (2021)
19. Tian, Y., Cheng, R., Zhang, X., Jin, Y.: PlatEMO: a MATLAB platform for evolutionary multi-objective optimization. IEEE Comput. Intell. Mag. **12**(4), 73–87 (2017)
20. Tanabe, R., Ishibuchi, H.: An easy-to-use real-world multi-objective optimization problem suite. Applied Soft Comput. **89**, 106078 (2020)
21. Ishibuchi, H., He, L., Shang, K.: Regular Pareto front shape is not realistic. In: Proc. IEEE Congr. Evol. Comput., pp. 2034–2041. Wellington, New Zealand (2019)
22. Nebro, A.J., Durillo, J.J., Vergne, M.: Redesigning the jMetal multi-objective optimization framework. In: Proc. Conf. Genet. Evol. Comput., pp. 1093–1100. Madrid Spain (2015)
23. Blank, J., Deb, K.: pymoo: multi-objective optimization in Python. IEEE Access **8**, 89497–89509 (2020)
24. He, C., Tian, Y., Wang, H., Jin, Y.: A repository of real-world datasets for data-driven evolutionary multiobjective optimization. Complex & Intelligent Syst. **6**(1), 189–197 (2019)
25. Ishibuchi, H., Imada, R., Setoguchi, Y., Nojima, Y.: Reference point specification in inverted generational distance for triangular linear Pareto front. IEEE Trans. Evol. Comput. **22**(6), 961–975 (2018)
26. Ishibuchi, H., Pang, L.M., Shang, K.: Difficulties in fair performance comparison of multi-objective evolutionary algorithms. IEEE Comput. Intell. Mag. **17**(1), 86–101 (2022)

27. Zhen, L., Li, M., Cheng, R., Peng, D., Yao, X.: Multiobjective test problems with degenerate Pareto fronts. https://arxiv.org/abs/1806.02706
28. Kerschke, P., Trautmann, H.: The R-Package FLACCO for exploratory landscape analysis with applications to multi-objective optimization problems. In: Proc. IEEE Congr. Evol. Comput., pp. 5262–5269. Vancouver, BC, Canada (2016)

Peak-A-Boo! Generating Multi-objective Multiple Peaks Benchmark Problems with Precise Pareto Sets

Lennart Schäpermeier[1(✉)], Pascal Kerschke[1], Christian Grimme[2], and Heike Trautmann[2]

[1] Big Data Analytics in Transportation, TU Dresden & ScaDS.AI, Dresden, Germany
{lennart.schaepermeier,pascal.kerschke}@tu-dresden.de
[2] Data Science: Statistics and Optimization, University of Münster, Münster, Germany
{christian.grimme,trautmann}@uni-muenster.de

Abstract. The design and choice of benchmark suites are ongoing topics of discussion in the multi-objective optimization community. Some suites provide a good understanding of their Pareto sets and fronts, such as the well-known DTLZ and ZDT problems. However, they lack diversity in their landscape properties and do not provide a mechanism for creating multiple distinct problem instances. Other suites, like bi-objective BBOB, possess diverse and challenging landscape properties, but their optima are not well understood and can only be approximated empirically without any guarantees.

This work proposes a methodology for creating complex continuous problem landscapes by concatenating single-objective functions from version 2 of the multiple peaks model (MPM2) generator. For the resulting problems, we can determine the distribution of optimal points with arbitrary precision w.r.t. a measure such as the dominated hypervolume. We show how the properties of the MPM2 generator influence the multi-objective problem landscapes and present an experimental proof-of-concept study demonstrating how our approach can drive well-founded benchmarking of MO algorithms.

Keywords: Multi-objective optimization · Multimodal optimization · Numeric optimization · Benchmarking · Problem generator

1 Introduction

In order to adequately understand problem hardness and to specifically tailor algorithmic approaches with respect to the criteria characterizing different facets and levels of difficulty in multi-objective (MO) optimization, comprehensive and carefully designed benchmark sets are an essential prerequisite [1].

An MO benchmark suite ideally should be a) comprehensive with regard to the representativeness of relevant real-world problems, b) scalable regarding

M. Emmerich et al. (Eds.): EMO 2023, LNCS 13970, pp. 291–304, 2023.
https://doi.org/10.1007/978-3-031-27250-9_21

both decision and objective space dimensionality, c) capable of covering the most important characteristics of MO landscape properties in relevant combinations, d) extendable in size by providing a means to specifically generate a desired number of problem instances with certain landscape characteristics, and most importantly e) well understood by providing analytical expressions of both Pareto front (PF) and Pareto sets (PS), ideally including local structures.

However, meeting all these requirements is an extremely challenging MO problem (MOP) by itself, and we have to aim for optimal trade-off solutions. In single-objective (SO) optimization, BBOB [13] presents a benchmark suite that ticks off many boxes of the previously stated wish list such that, e.g., it is well understood in terms of problem difficulties, optima are known, an arbitrary number of instances per problem type can be generated, and scalability regarding decision space dimensionality is provided. Moreover, a standardized algorithm evaluation procedure exists, which is widely accepted within the community. The community, however, is still largely debating on a).

In MO optimization, there is, unfortunately, no straightforward counterpart. While real-world representativeness is also an issue in this domain, we are specifically concerned about items d) and e) as critical issues. In our view, existing benchmark suites turn out to be either not challenging enough, if PF and PS are known analytically (e.g., ZDT [29] and DTLZ [7]), or extremely challenging if requirement e) is omitted as, e.g., in the bi-objective BBOB [25]. Specifically, algorithm performance evaluation is very challenging as no ground truth exists for comparison. Also, item c) cannot be assessed properly as MO landscape characteristics can only be empirically and heuristically approximated.

This paper concentrates on the MPM2 generator [26,27] and on an MO benchmark set creation concept based thereon [16]. We will show that we can largely contribute to understanding MOPs by providing a method for deriving both PS and PF analytically and allowing for the approximation of optimal Hypervolume (HV) up to an arbitrary precision. Thereby, a ground truth is provided in combination with MPM2 being flexible regarding the generation of different types of landscape structures and problem characteristics. So far, we concentrate on continuous bi-objective MOPs as proof-of-concept while on top simultaneously illustrating generalization and scalability potential.

We start by giving some background on MO benchmark suites and algorithms in Sect. 2. Then, in Sect. 3, we introduce our methodology first for individual pairs of unimodal functions and subsequently for multiple peaks. This is followed by a proof-of-concept experimental study showing problem properties and algorithm performances in Sect. 4. Section 5 concludes this work and comments on future research perspectives building on our methodology.

2 Background

Before diving into our methodology, we will start by introducing some background on MO benchmark suites, in particular the MPM2 generator, and the MO optimizers utilized in the experimental section later in this work.

MO Benchmark Suites. In a recent survey on continuous multimodal MO optimization [12], existing MO benchmark suites were discussed both in general as well as with a dedicated focus on multimodality. Therein, it has been concluded that the existing benchmark problems usually fall into one of two categories. The first group comprises hand-crafted MOPs with well-understood structural landscape properties, including analytically defined PSs. Typical representatives of this collection are historically well-established test suites such as DTLZ [7] or ZDT [29]. However, a severe limitation of these MOPs is, from a multimodal perspective, their lack of challenging landscape structures, as visually demonstrated in [19,20,22]. Likewise, the MOPs of the recently proposed MMF test suite [28] primarily exhibit extremely regular patterns; yet, structurally diverse problems are crucial for meaningful algorithm benchmarking [1]. Moreover, the scalability of the aforementioned MOPs is limited w.r.t. their number of optima and dimensionality of the decision space.

The second group of benchmark problems comprises suites based on concatenations of SO benchmark functions, like the bi-objective BBOB [25]. These problem collections are usually scalable in dimensionality and they contain more diverse MOPs with potentially complex landscapes due to the flexible concatenation of functions. Yet, these benefits come at the cost of poorly understood structural properties. For instance, although the global optima of the SO BBOB functions are analytically known [13], the PS of the corresponding bi-objective BBOB instance must be empirically approximated.

After comparing the strengths and weaknesses of the existing MO benchmark suites, it became evident that our community is currently lacking a sophisticated problem generator, which is capable of constructing a scalable, comprehensive, and diverse set of multimodal MOPs with known landscape structures.

To fill this gap, we herein utilize Wessing's *Multiple Peaks Model* (MPM2) generator [26,27] for generating SO functions with configurable topologies and scalable dimensionality. Each of these SO functions is essentially the minimum of a configurable number of individual peak functions, i.e., unimodal functions with ellipsoidal structure, aligned in one of two topologies: funnel or random. The *funnel* type contains a large funnel in which all optima are grouped around a global optimum, such that the depth per peak decreases with increasing distance from the global optimum. In contrast, the *random* type distributes the depths and locations of local optima uniformly across the search space. The implementation of the MPM2 generator allows us to extract valuable information about each of the underlying peaks, such as the covariance matrix, radius, and height properties, which we will later use as a basis for identifying the PS of the generated MOPs. Finally, note that the decision space of a d-dimensional MPM2 problem is usually $[0, 1]^d$ and that objective values are restricted to $[0, 1]$.

Similar to [16], we create bi-objective benchmark problems by concatenating functions generated with MPM2. Consequently, the resulting MOPs tend to be part of the second category of benchmark problems. Still, due to its modular composition of multiple peaks, we can generate configurable and scalable MOPs with *known* PSs (see Sect. 3). Despite their relatively simple components, the landscapes of the MOPs generated with this approach do not necessarily exhibit simple, regular patterns, which are easy to exploit by an algorithm. Instead, our approach enables the creation of diverse test functions with irregular, non-separable landscapes, convex, concave, and/or split (globally or locally efficient) PFs. The respective decision spaces also offer a variety of structural challenges (see, e.g., Fig. 3). Our approach thus provides a valuable framework for creating various configurable and scalable MOPs that will be useful for meaningfully benchmarking MOEAs and studying their strengths and weaknesses.

Algorithms. Next to classical MOEAs such as NSGA-II [6] and SMS-EMOA [10], which focus on the approximation of globally optimal solutions and convergence in objective space, specific MOEAs exist, which are explicitly suited for overcoming obstacles of multimodal MOPs and for exploiting local structures. A comprehensive overview of different MOEA categories, including multimodality aspects both from a multi-local and multi-global perspective, is given in [12,23].

Herein, we focus on specific MOEAs like Omni-Optimizer [8] and MOLE [21]. These approaches were shown to be more competitive than classical MOEAs w.r.t. convergence in objective space while simultaneously ensuring diversity in decision space [14,18]. Omni-Optimizer is conceptually similar to NSGA-II but additionally comprises a diversity preservation strategy in decision space. It is thus applicable to various types of MOPs. MOLE, however, is a gradient-based MO optimizer that actively explores and traverses locally efficient sets. Thereby, it exploits interactions between their respective basins of attraction for a directed descent towards dominating local (or even global) sets.

3 On the Pareto Set of Multiple Peaks Functions

As outlined in Sect. 2, MPM functions essentially take the minimum of multiple individual peak functions, whose shape and placement are up to the generator. Using the MPM2 generator, the individual peak functions obtain ellipsoidal level sets, which makes them approachable using analytical techniques. We will first illustrate how the PS of two unimodal peak functions can be derived and then discuss how this analysis scales with increasing numbers of peaks per function.

Bi-Objective Convex-Quadratic Problems. A theoretical analysis of the PS between individual peak functions stemming from the MPM2 generator has been conducted before in [16]. Here, however, we base our discussion on the results of [24], which focus on bi-objective convex-quadratic problems. Consider the following bi-objective convex-quadratic problem $F(x)$ with search space $\mathcal{X} = \mathbb{R}^d$:

$$F(x) = (f_1(x), f_2(x)) \rightarrow min! \quad \text{with } f_i(x) = \frac{1}{2}(x - x_i^*)^T H_i (x - x_i^*), i = 1, 2,$$

where $x_1^*, x_2^* \in \mathbb{R}^d$ are the global optima of f_1 and f_2, respectively. Likewise, $H_1, H_2 \in \mathbb{R}^{d \times d}$ are positive definite symmetric Hessian matrices determining the shape and orientation of the quadratic functions.

For convex quadratic functions, the position of the global optimum is determined by the singular point at which the gradient equals the zero vector. Further, the PF of two convex quadratic problems is convex [11], which enables us to express all globally optimal points by linear interpolation between the problems using a parameter $t \in [0,1]$:

$$\begin{aligned} F_t(x) &= (1-t)f_1(x) + tf_2(x) \to min! \\ \Rightarrow \nabla F_t(x) &= (1-t)\nabla f_1(x) + t\nabla f_2(x) = 0 \\ &(1-t)H_1(x - x_1^*) + tH_2(x - x_2^*) = 0 \\ &[(1-t)H_1 + tH_2]^{-1}[(1-t)H_1 x_1^* + tH_2 x_2^*] = x \end{aligned}$$

In the work of [24], some additional results are shown, e.g., monotone objective transformations do not impact the placement of the PS, while PF properties, such as its shape, can be adjusted. As the peaks resulting from MPM2 constitute a convex quadratic function to which a monotone transformation is applied in the objective space only, the above analysis also describes the PS of a bi-objective unimodal peak function. Finally, note that this does not guarantee that the PS is contained within the usual $[0,1]^d$ bounding box for MPM2 problems and the decision space boundaries may need slight adjustment when this unconstrained PS should still be reachable.

Multiple Peaks. While we can only describe the PS analytically given a combination of two unimodal peak functions, we can leverage this knowledge for constructing the PS of bi-objective MPM2 problems. The first necessary insight is that this specific PS has to be a subset of the theoretical PSs which are generated by an exhaustive combination of each pair of peaks, cf. Fig. 1.

This leaves us with considering all pairwise peak functions, i.e., one peak per objective, and their respective PSs, cf. Fig. 1a. While new locally efficient solutions cannot be generated, some of the analytical PSs may become partially or completely inactive because of being dominated by objective values of another peak combination. Additionally, they may become globally dominated by the PSs of other peak pairs, thereby creating complex local and global interactions. Considering the example in Fig. 1b, the blue set (corresponding to the bottom left image in Fig. 1a) is completely inactive, while the red and violet sets are partially inactive, and only the green set stays fully active. These dynamics are mirrored in the PLOT visualization (see bottom right image in Fig. 1b) [19], which only depicts the (local) PSs and PFs, respectively.

While these local dynamics can become very complex, it is not necessary to study them in detail, if one is only interested in deriving the PS and PF, i.e., the globally efficient points. We propose a simple, numerical procedure for deriving a set-based approximation building on the analytical description of the PSs of the peak pairs (see Fig. 2). As a target, we choose to approximate the dominated hypervolume (HV) w.r.t. the reference point $(1,1)$. We start by evaluating evenly spaced points w.r.t. the parameter t for each of the pairwise theoretical fronts. In our implementation, the minimum resolution is 4 points per set.

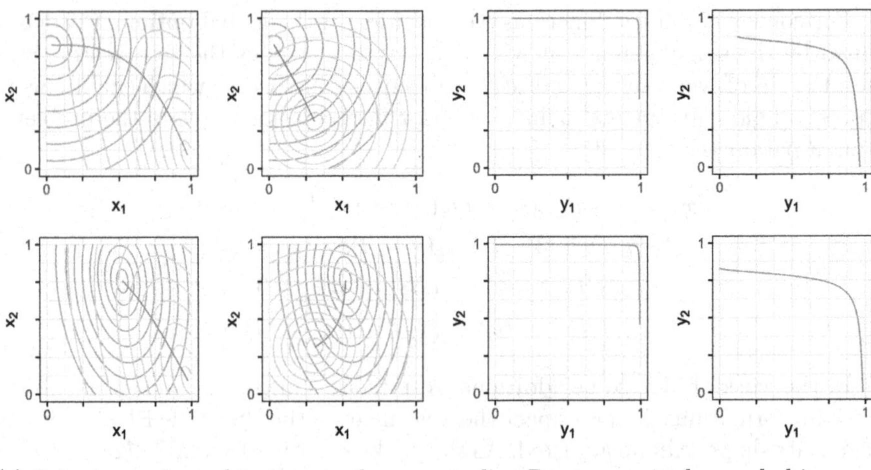

(a) Pairwise peak combinations and corresponding Pareto sets in dec. and obj. space.

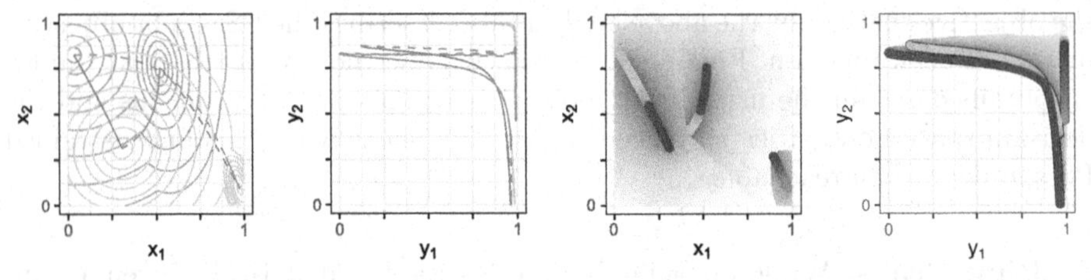

(b) Left: Active (solid) and inactive (dashed) sets of the combined MPM2 problems. Sets are inactive if they are dominated by another pair of peaks. In the objective space, the inactive theoretical sets are printed semi-transparently. Right: PLOT [19] visualization, showing locally efficient sets (in color) and attraction basins (in grayscale).

Fig. 1. Here, we demonstrate the effects of combining individual peak functions to one MPM2 problem on the Pareto set. The exemplary problem consists of two peaks per objective with a random topology and seeds 667172 and 540835.

For each set, we then individually compute the best possible intermediate points, i.e., the minimum of two consecutive points, that could still be contained within the set without dominating any other point in it. When interpolating on the same (theoretical) set, no more dominant points could be found, making the intermediates the most optimistic estimate between two adjacent points from a set. The (relative) HV gap is then given by the difference (percentage) of the HV dominated by i) the best possible intermediates, and ii) the actually evaluated points. This process is repeated with doubled resolution until the HV gap reaches a sufficiently small target value. To save on evaluations, sets whose best possible intermediates are fully dominated by the union of all evaluated points so far can be excluded in the respective iteration, as they cannot contribute to the PF.

Although we can pessimistically estimate the computational complexity proportional to the number of initial theoretical Pareto sets times the inverse of the

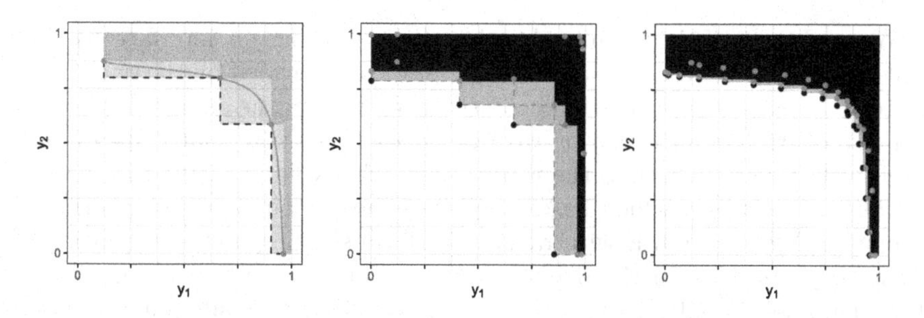

Fig. 2. Left: Illustration of the hypervolume gap (light shaded area) for the green front of the problem shown in Fig. 1. Middle: The hypervolume gap (gray area) for a resolution of 4 points per set, where the black points denote the nondominated points from the best possible intermediates for each set. Right: The same gap for a resolution of 16 points per set. Note that the red and blue sets could be excluded as they cannot contribute to the Pareto front.

target gap, practical computations become surprisingly efficient. This is due to the observation that many of the theoretical local sets can be excluded early.

4 Experimental Study

In the following proof-of-concept study, we (1) demonstrate the properties of the proposed test suite subject to its main parameters and (2) conduct first performance analyses of several standard and multimodality-affine solvers.

Setup. We created a total of 1,280 problems using the following configuration: We select the search space dimensionality $d \in \{2, 3, 5, 10\}$ to cover an increasing but still manageable decision space. The number of peaks $n_p \in \{1, 2, 4, 8, 16, 32, 64, 128\}$ per MPM2 function is exponentially scaled to enable analyzing the influence of SO multimodality on a log-scale. Note that n_p coincides for both constituent MPM2 functions to obtain a meaningful amount of unambiguous classes for further analyses. The same rationale applies to the topology, i.e., we select a specific topology parameter $t \in \{\texttt{funnel}, \texttt{random}\}$ for both problems simultaneously. Finally, we choose the random seeds $s \in \{1, \ldots, 20\}$ for the first objective, while the seed for the second objective is always set to $s + 1{,}000$. For all problems, we then approximate the HV of the PF to a relative gap of at most 10^{-4}, i.e., we have an uncertainty of less than 0.01% about the optimal HV of each problem. This approximation takes at most a few seconds per problem. We set the decision space boundaries to $[-0.2, 1.2]^d$, ensuring all Pareto sets are fully included.

To obtain the problem characteristics and to optimize the HV, we extract the parameters for all peaks of the MPM2 problems using an interface implemented in the R-package `smoof` [4]. Our experiments and analyses are also conducted in R and can be found at https://github.com/schaepermeier/2023-emo-mpm2.

We run the algorithms introduced in Sect. 2 using an interface to the C implementations of NSGA-II and Omni-Optimizer provided by the `mco` [17] and `omnioptr` [5] packages, respectively. Further, we rely on the default implementation of SMS-EMOA provided by the `ecr` package [3]. Note that it uses random parent selection rather than the tournament scheme implemented in NSGA-II and Omni-Optimizer, which has to be taken into account in the experimental evaluation. We use MOLE with random uniform starting points as provided by the `moleopt` package in the default configuration, but setting its internal HV target parameter to 10^{-3}, to reduce time spent refining already found solutions. All MOEAs have their population set to 100, which is a common default.

We run all algorithms for 10,000 evaluations, i.e., 100 generations for NSGA-II and Omni-Optimizer. As performance measure, we consider the dominated HV w.r.t. the reference point $(1, 1)$ provided by the archive of evaluated points. For performance reasons, we compute the achieved HV every 100 function evaluations. We perform 15 repetitions per combination of problem and algorithm to ensure statistically reliable results.

Analysis of Problem Properties. To illustrate interesting properties of the generated problems, we perform two separate analyses: We start by visualizing two-dimensional problem landscapes to visually show the impact of the degrees of multimodality and the two problem topologies. We then focus on problem characteristics in objective space when scaling the search space dimensionality.

Figure 3 shows PLOT [19] visualizations of two-dimensional bi-objective problems generated with funnel and random topologies for 2, 8, 32, and 128 peaks, respectively. Here, it can be observed that the degree of multimodality greatly increases with the number of peaks used in the individual problems, as demonstrated by the number and location of visualized locally efficient points. Although the problems with lower multimodality still seem somewhat similar between topologies, varying distribution of the locally and globally efficient sets becomes apparent with an increasing number of peaks. The funnel problems tend to cluster the globally (and to an extent locally) efficient points, while the random problems show a much higher dispersion in the decision space, with many smaller areas contributing to the (global) PS. These representative problems also show a distinctive property in objective space: In random topologies the PF rapidly approaches the ideal point $(0, 0)$ with increasing number of peaks, while respective closeness is limited in the funnel topology.

Figure 4 visualizes the influence of generator parameters on PS properties. On the one hand, across dimensions and topologies, we can see that the number of locally efficient sets contributing to the Pareto set increases with the number of peaks. However, this effect decreases with increasing dimensionality and differences between the topologies in this regard almost vanish. Further, the approximated HV increases with the number of peaks. While it approaches the maximum possible HV of 1 for the random topology, funnel problems demonstrate a much wider range of HV values and a slower increase with the number of peaks. Again, the effect diminishes with increasing dimensionality.

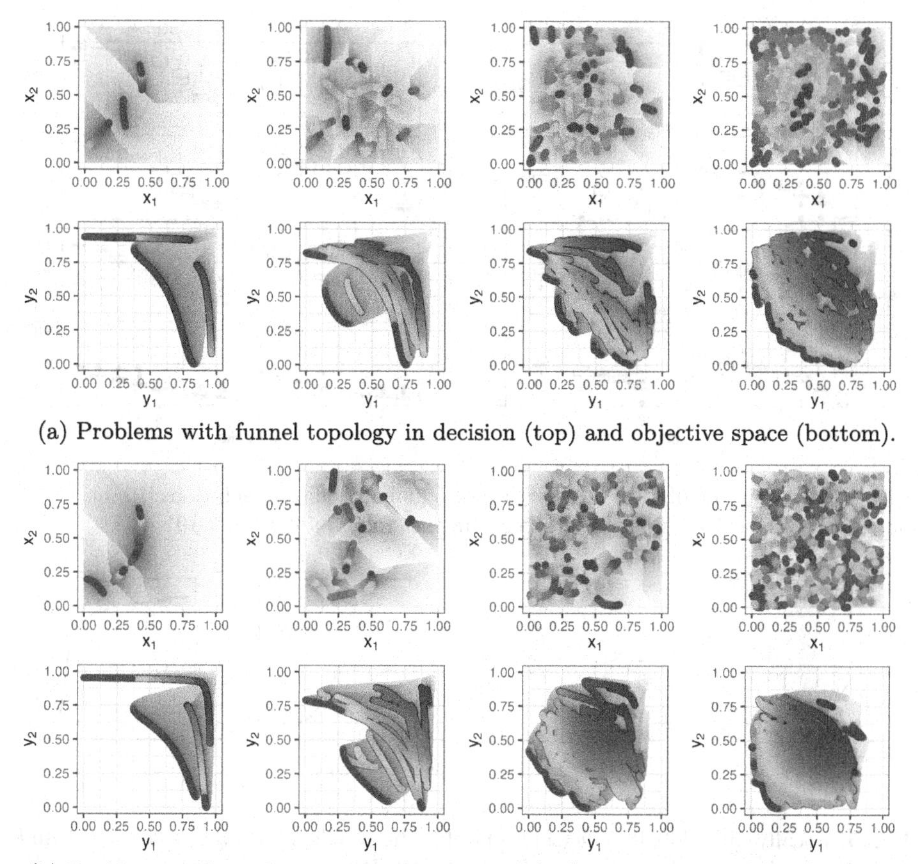

(a) Problems with funnel topology in decision (top) and objective space (bottom).

(b) Problems with random topology in decision (top) and objective space (bottom).

Fig. 3. PLOT visualizations of problems with 2, 8, 32 and 128 peaks per objective (left-to-right) for funnel and random topologies. All problems use seed $s = 1$.

Finally, Fig. 5 shows properties of the locally and globally efficient fronts, exemplarily for the random topology, with increasing dimension. Here, we can see that, using otherwise identical parameters, the PF is becoming more concave within our generator framework. This seems to be a property of the peak function, and should be investigated in detail in the future.

Algorithm Comparison. Based on our knowledge of optimal HV values and regularly conducted HV evaluations during the optimization process, we can generate convergence plots as exemplarily provided in Fig. 6 for 5D problems. It depicts the mean relative hypervolume gap, i.e., the mean percentage of hypervolume not yet covered, per problem w.r.t. function evaluations. Several insights can be gained: Firstly, problem hardness tends to increase with an increasing number of peaks per problem. This is particularly noticeable for the MOLE

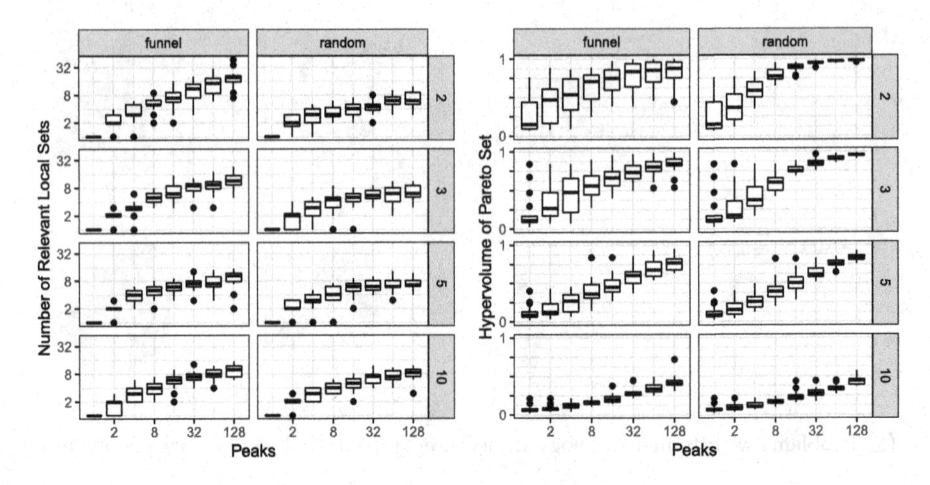

Fig. 4. Left: Number of local sets that contain globally efficient solutions. Right: Computed HV of PS. Rows indicate problem dimensionality, $d \in \{2, 3, 5, 10\}$.

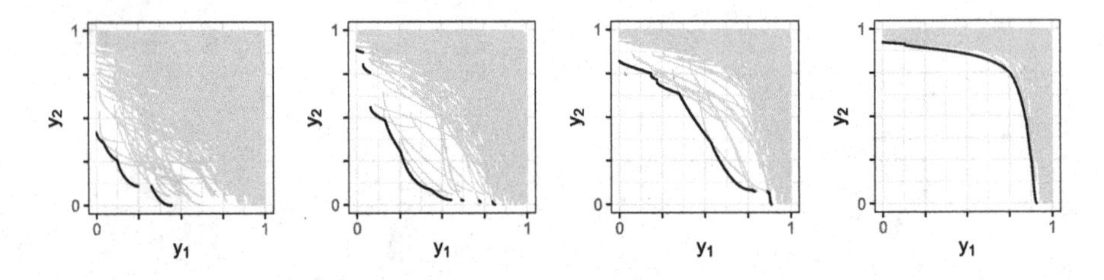

Fig. 5. Locally and globally efficient fronts for the problem with 32 peaks, $s = 1$ and random topology for dimensions 2, 3, 5, and 10. Higher-dimensional problems have fewer visible disconnects and overall a more concave front shape.

algorithm, which, as a purely local search approach, is slowed down by the amount of locally efficient points, while the performance of the evolutionary algorithms is less affected. Further, by comparing the achieved values at the end of the runs for the 32 and 128 peaks problems, we see that the random problems tend to be slightly harder to solve than the funnel problems for the EAs, while MOLE is less affected.

Figure 7 shows critical difference plots [9] for all problem dimensions and topologies. It validates that, in general, Omni-Optimizer and NSGA-II perform best, though only in 10D Omni-Optimizer is clearly superior. They are followed by SMS-EMOA and MOLE. For SMS-EMOA, the mentioned random parent selection scheme implemented in `ecr` might be the reason for its comparatively bad performance w.r.t. NSGA-II. Further, MOLE's performance relative to SMS-EMOA is improving with dimensionality, although they are always statistically tied. MOLE also tends to perform slightly better on random than on funnel topologies. Finally, while random search seems to have some merit in lower dimensions (2, 3), it is clearly the worst performer in higher dimensions (5, 10).

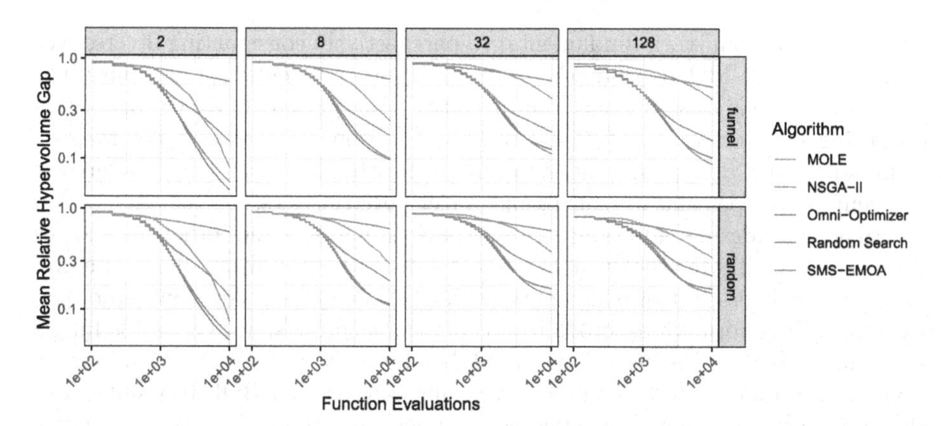

Fig. 6. Convergence plots of the aggregated algorithm performances for the 5D problems. Columns denote the number of peaks, while rows show the topology. Each algorithm was evaluated with 15 repetitions on 20 problems per group.

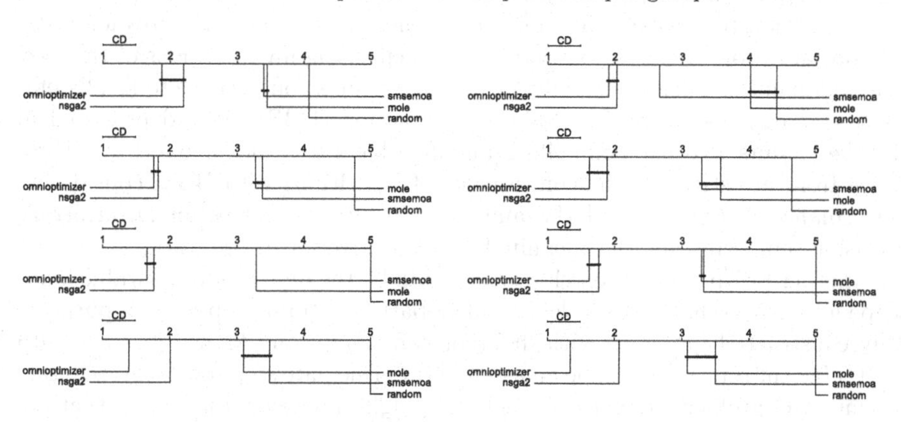

Fig. 7. Critical differences for the mean final HV gap per problem in the random (left) and funnel (right) topologies for dimensions 2, 3, 5, and 10 (top-to-bottom).

5 Conclusions

In this work, we introduced a new methodology for determining the globally optimal solutions of MOPs, which are created based on multiple peak problems, to an arbitrary precision in terms of dominated HV. We apply this methodology for developing tools that are able to generate a wide range of benchmark problems with ground truth regarding the PS and PF while simultaneously having complex landscape characteristics. The highly parametrizable generator facilitates the creation of problems with specific structural properties, which in turn is essential for conducting structured analyses of landscape properties of MOPs. Next to landscape analyses, the ground truth enables a systematic benchmarking of algorithms, which we demonstrated in a proof-of-concept study.

We consider our work fundamental to perspectively constructing diverse, yet well-understood, MO benchmark problems in order to enhance and meaningfully complement existing benchmark suites. Our presented framework offers promising perspectives for future research in various important areas ranging from benchmark design and understanding algorithm behavior to characterizing problem landscapes and measuring optimizer performances.

First, considering additional peak shape functions and other topologies enables and facilitates the straightforward construction of a broader scope and thus a more diverse set of benchmark problems. Integrating decision and objective space transformations provides an additional promising avenue for future extensions. Such transformations could be, for instance, the introduction of asymmetries (similar to those used in the single-objective BBOB test suite) into the previously constructed multiple peaks functions. We would also like to point out that the mathematical analysis of the PS can easily be extended to more objectives. However, in this case, it is not intuitively clear how to generalize the PS approximation.

In addition to constructing more comprehensive benchmark suites for *global* MO optimization, we are also interested in facilitating investigations of the *local* search dynamics. Therefore, analyzing an algorithm's convergence to locally efficient sets, e.g., using the Basin-Based Evaluation (BBE) method proposed in [14], represents another compelling and feasible extension of our framework. Aside from investigating the convergence of algorithms with BBE, considering performance metrics beyond HV and providing target values for an arbitrary precision represents another meaningful perspective for future work.

Another prospective research avenue could be the design of measurable landscape features to characterize (local and global) structural properties of purposefully constructed problems with different complexity and known ground truth w.r.t. efficient sets. This will be an essential intermediate step towards (i) characterizing MO problems in general (including high-dimensional problems that are not visualizable anymore), as well as (ii) developing feature-based approaches such as automated algorithm selection.

Finally, we emphasize that our experimental study is intended to illustrate first proof-of-concept takeaways. For future work, we envision our approach enabling a sound and reliable comparison of MO optimizers by evaluating them on a broader set of problems with known structural challenges and also configuring them via automated algorithm configuration methods [2,15,18]. This will ultimately lead to a better understanding of algorithmic components and pave the ground for better algorithm design.

Acknowledgment. The authors acknowledge support by the *European Research Center for Information Systems (ERCIS)*. Further, L. Schäpermeier and P. Kerschke acknowledge support by the *Center for Scalable Data Analytics and Artificial Intelligence (ScaDS.AI) Dresden/Leipzig*.

References

1. Bartz-Beielstein, T., et al.: Benchmarking in Optimization: Best Practice and Open Issues (2020). https://doi.org/10.48550/arxiv.2007.03488
2. Blot, A., Hoos, H.H., Jourdan, L., Kessaci-Marmion, M.É., Trautmann, H.: MO-ParamILS: a multi-objective automatic algorithm configuration framework. In: Festa, P., Sellmann, M., Vanschoren, J. (eds.) LION 2016. LNCS, vol. 10079, pp. 32–47. Springer, Cham (2016). https://doi.org/10.1007/978-3-319-50349-3_3
3. Bossek, J.: ECR 2.0: a modular framework for evolutionary computation in R. In: Proceedings of the Genetic and Evolutionary Computation Conference Companion, pp. 1187–1193 (2017)
4. Bossek, J.: smoof: single- and multi-objective optimization test functions. R J. **9**(1), 103 (2017)
5. Bossek, J., Deb, K.: omnioptr: omni-optimizer algorithm (2021). https://github.com/jakobbossek/omnioptr
6. Deb, K., Pratap, A., Agarwal, S., Meyarivan, T.: A fast and elitist multiobjective genetic algorithm: NSGA-II. IEEE Trans. Evol. Comput. (TEVC) **6**(2), 182–197 (2002)
7. Deb, K., Thiele, L., Laumanns, M., Zitzler, E.: Scalable test problems for evolutionary multiobjective optimization. In: Abraham, A., Jain, L., Goldberg, R. (eds.) Evolutionary Multiobjective Optimization, pp. 105–145. Springer, London (2005). https://doi.org/10.1007/1-84628-137-7_6
8. Deb, K., Tiwari, S.: Omni-optimizer: a generic evolutionary algorithm for single and multi-objective optimization. Eur. J. Oper. Res. (EJOR) **185**, 1062–1087 (2008)
9. Demšar, J.: Statistical comparisons of classifiers over multiple data sets. J. Mach. Learn. Res. **7**, 1–30 (2006)
10. Emmerich, M., Beume, N., Naujoks, B.: An EMO algorithm using the hypervolume measure as selection criterion. In: Coello Coello, C.A., Hernández Aguirre, A., Zitzler, E. (eds.) EMO 2005. LNCS, vol. 3410, pp. 62–76. Springer, Heidelberg (2005). https://doi.org/10.1007/978-3-540-31880-4_5
11. Glasmachers, T.: Challenges of convex quadratic bi-objective benchmark problems. In: Proceedings of the Genetic and Evolutionary Computation Conference, pp. 559–567 (2019)
12. Grimme, C., et al.: Peeking beyond peaks: challenges and research potentials of continuous multimodal multi-objective optimization. Comput. Oper. Res. **136**, 105489 (2021)
13. Hansen, N., Finck, S., Ros, R., Auger, A.: Real-parameter black-box optimization benchmarking 2009: noiseless functions definitions. Technical report, RR-6829, INRIA (2009)
14. Heins, J., Rook, J., Schäpermeier, L., Kerschke, P., Bossek, J., Trautmann, H.: BBE: basin-based evaluation of multimodal multi-objective optimization problems. In: Rudolph, G., Kononova, A.V., Aguirre, H., Kerschke, P., Ochoa, G., Tušar, T. (eds.) Parallel Problem Solving from Nature, pp. 192–206. Springer, Cham (2022). https://doi.org/10.1007/978-3-031-14714-2_14
15. Hoos, H.H.: Automated algorithm configuration and parameter tuning. In: Hamadi, Y., Monfroy, E., Saubion, F. (eds.) Autonomous Search, pp. 37–71. Springer, Heidelberg (2011). https://doi.org/10.1007/978-3-642-21434-9_3
16. Kerschke, P., et al.: Search dynamics on multimodal multiobjective problems. Evol. Comput. **27**(4), 577–609 (2019)

17. Mersmann, O., Trautmann, H., Steuer, D., Bischl, B., Deb, K.: MCO: multiple criteria optimization algorithms and related functions, R package, version 1.0-15.6 (2020). https://github.com/olafmersmann/mco
18. Rook, J., Trautmann, H., Bossek, J., Grimme, C.: On the potential of automated algorithm configuration on multi-modal multi-objective optimization problems. In: Proceedings of the Genetic and Evolutionary Computation Conference Companion, pp. 356–359. ACM, New York (2022)
19. Schäpermeier, L., Grimme, C., Kerschke, P.: One PLOT to show them all: visualization of efficient sets in multi-objective landscapes. In: Bäck, T., et al. (eds.) PPSN 2020. LNCS, vol. 12270, pp. 154–167. Springer, Cham (2020). https://doi.org/10.1007/978-3-030-58115-2_11
20. Schäpermeier, L., Grimme, C., Kerschke, P.: To boldly show what no one has seen before: a dashboard for visualizing multi-objective landscapes. In: Ishibuchi, H., et al. (eds.) EMO 2021. LNCS, vol. 12654, pp. 632–644. Springer, Cham (2021). https://doi.org/10.1007/978-3-030-72062-9_50
21. Schäpermeier, L., Grimme, C., Kerschke, P.: MOLE: digging tunnels through multimodal multi-objective landscapes. In: Proceedings of the Genetic and Evolutionary Computation Conference (GECCO), pp. 592–600. ACM (2022)
22. Schäpermeier, L., Grimme, C., Kerschke, P.: Plotting impossible? Surveying visualization methods for continuous multi-objective benchmark problems. IEEE Trans. Evol. Comput. **26**(6), 1306–1320 (2022)
23. Tanabe, R., Ishibuchi, H.: A review of evolutionary multi-modal multi-objective optimization. IEEE Trans. Evol. Comput. (TEVC) **24**(1), 193–200 (2020)
24. Toure, C., Auger, A., Brockhoff, D., Hansen, N.: On bi-objective convex-quadratic problems. In: Deb, K., et al. (eds.) EMO 2019. LNCS, vol. 11411, pp. 3–14. Springer, Cham (2019). https://doi.org/10.1007/978-3-030-12598-1_1
25. Tušar, T., Brockhoff, D., Hansen, N., Auger, A.: COCO: The Bi-Objective Black Box Optimization Benchmarking (bbob-biobj) Test Suite. arXiv preprint abs/1604.00359 (2016)
26. Wessing, S.: The multiple peaks model 2. Technical report, TR15-2-001, TU Dortmund University, Germany (2015)
27. Wessing, S.: Two-stage methods for multimodal optimization. Ph.D. thesis, University of Dortmund (2015). https://doi.org/10.17877/DE290R-7804
28. Yue, C., Qu, B., Yu, K., Liang, J., Li, X.: A novel scalable test problem suite for multimodal multiobjective optimization. Swarm Evol. Comput. **48**, 62–71 (2019)
29. Zitzler, E., Deb, K., Thiele, L.: Comparison of multiobjective evolutionary algorithms: empirical results. Evol. Comput. (ECJ) **8**(2), 173–195 (2000)

MACO: A Real-World Inspired Benchmark for Multi-objective Evolutionary Algorithms

Sebastian Mai, Tobias Benecke(✉), and Sanaz Mostaghim

Otto-von-Guericke University Magdeburg, Magdeburg, Germany
{sebastian.mai,tobias.benecke,sanaz.mostaghim}@ovgu.de
https://www.ci.ovgu.de

Abstract. The multi-agent coordination (MACO) problem is a real-world inspired multi-objective optimization problem for evolutionary algorithms. It recreates the challenges that are present in optimizing the real-world multi-objective multi-agent pathfinding (MOMAPF) problem. The MACO problem is a scalable, real-valued problem with two objective functions and a known optimal solution. Besides the base version, three variants are proposed, which are based on different properties of the real world MOMAPF problem. Independent sub-problems can be introduced using interaction classes, the multi-modality of the problem can be modified through a set of weights, and the interaction rate between the variables can be altered using the p-norm to approximate the min operator present in the second objective. In our experiments, we assess the performance of three popular multi-objective evolutionary algorithms, both for the basic version and all proposed variations.

Keywords: Multi-objective optimization · Benchmarking · Real-world problem · Multi-modality · Evolutionary algorithms

1 Introduction

Many real-world problems require coordinated planning for multiple agents sharing the same workspace, such as automated warehouses [9] or construction of large structures using multiple robots [14]. This type of planning is also useful in systems with multiple human operated vehicles that follow centrally planned routes, for example in harbors or airports. Often the problems are solved as a single-objective problems. However, in the real world many different objectives are relevant to the application, such as time, safety, travelled distance or energy use [13,18,19]. It is possible to apply multi-objective evolutionary algorithms (MOEAs) to the continuous and the discrete multi-objective multi-agent pathfinding (MOMAPF) problem [13,21]. However, testing the performance of different algorithms is difficult due to the high computational cost associated with the calculation of the fitness function and the fact that the optimal solution is unknown in the continuous version of the problem [13].

M. Emmerich et al. (Eds.): EMO 2023, LNCS 13970, pp. 305–318, 2023.
https://doi.org/10.1007/978-3-031-27250-9_22

In recent years, the established, artificially created benchmark problems traditionally used to evaluate multi-objective evolutionary algorithms (MOEAs) have been criticized, as their properties might not reflect the difficulties of real-world problems correctly [3,11,20]. To create a benchmark that reproduces the challenges found in the MOMAPF problem [13], we propose the multi-agent coordination (MACO) problem, a new benchmark problem for MOEAs. Besides the base version of the problem, three variations are introduced, which are also based on real-world applications of MOMAPF. The MACO problem is scalable in the number of dimensions and the Pareto-front and Pareto-sets are known. Furthermore, the variations of the problem provide interesting possibilities and challenges when benchmarking MOEAs. The multi-modality of the problem can be scaled using the weights variation, and independent sub-problems that need to be optimized at once can be created using interaction classes. This makes the MACO problem a great addition to benchmark and evaluate MOEAs.

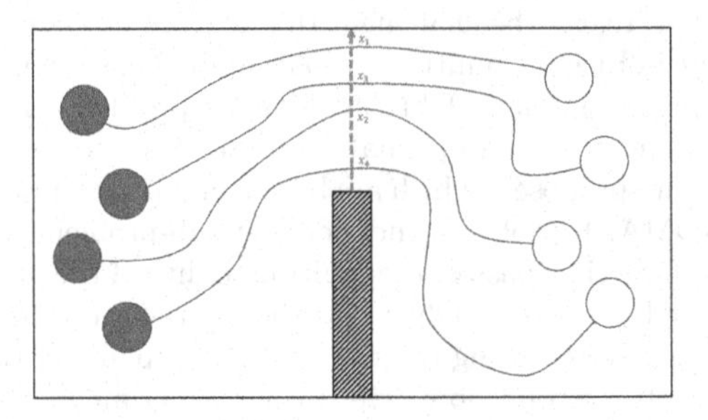

Fig. 1. Example plan (blue) for four agents that navigate around an obstacle. In our benchmark problem, we only consider the position of the agents in a single dimension at a critical location (Green Arrow). (Color figure online)

Figure 1 shows the idea of the benchmark problem presented in this paper and how it is related to the MOMAPF problem [13,21]. In a real application the whole trajectory needs to be optimized, which increases the solution space, as well as the computational effort needed to compute the objective values for each solution. In our test-problem, we optimize the position of the agents only in a critical scenario, i.e., when they pass a narrow passage. As such, we only optimize a single variable for each agent - the position at which the agent crosses the narrow passage (green arrow). As objectives, we use two functions: We assume that all agents need to take the shortest path, as it is associated with the lowest time and energy cost. We assume the shortest path to be as close as possible to the obstacle at position zero. The second objective function is the distance between the agents, which we want to maximize to increase the safety of all agents at the critical location.

The remainder of this paper is structured as follows: The next section describes related works. In Sect. 3 we provide a mathematical definition of the objective functions and the optimal solution to the benchmark problem. In Sect. 4 we show how state-of-the art algorithms cope with the problem in all four variations. In the last section, we conclude the paper.

2 Related Works

In recent years there has been increased criticism on artificially created multi-objective test functions, as they are the most common way to evaluate MOEAs, but contain artificial features that do not reflect the difficulty of real-world problems in practice [3,11,20]. Many of these test problems are designed around the Pareto-front shape. While for the performance assessment, knowledge about the Pareto-front is very beneficial, these design choices can lead to unrealistic properties. For example, the frequently used DTLZ [5] and WFG [10] problems contain position and distance variables. The distance variables are controlling the distance from an individual to the Pareto-front, while the position variables are controlling the position of the individual on the Pareto-front. Furthermore, the unusual properties of artificial test functions might be exploited by algorithms. For example, decomposition-based algorithms, like NSGA-III [8] or MOEA/D [23] are known to work well for problems with triangular shaped Pareto-fronts [11,20]. Using benchmark problems which overrepresent these kinds of unrealistic properties might lead to unfair performance advantages for algorithms exploiting them. Several works address these issues and try to find more realistic test suites. For example, Tanabe and Ishibuchi propose a benchmark suite consisting of 16 real-world test problems to achieve a more reliable evaluation [20]. In [3], Brockhoff et al. propose a multi-objective test suite based on well known single-objective test functions to generate a benchmark more representative of real-world problems.

In this work, we are introducing a new test problem, which tries to replicate the real-world MOMAPF problem [13,15,21]. There are two versions of this problem: a discrete planning problem based on a graph [15,21], which can be solved optimally [15] or suboptimally by meta-heuristic algorithms [21]. As such, for the discrete case, benchmark problems for evolutionary algorithms can be generated with search-based algorithms [15,22]. In the continuous version of the MOMAPF problem [13], no algorithm to find an optimal solution exists. A benchmark for single agent pathfinding with multiple objectives can be found in [22]. In addition, the evaluation of the fitness function in the real-world problem is expensive, as the pairwise distance in between multiple agents and obstacles need to be computed for all time-steps within the planning horizon [13]. To create a benchmark that is fast to compute but still contains the properties of the MOMAPF problem, we are simplifying the problem as described in the previous section.

3 Multi-agent Coordination Problem

The multi-agent coordination problem (MACO) is a real-valued optimization problem with two objectives. In this section, we define the base version of the problem and its optimal solution. In addition, we propose three modifications to the f_2 objective, that are also motivated by the real world MOMAPF problem [13]. Finally, we show how the optimal set is calculated when all modifications for the second objective are used.

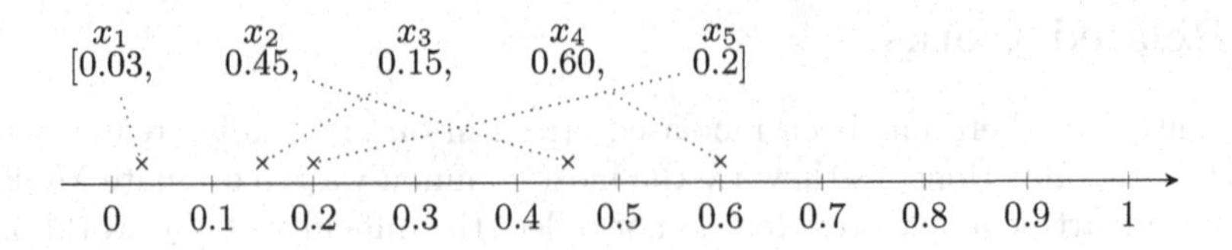

Fig. 2. Decoding a solution $\vec{x}$ for $N = 5$ agents

We assume each solution $\vec{x} \in \mathcal{R}^N$ is a N-dimensional vector, where each variable x_i can take values $0 < x_i \leq 1$. Each variable x_i in a solution $\vec{x}$ represents the position of an agent in a narrow passage. In Fig. 2 we show an example configuration with five agents at the positions $\vec{x} = [0.03, 0.45, 0.15, 0.60, 0.20]$. For our test-problem, we assume that it is better for all agents to pass the obstacle to closer to the left side, where $x_i = 0$. As such, we define f_1 as the average distance to the zero position (Eq. 1). Therefore, the f_1 objective represents the cost objective in a real-world application, as a longer distance needs more time and energy. The f_2 objective reflects the risk of collision. If the agents are closer to each other, risk is increased, while risk is decreased if agents are further from each other. Because we can not offset the risk for a collision between two agents by increasing the distance between a different pair of agents, we use the minimum distance between the agents. As such, f_2 is defined as described in Eq. 2. To formulate both objectives to be minimized, we subtract the minimum distance in f_2 from the largest possible distance between two agents, which is one. In our example (Fig. 2) the agents that relate to variables x_3 and x_5 are closest to each other and define the risk in this situation, which is $1 - |x_3 - x_5| = 1 - 0.05 = 0.95$.

$$f_1(\vec{x}) = \sum_{\forall x_i \in \vec{x}} \frac{x_i}{N} \tag{1}$$

$$f_2(\vec{x}) = 1 - \min_{\forall i \neq j \in \vec{x}} |x_i - x_j| \tag{2}$$

For the basic version, a solution is optimal when the minimal distances from each agent to all others are equal and one gene value is zero. We can systematically generate such a solution by setting the first gene value $x_0 = 0$ and all other genes equidistant, e.g. $\vec{x}_1 = (0.0, 0.1, 0.2, 0.3, 0.4)$. Other Pareto-optimal solutions can be generated by modifying the distance between the elements, for example $\vec{x}_2 = (0.0, 0.2, 0.4, 0.6, 0.8)$. Any permutation of these vectors is also an optimal solution.

Besides the base version of the problem, we propose three variations of the second objective that are also inspired by the MOMAPF problem. The variations can be applied individually or in any combination simultaneously.

3.1 Variation: P-Norm

A great difficulty in optimizing f_2 is, that only two variables interact at each time. Changes in values of the variables that are not involved in the minimum have no impact in the f_2 objective. To include the distances between all agents we redefine f_2 based on the p-norm $||\cdot||_p$ of a vector with all differences between the pairs of variables x_i, x_j (Eq. 3). One property of the p-norm is that for $p \rightarrow -\infty$ the p-norm is equal to the minimum of the entries. By minimizing f_2^P, with a negative value for p, e.g. $p = -10$, the solutions closely approximate those of f_2, but all variables have an impact on the objective (Eq. 4).

$$f_2^P(\vec{x}) = 1 - ||\{x_i - x_j | \forall i \neq j\}||_p \tag{3}$$

$$\lim_{p \rightarrow -\infty} f_2^P(\vec{x}) = f_2(\vec{x}) \tag{4}$$

The p-norm should make the problem easier to solve for an algorithm, as in the base problem only two genes in the genome affect the minimum. Using the p-norm, all genes are now influencing f_2^P. If p is further away from 0, some effects of the modification get lost in the floating-point precision. On the other hand, if p is closer to 0, the effects of the modification change the intended meaning of the base version of the f_2 objective, which has negative impacts in real-world applications (ideally, the p-norm only acts as a tiebreaker between otherwise equal solutions). In addition, the definition of f_2^P, using the p-norm, means that the objective has a known gradient (this property is not utilized in this paper).

While p can be chosen freely by the user, we propose $p = -10$ as a trade-off between accuracy and performance, and $p = -\infty$ as the base version of the problem, as the two $p - norm$ variations for the MACO problem.

3.2 Variation: Weights

The basic version of the problem is extremely multi-modal: We can apply any permutation to the N variables in a solution $\vec{x}$ while neither the f_1 nor f_2 objective values change. From the viewpoint of the multi-agent pathfinding application that motivates our test-problem, the order in which the agents are sorted at the narrow passage is not relevant in the objectives. In real applications, however, the order of agents in the narrow passages does have an impact. This leads to problems that are not multi-modal or have fewer true local solutions.

We modify the f_2 objective to addresses this issue: The minimum distance for each agent i (represented by variable x_i) is weighted by a specific weight w_i in the weight vector $\vec{w} \in R^N$ (Eq. 5).

$$f_2^W(\vec{x}) = 1 - \min_i \left(\frac{w_i}{\sum_{\forall j \neq L} w_j^{-1}} \min_{i \neq j} |x_i - x_j| \right) \tag{5}$$

Normalization of the weights is important for the computation of the Pareto-set (see Sect. 3.4), when we combine multiple interaction classes (see Sect. 3.3) with weight vectors. In this case, the normalization is performed for the weights of each class independently.

Weights affect the multi-modality of the problem. With equal weight, the order of the solutions in x_i is arbitrary. With all different weights, only the solution with the correct order, e.g. $x_1 < x_2 < x_3 ...$ is optimal. By choosing a set of weights, we can tune how much the local optimal solutions differ in their objective values. This means: If two weights in the weight vector $\vec{w}$ are the same ($w_i = w_j$) then there are $2! = 2$ Pareto-sets that map to the same globally optimal solution. If four weights are the same, we get $4! = 24$ Pareto-sets, and so on. Furthermore, we can choose weights very close to each other, which leads to local optima having near-optimal performance. In this situation, algorithms will struggle to find the global optimum, as they will likely get stuck in locally optimal solutions. In contrast, when we use dissimilar weights, algorithms are less likely to converge to a local optimum. However, if the true optimal solution is not found, the impact on the fitness is more severe. Because of this effect in the multi-modality of the problem, the weights are a key parameter in the benchmarks.

The weights can be chosen by the user of the benchmark, we propose to use the following three settings:

Equal: All weights being equal is equivalent to the base problem. In this version, the problem is extremely multi-modal, with $N!$ global optimal Pareto-sets.

Shallow: In this case, the weights are built linearly degrading from 1.0 for the first gene to 0.9 for the last gene. The weight vector for $N = 5$ genes would therefore be $\vec{w} = [1.0, 0.975, 0.95, 0.925, 0.9]$. We call it shallow because we degrade the weights only by a small amount. Using a different weight for each gene means the problem has only one global optimum. However, it is still very difficult to solve, as the weights being in close range to each other results in many close local optima of the problem.

Steep: Similar to the *shallow* configuration, the weights are linearly degraded, however with a more steep descent from 1.0 on the first gene to 0.1 on the last gene. In this case, the problem has only one global optimum and the resulting local optima are not as similar to the global optimum in their fitness value.

3.3 Variation: Interaction Classes

In many pathfinding problems, solutions are only partially coupled - i.e. some agents have to coordinate to find working solutions, other agents move in different areas of the workspace and do not affect each other. Ideally, the designer of a robotic system takes those partitions into account by creating separate plans for separated work spaces. Unfortunately, it is not always possible to know a-priori which plans affect each other and therefore need to be planned at once. While some algorithms (e.g. [16]), explicitly exploit the independence of those agents, black box approaches are not aware of the coupling between agents.

There are existing algorithms for large-scale optimization that aim to find groups of variables in a problem, which are coupled [24]. Those algorithms exploit the variable groupings to solve problems more efficiently. Variable groupings (and the automatic detection of those groupings) may help to solve the MOMAPF and MACO problem more efficiently.

In this benchmark, we create independent groups of agents by computing f_1 normally and modify the visibility of each solution in f_2. This also affects the multi-modality of the problem. To model this, we assign a class to each variable in the problem and modify f_2 to f_2^{IC}, as shown in Eq. (6). In practical terms, the interaction classes are stored in a vector $\vec{c}$ which has the same size as the genome of an individual. Each element of $\vec{c}$ assigns a class to the given genome at the same position. We consider the distance between two agents only if the agents belong to the same class.

$$f_2^{IC}(\vec{x}) = 1 - \min_{\substack{\forall i \neq j \in \vec{x} \\ c(i) = c(j)}} |x_i - x_j| \tag{6}$$

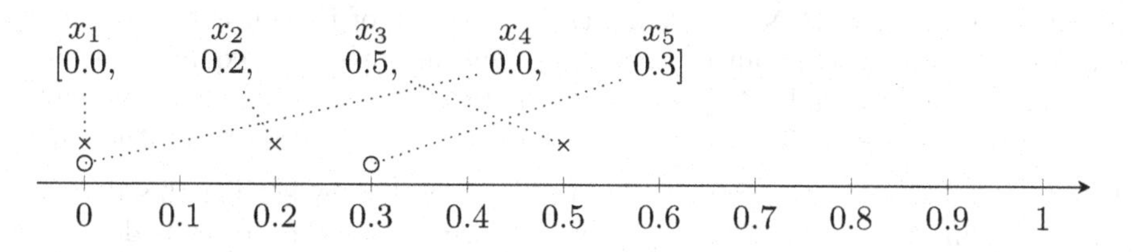

Fig. 3. Decoding a solution $\vec{x}$ for $N = 5$ agents with two different interaction classes. x_1, x_2, and x_3 belong to the same class (labelled with 'x'), while x_4 and x_5 belong to a different class (labelled with 'o')

An example for this modification with the class vector $\vec{c} = [1, 1, 1, 2, 2]$ is visualized in Fig. 3. The genes x_1, x_2, and x_3 belong to class 1, while the genes x_4, and x_5 belong to class 2. It can be imagined that the agents of class 1 need to navigate a different area than the agents of class 2, so there is no risk of collision between an agent of class 1 and 2. In the example of Fig. 3, the closest two genes are therefore x_1 and x_2. The distance between x_2 and x_5 is not considered, as they belong to different interaction classes.

The class vector $\vec{c}$ can be chosen freely by the user, we propose using the following four settings:

None. Every gene having the same class yields the same result as not using classes at all.

Half. In this configuration we assign two classes, with the first half getting class 1 and the second half class 2. If the vector length is odd, the middle element is assigned to class 1.

G3. In this configuration we assign the first three elements class 1, the second three elements class 2, and so on. The number of different classes assigned therefore depends on the number of genes. If we have $N = 7$ genes, the class vector would be $c = [1, 1, 1, 2, 2, 2, 3]$, with three different classes 1, 2 and 3, with class 1 and 2 having three elements and class 3 having one (for this element, the second objective is always perfect).

G4. This configuration is similar to *G3*, with the class is changed every 4th element this time. Again, the number of different classes depends on the number of genes. If we have $N = 7$ genes, the class vector would be $c = [1, 1, 1, 1, 2, 2, 2]$. As the amount of different classes is lower than for *G3*, the *G4* configuration might be easier to solve.

3.4 Optimal Solution

In this section, we will explain how to calculate the Pareto-set of the MACO problem. For the base version and the p-norm variation, the optimal solutions are the same. However, for the weight- and the interaction class variations we need to make slight adjustments.

Base Version and P-Norm Variation. Because of the *min* operator in f_2 and f_2^P, the minimum distance between all solutions have to be equally spaced to achieve an optimal trade-off between the first and second objective. Therefore, the distance between all adjacent agents $\mathbf{min}_{i \neq j} |x_i - x_j|$ needs to be the same. Different solutions in the Pareto-set can be generated by scaling this distance with a scaling factor $s \in [0, 1]$. The general form of the Pareto-set is described by Eq. (7). A special property of our benchmark is that all permutations of the solution vector $\vec{x}$ lead to the same objective values, i.e., we can change the indexing (i) of variables in an optimal solution $\vec{x}$ and get another optimal solution $\vec{x}'$.

$$x_i = s \cdot (i - 1) \cdot \frac{1}{n - 1} \tag{7}$$

Weights Variation. To explain the optimal solution for f_2^W, we are going to assume (without loss of generality) that the indices of all solutions are arranged, such that $w_1 \leq w_2 \leq w_3 \cdots \leq w_L$. We could transform the weighted problem into the base problem by optimizing $x_i \frac{1}{w_i}$ instead of x_i. Hence, in the weighted problem $\frac{1}{w_i} \cdot \mathbf{min}_{i \neq j} |x_i - x_j|$ takes the same value for all agents (with the exception of w_L). This relationship can be used to construct the optimal solution: The x_i in the optimal solution are sorted to best satisfy f_1, i.e., larger weights lead to smaller pairwise distances between the agents. Thus, $x_L = 0$ can be used as a starting point to iteratively add the agent with the next largest weight using the pairwise distance relationship. A solution obtained in this way can be scaled by a scaling factor s in order to generate more solutions. In case some weights are equal, we also get ambiguity for the permutation of those values in the solution, leading to multiple Pareto-optimal solutions.

A problem with this method of constructing the ideal solution arises, when we use interaction classes, because different weights would couple two subproblems that should be independent by our definition. Therefore, we include the normalization of the weights in Eq. (5), which leads to fixed scaling of each (sub)problem.

Interaction Classes Variation. In case we have multiple interaction classes (3.3), we simply compute one optimal solution for each class in the vector separately and append the partial solutions to the full solution vector in the correct order of variables according to the weights.

4 Experiments

To see how different algorithms perform in the new benchmark, we implemented the MACO problem in the pymoo framework [2]. We performed several experiments, both on the base version of the problem and the variations, using their proposed settings.

The problem was run using three common MOEAs, NSGA-II [7], NSGA-III [8], and MOEA/D [23]. Each algorithm was configured with a population size of 100 and was run with 30.000 function evaluations, resulting in 300 generations. For crossover, simulated binary crossover [4] was used with an $\eta = 20$, and for mutation the polynomial mutation operator [6] was used, also with an $\eta = 20$.

For each variation, we use the proposed settings. For the p-norm variation, we use $p = -\infty$ as the base version and $p = -10$ as the approximation. For the weights, we use the settings described in Sect. 3.2 (*equal*, *shallow*, and *steep*). For the interaction classes, we use *none*, $half$, $G3$, and $G4$. Four different genome sizes N were used (3, 5, 10, and 20). Each configuration was repeated 31 times and was initialized with a uniformly generated population.

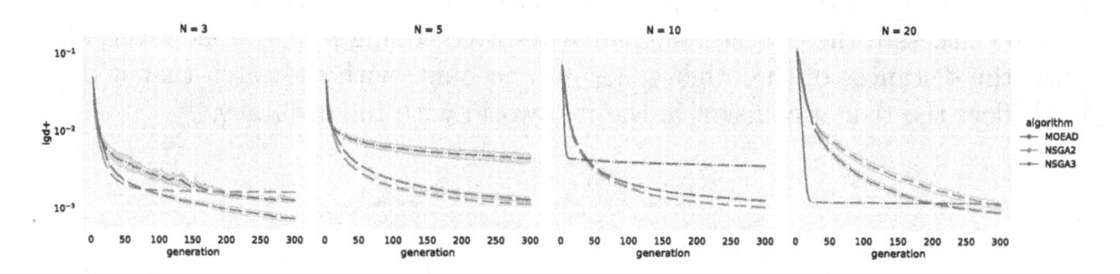

Fig. 4. Median IGD+ results for the base version of the MACO problem using MOEA/D, NSGA-II, and NSGA-III. Each Graph shows a different genome size N.

Base Version. Figure 4 shows the median IGD+ results for 300 generations of the base version of the MACO problem. Each graph shows a different genome size N.

As it can be expected, the lower genome sizes are less difficult, starting with a lower IGD+ value. Also, the NSGA-II and NSGA-III algorithms show a slower convergence for larger genome sizes. Interestingly, this effect was not found with the MOEA/D algorithm. While it also does start with a higher IGD+ value, larger genome sizes show lower IGD+ values in the later generations, which is unusual. In terms of convergence speed, MOEA/D is also fairly consistent, having a fast convergence until generation 25 to 50. After this, the MOEA/D seems to get stuck on the same IGD+ value. This is probably due to the high multi-modality of the MACO problem, which we can observe later in the results of the weights variation. NSGA-II and NSGA-III on the other hand seem to consistently find better solutions in higher generations.

In terms of the observed median IGD+ results shown in Fig. 4, there appears to be no algorithm that is best suited to solve the base version of the MACO problem when considering different genome sizes. After 300 generations, NSGA-III performed the best for the genome sizes $N = 3$ and $N = 20$ and NSGA-II for $N = 5$ and $N = 10$. When considering the larger genome sizes of $N = 10$ and $N = 20$, MOEA/D showed a very fast convergence, however getting stuck in local optima and being overtaken by the other two algorithms eventually.

Variation: P-Norm. To compare the performance for the two $p-norm$ variations of $p = -\infty$ and $p = -10$, Fig. 5 shows the IGD+ values of generation 300 for the genome size of $N = 10$ as a box plot. Similar to the previously described base version of the problem, we see NSGA-II to perform the best for this genome size, closely followed by NSGA-III, MOEA/D being last by a more significant margin. Using the p-norm variation of $p = -10$ shows better IGD+ values for all three problems, indicating a decrease in problem complexity. This is to be expected, as $p = -\infty$ is equivalent to the base problem, meaning only the closest two values are influencing the second objective. Two different individuals that have the same distance on their closest genes, but different distances on the second-closest gene pair, will still show the same value. This increases the difficulty for the algorithm, as there is no indication that one individual will have a better fitness if the closest gene pair is resolved. Using $p = -10$, however, also takes the distances of the other genes into account, with the catch that it does not reflect the risk of collision in the real world with full accuracy.

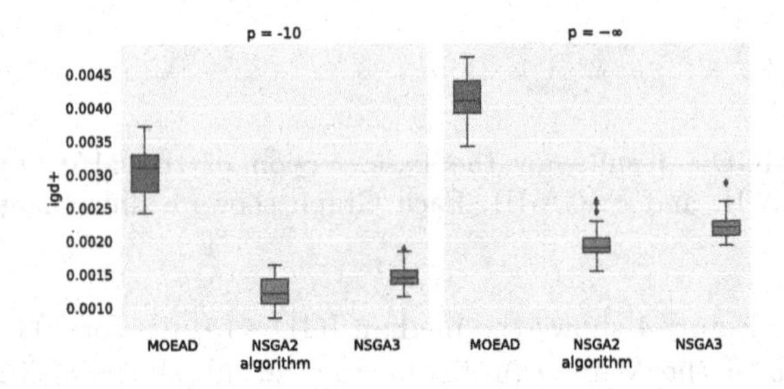

Fig. 5. Box plots of the IGD+ values for the two p-value variations for the genome size of $N = 10$ in the last generation (300).

Variation: Interaction Classes. The results for the interaction classes variation can be found in Fig. 6. The respective class vectors for the genome size of $N = 10$ are $G3 = [1, 1, 1, 2, 2, 2, 3, 3, 3, 4]$, $G4 = [1, 1, 1, 1, 2, 2, 2, 2, 3, 3]$, $half = [1, 1, 1, 1, 1, 2, 2, 2, 2, 2]$, and $none = [1, 1, 1, 1, 1, 1, 1, 1, 1, 1]$.

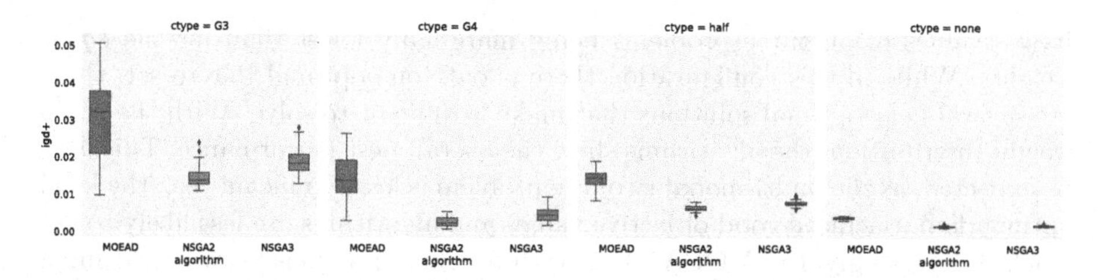

Fig. 6. Box plots of the IGD+ values for the four interaction class variations in the last generation (300).

It can be observed that the interaction classes increase the difficulty of the problem. All three classes increase the difficulty of the problem, with the *none* configuration, representing the base version of the problem, performing the best. However, we do not find the number of different interaction classes to be the only factor when determining the difficulty of the problem. The configuration with the highest number of different interaction classes, $G3$, was also the hardest to solve. Interestingly though, the $G4$ configuration was on average performing better than the $half$ configuration, even though it has a higher number of interaction classes. However, we can observe a higher spread in the solutions of the $G4$ problem, especially for the MOEA/D algorithm. The size or even distribution of the classes could also be a factor, as both the $G4$ and $G3$ configurations show a higher spread in their solutions, also having fewer elements in the last class.

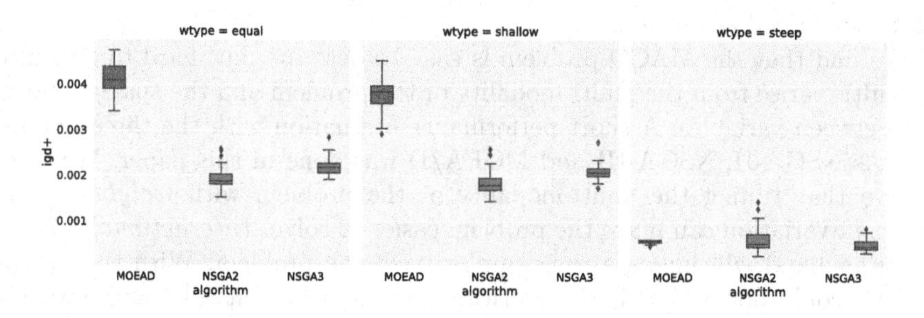

Fig. 7. Box plots of the IGD+ values for the three weight variations in the last generation (300).

Variation: Weights. Finally, Fig. 7 shows the IGD+ results of the three weight types in the final generation (300) for the genome size of $N = 10$. The *equal* weights, which represent the base version of the problem, show the highest IGD+ values. This was to be expected, as the problem is highly multi-modal, with $N!$ optimal Pareto-sets. As already seen in the evaluation of different genome sizes in Fig. 4, MOEA/D performs the worst with the highest IGD+ value, with NSGA-III and NSGA-III performing notably better. Results for the *shallow* weights show this problem to be just slightly less complex to solve, with the

IGD+ values of all three problems being marginally lower than for the *equal* weights. While in this configuration there is only one optimal Pareto-set, there are several near optimal solutions that make it difficult to solve. With the *steep* weight distribution, the algorithms show the overall best performance. This is to be expected, as the multi-modality of the problem is less significant, i.e., the local optima do not achieve good objective values and algorithms are less likely to get stuck. Interestingly, the MOEA/D algorithm, which previously was performing the worst, now shows way better values, similar to NSGA-II and NSGA-III. This indicates MOEA/D to struggle more with the high multi-modality when compared to NSGA-II and NSGA-III.

5 Conclusion

This work presents the MACO problem, a multi-objective benchmark for evolutionary algorithms based on the real-world MOMAPF problem. The problem has two objectives, where f_1 is to minimize the average value of all variables and f_2 relates to the minimum pairwise distances between variables. Besides the base problem, three variations of the second objective were proposed, that are also inspired by properties of the MOMAPF problem. The $p - norm$ variation increases interaction between variables. Multiple sub-problems at once (interaction classes) can appear in the real-world and make the problem even more difficult to solve. Finally, the weighted variation allows us to tune the multi-modality of the problem. While we provide multiple configurations for each variation, they can also be freely configured by the user. Furthermore, the variations can be combined as desired.

We find that the MACO problem is easy to describe, but hard to optimize. Difficulties arise from the multi-modality of the problem and the sparse interaction between variables. A short performance evaluation with the three popular MOEAs NSGA-II, NSGA-III and MOEA/D was done in this paper. We could observe that tuning the multi-modality of the problem with *weights* or the $p - norm$ variation can make the problem easier to solve. $Interactionclasses$ on the other hand will increase the complexity of the problem. With the evaluation, we could not find a single algorithm to be the best at solving the MACO problem.

In future works, we plan to further evaluate different combinations of variations. Using the problem to evaluate the performance of different algorithms, especially ones designed to solve multi-modal problems, like [12], or independent sub-problems, e.g. [24], will also be interesting. Furthermore, more information about the MACO problem and especially the proposed variants can be gained by evaluating beyond the performance in the objective space, for example in terms of population dynamics [1] or influence of variables [17]. Finally, using constraints like a minimum distance between the agents or the obstacles would be an interesting addition to the current versions of the MACO problem.

References

1. Benecke, T., Mostaghim, S.: Tracking the heritage of genes in evolutionary algorithms. In: 2021 IEEE Congress on Evolutionary Computation (CEC), pp. 1800–1807 (2021). https://doi.org/10.1109/CEC45853.2021.9504916
2. Blank, J., Deb, K.: Pymoo: multi-objective optimization in python. IEEE Access **8**, 89497–89509 (2020). https://doi.org/10.1109/ACCESS.2020.2990567. arXiv: 2002.04504
3. Brockhoff, D., Auger, A., Hansen, N., Tušar, T.: Using well-understood single-objective functions in multiobjective black-box optimization test suites. Evol. Comput. **30**(2), 165–193 (2022)
4. Deb, K., Agrawal, R.: Simulated binary crossover for continuous search space. Complex Syst. (1995). https://www.semanticscholar.org/paper/Simulated-Binary-Crossover-for-Continuous-Search-Deb-Agrawal/b8ee6b68520ae0291075cb1408046a7dff9dd9ad
5. Deb, K., Thiele, L., Laumanns, M., Zitzler, E.: Scalable multi-objective optimization test problems. In: Proceedings of the 2002 Congress on Evolutionary Computation. CEC 2002 (Cat. No. 02TH8600), vol. 1, pp. 825–830 (2002). https://doi.org/10.1109/CEC.2002.1007032
6. Deb, K., Agrawal, S.: A niched-penalty approach for constraint handling in genetic algorithms. In: Dobnikar, A., Steele, N.C., Pearson, D.W., Albrecht, R.F. (eds.) Artificial Neural Nets and Genetic Algorithms, pp. 235–243. Springer, Vienna (1999). https://doi.org/10.1007/978-3-7091-6384-9_40
7. Deb, K., Agrawal, S., Pratap, A., Meyarivan, T.: A fast elitist non-dominated sorting genetic algorithm for multi-objective optimization: NSGA-II. In: Schoenauer, M., et al. (eds.) PPSN 2000. LNCS, vol. 1917, pp. 849–858. Springer, Heidelberg (2000). https://doi.org/10.1007/3-540-45356-3_83
8. Deb, K., Jain, H.: An evolutionary many-objective optimization algorithm using reference-point-based nondominated sorting approach, part I: solving problems with box constraints. IEEE Trans. Evol. Comput. **18**(4), 577–601 (2014). https://doi.org/10.1109/TEVC.2013.2281535
9. Honig, W., Kiesel, S., Tinka, A., Durham, J.W., Ayanian, N.: Persistent and robust execution of MAPF schedules in warehouses. IEEE Robot. Autom. Lett. **4**(2), 1125–1131 (2019). https://doi.org/10.1109/LRA.2019.2894217
10. Huband, S., Barone, L., While, L., Hingston, P.: A scalable multi-objective test problem toolkit. In: Coello Coello, C.A., Hernández Aguirre, A., Zitzler, E. (eds.) EMO 2005. LNCS, vol. 3410, pp. 280–295. Springer, Heidelberg (2005). https://doi.org/10.1007/978-3-540-31880-4_20
11. Ishibuchi, H., He, L., Shang, K.: Regular pareto front shape is not realistic. In: 2019 IEEE Congress on Evolutionary Computation (CEC), pp. 2034–2041 (2019). https://doi.org/10.1109/CEC.2019.8790342
12. Javadi, M., Mostaghim, S.: Using neighborhood-based density measures for multimodal multi-objective optimization. In: Ishibuchi, H., et al. (eds.) EMO 2021. LNCS, vol. 12654, pp. 335–345. Springer, Cham (2021). https://doi.org/10.1007/978-3-030-72062-9_27
13. Mai, S., Mostaghim, S.: Modeling pathfinding for swarm robotics. In: Dorigo, M., Stützle, T., Blesa, M.J., Blum, C., Hamann, H., Heinrich, M.K., Strobel, V. (eds.) ANTS 2020. LNCS, vol. 12421, pp. 190–202. Springer, Cham (2020). https://doi.org/10.1007/978-3-030-60376-2_15

14. Ochalek, M., Jenett, B., Formoso, O., Gregg, C., Trinh, G., Cheung, K.: Geometry systems for lattice-based reconfigurable space structures. In: 2019 IEEE Aerospace Conference, pp. 1–10 (2019). https://doi.org/10.1109/AERO.2019.8742178. ISSN 1095-323X
15. Ren, Z., Zhan, R., Rathinam, S., Likhachev, M., Choset, H.: Enhanced multi-objective A * using balanced binary search trees. Technical report (2022)
16. Sharon, G., Stern, R., Felner, A., Sturtevant, N.: Meta-agent conflict-based search for optimal multi-agent path finding. In: Proceedings of the 5th Annual Symposium on Combinatorial Search, SoCS 2012, pp. 97–104 (2012). ISBN 9781577355847
17. Smedberg, H., Bandaru, S.: Finding influential variables in multi-objective optimization problems. In: 2020 IEEE Symposium Series on Computational Intelligence (SSCI), pp. 173–180 (2020). https://doi.org/10.1109/SSCI47803.2020.9308383
18. Steup, C., Parlow, S., Mai, S.: Generic component-based mission-centric energy model for micro-scale unmanned aerial vehicles. Drones **4**(63), 1–17 (2020). https://doi.org/10.3390/drones4040063
19. Surynek, P., Felner, A., Stern, R., Boyarski, E.: An empirical comparison of the hardness of multi-agent path finding under the makespan and the sum of costs objectives. In: Proceedings of the 9th Annual Symposium on Combinatorial Search, SoCS 2016, pp. 145–146 (2016). ISBN 9781577357698
20. Tanabe, R., Ishibuchi, H.: An easy-to-use real-world multi-objective optimization problem suite. Appl. Soft Comput. **89**, 106078 (2020)
21. Weise, J., Mai, S., Zille, H., Mostaghim, S.: On the scalable multi-objective multi-agent pathfinding problem. In: 2020 IEEE Congress on Evolutionary Computation, CEC 2020 - Conference Proceedings (2020). https://doi.org/10.1109/CEC48606.2020.9185585
22. Weise, J., Mostaghim, S.: Scalable Many-Objective Pathfinding Benchmark Suite, pp. 1–10 (2020). https://arxiv.org/abs/2010.04501. arXiv: 2010.04501
23. Zhang, Q., Li, H.: MOEA/D: a multiobjective evolutionary algorithm based on decomposition. IEEE Trans. Evol. Comput. **11**(6), 712–731 (2007). https://doi.org/10.1109/TEVC.2007.892759
24. Zille, H., Ishibuchi, H., Mostaghim, S., Nojima, Y.: Mutation operators based on variable grouping for multi-objective large-scale optimization. In: 2016 IEEE Symposium Series on Computational Intelligence, SSCI 2016 (2017). https://doi.org/10.1109/SSCI.2016.7850214. ISBN 9781509042401

A Scalable Test Suite for Bi-objective Multidisciplinary Optimization

Victoria Johnson(✉), João A. Duro, Visakan Kadirkamanathan, and Robin C. Purshouse

University of Sheffield, Sheffield, UK
vjohnson2@sheffield.ac.uk

Abstract. Multidisciplinary design optimization (MDO) involves solving problems that feature multiple subsystems or disciplines, which is an important characteristic of many complex real-world problems. Whilst a range of single-objective benchmark problems have been proposed for MDO, there exists only a limited selection of multi-objective benchmarks, with only one of these problems being scalable in the number of disciplines. In this paper, we propose a new multi-objective MDO test suite, based on the popular ZDT bi-objective benchmark problems, which is scalable in the number of disciplines and design variables. Dependencies between disciplines can be defined directly in the problem formulation, enabling a diverse set of multidisciplinary topologies to be constructed that can resemble more realistic MDO problems. The new problems are solved using a multidisciplinary feasible architecture which combines a conventional multi-objective optimizer (NSGA-II) with a Newton-based multidisciplinary analysis solver. Empirical findings show that it is possible to solve the proposed ZDT-MDO problems but that multimodal problem landscapes can pose a significant challenge to the optimizer. The proposed test suite can help stimulate more research into the neglected but important topic of multi-objective multidisciplinary optimization.

Keywords: Multidisciplinary design optimization · Multi-objective optimization · Benchmark problems · Scalability

1 Introduction

Multidisciplinary design optimization (MDO) is an area of research that handles optimization problems involving multiple disciplines, subsystems or components. MDO recognises that large, complex or interwoven engineered systems are often partitioned into smaller subsystems. This decomposition can arise for a

VJ was supported by the UK Engineering and Physical Sciences Research Council. JAD, VK and RCP were supported by SIPHER (MR/S037578/1), a UK Prevention Research Partnership funded by the UK Research and Innovation Councils, the Department of Health and Social Care (England) and the UK devolved administrations, and leading health research charities https://ukprp.org/..

M. Emmerich et al. (Eds.): EMO 2023, LNCS 13970, pp. 319–332, 2023.
https://doi.org/10.1007/978-3-031-27250-9_23

number of interrelated reasons: from engineering practitioners taking a 'divide-and-conquer' approach to solving complex design problems, to the way in which engineering disciplines have emerged over time as discrete entities, to the functional organisation of teams in large engineering companies and institutions. Whilst MDO arose in the design of complex engineered products, such as exist in the aerospace and automotive sectors, its application is not limited to engineering, but is equally applicable to other complex systems contexts such as environmental and public policy [14,18].

One important consideration in MDO is the need to model the interactions between subsystems, because the performance of a system is not necessarily defined just by its components, but also by the interactions between those components. It is common to model the interactions by using *linking* (or *coupling*) variables that are exchanged between the subsystems. However, when the subsystems have circular dependencies, it is not trivial to determine the values of the linking variables, and it might be necessary to use numerical approximation techniques, such as a multidisciplinary analysis (MDA) solver.

Several architectures have been proposed for dealing with MDO problems—see, for example, the seminal survey paper by Martins & Lambe [17]. These MDO architectures specify how to organize the discipline analysis models (and other types of models) within the problem formulation, in order to facilitate the process of finding the optimal design for the entire system. Some typical examples are the *individual discipline feasible* and the *multidisciplinary feasible* (MDF) architectures [4]. However, the focus of the MDO literature is primarily on single-objective problems. Multiple conflicting objectives are often found in real-world applications and, given that MDO problems are traditionally aimed at engineering applications, it is perhaps surprising that, to our knowledge, no multi-objective multidisciplinary optimization (MO-MDO) test suite has yet been proposed. Such a test suite would provide an opportunity for researchers and others to develop and test new optimization algorithms making them better equipped for dealing with multi-objective MDO problems.

In earlier work, we proposed an MDO version of a bi-objective benchmark problem known as ZDT1 [13]. This problem was then solved using an MDF architecture, encompassing a conventional multi-objective evolutionary optimization algorithm, NSGA-II [5], as the system optimizer, and a Newton-based method as the MDA solver. The present paper builds upon [13] and its distinctive contribution is as follows:

1. the approach used to transform the original ZDT1 problem into an MDO variant is extended to the remaining continuous ZDT problems;
2. the way the linking variables are integrated into the optimization problem is improved, in that the deviation of the linking variables from their optimal values is used to perturb the decision variables; and
3. two new topologies for connecting the disciplines via their linking variables are proposed, and we show how it is possible to create arbitrary problem structures.

The scope of this work encompasses bi-objective MDO problems with both varying number of variables and number of disciplines. The discipline analysis models are mutually interdependent. Although all optimization problems in this work contain only continuous variables, discrete variables are also within scope as long as they are supported by the optimization algorithm. We have only considered the MDF architecture in this work, but other architectures could be used instead as long they are compatible with the problem formulation.

The remainder of this paper is organised as follows. Section 2 discusses and analyses the current state of multidisciplinary and multi-objective benchmark problems. Section 3 introduces the proposed MO-MDO test suite. Several topologies for connecting the disciplines via linking variables are proposed in Sect. 4. The experimental setup is in Sect. 5, while the obtained experimental results are in Sect. 6. Section 7 gives a short summary of the work undertaken and proposes directions for future work.

2 Related Literature

The MDO paradigm originated in industrial settings, where different parts of complex engineered products are designed or optimised by different disciplinary teams. MDO codifies this arrangement via the structure of the optimization problem, including concepts such as: global variables, which are accessible by more than one discipline; local variables, which are used only within one discipline; and linking variables that are exchanged between disciplines as a way to model disciplinary interdependencies. The MDO literature is extensive [17], and we therefore focus our review on the benchmark problems that have been proposed for testing MDO approaches, since this is the area most pertinent to our paper's aims, and contrast these to popular benchmarks for multi-objective optimization.

2.1 Multi-disciplinary Benchmarks

There are comparatively few MDO benchmark problems compared with multi-objective benchmarks. Many of these derive from the NASA MDO test suite [19], which contains 14 problems, including the Golinski speed reducer problem, propane combustion and aerospike nozzle design. While some of the benchmark problems have been expanded, such as the speed reducer problem, other problems are outdated and do not fulfil the needs of current MDO research in terms of complexity and scalability. Further, the original test suite is no longer available from its primary source, with the suite now distributed across a number of secondary sources, e.g. [21].

Another popular MDO benchmark problem is the Sellar (also known as the 'analytical') problem [20]. This problem is small, consisting of only two disciplines, each containing one equation for the multidisciplinary analysis, one local variable, two global variables and two linking variables. As such, the problem cannot provide an indication of how a complex MDO architecture will perform. Further MDO problems are esoteric, having been proposed for specific applications and typically solved only by the problem proposers; examples include building envelope design [23], robotic fish [2], automotive design [1] and wing design [3]. These problems are unsuitable as benchmarks because of the narrowness of the application and/or lack of availability of the MDA equations in the public domain.

2.2 Multi-objective Benchmarks

The literature on multi-objective benchmark problems is very extensive and we focus only on some popular examples in this section. The ZDT test suite, proposed by Zitzler et al. [24], consists of six two-objective test problems, five of which are continuous and one of which is discrete. For the purposes of this paper, we will only discuss the continuous problems. In each problem, the first objective f_1 is a function of the first design variable, and the second objective f_2 comprises the product of a so-called $g(.)$ function, which is a variation of the sum of all design variables except the one found in the first objective, and an $h(.)$ function which defines the relationship between the first design variable (and, by extension, f_1) and the remainder. The ZDT test suite can be criticised as unrealistic or incomparable with real-world problems, with structures that provide only a limited reflection of the challenges posed by the current state of research in multi-objective optimization. However, the problems are also simple to modify and are scalable in the number of design variables.

Other test suites include those with similar $g(.)$ functions, such as the DTLZ problems which are scalable in the number of objectives [6], modular problems such as WFG [12], and problems with varied constraints such as those provided by DAS-CMOP [7] and MW [16].

2.3 Multi-objective Multidisciplinary Benchmarks

All the MDO problems mentioned above contain a single objective. Existing multi-objective multidisciplinary optimization problems are derived from single-objective MDO benchmarks which are not scalable, such as the Golinski speed reducer problem [8,11,15]. More recently, we proposed a MO-MDO problem based on the bi-objective ZDT1 problem [13], which is scalable in the number of variables and disciplines but has a cost landscape that is not otherwise challenging to an optimizer.

3 Proposed MO-MDO Test Suite: ZDT-MDO

The proposed MO-MDO test suite is based on the ZDT benchmark problems and we therefore label it *ZDT-MDO*. Despite the limitations of ZDT as a test set, the original structure of the problems makes them amenable to restructuring into MDO problems in which the original Pareto front is recoverable (which is highly advantageous from an analysis perspective). Here, we consider the five continuous ZDT problems, with ZDT5 omitted because it is binary encoded. For all problems, the first decision variable controls the position across the Pareto front, while the others are called distance decision variables because they control the convergence towards the Pareto front.

The multidisciplinary system contains global variables that are shared between the disciplines, and each discipline has its own set of local variables. The decision variables of the original ZDT problem are partitioned into global and local ones, where the first n_z are global and are represented by the vector $\mathbf{z} = (z_1, \ldots, z_{n_z})^T$. The remaining ones are local variables and are distributed across N disciplines as given by the vector $\mathbf{x} = (\mathbf{x}_1, \ldots, \mathbf{x}_N)^T$. Each $\mathbf{x}_i = (x_{i,1}, \ldots, x_{i,n_{x_i}})^T$ contains a total of n_{x_i} local variables at the ith discipline where $i \in \{1, \ldots, N\}$.

The disciplines exchange linking variables to model the interactions of the overall system. These linking variables are the output of an analysis conducted by each discipline that simulates the behaviour of a particular component of the multidisciplinary system. There is a total of n_{y_i} output linking variables at the ith discipline, given by the vector $\mathbf{y}_i = (y_{i,1}, \ldots, y_{i,n_{y_i}})^T$, and $\mathbf{y} = (\mathbf{y}_1, \ldots, \mathbf{y}_N)^T$ contains the output linking variables of all disciplines. Each discipline may require one or more linking variables from other disciplines to conduct its own disciplinary analysis. To keep track of the linking variable connections in the system consider the following:

1. let n_{p_i} $(1 \leq n_{p_i} < N)$ denote the number of disciplines that provide linking variables to the ith discipline;
2. the indices of the disciplines that provide linking variables to the ith discipline are stored in the set $\mathbf{p}_i = \{p_{i,1}, \ldots, p_{i,n_{p_i}}\}$ where $p_{i,j} \in \{1, \ldots, N\} \backslash \{i\}$ $\forall_{j=1,\ldots,n_{p_i}}$.

For instance, for a hypothetical four-discipline system, if the second and fourth disciplines provide linking variables to the first discipline, then $\mathbf{p}_1 = \{2, 4\}$. The discipline analysis at the ith discipline is to find $\mathbf{y}_i$ that satisfies the following expression:

$$A_{i,i}\mathbf{y}_i + \sum_{j=1}^{n_{p_i}} (A_{i,p_{i,j}}\mathbf{y}_{p_{i,j}}) = -C_i\bar{\mathbf{z}} - D_i\mathbf{x}_i, \tag{1}$$

where $\bar{\mathbf{z}} = (z_2, \ldots, z_{n_z})^T$ excludes the first decision variable of the original ZDT problem. The above expression only relies on the decision variables of the distance type, implying that the positional decision variable (z_1) is not included to ensure that there is a single solution to the systems of equations. The matrices in Eq. 1 are defined as follows:

1. $A_{i,i} \in \mathbb{R}^{n_{y_i} \times n_{y_i}}$, $C_i \in \mathbb{R}^{n_{y_i} \times (n_z - 1)}$, and $D_i \in \mathbb{R}^{n_{y_i} \times n_{x_i}}$ $\forall_{i=1\ldots,N}$,
2. $A_{i,p_{i,j}} \in \mathbb{R}^{n_{y_i} \times n_{y_j}}$ $\forall_{i=1\ldots,N}$ and $\forall_{j=1,\ldots,n_{p_i}}$.

An important aspect of Eq. 1 is that, depending on how the disciplines are connected, determining the linking variables for one discipline may require knowing the values of the linking variables from the other disciplines. It can become even harder to solve in case there are cyclic connections in the system. The complete set of equations across disciplines can form a full system of equations as given by:

$$\begin{bmatrix} A_{1,1} & A_{1,2} & \dots & A_{1,N} \\ A_{2,1} & A_{2,2} & \dots & A_{2,N} \\ \vdots & \vdots & \ddots & \vdots \\ A_{N,1} & A_{N,2} & \dots & A_{N,N} \end{bmatrix} \begin{bmatrix} \mathbf{y}_1 \\ \mathbf{y}_2 \\ \vdots \\ \mathbf{y}_N \end{bmatrix} = - \begin{bmatrix} C_1 \\ C_2 \\ \vdots \\ C_N \end{bmatrix} \bar{\mathbf{z}} - \begin{bmatrix} D_1 & 0 & \dots & 0 \\ 0 & D_2 & \dots & 0 \\ \vdots & \vdots & \ddots & \vdots \\ 0 & 0 & \dots & D_N \end{bmatrix} \begin{bmatrix} \mathbf{x}_1 \\ \mathbf{x}_2 \\ \vdots \\ \mathbf{x}_N \end{bmatrix} \tag{2}$$

or equivalently given by:

$$\boldsymbol{A}\mathbf{y} = -\boldsymbol{C}\bar{\mathbf{z}} - \boldsymbol{D}\mathbf{x}. \tag{3}$$

To ensure that the full system of equations has a unique solution, $\boldsymbol{A}$ needs to be invertible. Additionally, any column in $\boldsymbol{A}^{-1}\boldsymbol{C}$ or $\boldsymbol{A}^{-1}\boldsymbol{D}$ cannot be all zeros to ensure that there are no redundant design variables. Finding all $\mathbf{y}_i$s for the entire system requires the use of numerical techniques, such as Gauss–Seidel and Newton-based methods that are often called multidisciplinary analysis solvers in the MDO literature [17].

The linking variables are incorporated into the optimization problem by penalising the local variables as given by the function:

$$\xi(\mathbf{x}_i, \mathbf{y}_i) = \mathbf{x}_i + \|\mathbf{y}_i - \mathbf{y}_i^*\|_1, \tag{4}$$

where $\mathbf{y}_i^*$ are the linking variable optimal values for the ith discipline, and the operator $\|\bullet\|_1$ is the L^1-norm. Let the output of Eq. 4 be the vector $\hat{\mathbf{x}}_i = (\hat{x}_{i,1}, \dots, \hat{x}_{i,n_{x_i}})^T$, and the function that applies the same transformation to all $\mathbf{x}_i$s is denoted by $\boldsymbol{\xi}(\mathbf{x}, \mathbf{y})$. The proposed MO-MDO problem formulation based on ZDT1 is given by:

$$\begin{aligned} \min \quad & f_1(\mathbf{z}) = z_1 \\ \min \quad & f_2(\mathbf{z}, \boldsymbol{\xi}(\mathbf{x}, \mathbf{y})) = g(\mathbf{z}, \boldsymbol{\xi}(\mathbf{x}, \mathbf{y}))h(\mathbf{z}, \boldsymbol{\xi}(\mathbf{x}, \mathbf{y})) \\ \text{s.t.} \quad & g(\mathbf{z}, \boldsymbol{\xi}(\mathbf{x}, \mathbf{y})) = 1 + \frac{9}{n_v - 1} \left(\sum_{i=2}^{n_z} z_i + \sum_{i=1}^{N} \sum_{j=1}^{n_{x_i}} \hat{x}_{i,j} \right) \\ & h(\mathbf{z}, \boldsymbol{\xi}(\mathbf{x}, \mathbf{y})) = 1 - \sqrt{\frac{f_1(\mathbf{z})}{g(\mathbf{z}, \boldsymbol{\xi}(\mathbf{x}, \mathbf{y}))}} \end{aligned} \tag{5}$$

where $n_v = n_z + \sum_{i=1}^{N} n_{x_i}$. For the remaining ZDT problems, $f_1(\mathbf{z}) = z_1$, with the exception of ZDT6 which is $f_1(\mathbf{z}) = 1 - \exp(-4z_1)\sin^6(6\pi z_1)$, while the g and h functions are shown in Table 1. For optimality, all decision variables (global and local) with the exception of z_1 have to be zero for the given $g(\cdot)$ functions, unless transformations are applied. This means that Eq. 2 becomes an homogeneous system of linear equations which is solved when all the $\mathbf{y}_i$s are zero vectors. The benchmarks established in this section can be found in the project's github repository[1].

[1] https://github.com/vj2Sheffield/mdo_zdt.

Table 1. The g and h functions of the proposed MO-MDO test suite

	g	h
ZDT2	$1+\frac{9}{n_v-1}\left(\sum_{i=2}^{n_z} z_i+\sum_{i=1}^{N}\sum_{j=1}^{n_{x_i}}\hat{x}_{i,j}\right)$	$1-(f_1/g)^2$
ZDT3	$1+10(n_v-1)+\sum_{i=2}^{n_z}(z_i-10\cos(4\pi z_i)) + \sum_{i=1}^{N}\sum_{j=1}^{n_{x_i}}(\hat{x}_{i,j}-10\cos(4\pi\hat{x}_{i,j}))$	$1-\sqrt{f_1/g}$
ZDT4	$1+\frac{9}{n_v-1}\left(\sum_{i=2}^{n_z} z_i+\sum_{i=1}^{N}\sum_{j=1}^{n_{x_i}}\hat{x}_{i,j}\right)$	$1-\sqrt{f_1/g}$
ZDT6	$1+\frac{9}{n_v-1}\left(\sum_{i=2}^{n_z} z_i+\sum_{i=1}^{N}\sum_{j=1}^{n_{x_i}}\hat{x}_{i,j}\right)^{0.25}$	$1-(f_1/g)^2$

4 Defining Dependencies Between Disciplines

The proposed formulation in Eq. 2 offers the flexibility to connect the disciplines in different ways via linking variables. For instance, for a three-discipline system, in case the second and third disciplines receive linking variables from the first discipline, then $A_{2,1}$ and $A_{3,1}$ have non-zero elements. If there are no more connections between the disciplines (except $A_{i,i}$ $\forall_{i=1...N}$ which are set to the identity matrix), then the remaining matrices in $\boldsymbol{A}$ are set to zero. On the other hand, in case the first discipline receives linking variables from either the second or third discipline (implying that $A_{1,2}$ and/or $A_{1,3}$ have non-zero elements), then it can be said that the topology contains cyclic connections.

Figure 1a shows a five-discipline system where each discipline is only connected to the next one, and a cyclic connection is created by connecting the last discipline to the first one. The same topology is depicted by an extended design structure matrix (XDSM) as shown in Fig. 1b. This technique has been popularised by [17] to visualise the interconnections between the components of a complex system. It is useful in particular to visualise both data dependencies and process flow. The discipline analysis are represented in a diagonal, the input data flows along the vertical direction, while the output data flows along the horizontal direction. The data is labelled inside parallelograms, and the way the data flows is shown as thick grey lines. Other possible ways of connecting the disciplines are shown in the remaining subfigures in Fig. 1. We now propose the following three topologies for connecting the disciplines:

1. OIOO: stands for "one-in-one-out" since each discipline only receives and sends linking variables to a single discipline. We adopt a circular topology where the first discipline receives linking variables from the last discipline. This is given by $\mathbf{p}_1 = \{N\}$ and $\mathbf{p}_i = \{i-1\}$ $\forall_{i=2,...,N}$, and the XDSM is shown in Fig. 1b.
2. TITO: stands for "two-in-two-out" since each discipline sends and receives linking variables to two disciplines. This is given by $\mathbf{p}_1 = \{N, i+1\}$, $\mathbf{p}_i = \{i-1, i+1\}$ and $\mathbf{p}_N = \{i-1, 1\}$, and the XDSM is shown in Fig. 1d.
3. AIAO: stands for "all-in-all-out" since each discipline sends and receives linking variables to all disciplines. This is given by $\mathbf{p}_i = \{1, \ldots, N\}\backslash\{i\}$ $\forall_{i=1,...,N}$, and the XDSM is shown in Fig. 1f.

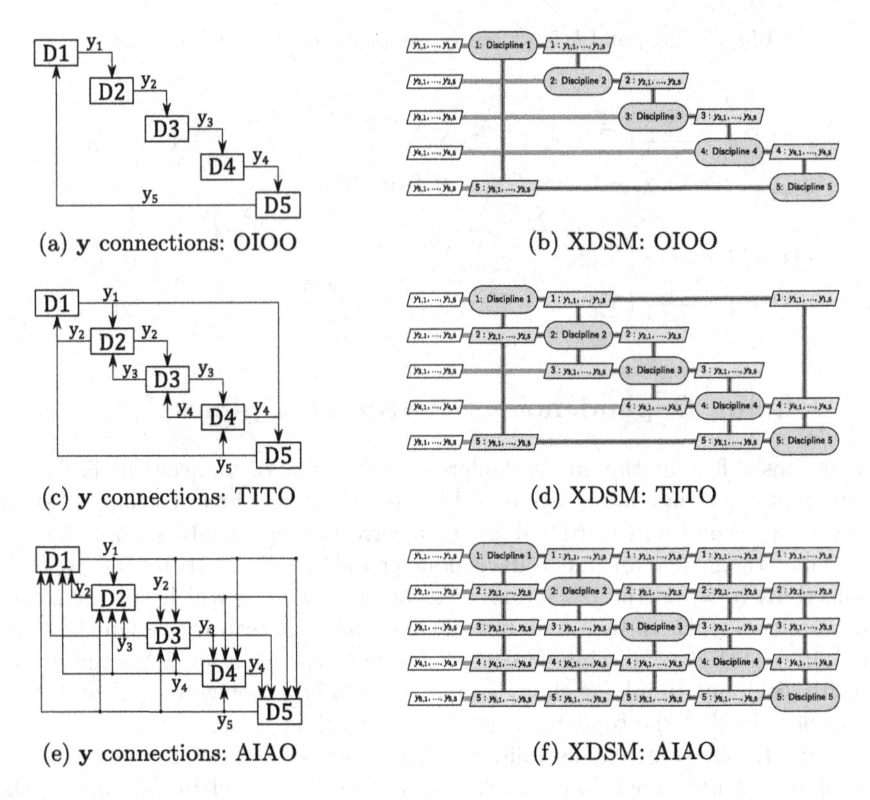

(a) **y** connections: OIOO (b) XDSM: OIOO

(c) **y** connections: TITO (d) XDSM: TITO

(e) **y** connections: AIAO (f) XDSM: AIAO

Fig. 1. Different topologies showcasing the dependencies between disciplines on a five-discipline system.

5 Experimental Setup

The matrices in Eq. 1 are randomly generated and then row-normalised. The only exception is $A_{i,i}\ \forall_{i=1,\dots,N}$ which is set to the identity matrix. The number of global variables are set to 10 ($n_z = 10$) and for all disciplines the number of local variables and the size of the linking variables vector is set to 5 (i.e. $n_{x_i} = 5$ and $n_{y_i} = 5\ \forall_{i=1,\dots,N}$). The lower and upper bounds for the decision variables of all the problems are set to 0 and 1, respectively. The only exception is ZDT4 where the lower bounds are -5 and 5 for all decision variables with the exception of z_1 which takes values in the range $[0, 1]$.

For dealing with the MO-MDO problems, we adopt an MDF architecture involving a system optimizer and a MDA solver that conducts the disciplinary analysis one discipline at a time. For the system optimizer we use a popular multi-objective optimization algorithm known as NSGA-II [5]. The crossover and mutation probabilities are set to 90% and $1/n_v$, respectively, while the crossover and mutation index are both set to 20. The number of generations is set to 1000 with a population size of 100. The initial population is randomly initialised. The MDF architecture is provided by the OpenMDAO package in Python [10], and

NSGA-II implementation by PyOptSparse [22]. The MDA solver is also provided by OpenMDAO and uses a combination of a nonlinear and linear solver. The nonlinear solver is a Newton method while the linear solver relies on linear algebra techniques such as LU decomposition. The MDA solver runs for a maximum of 1000 iterations. For comparing different problem instances the hypervolume indicator is used. To compute the hypervolume we have used a dimension-sweep algorithm, taken from [9]. The reference point used in the hypervolume computations is $\{1.1, 14\}$ for ZDT1, $\{1.1, 13\}$ for ZDT3, $\{1.1, 1620\}$ for ZDT4, and $\{1.1, 17\}$ for ZDT6.

6 Experimental Results

In this section we show the obtained results for the MDO version of ZDT1, ZDT3, ZDT4 and ZDT6 problems. Due to space limitations, ZDT2 results are omitted, since they are very similar to those obtained for ZDT1. For all cases the MDA solver has run for sufficient number of iterations to guarantee convergence, implying that the correct linking variables were obtained for the given global ($\bar{\mathbf{z}}$) and local variables ($\mathbf{x}$). Therefore our analysis will be mostly focused on the performance of the system optimizer (NSGA-II) in dealing with these problems.

The convergence across generations is captured by the hypervolume metric in Fig. 2 for five and 10 discipline problems with different linking variable topologies. Figure 3 depicts the non-dominated solutions obtained at the end of the optimization run shown alongside the Pareto optimal front (POF). In all plots, the notation D5 and D10 denotes the number of disciplines. Good convergence is achieved for all problems instances involving ZDT1, ZDT3 and ZDT6, although not all solutions are co-located on the POF for ZDT6. ZDT4 shows constant improvement in terms of hypervolume across the generations, but achieves poor convergence overall within the given computational budget.

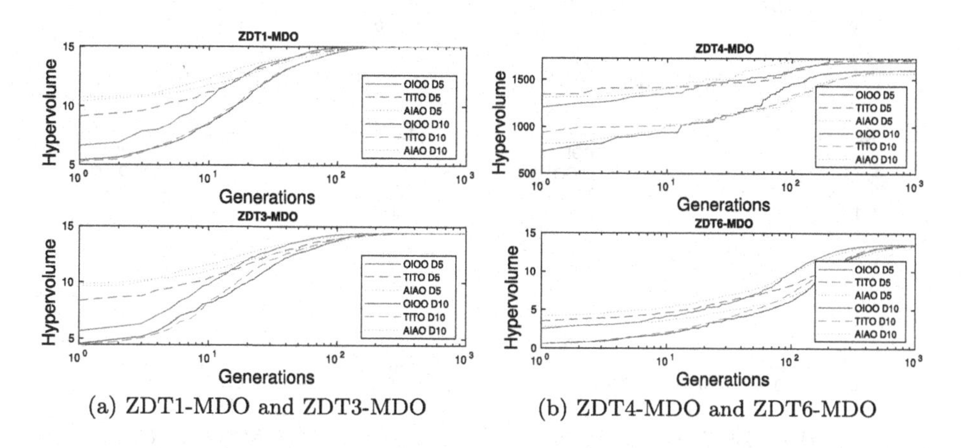

(a) ZDT1-MDO and ZDT3-MDO (b) ZDT4-MDO and ZDT6-MDO

Fig. 2. Hypervolume across generations.

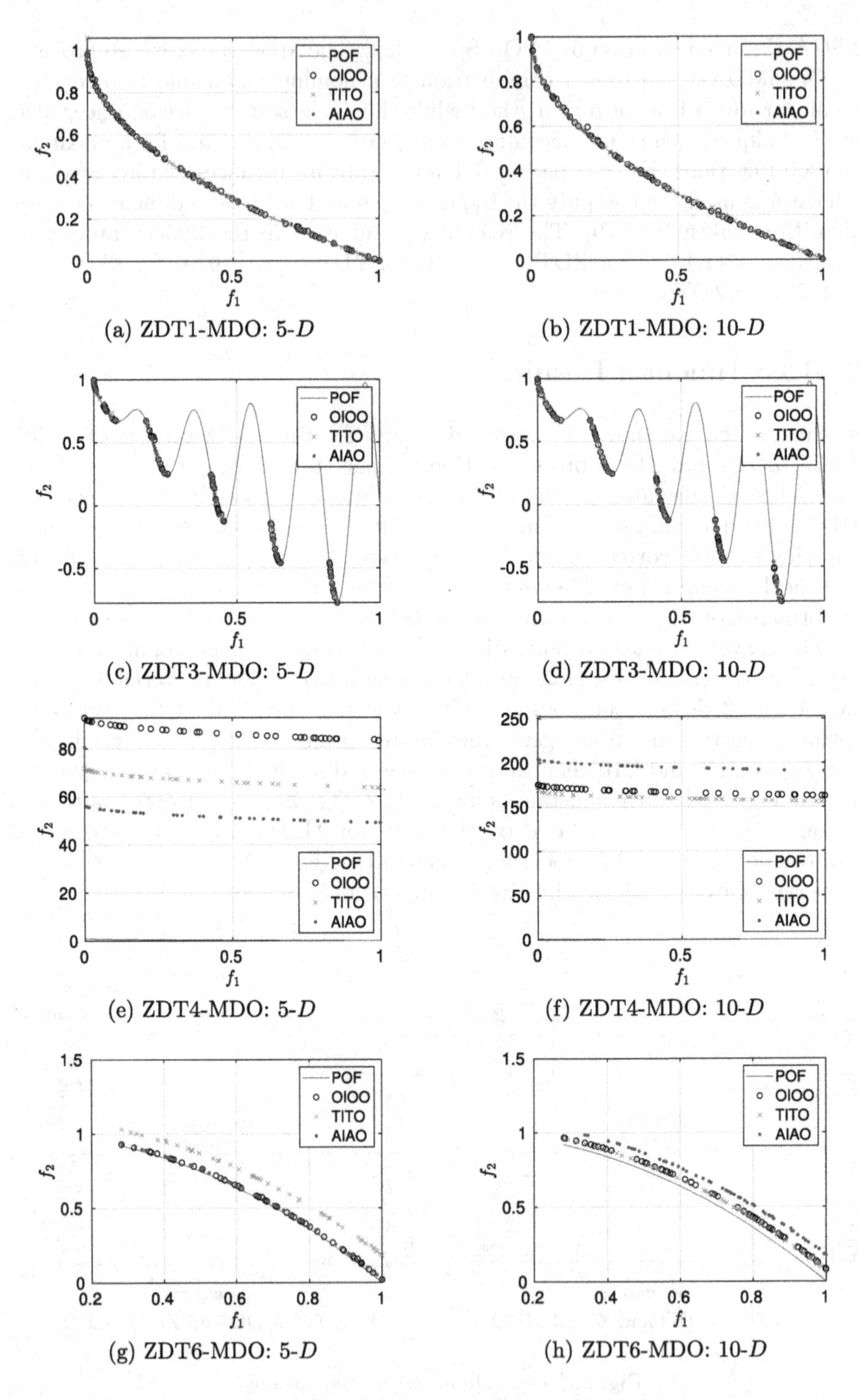

Fig. 3. Non-dominated solutions at the end of the optimization run.

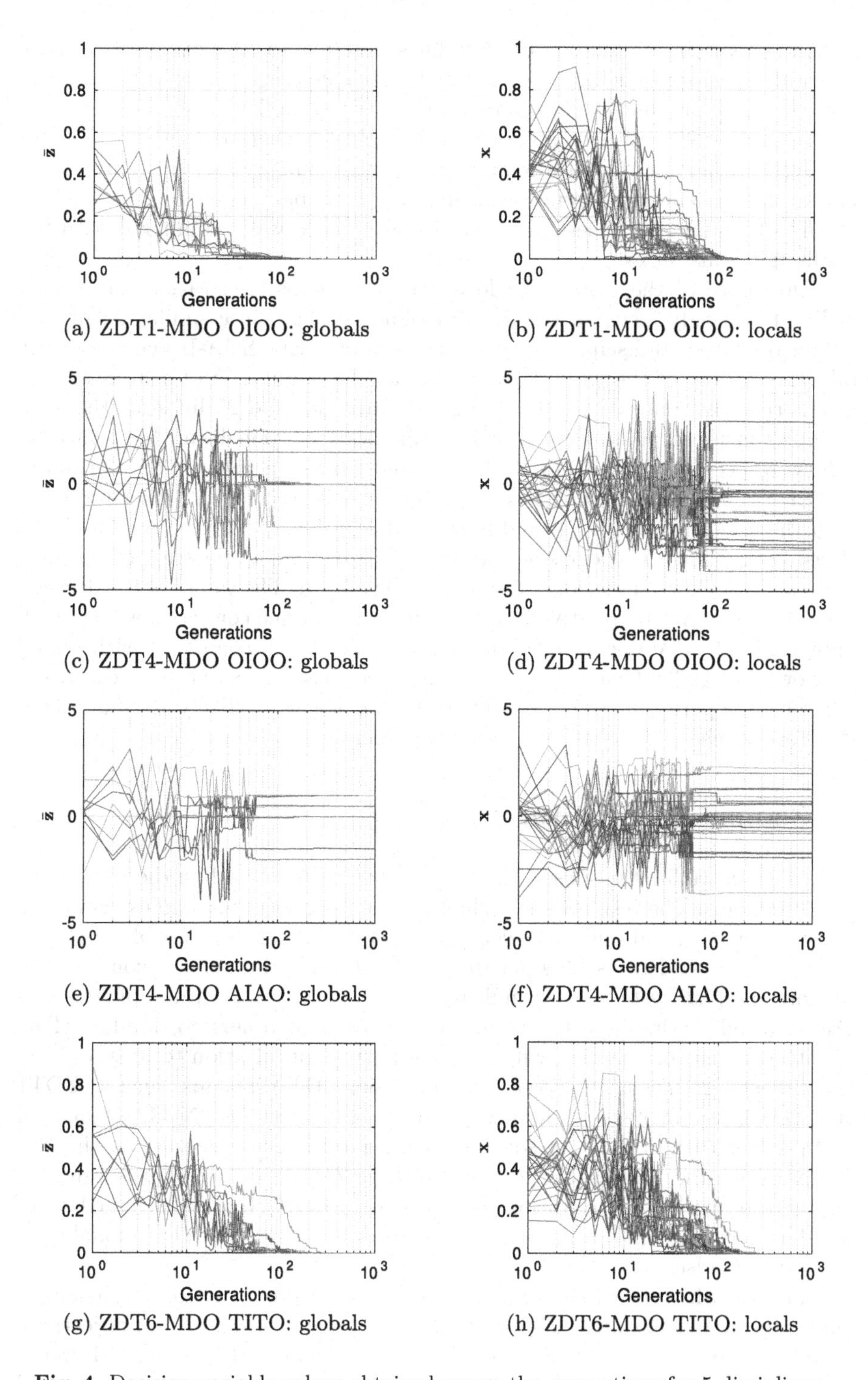

(a) ZDT1-MDO OIOO: globals

(b) ZDT1-MDO OIOO: locals

(c) ZDT4-MDO OIOO: globals

(d) ZDT4-MDO OIOO: locals

(e) ZDT4-MDO AIAO: globals

(f) ZDT4-MDO AIAO: locals

(g) ZDT6-MDO TITO: globals

(h) ZDT6-MDO TITO: locals

Fig. 4. Decision variable values obtained across the generations for 5 disciplines.

An increase in the number of disciplines from five to 10 is expected to make the problem more difficult, since it implies an increase in the number of decision variables (30 and 60 decision variables for five and 10 disciplines, respectively). This difficulty is mostly reflected on ZDT4 where the values of f_2 are relatively higher for the 10 discipline case when compared with the five discipline problem. The same trend is captured by the hypervolume for ZDT4, where the five-discipline instances show better convergence when compared with the 10-discipline instances.

The values of the global and local variables across generations are shown in Fig. 4. We only focus on the five discipline problem since similar results are obtained for the 10 discipline case. At the end of each NSGA-II generation, we take the median of the variable values across the population of solutions. This means that there are 9 lines for the global variables and 25 lines for the local variables in these plots. Figures 4a and 4b show the global and local variable values, respectively, for ZDT1. Both variables converge towards the optima in less than 200 generations. The same pattern is observed for the other problems with the exception of ZDT4, and it took slightly longer to converge for ZDT6 (Figs. 4g and 4h). The decision variables for ZDT4 become trapped in local optima after a few generations as shown in Figs. 4c–4f. The values of the decision variables for AIAO are relatively close to the optima when compared with OIOO, implying that AIAO achieves better performance when compared with OIOO as shown in Fig. 3e. Given that the MDA solver has converged in all cases, the differences in performance observed between topologies are likely attributable to the stochasticity of the optimizer at the system level.

7 Summary and Future Work

In this paper we have proposed an MO-MDO test suite based on the continuous ZDT problems. The test suite is scalable in the number of disciplines, as well as the number of global and local decision variables. It offers a flexible approach to defining dependencies between the disciplines, allowing for the construction of more complex systems with multiple dependencies between disciplines. This test suite offers the opportunity for researchers and others to develop MDO architectures in combination with multi-objective optimization techniques. The experimental results have shown that for easier ZDT problems, such as ZDT1 and ZDT3, it can be straightforward for an optimizer like NSGA-II and an MDA solver to find a set of solutions with good convergence across the PF. For problems that are harder to solve, such as ZDT4, it may require using an impractical number of generations (beyond 1000) to find a well-converged set of solutions, or a system optimizer more capable of dealing with multimodality in the fitness landscape.

Future work will include an expansion of MDO problems to more complex multi-objective test suites. This will allow for greater scalability in objectives, as well as being potentially more representative of real-world problems. Extending some of these problems to MDO formulations is not straightforward and will require revisions to the present architecture. Additionally, alternative MDO architectures will be considered for application to MO-MDO benchmarks.

References

1. Bäckryd, R.D., Ryberg, A.B., Nilsson, L.: Multidisciplinary design optimisation methods for automotive structures. Int. J. Autom. Mech. Eng. **14**(1), 4050–4067 (2017). https://doi.org/10.15282/ijame.14.1.2017.17.0327
2. Chen, H., Li, W., Cui, W., Yang, P., Chen, L.: Multi-objective multidisciplinary design optimization of a robotic fish system. J. Marine Sci. Eng. (5) (2021). https://doi.org/10.3390/jmse9050478
3. Mas Colomer, J., Bartoli, N., Lefebvre, T., Martins, J.R.R.A., Morlier, J.: An MDO-based methodology for static aeroelastic scaling of wings under non-similar flow. Struct. Multidiscip. Optim. **63**(3), 1045–1061 (2021). https://doi.org/10.1007/s00158-020-02804-z
4. Cramer, E.J., Dennis, J.E., Jr., Frank, P.D., Lewis, R.M., Shubin, G.R.: Problem formulation for multidisciplinary optimization. SIAM J. Optim. **4**(4), 754–776 (1994). https://doi.org/10.1137/0804044
5. Deb, K., Pratap, A., Agarwal, S., Meyarivan, T.: A fast and elitist multiobjective genetic algorithm: NSGA-II. IEEE Trans. Evol. Comput. **6**(2), 182–197 (2002). https://doi.org/10.1109/4235.996017
6. Deb, K., Thiele, L., Laumanns, M., Zitzler, E.: Scalable test problems for evolutionary multiobjective optimization. In: Abraham, A., Jain, L., Goldberg, R. (eds.) Evolutionary Multiobjective Optimization. Advanced Information and Knowledge Processing, pp. 105–145. Springer, London (2005). https://doi.org/10.1007/1-84628-137-7_6
7. Fan, Z., et al.: Difficulty adjustable and scalable constrained multiobjective test problem toolkit. Evol. Comput. **28**(3), 339–378 (2020). https://doi.org/10.1162/evco_a_00259
8. Farnsworth, M., Tiwari, A., Zhu, M., Benkhelifa, E.: A multi-objective and multidisciplinary optimisation algorithm for microelectromechanical systems. In: Maldonado, Y., Trujillo, L., Schütze, O., Riccardi, A., Vasile, M. (eds.) NEO 2016. SCI, vol. 731, pp. 205–238. Springer, Cham (2018). https://doi.org/10.1007/978-3-319-64063-1_9
9. Fonseca, C.M., Paquete, L., López-Ibáñez, M.: An improved dimension-sweep algorithm for the hypervolume indicator. In: International Conference on Evolutionary Computation, pp. 1157–1163. IEEE, Vancouver (2006)
10. Gray, J.S., Hwang, J.T., Martins, J.R.R.A., Moore, K.T., Naylor, B.A.: OpenMDAO: an open-source framework for multidisciplinary design, analysis, and optimization. Struct. Multidiscip. Optim. **59**(4), 1075–1104 (2019). https://doi.org/10.1007/s00158-019-02211-z
11. Gunawan, S., Farhang-Mehr, A., Azarm, S.: On maximizing solution diversity in a multiobjective multidisciplinary genetic algorithm for design optimization. Mech. Based Des. Struct. Mach. **32**(4), 491–514 (2004). https://doi.org/10.1081/SME-200034164
12. Huband, S., Hingston, P., Barone, L., While, L.: A review of multiobjective test problems and a scalable test problem toolkit. IEEE Trans. Evol. Comput. **10**(5), 477–506 (2006). https://doi.org/10.1109/TEVC.2005.861417

13. Johnson, V., Duro, J.A., Kadirkamanathan, V., Purhouse, R.C.: Toward scalable benchmark problems for multi-objective multidisciplinary optimization. In: Proceedings of the 2022 IEEE Symposium Series on Computational Intelligence (IEEE SSCI) (2022)
14. Klamroth, K., et al.: Multiobjective optimization for interwoven systems. J. Multi-Criteria Decis. Anal. **24**(1–2), 71–81 (2017). https://doi.org/10.1002/mcda.1598
15. Kurapati, A., Azarm, S.: Immune network simulation with multiobjective genetic algorithms for multidisciplinary design optimization. Eng. Optim. **33**(2), 245–260 (2000). https://doi.org/10.1080/03052150008940919
16. Ma, Z., Wang, Y.: Evolutionary constrained multiobjective optimization: test suite construction and performance comparisons. IEEE Trans. Evol. Comput. **23**(6), 972–986 (2019). https://doi.org/10.1109/TEVC.2019.2896967
17. Martins, J.R.R.A., Lambe, A.B.: Multidisciplinary design optimization: a survey of architectures. AIAA J. **51**(9), 2049–2075 (2013). https://doi.org/10.2514/1.J051895
18. Meier, P., et al.: The SIPHER consortium: introducing the new UK hub for systems science in public health and health economic research. Wellcome Open Res. **4**(174), 174 (2019). https://doi.org/10.12688/wellcomeopenres.15534.1
19. Padula, S., Alexandrov, N., Green, L.: MDO test suite at NASA langley research center. In: 6th Symposium on Multidisciplinary Analysis and Optimization, pp. 410–420 (1996). https://doi.org/10.2514/6.1996-4028
20. Sellar, R., Batill, S., Renaud, J.: Response surface based, concurrent subspace optimization for multidisciplinary system design. In: 34th Aerospace Sciences Meeting and Exhibit (1996). https://doi.org/10.2514/6.1996-714
21. Tedford, N., Martins, J.: Benchmarking multidisciplinary design optimization algorithms. Optim. Eng. **11**, 159–183 (2010). https://doi.org/10.1007/s11081-009-9082-6
22. Wu, N., Kenway, G., Mader, C.A., Jasa, J., Martins, J.R.R.A.: pyOptSparse: a python framework for large-scale constrained nonlinear optimization of sparse systems. J. Open Sour. Softw. **5**(54), 2564 (2020). https://doi.org/10.21105/joss.02564
23. Yang, D., Turrin, M., Sariyildiz, S., Sun, Y.: Sports building envelope optimization using multi-objective multidisciplinary design optimization (M-MDO) techniques: case of indoor sports building project in China. In: 2015 IEEE Congress on Evolutionary Computation (CEC), pp. 2269–2278 (2015). https://doi.org/10.1109/CEC.2015.7257165
24. Zitzler, E., Deb, K., Thiele, L.: Comparison of multiobjective evolutionary algorithms: empirical results. Evol. Comput. **8**(2), 173–195 (2000). https://doi.org/10.1162/106365600568202

Performance Evaluation of Multi-objective Evolutionary Algorithms Using Artificial and Real-world Problems

Hisao Ishibuchi(✉), Yang Nan, and Lie Meng Pang

Guangdong Provincial Key Laboratory of Brain-Inspired Intelligent Computation, Department of Computer Science and Engineering, Southern University of Science and Technology, Shenzhen 518055, China
{hisao,panglm}@sustech.edu.cn, nany@mail.sustech.edu.cn

Abstract. Performance of evolutionary multi-objective optimization (EMO) algorithms is usually evaluated using artificial test problems such as DTLZ and WFG. Every year, new EMO algorithms with high performance on those test problems are proposed. One question is whether they also work well on real-world problems. In this paper, we try to find an answer to this question by examining the performance of ten EMO algorithms including both well-known representative algorithms and recently-proposed new algorithms. First, those algorithms are applied to five artificial test suites (DTLZ, WFG, Minus-DTLZ, Minus-WFG and MaF) and three real-world problem suites. The performance of each algorithm is evaluated by the hypervolume indicator. Next, the ranking of the ten EMO algorithms is created for each problem suite. That is, eight different rankings are obtained (each ranking is for each problem suite). Then, the eight different rankings are visually compared to answer our research question. The distance between two rankings is also calculated to support visual comparison results. Our experimental results show that similar rankings of the ten EMO algorithms are obtained for the three real-world problem suites and Minus-WFG. It is also shown that the ranking for each of the three real-world problem suites is clearly different from their ranking for DTLZ.

Keywords: Evolutionary multi-objective optimization · Performance evaluation · Artificial test problems · Real-world test problems · Many-objective optimization

1 Introduction

An m-objective minimization problem is written as

$$\text{Minimize } \boldsymbol{f}(\boldsymbol{x}) = (f_1(\boldsymbol{x}), f_2(\boldsymbol{x}), \ldots, f_m(\boldsymbol{x})), \tag{1}$$

$$\text{subject to } \boldsymbol{x} \in \boldsymbol{X}, \tag{2}$$

M. Emmerich et al. (Eds.): EMO 2023, LNCS 13970, pp. 333–347, 2023.
https://doi.org/10.1007/978-3-031-27250-9_24

where $f_i(\boldsymbol{x})$ is the ith objective to be minimized (i = 1, 2,..., m), $\boldsymbol{x}$ is a decision vector, and $\boldsymbol{X}$ is the feasible region of $\boldsymbol{x}$. Two solutions $\boldsymbol{x}^{A}$ and $\boldsymbol{x}^{B}$ are compared by the Pareto dominance relation as follows: When $f_i(\boldsymbol{x}^{A}) \leq f_i(\boldsymbol{x}^{B})$ for all i and $f_j(\boldsymbol{x}^{A}) < f_j(\boldsymbol{x}^{B})$ for at least one j, $\boldsymbol{x}^{A}$ dominates $\boldsymbol{x}^{B}$ (i.e., $\boldsymbol{x}^{A}$ is better than $\boldsymbol{x}^{B}$). If $\boldsymbol{x}^{B}$ is not dominated by any other solutions in $\boldsymbol{X}$, $\boldsymbol{x}^{B}$ is a Pareto optimal solution. The set of all Pareto optimal solutions is the Pareto optimal solution set. The image of the Pareto optimal solution set on the objective space is the Pareto front. If all solutions in a solution set are non-dominated with each other, the set is referred to as a non-dominated solution set. Various evolutionary multi-objective optimization (EMO) algorithms have been proposed to search for a non-dominated solution set which approximates the entire Pareto front.

When a new EMO algorithm is proposed, its high performance is usually demonstrated through computational experiments in comparison with other algorithms using well-known scalable artificial test problems such as DTLZ [1] and WFG [2]. Every year, a number of new EMO algorithms are proposed in this manner. Recently, it is pointed out in some studies [3–5] that test problems in the DTLZ and WFG suites have somewhat unrealistic features. One is the use of a single common distance function in all objectives. As a result, we can find a Pareto optimal solution by minimizing the single distance function independent of the number of objectives. Another feature is that any $(m-1)$ objectives of an m- objective test problem can be simultaneously optimized. However, these two artificial test suites are still frequently used for evaluating EMO algorithms in many recent papers (e.g., [6–17]).

One question is whether high-performance EMO algorithms on artificial test problems also work well on real-world problems or not. In this paper, we try to find an answer to this question. The performance of ten EMO algorithms is examined: NSGA-II [18], MOEA/D-PBI [19], SMS-EMOA [20] (HypE [21] for problems with more than five objectives), NSGA-III [22], MOEA/DD [23], RVEA [24], SparseEA [25], DEA-GNG [26], R2HCA-EMOA [27], and PREA [28]. They are classified into three categories in Table 1: Pareto dominance-based, decomposition-based and indicator-based algorithms. In Table 1, we also show the test problems used in computational experiments for performance evaluation in the paper where each algorithm was proposed. Constrained test problems which were used for performance evaluation of constrained versions of these algorithms are not shown in Table 1 for the sake of conciseness (since multi-objective problems with only box constraints are used in this paper). In Table 1, each algorithm is numbered based on the publication year of the related paper. The ID number of each algorithm is used in all figures in this paper for concise presentation. Except for SparseEA (which is a large-scale sparse multi-objective optimization algorithm proposed in 2020), EMO algorithms in Table 1 proposed in the 2010s and 2020s were evaluated by many-objective test problems.

The ranking of these ten EMO algorithms is created for each of five artificial test suites (DTLZ1–7 [1], WFG1–9 [2], Minus-DTLZ1–7 [3], Minus-WFG1–9 [3], and MaF1–15 [4]) through computational experiments on problem instances with 3, 5 and 8 objectives. Thus, five different rankings are created. The ranking of the ten EMO algorithms is also created for each of three real-world problem suites: Tanabe & Ishibuchi [29], He et al. [30], and Kumar et al. [31]. As a result, eight different rankings are created in total. Then, they are compared with each other in order to analyze the difference between artificial test problem-based and real-world problem-based comparison results.

Table 1. Ten algorithms examined in this paper and some related information. Each algorithm is numbered based on the publication year of the related paper. The ID number of each algorithm is used in all figures in this paper.

ID	Algorithm	Year	Algorithm category	Test problems used in algorithm's original paper	Number of objectives
1	NSGA-II [18]	2002	Pareto dominance-based algorithm	SCH, FON, POL, KUR ZDT1–4, 6	2 2
2	MOEA/D-PBI [19]	2007	Decomposition-based algorithm	ZDT1–4, 6 DTLZ1–2 Knapsack	2 3 2, 3, 4
3a	SMS-EMOA [20]	2007	Indicator-based algorithm	ZDT1–4, 6 DTLZ1–4, Airfoil design	2 3
3b	HypE [21]	2011	Indicator-based algorithm	DTLZ1–7, WFG1–9, Knapsack	2, 3, 5, 7, 10, 25, 50
4	NSGA-III [22]	2014	Decomposition-based algorithm	DTLZ1–4, Scaled DTLZ1–2 Convex DTLZ2 Crash-worthiness problem Car cab design	3, 5, 8, 10, 15 3, 5, 8, 10, 15 3 9
5	MOEA/DD [23]	2015	Decomposition-based algorithm	DTLZ1–4, WFG1–9	3, 5, 8, 10, 15
6	RVEA [24]	2016	Decomposition-based algorithm	DTLZ1–7, WFG1–9 Scaled DTLZ1, 3	3, 6, 8, 10 3, 6, 8, 10
7	SparseEA [25]	2020	Pareto dominance-based algorithm	Sparse MOPs (MOP1–8) ML application problems	2 2
8	DEA-GNG [26]	2020	Decomposition-based algorithm	Scaled DTLZ2, Convex DTLZ2 Minus-DTLZ2, DTLZ2BZ DTLZ7, WFG1–2 DTLZ5, Polygon problems	2, 3, 5, 8 2, 3, 5, 8 2, 3, 5, 8 3, 5, 8
9	R2HCA-EMOA [27]	2020	Indicator-based algorithm	DTLZ1–4, Minus-DTLZ1–4 WFG1–9, Minus-WFG1–9	5, 10, 15 5, 10, 15
10	PREA [28]	2021	Indicator-based algorithm	DTLZ1–4, WFG4–9 WFG1–3, MaF1, 3, 4, 6, 7	3, 5, 10 3, 5, 10, 15, 20

This paper is organized as follows. In Sect. 2, we briefly explain five artificial test suites and three real-world problem suites used in this paper. In Sect. 3, the ranking of the ten EMO algorithms for each problem suite is shown and compared. Experimental results demonstrate that the obtained ranking of the ten EMO algorithms is strongly problem suite dependent. In Sect. 4, this paper is concluded and some future research topics are explained.

2 Artificial Test Problems and Real-World Problems

In the 1990s, simple two-objective artificial test problems were often utilized for explaining the usefulness of non-elitist EMO algorithms (e.g., Schaffer's test problem in the NPGA [32] and NSGA [33] papers, and F3 in [33]). As shown in [34], randomly generated initial solutions for those test problems are on the Pareto front or very close to the Pareto front. Thus, no strong convergence ability to push the population towards the Pareto front is needed in EMO algorithms for those test problems.

In 2000, the first well-known artificial test suite ZDT [35] with six two-objective problems (ZDT1–6) was proposed. One challenge in solving ZDT is that no randomly generated initial solutions are close to the Pareto front [34]. That is, the initial population is not close to the Pareto front. Thus, strong convergence ability is needed to push the population towards the Pareto front. Slow convergence of the population by non-elitist EMO algorithms was clearly demonstrated in the ZDT paper [35]. As shown in Table 1, ZDT was often used for performance evaluation of elitist EMO algorithms such as NSGA-II [18], MOEA/D [19] and SMS-EMOA [20] in the 2000s.

In 2002, the first well-known scalable artificial test suite DTLZ [1] with seven problems (DTLZ1–7) was proposed. One advantage of DTLZ over ZDT is that the number of objectives can be arbitrarily specified. Thus, DTLZ has been frequently used for performance evaluation of EMO algorithms for many-objective optimization (see Table 1). One feature of DTLZ is that the feasible region in the objective space spreads out from the Pareto front towards all non-negative directions as shown in Fig. 1 (a). As a result, the diversity maximization of the population totally conflicts with its convergence towards the Pareto front. Moreover, when a Pareto dominance-based EMO algorithm is applied to DTLZ, the current population often includes dominance resistant solutions (DRSs [36, 37]) which have near optimal values for some objectives and very bad values for other objectives. Thanks to some near optimal values, DRSs are hardly dominated by other solutions in the population. Due to some very bad values, they are evaluated as having large diversity. As a result, they have high fitness in Pareto dominance-based EMO algorithms (i.e., they can survive over many generations and create other DRSs). DTLZ was often used to demonstrate poor performance of dominance-based EMO algorithms (and good performance of other EMO algorithms) on many-objective test problems with four or more objectives [38, 39].

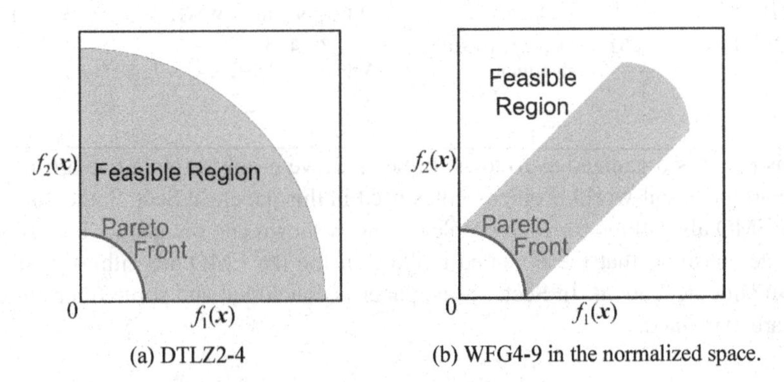

(a) DTLZ2-4 (b) WFG4-9 in the normalized space.

Fig. 1. Feasible region of each test problem in the objective space.

Another feature of DTLZ is that the Pareto front shape of DTLZ1–4 is triangular. Decomposition-based EMO algorithms work well on these test problems since the weight vector distribution is also triangular (i.e., since the weight vector distribution is consistent with the Pareto front shape [3]). In this paper, we use DTLZ1–7 with 3, 5 and 8 objectives (i.e., 21 problem instances in total). Since its proposal in 2002, some modifications have been made on DTLZ (i.e., there exist different versions of DTLZ). In this paper, we use DTLZ1–7 in PlatEMO [40].

In 2006, another well-known scalable artificial test suite WFG [2] with nine problems (WFG1–9) was proposed. One additional challenge in solving WFG in comparison with DTLZ is that each objective has a different scale. More specifically, the ideal and nadir points of the m-objective WFG problems are (0, 0,..., 0) and (2, 4,..., $2m$), respectively (whereas they are (0, 0,..., 0) and (1, 1,..., 1), respectively, in most DTLZ problems). This means that a uniformly distributed weight vector set cannot find a uniformly distributed solution set on the Pareto front. As a result, decomposition-based EMO algorithms with no normalization mechanism of the objective space such as MOEA/D [19] and MOEA/DD [23] do not work well. The feasible region of WFG spreads out along the center vector (1, 1,..., 1) in the normalized objective space as shown in Fig. 1 (b). Thus, there exist no dominance resistant solutions (DRSs) in WFG since very bad values of some objectives mean very bad values of the other objectives in Fig. 1 (b). As a result, many-objective WFG is easier for Pareto dominance-based EMO algorithms than many-objective DTLZ [41]. In this paper, we use WFG1–9 with 3, 5 and 8 objectives (i.e., 27 problem instances in total).

In 2017, the minus versions of DTLZ and WFG were formulated by simply assigning a minus sign to all objectives in DTLZ and WFG [3]. The assignment of a minus sign is equivalent to the change from "minimization" to "maximization". Thus, the Pareto front is the opposite side of the feasible region in comparison with DTLZ and WFG (i.e., the top-right boundary of the feasible region in each figure in Fig. 1). This simple change has large effects on the problem features of DTLZ and WFG. For example, whereas DTLZ2–4 and WFG4–9 have concave triangular Pareto fronts, their minus versions have convex inverted triangular Pareto fronts. Since inverted triangular Pareto fronts are not consistent with the weight vector distribution [3], Minus-DTLZ and Minus-WFG are difficult for decomposition-based EMO algorithms with the fixed weight vectors such as MOEA/D [19], NSGA-III [22] and MOEA/DD [23]. They are also difficult for hypervolume-based EMO algorithms (e.g., SMS-EMOA [20] and HypE [21]) since the reference point specification for hypervolume calculation has a large effect on their performance when test problems have inverted triangular Pareto fronts [42]. For Pareto dominance-based EMO algorithms, Minus-DTLZ is easier than DTLZ since (i) the population can be driven towards the Pareto front by diversity maximization and (ii) Minus-DTLZ has no DRSs [41]. In this paper, we use Minus-DTLZ1–7 with 3, 5 and 8 objectives (21 problem instances in total) and Minus-WFG1–9 with 3, 5 and 8 objectives (27 problem instances in total).

In 2017, MaF was proposed as a new many-objective test suite [4]. Its main feature is that all of its 15 test problems (i.e., MaF1–15) are collected from the literature instead of designing new test problems. For example, MaF3 is the convex DTLZ3 problem in [22], and MaF10, 11, 12 are the same as WFG1, 2, 9, respectively. MaF includes test problems with various Pareto front shapes (e.g., concave, convex, disconnected, triangular, and inverted triangular Pareto fronts). In this paper, we use MaF1–15 with 3, 5, 8 objectives (45 problem instances in total).

Recently, some real-world problem suites have been proposed in the literature. In this paper, three suites in Table 2 are used: RE problems in Tanabe & Ishibuchi [29], DDMOP problems in He et al. [30], and RCM problems in Kumar et al. [31]. Tanabe & Ishibuchi [29] has 16 problems (RE2-4-1, RE2-3-2,..., RE9-7-1). He et al. [30] has seven problems (DDMOP1,..., DDMOP7). Among them, DDMOP2 is the same as RE3-5-4. Kumar et al. [31] has 50 problems (RCM01,..., RCM50) which are classified into five categories: (i) mechanical design problems, (ii) chemical engineering problems, (iii) process, design and synthesis problems, (iv) power electronics problems, and (v) power system optimization problems. Among them, six problems are also included in the RE suite [29]. Most problems in the original RCM suite [31] are constrained multi-objective problems. In the RE suite [29], all constraint conditions of each problem were combined into an additional objective to minimize the total constraint violation. In this paper, we apply the same reformulation method to the original RCM problems. As a result, except for some problems with no constraint conditions (e.g., RCM09), the number of objectives in RCM in Table 2 is larger than that in the original RCM suite [31]. The MATLAB code of all reformulated problems is available from https://github.com/Nany12345/EMO-2023-Performance-Comparison.

We can see that many problems in Table 2 have only two or three objectives. Only five problems have six or more objectives. These observations suggest that Pareto dominance-based algorithms (e.g., NSGA-II [18]) will work well on most problems. We can also see that no problems in Table 2 have more than 34 decision variables. This observation suggests that no special mechanisms to handle large-scale problems (e.g., SparseEA [25]) are needed. As shown in Tanabe & Ishibuchi [29] and He et al. [30], many real-world problems have complicated Pareto front shapes. This suggests that decomposition-based EMO algorithms with the fixed weight vectors (e.g., MOEA/D-PBI [19], NSGA-III [22], MOEA/DD [23]) will have some difficulties. It was also shown in [29] and [30] that each objective in many real-world problems has a totally different scale. For example, the range of each objective in the approximated Pareto front of RE3-4-2 [29] is f_1: [0.01, 35], f_2: [0.004, 20,000], and f_3: [0.02, 1,000,000,000]. This observation suggests the necessity of objective space normalization. Among the examined algorithms, the following five do not have any objective space normalization mechanism: MOEA/D-PBI [19], SMS-EMOA [20], HypE [21], MOEA/DD [23], and RVEA [24]. These algorithms will have some difficulties in handling badly scaled real-world problems. Objective space normalization will improve the performance of decomposition-based EMO algorithms (i.e., MOEA/D-PBI [19], MOEA/DD [23] and RVEA [24]) since it helps to find uniformly distributed solutions using uniformly distributed weight vectors [43]. In hypervolume-based EMO algorithms (i.e., SMS-EMOA [20] and HypE [21]), objective space normalization is needed to appropriately specify the reference point for hypervolume calculation [44]. For example, in SMS-EMOA, the reference point is specified by adding (1, 1,..., 1) to the estimated nadir point. In this specification, the effect of "+1" on hypervolume calculation can be totally different depending on the scale of each objective (e.g., the effect of "+1" is totally different between "1 + 1" and "1,000,000 + 1").

Table 2. Three real-world problem suites used in this paper. In this table, m is the number of objectives, and n is the number of decision variables.

RE [29]			DDMOP [30]			RCM [31]					
Problem	m	n	Problem	m	n	Problem	m	n	Problem	m	n
RE2-4-1	2	4	DDMOP1	9	11	RCM1	3	4	RCM26	3	3
RE2-3-2	2	3	DDMOP2	3	5	RCM2	3	5	RCM27	3	3
RE2-4-3	2	4	DDMOP3	3	6	RCM3	3	3	RCM28	3	7
RE2-2-4	2	2	DDMOP4	10	13	RCM4	3	4	RCM29	3	7
RE2-3-5	2	3	DDMOP5	3	11	RCM5	3	4	RCM30	3	25
RE3-3-1	3	3	DDMOP6	2	10	RCM6	3	7	RCM31	3	25
RE3-4-2	3	4	DDMOP7	2	17	RCM7	3	4	RCM32	3	25
RE3-4-3	3	4				RCM8	4	7	RCM33	3	30
RE3-5-4	3	5				RCM9	2	4	RCM34	3	30
RE3-7-5	3	7				RCM10	3	2	RCM35	3	30
RE3-4-6	3	4				RCM11	6	3	RCM36	3	28
RE3-4-7	3	4				RCM12	3	4	RCM37	3	28
RE4-7-1	4	7				RCM13	4	7	RCM38	3	28
RE4-6-2	4	6				RCM14	3	5	RCM39	4	28
RE6-3-1	6	3				RCM15	3	3	RCM40	3	34
RE9-7-1	9	7				RCM16	3	2	RCM41	4	34
						RCM17	4	6	RCM42	3	34
						RCM18	3	3	RCM43	3	34
						RCM19	4	10	RCM44	4	34
						RCM20	3	4	RCM45	4	34
						RCM21	3	6	RCM46	5	34
						RCM22	3	9	RCM47	3	18
						RCM23	3	6	RCM48	3	18
						RCM24	4	9	RCM49	4	18
						RCM25	3	2	RCM50	3	6

3 Results of Computational Experiments

In our computational experiments, the ten EMO algorithms in Table 1 are compared using the eight problem suites explained in Sect. 2. We use the standard specification with respect to the number of decision variables in each artificial test problem. More specifically, we use the same specification as in PlatEMO [40] for DTLZ, WFG and MaF. The specifications for Minus-DTLZ and Minus-WFG are the same as those for DTLZ and WFG, respectively. All algorithms are applied to all problems under the same termination condition: 200 generations (in the case of steady state algorithms, the number of generations is adjusted so that the total number of examined solutions is the same). Different population size is used depending on the number of objectives as shown in Table 3. This is because the population size cannot be arbitrarily specified in most decomposition-based EMO algorithms.

Table 3. Population size specifications.

Number of objectives (m)	2	3	4	5	6	8	9	10
Population size	100	91	120	210	112	156	174	275

In all EMO algorithms except for SparseEA [25], the same crossover and mutation operators are used with the same crossover and mutation probabilities as follows: The simulated binary crossover (SBX) with the distribution index 20 and the crossover probability 1.0, and the polynomial mutation with the distribution index 20 and the mutation probability $1/n$ where n is the number of decision variables. SparseEA uses its own method to generate offspring. All other parameters in each algorithm are the same as in the related paper and the authors' code. For example, the penalty parameter θ in MOEA/D-PBI is specified as $\theta = 5$ as in the MOEA/D paper [19]. The number of samples in HypE for approximating hypervolume contribution is 10,000, and the number of direction vectors in R2HCA-EMOA [27] is 100.

Each algorithm is applied to each problem instance (e.g., 3-objective DTLZ1) 31 times. Then the average hypervolume value is calculated over 31 runs. The reference point is specified as $(1 + 1/H)(1, 1,..., 1)$ in the normalized objective space with the ideal point (0, 0,..., 0) and the nadir point (1, 1,..., 1) as suggested in [42] where $H = 99$, 12, 7, 6, 3, 3, 3, 3 for $m = 2, 3, 4, 5, 6, 8, 9, 10$, respectively. The true ideal and nadir points are unknown in the real-world problem suites (i.e., RE, DDMOP and RCM) and some Minus problems (i.e., Minus-DTLZ5–7 and Minus-WFG1–3). For those problems, the ideal and nadir points are estimated using all non-dominated solutions obtained from all runs of all algorithms in our computational experiments.

The ranking of the ten algorithms is created for each problem instance simply using the average hypervolume value over 31 runs of each algorithm. Such a ranking is created for all problem instances in each problem suite. For example, 27 different rankings are created for the WFG test suite (each ranking is for each WFG problem instance). Then the average ranking is calculated for each problem suite simply by calculating the average of the rankings over the problem instances. For example, the average ranking for WFG is calculated as the average over its 27 problem instances. In this manner, we have eight different average rankings in Fig. 2, each of which is for each problem suite. In each figure in Fig. 2, the best rank is 1, and the worst rank is 10. If one algorithm is always the best for all problem instances, its average rank is 1.0. If another algorithm is always the worst for all problem instances, its average rank is 10.0. When different algorithms work well on different problem instances in a problem suite, their average ranks tend to be around 5.5 as in the ranking for MaF in Fig. 2. From a quick visual comparison among the eight rankings, we can see that the three rankings for the real-world problems are similar to the ranking for Minus-WFG and clearly different from the ranking for DTLZ.

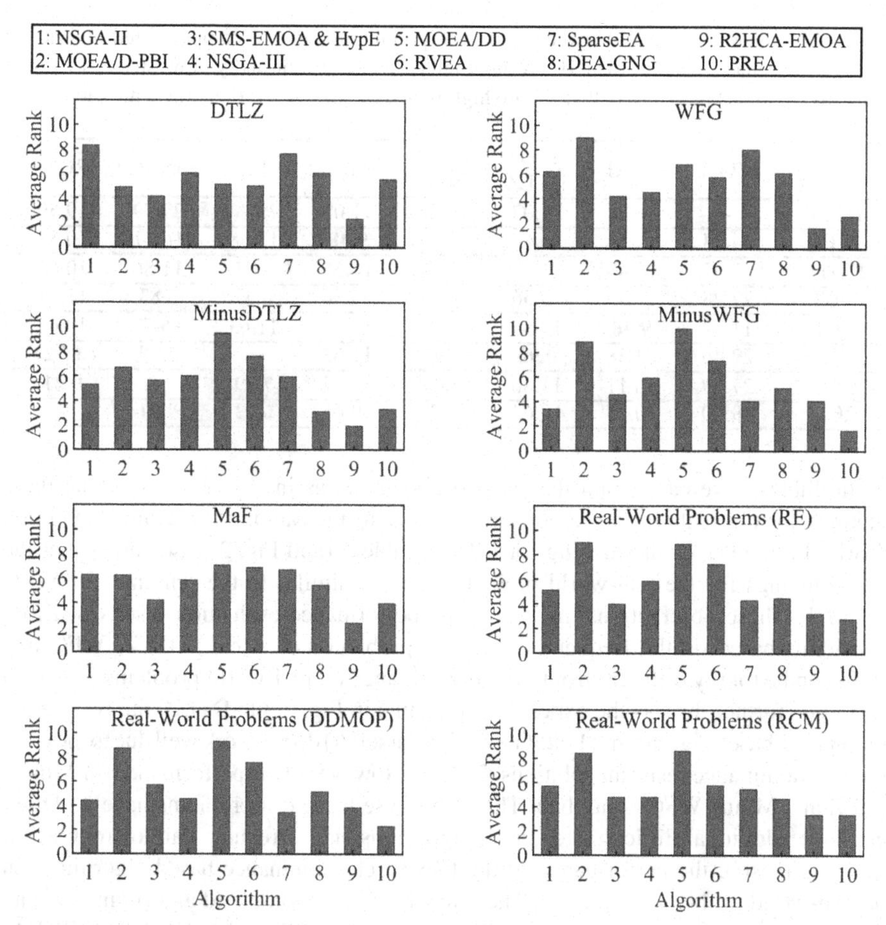

Fig. 2. Average ranking of the examined ten algorithms on each problem suite. Smaller ranks mean better algorithms. The final population of each run is used for comparison.

To confirm these visual observations, we calculate the distance between two average rankings using the Manhattan distance by handling each average ranking in Fig. 2 as a 10-dimensional vector. That is, the distance of two average rankings $\boldsymbol{R}_{\mathrm{A}} = (R_{\mathrm{A}1}, R_{\mathrm{A}2},..., R_{\mathrm{A}10})$ and $\boldsymbol{R}_{\mathrm{B}} = (R_{\mathrm{B}1}, R_{\mathrm{B}2},..., R_{\mathrm{B}10})$ is calculated as follows:

$$Distance(\boldsymbol{R}_{\mathrm{A}}, \boldsymbol{R}_{\mathrm{B}}) = |R_{\mathrm{A}1} - R_{\mathrm{B}1}| + |R_{\mathrm{A}2} - R_{\mathrm{B}2}| + \ldots + |R_{\mathrm{A}10} - R_{\mathrm{B}10}|. \quad (3)$$

The calculated distance is summarized in Table 4. Similar pairs of average rankings with the Manhattan distance smaller than 10 are highlighted by bold font and yellow shade (e.g., between WFG and MaF). Dissimilar pairs of average rankings with the Manhattan distance larger than 15 are highlighted by gray shade.

Table 4. Manhattan distance between two average rankings of the compared ten algorithms in Fig. 2. Distances smaller than 10 are highlighted by bold font and yellow shade (i.e., high similarity), and distances larger than 15 are highlighted by gray shade (i.e., low similarity).

	DTLZ	WFG	Minus-DTLZ	Minus-WFG	MaF	RE	DDMOP	RCM
DTLZ	-	14.34	21.07	27.28	11.02	22.46	27.79	19.50
WFG	14.34	-	16.48	17.83	**9.48**	13.38	16.11	10.15
M-DTLZ	21.07	16.48	-	13.56	10.55	**9.93**	11.86	10.63
M-WFG	27.28	17.83	13.56	-	16.36	**6.67**	**5.83**	11.31
MaF	11.02	**9.48**	10.55	16.36	-	11.63	16.77	10.65
RE	22.46	13.38	**9.93**	**6.67**	11.63	-	**7.39**	**6.72**
DDMOP	27.79	16.11	11.86	**5.83**	16.77	**7.39**	-	**9.21**
RCM	19.50	10.15	10.63	11.31	10.65	**6.72**	**9.21**	-

In Table 4, we can obtain the same observations as in Fig. 2. That is, the three rankings for the real-world problems are similar to the ranking for Minus-WFG and clearly different from the ranking for DTLZ. Table 4 (and Fig. 2) also shows that the three rankings for the real-world problems are not similar to the rankings for WFG and MaF. These observations imply that the performance evaluation results of EMO algorithms based on the frequently-used test problem suites (e.g., DTLZ, WFG and MaF) can be totally different from their performance on real-world problems.

Let us further discuss the experimental results in Fig. 2. On DTLZ, the two Pareto dominance-based algorithms (1: NSGA-II, 7: SparseEA) do not work well due to the existence of dominance resistant solutions (DRSs). However, their performance on Minus-DTLZ and Minus-WFG is not bad. This is because these test problems have no DRSs and have additional difficulty for other algorithms (i.e., irregular Pareto fronts). For the same reason, the performance of the two Pareto dominance-based algorithms on the real-world problems is not bad. The three decomposition-based algorithms with no objective space normalization mechanisms (2: MOEA/D-PBI, 5: MOEA/DD, 6: RVEA) do not work well on the real-world problems in Fig. 2 whereas their performance on DTLZ is not bad. This is because each objective in real-world problems has a totally different scale. For the same reason, the other decomposition-based algorithms with normalization mechanisms (4: NSGA-III, 8: DEA-GNG) show better performance on the real-world problems. The difference in the performance between these two algorithms is not clear whereas NSGA-III has a fixed reference vector set and DEA-GNG has a reference vector adaptation mechanism. This is due to the termination condition specification: 200 generations. The number of generations is not enough for carefully adjusting the reference vector in DEA-GNG (and also in RVEA).

In Fig. 2, the hypervolume-based EMO algorithms (3: SMS-EMOA & HypE, 9: R2HCA-EMOA) show the best results on DTLZ and WFG. However, they do not always show the best results on the real-world problems. For example, in Fig. 2, algorithm 3 (SMS-EMOA & HypE) is outperformed by some other algorithms (e.g., algorithm 1: NSGA-II) on DDMOP. This is because the reference point specification mechanisms in SMS-EMOA & HypE are not suitable for real-world problems with complicated Pareto fronts and totally different scales in objective spaces. By specifying the reference point in the normalized objective space, their performance can be improved [44]. PREA (algorithm 10) is based on a distance indicator (instead of the hypervolume indicator). It shows the best performance on the real-world problems (since the other algorithms have some difficulties) whereas its performance is not good on DTLZ. From the experimental results in Fig. 2, we can say that high-performance algorithms on the frequently-used artificial test suites DTLZ and WFG do not always work well on real-world problems. We can also see from Fig. 2 that a wide variety of test problems are needed for reliable performance comparison [45].

One possible reason for the large differences in the average rankings between the artificial test suites and the real-world problem suites is the difference in the number of objectives. Whereas almost all the examined real-world problems have only 2–4 objectives, the artificial test problems have 3, 5 and 8 objectives in Fig. 2 and Table 4. We recalculate the average ranking for each of the five artificial test suites using only the three-objective problem instances. The results are shown in Fig. 3 and Table 5.

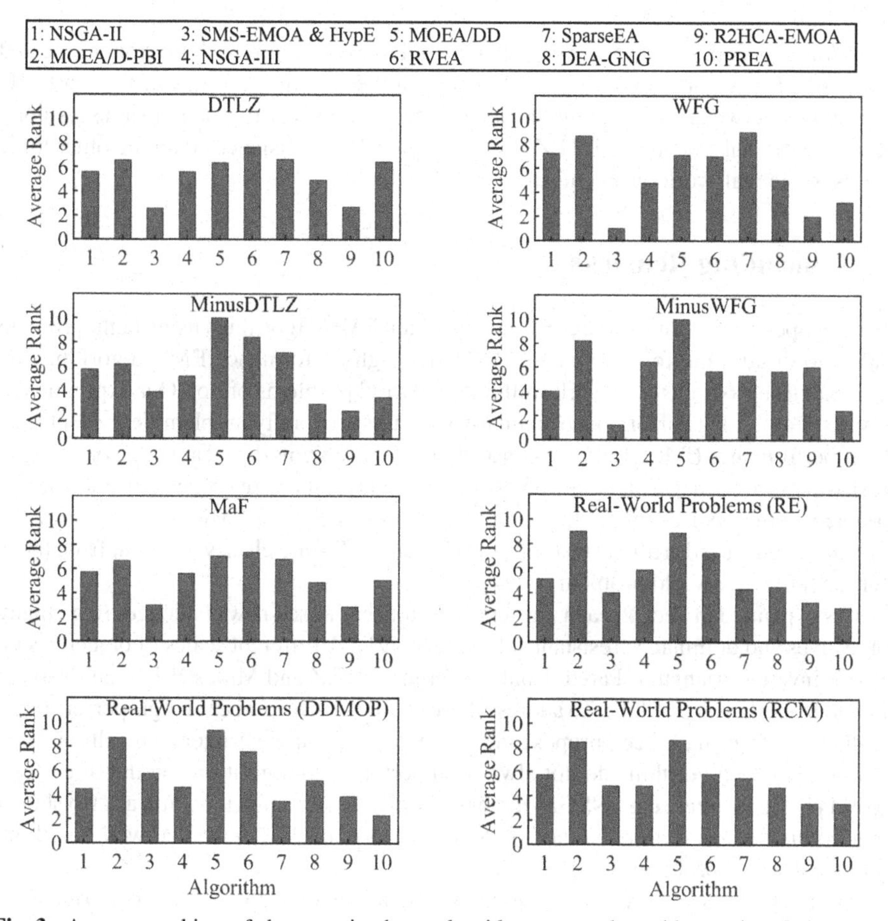

Fig. 3. Average ranking of the examined ten algorithms on each problem suite. Only three-objective artificial test problems are used for ranking whereas all real-world problems are used.

Table 5. Manhattan distance between two rankings of the compared ten algorithms in Fig. 3. Distances smaller than 10 are highlighted by bold font and yellow shade (i.e., high similarity), and distances larger than 15 are highlighted by gray shade (i.e., low similarity).

	DTLZ	WFG	Minus-DTLZ	Minus-WFG	MaF	RE	DDMOP	RCM
DTLZ	-	13.86	12.71	20.38	**3.40**	14.43	19.14	14.23
WFG	13.86	-	16.06	18.89	11.80	15.39	19.02	13.64
M-DTLZ	12.71	16.06	-	19.81	10.46	12.68	16.57	12.25
M-WFG	20.38	18.89	19.81	-	18.60	13.26	14.63	13.32
MaF	**3.40**	11.80	10.46	18.60	-	11.67	16.97	11.27
RE	14.43	15.39	12.68	13.26	11.67	-	**7.39**	**6.72**
DDMOP	19.14	19.02	16.57	14.63	16.97	**7.39**	-	**9.21**
RCM	14.23	13.64	12.25	13.32	11.27	**6.72**	**9.21**	-

From Fig. 3 and Table 5, we can obtain similar observations to those from Fig. 2 and Table 4. That is, there exist large differences in the average rankings of the ten EMO algorithms between the artificial test suites and the real-world problem suites. Among the five artificial test suites, DTLZ and WFG have larger distances than the other three suites to the real-world problem suites in Table 4.

4 Concluding Remarks

In this paper, we examined the performance of ten EMO algorithms using eight problem suites to discuss the following issue: Whether high-performance EMO algorithms on artificial test problems also work well on real-world problems or not. Our experimental results showed that performance comparison results are totally problem-dependent (i.e., high-performance EMO algorithms on artificial test problems do not always work well on real-world problems). Totally different comparison results were obtained from different problem suites. For example, the average rankings of the ten EMO algorithms for the two frequently-used artificial test suites DTLZ and WFG are clearly different from those for the three real-world problem suites.

As explained in Sect. 2, each artificial test suite has posed new challenges (e.g., many objectives and dominance resistant solutions by DTLZ, different scales of objectives by WFG, inverted triangular Pareto fronts by Minus-DTLZ and Minus-WFG, and various Pareto front shapes by MaF). To address these challenges, a number of high-performance EMO algorithms have been proposed in the literature. Our experimental results showed that new EMO algorithms do not always outperform old algorithms on the three real-world problem suites (e.g., NSGA-II worked well). At the same time, it was also shown that all the four algorithms in the 2020s (i.e., algorithms 7–10) worked well on all the three real-world problem suites.

In this paper, we used a very simple method to calculate the ranking of EMO algorithms for each problem instance, which was simply based on the average hypervolume values. More rigorous ranking mechanisms may be needed to evaluate their performance in future studies. Another future research direction is to use various settings in computational experiments for performance comparison of EMO algorithms. This direction includes the use of other termination conditions (e.g., 100, 500 and 1000 generations in addition to 200 generations), multiple performance indicators, and other problem suites such as a traditional artificial test suite ZDT [35] and a large-scale artificial test suite LSMOP [46, 47].

Acknowledgement. This work was supported by National Natural Science Foundation of China (Grant No. 61876075), Guangdong Provincial Key Laboratory (Grant No. 2020B121201001), the Program for Guangdong Introducing Innovative and Enterpreneurial Teams (Grant No. 2017ZT07X386), The Stable Support Plan Program of Shenzhen Natural Science Fund (Grant No. 20200925174447003), Shenzhen Science and Technology Program (Grant No. KQTD2016112514355531).

References

1. Deb, K., Thiele, L., Laumanns, M., Zitzler, E.: Scalable multi-objective optimization test problems. Proc. of IEEE CEC **2002**, 825–830 (2002)
2. Huband, S., Hingston, P., Barone, L., While, L.: A review of multiobjective test problems and a scalable test problem toolkit. IEEE Trans. on Evolutionary Computation **10**(5), 477–506 (2006)
3. Ishibuchi, H., Setoguchi, Y., Masuda, H., Nojima, Y.: Performance of decomposition-based many-objective algorithms strongly depends on Pareto front shapes. IEEE Trans. on Evolutionary Computation **21**(2), 169–190 (2017)
4. Cheng, R., et al.: A benchmark test suite for evolutionary many-objective optimization. Complex & Intelligent Syst. **3**(1), 67–81 (2017)
5. Ishibuchi, H., He, L., Shang, K.: Regular Pareto front shape is not realistic. Proc. of IEEE CEC **2019**, 2035–2042 (2019)
6. Zhu, Q., et al.: An elite gene guided reproduction operator for many-objective optimization. IEEE Trans. on Cybernetics **51**(2), 765–778 (2021)
7. Liang, Z., Hu, K., Ma, X., Zhu, Z.: A many-objective evolutionary algorithm based on a two-round selection strategy. IEEE Trans. on Cybernetics **51**(3), 1417–1429 (2021)
8. Li, L., Chang, L., Gu, T., Sheng, W., Wang, W.: On the norm of dominant difference for many-objective particle swarm optimization. IEEE Trans. on Cybernetics **51**(4), 2055–2067 (2021)
9. Liang, Z., Luo, T., Hu, K., Ma, X., Zhu, Z.: An indicator-based many-objective evolutionary algorithm with boundary protection. IEEE Trans. on Cybernetics **51**(9), 4553–4566 (2021)
10. Liu, Y., Zhu, N., Li, M.: Solving many-objective optimization problems by a Pareto-based evolutionary algorithm with preprocessing and a penalty mechanism. IEEE Trans. on Cybernetics **51**(11), 5585–5594 (2021)
11. Zhang, K., Xu, Z., Xie, S., Yen, G.G.: Evolution strategy-based many-objective evolutionary algorithm through vector equilibrium. IEEE Trans. on Cybernetics **51**(11), 5455–5467 (2021)
12. Moreno, J., Rodriguez, D., Nebro, A.J., Lozano, J.A.: Merge nondominated sorting algorithm for many-objective optimization. IEEE Trans. on Cybernetics **51**(12), 6154–6164 (2021)
13. Chen, L., Deb, K., Liu, H.-L., Zhang, Q.: Effect of objective normalization and penalty parameter on penalty boundary intersection decomposition-based evolutionary many-objective optimization algorithms. Evol. Comput. **29**(1), 157–186 (2021)

14. Song, Z., Wang, H., He, C., Jin, Y.: A Kriging-assisted two-archive evolutionary algorithm for expensive many-objective optimization. IEEE Trans. on Evolutionary Computation **25**(6), 1013–1027 (2021)
15. Liu, S.-C., Zhan, Z.-H., Tan, K.C., Zhang, J.: A multiobjective framework for many-objective optimization. IEEE Trans. on Cybernetics **52**(12), 13654–13668 (2022)
16. Wang, Z., Zhang, Q., Ong, Y.-S., Yao, S., Liu, H., Luo, J.: Choose appropriate subproblems for collaborative modeling in expensive multiobjective optimization. IEEE Trans. on Cybernetics (Early Access Paper)
17. Sonoda, T., Nakata, M.: Multiple classifiers-assisted evolutionary algorithm based on decomposition for high-dimensional multi-objective problems. IEEE Trans. on Evolutionary Computation **26**(6), 1581–1595 (2022)
18. Deb, K., Pratap, A., Agarwal, S., Meyarivan, T.: A fast and elitist multiobjective genetic algorithm: NSGA-II. IEEE Trans. on Evolutionary Computation **6**(2), 182–197 (2002)
19. Zhang, Q., Li, H.: MOEA/D: a multiobjective evolutionary algorithm based on decomposition. IEEE Trans. on Evolutionary Computation **11**(6), 712–731 (2007)
20. Beume, N., Naujoks, B., Emmerich, M.: SMS-EMOA: multiobjective selection based on dominated hypervolume. European J. of Operational Res. **181**(3), 1653–1669 (2007)
21. Bader, J., Zitzler, E.: HypE: an algorithm for fast hypervolume-based many-objective optimization. Evol. Comput. **19**(1), 45–76 (2011)
22. Deb, K., Jain, H.: An evolutionary many-objective optimization algorithm using reference-point-based non-dominated sorting approach, part I: solving problems with box constraints. IEEE Trans. on Evolutionary Computation **18**(4), 577–601 (2014)
23. Li, K., Deb, K., Zhang, Q., Kwong, S.: An evolutionary many-objective optimization algorithm based on dominance and decomposition. IEEE Trans. on Evolutionary Computation **19**(5), 694–716 (2015)
24. Cheng, R., Jin, Y., Olhofer, M., Sendhoff, B.: A reference vector guided evolutionary algorithm for many-objective optimization. IEEE Trans. on Evolutionary Computation **20**(5), 773–791 (2016)
25. Tian, Y., Zhang, X., Wang, C., Jin, Y.: An evolutionary algorithm for large-scale sparse multi-objective optimization problems. IEEE Trans. on Evolutionary Computation **24**(2), 380–393 (2020)
26. Liu, Y., Ishibuchi, H., Masuyama, N., Nojima, Y.: Adapting reference vectors and scalarizing functions by growing neural gas to handle irregular Pareto fronts. IEEE Trans. on Evolutionary Computation **24**(3), 439–453 (2020)
27. Shang, K., Ishibuchi, H.: A new hypervolume-based evolutionary algorithm for many-objective optimization. IEEE Trans. on Evolutionary Computation **24**(5), 839–852 (2020)
28. Yuan, J., Liu, H.-L., Gu, F., Zhang, Q., He, Z.: Investigating the properties of indicators and an evolutionary many-objective algorithm using promising regions. IEEE Trans. on Evolutionary Computation **25**(1), 75–86 (2021)
29. Tanabe, R., Ishibuchi, H.: An easy-to-use real-world multi-objective optimization problem suite. Applied Soft Computing **89**, 106078 (2020)
30. He, C., Tian, Y., Wang, H., Jin, Y.: A repository of real-world datasets for data-driven evolutionary multiobjective optimization. Complex & Intelligent Systems **6**(1), 189–197 (2019)
31. Kumar, A., et al.: A benchmark-suite of real-world constrained multi-objective optimization problems and some baseline results. Swarm and Evolutionary Computation **67**, 100961 (2021)
32. Horn, J., Nafpliotis, N., Goldberg, D.E.: A niched Pareto genetic algorithm for multi-objective optimization. Proc. of IEEE ICEC **1994**, 82–87 (1994)
33. Srinivas, N., Deb, K.: Multiobjective optimization using nondominated sorting in genetic algorithms. Evol. Comput. **2**(3), 221–248 (1994)

34. Ishibuchi, H., Masuda, H., Tanigaki, Y., Nojima, Y.: Review of coevolutionary developments of evolutionary multi-objective and many-objective algorithms and test problems. Proc. of IEEE SSCI **2014**, 178–185 (2014)
35. Zitzler, E., Deb, K., Thiele, L.: Comparison of multiobjective evolutionary algorithms: empirical results. Evol. Comput. **8**(2), 173–195 (2000)
36. Ikeda, K., Kita, H., Kobayashi, S.: Failure of Pareto-based MOEAs: does non-dominated really mean near to optimal? Proc. of CEC **2001**, 957–962 (2001)
37. Ishibuchi, H., Matsumoto, T., Masuyama, N., Nojima, Y.: Effects of dominance resistant solutions on the performance of evolutionary multi-objective and many-objective algorithms. Proc. of GECCO **2020**, 507–515 (2020)
38. Wagner, T., Beume, N., Naujoks, B.: Pareto-, aggregation-, and indicator-based methods in many-objective optimization. In: Obayashi, S., Deb, K., Poloni, C., Hiroyasu, T., Murata, T. (eds.) EMO 2007. LNCS, vol. 4403, pp. 742–756. Springer, Heidelberg (2007). https://doi.org/10.1007/978-3-540-70928-2_56
39. Mostaghim, S., Schmeck, H.: Distance based ranking in many-objective particle swarm optimization. In: Rudolph, G., Jansen, T., Beume, N., Lucas, S., Poloni, C. (eds.) PPSN 2008. LNCS, vol. 5199, pp. 753–762. Springer, Heidelberg (2008). https://doi.org/10.1007/978-3-540-87700-4_75
40. Tian, Y., Cheng, R., Zhang, X., Jin, Y.: PlatEMO: a MATLAB platform for evolutionary multi-objective optimization. IEEE Comput. Intell. Mag. **12**(4), 73–87 (2017)
41. Ishibuchi, H., Matsumoto, T., Masuyama, N., Nojima, Y.: Many-objective problems are not always difficult for Pareto dominance-based evolutionary algorithms. Proc. of ECAI **2020**, 291–299 (2020)
42. Ishibuchi, H., Imada, R., Setoguchi, Y., Nojima, Y.: How to specify a reference point in hypervolume calculation for fair performance comparison. Evol. Comput. **26**(3), 411–440 (2018)
43. He, L., Shang, K., Nan, Y., Ishibuchi, H., Srinivasan, D.: Relation between objective space normalization and weight vector scaling in decomposition-based multi-objective evolutionary algorithms. IEEE Trans. on Evolutionary Computation (Early Access Paper)
44. Ishibuchi, H., Imada, R., Masuyama, N., Nojima, Y.: Use of two reference points in hypervolume-based evolutionary multiobjective optimization algorithms. In: Auger, A., Fonseca, C.M., Lourenço, N., Machado, P., Paquete, L., Whitley, D. (eds.) PPSN 2018. LNCS, vol. 11101, pp. 384–396. Springer, Cham (2018). https://doi.org/10.1007/978-3-319-99253-2_31
45. Ishibuchi, H., Pang, L.M., Shang, K.: Difficulties in fair performance comparison of multi-objective evolutionary algorithms. IEEE Comput. Intell. Mag. **17**(1), 86–101 (2022)
46. Cheng, R., Jin, Y., Olhofer, M., Sendhoff, B.: Test problems for large-scale multiobjective and many-objective optimization. IEEE Trans. on Cybernetics **47**(12), 4108–4121 (2017)
47. Pang, L.M., Ishibuchi, H., Shang, K.: Counterintuitive experimental results in evolutionary large-scale multi-objective optimization. IEEE Trans. on Evolutionary Computation **26**(6), 1609–1616 (2022)

A Novel Performance Indicator Based on the Linear Assignment Problem

Diana Cristina Valencia-Rodríguez(✉) and Carlos A. Coello Coello

CINVESTAV-IPN (Evolutionary Computation Group),
Av. IPN 2508, San Pedro Zacatenco, 07360 Ciudad de México, Mexico
diana.valencia@cinvestav.mx, ccoello@cs.cinvestav.mx

Abstract. Evaluating the performance of Multi-Objective Evolutionary Algorithms is complex since we have to assess different characteristics of the approximation sets that they generate. Over the years, a variety of performance indicators have been proposed to fulfill this task. One of the most popular performance indicators has been the hypervolume because it can assess both convergence and spread of a set of solutions and it is fully Pareto compliant. However, its computational cost grows exponentially with the number of objectives. A good alternative is the $R2$ indicator which has a similar behavior but a much lower computational cost. Nevertheless, $R2$ sometimes is unable to differentiate two sets with different distributions. In this work, we propose a novel performance indicator based on the linear assignment problem called "ILAP", which offers advantages over $R2$. To illustrate this, we include an example in which the ILAP can differentiate two sets when the $R2$ indicator cannot do it. Furthermore, our experimental analysis shows that our proposed indicator correctly ranks solution sets with different distributions and shapes.

Keywords: Indicator · Linear assignment problem · Multi-objective optimization

1 Introduction

The solution to a multi-objective optimization problem consists of a set of non-dominated solutions which can not be easily evaluated as in the case of single-objective problems. Therefore, the performance assessment of Multi-Objective Evolutionary Algorithms (MOEAs) is an essential research topic. Over the years, a variety of indicators have been proposed to assess different characteristics of the Pareto Front approximations [1,7,12]. One of the most popular indicators has been the hypervolume [12], which measures the space covered by an approximation set given a reference point. This indicator is Pareto compliant and can assess both convergence and spread of the approximations produced by a MOEA. However, its computational cost becomes unaffordable as the number of objectives increases.

M. Emmerich et al. (Eds.): EMO 2023, LNCS 13970, pp. 348–360, 2023.
https://doi.org/10.1007/978-3-031-27250-9_25

Another commonly used performance indicator is $R2$ [1]. This indicator can assess the convergence and diversity of the solutions by using a set of weight vectors and a scalarizing function. Moreover, the behavior of the $R2$ indicator is similar to that of the hypervolume (although $R2$ is weakly Pareto compliant) but has a significantly lower computational cost [1]. Nevertheless, as we will see later on, the $R2$ indicator may obtain the same value for approximation sets with different distributions.

This work introduces a performance indicator based on the linear assignment problem [2]: ILAP. In a linear assignment problem, we have to assign a set of agents to a set of tasks. The assignment of an agent to a task corresponds to a cost. Therefore, the aim is to find an assignment with the lowest cost. In the case of the ILAP, we use a set of weight vectors and a set of individuals, where the assignment cost is computed using a scalarizing function. Thus, we use the cost of the best assignment as the indicator value.

Our experimental results show that ILAP correctly ranks solution sets with different distributions and shapes. Moreover, we present an example in which ILAP distinguishes two approximation sets in a better way than the $R2$ indicator.

The remainder of the paper is organized in the following way. First, we introduce the necessary concepts to understand this paper in Sect. 2. Then, we present in Sect. 3 our proposed indicator. After that, in Sect. 4, we discuss the differences and similarities between our approach and the $R2$ indicator. In Sect. 5, we evaluate our proposed ILAP. Finally, we present the conclusions and some possible paths for future research in Sect. 6.

2 Background

2.1 Multi-objective Optimization

A Multi-objective Optimization Problem (MOP) is defined as follows[1]:

$$\text{minimize} \quad F(\boldsymbol{x}) = [f_1(\boldsymbol{x}), \ldots, f_m(\boldsymbol{x})]^T \tag{1}$$

$$\text{subject to} \quad g_i(\boldsymbol{x}) \leq 0 \quad i = 1, \ldots, p, \tag{2}$$

$$h_j(\boldsymbol{x}) = 0 \quad j = 1, \ldots, q \tag{3}$$

where $\boldsymbol{x} = [x_1, \ldots, x_n]$ is the vector of decision variables, $f_i : \mathbb{R}^n \to \mathbb{R}$ for $i = 1, \ldots, m$ are the objective functions, and $g_i, h_i : \mathbb{R}^n \to \mathbb{R}$ for $i = 1, \ldots, p$, $j = 1, \ldots, q$ are the constraints of the problem. We denote Ω as the decision space and $\mathcal{F}$ as the feasible region.

In a MOP, we cannot easily compare the solutions because the objective functions are usually in conflict with each other. Therefore, we use the Pareto dominance relation to define a partial order of the solutions:

Definition 1. *A vector* $\boldsymbol{x} \in \Omega$ *is said to dominate* $\boldsymbol{y} \in \Omega$ *(denoted as* $\boldsymbol{x} \prec \boldsymbol{y}$*), if* $f_i(\boldsymbol{x}) \leq f_i(\boldsymbol{y})$ *for all* $i = 1, \ldots, m$*, and* $f_j(\boldsymbol{x}) < f_j(\boldsymbol{y})$ *in at least one* j.

[1] without loss of generality, we assume minimization problems.

Pareto optimality, which is the most commonly used notion of optimality adopted in multi-objective optimization, is formally defined as follows:

Definition 2. *A vector $\boldsymbol{x} \in \mathcal{F}$ is Pareto optimal if there does not exist another vector $\boldsymbol{y} \in \mathcal{F}$ such that $\boldsymbol{y} \prec \boldsymbol{x}$.*

Moreover, we also adopt the following definitions commonly used in multi-objective optimization:

Definition 3. *The Pareto Optimal Set $\mathcal{P}^*$ is defined as:*

$$\mathcal{P}^* := \{\boldsymbol{x} \mid \boldsymbol{x} \text{ is Pareto optimal}\}$$

Definition 4. *The Pareto Optimal Front $\mathcal{PF}^*$ is defined as:*

$$\mathcal{PF}^* := \{\boldsymbol{F}(\boldsymbol{x}) \in \mathbb{R}^m \mid \boldsymbol{x} \in \mathcal{P}^*\}$$

Definition 5. *Given a predefined weight vector $\boldsymbol{w} \in \mathbb{R}^m$, a scalarizing function s transforms a multi-objective problem into a single-objective problem of the following form:*

$$\begin{aligned} \text{minimize} \quad & s(\boldsymbol{f}'(\boldsymbol{x}), \boldsymbol{w}) && (4) \\ \text{subject to} \quad & \boldsymbol{x} \in \mathcal{F}, && (5) \end{aligned}$$

where $\boldsymbol{x}$ is the decision vector, $\mathcal{F} \in \mathbb{R}^n$ is the feasible region, $\boldsymbol{f} \in \mathbb{R}^m$ is the vector of m objective functions, $\boldsymbol{f}'(\boldsymbol{x}) := \boldsymbol{f}(\boldsymbol{x}) - \boldsymbol{z}$, and $\boldsymbol{z} \in \mathbb{R}^m$ is a reference point.

2.2 Linear Assignment Problem

A Linear Assignment Problem (LAP) comprises a set of agents and a set of tasks where assigning an agent to a task involves a cost. Therefore, the aim is to find an assignment that minimizes the overall cost. Formally, a LAP can be formulated as follows.

Definition 6. *Given a set of agents $A = \{a_1, \ldots, a_n\}$, a set of tasks $T = \{t_1, \ldots, t_n\}$, and the cost function $C : A \times T \to \mathbb{R}$. Let $\Phi : A \to T$ the set of all possible bijections between A and T, the linear assignment problem (LAP) can then be stated as*

$$\underset{\phi \in \Phi}{\text{minimize}} \sum_{a \in A} C(a, \phi(a)) \tag{6}$$

Usually, the cost function can be expressed as a real-valued matrix C with elements $C_{ij} = C(a_i, t_j)$. Moreover, the set Φ of all possible bijections can be viewed as a set of permutation matrices $\mathcal{X}$ where each matrix $x \in \mathcal{X}$ holds

$$x_{ij} = \begin{cases} 1 & \text{if agent } i \text{ is assigned to task } j, \\ 0 & \text{otherwise.} \end{cases}$$

Therefore, a LAP can be modeled as [2]:

$$\underset{x \in \mathcal{X}}{\text{minimize}} \quad \sum_{i=1}^{n} \sum_{j=1}^{n} C_{ij} x_{ij} \tag{7}$$

$$\text{such that} \quad \sum_{j=1}^{n} x_{ij} = 1 \quad (i = 1, 2, \ldots, n), \tag{8}$$

$$\sum_{i=1}^{n} x_{ij} = 1 \quad (j = 1, 2, \ldots, n), \tag{9}$$

$$x_{ij} \in \{0, 1\} \quad (i, j = 1, 2, \ldots, n). \tag{10}$$

This problem can be solved using the so-called Hungarian algorithm, which has an $O(n^3)$ computational complexity [2].

3 Our Proposed Indicator

Molinet Berenguer and Coello Coello [9] transformed the selection process of a MOEA into a LAP. In this proposal, the authors consider a set of individuals and a set of weight vectors representing different regions of the Pareto front. Moreover, the cost of assigning an individual to a weight vector is computed using a scalarizing function. Therefore, after solving the LAP, the individuals assigned to a weight vector are selected for the next generation. The authors proposed an algorithm called Hungarian Differential Evolution (HDE) that uses the LAP selection scheme. The experimental results show that the HDE is very competitive with respect to state-of-the-art algorithms.

In the case of the LAP selection process, the size of the set of individuals is bigger than the set of weight vectors. Therefore, the Hungarian algorithm finds the subset of individuals that minimizes the overall assignment cost and discards the subsets with the worst values. Hence, we can deduce that the minimum overall assignment cost gives us an estimation of how good or bad a set is. Using this idea, we propose an indicator based on the LAP. The ILAP indicator is defined in the following.

Definition 7. *Given a set of uniformly distributed weight vectors* $W = \{\boldsymbol{w}_1, \ldots, \boldsymbol{w}_n\}$, *an approximation set* $A = \{\boldsymbol{a}_1, \ldots, \boldsymbol{a}_n\}$, *and a cost matrix* C *such that* $C_{ij} = s(\boldsymbol{w}_i, \boldsymbol{a}_j)$ *where* s *is a scalarizing function. Then, the ILAP is defined as:*

$$I_{LAP} = \frac{1}{n} \min_{x \in \mathcal{X}} \left\{ \sum_{i=1}^{n} \sum_{j=1}^{n} C_{ij} x_{ij} \right\} \tag{11}$$

where $\mathcal{X}$ *is the set of permutation matrices.*

We compute the ILAP by obtaining a cost matrix C using s, A, and W. Then, we solve the LAP defined by C employing the Hungarian algorithm. Finally, the indicator value is the best assignment's cost divided by n. The cost matrix computation is performed in $O(mn^2)$, where m is the number of objectives. Moreover, the LAP problem is solved in $O(n^3)$. Therefore the computational complexity of computing the ILAP is $O(mn^2 + n^3)$.

In the ILAP, each weight vector must be assigned to a currently unassigned solution while minimizing the cost. In the ideal case, each weight vector is assigned to a solution where it obtains its lowest cost. However, let's assume that more than one weight vector obtains its lowest cost with the same solution. In that case, the indicator will assign the solution to the vector with the lowest value and will use the second-best solutions for the remaining vectors.

This process allows the ILAP to assess convergence and diversity at the same time. On the one hand, it measures convergence by always considering the best values of the scalarizing functions. On the other hand, it measures diversity because it tries to quantify how much the solutions cover the regions of the weight vectors. Examples of these two cases are shown in Fig. 1a and Fig. 1b, where the ILAP successfully ranks the sets. We used the Achievement Scalarizing Function (ASF) [10] for these examples and for the rest of the paper.

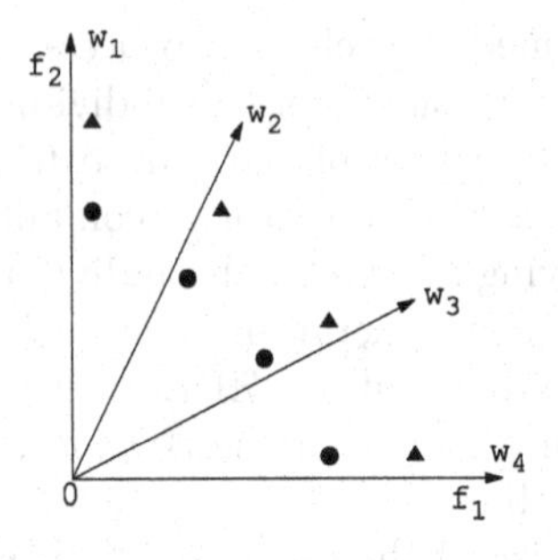

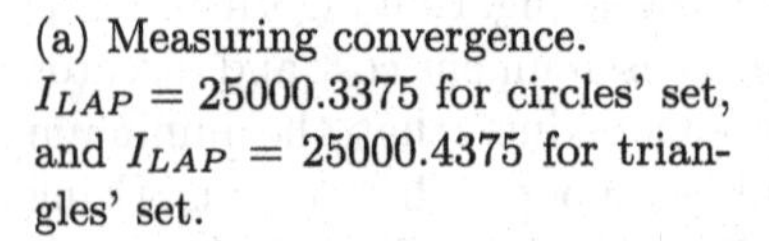
(a) Measuring convergence. $I_{LAP} = 25000.3375$ for circles' set, and $I_{LAP} = 25000.4375$ for triangles' set.

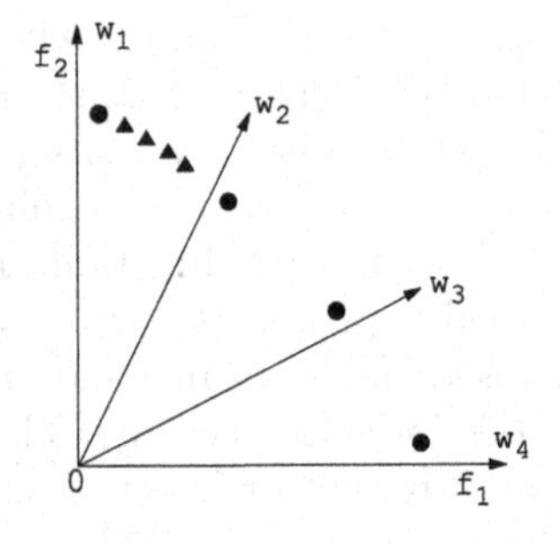

(b) Measuring diversity. $I_{LAP} = 25000.4375$ for circles' set, and $I_{LAP} = 197500.675$ for triangles' set.

Fig. 1. Examples where the ILAP assesses both convergence and diversity. A lower value is preferred; therefore, the ILAP ranks the sets correctly in both cases.

4 Comparison Between Our Approach and the $R2$-indicator

The $R2$ indicator is a performance indicator that assesses the convergence and the diversity of a solution set. It assesses performance by mapping the candidate solutions from objective space into utility space. Given a reference set A, a set of reference vectors V, and a scalarizing function[2] s. The $R2$ indicator is defined as follows [1]:

$$R2(A, V) = \frac{1}{|V|} \sum_{v \in V} \min_{a \in A}\{s(\boldsymbol{a}, \boldsymbol{v})\}$$

The ILAP and $R2$ indicators have some similarities. Both use scalarizing functions and weight vectors to assess the performance of an approximation set. Moreover, the indicators will obtain the same value when the regions given by the weight vectors are equally covered (as shown in Fig. 2).

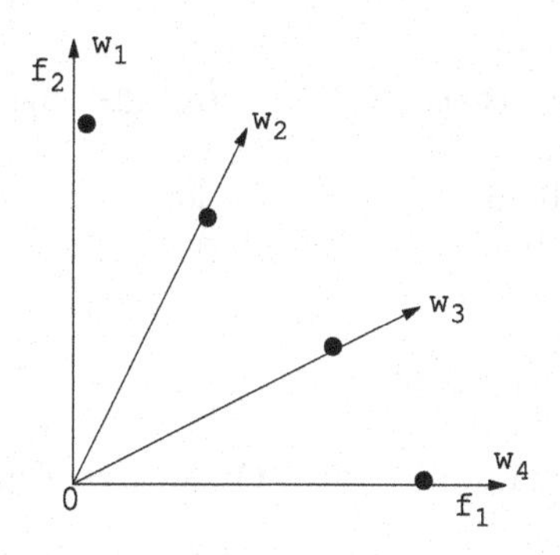

Fig. 2. Example of a case where ILAP and $R2$ obtain the same values: $I_{LAP} = R2 = 10000.3875$

However, the $R2$ indicator only considers the solutions with the best values of the scalarizing function, discarding the information provided by the solutions with the worst values. Therefore, the $R2$ indicator may not evaluate the performance of the whole set and may obtain the same value for two different approximations. On the other hand, the ILAP indicator considers the whole set since it assigns each weight vector with a different solution and obtains the indicator's value from this assignment.

An example of the previous situation is shown in Fig. 3a and Fig. 3b. Given two different approximation sets, the $R2$ indicator obtains the same value, while the ILAP obtains distinct values. Furthermore, the ILAP prefers the approximation set with a solution nearer an uncovered vector.

[2] also known as utility function.

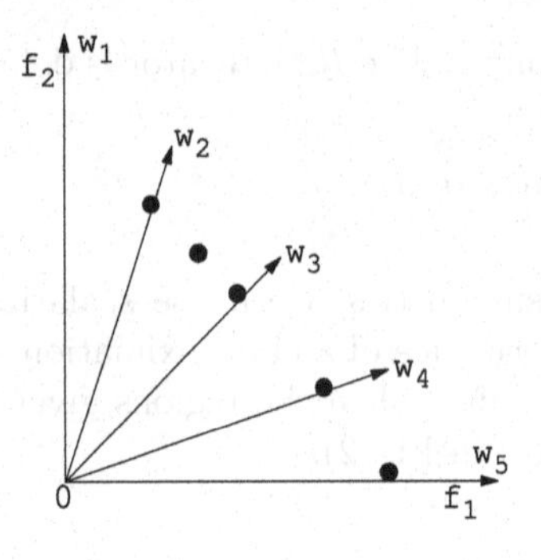

(a) R2=44000.50133, I_{LAP}=44000.584

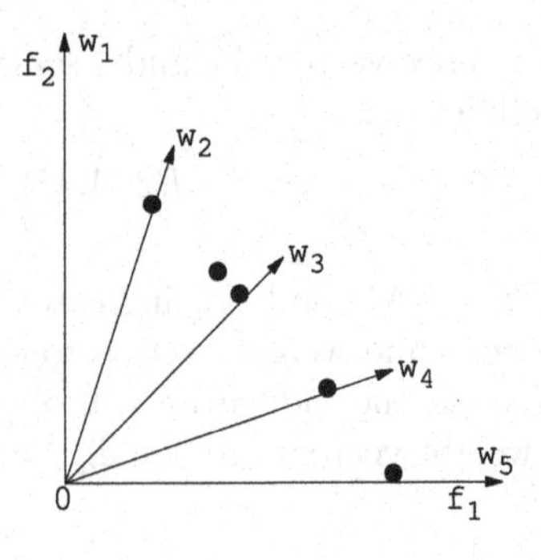

(b) R2=44000.50133, I_{LAP}=44000.616

Fig. 3. The $R2$ indicator obtains the same value for two sets with distinct distributions, while the ILAP indicator obtains different values.

5 Experimental Analysis

5.1 Evaluation in Artificial Many-Objective Pareto Fronts

In this section, we study the performance of the ILAP in artificial Pareto fronts. We employed three types of solutions sets generated in a unit m-simplex:

- C1. The solutions are concentrated in one corner of the simplex.
- C2. The solutions are randomly generated.
- C3. The solutions are uniformly distributed. We employ the method proposed in [5] for this type of set.

Moreover, the set size for each dimension is shown in Table 1, and Fig. 4a to Fig. 5l show the parallel coordinates graphs of the sets. Regarding the ILAP, we use the ASF, and the Uniform Design with the Hammersley method (UDH) [9] for generating the weight vectors.

Table 1. Set size for each dimension

m	3	4	5	6	7	8	9	10
Set size	100	110	120	130	140	150	160	170

The results are shown in Tables 2a and 2b. Moreover, we include the results of the hypervolume indicator (HV) [12] as a reference. We can observe that the ILAP consistently ranks the C3 sets in first place, the C2 sets in second place, and the C1 sets in last place. Furthermore, HV obtains the same ranking. Therefore, the ILAP can correctly rank a set of solutions in 3 to 10 dimensions.

Table 2. ILAP and HV values of the sets C1, C2, and C3 for each dimension m. Darker cells imply better values.

m	C1	C2	C3
3	5.0453	1.5322	1.1263
4	5.6031	1.8407	1.3157
5	5.9506	2.3113	1.5775
6	5.8495	2.6022	1.9827
7	6.0938	2.9235	2.1356
8	5.9524	3.0488	2.4045
9	5.8106	3.2949	2.9855
10	5.5003	3.4681	3.1862

(a) ILAP

m	C1	C2	C3
3	0.77462	1.076862	1.11977
4	0.906105	1.32058	1.369026
5	1.036677	1.49916	1.560266
6	1.197589	1.659354	1.737507
7	1.284942	1.862622	1.920832
8	1.400768	2.057528	2.111709
9	1.586619	2.252491	2.328822
10	1.805137	2.501373	2.563038

(b) HV

5.2 Evaluation in Pareto Front Approximations

In this section, we use the ILAP, the hypervolume, and the $R2$ indicator to evaluate the performance of two well-known MOEAs: the NSGA-II [3] and the MOEA/D [11]. For this purpose, we ran each algorithm 30 times using different problems. We adopted the DTLZ1, DTLZ2, and DTLZ7 problems from the Deb-Thiele-Laumanns-Zitzler (DTLZ) [4] test suite, the DTLZ1^{-1} from the Minus-DTLZ test problems [8], and the WFG1-WFG3 from the Walking-Fish-Group (WFG) [6] test suite with m = 3, 5, 8, and 10 objectives. Regarding the DTLZ problems, we set the number of decision variables to $n = m + k - 1$, where $k = 5$ for DTLZ1, $k = 10$ for DTLZ2, and $k = 20$ for DTLZ7. In the case of the WFG problems, we set the position-related parameters to $2 \times (m - 1)$ and the distance-related parameters to 20. Finally, we use the same configuration of DTLZ1 for DTLZ1^{-1}.

In the case of the algorithm's parameters, we set the population sizes to 100 for three objectives, 120 for five, 140 for eight, and 160 for ten. We set the crossover and mutation parameters to $pc = 1.0$, $pm = 1/\text{number of variables}$, $n_c = 20$, and $n_m = 20$. Regarding the MOEA/D parameters, we used a neighborhood size $T = 20$, the ASF function, and the UDH weight vectors. Finally, the ILAP and the $R2$ indicators adopted the ASF function and UDH weight vectors.

Tables 3a, 3b, and 3c display the average and the standard deviation of each indicator. We can observe that the three indicators obtain the same results for DTLZ1, DTLZ2, DTLZ7, WFG1, WFG2, and DTLZ1^{-1}. In the case of the WFG3 problem, the $R2$ and the hypervolume get the same rank in 5 and 8 objectives. In contrast, the ILAP and the hypervolume get the same rank in 3 and 5 objectives. This situation could happen because the WFG3 is a linear problem that hardly fits the shape of a simplex. Therefore, the R2 and ILAP indicators may have some trouble with the performance assessment because they employ reference vectors sampled in a simplex.

Table 3. Average and standard deviation of the hypervolume, $R2$, and ILAP indicators. Gray cells imply better values. Moreover, the symbol "*" represents that the algorithm is statistically better according to the Wilcoxon rank sum test.

(a) HV

	M	MOEAD	NSGA-II
DTLZ1	3	1.324e+0 (8.5e-4)	*1.331e+0 (3.1e-6)
	5	*1.610e+0 (7.5e-6)	1.610e+0 (1.4e-4)
	8	*2.144e+0 (6.2e-6)	2.143e+0 (4.4e-4)
	10	*2.594e+0 (8.3e-6)	2.593e+0 (1.4e-4)
DTLZ2	3	*8.330e-1 (9.3e-4)	8.169e-1 (5.1e-3)
	5	*1.593e+0 (2.1e-3)	1.584e+0 (5.5e-3)
	8	*2.139e+0 (1.2e-3)	2.002e+0 (4.2e-2)
	10	*2.589e+0 (1.2e-3)	2.454e+0 (3.9e-2)
DTLZ7	3	6.436e-1 (5.e-2)	*6.908e-1 (2.6e-2)
	5	6.136e-1 (5.6e-2)	*8.222e-1 (1.9e-2)
	8	2.e-1 (1.2e-1)	*7.608e-1 (9.4e-2)
	10	1.102e-1 (9.4e-2)	*5.378e-1 (1.3e-1)
WFG1	3	*1.197e+0 (2.7e-2)	1.108e+0 (2.5e-2)
	5	*1.526e+0 (3.6e-2)	1.273e+0 (2.9e-2)
	8	*2.055e+0 (5.0e-2)	1.495e+0 (3.6e-2)
	10	*2.465e+0 (3.5e-2)	1.378e+0 (3.8e-2)
WFG2	3	1.072e+0 (8.e-2)	*1.155e+0 (8.2e-2)
	5	1.328e+0 (1.1e-1)	*1.58e+0 (6.9e-3)
	8	1.711e+0 (1.4e-1)	*2.129e+0 (6.3e-3)
	10	2.074e+0 (2.1e-1)	*2.576e+0 (7.8e-3)
WFG3	3	8.132e-1 (7.8e-3)	8.162e-1 (4.e-3)
	5	1.140e+0 (1.4e-2)	1.134e+0 (1.4e-2)
	8	1.472e+0 (2.9e-2)	*1.501e+0 (2.7e-2)
	10	1.518e+0 (5.1e-2)	*1.798e+0 (3.5e-2)
DTLZ1^{-1}	3	*2.775e-1 (2.9e-5)	2.733e-1 (2.2e-3)
	5	1.224e-2 (9.e-5)	1.215e-2 (5.3e-4)
	8	*3.472e-5 (1.3e-6)	3.167e-5 (2.3e-6)
	10	4.988e-7 (3.8e-8)	4.924e-7 (3.6e-8)

(b) R2

	M	MOEAD	NSGA-II
DTLZ1	3	7.692e-2 (7.e-3)	*2.856e-2 (6.e-4)
	5	*1.155e-3 (3.1e-5)	2.388e-1 (1.2e-1)
	8	*1.318e-3 (5.5e-5)	8.6e-1 (3.1e-1)
	10	*1.574e-3 (9.5e-5)	1.165e+0 (2.6e-1)
DTLZ2	3	*1.360e+0 (2.5e-4)	1.446e+0 (3.8e-2)
	5	*8.176e-1 (5.2e-3)	1.196e+0 (5.9e-2)
	8	*6.764e-1 (2.4e-2)	2.832e+0 (2.4e-1)
	10	*8.836e-1 (1.0e-1)	3.216e+0 (2.0e-1)
DTLZ7	3	3.517e+0 (1.1e+0)	*3.074e+0 (5.8e-1)
	5	7.344e+0 (9.4e-1)	*6.085e+0 (1.7e-1)
	8	1.702e+1 (2.2e+0)	*1.098e+1 (4.8e-1)
	10	2.339e+1 (3.5e+0)	*1.487e+1 (9.1e-1)
WFG1	3	*1.076e+0 (1.6e-1)	1.493e+0 (2.7e-1)
	5	*1.375e+0 (2.2e-1)	2.922e+0 (2.6e-1)
	8	*1.412e+0 (2.3e-1)	4.859e+0 (2.3e-1)
	10	*1.797e+0 (2.1e-1)	8.452e+0 (2.7e-1)
WFG2	3	2.126e+0 (7.0e-1)	*1.488e+0 (7.6e-1)
	5	3.305e+0 (1.1e+0)	*1.174e+0 (5.4e-2)
	8	5.121e+0 (1.5e+0)	*1.425e+0 (5.4e-2)
	10	5.187e+0 (2.3e+0)	*1.533e+0 (6.7e-2)
WFG3	3	*2.667e+0 (2.5e-2)	2.673e+0 (2.0e-2)
	5	*3.724e+0 (6.2e-2)	3.796e+0 (6.2e-2)
	8	*5.617e+0 (8.1e-2)	5.711e+0 (7.7e-2)
	10	6.985e+0 (2.2e-1)	*6.594e+0 (1.1e-1)
DTLZ1^{-1}	3	*5.014e+0 (2.0e-4)	5.46e+0 (6.4e-2)
	5	*1.349e+1 (3.2e-2)	1.581e+1 (1.4e-1)
	8	*3.073e+1 (8.9e-2)	3.825e+1 (3.5e-1)
	10	*4.034e+1 (1.2e-1)	5.187e+1 (4.4e-1)

(c) ILAP

	M	MOEAD	NSGA-II
DTLZ1	3	6.127e-1 (1.0e-3)	*5.061e-1 (5.9e-2)
	5	*1.155e-3 (3.0e-5)	9.361e-1 (1.0e-1)
	8	*1.318e-3 (5.6e-5)	1.800e+0 (1.1e-1)
	10	*1.577e-3 (9.8e-5)	2.105e+0 (6.0e-2)
DTLZ2	3	*1.363e+0 (3.4e-4)	1.958e+0 (7.2e-2)
	5	*8.181e-1 (5.2e-3)	1.478e+0 (6.5e-2)
	8	*6.765e-1 (2.5e-2)	4.292e+0 (2.2e-1)
	10	*8.851e-1 (1.1e-1)	5.212e+0 (1.8e-1)
DTLZ7	3	4.797e+0 (1.1e+0)	*3.298e+0 (6.3e-1)
	5	1.257e+1 (1.2e+0)	*6.904e+0 (1.7e-1)
	8	2.834e+1 (3.0e+0)	*1.333e+1 (4.4e-1)
	10	3.711e+1 (6.0e+0)	*1.825e+1 (6.9e-1)
WFG1	3	*1.359e+0 (1.7e-1)	1.721e+0 (3.8e-1)
	5	*1.62e+0 (3.1e-1)	3.338e+0 (3.3e-1)
	8	*1.727e+0 (2.9e-1)	5.834e+0 (4.5e-1)
	10	*2.213e+0 (2.8e-1)	1.021e+1 (4.6e-1)
WFG2	3	2.377e+0 (6.3e-1)	*1.75e+0 (8.5e-1)
	5	3.568e+0 (1.1e+0)	*1.954e+0 (2.3e-1)
	8	5.302e+0 (1.5e+0)	*2.953e+0 (2.5e-1)
	10	5.314e+0 (2.2e+0)	*3.271e+0 (3.0e-1)
WFG3	3	3.276e+0 (3.7e-2)	*2.929e+0 (5.6e-2)
	5	*4.172e+0 (1.2e-1)	4.804e+0 (1.3e-1)
	8	*6.833e+0 (1.6e-1)	7.874e+0 (2.3e-1)
	10	*9.165e+0 (2.5e-1)	9.488e+0 (2.6e-1)
DTLZ1^{-1}	3	*5.014e+0 (2.0e-4)	5.627e+0 (6.7e-2)
	5	*1.349e+1 (3.2e-2)	1.637e+1 (1.9e-1)
	8	*3.076e+1 (7.4e-2)	3.915e+1 (3.e-1)
	10	*4.037e+1 (1.2e-1)	5.325e+1 (3.7e-1)

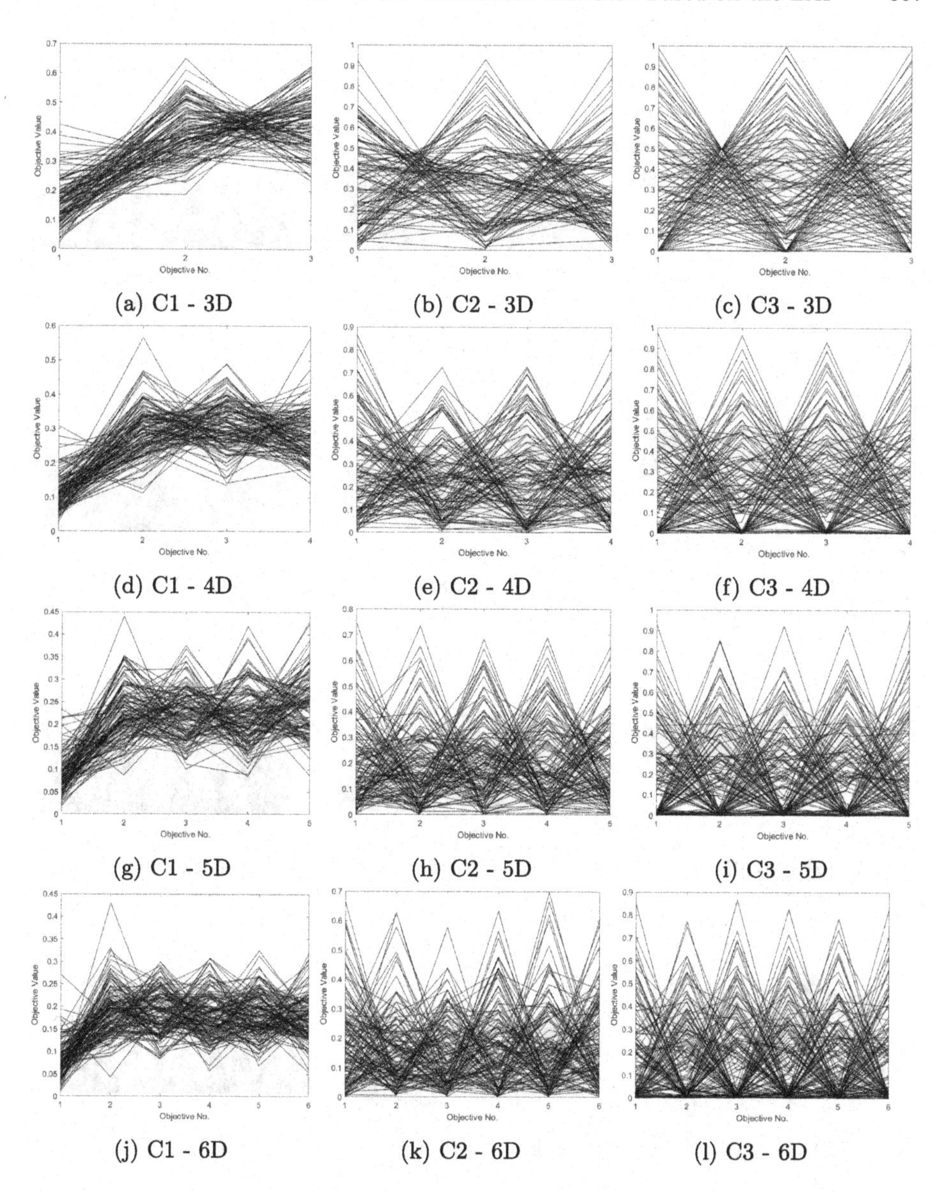

Fig. 4. Artificial solution sets generated in a unit simplex. Solutions in C1 are concentrated in a corner, in C2 are randomly generated, and in C3 are uniformly distributed.

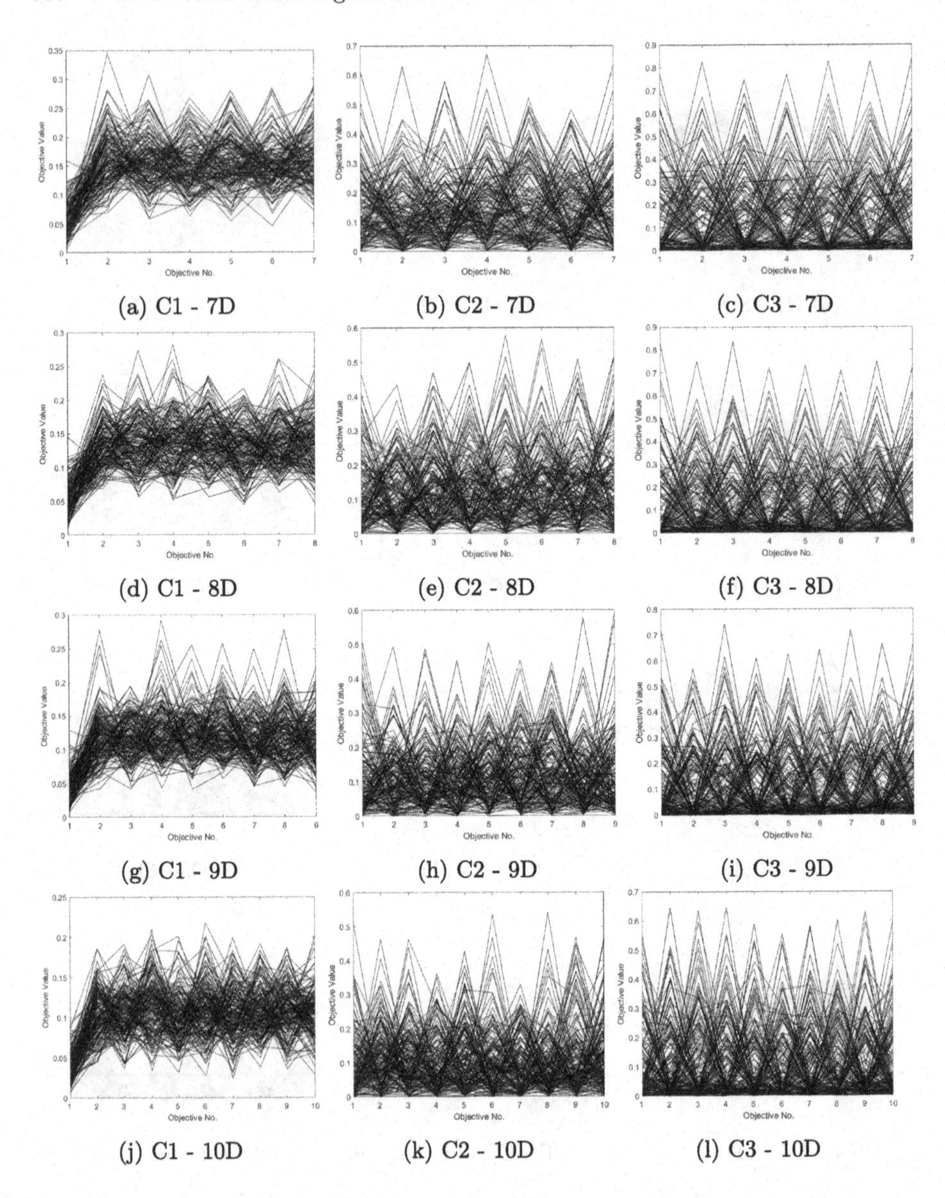

(a) C1 - 7D (b) C2 - 7D (c) C3 - 7D

(d) C1 - 8D (e) C2 - 8D (f) C3 - 8D

(g) C1 - 9D (h) C2 - 9D (i) C3 - 9D

(j) C1 - 10D (k) C2 - 10D (l) C3 - 10D

Fig. 5. Artificial solution sets generated in a unit simplex (continuation)

6 Conclusions and Future Work

We proposed a novel performance indicator based on the Linear Assignment Problem, called ILAP. The experimental results showed that our proposed ILAP could successfully rank the solutions sets using different distributions and Pareto Front shapes of many-objective problems. Moreover, we described an example where the $R2$ indicator (the performance indicator with the most significant similarity with the ILAP) can not distinguish between two different approximation sets. And our proposed ILAP can differentiate them and prefers the one with a solution near an uncovered region. As part of our future work, we would like to analyze the mathematical properties of our proposed indicator.

Acknowledgements. The first author acknowledges support from CONACyT to pursue graduate studies at the Department of Computer Science of CINVESTAV-IPN. The second author gratefully acknowledges support from CONACyT grant no. 2016-01-1920 (Investigación en Fronteras de la Ciencia 2016).

References

1. Brockhoff, D., Wagner, T., Trautmann, H.: On the properties of the R2 indicator. In: Proceedings of the 14th Annual Conference on Genetic and Evolutionary Computation, pp. 465–472. Association for Computing Machinery, New York (2012)
2. Burkard, R.E., Dell'Amico, M., Martello, S.: Assignment Problems, Revised Reprint. Other Titles in Applied Mathematics. Society for Industrial and Applied Mathematics (SIAM) (2012)
3. Deb, K., Pratap, A., Agarwal, S., Meyarivan, T.: A fast and elitist multiobjective genetic algorithm: NSGA-II. IEEE Trans. Evol. Comput. **6**(2), 182–197 (2002)
4. Deb, K., Thiele, L., Laumanns, M., Zitzler, E.: Scalable test problems for evolutionary multiobjective optimization. In: Abraham, A., Jain, L., Goldberg, R. (eds.) Evolutionary Multiobjective Optimization: Theoretical Advances and Applications. Advanced Information and Knowledge Processing, pp. 105–145. Springer, London, London (2005). https://doi.org/10.1007/1-84628-137-7_6
5. Falcón-Cardona, J.G., Ishibuchi, H., Coello Coello, C.A.: Riesz s-energy-based reference sets for multi-objective optimization. In: 2020 IEEE Congress on Evolutionary Computation (CEC), pp. 1–8 (2020)
6. Huband, S., Barone, L., While, L., Hingston, P.: A scalable multi-objective test problem toolkit. In: Coello Coello, C.A., Hernández Aguirre, A., Zitzler, E. (eds.) EMO 2005. LNCS, vol. 3410, pp. 280–295. Springer, Heidelberg (2005). https://doi.org/10.1007/978-3-540-31880-4_20
7. Ishibuchi, H., Masuda, H., Tanigaki, Y., Nojima, Y.: Modified distance calculation in generational distance and inverted generational distance. In: Gaspar-Cunha, A., Henggeler Antunes, C., Coello, C.C. (eds.) EMO 2015. LNCS, vol. 9019, pp. 110–125. Springer, Cham (2015). https://doi.org/10.1007/978-3-319-15892-1_8
8. Ishibuchi, H., Setoguchi, Y., Masuda, H., Nojima, Y.: Performance of decomposition-based many-objective algorithms strongly depends on pareto front shapes. IEEE Trans. Evol. Comput. **21**(2), 169–190 (2017)
9. Molinet Berenguer, J.A., Coello Coello, C.A.: Evolutionary many-objective optimization based on Kuhn-Munkres' algorithm. In: Gaspar-Cunha, A., Henggeler Antunes, C., Coello, C.C. (eds.) EMO 2015. LNCS, vol. 9019, pp. 3–17. Springer, Cham (2015). https://doi.org/10.1007/978-3-319-15892-1_1
10. Pescador-Rojas, M., Hernández Gómez, R., Montero, E., Rojas-Morales, N., Riff, M.-C., Coello Coello, C.A.: An overview of weighted and unconstrained scalarizing functions. In: Trautmann, H., et al. (eds.) EMO 2017. LNCS, vol. 10173, pp. 499–513. Springer, Cham (2017). https://doi.org/10.1007/978-3-319-54157-0_34

11. Zhang, Q., Li, H.: MOEA/D: a multiobjective evolutionary algorithm based on decomposition. IEEE Trans. Evol. Comput. **11**(6), 712–731 (2007)
12. Zitzler, E.: Evolutionary algorithms for multiobjective optimization: methods and applications. Ph.D. thesis, Swiss Federal Institute of Technology (ETH), Zurich, Suiza (1999)

A Test Suite for Multi-objective Multi-fidelity Optimization

Angus Kenny[1], Tapabrata Ray[1], Hemant Kumar Singh[1(✉)], and Xiaodong Li[2]

[1] School of Engineering and IT, The University of New South Wales, Canberra, ACT, Australia
{angus.kenny,t.ray,h.singh}@adfa.edu.au

[2] School of Computing Technologies, RMIT University, Melbourne, VIC, Australia
xiaodong.li@rmit.edu.au

Abstract. Multi-objective optimization problems (MOP) are frequently encountered in practice. In some cases, different computationally expensive analyses may be independently used for computing different objectives of the MOP. Additionally, the analyses may be executed to obtain estimates with different fidelity, typically higher fidelity requiring a longer run-time. For instance, in automotive design, the aerodynamic drag is computed using computational fluid dynamic (CFD) analysis and its crashworthiness/strength is assessed using finite element analysis (FEA). Both the objectives can be independently computed and the underlying fidelity of each analysis can also be controlled using different mesh sizes/thresholds on the residual errors. While there exist a number of generic MOP benchmark problems in the literature, there is scarce work on constructing MOPs with multi-fidelity (MF) analyses to support the development of multi-fidelity, multi-objective optimization algorithms. The existing MF benchmarks are limited to unconstrained, single-objective optimization problems only. Towards addressing this gap, in this paper, we introduce a test suite for multi-objective, multi-fidelity optimization (MOMF). The problems are derived by combining existing unconstrained, multi-objective design optimization problems with resolution/stochastic/instability errors that are common manifestations of MF simulations. The method allows for the construction of any number of low-fidelity functions with desired level of correlations for a given high-fidelity objective function. We hope that the test suite would motivate novel algorithmic developments to support optimization involving computationally expensive and independently evaluable objectives.

1 Background

Evolutionary multi-objective optimization has been an active area of research with a number of efficient algorithms developed over the last three decades. The fundamental development of the algorithms is often supplemented by numerical benchmark problems. Benchmark problems are designed to pose specific challenges to the algorithms and provide a systematic documentation of the strengths and weaknesses of the algorithms. The benchmarks have themselves evolved over time to capture new and previously unsolved challenges. For example, among the older generation of benchmarks, the ZDT test suite was introduced in 2000 [17], followed by scalable DTLZ and WFG test

M. Emmerich et al. (Eds.): EMO 2023, LNCS 13970, pp. 361–373, 2023.
https://doi.org/10.1007/978-3-031-27250-9_26

suites in 2002 [4,5] and 2005 [7], respectively. Further challenges were introduced later by designing problems with more irregular Pareto front shapes, such as inverted [8] and those with extremely convex/non-convex fronts [15,16]. While all of the above benchmarks have led to improved optimization algorithms over time, one of the criticisms has been regarding whether these synthetic benchmarks capture important aspects of the optimization problems found in real-world settings. To address this issue, a benchmark suite derived from real-world multi-objective optimization problems (MOP) comprising 16 real-world, mixed-variable optimization problems was presented in [13]. The problems in the set are computationally cheap to evaluate and capture some of the complexity and nuance that is missed by the synthetic test functions. Of the 16 test problems, 3 of them are unconstrained multi-objective optimization problems, while for the remaining 13, the sum of constraint violation has been considered as the second objective. The unconstrained problems (four-bar truss design, vehicle crashworthiness design, and rocket injector design) from this set are later built upon in this study to construct multi-fidelity versions of these test problems.

In a number of practical applications, computational cost of evaluating a design is a significant consideration. A high evaluation cost translates to fewer design evaluations available to conduct the optimization in a reasonable time-frame, requiring innovative and efficient strategies to sample designs during optimization. One of the strategies relies on evaluating the designs in different fidelities, so called *multi-fidelity* (MF) optimization [3,10], which attempts to prudently balance the computational effort invested in evaluating candidate designs in low-fidelity (LF) vs high-fidelity (HF). A higher fidelity analysis is more accurate and comes with a higher computational cost. A lower fidelity analysis on the other hand is less accurate albeit computationally cheap. MF approaches are particularly attractive when iterative numerical simulations such as computational fluid dynamics (CFD) or finite element analysis (FEA) are involved in evaluating the objectives, since there is a potential to evaluate the designs in different fidelities by, e.g., using different mesh sizes (coarse vs fine), or using different thresholds on residual errors for terminating the simulations [3].

Most studies on MF optimization have focused on two levels of fidelity (high and low) although in principle there could potentially be many more levels of fidelity. In an attempt to develop a test suite to assess the performance of evolutionary multi-fidelity optimization algorithms, Wang et al. [14] analyzed a number of CFD simulations and identified three main types of error that manifest in different fidelity settings: resolution errors, when there is a discrepancy between local and global errors, e.g., due to different FEA mesh-sizes; stochastic errors, where the fitness of a given sample can vary from simulation to simulation; and instability errors, which represent failed simulations. These behaviors, referred to as *error functions*, were used to generate a MF test suite for unconstrained, single-objective optimization [14]. However, there do not exist similar generalizable instances for multi-objective scenarios that can be used to benchmark MF algorithms. In the few such works that currently exist, e.g., [6], a few specific synthetic instances are generated to demonstrate the performance. For such cases, there is low flexibility in terms of modifying the instances to match desirable correlations between different objectives; unlike the test suite proposed later in this paper.

Another related research gap in the literature is the lack of optimization algorithms that allow for selective evaluation of specific objective(s) or constraint(s) during the course of search, without the need to evaluate all of them [1,11,12]. For overwhelming majority of existing studies, it is implicitly assumed that a call to the design evaluation yields the values of all objectives and constraints concurrently. However, the paradigm of selective evaluation is relevant to study in practice since there may exist problems where different analyses independently compute the different objectives/constraints. This provides an opportunity to make an algorithmic choice, based on the cost vs benefit in evaluating an additional objective/constraint when the additional information is not likely to change the ranking of the solution. For example, for the design of a wing, the value of drag (computed via CFD) may be irrelevant to the ranking of the design if it violates structural constraints (computed via FEA). While design of such algorithms, as well as those that deal with heterogeneous cost of objectives [2] are increasingly gaining attention, there currently do not exist benchmarks targeted towards such developments.

In this paper, we attempt to make a contribution at the confluence of the above two research gaps. Towards this end, we construct an improvised multi-objective test suite where each of the objectives can be evaluated in a different fidelity, independent of others. To construct the proposed multi-objective, multi-fidelity (MOMF) problems, we combine a subset of the problems presented in [13] with a subset of the error functions introduced in [14]. The proposed MOMF test problems are presented in Sect. 2. Furthermore, some baseline results obtained using different fidelities of objective evaluation are presented in Sect. 3. Some key directions for future development of efficient MOMF algorithms are presented in Sect. 4. The problem suite itself is available for download at http://www.mdolab.net/research_resources.html.

2 The Proposed MOMF Problem Suite

Many of the problems presented by Tanabe and Ishibuchi in [13] are originally constrained, single-objective optimization problems which have been converted into an unconstrained form by adding an extra objective which minimises the sum of all constraint violations. This would effectively mean that the additional objective would have a value of 0 for all feasible solutions and a value greater than 0 for all infeasible solutions. Furthermore, it can be argued that the additional objective does not reflect the properties of objectives in real life settings. We wanted to keep this method as problem-independent as possible and thus have opted to treat the objective functions as "black-boxes", the output of which is modulated by adding error to produce a low-fidelity versions. For these two reasons, we selected three problems RE2-4-1, RE3-4-7 and RE3-5-4, from [13] as the baseline, which are natively unconstrained, multi-objective problems. The functions are reproduced here for convenience:

Four Bar Truss: The four bar truss design problem (RE2-4-1) involves minimization of the structural volume f_1 and the joint displacement f_2. The variables x_1, x_2, x_3 and x_4 are the lengths of the four bars.

$$f_1(\boldsymbol{x}) = L\left(2x_1 + \sqrt{2}x_2 + \sqrt{x_3} + x_4\right), \quad f_2(\boldsymbol{x}) = \frac{FL}{E}\left(\frac{2}{x_1} + \frac{2\sqrt{2}}{x_2} - \frac{2\sqrt{2}}{x_3} + \frac{2}{x_4}\right), \tag{1}$$

where, $x_1, x_4 \in [a, 3a]$, $x_2, x_3 \in [\sqrt{2}, 3a]$, and $a = \frac{F}{\sigma}$. The parameters are given as: $F = 10$ kN, $E = 2 \times 10^5$ kN, $L = 200$ cm and $\sigma = 10$ kN/cm^2.

Rocket Injector: The rocket injector design problem (RE3-4-7) involves minimization of the maximum temperature of the injector face f_1, the distance from the inlet f_2 and the maximum temperature of the post tip f_3. The variables x_1, x_2, x_3 and x_4 denote the hydrogen flow angle (α), the hydrogen area (Δ_{HA}), the oxygen area (Δ_{OA}) and the oxidiser post tip thickness ($OPTT$).

$$\begin{aligned} f_1(\boldsymbol{x}) = {} & 0.692 + 0.477x_1 - 0.687x_2 - 0.080x_3 - 0.0650x_4 - 0.167{x_1}^2 \\ & - 0.0129x_2x_1 + 0.0796{x_2}^2 - 0.0634x_3x_1 - 0.0257x_3x_2 + 0.0877{x_3}^2 \\ & - 0.0521x_4x_1 + 0.00156x_4x_2 + 0.00198x_4x_3 + 0.0184{x_4}^2, \end{aligned} \tag{2}$$

$$\begin{aligned} f_2(\boldsymbol{x}) = {} & 0.153 - 0.322x_1 + 0.396x_2 + 0.424x_3 + 0.0226x_4 + 0.175{x_1}^2 + 0.0185x_2x_1 \\ & - 0.0701{x_2}^2 - 0.251x_3x_1 + 0.179x_3x_2 + 0.0150{x_3}^2 + 0.0134x_4x_1 \\ & + 0.0296x_4x_2 + 0.0752x_4x_3 + 0.0192{x_4}^2, \end{aligned} \tag{3}$$

$$\begin{aligned} f_3(\boldsymbol{x}) = {} & 0.370 - 0.205x_1 + 0.0307x_2 + 0.108x_3 + 1.019x_4 - 0.135{x_1}^2 + 0.0141x_2x_1 \\ & + 0.0998{x_2}^2 + 0.208x_3x_1 - 0.0301x_3x_2 - 0.226{x_3}^2 + 0.353x_4x_1 \\ & - 0.0497x_4x_3 - 0.423{x_4}^2 + 0.202x_2{x_1}^2 - 0.281x_3{x_1}^2 - 0.342x_1{x_2}^2 \\ & - 0.245{x_2}^2x_3 + 0.281x_2{x_3}^2 - 0.184x_1{x_4}^2 - 0.281x_1x_2x_3, \end{aligned} \tag{4}$$

where $x_i \in [0, 1]$ for each $i \in \{1, \ldots, 4\}$.

Vehicle Crashworthiness: The vehicle crashworthiness design problem (RE3-5-4) involves minimization of the weight f_1, acceleration characteristics f_2 and toe-board instruction of the vehicle f_3. The variables x_1, x_2, x_3, x_4 and x_5 are all real-valued and specify the thickness of the five reinforced members around the frontal structure of the vehicle.

$$\begin{aligned} f_1(\boldsymbol{x}) = {} & 1640.2823 + 2.3573285x_1 + 2.3220035x_2 \\ & + 4.5688768x_3 + 7.7213633x_4 + 4.4559504x_5, \end{aligned} \tag{5}$$

$$\begin{aligned} f_2(\boldsymbol{x}) = {} & 6.5856 + 1.15x_1 - 1.0427x_2 + 0.9738x_3 + 0.8364x_4 - 0.3695x_1x_4 \\ & + 0.0861x_1x_5 + 0.3628x_2x_4 - 0.1106{x_1}^2 - 0.3437{x_3}^2 + 0.1764{x_4}^2, \end{aligned} \tag{6}$$

$$\begin{aligned} f_3(\boldsymbol{x}) = {} & -0.0551 + 0.0181x_1 + 0.1024x_2 + 0.0421x_3 - 0.0073x_1x_2 + 0.024x_2x_3 \\ & - 0.0118x_2x_4 - 0.0204x_3x_4 - 0.008x_3x_5 - 0.0241{x_2}^2 + 0.0109{x_4}^2, \end{aligned} \tag{7}$$

where $x_i \in [1, 3]$ for each $i \in \{1, \ldots, 5\}$.

In order to transform the above standard formulations to MF instances, we add error functions to the objective functions to yield various low fidelity estimates. Each error function is problem-independent and is a function of the decision variables, $\boldsymbol{x}$, and a user-specified parameter ($\phi \in [0, 10000]$) which represents the fidelity. Because of this problem-independent property, any test function f can be transformed into its lower-fidelity counterpart $\tilde{f}$:

$$\tilde{f}(\boldsymbol{x}, \phi) = f(\boldsymbol{x}) + e(\boldsymbol{x}, \phi), \tag{8}$$

where $\boldsymbol{x}$ is a vector with D decision variables, such that $l_i \leq x_i \leq u_i \quad \forall i \in [D]$, where $\boldsymbol{l}, \boldsymbol{u} \in \mathbb{R}^D$ are the lower- and upper-bounds, respectively[1] and ϕ is the *fidelity factor*, with $\tilde{f}(\boldsymbol{x}, 10000) = f(\boldsymbol{x})$ and $\tilde{f}(\boldsymbol{x}, 0)$ having the worst possible correlation to $f(\boldsymbol{x})$.

When transforming these multi-objective problems into multi-fidelity ones, noise is introduced using the error functions given in [14]. Here, ten functions were presented in [14] modelling different characteristic errors that can occur in CFD simulations. The first set models so-called resolution errors e_r, when there is an inconsistency between local and global errors; the second set models stochastic errors e_s, where the fitness value of the same solution can vary across different simulations; and the final set models instability errors e_{ins}, when failure occurs during a simulation. For this set of problems we use seven of this ten, having excluded e_{r4}, which requires prior knowledge of the global optimum, and both e_{ins1} and e_{ins2} as we are not modelling instability errors.

These error functions depend on the decision variables $\boldsymbol{x}$ and a user-specified parameter $\phi \in [0, 10000]$, which determines *how much* noise is introduced. This gives the user a fine-grained control over the fidelity of a given analysis. However, from an algorithmic design standpoint, it is more practical to discretize these continuous fidelity values to several so-called *fidelity levels*. The fidelity level Φ is a given value of ϕ, such that the correlation $r_f = \text{cor}\left(f(\boldsymbol{x}), \tilde{f}(\boldsymbol{x}, \phi)\right)$ is some specific value. Here we present the problems where each objective can be evaluated in four fidelity levels, with $\Phi_0 : r_f = 1$ indicating the highest fidelity, and the rest being: $\Phi_1 : r_f = 0.966$, $\Phi_2 : r_f = 0.933$ and $\Phi_3 : r_f = 0.9$, when all the functions are considered to be perfectly correlated across the entire search space ($\Phi_0 : r_f = 1.0$); and $\Phi_1 : r_f = 0.9$, $\Phi_2 : r_f = 0.8$ and $\Phi_3 : r_f = 0.7$, otherwise[2].

In [14], error functions were applied to a modified Rastrigin function with variable bounds $\boldsymbol{x} \in [-1, 1]^D$. Due to the intended problem-independent nature of these functions, it is important to ensure the scale of the noise being added is appropriate to produce the desired correlation between estimates in various fidelities. This is achieved through the use of a scaling factor, s_f, which is applied to the output of the error function such that Eq. (8) becomes:

$$\tilde{f}(\boldsymbol{x}, \phi, s_f) = f(\boldsymbol{x}) + s_f \cdot e(\boldsymbol{x}', \phi), \tag{9}$$

where $\boldsymbol{x}' \in [-1, 1]^D$ is a scaled version of $\boldsymbol{x}$. Each error and objective function pair must have its own scaling factor s_f, which is determined experimentally.

Irrespective of the choice of fidelity levels and their respective correlation values, in order for each fidelity level Φ to have the desired correlation, the values of s_f and ϕ must be computed. For each problem, we set $\Phi_3 = 5000$ as a base line[3] We then generate 100,000 samples using Latin hypercube sampling and for any given error and objective function pair, we solve a minimization problem to find the scaling factor s_f

[1] To conserve space, the following shorthand is used in this paper: $[k] = \{1, 2, \ldots, k\}$ and $[k^*] = \{0, 1, \ldots, k-1\}$. As is conventional, $[i, j]$ indicates the real interval between (and including) endpoints i and j.

[2] It is helpful to think of the index to Φ as indicating the degree of noise introduced, e.g., Φ_0 has no noise and thus the highest fidelity, whereas Φ_3 has a more noise and thus lower fidelity.

[3] We use Φ_3 as a baseline here, because we have 4 levels of fidelity. If 5 levels of fidelity were required, the baseline would be $\Phi_4 = 5000$.

Table 1. Table of correlation values. Given are the fidelity factors ϕ and their resultant Kendall-Tau correlations r_f for each objective function of each problem in the MOMF suite, for each error function e and prescribed fidelity level Φ.

	Φ	MOMF2-4-1a			MOMF2-4-1b			MOMF2-4-1c			MOMF2-4-1d			MOMF3-4-1				MOMF3-5-1			
		ϕ	r_{f_1}	r_{f_2}	ϕ	r_{f_1}	r_{f_2}	ϕ	r_{f_1}	r_{f_2}	ϕ	r_{f_1}	r_{f_2}	ϕ	r_{f_1}	r_{f_2}	r_{f_3}	ϕ	r_{f_1}	r_{f_2}	r_{f_3}
e_{r1}	1	8436	0.90	0.90	8426	0.97	0.90	8436	0.90	0.97	8286	0.97	0.97	8286	0.90	0.90	0.89	8436	0.92	0.90	0.91
	2	7064	0.83	0.80	7054	0.94	0.80	6663	0.80	0.93	6653	0.93	0.93	6823	0.80	0.80	0.80	6583	0.80	0.80	0.80
	3	5000	0.70	0.70	5000	0.90	0.70	5000	0.70	0.90	5000	0.90	0.90	5000	0.70	0.70	0.70	5000	0.70	0.70	0.70
e_{r2}	1	9178	0.92	0.87	9879	0.99	0.90	9028	0.90	0.97	9028	0.97	0.97	9368	0.90	0.90	0.90	7995	0.90	0.90	0.90
	2	6893	0.78	0.80	6913	0.94	0.80	7304	0.80	0.95	6673	0.93	0.93	6643	0.80	0.81	0.79	6012	0.80	0.77	0.79
	3	5000	0.70	0.70	5000	0.90	0.70	5000	0.70	0.90	5000	0.90	0.90	5000	0.70	0.70	0.70	5000	0.70	0.70	0.70
e_{r3}	1	8266	0.90	0.89	8476	0.98	0.90	8266	0.90	0.97	8256	0.97	0.97	8386	0.90	0.90	0.90	8246	0.91	0.90	0.90
	2	6583	0.80	0.76	6993	0.95	0.80	6673	0.80	0.93	6683	0.93	0.93	6733	0.80	0.80	0.81	6633	0.80	0.80	0.80
	3	5000	0.70	0.70	5000	0.90	0.70	5000	0.70	0.90	5000	0.90	0.90	5000	0.70	0.70	0.70	5000	0.70	0.70	0.70
e_{s1}	1	8697	0.92	0.90	8757	0.97	0.90	8416	0.90	0.97	8396	0.97	0.97	8436	0.90	0.90	0.90	8426	0.90	0.90	0.90
	2	6853	0.80	0.79	7224	0.94	0.80	6773	0.80	0.93	6683	0.93	0.93	6803	0.80	0.80	0.80	6783	0.80	0.80	0.80
	3	5000	0.70	0.70	5000	0.90	0.69	5000	0.70	0.90	5000	0.90	0.90	5000	0.70	0.70	0.70	5000	0.70	0.70	0.70
e_{s2}	1	7715	0.92	0.90	7725	0.97	0.90	7304	0.90	0.97	7254	0.97	0.97	7324	0.90	0.90	0.90	7304	0.90	0.90	0.90
	2	5891	0.80	0.78	6122	0.94	0.80	5881	0.80	0.93	5851	0.93	0.93	5891	0.80	0.80	0.80	5881	0.80	0.80	0.80
	3	5000	0.70	0.70	5000	0.90	0.70	5000	0.70	0.90	5000	0.90	0.90	5000	0.70	0.70	0.70	5000	0.70	0.70	0.70
e_{s3}	1	8697	0.90	0.91	8577	0.97	0.90	8717	0.90	0.97	8376	0.97	0.97	8426	0.90	0.90	0.90	8567	0.89	0.90	0.90
	2	7014	0.79	0.80	6983	0.94	0.80	7134	0.80	0.94	6723	0.93	0.93	6793	0.80	0.80	0.80	6943	0.79	0.80	0.80
	3	5000	0.70	0.70	5000	0.90	0.70	5000	0.70	0.90	5000	0.90	0.90	5000	0.70	0.70	0.70	5000	0.70	0.70	0.70
e_{s4}	1	7605	0.89	0.90	7525	0.97	0.90	7715	0.90	0.97	7254	0.97	0.97	7314	0.90	0.90	0.90	7525	0.89	0.90	0.90
	2	6102	0.80	0.81	6022	0.94	0.80	6112	0.80	0.94	5851	0.93	0.93	5891	0.80	0.80	0.80	5991	0.79	0.80	0.80
	3	5000	0.70	0.70	5000	0.90	0.70	5000	0.70	0.90	5000	0.90	0.90	5000	0.70	0.70	0.70	5000	0.70	0.70	0.70

which gives the smallest absolute difference between the Kendall-Tau correlation [9] $r_f = \text{cor}\left(f(\boldsymbol{x}), \tilde{f}(\boldsymbol{x}, \phi, s_f)\right)$ and the required correlation (0.9 for highly correlated and 0.7 otherwise) at $\phi = 5000$. The remaining values for ϕ were chosen as those that minimize the distance between the correlation values for all objectives and the target correlation value for a given level Φ. Table 1 presents these ϕ values and their associated correlations for each fidelity level Φ, and Table 2 lists the scaling factors for each combination of problem and error function.

In this suite, the problems MOMF2-4-1a to MOMF2-4-1d are based on RE2-4-1. The variants present different scenarios: in MOMF2-4-1a, f_1 and f_2 have positive correlations from 0.7 through to 1; in MOMF2-4-1b, f_1 is highly correlated across all fidelity levels and f_2 correlations vary between 0.7 and 1; in MOMF2-4-1c, f_1 correlations vary between 0.7 and 1 and f_2 is highly correlated across all fidelity levels; and in MOMF2-4-1d, both f_1 and f_2 are highly correlated across all fidelity levels. The other two problems, i.e., MOMF3-4-1 and MOMF3-5-1 involve three objectives and are based on RE3-4-7 and RE3-5-4 respectively. They have been designed, along with MOMF2-4-1a to have a correlation between 0.7 and 1.0 for all the objectives.

As discussed earlier, we have used 7 error functions in this study. Thus, for the 2-objective four-bar truss design problem, 3 low fidelity functions have been constructed for each of the objectives. One can observe from Table 2 (MOMF2-4-1a), that the correlations of the lowest fidelity i.e., Φ_3 is set to 0.7 for both objectives. Subsequent higher fidelities i.e., Φ_2 and Φ_1 have correlations close to 0.8 and 0.9 with the corresponding

Table 2. Table of scaling factors. Given are the scalar values s_f which produce the correlations in Table 1, where $r_f = \text{cor}\left(f(\boldsymbol{x}), \tilde{f}(\boldsymbol{x}, \phi, s_f)\right)$. This value is given for each combination of objective and error functions, respectively.

	MOMF2-4-1a		MOMF2-4-1b		MOMF2-4-1c		MOMF2-4-1d	
	s_{f_1}	s_{f_2}	s_{f_1}	s_{f_2}	s_{f_1}	s_{f_2}	s_{f_1}	s_{f_2}
e_{r1}	1.10e+02	1.64e-01	3.51e+01	1.64e-01	1.10e+02	3.91e-02	3.52e+01	3.90e-02
e_{r2}	2.01e+02	3.09e-01	6.34e+01	3.12e-01	2.02e+02	7.20e-02	6.36e+01	7.14e-02
e_{r3}	1.31e+02	2.17e-01	4.30e+01	2.17e-01	1.31e+02	5.09e-02	4.32e+01	5.07e-02
e_{s1}	5.39e+03	7.25e+00	1.74e+03	7.45e+00	5.35e+03	1.87e+00	1.69e+03	1.86e+00
e_{s2}	3.26e+04	4.42e+01	1.03e+04	4.41e+01	3.26e+04	1.14e+01	1.03e+04	1.13e+01
e_{s3}	6.93e+03	5.61e+00	1.77e+03	5.59e+00	6.92e+03	1.60e+00	1.78e+03	1.59e+00
e_{s4}	4.23e+04	3.44e+01	1.08e+04	3.42e+01	4.22e+04	9.70e+00	1.08e+04	9.61e+00

	MOMF3-4-1			MOMF3-5-1		
	s_{f_1}	s_{f_2}	s_{f_3}	s_{f_1}	s_{f_2}	s_{f_3}
e_{r1}	1.44e-01	1.50e-01	1.67e-01	2.03e+00	3.30e-01	1.78e-02
e_{r2}	2.54e-01	2.64e-01	2.96e-01	3.54e+00	5.84e-01	3.08e-02
e_{r3}	1.87e-01	1.99e-01	2.05e-01	2.31e+00	3.86e-01	2.09e-02
e_{s1}	7.04e+00	7.40e+00	7.94e+00	1.07e+02	1.74e+01	9.50e-01
e_{s2}	4.30e+01	4.51e+01	4.84e+01	6.49e+02	1.05e+02	5.75e+00
e_{s3}	6.41e+00	6.55e+00	7.22e+00	1.31e+02	1.79e+01	1.03e+00
e_{s4}	3.89e+01	4.00e+01	4.39e+01	8.01e+02	1.09e+02	6.29e+00

scaling factors listed in Table 2. The correlations for different values of Φ using each of the error functions for MOMF2-4-1a are presented in Fig. 1. It is perfectly possible to use this information to choose different Φ_3, Φ_2 and Φ_1 values for each of the objectives to reflect different scenarios. For example, as shown in Fig. 2, low fidelity estimates of f_1 have a higher correlation overall (say, varying between 0.9 and 1) than those of f_2 (say, varying between 0.7 and 1).

3 Numerical Experiments and Discussion

In all, there are 6 problems MOMF2-4-1a to MOMF2-4-1d, MOMF3-4-1a and MOMF2-5-1a, each of which can be modified using error functions e_{r1}, e_{r2}, e_{r3}, e_{r4}, e_{s1}, e_{s2} and e_{s3} resulting in 42 possible instances. For a preliminary assessment of problem behavior and algorithm performance, we use the well established non-dominated sorting genetic algorithm (NSGA-II) [4] as a canonical optimization method to solve these instances. A population of 200 individuals is evolved over 1000 generations. The probability of crossover and mutation was set to 0.9 $1/|\boldsymbol{x}|$, respectively, with the distribution index of both the simulated binary crossover (SBX) and polynomial mutation (PM) set as 20.

We first focus on MOMF2-4-1a with the error function e_{r1} with correlations of both objectives varying between 0.7 and 1. The non-dominated set of solutions obtained

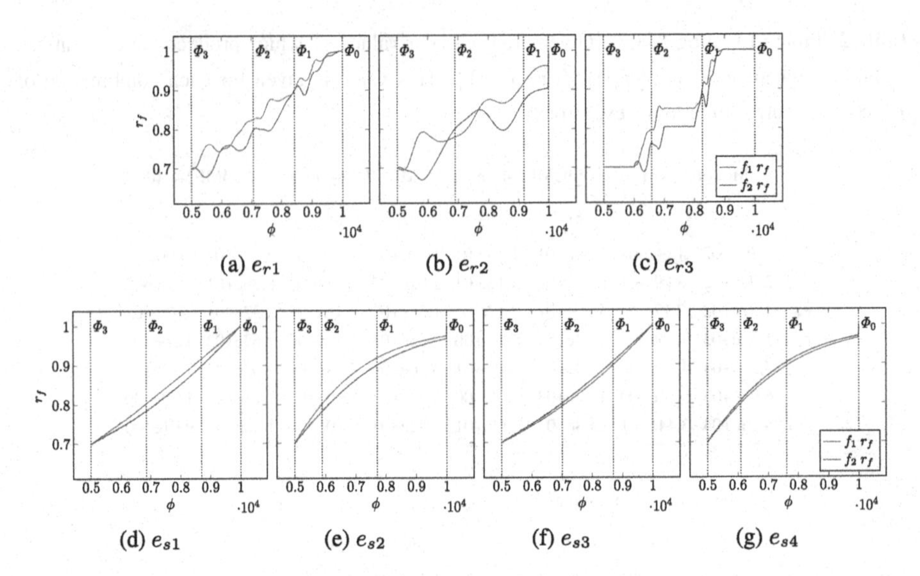

Fig. 1. Plots of ϕ value against $r_f = \text{cor}\left(f(\boldsymbol{x}), \tilde{f}(\boldsymbol{x}, \phi, s_f)\right)$ for problem MOMF2-4-1a. Values for f_1 and f_2 are shown in red and blue, respectively. The ϕ values for each fidelity level Φ are indicated by vertical dotted lines. (Color figure online)

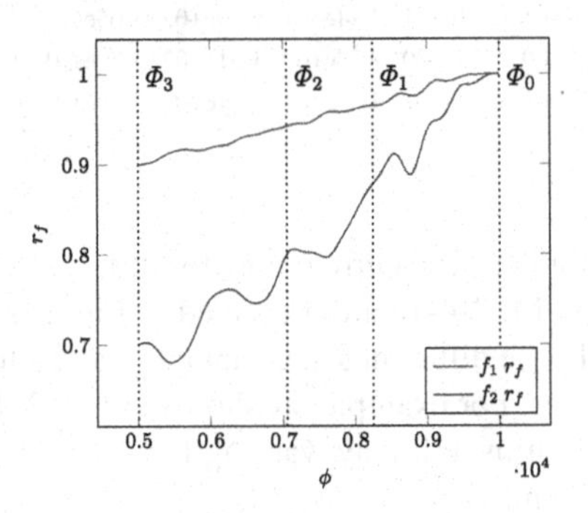

Fig. 2. Correlation plot of problem MOMF2-4-1b with error function e_{r1}. This variant has correlation ranges of $r_f \in [0.9, 1]$ for f_1 and $r_f \in [0.7, 1]$ for f_2.

using Φ_0/Φ_0 i.e., highest fidelity analysis for both objectives is marked in blue in Fig. 3. The normalized variable values of the non-dominated solutions are in blue and are presented next to it. It can be observed that the non-dominated solutions span a continuous convex curve. The variables x_1, x_2 and x_4 span the entire variable range, while variable x_3 is at its lower bound. If one opts to solve the above problem using Φ_0/Φ_1, i.e., using a lower fidelity analysis for the second objective, the obtained solutions would be far from optimal (Fig. 3a). On the other hand, if one uses Φ_1/Φ_0, i.e., using a lower fidelity analysis for the first objective, the obtained set of solutions would be close to optimal as evidenced by Fig. 3b. The results obtained using Φ_0/Φ_2, Φ_0/Φ_3, Φ_2/Φ_0, Φ_3/Φ_0 are also presented in Fig. 3 which suggests that for this problem, the second objective should be

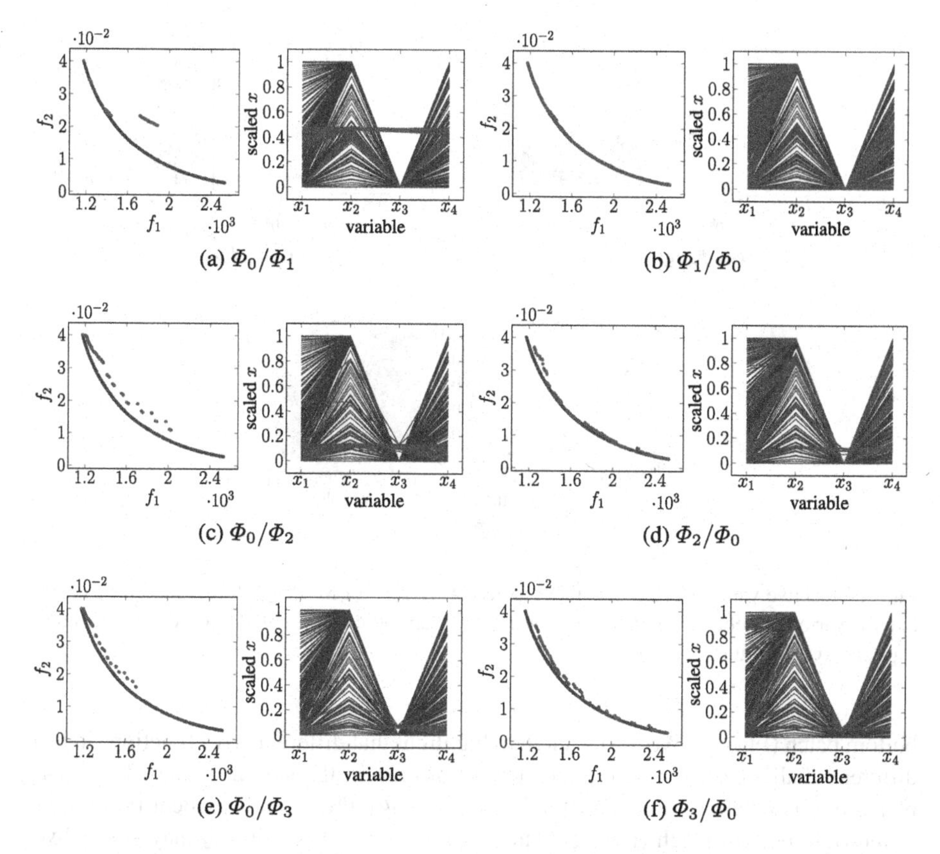

Fig. 3. Objective value (left) and parallel axis (right) plots for MOMF2-4-1a with error function e_{r1}. For f_1/f_2, blue indicates Φ_0/Φ_0 and red is given in the captions. (Color figure online)

always be evaluated in highest fidelity, while the choice of lower fidelity analysis for the first objective will have low impact on the convergence and diversity.

To observe a case where the second objective needs to be evaluated in HF for a better performance, one may consider MOMF2-4-1c where the correlations of the second objective across the design space is between 0.9 and 1. Results of Φ_0/Φ_1, Φ_0/Φ_2, and Φ_0/Φ_3 are presented in Fig. 4 which does show some improvement, although only parts of the PF are uncovered or pre-converged solutions are delivered with the variables spanning limited ranges. It is important to note that although correlations of f_2 are high and vary between 0.9 and 1, they are computed based on 100,000 LHS samples over the variable range. A denser sampling may be considered for more accurate estimates.

Next, we observe the three-objective problems MOMF3-4-1 and MOMF3-5-1. We focus on cases when all objectives are evaluated in the same fidelity level, i.e., $\Phi_1/\Phi_1/\Phi_1$, $\Phi_2/\Phi_2/\Phi_2$, and $\Phi_3/\Phi_3/\Phi_3$ with error function e_{r1}. One can observe that the choice of lower fidelity levels can lead to (i) identification of subsets of PF instead of a good spread over whole PF, and (ii) poorly converged solutions. As an easily observable example, use of $\Phi_3/\Phi_3/\Phi_3$ for MOMF3-5-1 fails to identify solutions in the right

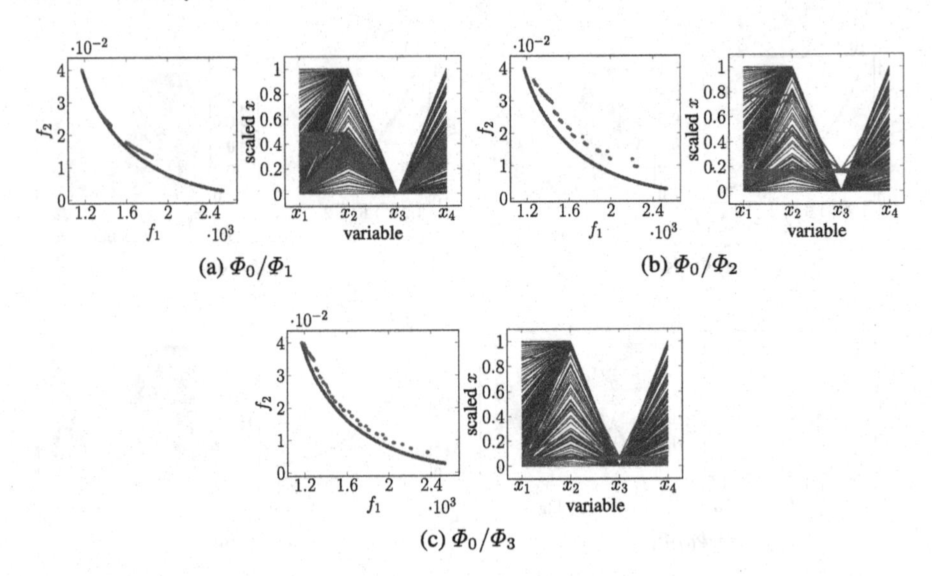

Fig. 4. Objective values (left) and parallel axis plots (right) for MOMF2-4-1c with error function e_{r1} for various fidelity combinations. For f_1/f_2, blue indicates Φ_0/Φ_0 and red is given in the captions. (Color figure online)

bottom patch (Fig. 5f). We also wanted to highlight that different error functions induce different challenges even where the correlations are in the similar ranges. For example, as discussed above, for MOMF3-5-1 with e_{r1}, $\Phi_3/\Phi_3/\Phi_3$ fails to identify solutions in the right bottom patch (Fig. 5f), whereas $\Phi_3/\Phi_3/\Phi_3$ based on e_{s2} may still deliver competitive results as can be observed from Fig. 6). In all these examples, we have considered 4 levels of fidelity for the objective functions and the correlations increased in an uniform manner either between 0.7 and 1 or between 0.9 and 1. However, it is worth mentioning that the flexibility of the presented approach allows for creation of any number of fidelities for any of the objectives.

Since the problems presented above are derived from real world problems, the theoretical Pareto Front (PF) or the Pareto Set (PS) are unknown, which makes it difficult to ascertain the performance relative to the theoretical optimum. Therefore, as a last experiment, and also to emphasize the generalizability of the proposed approach, we present here the cases using some synthetic benchmarks with known PF. For this exercise, we selected two problems, namely, DTLZ2 [5] and minus DTLZ2 [8]. We consider 3-objective formulations of both problems involving 7 variables. Furthermore, we consider 5 fidelity levels with non-uniform increment in correlation values i.e., $r_f \in \{0.7, 0.8, 0.9, 0.95, 1.0\}$ based on error function e_{s3} as shown in Fig. 7.

The results obtained using for DTLZ2 with $\Phi_1/\Phi_1/\Phi_1$, $\Phi_2/\Phi_2/\Phi_2$, and $\Phi_3/\Phi_3/\Phi_3$, and $\Phi_4/\Phi_4/\Phi_4$ are presented in (a) to (d) in Fig. 8. One can clearly observe that the obtained PF approximation becomes worse in quality as lower fidelity analysis is used. For example, $\Phi_1/\Phi_1/\Phi_1$ still has some solutions close to the optimal albeit a few solutions in dominance resistant regions (axial directions). When the lowest fidelity analysis

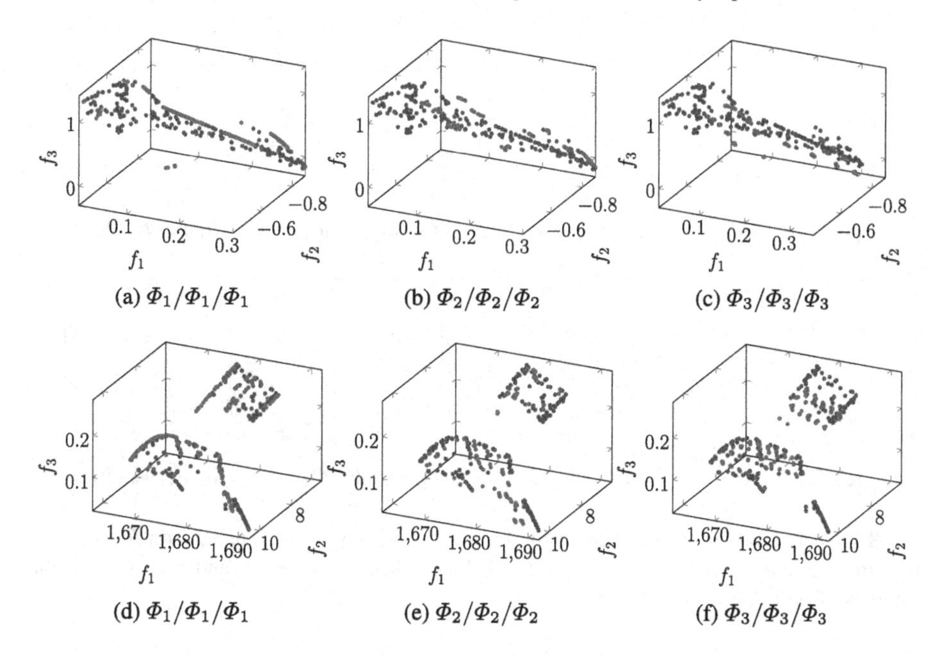

Fig. 5. Objective values for MOMF3-4-1 (top row) and MOMF3-5-1 (bottom row) using error function e_{r1} for various fidelity combinations. For $f_1/f_2/f_3$, blue indicates $\Phi_0/\Phi_0/\Phi_0$ and red is given in the captions. (Color figure online)

is used for all the objective functions, only solutions along axial directions are obtained, with none in the vicinity of the PF.

For the inverted DTLZ2 problem with e_{s3} as the error function, the use of lower fidelity functions yield poorly converged solutions with the distribution biased towards the edges (and far from the PF) as can be observed from (e) to (h) of Fig. 8. This again reinforces the poorer performance obtained when lower fidelities are used.

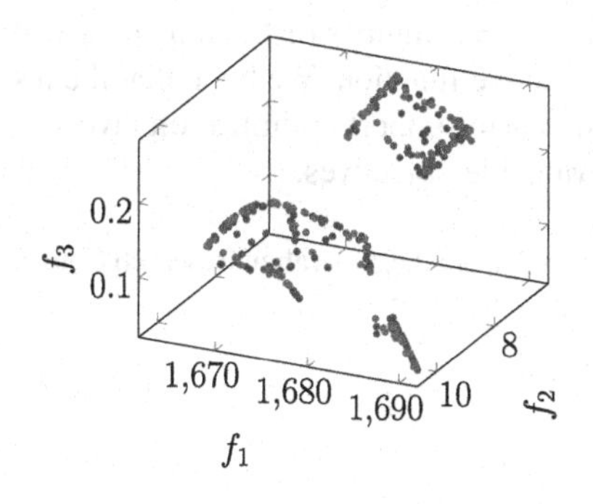

Fig. 6. Objective value for MOMF3-5-1. For $f_1/f_2/f_3$, blue indicates $\Phi_0/\Phi_0/\Phi_0$ and red indicates $\Phi_3/\Phi_3/\Phi_3$. (Color figure online)

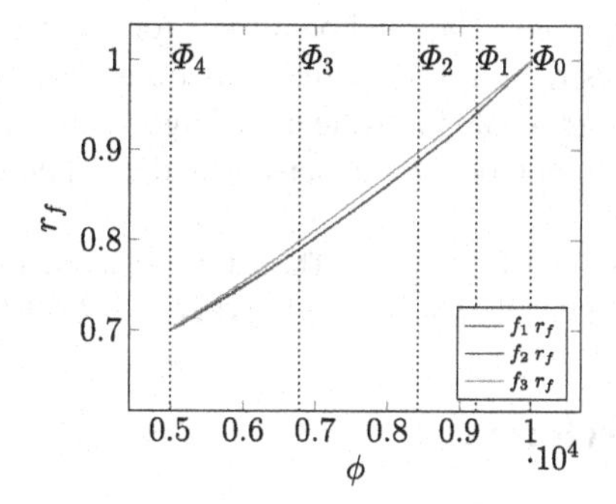

Fig. 7. Correlation of problem MOMF3-7-1 with error function e_{s3}. Here we have used 5 fidelity levels such that $r_f \in \{0.7, 0.8, 0.9, 0.95, 1.0\}$.

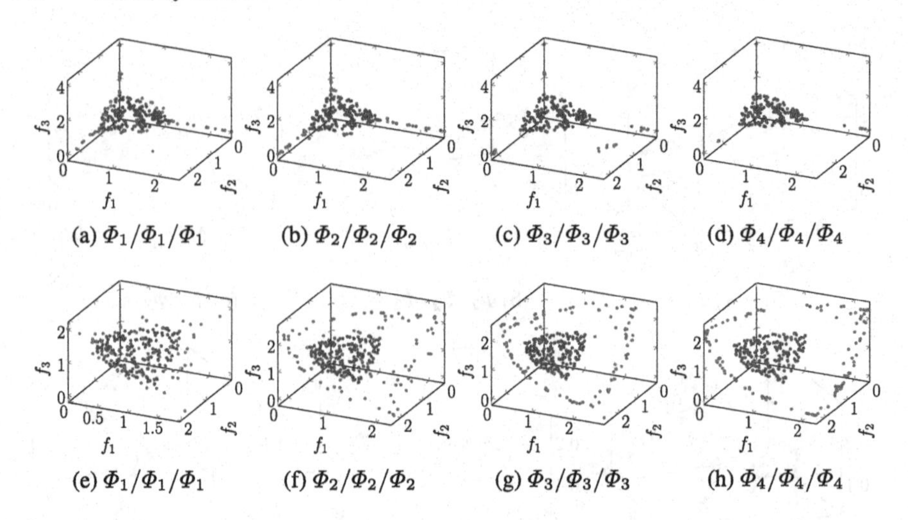

Fig. 8. Objective values for DTLZ2 (top row) and DTLZ2^{-1} (bottom row) using error function e_{s3} for various fidelity combinations. For $f_1/f_2/f_3$, blue indicates $\Phi_0/\Phi_0/\Phi_0$ and red is given in the captions. (Color figure online)

4 Conclusion and Future Work

Multi-objective optimization problems involving independently evaluable objectives are often encountered in practice. The objective computations in such problems may involve different iterative numerical solvers such as FEA, CFD etc. that can be invoked with various levels of fidelity. The lack of benchmarks has been one of the factors limiting the fundamental development of efficient optimization algorithms to deal with such classes of problems. In this paper, we present an approach and a test suite for multi-objective multi-fidelity optimization problems. The problems are derived by combining existing unconstrained, multi-objective design optimization problems with resolution/stochastic/instability errors that are common manifestations of multi-fidelity simulations. The method allows for the construction of any number of LF functions with desired level of correlations for any given HF objective function. We hope that the test suite would motivate novel algorithmic developments to support optimization involving computationally expensive and independently evaluable objectives.

Acknowledgement. The authors acknowledge the support from the Australian Research Council through the Discovery Project grant DP190101271.

References

1. Allmendinger, R., Handl, J., Knowles, J.: Multiobjective optimization: when objectives exhibit non-uniform latencies. Eur. J. Oper. Res. **243**(2), 497–513 (2015)
2. Blank, J., Deb, K.: Handling constrained multi-objective optimization problems with heterogeneous evaluation times: proof-of-principle results. Memetic Comput. 1–16 (2022)

3. Branke, J., Asafuddoula, M., Bhattacharjee, K.S., Ray, T.: Efficient use of partially converged simulations in evolutionary optimization. IEEE Trans. Evol. Comput. **21**(1), 52–64 (2016)
4. Deb, K., Thiele, L., Laumanns, M., Zitzler, E.: Scalable multi-objective optimization test problems. In: Proceedings of the 2002 Congress on Evolutionary Computation, CEC 2002 (Cat. No. 02TH8600), vol. 1, pp. 825–830. IEEE (2002)
5. Deb, K., Thiele, L., Laumanns, M., Zitzler, E.: Scalable test problems for evolutionary multiobjective optimization. In: Abraham, A., Jain, L., Goldberg, R. (eds.) Evolutionary Multiobjective Optimization, pp. 105–145. Springer, Heidelberg (2005). https://doi.org/10.1007/1-84628-137-7_6
6. Habib, A., Singh, H.K., Ray, T.: A multiple surrogate assisted multi/many-objective multifidelity evolutionary algorithm. Inf. Sci. **502**, 537–557 (2019)
7. Huband, S., Barone, L., While, L., Hingston, P.: A scalable multi-objective test problem toolkit. In: Coello Coello, C.A., Hernández Aguirre, A., Zitzler, E. (eds.) EMO 2005. LNCS, vol. 3410, pp. 280–295. Springer, Heidelberg (2005). https://doi.org/10.1007/978-3-540-31880-4_20
8. Ishibuchi, H., Setoguchi, Y., Masuda, H., Nojima, Y.: Performance of decomposition-based many-objective algorithms strongly depends on pareto front shapes. IEEE Trans. Evol. Comput. **21**(2), 169–190 (2016)
9. Kendall, M.G.: A new measure of rank correlation. Biometrika **30**(1/2), 81–93 (1938)
10. Kenny, A., Ray, T., Singh, H.K.: An iterative two-stage multi-fidelity optimization algorithm for computationally expensive problems. IEEE Trans. Evol. Comput. (2022)
11. Mamun, M., Singh, H., Ray, T.: An approach for computationally expensive multi-objective optimization problems with independently evaluable objectives. Swarm Evol. Comput. (2022)
12. Rahi, K.H., Singh, H.K., Ray, T.: Partial evaluation strategies for expensive evolutionary constrained optimization. IEEE Trans. Evol. Comput. **25**(6), 1103–1117 (2021)
13. Tanabe, R., Ishibuchi, H.: An easy-to-use real-world multi-objective optimization problem suite. Appl. Soft Comput. **89**, 106078 (2020)
14. Wang, H., Jin, Y., Doherty, J.: A generic test suite for evolutionary multifidelity optimization. IEEE Trans. Evol. Comput. **22**(6), 836–850 (2017)
15. Wang, Z., Li, Q., Yang, Q., Ishibuchi, H.: The dilemma between eliminating dominance-resistant solutions and preserving boundary solutions of extremely convex pareto fronts. Complex Intell. Syst. 1–10 (2021)
16. Wang, Z., Ong, Y.S., Ishibuchi, H.: On scalable multiobjective test problems with hardly dominated boundaries. IEEE Trans. Evol. Comput. **23**(2), 217–231 (2018)
17. Zitzler, E., Deb, K., Thiele, L.: Comparison of multiobjective evolutionary algorithms: empirical results. Evol. Comput. **8**(2), 173–195 (2000)

Indicator Design and Complexity Analysis

Diversity Enhancement via Magnitude

Steve Huntsman(✉)

STR, Arlington, VA 22203, USA
steve.huntsman@str.us

Abstract. Promoting and maintaining diversity of candidate solutions is a key requirement of evolutionary algorithms. In this paper, we use the recently developed theory of *magnitude* to construct a gradient flow that systematically manipulates finite subsets of Euclidean space to enhance their diversity, and we apply the ideas in service of multi-objective evolutionary algorithms. We demonstrate diversity enhancement on benchmark problems using leading algorithms.

1 Introduction

Promoting and maintaining diversity of candidate solutions is a key requirement of *evolutionary algorithms* (EAs) in general and *multi-objective EAs* (MOEAs) in particular [1,2]. Many ways of measuring diversity have been considered, and many shortcomings identified [3]. Perhaps the most theoretically attractive diversity measure, used by [4,5], is the *Solow-Polasky diversity* [6]. It turns out that a recently systematized theory of diversity in generalized metric spaces [7] singles out the Solow-Polasky diversity or *magnitude* of a (certain frequently total subset of a) finite metric space as equal to the maximum value of the "correct" definition (1) of diversity that uniquely satisfies various natural desiderata. While the notion of magnitude was implicit in the mathematical ecology literature over 25 years ago, an underlying notion of a diversity-maximizing probability distribution is much more recent and has not yet been applied to EAs.

In the context of MOEAs, a practical shortcoming associated with magnitude is its $O(n^3)$ algorithmic cost. To avoid this, [4,5] use an efficient approximation to merely *measure* diversity rather than attempting to *enhance* it.

However, it can be profitable to incur the marginal cost of computing a so-called weighting *en route* to the magnitude, since we can use it to enhance diversity near the boundary of the image of the candidate solution set under the objective functions. The nondominated part of this image is the current approximation to the Pareto front; the ability of weightings to couple both diversity and convergence to the Pareto front dovetails with recent indicator-based EA approaches to Pareto-dominance based MOEAs [8,9]. Moreover, the agnosticism of weightings to dimension further enhances their suitability for such applications.

In this paper, we construct a gradient flow that systematically manipulates finite subsets of Euclidean space to enhance their diversity, which provides a

M. Emmerich et al. (Eds.): EMO 2023, LNCS 13970, pp. 377–390, 2023.
https://doi.org/10.1007/978-3-031-27250-9_27

useful primitive for quality diversity [10]. We then apply this primitive in service of MOEAs by diversifying solution data through local mutations. For the sake of illustration, we only perform these mutations on the results already obtained by a MOEA, though they can be performed during evolution.

The paper is organized as follows. In Sect. 2, we sketch the concepts of weightings, magnitude, and diversity, and describe an efficiently computable scale above which a weighting is guaranteed to be proportional to the unique diversity-maximizing distribution. In Sect. 3, we develop a notion of a weighting gradient (estimate) and an associated flow. In Sect. 4, we use this gradient flow to demonstrate diversity enhancement on a toy problem before turning to benchmark problems in Sect. 5. Finally, we discuss algorithmic extensions in Sect. 6 before remarks in Sect. 7.

2 Weightings, Magnitude, and Diversity

For details on the ideas in this section, see §6 of [7] and also [16,30].

Call a square matrix $Z \geq 0$ a *similarity matrix* if $\text{diag}(Z) > 0$. A motivating class of examples is $Z = \exp[-td]$ where square brackets indicate entrywise function application, $t \in (0, \infty)$, and d is a square *dissimilarity matrix* (e.g., the matrix encoding a finite metric space). A *weighting* w is a column vector satisfying $Zw = 1$, where the vector of all ones is indicated on the right. A *coweighting* is the transpose of a weighting for Z^T. If Z admits both a weighting w and a coweighting, then its *magnitude* is defined via $1^T w = \sum_j w_j$, which also turns out to equal the sum of the coweighting components.

In the case $Z = \exp[-td]$ and d is the distance matrix corresponding to a finite subset of Euclidean space, Z is positive definite [11], hence invertible, and so its weighting and magnitude are well-defined and unique. More generally, if Z is invertible then its magnitude is $1^T Z^{-1} 1$. For d as specified above, the *magnitude function* is defined as the map $t \mapsto 1^T (\exp[-td])^{-1} 1$.

Weightings are excellent scale-dependent boundary detectors in Euclidean space (see, e.g., Fig. 2 and [12,13]). Meanwhile, magnitude is a very general notion of size that encompasses rich scale-dependent geometrical data [14].

Example 1. Consider a three-point space with $d_{12} = d_{13} = 1 = d_{21} = d_{31}$ and $d_{23} = \delta = d_{32}$. A routine calculation yields that

$$w_1 = \frac{e^{(\delta+2)t} - 2e^{(\delta+1)t} + e^{2t}}{e^{(\delta+2)t} - 2e^{\delta t} + e^{2t}}; \quad w_2 = w_3 = \frac{e^{(\delta+2)t} - e^{(\delta+1)t}}{e^{(\delta+2)t} - 2e^{\delta t} + e^{2t}}.$$

For $\delta = 10^{-3}$, Fig. 1 shows that at $t = 10^{-2}$, the "effective size" of the nearby points is ≈ 0.25; that of the distal point is ≈ 0.5, so the "effective number of points" is ≈ 1. At $t = 10$, these effective sizes are respectively ≈ 0.5 and ≈ 1, so the effective number of points is ≈ 2. Finally, at $t = 10^4$, the effective sizes are all ≈ 1, so the effective number of points is ≈ 3.

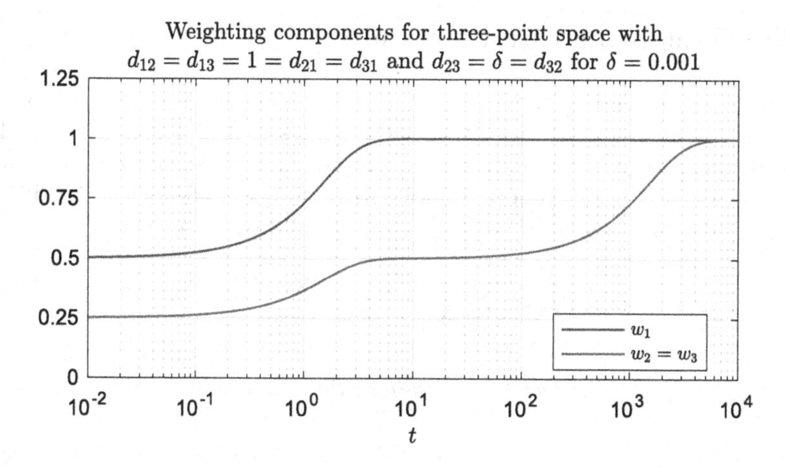

Fig. 1. Weighting components for an "isoceles" metric space. The magnitude function $w_1 + w_2 + w_3$ gives a scale-dependent "effective number of points."

For a probability distribution p and similarity matrix Z, the *diversity* $D_q^Z(p)$ is defined for $1 < q < \infty$ (and via limits for $q = 1, \infty$) via

$$\log D_q^Z(p) := \frac{1}{1-q} \log \sum_{j:p_j>0} p_j (Zp)_j^{q-1}. \tag{1}$$

This is the "correct" measure of diversity in essentially the same way that Shannon entropy is the "correct" measure of information [7]. (In fact, the expression (1) is a generalization of the Rényi entropy of order q. In the event $Z = I$, the usual Rényi entropy is recovered, with Shannon entropy as the case $q = 1$.) We therefore restrict our attention to it versus other measures such as those discussed in [2,3].

Recent mathematical developments [7,15] have clarified the role of magnitude in *maximizing* (1) versus merely computing it. Specifically, if $Z = \exp[-td]$ is positive definite with d symmetric, nonnegative, and with zero diagonal, and if Z admits a positive weighting $w = Z^{-1}1$, then this (unique) weighting is proportional to the diversity-maximizing distribution. This situation holds automatically if d is the distance matrix of a finite subset of Euclidean space and if Z is diagonally dominant (i.e., $Z_{jj} > \sum_{k \neq j} Z_{jk}$).

For d with zero diagonal and all other entries positive, there is a least $t_d > 0$ such that $\exp[-td]$ is diagonally dominant for any $t > t_d$. Because $\exp[-td]$ is diagonally dominant iff $1 > \max_j \sum_{k \neq j} \exp(-td_{jk})$, we can efficiently estimate t_d using the following elementary bounds and a binary search:

Lemma 1. *For $d \in M_n$ as above,* $\frac{\log(n-1)}{\min_j \max_k d_{jk}} \leq t_d \leq \frac{\log(n-1)}{\min_j \min_{k \neq j} d_{jk}}$.

More importantly, we can also use Lemma 1 to find the least $t_+ < t_d$ such that $\exp[-td]$ admits a positive weighting for $t > t_+$.

3 The Weighting Gradient Flow

We now define a gradient flow that (for $t \geq t_+$) increases the diversity of finite subsets of Euclidean space and thereby provides a useful primitive for EAs. Although there are various sophisticated approaches to estimating gradients on point clouds (see, e.g., [17]), a reasonable heuristic estimate for the specific case of the gradient of a weighting w on $\{x_j\}_j$ in Euclidean space is

$$(\hat{\nabla} w)_j := \sum_{k \neq j} \frac{Z_{jk}}{\sum_{k' \neq j} Z_{jk'}} \frac{w_k - w_j}{d_{jk}} e_{jk}, \tag{2}$$

where $e_{jk} := \frac{x_k - x_j}{d_{jk}}$. The *weighting gradient flow* induced by (2) is

$$\dot{x} = \hat{\nabla} w. \tag{3}$$

Example 2. Figure 2 illustrates how weightings identify boundaries at various scales, and the corresponding weighting gradient estimates (2).

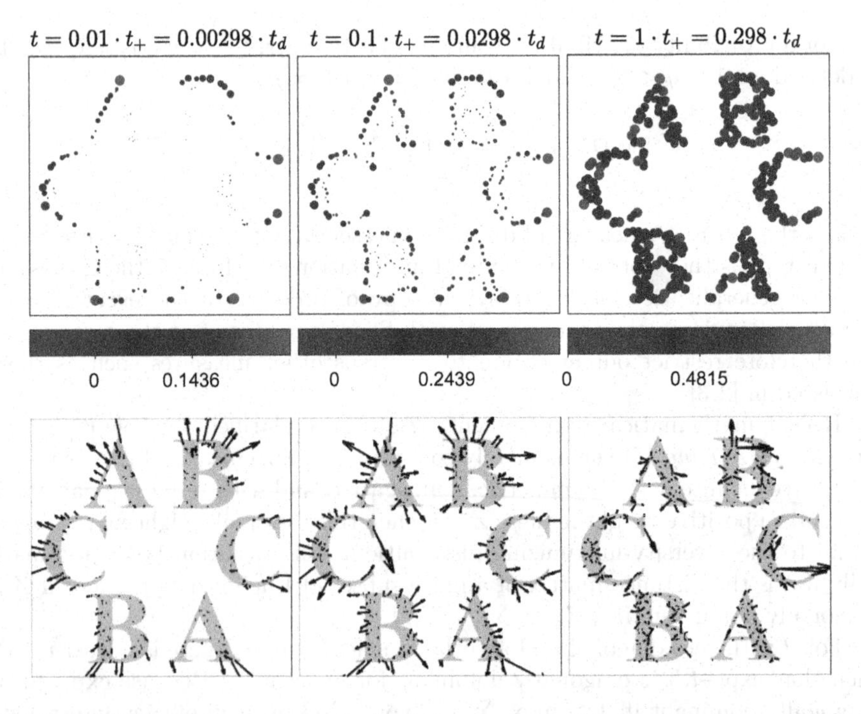

Fig. 2. (Top) Weighting components for 500 points sampled without replacement from a probability distribution on $\mathbb{Z}^2$ that is approximately uniform on its support. From left to right, various scale factors t defining $Z = \exp[-td]$ (with d = Euclidean distance) are shown in terms of the intrinsic scales t_d and t_+. Both the color and size of a point indicate its weighting component; the nonzero color axis tick mark is at half maximum. (Bottom) Weighting gradient estimate (2) for the data above. The gradient vectors are scaled uniformly in each panel for visualization purposes. Note that for the largest value of t the large gradient vectors have basepoints near other large gradient vectors.

4 Enhancing Diversity

Following [18], we apply the ideas sketched above to a toy problem where the objective function f has three components, each measuring the distance to a vertex of a regular triangle with vertices in S^1. The application is mostly conceptually straightforward, but we mention a few implementation details:

- We begin with a uniformly distributed sample of $n_0 = 10^3$ points in the disk of radius 1.25, and retain n points that are dominated by $\leq \delta = 0.1$;
- Replace misbehaving points (e.g., out of bounds or `NaN`s) with predecessors;
- Set $S_j := 1 - 2\frac{\text{dom}_j}{\max_k \text{dom}_k}$, where $\text{dom}_j = |\{\text{points dominating the } j\text{th point}\}|$;
- Evolve the n points under a modulated version of (3) *on the objective space* with $t = t_+$ as $dy_j = ds \cdot S_j(\hat{\nabla} w)_j$ for only $N = 10$ steps and step size $ds = \sqrt{\langle \min_{k \neq j}(d_f)_{jk} \rangle / n}$, where the pullback metric is $d_f(x, x') := d(f(x), f(x'))$;
- Pull back the weighting gradient flow from objective to solution space using the Jacobian's pseudoinverse, then recompute points in objective space.

The result of this experiment is depicted in Figs. 3 and 4. The salutary effect on diversity in objective (and solution) space is apparent. This can be quantified via the objective space magnitude functions, as shown in Fig. 5.

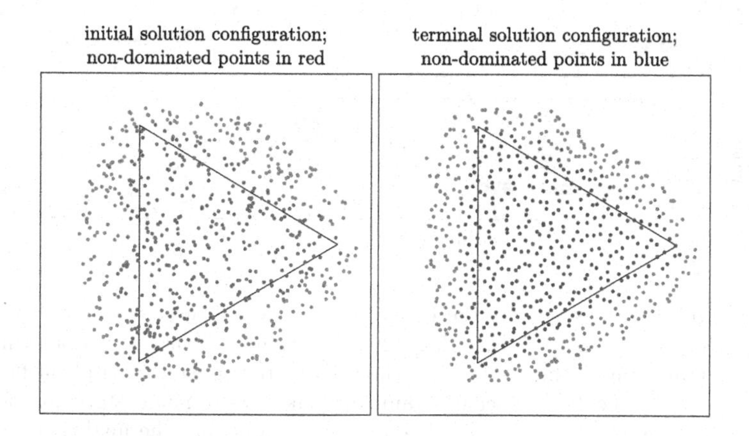

initial solution configuration;
non-dominated points in red
terminal solution configuration;
non-dominated points in blue

Fig. 3. Comparision of initial (red; left) and terminal (blue; right) locations of points in the solution space. The weighting gradient flow produces more evenly distributed terminal points. The triangle defining objective components (by distance to vertices) is shown. The actual Pareto front is the interior of the triangle; the area displayed is $[-1, 1]^2$. Bottom: comparision of initial (red; left) and terminal (blue; right) points in the objective space. The terminal points are more evenly distributed. (Color figure online)

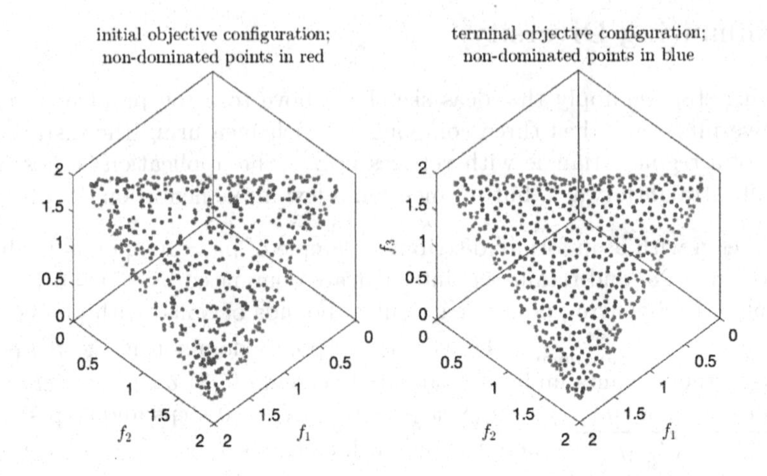

Fig. 4. Comparision of initial (red; left) and terminal (blue; right) points in the objective space. The terminal points are more evenly distributed. (Color figure online)

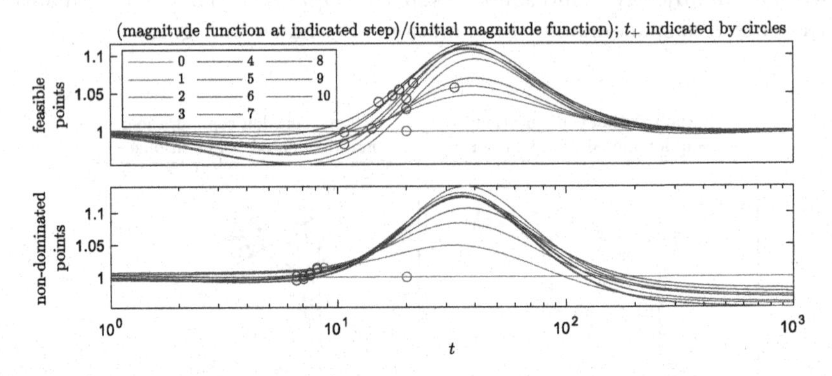

Fig. 5. Magnitude increases for the experiment of Sect. 4 at scales above t_+, where magnitude equals diversity. (Top) Magnitude function quotients at various timesteps for feasible points under the evolution of the (modulated) weighting gradient flow. The horizontal axis t indicates the scale parameter; timesteps of numerators are indicated via color, going from red at the initial timestep (0) to blue at the final timestep (10): the denominator is the function at the initial timestep. Circles indicate the scales t_+. (Bottom) As above, but for non-dominated points. (Color figure online)

5 Performance on Benchmarks

The effectiveness of the (modulated) weighting gradient flow approach hinges on the ability to cover and thereby "keep pressure on" the Pareto front. A straightforward way to do this is to use a MOEA to produce an initial overapproximation of the Pareto front as in [19], and then improve the diversity of the overapproximation via the weighting gradient flow. We proceed to detail the results of an experiment along these lines. For the experiment we considered two leading MOEAs (NSGA-II [20] and SPEA2 [21]) and two leading benchmark problem sets (DTLZ [22][1] and WFG [24]), all implemented in PlatEMO version 2.9 [25]. For each problem, we used 10 decision variables, three objectives (to enable visualization), and performed 10 runs (which appears quite adequate for characterization purposes) with population size 250 and 10^4 fitness evaluations. We then took $N = 10$ timesteps for the weighting gradient flow as before.

Figure 6 (cf. Fig. 5) shows magnitude functions at various timesteps of the (modulated) weighting gradient flow applied to the results of NSGA-II on the WFG2 benchmark. Feasible points show a diversity (as measured by magnitude at scale t_+ for the feasible objective points) increase of about 10%, whereas non-dominated points show a diversity increase of several percent as well, even as the total number of non-dominated points decreases by about 15%.

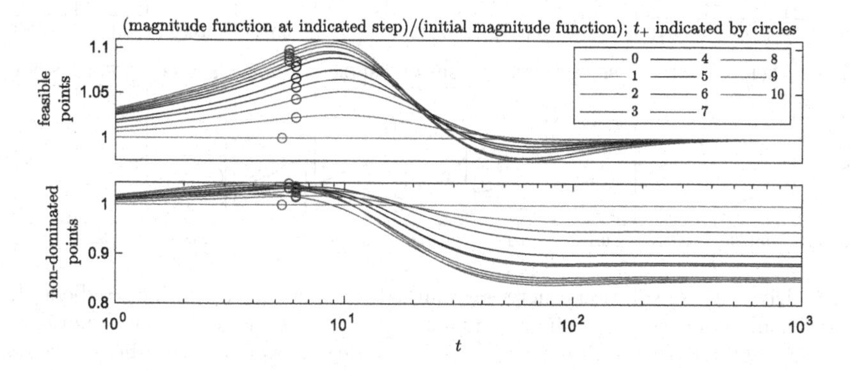

Fig. 6. As in Fig. 5, but for a solution of the WFG2 benchmark via NSGA-II.

We produce an ensemble characterization in Fig. 7. The figure shows that the number of non-dominated points decreases since the weighting gradient flow pushes some points a short distance away from the Pareto front (as illustrated in Fig. 9) before they are halted or reversed. The figure also shows that the diversity of non-dominated points generally increases slightly, and the diversity of feasible points increases significantly. As a consequence, the diversity contributions of individual solutions (as measured by the average weighting, i.e., the magnitude of non-dominated points divided by their cardinality) also increases significantly. For less challenging problems such as in Fig. 3, the number of non-dominated points will decrease less, and the diversity gains will be enhanced.

[1] For DTLZ, we considered only the two most relevant problems, viz. DTLZ4 and DTLZ7. DTLZ4 was formulated "to investigate an MOEA's ability to maintain a good distribution of solutions" and DTLZ7 was formulated to "test an algorithm's ability to maintain subpopulation in different Pareto-optimal regions" [22]. (NB. One approach for the latter, not pursued here, is to resample points so that the diversity per point in each connected component of the Pareto front is approximately equal. For the application of topological data analysis to Pareto fronts, see [23].).

On the other hand, the effects of the weighting gradient flow are considerably reduced in the case of SPEA2, which produces a visibly more uniform distribution in objective space than NSGA-II: see Fig. 11. The weighting gradient flow appears to *decrease* this uniformity; the formation of a gap just behind the boundary along with a slight increase in the population near the boundary are the main visible indicators that something useful (at least for DTLZ4, WFG2, WFG3, WFG6, and WFG8, per Fig. 7) is actually happening.

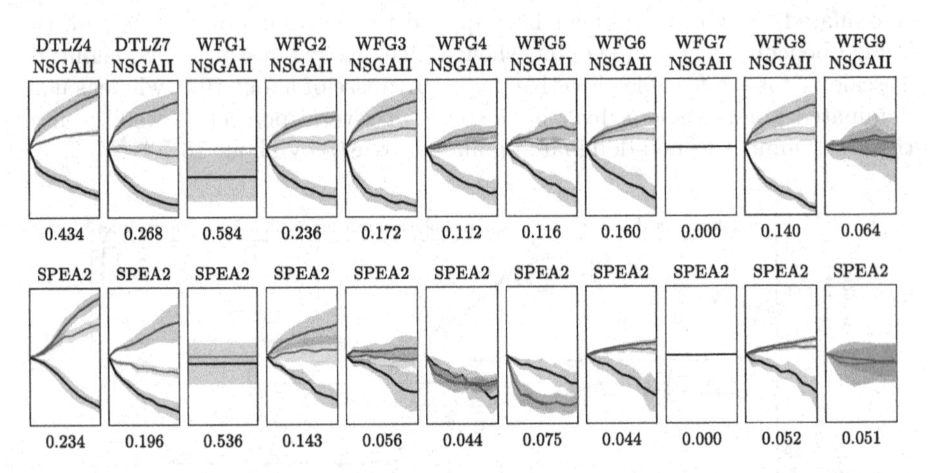

Fig. 7. Diversity of solutions increases markedly under the weighting gradient flow, even as some points become slightly dominated. (Top) Average diversity quotients of feasible (blue) and non-dominated (red) points under the weighting gradient flow along with proportion of population that remains non-dominated (black). Here the diversity is the magnitude at scale t_+. Shaded bands indicate one standard deviation. All panels have the same horizontal axis, viz., the number of timesteps (from 0 to $N = 10$). The vertical axes are $[1 - \Delta, 1 + \Delta]$, with Δ shown below each panel. Not shown explicitly is the average weighting of non-dominated points, i.e., the red curve divided by the black one, but so long as the colored bands already shown are visibly separate, this consistently lies above the blue band. (Bottom) As for the top panels, but for SPEA2. (Color figure online)

Although Figs. 7 and 8 shows that the weighting gradient flow causes a significant proportion of points to become dominated, Fig. 9 uses the *inverted generational distance* (IGD) relative to uniformly distributed reference points on Pareto fronts [26] to show that this qualitative change in dominance is belied by only minor quantitative changes in the distance to Pareto fronts.[2] (Note that the relatively large increases in IGD for DTLZ4 and DTLZ7 are consequences of starting from a low baseline.) That is, feasible points give a better quantitative sense of diversification performance than nondominated points, especially in light of use cases in which the weighting gradient flow is not limited to postprocessing.

[2] Recall that the IGD for X relative to reference set R is $\frac{1}{|R|} \sum_{r \in R} \min_{x \in X} d(x, r)$.

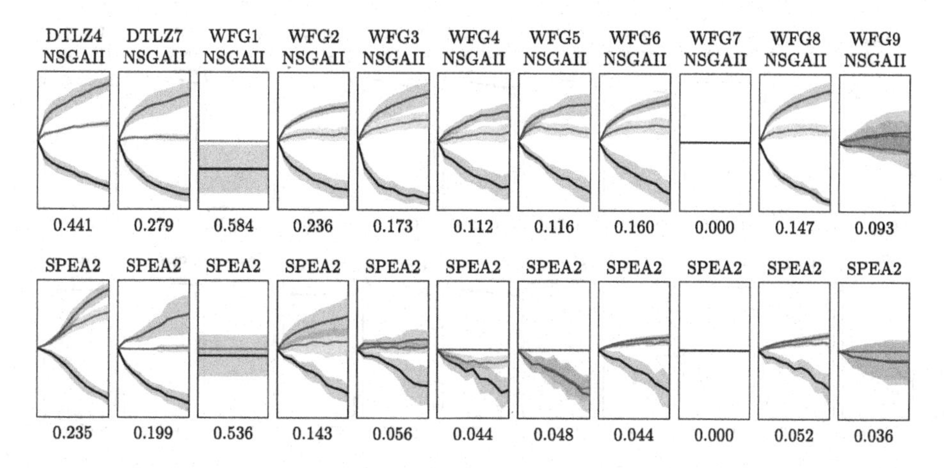

Fig. 8. As in Fig. 7, but for diversity taken as the magnitude at the scale maximizing the quotient by the initial timestep.

Rather than relying solely on a delicate characterization of diversity, we also visualize some of the results directly: this is the rationale for three-objective problems. Figure 10 shows how diversity in objective space is promoted for WFG2-3. Figure 11 shows analogous results for SPEA2.

Careful inspection reveals that the weighting gradient flow tends to induce a gap between the boundary of the non-dominated region and its interior, which is consistent with the generally observed phenomenon that the largest weights in finite subsets of Euclidean space tend to occur on boundaries and the smallest weights immediately "behind" the boundary. Meanwhile, the boundary region tends to become slightly more populated.[3] From the perspective of a MOEA, this is frequently a benefit, since extremal and non-extremal points on the non-dominated approximation of the Pareto front differ in practical significance.[4]

6 Algorithmic Extensions

6.1 Multi-objective Weighting Gradient Flow

We can combine the weighting gradient flow with a multi-gradient descent strategy in a way somewhat akin to [28]. The basic additional ideas are:

- Introduce variable regularizing terms λ_w and λ_f for the weighting and function gradient flows, respectively;

[3] This highlights the need to distinguish between diversity and uniformity. The maximally diverse probability distribution on the interval $[0, L]$ is $\frac{1}{2+L}(\delta_0 + \lambda|_{[0,L]} + \delta_L)$, where Dirac and restricted Lebesgue measures are indicated on the right hand side [27]. Only in a suitable limit can boundary effects be ignored in relation to diversity.

[4] Using a scale $t > t_+$ for the weighting gradient flow would tend to diminish the distinction between uniformity (which is not a function of scale) and diversity (which is). That is, our experiments make this distinction to the greatest possible extent.

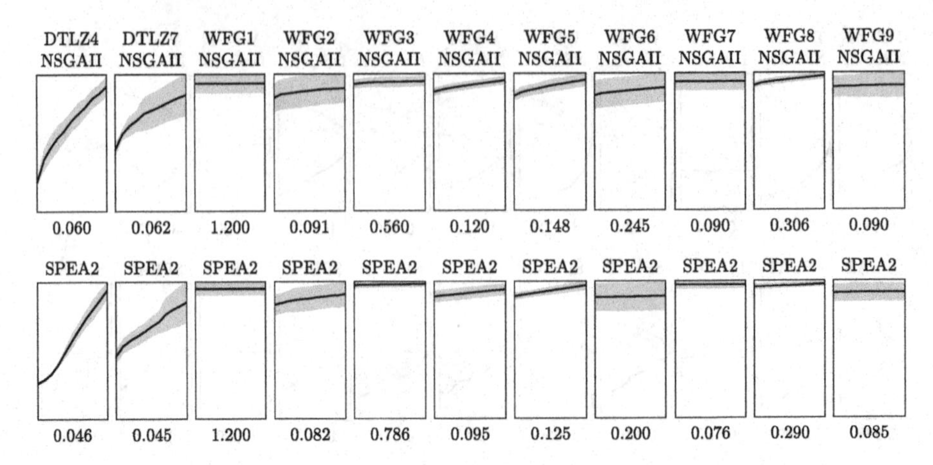

Fig. 9. The weighting gradient flow only slightly affects the quantitative dominance behavior of points, as measured by IGD. (Top) IGD under the weighting gradient flow starting from the results of NSGA-II runs, using uniformly distributed reference points on Pareto fronts. Shaded bands indicate one standard deviation. All panels have the same horizontal axis, viz., the number of timesteps (from 0 to $N = 10$). The vertical axes are $[0, y]$, where y is shown below each panel. (Bottom) As above, but for SPEA2.

- Form the objective-space differentials $dy_j = ds \cdot [\lambda_w S_j(\hat{\nabla} w)_j + \lambda_f \sum_\ell (\hat{\nabla} f)_\ell]$, where the sum is over ℓ such that $\langle (\hat{\nabla} w)_j, (\hat{\nabla} f_\ell)_j \rangle > 0$.

While we have tried this technique in isolation on MOEA benchmarks, the results are poor. However, this is unsurprising: the benchmarks are designed to frustrate MOEAs, much less multi-objective techniques relying on gradients.

6.2 Recycling Function Evaluations

In our experiments with post-processing the output of MOEAs, the weighting gradient flow evolution took time comparable to (and in the case of NSGA-II, slightly more than) the MOEA itself. Most of the time is spent evaluating the fitness function: apart from an initialization step, the evaluations are performed to compute Jacobians in service of pullback operations, and a lesser number are performed to compute pushforwards to maintain consistency.

However, our motivating problems require significant time (on the order of a second) for function evaluations. This demands a more efficient pullback scheme that minimizes or avoids function evaluations, even if the results are substantially worse. A reasonable idea is "recycling" in a sense similar to that employed in some modern Monte Carlo algorithms [29]. Specifically, rather than computing a good approximation to the Jacobian by evaluating functions afresh at very close points along coordinate axes, we settle instead for an approximation of lesser quality that exploits existing function evaluations. We have implemented this in concert with a *de novo* computation of the Jacobian in the event that this

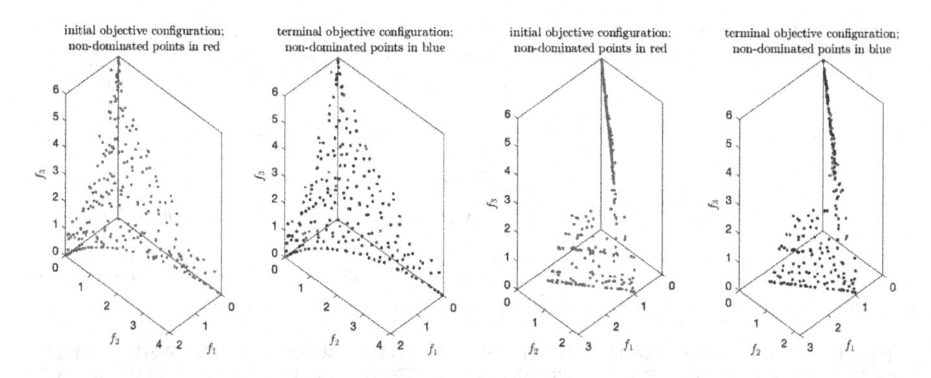

Fig. 10. (Far left) Initial configuration in objective space for WFG2 after a NSGA-II run. (Center left) Configuration after subsequently evolving under the weighting gradient flow. Dominated points are gray. (Right panels) As on the left, but for WFG3.

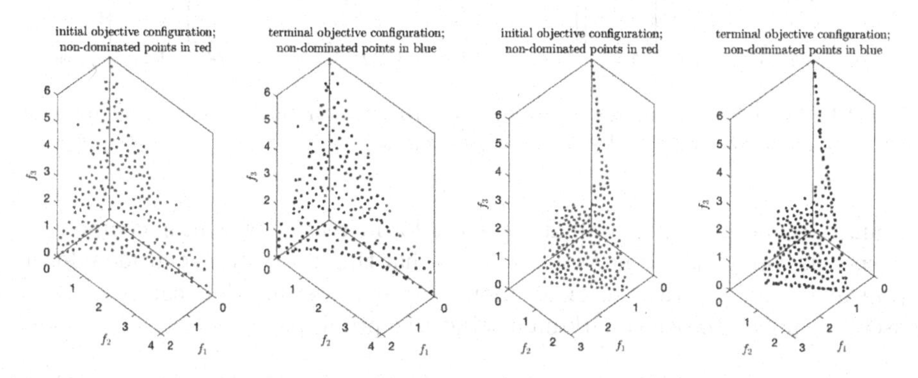

Fig. 11. As in Fig. 10, but for SPEA2. Note the formation of gaps behind the boundary.

initial Jacobian estimate does not have full rank. Our experiments suggest that this works reasonably well: for a typical run from Sect. 5, the number of function evaluations is reduced from 30250 to 2750, and the actual results are broadly comparable (sometimes better, sometimes worse): see Figs. 12 and 13.

This strategy will work poorly if evaluation points lie on a manifold of nonzero codimension or low curvature, because in such cases a matrix that transforms vectors from a base point to evaluation points into (a small multiple of) the standard basis will have a large condition number. However, these situations are relatively unlikely to present major problems in practice, and the recycling approach is likely to be useful when function evaluations are expensive.

7 Remarks

Although our experiments have focused on the results of applying the weighting gradient flow and related constructions after a MOEA has been applied, the more natural application is in the course of a MOEA. As mentioned in Sect. 6, there

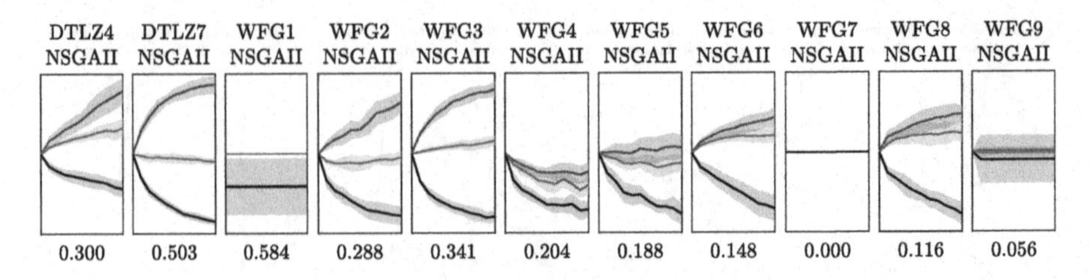

Fig. 12. As in the top panel of Fig. 7, but for a Jacobian approximation that uses existing function evaluations, increasing speed at the cost of accuracy.

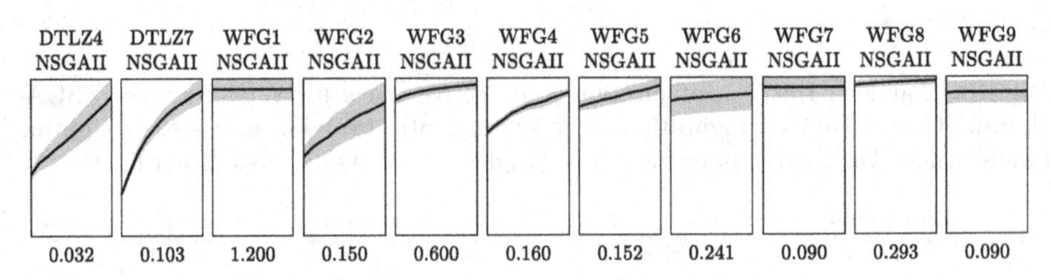

Fig. 13. As in Fig. 9, but for a Jacobian approximation that uses existing function evaluations, increasing speed at the cost of accuracy.

is ample scope to refine and build on ideas for increasing weighting components in specific contexts. It is nevertheless clear that the theory of magnitude informs principled and practical diversity-promoting mechanisms that can already be usefully applied to benchmark multi-objective problems.

Acknowledgment. Thanks to Andy Copeland, Megan Fuller, Zac Hoffman, Rachelle Horwitz-Martin, and Daryl St. Laurent for many patient questions and observations that clarified and simplified the ideas herein. This research was developed with funding from the Defense Advanced Research Projects Agency (DARPA). The views, opinions and/or findings expressed are those of the author and should not be interpreted as representing the official views or policies of the Department of Defense or the U.S. Government.

References

1. Eiben, A.E., Smith, J.E.: Introduction to Evolutionary Computing. Springer, Heidelberg (2015). https://doi.org/10.1007/978-3-662-44874-8
2. Basto-Fernandes, V., Yevseyeva, I., Deutz, A., Emmerich, M.: A survey of diversity oriented optimization: problems, indicators, and algorithms. In: Emmerich, M., Deutz, A., Schütze, O., Legrand, P., Tantar, E., Tantar, A.-A. (eds.) EVOLVE – A Bridge between Probability, Set Oriented Numerics and Evolutionary Computation VII. SCI, vol. 662, pp. 3–23. Springer, Cham (2017). https://doi.org/10.1007/978-3-319-49325-1_1
3. Yan, J., Li, C., Wang, Z., Deng, L., Sun, D.: Diversity metrics in multi-objective optimization: review and perspective. In: 2007 IEEE International Conference on Integration Technology, pp. 553–557. IEEE (2007)

4. Ulrich, T., Bader, J., Thiele, L.: Defining and optimizing indicator-based diversity measures in multiobjective search. In: Schaefer, R., Cotta, C., Kołodziej, J., Rudolph, G. (eds.) PPSN 2010. LNCS, vol. 6238, pp. 707–717. Springer, Heidelberg (2010). https://doi.org/10.1007/978-3-642-15844-5_71
5. Ulrich, T., Thiele, L.: Maximizing population diversity in single-objective optimization. In: Proceedings of the 13th Annual Conference on Genetic and Evolutionary Computation, pp. 641–648 (2011)
6. Solow, A.R., Polasky, S.: Measuring biological diversity. Environ. Ecol. Stat. **1**(2), 95–103 (1994)
7. Leinster, T.: Entropy and Diversity: the Axiomatic Approach. Cambridge (2021)
8. Zitzler, E., Künzli, S.: Indicator-based selection in multiobjective search. In: Yao, X., et al. (eds.) PPSN 2004. LNCS, vol. 3242, pp. 832–842. Springer, Heidelberg (2004). https://doi.org/10.1007/978-3-540-30217-9_84
9. Wang, Y., Emmerich, M., Deutz, A., Bäck, T.: Diversity-indicator based multi-objective evolutionary algorithm: DI-MOEA. In: Deb, K., et al. (eds.) EMO 2019. LNCS, vol. 11411, pp. 346–358. Springer, Cham (2019). https://doi.org/10.1007/978-3-030-12598-1_28
10. Pugh, J.K., Soros, L.B., Stanley, K.O.: Quality diversity: a new frontier for evolutionary computation. Front. Robot. AI **3**, 40 (2016)
11. Steinwart, I., Christmann, A.: Support Vector Machines. Springer, Heidelberg (2008). https://doi.org/10.1007/978-0-387-77242-4
12. Willerton, S.: Heuristic and computer calculations for the magnitude of metric spaces. arXiv preprint arXiv:0910.5500 (2009)
13. Bunch, E., Dickinson, D., Kline, J., Fung, G.: Practical applications of metric space magnitude and weighting vectors. arXiv preprint arXiv:2006.14063 (2020)
14. Leinster, T., Meckes, M.W.: The magnitude of a metric space: from category theory to geometric measure theory. In: Gigli, N. (ed.) Measure Theory in Non-Smooth Spaces (2017)
15. Leinster, T., Meckes, M.W.: Maximizing diversity in biology and beyond. Entropy **18**(3), 88 (2016)
16. Leinster, T.: The magnitude of metric spaces. Doc. Math. **18**, 857–905 (2013)
17. Luo, C., Safa, I., Wang, Y.: Approximating gradients for meshes and point clouds via diffusion metric. In: Computer Graphics Forum, vol. 28, no. 5, pp. 1497–1508. Wiley Online Library (2009)
18. Ishibuchi, H., Yamane, M., Akedo, N., Nojima, Y.: Two-objective solution set optimization to maximize hypervolume and decision space diversity in multiobjective optimization. In: The 6th International Conference on Soft Computing and Intelligent Systems, and The 13th International Symposium on Advanced Intelligence Systems, pp. 1871–1876. IEEE (2012)
19. Guariso, G., Sangiorgio, M.: Improving the performance of multiobjective genetic algorithms: an elitism-based approach. Information **11**(12), 587 (2020)
20. Deb, K., Pratap, A., Agarwal, S., Meyarivan, T.: A fast and elitist multiobjective genetic algorithm: NSGA-II. IEEE Trans. Evol. Comput. **6**(2), 182–197 (2002)
21. Zitzler, E., Laumanns, M., Thiele, L.: SPEA2: improving the strength Pareto evolutionary algorithm. TIK-report, vol. 103 (2001)
22. Deb, K., Thiele, L., Laumanns, M., Zitzler, E.: Scalable test problems for evolutionary multi-objective optimization. TIK-report, vol. 112 (2001)
23. Hamada, N., Goto, K.: Data-driven analysis of Pareto set topology. In: Proceedings of the Genetic and Evolutionary Computation Conference, pp. 657–664 (2018)

24. Huband, S., Hingston, P., Barone, L., While, L.: A review of multiobjective test problems and a scalable test problem toolkit. IEEE Trans. Evol. Comput. **10**(5), 477–506 (2006)
25. Tian, Y., Cheng, R., Zhang, X., Jin, Y.: PlatEMO: a MATLAB platform for evolutionary multi-objective optimization [educational forum]. IEEE Comput. Intell. Mag. **12**(4), 73–87 (2017)
26. Tian, Y., Xiang, X., Zhang, X., Cheng, R., Jin, Y.: Sampling reference points on the Pareto fronts of benchmark multi-objective optimization problems. In: 2018 IEEE congress on evolutionary computation (CEC), pp. 1–6. IEEE (2018)
27. Leinster, T., Roff, E.: The maximum entropy of a metric space. arXiv preprint arXiv:1908.11184 (2019)
28. Désidéri, J.-A.: Multiple-gradient descent algorithm (MGDA) for multiobjective optimization. C.R. Math. **350**(5–6), 313–318 (2012)
29. Frenkel, D.: Speed-up of Monte Carlo simulations by sampling of rejected states. Proc. Natl. Acad. Sci. **101**(51), 17 571–17 575 (2004)
30. Leinster, T., Cobbold, C.A.: Measuring diversity: the importance of species similarity. Ecology **93**(3), 477–489 (2012)

Two-Stage Greedy Approximated Hypervolume Subset Selection for Large-Scale Problems

Yang Nan, Hisao Ishibuchi(✉), Tianye Shu, and Ke Shang

Guangdong Provincial Key Laboratory of Brain-Inspired Intelligent Computation, Department of Computer Science and Engineering, Southern University of Science and Technology, Shenzhen 518055, China
{nany,12132356}@mail.sustech.edu.cn, hisao@sustech.edu.cn

Abstract. Recently, it has been demonstrated that a solution set that is better than the final population can be obtained by subset selection in some studies on evolutionary multi-objective optimization. The main challenge in this type of subset selection is how to efficiently handle a huge candidate solution set, especially when the hypervolume-based subset selection is used for many-objective optimization. In this paper, we propose an efficient two-stage greedy algorithm for hypervolume-based subset selection. In each iteration of the proposed greedy algorithm, a small number of promising candidate solutions are selected in the first stage using the rough hypervolume contribution approximation. In the second stage, a single solution among them is selected using the more precise approximation. Experimental results show that the proposed algorithm is much faster than state-of-the-art hypervolume-based greedy subset selection algorithms at the cost of a slight deterioration of the selected subset quality.

Keywords: Evolutionary multi-objective optimization · Hypervolume subset selection · Two-stage hypervolume contribution approximation

1 Introduction

Recently, the use of an unbounded external archive was examined in many studies [1–10] in the evolutionary multi-objective optimization (EMO) community. It was shown in [6–8] that the selected subset from the unbounded external archive is usually better than the final population. This is because in general the final population is not the best subset of all examined solutions. For example, final solutions can be dominated by other generated and discarded solutions in previous generations [10]. The main difficulty in subset selection from the unbounded external archive is a huge candidate solution set. For example, more than two million non-dominated solutions are included in a candidate solution set in [7]. Thus, efficient subset selection algorithms are needed.

Hypervolume subset selection (HSS) is a popular research topic since the hypervolume indicator [11] is the only performance indicator with Pareto compliant property [12]. In general, the HSS problem is to select a subset S_{sub} from

M. Emmerich et al. (Eds.): EMO 2023, LNCS 13970, pp. 391–404, 2023.
https://doi.org/10.1007/978-3-031-27250-9_28

a candidate set S_c so that the hypervolume of S_{sub} is maximized. Thus, the HSS problem can be formalized as follows:

$$S^*_{sub} = \underset{S_{sub} \subset S_c, |S_{sub}|=k}{\arg\max} HV(S_{sub}), \tag{1}$$

where S^*_{sub} is the optimal subset with k solutions, S_c is the given set of n candidate solutions (i.e., $|S_c| = n$), and $HV(S_{sub})$ is the hypervolume of S_{sub}. HSS methods can be classified into four categories [13]: exact methods, evolutionary methods, local search methods, and greedy methods.

Among them, the greedy methods [14–18] are the most well-known since they can obtain a near optimal subset (i.e., $(1 - 1/e)$ to the optimal subset is guaranteed [19]) within a reasonable computation time. Bradstreet *et al.* [14,15] proposed two basic greedy HSS methods: the greedy reduction and the greedy inclusion. However, the efficiency of the basic greedy HSS methods is low. Jiang *et al.* [16] proposed a hypervolume contribution update strategy that can significantly reduce the computation time of the basic greedy HSS methods. To further improve the efficiency of the greedy HSS method, Chen *et al.* [17] proposed a lazy greedy inclusion HSS method. In the lazy greedy HSS method, the submodular property of the hypervolume indicator [20] is utilized to avoid unnecessary hypervolume contribution calculations. Currently, the lazy greedy HSS method is the most efficient greedy HSS method.

Since the above-mentioned greedy HSS methods use the exact hypervolume contribution calculation, their computation time is very large in high-dimensional cases (i.e., many-objective problems). This is because the time complexity of the exact hypervolume contribution calculation is $O(k^{m-1})$ [13] where k is the number of solutions involved in hypervolume contribution calculation and m is the dimension of solutions (i.e., the number of objectives). To address this issue, Shang *et al.* [18] proposed a greedy approximated HSS method which uses an R2-based hypervolume contribution approximation method [21] instead of the exact calculation. As a result, the greedy approximated HSS method is much faster than the exact greedy HSS methods (including the lazy greedy HSS method). The selected subset quality (i.e., hypervolume value) by the approximated method is slightly worse than that by the exact HSS methods.

As pointed out in [22], the computation time of the greedy HSS methods (both the exact and approximated methods) severely increases as the number of candidate solutions increases. This is because the greedy methods need to examine all the unselected candidate solutions in each iteration. Thus, the above-mentioned greedy methods are not applicable to the large-scale subset selection where the number of candidate solutions is huge (e.g., more than 2,000,000 non-dominated solutions in the unbounded external archive [7]).

In this paper, we propose a two-stage greedy approximated HSS method to efficiently select a subset from a huge candidate solution set. In the proposed method, we use a two-stage hypervolume contribution approximation method [23] for a hypervolume-based EMO algorithm to select a good solution in each iteration of greedy inclusion. Each iteration is divided into two stages. In the first stage, we roughly approximate the hypervolume contribution of each unselected candidate solution. Then, we preselect only a small number of promising candidate solutions. In the second stage, the hypervolume contribution of each preselected promising solution is more precisely approximated and one of them is selected. In this manner, we do not have to precisely approximate the hypervolume contribution of each candidate solution. Thus, the proposed two-stage method is much faster than other greedy methods for large-scale subset selection.

The remainder of this paper is organized as follows. In Sect. 2, the existing greedy HSS methods are briefly reviewed. The proposed two-stage greedy approximated HSS method is described in Sect. 3. Experimental results are reported in Sect. 4 to demonstrate the superiority of the proposed method. Finally, the conclusion is given in Sect. 5.

2 Greedy Hypervolume Subset Selection

For a large candidate set (i.e., $k \ll n$), the use of greedy reduction is unrealistic. Thus, in this paper, we focus only on greedy inclusion HSS methods where k solutions are selected from the candidate set S_c with n solutions one by one. In this section, we explain greedy exact and greedy approximated HSS methods.

2.1 Greedy Exact HSS Methods

Basic Greedy Inclusion HSS (GI-HSS [14]**).** The framework of the basic greedy inclusion HSS method (GI-HSS) is shown in Algorithm 1, which is also the framework of all greedy HSS methods. As explained in Sect. 1, the size of the candidate solution set S_c is n (i.e., $|S_c| = n$) and the size of the subset S_{sub} to be selected is k (i.e., $|S_{sub}| = k$). First, S_{sub} is empty. Then, a candidate solution is selected from S_c and added to S_{sub} in each iteration one by one. GI-HSS is terminated when $|S_{sub}|$ reaches k. In each iteration, GI-HSS calculates the hypervolume contribution of each unselected candidate solution. Then, the candidate solution with the largest hypervolume contribution is added to S_{sub}. Since the hypervolume contribution calculation is time-consuming, the efficiency of GI-HSS is low, especially when n is large (i.e., $k \ll n$).

Algorithm 1: Basic framework of greedy methods

input : S_c (candidate solution set), k (selected subset size)
output: S_{sub} (selected subset)
begin
1. $S_{sub} \longleftarrow \emptyset$;
2. **while** $|S_{sub}| < k$ **do**
3. Calculate the hypervolume contribution to S_{sub} of each solution $\boldsymbol{s}$ in S_c;
4. Select the best candidate solution $\boldsymbol{a}$ with the largest hypervolume contribution to S_{sub} from S_c;
5. $S_{sub} \longleftarrow S_{sub} \cup \{\boldsymbol{a}\}$; $S_c \longleftarrow S_c \backslash \{\boldsymbol{a}\}$;

Greedy Inclusion HSS with Hypervolume Contribution Updating (UGI-HSS [16]). To improve the efficiency of the basic greedy HSS method, Jiang *et al.* [16] proposed a hypervolume update strategy. In each iteration, GI-HSS calculates the hypervolume contribution of each candidate solution whereas UGI-HSS updates their hypervolume contribution more efficiently. As a result, UGI-HSS is much faster than GI-HSS.

Lazy Greedy Inclusion HSS (LGI-HSS [17]). The UGI-HSS method improves the efficiency of the basic greedy HSS method by reducing the computation time of the hypervolume contribution calculation for each candidate solution. In contrast, the LGI-HSS method improves the efficiency of GI-HSS by reducing the number of hypervolume contribution calculations. That is, LGI-HSS uses the submodular property of the hypervolume indicator [20] to avoid unnecessary hypervolume contribution calculations. Currently, the LGI-HSS method is the most efficient greedy exact HSS method.

2.2 Greedy Approximated HSS Method

Since the time complexity of hypervolume contribution calculation increases exponentially as the number of objectives m increases [13], the greedy exact HSS methods are impractical in high-dimensional cases (e.g., $m > 10$). To overcome this issue, a greedy approximated HSS method (GAHSS [18]) was proposed. In GAHSS, the hypervolume contribution of each candidate solution is approximated using the R2-based hypervolume contribution approximation method [21]. In this subsection, we briefly explain the mechanism of the R2-based hypervolume contribution approximation method and the framework of GAHSS.

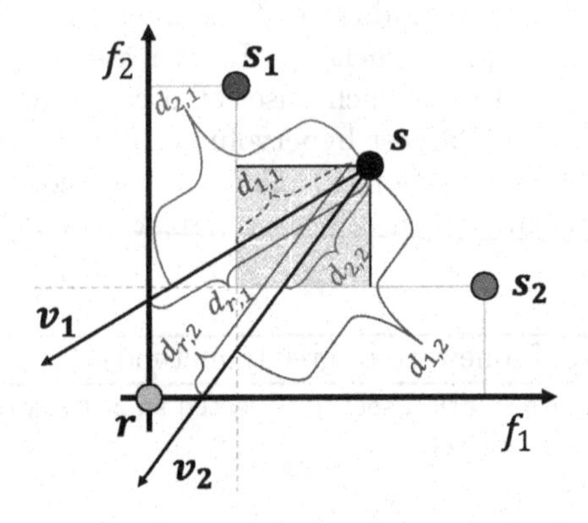

Fig. 1. Illustration of R2-based hypervolume contribution approximation

R2-Based Hypervolume Contribution Approximation (R2-HVC [21]). Figure 1 illustrates the R2-based hypervolume contribution approximation method. Solutions $\boldsymbol{s}_1$, $\boldsymbol{s}_2$ and $\boldsymbol{s}$ form a solution set S_c, and $\boldsymbol{r}$ is the reference point. To approximate the hypervolume contribution of the solution $\boldsymbol{s}$, R2-HVC uses a vector set Λ (i.e., $\Lambda = \{\boldsymbol{v}_1, \boldsymbol{v}_2\}$ in Fig. 1) to detect the boundary of the hypervolume contribution region of $\boldsymbol{s}$ (i.e., shaded region in Fig. 1). The average length of the line segment for each vector from $\boldsymbol{s}$ to the boundary (e.g., $d_{1,1}$ for $\boldsymbol{v}_1$ and $d_{2,2}$ for $\boldsymbol{v}_2$) is used as the hypervolume contribution approximation for $\boldsymbol{s}$. Thus, the computation time and the approximation quality directly depends on the number of vectors in Λ.

For each vector, to obtain the corresponding line segment (e.g., $d_{1,1}$ for $\boldsymbol{v}_1$ in Fig. 1), we calculate the length of each of all line segments determined by other solutions in S_c and the reference point (e.g., $d_{1,1}$, $d_{2,1}$ and $d_{r,1}$ in Fig. 1). The line segment with the minimum length (e.g., $d_{1,1}$ for v_1) is the corresponding line segment and is used for hypervolume contribution approximation.

To approximate the hypervolume contribution of each candidate solution $\boldsymbol{s}_i \in S_c$ to the subset S_{sub} (i.e., $HVC(\boldsymbol{s}_i, S_{sub}, \boldsymbol{r})$), we need to calculate the length of each of all related line segments and store them in a matrix M_i as

$$M_i = \begin{bmatrix} d^i_{1,1} & d^i_{1,2} & \cdots & d^i_{1,|\Lambda|} \\ d^i_{2,1} & d^i_{2,2} & \cdots & d^i_{2,|\Lambda|} \\ \vdots & \vdots & \ddots & \vdots \\ d^i_{|S_{sub}|,1} & d^i_{|S_{sub}|,2} & \cdots & d^i_{|S_{sub}|,|\Lambda|} \\ d^i_{r,1} & d^i_{r,2} & \cdots & d^i_{r,|\Lambda|} \end{bmatrix}. \tag{2}$$

In this matrix M_i, each row refers to the corresponding solution in S_{sub} or the reference point $\boldsymbol{r}$, and each column refers to the corresponding vector in the vector set Λ. The hypervolume contribution of $\boldsymbol{s}_i$ is approximated by the average of the minimum value in each column (i.e., $\min(d^i_j) = \min\{d^i_{1,j}, ..., d^i_{|S_{sub}|,j}, d^i_{r,j}\}$) as $(\min(d^i_1) + \ldots + \min(d^i_{|\Lambda|}))/|\Lambda|$. For more details, see [21].

Greedy Approximated HSS (GAHSS [18]). In GAHSS, a tensor $T_{\min}$ is used to calculate the approximated hypervolume contribution of each candidate solution. Its structure is

$$T_{\min} = \begin{bmatrix} \min(d^1_1) & \min(d^1_2) & \cdots & \min(d^1_{|\Lambda|}) \\ \vdots & \vdots & \ddots & \vdots \\ \min(d^{|S_c|}_1) & \min(d^{|S_c|}_2) & \cdots & \min(d^{|S_c|}_{|\Lambda|}) \end{bmatrix}. \tag{3}$$

Each row of $T_{\min}$ refers to the corresponding candidate solution (from M_i in Eq. (2)), and each column of $T_{\min}$ refers to the corresponding vector. For a candidate solution $\boldsymbol{s}_i \in S_c$, each value $\min(d^i_j)$ in $T_{\min}$ is the minimum value of the j-th column of M_i (i.e., $\min(d^i_j) = \min\{d^i_{1,j}, \ldots, d^i_{|S_{sub}|,j}, d^i_{r,j}\}$). Then, each row of $T_{\min}$ is used to approximate the hypervolume contribution of each candidate solution in S_c.

Algorithm 2: Greedy approximated HSS

input : S_c (candidate solution set), k (selected subset size), Λ (vector set)
output: S_{sub} (selected subset)
begin
1 $S_{sub} \longleftarrow \emptyset$;
2 Calculate the tensor $T_{\min}$ in (3) using the reference point $\boldsymbol{r}$ for all candidate solutions in S_c and all vectors in Λ;
3 **while** $|S_{sub}| < k$ **do**
4 Approximate the hypervolume contribution to S_{sub} of each candidate solution in S_c using $T_{\min}$;
5 Select the best candidate solution $\boldsymbol{a}$ with the largest hypervolume contribution from S_c;
6 $S_{sub} \longleftarrow S_{sub} \cup \{\boldsymbol{a}\}$; $S_c \longleftarrow S_c \backslash \{\boldsymbol{a}\}$;
7 Calculate the tensor T in (4) using $\boldsymbol{a}$ for all candidate solutions in S_c and all vectors in Λ;
8 Update $T_{\min}$ using $T_{\min}$ and T;

When a new solution $\boldsymbol{a}$ is added to the subset S_{sub}, we first calculate a tensor T using $\boldsymbol{a}$ for all candidate solutions in S_c and all vectors in Λ as:

$$T = \begin{bmatrix} d^1_{\boldsymbol{a},1} & d^1_{\boldsymbol{a},2} & \cdots & d^1_{\boldsymbol{a},|\Lambda|} \\ \vdots & \vdots & \ddots & \vdots \\ d^{|S_c|}_{\boldsymbol{a},1} & d^{|S_c|}_{\boldsymbol{a},2} & \cdots & d^{|S_c|}_{\boldsymbol{a},|\Lambda|} \end{bmatrix}. \quad (4)$$

Similar to $T_{\min}$, each row of T refers to the corresponding candidate solution, and each column of T refers to the corresponding vector. The i-th row of T is a new row of M_i: $(d^i_{\boldsymbol{a},1}, \dots, d^i_{\boldsymbol{a},|\Lambda|})$. In this manner, each element of $T_{\min}$ is updated as $\min(d^i_j) = \min\{\min(d^i_j), d^i_{\boldsymbol{a},j}\}$.

The framework of GAHSS is shown in Algorithm 2. As in GI-HSS, we first initialize the subset S_{sub} as an empty set. Then, the tensor $T_{\min}$ is initialized as the tensor using the reference point $\boldsymbol{r}$ for each candidate solution i and each vector j (i.e., $\min(d^i_j) = d^i_{\boldsymbol{r},j}$, where $d^i_{\boldsymbol{r},j}$ is an element of M_i). In each iteration, we first use $T_{\min}$ to calculate the approximated hypervolume contribution of each candidate solution. Then, the best candidate solution with the largest approximated hypervolume contribution is added to S_{sub}, and the tensor T is calculated based on the newly added solution. Finally, the tensor $T_{\min}$ is updated using $T_{\min}$ and T.

3 Proposed Two-Stage Greedy Approximated HSS

As shown in Algorithm 2, GAHSS calculates the tensor T using the newly added solution $\boldsymbol{a}$ for all candidate solutions in S_c and all vectors in Λ in each iteration. Thus, its computation time can be unacceptably large when the size of the candidate set is huge. To address this issue, we propose a two-stage GAHSS (TGAHSS) method in this paper. In each iteration of TGAHSS, we select a small number of promising candidate solutions from S_c in the first stage. Then, we select a single candidate solution from them in the second stage. The basic idea of the proposed two-stage method is to use a different vector set for hypervolume contribution approximation in each stage. By using a small vector set (i.e., only a small number of vectors) in the first stage, we can significantly decrease the computation time without severely degrading the quality of the selected subset.

We need two tensors $T^1_{\min}$ and $T^2_{\min}$ in TGAHSS for the first and second stages, respectively. Let the two vector sets for the first and second stages be Λ_1 and Λ_2 (where $|\Lambda_1| > |\Lambda_2|$), respectively. The number of candidate solutions used in the second stage is n_2. $T^1_{\min}$ and $T^2_{\min}$ in TGAHSS are described as:

$$T^1_{\min} = \begin{bmatrix} \min(d^1_1) & \cdots & \min(d^1_{|\Lambda_1|}) \\ \vdots & \ddots & \vdots \\ \min(d^{|S_c|}_1) & \cdots & \min(d^{|S_c|}_{|\Lambda_1|}) \end{bmatrix}, \quad T^2_{\min} = \begin{bmatrix} \min(d^1_1) & \cdots & \min(d^1_{|\Lambda_2|}) \\ \vdots & \ddots & \vdots \\ \min(d^{|S_c|}_1) & \cdots & \min(d^{|S_c|}_{|\Lambda_2|}) \end{bmatrix}. \quad (5)$$

Each row of $T^1_{\min}$ and $T^2_{\min}$ refers to the corresponding candidate solution, and each column of $T^1_{\min}$ and $T^2_{\min}$ refers to the corresponding vector in Λ_1 and Λ_2, respectively. In each iteration of TGAHSS, we update the entire $T^1_{\min}$ and only a small part of $T^2_{\min}$.

The framework of TGAHSS is shown in Algorithm 3. Different from GAHSS, we need to initialize two tensors $T^1_{\min}$ and $T^2_{\min}$ at the beginning (Line 2 in Algorithm 3). In each iteration, we first use $T^1_{\min}$ to roughly approximate the hypervolume contribution of each candidate solution and select n_2 promising candidate solutions (Lines 4–5 in Algorithm 3). In the second stage, we only need to update a small part of $T^2_{\min}$, instead of the entire $T^2_{\min}$. That is, only n_2 rows of $T^2_{\min}$, which are related to the n_2 promising candidate solutions, need to be updated. After that, the hypervolume contribution of each of the n_2 candidate solutions is approximated (using much more vectors in the second stage than those in the first stage: $|\Lambda_1| > |\Lambda_2|$) from the n_2 rows of $T^2_{\min}$ (Lines 6–7 in Algorithm 3). Then, the best solution $\boldsymbol{a}$ with the largest approximated hypervolume contribution is selected from the n_2 promising candidate solutions, and added to the subset S_{sub} (Line 9 in Algorithm 3). Finally, the first tensor $T^1_{\min}$ is updated using the newly added solution $\boldsymbol{a}$ (Lines 10, 11 in Algorithm 3).

4 Experimental Results

In this section, we first examine the proposed two-stage greedy approximated HSS (TGAHSS) method under different parameter settings. Then, the proposed method is compared with two state-of-the-art methods: the greedy approximated HSS (GAHSS [18]) and the lazy greedy inclusion HSS (LGI-HSS [17]).

Algorithm 3: Two-stage greedy approximated HSS

input : S_c (candidate solution set), k (selected subset size), Λ_1 (first-stage vector set), Λ_2 (second-stage vector set)
output: S_{sub} (selected subset)
begin
1. $S_{sub} \longleftarrow \emptyset$;
2. Calculate the tensor $T^1_{\min}$ and $T^2_{\min}$ in (5) using the reference point $\boldsymbol{r}$ for all candidate solutions in S_c and all vectors in Λ_1 and Λ_2, respectively;
3. **while** $|S_{sub}| < k$ **do**
4. /* First stage */
5. Approximate the hypervolume contribution to S_{sub} of each candidate solution in S_c using $T^1_{\min}$;
6. Select n_2 candidate solutions with the largest hypervolume contributions from S_c;
7. /* Second stage */
8. Update the corresponding n_2 rows of T^2_{min} which are related to the n2 candidate solutions;
9. Approximate the hypervolume contribution of each of the n_2 candidate solutions to S_{sub} using the corresponding rows of T^2_{min};
10. Select the best solution $\boldsymbol{a}$ with the largest approximated hypervolume contribution from the n_2 candidate solutions;
11. $S_{sub} \longleftarrow S_{sub} \cup \{\boldsymbol{a}\}$; $S_c \longleftarrow S_c \backslash \{\boldsymbol{a}\}$;
12. /* Update of T^1_{min} */
13. Calculate the tensor T^1 using $\boldsymbol{a}$ for all the candidate solutions and all vectors in Λ_1;
14. Update $T^1_{\min}$ using $T^1_{\min}$ and T^1;

4.1 Experimental Settings

We use eight candidate solution sets in [24][1], which are generated in the following manner. First, under the termination condition of 100,000 solution evaluations, NSGA-III [25] is applied to eight test problems: DTLZ1-2 [26] and Minus-DTLZ1-2 [27] with five and ten objectives (i.e., $m = 5, 10$). Next, all examined solutions (i.e., 100,000 solutions) are stored for each test problem. Then, all non-dominated solutions among the stored solutions are used as a candidate solution set for each test problem. The size of each candidate solution set (i.e., n) is shown in Table 1.

The selected subset size k is set to 100. As suggested in [28], the reference point for hypervolume subset selection is specified as $(1 + 1/H) \times nadir$ where $nadir$ is the estimated nadir point of the candidate set and $H = 4, 2$ for $m = 5, 10$, respectively. For performance evaluation, the reference point is set to $(1 + 1/H) \times trueNadir$ where $trueNadir$ is the true nadir point of each test problem.

In the first stage of the proposed TGAHSS method, the first vector in the vector set Λ_1 is specified as $(1/\sqrt{m}, \ldots, 1/\sqrt{m})$. Other vectors in Λ_1 are gen-

[1] https://github.com/HisaoLabSUSTC/BenchSS.

Table 1. Size of each candidate solution set

Problem	$m = 5$	$m = 10$
DTLZ1	29,194	30,194
DTLZ2	45,605	62,601
Minus-DTLZ1	35,798	76,701
Minus-DTLZ2	48,741	85,631

erated using the UNV method [30]. In the second stage, the vector set Λ_2 is generated using the UNV method as suggested in [29]. In the GAHSS method, the vector set is also generated by the UNV method as in its original paper [18]. Each HSS method is executed 21 times independently.

All experiments are performed on a machine with AMD Ryzen Threadripper 3990X 64-Core Processor 2.90 GHz and Windows 10 Pro.

4.2 Performance of TGAHSS Under Different Parameter Settings

In the proposed TGAHSS, there are three parameters: the number of first-stage vectors $|\Lambda_1|$, the number of second-stage vectors $|\Lambda_2|$, and the number of second-stage solutions n_2. We set the number of the second-stage vectors $|\Lambda_2|$ as $|\Lambda_2| = 100$, which is the same setting as in GAHSS. In this subsection, we examine the sensitivity of the performance of TGAHSS to the other two parameters. For the number of first-stage vectors $|\Lambda_1|$, we examine the settings of $|\Lambda_1| = \{1, 2, 10, 20, 30, 40, 50\}$. For the number of second-stage solutions n_2, we examine the settings of $n_2 = \{1, 2, 5, 10, 20, 50, 100, 500\}$.

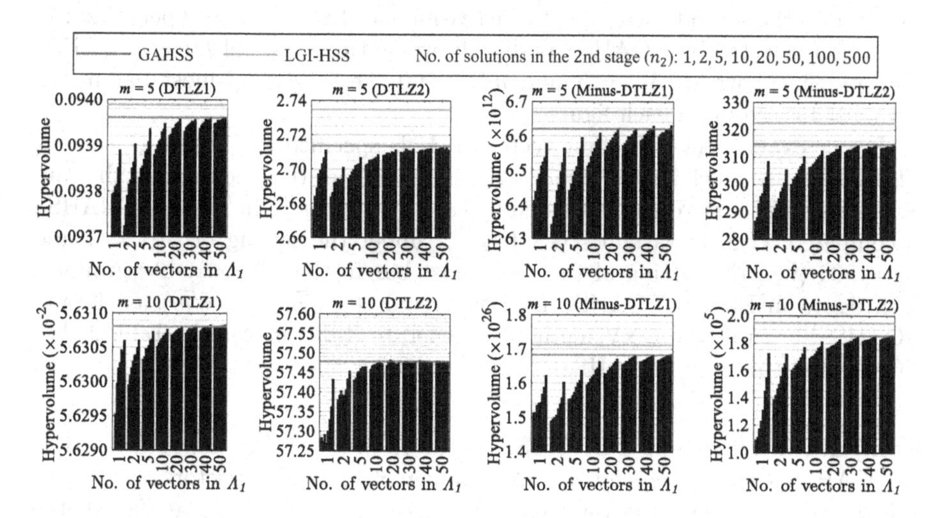

Fig. 2. Hypervolume of TGAHSS with different parameter settings compared to GAHSS and LGI-HSS. (Color figure online)

Table 2. Hypervolume of the subsets selected by TGAHSS^{UNV} and TGAHSS. Λ_1 in TGAHSS^{UNV} has a single vector generated by the UNV method.

Data Shape	m	TGAHSS^{UNV}	TGAHSS	TGAHSS/TGAHSS^{UNV}
DTLZ1	5	9.3800E-2	9.3888E-2 (+)	1.0009
	10	5.6300E-2	5.6306E-2 (+)	1.0000
DTLZ2	5	2.6693E+0	2.7109E+0 (+)	1.0156
	10	5.7368E+1	5.7431E+1 (+)	1.0011
Minus-DTLZ1	5	6.2018E+12	6.5753E+12 (+)	1.0602
	10	1.4123E+26	1.6209E+26 (+)	1.1477
Minus-DTLZ2	5	2.9341E+2	3.0824E+2 (+)	1.0506
	10	1.5589E+5	1.7201E+5 (+)	1.1034
(+/-/≈)			(8/0/0)	

Hypervolume of the Selected Subset. Figure 2 shows the average hypervolume of the subsets selected by TGAHSS with different parameter settings, which are compared with the results selected by GAHSS (i.e., red lines) and LGI-HSS (i.e., green lines). The horizontal axis of each figure is the number of first-stage vectors (i.e., $|\Lambda_1|$). For each specification of $|\Lambda_1|$, experimental results obtained by various specifications of the number of second-stage solutions (i.e., $n_2 = \{1, 2, 5, 10, 20, 50, 100, 500\}$) are shown as a group of bars.

As shown in Fig. 2, the hypervolume of the selected subset clearly increases as the number of first-stage vectors increases (i.e., as $|\Lambda_1|$ increases). When $|\Lambda_1|$ is small (e.g., the left-most group of bars for $|\Lambda_1| = 1$), the hypervolume of the subset selected by TGAHSS is significantly improved by increasing the number of solutions in the second stage (i.e., by increasing n_2). In Fig. 2, the hypervolume of the subset selected by TGAHSS is slightly worse than that selected by GAHSS (i.e., red lines) when $|\Lambda_1| = 1$ and $n_2 = 500$ (i.e., the right-most bar in the left-most bar group in each figure).

In our experiments, the first vector of Λ_1 is specified as $(1/\sqrt{m}, ..., 1/\sqrt{m})$. This is because much better results are obtained from this vector than the randomly specified first vector generated by the UNV method. In Table 2, TGAHSS is compared with its variant TGAHSS^{UNV} under the setting of $|\Lambda_1| = 1$ and $n_2 = 500$ (i.e., the setting of the right-most bar in the left-most bar group in each figure in Fig. 2). TGAHSS uses the vector $(1/\sqrt{m}, ..., 1/\sqrt{m})$ as Λ_1 and TGAHSS^{UNV} uses the UNV method to generate Λ_1. It is clear in Table 2 that TGAHSS outperforms TGAHSS^{UNV}.

Computation Time of TGAHSS. As shown in Fig. 3, the computation time of TGAHSS strongly depends on the number of first-stage vectors (i.e., $|\Lambda_1|$). When $|\Lambda_1| = 50$ (i.e., the right-most bar group in each figure in Fig. 3), the computation time of TGAHSS is about 1/3 less than that of GAHSS. However, when $|\Lambda_1| = 1$ (i.e., the left-most bar group in each figure in Fig. 3), TGAHSS is about ten times faster than GAHSS. This is because the computation time of TGAHSS in the first stage is much larger than that in the second stage when the number of candidate solutions is very large. We need to approximate the hypervolume contribution of all candidate solutions (e.g., 85,631 solutions for Minus-DTLZ1) in the first stage whereas we handle only a small number of candidate solutions in the second stage (i.e., up to 500 solutions in Fig. 3). Thus, the computation time of TGAHSS strongly depends on the number of first-stage vectors (i.e., $|\Lambda_1|$).

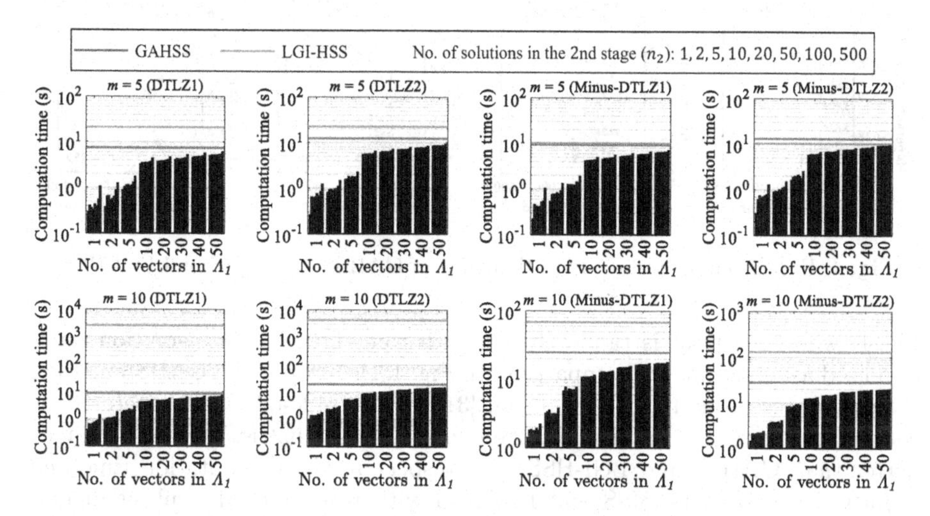

Fig. 3. Computation time of TGAHSS with different parameter settings compared to GAHSS and LGI-HSS.

It is also shown in Fig. 3 that the specifications of the number of the second-stage solutions (i.e., different bars in each bar group in each figure of Fig. 3) have no strong effect on the computation time of TGAHSS. This is because the computation time of the first stage is much larger than that of the second stage. In Fig. 2 and Fig. 3, we can observe that TGAHSS (with $|\Lambda_1| = 1$ and $n_2 = 500$) can obtain a slightly worse subset using a much smaller computation time compared to GAHSS. We use this setting in the next subsection.

4.3 Comparison with State-of-the-Art Methods

In this section, we compare the proposed method with the two most efficient greedy HSS methods: the greedy approximated HSS method (GAHSS [18]) and the lazy greedy inclusion HSS method (LGI-HSS [17]). LGI-HSS is the most efficient greedy HSS method with exact hypervolume contribution calculation. Since GAHSS uses the approximate calculation, GAHSS is faster than LGI-HSS but its selected subset is worse than the subset selected by LGI-HSS.

Each algorithm is applied to each test problem 21 times. Average results over 21 runs are summarized in Fig. 4. The random method (black point in

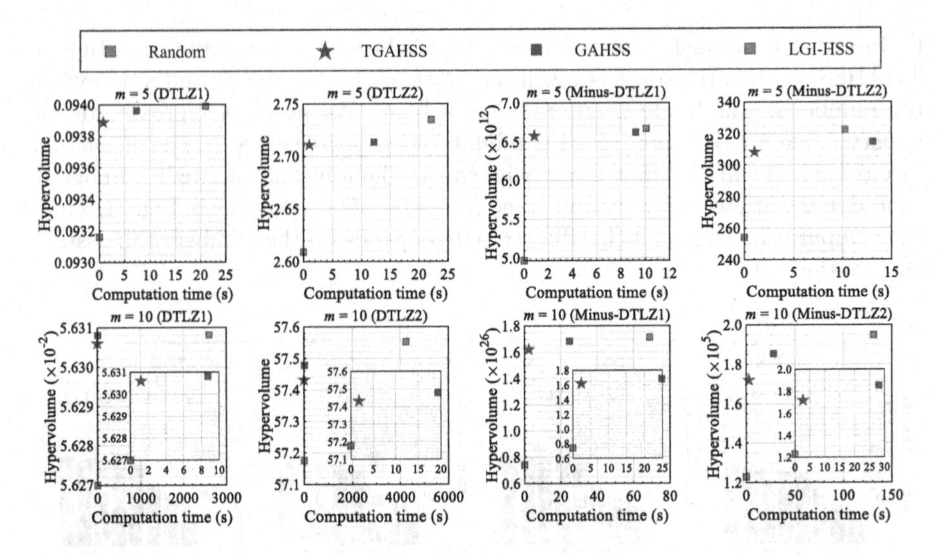

Fig. 4. Hypervolume and computation time of GAHSS, TGAHSS and LGI-HSS.

Fig. 4) where a subset is randomly selected from the candidate solution set is also used as a baseline for comparison. In Fig. 4, the proposed TGAHSS method always locates around the knee region [31] of the trade-off curve generated by connecting the results obtained by the four methods. That is, TGAHSS is much better than GAHSS and LGI-HSS with respect to the computation time and slightly worse than GAHSS and LGI-HSS with respect to the subset quality in Fig. 4. For example, for 5-objective DTLZ2, the hypervolume of the subset selected by TGAHSS is slightly worse than that of GAHSS but TGAHSS is about ten times faster than GAHSS.

5 Conclusion

In this paper, we proposed a two-stage greedy approximated hypervolume subset selection method (TGAHSS) for large-scale candidate solution sets (e.g., 50,000 solutions). Experimental results showed that the proposed TGAHSS method is much faster than the two state-of-the-art greedy HSS methods. The quality of the subset selected by TGAHSS is slightly worse than those selected by GAHSS and LGI-HSS in terms of the hypervolume. In the future, we will examine the use of TGAHSS to generate an initial subset for local search HSS.

Acknowledgements. This work was supported by National Natural Science Foundation of China (Grant No. 62002152, 61876075), Guangdong Provincial Key Laboratory (Grant No. 2020B121201001), the Program for Guangdong Introducing Innovative and Enterpreneurial Teams (Grant No. 2017ZT07X386), The Stable Support Plan Program of Shenzhen Natural Science Fund (Grant No. 20200925174447003), Shenzhen Science and Technology Program (Grant No. KQTD2016112514355531).

References

1. Ishibuchi, H., Pang, L.M., Shang, K.: A new framework of evolutionary multi-objective algorithms with an unbounded external archive. In: Proceedings of the European Conference on Artificial Intelligence, Santiago de Compostela, Spain, pp. 283–290 (2020)
2. Fieldsend, J.E., Everson, R.M., Singh, S.: Using unconstrained elite archives for multiobjective optimization. IEEE Trans. Evol. Comput. **7**(3), 305–323 (2003)
3. Schütze, O., Coello Coello, C.A., Mostaghim, S., Talbi, E.-G., Dellnitz, M.: Hybridizing evolutionary strategies with continuation methods for solving multi-objective problems. J. Eng. Optim. **40**(5), 383–402 (2008)
4. Brockhoff, D., Tran, T.-D., Hansen, N.: Benchmarking numerical multiobjective optimizers revisited. In: Proceedings of the Conference on Genetic and Evolutionary Computation, Madrid Spain, pp. 639–646 (2015)
5. Tanabe, R., Ishibuchi, H., Oyama, A.: Benchmarking multi- and many-objective evolutionary algorithms under two optimization scenarios. IEEE Access **5**, 19597–19619 (2017)
6. Ishibuchi, H., Setoguchi, Y., Masuda, H., Nojima, Y.: How to compare many-objective algorithms under different settings of population and archive sizes. In: Proceedings of the IEEE Congress on Evolutionary Computation, Vancouver, BC, Canada, pp. 1149–1156 (2016)
7. Ishibuchi, H., Pang, L.M., Shang, K.: Difficulties in fair performance comparison of multi-objective evolutionary algorithms. IEEE Comput. Intell. Mag. **17**(1), 86–101 (2022)
8. Bringmann, K., Friedrich, T., Klitzke, P.: Generic postprocessing via subset selection for hypervolume and epsilon-indicator. In: Bartz-Beielstein, T., Branke, J., Filipič, B., Smith, J. (eds.) PPSN 2014. LNCS, vol. 8672, pp. 518–527. Springer, Cham (2014). https://doi.org/10.1007/978-3-319-10762-2_51
9. Bezerra, L.C.T., López-Ibáñez, M., Stützle, T.: Archiver effects on the performance of state-of-the-art multi- and many-objective evolutionary algorithms. In: Proceedings of the Conference on Genetic and Evolutionary Computation, Prague Czech Republic, pp. 620–628 (2019)
10. Li, M., Yao, X.: An empirical investigation of the optimality and monotonicity properties of multiobjective archiving methods. In: Deb, K., et al. (eds.) EMO 2019. LNCS, vol. 11411, pp. 15–26. Springer, Cham (2019). https://doi.org/10.1007/978-3-030-12598-1_2
11. Zitzler, E., Thiele, L.: Multiobjective evolutionary algorithms: a comparative case study and the strength Pareto approach. IEEE Trans. Evol. Comput. **3**(4), 257–271 (1999)
12. Zitzler, E., Thiele, L., Laumanns, M., Fonseca, C.M., da Fonseca, V.G.: Performance assessment of multiobjective optimizers: an analysis and review. IEEE Trans. Evol. Comput. **7**(2), 117–132 (2003)
13. Shang, K., Ishibuchi, H., He, L., Pang, L.M.: A survey on the hypervolume indicator in evolutionary multiobjective optimization. IEEE Trans. Evol. Comput. **25**(1), 1–20 (2021)
14. Bradstreet, L., Barone, L., While, L.: Maximising hypervolume for selection in multi-objective evolutionary algorithms. In: Proceedings of the IEEE Congress on Evolutionary Computation, Vancouver, BC, Canada, pp. 1744–1751 (2006)
15. Bradstreet, L., While, L., Barone, L.: Incrementally maximising hypervolume for selection in multi-objective evolutionary algorithms. In: Proceedings of the IEEE Congress on Evolutionary Computation, Singapore, pp. 3203–3210 (2007)

16. Jiang, S., Zhang, J., Ong, Y., Zhang, A.N., Tan, P.S.: A simple and fast hypervolume indicator-based multiobjective evolutionary algorithm. IEEE Trans. Cybern. **45**(10), 2202–2213 (2015)
17. Chen, W., Ishibuchi, H., Shang, K.: Fast greedy subset selection from large candidate solution sets in evolutionary multi-objective optimization. IEEE Trans. Evol. Comput. **26**(4), 750–764 (2022)
18. Shang, K., Ishibuchi, H., Chen, W.: Greedy approximated hypervolume subset selection for many-objective optimization. In: Proceedings of the Genetic and Evolutionary Computation Conference, Lille, France, pp. 448–456 (2021)
19. Nemhauser, G.L., Wolsey, L.A., Fisher, M.L.: An analysis of approximations for maximizing submodular set functions-I. Math. Program. **14**(1), 265–294 (1978)
20. Ulrich, T., Thiele, L.: Bounding the effectiveness of hypervolume-based $(\mu + \lambda)$-archiving algorithms. In: Hamadi, Y., Schoenauer, M. (eds.) LION 2012. LNCS, pp. 235–249. Springer, Heidelberg (2012). https://doi.org/10.1007/978-3-642-34413-8_17
21. Shang, K., Ishibuchi, H., Ni, X.: R2-based hypervolume contribution approximation. IEEE Trans. Evol. Comput. **24**(1), 185–192 (2020)
22. Nan, Y., Shang, K., Ishibuchi, H., He, L.: Improving local search hypervolume subset selection in evolutionary multi-objective optimization. In: Proceedings of the IEEE International Conference on Systems, Man and Cybernetics, Melbourne, Australia, pp. 751–757 (2021)
23. Nan, Y., Shang, K., Ishibuchi, H., He, L.: A two-stage hypervolume contribution approximation method based on R2 indicator. In: Proceedings of the IEEE Congress on Evolutionary Computation, Kraków, Poland, pp. 2468–2475 (2021)
24. Shang, K., Shu, T., Ishibuchi, H., Nan, Y., Pang, L.M.: Benchmarking subset selection from large candidate solution sets in evolutionary multi-objective optimization. arXiv:2201.06700
25. Deb, K., Jain, H.: An evolutionary many-objective optimization algorithm using reference-point-based nondominated sorting approach, Part I: solving problems with box constraints. IEEE Trans. Evol. Comput. **18**(4), 577–601 (2013)
26. Deb, K., Thiele, L., Laumanns, M., Zitzler, E.: Scalable test problems for evolutionary multiobjective optimization. In: Abraham, A., Jain, L., Goldberg, R. (eds.) Evolutionary Multiobjective Optimization. AI&KP, pp. 105–145. Springer, London (2005). https://doi.org/10.1007/1-84628-137-7_6
27. Ishibuchi, H., Setoguchi, Y., Masuda, H., Nojima, Y.: Performance of decomposition-based many-objective algorithms strongly depends on Pareto front shapes. IEEE Trans. Evol. Comput. **21**(2), 169–190 (2017)
28. Ishibuchi, H., Imada, R., Setoguchi, Y., Nojima, Y.: How to specify a reference point in hypervolume calculation for fair performance comparison. Evol. Comput. **26**(3), 411–440 (2018)
29. Nan, Y., Shang, K., Ishibuchi, H.: What is a good direction vector set for the R2-based hypervolume contribution approximation. In: Proceedings of the Conference on Genetic and Evolutionary Computation, Lisbon, Portugal, pp. 524–532 (2020)
30. Deng, J., Zhang, Q.: Approximating hypervolume and hypervolume contributions using polar coordinate. IEEE Trans. Evol. Comput. **23**(5), 913–918 (2019)
31. Zhang, X., Tian, Y., Jin, Y.: A knee point-driven evolutionary algorithm for many-objective optimization. IEEE Trans. Evol. Comput. **19**(6), 761–776 (2015)

The Hypervolume Indicator Hessian Matrix: Analytical Expression, Computational Time Complexity, and Sparsity

André Deutz[1], Michael Emmerich[1], and Hao Wang[1,2](✉)

[1] Leiden Institute of Advanced Computer Science, Leiden University, Leiden, The Netherlands
{a.h.deutz,m.t.m.emmerich}@liacs.leidenuniv.nl

[2] applied Quantum algorithms (aQa), Leiden University, Leiden, The Netherlands
h.wang@liacs.leidenuniv.nl

Abstract. The problem of approximating the Pareto front of a multi-objective optimization problem can be reformulated as the problem of finding a set that maximizes the hypervolume indicator. This paper establishes the analytical expression of the Hessian matrix of the mapping from a (fixed size) collection of n points in the d-dimensional decision space (or m dimensional objective space) to the scalar hypervolume indicator value. To define the Hessian matrix, the input set is vectorized, and the matrix is derived by analytical differentiation of the mapping from a vectorized set to the hypervolume indicator. The Hessian matrix plays a crucial role in second-order methods, such as the Newton-Raphson optimization method, and it can be used for the verification of local optimal sets. So far, the full analytical expression was only established and analyzed for the relatively simple bi-objective case. This paper will derive the full expression for arbitrary dimensions ($m \geq 2$ objective functions). For the practically important three-dimensional case, we also provide an asymptotically efficient algorithm with time complexity in $O(n \log n)$ for the exact computation of the Hessian Matrix' non-zero entries. We establish a sharp bound of $12m-6$ for the number of non-zero entries. Also, for the general m-dimensional case, a compact recursive analytical expression is established, and its algorithmic implementation is discussed. Also, for the general case, some sparsity results can be established; these results are implied by the recursive expression. To validate and illustrate the analytically derived algorithms and results, we provide a few numerical examples using Python and Mathematica implementations. Open-source implementations of the algorithms and testing data are made available as a supplement to this paper.

Keywords: Multi-objective optimization · Hypervolume indicator · Hessian matrix

M. Emmerich et al. (Eds.): EMO 2023, LNCS 13970, pp. 405–418, 2023.
https://doi.org/10.1007/978-3-031-27250-9_29

1 Introduction

In this paper, we delve into continuous m-dimensional multi-objective optimization problems (MOPs), where multiple objective functions, e.g., $\mathbf{f} = (f_1, \ldots, f_m) : \mathcal{X} \subseteq \mathbb{R}^d \to \mathbb{R}^m$ are subject to minimization. Also, we assume $\mathbf{f}$ is at least twice continuously differentiable. When solving such problems, it is a common strategy to approximate the Pareto front for m-objective functions mapping from a continuous decision space $\mathbb{R}^d$ to the $\mathbb{R}$ (or as a vector-valued function from $\mathbb{R}^d$ to $\mathbb{R}^m$. MOPs can be accomplished by means of a finite set of points that distributes across the at most $m-1$-dimensional manifold of the Pareto front. The hypervolume indicator of a set of points is the m dimensional Lebesgue measure of the space that is jointly dominated by a set of objective function vectors in $\mathbb{R}^m$ and bound from above by a reference point $\mathbf{r} \in \mathbb{R}^m$. More precisely, for minimization problems, the hypervolume indicator (HV) [16,17] is defined as the Lebesgue measure of the compact set dominated by a Pareto approximation set $Y \subset \mathbb{R}^m$ and cut from above by a reference point $\mathbf{r}$:

$$\mathrm{HV}(Y; \mathbf{r}) = \lambda_m \left(\{ \mathbf{p} \colon \exists \mathbf{y} \in Y (\mathbf{y} \prec \mathbf{p} \wedge \mathbf{p} \prec \mathbf{r}) \} \right),$$

where λ_m denotes the Lebesgue measure on $\mathbb{R}^m$. We will omit the reference point for simplicity henceforth. HV is Pareto compliant, i.e., for all $Y \prec Y'$, $\mathrm{HV}(Y) > \mathrm{HV}(Y')$, and is extensively used to assess the quality of approximation sets to the Pareto front, e.g., in SMS-EMOA [1] and multiobjective Bayesian optimization [5]. Being a set function, it is cumbersome to define the derivative of HV. Therefore, we follow the generic set-based approach for MOPs [4], which considers a finite set of objective points (of size n) as a single point in $\mathbb{R}^{nm}$, obtained via the following concatenation map (and its inverse):

$$\mathrm{concat} \colon (\mathbb{R}^m)^n \to \mathbb{R}^{nm}, \quad Y \mapsto \left(y_1^{(1)}, \ldots, y_m^{(1)}, \ldots, y_1^{(n)}, \ldots, y_m^{(n)} \right)^\top,$$

$$\mathrm{concat}^{-1} \colon \mathbb{R}^{nm} \to (\mathbb{R}^m)^n, \quad \mathbf{Y} \mapsto \left\{ (Y_1, \ldots, Y_m)^\top, \ldots, (Y_{(n-1)m+1}, \ldots, Y_{nm})^\top \right\}.$$

The concatenation map gives rise to a hypervolume function that takes vectors in $\mathbb{R}^{nm}$ as input (Table 1):

$$\mathcal{H} \colon \mathbb{R}^{nm} \to \mathbb{R}_{\geq 0}, \quad \mathbf{Y} \mapsto \left[\mathrm{HV} \circ \mathrm{concat}^{-1} \right] (\mathbf{Y}), \tag{1}$$

Table 1. Basic notation used throughout the paper. HVI stands for "Hypervolume Indicator".

Symbol	Domain	Description
m	N	Number of objective functions
d	N	Number of decision variables
n	N	Number of points in the approximation set
$\mathbf{f} = (f_1, \ldots, f_m)$	$\mathbb{R}^d \to \mathbb{R}^m$	Vector-valued objective function
$\mathbf{X} = (\mathbf{x}^{(1)\top}, \ldots, \mathbf{x}^{(n)\top})^\top$	$\mathbb{R}^{nd}$	Concatenation of n points in the decision space
$\mathbf{Y} = (\mathbf{y}^{(1)\top}, \ldots, \mathbf{y}^{(n)\top})^\top$	$\mathbb{R}^{nm}$	Concatenation of n points in the objective space
HV	$\mathbb{R}^{nm} \to \mathbb{R}$	HVI for subsets of the objective space
$\mathcal{H}$	$\mathbb{R}^{nm} \to \mathbb{R}_{\geq 0}$	HVI on the product of n objective spaces
$\mathcal{H}_\mathbf{F}$	$\mathbb{R}^{nd} \to \mathbb{R}_{\geq 0}$	HVI on the product of n decision spaces

Similarly, we also consider a finite set of decision points (of size n) as a single point in $\mathbb{R}^{nd}$, i.e., $\mathbf{X} = [\mathbf{x}^{(1)\top}, \mathbf{x}^{(2)\top}, \ldots, \mathbf{x}^{(n)\top}]^\top \in \mathbb{R}^{nd}$. In this sense, the objective function $\mathbf{f}$ is also extended to: $\mathbf{F}\colon \mathbf{X} \mapsto [\mathbf{f}(X_1, \ldots, X_d), \ldots, \mathbf{f}(X_{(n-1)d+1}, \ldots, X_{nd})]^\top$. Taking $\mathbf{F}$, we can express hypervolume indicator as a function supported on $\mathbb{R}^{nd}$:

$$\mathcal{H}_\mathbf{F}\colon \mathbb{R}^{nd} \to \mathbb{R}_{\geq 0}, \quad \mathbf{X} \mapsto \left[\mathrm{HV} \circ \mathrm{concat}^{-1} \circ \mathbf{F}\right](\mathbf{X}). \tag{2}$$

Notably, assuming $\mathbf{f}$ is twice differentiable, the above hypervolume functions $\mathcal{H}$ and $\mathcal{H}_\mathbf{F}$ are twice differentiable almost everywhere in their domains, respectively. In our previous works [4,14], we have provided the gradient of $\mathcal{H}_\mathbf{F}$ w.r.t. $\mathbf{X}$ using the chain rule: $\nabla\mathcal{H}_\mathbf{F}(\mathbf{X}) = (\partial\mathcal{H}/\partial\mathbf{F})(\partial\mathbf{F}/\partial\mathbf{X})$.

In this work, we investigate the hypervolume indicator Hessian matrix for more than two objectives and propose an algorithm to compute it efficiently. The work is structured as follows: In Sect. 2, we briefly review the general construction of the hypervolume gradient and hypervolume Hessian $\mathcal{H}_\mathbf{F}$ via the chain-rule as it has been outlined previously in [4] and, respectively, in [12]. Section 3 provides a discussion of the 3-D hypervolume indicator Hessian matrix HV, including a $O(n \log n)$ dimension sweep algorithm for its asymptotically optimal computation and an analysis of its sparsity, i.e., the number of its non-zero components. Furthermore, it is argued that in the 3-D case, the hypervolume Hessian matrix is sparse and has at most $O(n)$ non-zero components. We provide a `Mathematica` implementation for computing the 3-D Hessian matrix.

In Sect. 4, we discuss the analytical formulations of the hypervolume Hessian for the general case of $m > 1$ objective functions. The result reduces the computation of the Hessian of m objective functions to the repeated computation of the hypervolume indicator gradient for collections of vectors in $\mathbb{R}^{m-1}$. A `Python` implementation is provided for the general cases.

In Sect. 5, we provide numerical examples. In Sect. 6, we finish the paper with a discussion of some basic properties of the hypervolume Hessian matrix, such as its continuity, one-sidedness, and rank, and point out interesting open questions for its further analysis. Due to the space limitation, we have excluded all proofs to theorems in this work and point the interested readers to [2], which is the arXiv version of this paper.

2 General Construction of Hypervolume Hessian and Gradient via the Chain Rule

In general, the Hessian matrix of the hypervolume indicator can be expressed as follows:

$$\nabla^2\mathcal{H}_\mathbf{F} = \frac{\partial}{\partial\mathbf{X}}\left(\frac{\partial\mathcal{H}}{\partial\mathbf{F}}\frac{\partial\mathbf{F}}{\partial\mathbf{X}}\right) = \nabla\mathbf{F}^\top\frac{\partial^2\mathcal{H}}{\partial\mathbf{F}\partial\mathbf{F}^\top}\nabla\mathbf{F} + \frac{\partial\mathcal{H}}{\partial\mathbf{F}}\frac{\partial^2\mathbf{F}}{\partial\mathbf{X}\partial\mathbf{X}^\top}. \tag{3}$$

The Hessian of vector-valued objective function $\mathbf{F}$, i.e., $\partial^2\mathbf{F}/\partial\mathbf{X}\partial\mathbf{X}^\top\colon \mathbb{R}^{nd} \to \mathrm{Hom}(\mathbb{R}^{nd}, \mathrm{Hom}(\mathbb{R}^{nd}, \mathbb{R}^{nm}))$, is a tensor of $(1,2)$ type. Let $T^k_{i,j} = \partial^2 F_k/\partial \mathrm{X}_i \partial \mathrm{X}_j$, $i, j \in [1..nd], k \in [1..nm]$, we specify the entries of T as follows:

$$T^k_{i,j} = \begin{cases} \partial^2 f_\beta(\mathbf{x}^{(\alpha)})/\partial x^{(\alpha)}_{i'} \partial x^{(\alpha)}_{j'}, & \text{if } (\alpha-1)d+1 \leq i,j \leq \alpha d, \\ 0, & \text{otherwise.} \end{cases}$$

$$\alpha = \lceil k/m \rceil, \beta = k - (\alpha-1)m, i' = i - (\alpha-1)d, j' = j - (\alpha-1)d.$$

Since the above map from k to (α, β) is bijective (its inverse is $k = \alpha\beta$), we will equivalently use $\alpha\beta$ for the contravariant index k. It is obvious that tensor T is sparse, where for each k, only d^2 entries are nonzero, giving up to nmd^2 nonzero entries in total. Using the Einstein summation convention, we can expand the second term in Eq. (3) as:

$$\left(\frac{\partial \mathcal{H}}{\partial \mathbf{F}} \frac{\partial^2 \mathbf{F}}{\partial \mathbf{X} \partial \mathbf{X}^\top}\right)_{i,j} = \left(\frac{\partial \mathcal{H}}{\partial \mathbf{F}}\right)_k T^k_{i,j} = \frac{\partial \mathcal{H}}{\partial f_\beta(\mathbf{x}^{(\alpha)})} T^{\alpha\beta}_{i,j} \tag{4}$$

Notably, the above expression leads to a *block-diagonal matrix* containing n matrices of shape $d \times d$ on its diagonal. Therefore, we observe a high sparsity of the second term in Eq. (3). As for the first term, $\partial^2 \mathcal{H}/\partial \mathbf{F} \partial \mathbf{F}^\top$ denotes the Hessian of the hypervolume indicator w.r.t. objective vectors, whose computation and sparsity will be discussed in the following sections.

In our previous work [12], we have derived the analytical expression of $\nabla^2 \mathcal{H}_\mathbf{F}$ for bi-objective cases and analyzed the structure and properties of the hypervolume Hessian matrix. We have shown that the Hessian $\nabla^2 \mathcal{H}_\mathbf{F}$ is a tri-diagonal block matrix in bi-objective cases and provided the non-singularity condition thereof, which states the Hessian is only singular on a null subset of $\mathbb{R}^{nd}$ [12].

3 Hypervolume Indicator Hessian Matrix in 3-D

As with many problems related to Pareto dominance, the 2-D and 3-D cases have a special structure that can be exploited for formulating asymptotically efficient dimension sweep algorithms [8,9]. Next, the dimension sweep technique will be applied to yield an asymptotically efficient algorithm for the problem of computing the Hessian Matrix of the 3-D hypervolume indicator HV. The basic idea is sketched in Fig. 1 and consists of computing first the facets of the polyhedron that is measured by the hypervolume indicator by lowering a sweeping plane along each one of the axes. The gradient components are given by the areas of the facets (e.g., the yellow shaded area in Fig. 1, and the *length of the line segments of the ortho-convex polygon that surrounds this facet determines the components of the Hessian matrix of* HV. This can be easily verified by studying geometrically the effect of small perturbations of the coordinates of points in $\mathbf{Y}$ along the coordinate axis on the value of the hypervolume indicator (gradient components) HV.

Without loss of generality, we first compute the derivatives with respect to y_3. By permuting the roles of y_1, y_2, and y_3, we can get all derivatives. In the context of the dimension sweep algorithm, we assume that points in Fig. 1 are sorted by the 3rd coordinate y_3, that is, Y is represented in such a way that

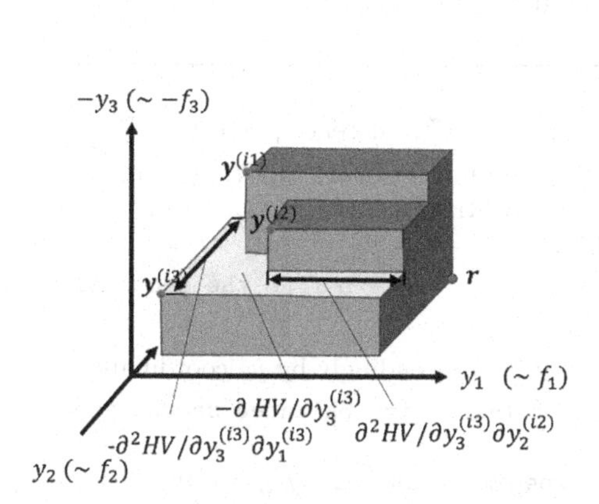

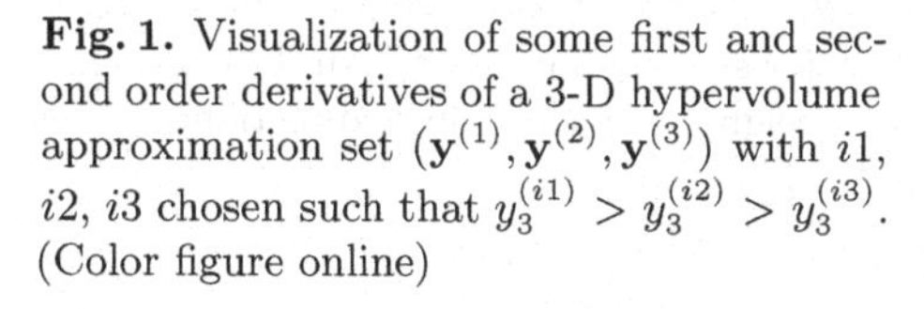

Fig. 1. Visualization of some first and second order derivatives of a 3-D hypervolume approximation set $(\mathbf{y}^{(1)}, \mathbf{y}^{(2)}, \mathbf{y}^{(3)})$ with $i1$, $i2$, $i3$ chosen such that $y_3^{(i1)} > y_3^{(i2)} > y_3^{(i3)}$. (Color figure online)

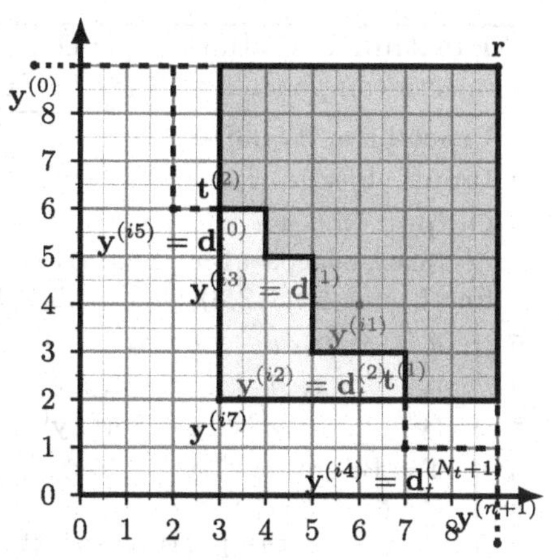

Fig. 2. Snapshot of the components of the polygon (yellow shaded), in which the length of each edge constitutes the non-zero components of the Hessian matrix of HV in a single iteration (lowering of sweeping plane) t of Algorithm 1. (Color figure online)

$y_3^{(i3)} < y_3^{(i2)} < y_3^{(i1)}$. We assume that the points in $\mathbf{Y}$ are in general position (otherwise, one-sided derivatives can occur, which will be discussed later).

Algorithm 1 computes all positive entries of the Hessian matrix of HV at a point. The algorithm proceeds in three sweeps. The sweep coordinate has index h (like height), and the other two coordinates are termed l and w (like length and width). The first sweep sets $h = 3$, the second sweep $h = 2$, and the third sweep $h = 1$. The values of l and w are set to the remaining two coordinates. The roles of l and w are interchangeable, but here we set them to $l = 1$, $w = 2$ in the first sweep, to $l = 1, w = 3$ in the second sweep, and to $l = 3, w = 2$ in the third sweep. Next, we describe a single sweep in detail. Without loss of generality, let us choose the sweep along the 3-rd coordinate, e.g., $l = 1, w = 2, h = 3$: First, we introduce sentinels $\mathbf{y}^{(0)}$ and $\mathbf{y}^{(n+1)}$ that make sure that every point has always a left and a lower neighbor. We use a balanced binary search tree T to efficiently maintain a list of all non-dominated points in the lw-plane among the points that have been visited in a single sweep so far. The sorted list represented by tree T is initialized by the sentinels; note that the sentinels cannot become dominated in the lw-plane because one of their coordinates is $-\infty$. The other coordinates are from the reference point. Next, start the loop that starts from the highest y_h coordinate and lowers the sweeping plane to the next highest y_h coordinate in each iteration t. The value of t is thus the index of the point that is currently processed, and points are sorted in the h-direction using the index transformation $a[t], t = 1, \ldots, n$. In each iteration, we determine from the sorted list the sublist starting from $\mathbf{y}^{d[t][0]}$ and terminating with $\mathbf{y}^{d[t][N_t]}$. We

Algorithm 1: HVH3D Triple Dimension Sweep algorithm to Compute all non-zero components of $\partial\mathcal{H}^2/\partial\mathbf{Y}\partial\mathbf{Y}^\top$

1 **Procedure:** HVH3D-TriSweep($\mathbf{Y}$);
2 **Input:** objective vectors $\mathbf{Y} = (\mathbf{y}^{(1)}, \ldots, \mathbf{y}^{(n)}) \in \mathbb{R}^m$, reference point $\mathbf{r} \in \mathbb{R}^m$;
3 **Output:** Non-zero components of the Hessian matrix of HV: $\partial^2\mathcal{H}/\partial y_m^{(i)} y_k^{(j)}$;
`// Three sweeps along the different coordinate axis.`
4 **for** $(l, w, h) \in ((1,2,3),(1,3,2),(3,2,1))$ **do**
5 $y_l^{(0)} = -\infty, y_w^{(0)} = r_w, y_h^{(0)} = r_h$; ▷ `Define the sentinels`
6 $y_l^{(n+1)} = r_l, y_l^{(n+1)} = -\infty, y_l^{(n+1)} = r_h$;
7 $(\mathbf{y}^{(a[1])}, \ldots, \mathbf{y}^{(a[n])}) \leftarrow$ Sort $(\mathbf{y}^{(1)}, \ldots, \mathbf{y}^{(n)})$ descendingly by y_h coordinate;
8 T = BSTree(($\mathbf{y}^{(0)}, \mathbf{y}^{(n+1)}$)) ; ▷ `Initialize balanced search tree`
9 **for** $t \in (1, \ldots, n)$ **do**
▷ `t is the position of the sweeping plane along` y_h `axis`
10 Determine $N[t]$ and $(\mathbf{y}^{(d[t][0])}, \ldots, \mathbf{y}^{(d[t][N[t]+1])})$ based on $\mathbf{y}^{(a[t])}$ as sublist of T starting from nearest lower neighbor to $\mathbf{y}^{(a[t])}$ in l direction and terminating at nearest lower neighbor of $\mathbf{y}^{(a[t])}$ in w-direction;
11 $\partial^2 \text{HV} / \partial y_h^{(a[t])} y_l^{(a[t])} = -(y_w^{(d[t][0])} - y_w^{(a[t])})$;
12 $\partial^2 \text{HV} / \partial y_h^{(a[t])} y_w^{(a[t])} = -(y_l^{(d[t][N_t])} - y_l^{(a[t])})$;
13 **if** $N[t] > 0$ **then**
14 $\partial^2 \text{HV} / \partial y_h^{(a[t])} y_w^{(d[t][0])} = y_l^{(d[t][1])} - y_l^{(a[t])}$;
15 $\partial^2 \text{HV} / \partial y_h^{(a[t])} y_l^{(d[t][N_t])} = y_w^{(d[t][N_t-1])} - y_w^{(a[t])}$;
16 **for** $j = 1, \ldots, N_t$ **do**
17 $\partial^2 \text{HV} / \partial y_h^{(a[t])} y_l^{(d[t][N_t])} = y_w^{(d[t][j-1])} - y_w^{(d[t][j])}$;
18 $\partial^2 \text{HV} / \partial y_h^{(a[t])} y_w^{(d[t][N_t])} = y_l^{(d[t][j+1])} - y_l^{(d[t][j])}$;
19 Discard $(\mathbf{y}^{(d[t][1])}, \ldots, \mathbf{y}^{(d[t][N[t]])})$ from T;
20 Add $\mathbf{y}^{a[t]}$ to tree T;
21 **return** $\partial^2\mathcal{H}/\partial y_m^{(i)} y_k^{(j)}$; ▷ `return only the` $O(n)$ `computed elements`

assume the list is sorted ascendingly in the l-coordinate. The point in Twith the highest y_l coordinate that does not exceed $y_l^{a[t]}$ is chosen as $\mathbf{t}^{\mathbf{d[t][0]}}$ and the point with the highest y_w coordinate that does not yet exceed $y_w^{(a[t])}$ is chosen as $\mathbf{t}^{(\mathbf{d[t][N_t+1]})}$. These two points always exist because of the sentinels we set initially. The points between these points in the list, given there exist such points, are referred to by $\mathbf{y}^{(d[t][1])}$, ..., $\mathbf{y}^{(d[t][N_t])}$. If no such points exist, then N_t is set to 0. Note, that the points $\mathbf{y}^{(d[t][1])}, \ldots, \mathbf{y}^{(d[t][N_t])}$ are points that become dominated by $\mathbf{y}^{(a[t])}$ in the lw-projection. They will be discarded from T at the end of the iteration, and $\mathbf{y}^{(a[t])}$ will be inserted to T thereafter so that the list represented by T remains a list of mutually non-dominated points in the lw-projection. Before discarding the points from T, the new positive components of the hypervolume Hessian are computed. This is done by computing the line segments of the polygonal area that is marked by the points $\mathbf{y}^{(a[t])}$ and $\mathbf{y}^{(d[t][0])}$,

$\ldots, \mathbf{y}^{(d[t][N_t+1])}$ as it is indicated graphically in Fig. 2 for a single iteration of the algorithm. This is the polygonal region that marks the area of the hypervolume gradient $\partial\mathcal{H}/y_h^{(a[t])}$. Changing infinitesimally the coordinates of the corners of this polygon adds a differential change to the area, which is the aforementioned hypervolume indicator gradient component. The details of the assignment to the hypervolume Hessian can be determined by computing the side-lengths of the edges of the polygon and carefully tracing which coordinates of points in $\mathbf{Y}$ determine the coordinates of the region the area of which determines the gradient component (the yellow area in Fig. 1 and in Fig. 2).

Theorem 1 (Computation of nonzero components of the Hessian). *Assume that n mutually non-dominated points in $\mathbb{R}^3$ are given by a collection $\mathbf{Y}$, and assume they are in general position (no duplicate coordinates). Furthermore, assume that all points in $\mathbf{Y}$ dominate the reference point $\mathbf{r}$. Then Algorithm 1 that we will term* HV3D-TriSweep *computes all non-zero components of the Hessian matrix* HV.

Proof: see [2, Theorem 1]. Next, we analyze the time complexity of Algorithm 1 and study the number of non-zero components it computes, which corresponds exactly to the non-zero components of the Hessian matrix of HV.

Lemma 1. *In Algorithm 1 it holds that $\sum_t^n N_t = n - 1$.*

Proof: see [2, Lemma 1].

Theorem 2 (Time complexity of 3-D Hessian matrix of HV**).** *The computation of all non-zero components of the Hessian matrix of the mapping from a set $\mathbf{Y} \in \mathbb{R}^3$ in the objective space to the hypervolume indicator takes computational time $\Theta(n \log n)$.*

Proof: see [2, Theorem 2].

Theorem 3 (Sparsity and space of 3-D Hessian matrix of HV**).** *The number of all non-zero components of the Hessian matrix of the mapping from a set $\mathbf{Y} \in \mathbb{R}^3$ in the objective space to the hypervolume indicator does never exceed* $12n - 6$.

Proof: see [2, Theorem 3].

4 General N-Dimensional Expression of the Hypervolume Hessian Matrix

For the general cases ($m > 3$), it suffices to compute the term $\partial^2\mathcal{H}/\partial\mathbf{F}\partial\mathbf{F}^\top$ and utilize Eq. (3) for computing the hypervolume Hessian. We summarize the computation in Algorithm 2 and explain the details as follows. Let $\mathbf{A} = \partial^2\mathcal{H}/\partial\mathbf{F}\partial\mathbf{F}^\top \in \mathbb{R}^{nm\times nm}$. Without loss of generality, we calculate the entries of $\mathbf{A}$ in a column-wise manner - given indices $i \in [1..n]$ and $k \in [1..m]$, column ik of $\mathbf{A}$ takes the following form:

$[\partial(\partial\mathcal{H}\partial y_k^{(i)})/\partial\mathbf{y}^{(1)}, \ldots, \partial(\partial\mathcal{H}\partial y_k^{(i)})/\partial\mathbf{y}^{(i)}, \ldots, \partial(\partial\mathcal{H}\partial y_k^{(i)})/\partial\mathbf{y}^{(n)}]^\top$, which will be discussed in two scenarios: (1) $\partial/\partial\mathbf{y}^{(i)}$ and (2) $\partial/\partial\mathbf{y}^{(j)}, i \neq j$. The dominated space of $S \subseteq \mathbb{R}^m$, that is the subset of $\mathbb{R}^m$ dominated by $\mathbf{y}$, is denoted by $\mathrm{Dom}_m(S) = \{\mathbf{p} \in \mathbb{R}^m \colon \exists \mathbf{y} \in S(\mathbf{y} \prec \mathbf{p}) \wedge \mathbf{p} \prec \mathbf{r}\}$. Also, we take the canonical basis $\{\mathbf{e}_i\}_i$ for $\mathbb{R}^m$ in our elaboration.

4.1 Partial Derivative $\partial(\partial\mathcal{H}/\partial y_k^{(i)})/\partial\mathbf{y}^{(i)}$

Intuitively, from Fig. 1, we observe that $\partial^2\mathcal{H}/\partial y_k^{(i)}\partial y_k^{(i)}$ is always zero for the 3D case since $\partial\mathcal{H}/\partial y_k^{(i)}$ is essentially the hypervolume improvement of the projection of $\mathbf{y}^{(i)}$ along axis $\mathbf{e}_k$ (bright yellow area in Fig. 1), ignoring the points that dominates $\mathbf{y}^{(i)}$ after the projection. In addition, $\partial^2\mathcal{H}/\partial y_k^{(i)}\partial y_\alpha^{(i)}, \alpha \neq i$ equals the negation of partial derivative of the hypervolume indicator w.r.t. $y_\alpha^{(i)}$ in the $m-1$-dimensional space, resulted from the projection. The computation of this quantity has been investigated in great detail previously [4]. We prove this argument for $m > 3$ as follows. First, we define the orthogonal projection operator $\mathrm{proj}_k \colon \mathbf{y} \mapsto (\ldots, y_{k-1}, y_{k+1}, \ldots)^\top$, which drops the k-th components of the input point $\mathbf{y}$. When applied to a subset $S \subseteq \mathbb{R}^m$, $\mathrm{proj}_k(S)$ operates on each element of S, resulting in a subset of $\mathbb{R}^{m-1}$.

Theorem 4 (Partial derivative of $\mathcal{H}$). *Assume a finite approximation set* $\mathbf{Y} = \{\mathbf{y}^{(1)}, \ldots \mathbf{y}^{(n)}\} \in (\mathbb{R}^m)^n$, *where, without loss of generality, the points are sorted in the ascending order w.r.t. the k-component, i.e.,* $y_k^{(1)} < \cdots y_k^{(i-1)} < y_k^{(i)} < y_k^{(i+1)} \cdots < y_k^{(n)}$. *Let* $\mathbf{y}_{\text{-}k}^{(\alpha)} = \mathrm{proj}_k(\mathbf{y}^{(\alpha)}), \alpha \in [1..n]$. *The partial derivative of* $\mathcal{H}$ *w.r.t.* $y_k^{(i)}$ *for all* $i \in [1..n]$ *and* $k \in [1..m]$ *is:*

$$\frac{\partial\mathcal{H}(\mathbf{Y})}{\partial y_k^{(i)}} = -\,\mathrm{HVC}\left(\mathbf{y}_{\text{-}k}^{(i)}, \left\{\mathbf{y}_{\text{-}k}^{(\alpha)} : \alpha \in [1..i-1]\right\}\right),$$

where $\mathrm{HVC}(\mathbf{y}, P)$ *is the hypervolume contribution of point* $\mathbf{y} \in \mathbb{R}^{m-1}$ *to a finite subset* $P \subset \mathbb{R}^{m-1}$, *i.e.,*

$$\mathrm{HVC}(\mathbf{y}, P) = \mathrm{HV}(P \cup \{\mathbf{y}\}) - \mathrm{HV}(P) = \lambda_{m-1}\left(\mathrm{Dom}_{m-1}(\mathbf{y}) \setminus \cup_{\mathbf{p}\in P}\mathrm{Dom}_{m-1}(\mathbf{p})\right).$$

Proof: see [2, Theorem 4].

Corollary 1. *It follows immediately from Theorem 4 that*

$$\partial^2\mathcal{H}/\partial y_k^{(i)}\partial y_k^{(i)} = 0, \quad i \in [1..n], k \in [1..m], m \in \mathbb{N}_{>0}. \tag{5}$$

For computing $\partial(\partial\mathcal{H}/\partial y_k^{(i)})/\partial y_l^{(i)}, l \neq i$, *it suffices to calculate the partial derivatives of the hypervolume contribution, i.e., for* $p = l$ *if* $l < k$; *otherwise* $p = l-1$,

$$\frac{\partial^2\mathcal{H}}{\partial y_l^{(i)}\partial y_k^{(i)}} = -\frac{\partial}{\partial y_p^{(i)}}\left(\mathrm{HVC}\left(\mathbf{y}_{\text{-}k}^{(i)}, \left\{\mathbf{y}_{\text{-}k}^{(\alpha)} : \alpha \in [1..i-1]\right\}\right)\right).$$

Algorithm 2: General algorithm for the hypervolume Hessian matrix

1 **Input:** $\mathbf{X} = \{\mathbf{x}^{(1)}, \dots, \mathbf{x}^{(n)}\}$: decision points, $\mathbf{Y} = \{\mathbf{y}^{(1)}, \dots, \mathbf{y}^{(n)}\}$: objective points, $\nabla\mathbf{F}, \nabla^2\mathbf{F}$: Jacobian and Hessian of the objective function;
2 $\mathbf{A} \leftarrow \mathbf{0}_{nm \times nm}$; ▷ $\mathbf{A} := \partial^2\mathcal{H}/\partial\mathbf{F}\partial\mathbf{F}^\top$
3 **for** $i = 1, \dots, n$ **do**
4   **if** $\mathbf{x}^{(i)}$ *is dominated* **then continue**;
5   **for** $k = 1, \dots, m$ **do**
6     $\mathbf{r}_{\text{-}k}, \mathbf{y}^{(i)}_{\text{-}k} \leftarrow \text{proj}_k(\mathbf{r}), \text{proj}_k(\mathbf{y}^{(i)})$;
7     $\mathbf{Y}' \leftarrow \text{proj}_k\left(\left\{\mathbf{y} \in \mathbf{Y} : y_k < y^{(i)}_k\right\}\right)$;
8     $Q \leftarrow \left\{\alpha \in [1..n] : \forall \mathbf{y}^{(\alpha)}_{\text{-}k} \in \mathbf{Y}' \nexists \mathbf{p} \in \mathbf{Y}'(\mathbf{p} \prec \mathbf{y}^{(\alpha)}_{\text{-}k})\right\}$;
9     $(v_1, \dots, v_{m-1}) \leftarrow \text{HVC}\left(\mathbf{y}^{(i)}_{\text{-}k}, \mathbf{Y}', \mathbf{r}_{\text{-}k}\right)$; ▷ `Apply Eq. (6)`
10     $\partial\left(\partial^2\mathcal{H}/\partial y^{(i)}_k\right)/\partial\mathbf{y}^{(i)} \leftarrow (v_1, \dots, v_{k-1}, 0, v_k, \dots, v_{m-1})^\top$; ▷ `Eq. (5)`
11   **for** $\alpha = im + 1, \dots, (i+1)m$ **do**
12       $\mathbf{A}_{\alpha, im+k} \leftarrow \left[\partial\left(\partial\,\text{HV}/\partial y^{(i)}_k\right)/\partial\mathbf{y}^{(i)}\right]_\alpha$;
      `// compute` $\partial(\partial\mathcal{H}/\partial y^{(i)}_k)/\partial\mathbf{y}^{(j)}$
13     $\mathbf{Y}' \leftarrow \left\{\text{CLIP}(\mathbf{p}; \mathbf{y}^{(i)}_{\text{-}k}) : \mathbf{p} \in \mathbf{Y}'\right\}$;
14   **for** $j \in Q$ **do**
15       $\mathbf{y}^{(j)}_{\text{-}k} \leftarrow \mathbf{Y}'[j]$; ▷ `take element` j `from set` $\mathbf{Y}'$
16       $(w_1, \dots, w_{m-1}) \leftarrow \text{HVC}\left(\mathbf{y}^{(j)}_{\text{-}k}, \mathbf{Y}', \mathbf{r}_{\text{-}k}\right)$; ▷ `Apply Eq. (8)`
17       $\partial\left(\partial\mathcal{H}/\partial y^{(i)}_k\right)/\partial\mathbf{y}^{(j)} \leftarrow (w_1, \dots, w_{k-1}, 0, w_k, \dots, w_{m-1})^\top$;
18   **for** $\alpha = jm + 1, \dots, (j+1)m$ **do**
19         $\mathbf{A}_{\alpha, im+k} \leftarrow \left[\partial\left(\partial\mathcal{H}/\partial y^{(i)}_k\right)/\partial\mathbf{y}^{(j)}\right]_\alpha$;
20 $T \leftarrow \partial^2\mathbf{F}/\partial\mathbf{X}\partial\mathbf{X}^\top \leftarrow \nabla^2\mathbf{F}(\mathbf{X})$;
21 $\mathbf{H} \leftarrow \nabla\mathbf{F}(\mathbf{X})^\top\mathbf{A}\nabla\mathbf{F}(\mathbf{X}) + \sum_{\alpha=1}^{n}\sum_{\beta=1}^{m}\left(\partial\mathcal{H}_\mathbf{F}/\partial f_\beta(\mathbf{x}^{(\alpha)})\right)T^{\alpha\beta}$; ▷ `Eq. (3)`
22 **return** $\mathbf{H}$;

Note that, when the perturbation on $y^{(i)}_p$ is sufficiently small, the resulting change on HV *of the approximation set $\{\mathbf{y}^{(i)}_{\text{-}k}\} \cup \{\mathbf{y}^{(\alpha)}_{\text{-}k} : \alpha \in [1..i-1]\}$ is equivalent to that on* HVC *of $\mathbf{y}^{(i)}_{\text{-}k}$. Hence, we could apply Theorem 4 again to the above equation, which involves projecting the objective points in $\mathbb{R}^{m-1}$ along axis $\mathbf{e}_p$, i.e., dropping the p-th component of $\{\mathbf{y}^{(\alpha)}_{\text{-}k}\}$. Let $I = \{\alpha \in [1..i-1] : y^{(\alpha)}_l < y^{(i)}_l\}$, we have:*

$$\frac{\partial^2\mathcal{H}}{\partial y^{(i)}_l \partial y^{(i)}_k} = \text{HVC}\left(\text{proj}_p(\mathbf{y}^{(i)}_{\text{-}k}), \left\{\text{proj}_p(\mathbf{y}^{(\alpha)}_{\text{-}k}) : \alpha \in I\right\}\right), \tag{6}$$

which is $m-2$-dimensional Lebesgue measure. This recursive computation is employed by sub-procedure HVC *at line 9 of Algorithm 2.*

In all, we have elaborated the method to compute $\partial(\partial\mathcal{H}/\partial y_k^{(i)})/\partial\mathbf{y}^{(i)}$, which constitutes d entries of column ik in matrix $\mathbf{A}$.

4.2 Partial Derivative $\partial\left(\partial\mathcal{H}/\partial y_k^{(i)}\right)/\partial\mathbf{y}^{(j)}, i \neq j$

Based on Theorem 4, we conclude that $\partial(\partial\mathcal{H}/\partial y_k^{(i)})/\partial\mathbf{y}^{(j)} = 0 \iff j \notin [1..i-1]$, since $\partial\mathcal{H}/\partial y_k^{(i)}$ only depends on $\mathbf{y}_{\text{-}k}^{(i)}$ and $\{\mathbf{y}_{\text{-}k}^{(\alpha)} : \alpha \in [1..i-1]\}$. Also, for all $j \in [1..i-1]$, we have

$$\partial^2\mathcal{H}/\partial y_k^{(j)}\partial y_k^{(i)} = 0, \tag{7}$$

due to the projection operation. Note that, $\partial(\partial\mathcal{H}/\partial y_k^{(i)})/\partial\mathbf{y}^{(j)}$ constitutes the remaining $(n-1)$ entries of column ik of matrix $\mathbf{A}$, containing at most $(i-1)(d-1)$ nonzero values, where i depends on the number of points which have a smaller kth-component than that of $\mathbf{y}^{(i)}$. Hence, we can bound the nonzero elements in $\mathbf{A}$ by $O(n(n-1)/2(d-1)) = O(n^2d)$. Note that possible a sharper bound can be formulated by considering the technique used to prove Theorem 3 in more than three dimensions. The remaining partial derivatives can be computed by first clipping points in $\{\mathbf{y}_{\text{-}k}^{(\alpha)} : \alpha \in [1..i-1]\}$ by $\mathbf{y}_{\text{-}k}^{(i)}$ from below (line 13 in Algorithm 2; sub-procedure CLIP): $\widehat{\mathbf{y}}_{\text{-}k}^{(\alpha)} = \text{CLIP}(\mathbf{y}_{\text{-}k}^{(\alpha)}; \mathbf{y}_{\text{-}k}^{(i)})$, where for $\mathbf{a}, \mathbf{b} \in \mathbb{R}^{m-1}$, $\text{CLIP}(\mathbf{a}; \mathbf{b}) = (a_1 + \min\{0, a_1 - b_1\}, \ldots, a_{m-1} + \min\{0, a_{m-1} - b_{m-1}\})^\top$. Note that this clipping operation does not change the volume of $\text{Dom}_{m-1}(\mathbf{y}_{\text{-}k}^{(i)}) \setminus (\cup_{\alpha\in I}\text{Dom}_{m-1}(\mathbf{y}_{\text{-}k}^{(\alpha)}))$ since the points that clipped out are not in $\text{Dom}_{m-1}(\mathbf{y}_{\text{-}k}^{(i)})$. Taking the clipping operation, we have the following relation:

$$\begin{aligned}
&\text{HVC}\left(\mathbf{y}_{\text{-}k}^{(i)}, \{\mathbf{y}_{\text{-}k}^{(\alpha)} : \alpha \in [1..i-1]\}\right) \\
&= \lambda_{m-1}\left(\text{Dom}_{m-1}(\mathbf{y}_{\text{-}k}^{(i)}) \setminus \left(\cup_{\alpha\in I}\text{Dom}_{m-1}(\widehat{\mathbf{y}}_{\text{-}k}^{(\alpha)})\right)\right) \\
&= \lambda_{m-1}\left(\text{Dom}_{m-1}(\mathbf{y}_{\text{-}k}^{(i)}) \setminus \text{Dom}_{m-1}(\{\widehat{\mathbf{y}}_{\text{-}k}^{(\alpha)} : \alpha \in [1..i-1]\})\right) \\
&= \lambda_{m-1}\left(\text{Dom}_{m-1}(\mathbf{y}_{\text{-}k}^{(i)})\right) - \lambda_{m-1}\left(\text{Dom}_{m-1}(\{\widehat{\mathbf{y}}_{\text{-}k}^{(\alpha)} : \alpha \in [1..i-1]\})\right).
\end{aligned}$$

The last step in the above equation is due to the fact that after clipping, $\text{Dom}_{m-1}(\{\widehat{\mathbf{y}}_{\text{-}k}^{(\alpha)} : \alpha \in [1..i-1]\}) \subset \text{Dom}_{m-1}(\mathbf{y}_{\text{-}k}^{(i)})$. For $j \neq i$, $l \neq k$, let $p = l$, if $l < k$; otherwise $p = l - 1$, we have:

$$\begin{aligned}
\frac{\partial^2\mathcal{H}}{\partial y_l^{(j)}\partial y_k^{(i)}} &= -\frac{\partial}{\partial y_p^{(j)}}\left(\text{HVC}\left(\mathbf{y}_{\text{-}k}^{(i)}, \{\mathbf{y}_{\text{-}k}^{(\alpha)} : \alpha \in [1..i-1]\}\right)\right) \\
&= \frac{\partial}{\partial y_p^{(j)}}\lambda_{m-1}\left(\text{Dom}_{m-1}(\{\widehat{\mathbf{y}}_{\text{-}k}^{(\alpha)} : \alpha \in I\})\right) \\
&= \frac{\partial}{\partial y_p^{(j)}}\text{HV}\left(\left\{\widehat{\mathbf{y}}_{\text{-}k}^{(\alpha)} : \alpha \in [1..i-1]\right\}\right).
\end{aligned}$$

We can apply the result of Theorem 4 to compute the above expression.

Corollary 2. *For $j \neq i$, $l \neq k$, let $p = l$, if $l < k$; otherwise $p = l - 1$. The partial derivative $\partial(\partial \mathcal{H}/\partial y_k^{(i)})/\partial y_l^{(j)}$ admits the following expression:*

$$\frac{\partial^2 \mathcal{H}}{\partial y_l^{(j)} \partial y_k^{(i)}} = -\operatorname{HVC}\left(\operatorname{proj}_p(\widehat{\mathbf{y}}^{(j)}_{\text{-}k}), \{\operatorname{proj}_p(\widehat{\mathbf{y}}^{(\alpha)}_{\text{-}k}) \colon \alpha \in I\}\right), \tag{8}$$

where $I = \{\alpha \in [1..i-1] \colon [\widehat{\mathbf{y}}^{(\alpha)}_{\text{-}k}]_p < [\widehat{\mathbf{y}}^{(j)}_{\text{-}k}]_p\}$ ($[\cdot]_p$ denotes taking the p-th component of a vector). This recursive computation is employed by sub-procedure HVC *at line 16 of Algorithm 2.*

5 Numerical Examples

In this section, we showcase some numerical examples of the computation of the hypervolume Hessian matrix. For the sake of comprehensibility, we only compute the Hessian w.r.t. the objective points. We specify the objective points for the numerical problem below.

- Example 1: $m = 3$, $n = 2$, $\mathbf{Y} = [(5, 3, 7)^\top, (2, 1, 10)^\top]^\top$, and $\mathbf{r} = (9, 10, 12)^\top$.
- Example 2: $m = 3$, $n = 3$, $\mathbf{Y} = [(8, 7, 10)^\top, (4, 11, 17)^\top, (2, 9, 21)^\top]^\top$, and $\mathbf{r} = (10, 13, 23)^\top$.
- Example 3: $m = 3$, $n = 6$, $\mathbf{Y} = [(16, 23, 1)^\top, (14, 32, 2)^\top, (12, 27, 3)^\top, (10, 21, 4)^\top, (8, 33, 5)^\top, (6.5, 31, 6)^\top]^\top$, and $\mathbf{r} = (17, 35, 7)^\top$.

We illustrate the corresponding Hessian matrices as heatmaps in Fig. 3 of [2], from which we see clearly a high sparsity in all cases. Moreover, for the second example with $n = 3$ points in 3-D objective space, as predicted by Theorem 3, we obtain exactly $12n - 6$ ($= 30$) positive components. Also, we have verified the above computation by comparing the results obtained from `Python`, `Mathematica`, and automatic differentiation performed in our previous work [15].

6 Discussion and Outlook

This paper highlights two approaches for computing the components of the Hessian matrix of the hypervolume indicator HV of a multi-set of points in objective space and of a multi-set of points in the decision space $\mathcal{H}_\mathbf{F}$. The approach of set-scalarization, as originated in [3,4] for the gradient of the hypervolume indicator. The main results of the paper are as follows:

1. the hypervolume indicator of $\mathcal{H}_\mathbf{F}$ can now be computed analytically not only for the bi-objective case as in [12], but also for more than two objective functions (Theorem 1).
2. the time complexity of computing all non-zero components of the 3-D hypervolume indicator HV for vectorized sets $\mathbf{Y}$ with n points in general position is in $\Theta(n \log n)$. (Theorem 2)

3. The number of non-zero components of the Hessian matrix is at most $12n - 6$. The space complexity of the Hessian matrix computation and the space required to store all components is in $O(n)$. (Theorem 3).
4. it holds that $\partial \mathcal{H} / \partial y_k$ is always the hypervolume contribution of the projection of y_k along axis k. (Theorem 4).
5. the analytical computation of the higher derivatives of the m-dimensional hypervolume indicator HV for $m > 1$can be formulated by computing the gradient of the gradient, which can be essentially achieved by the recursive application of Theorem (Theorem 4) (computing the $m-2$-dimensional projection's contributions along the y_k axis, of the $m-1$ dimensional projection's hypervolume contributions along the y_k); and taking special care of the role of the reference points and signs of non-zero components as detailed in Algorithm 2.

Some interesting next steps would be to

1. investigate the rank of the hypervolume Hessian matrix and its numerical stability of second-order methods that use the Hessian matrix of $\mathcal{H}_{\mathbf{F}}$ (or its inverse) in their iteration, such as the hypervolume Newton method [11,12].
2. find (asymptotically) efficient algorithms for the computation of the higher-order derivative tensors and for more than three-dimensional cases. The latter might, however, find the asymptotical time complexity of the N-D Hessian matrix computation might turn out to be a difficult endeavor, as it is not even known what the asymptotical time complexity of HV is. What is more promising is to bound the number of the non-zero components in the Hessian matrix, which is related to the number of $n-2$ dimensional facets in the ortho-convex polyhedron that marks the measured region of HV. It is conjectured that in the m-dimensional case, it also grows linearly in n (the number of points in the approximation set) but exponentially in m (the number of objectives). However, dimension sweep algorithms also yield high efficiency in the 4-D case and can probably be easily adapted [6].

More generally, it is remarked that besides the hypervolume indicator, other measures have been proposed for the quality of Pareto front approximations, such as the inverted generational distance [7] or the averaged Hausdorff distance [10,13], with sometimes advantageous properties regarding the uniformity the point distributions in their maximum. We believe that for such measures, vectorization of the input set is promising, and the analytical computation and subsequent analysis of the Hessian matrix is worthwhile to delve into. The code for computing analytically the Hessian matrix of HV has been validated on example data and made available in a GitHub repository.[1] The repository includes an implementation based on Algorithm 2 in `Python` and in `Mathematica`, as well as the data of the examples.

Remark: The authors have been listed alphabetically in this paper, and all authors have contributed to the completion of the manuscript.

[1] https://github.com/wangronin/HypervolumeDerivatives.

Author contributions. HW+ME+AD: Concept and formulation of the general analytical expression for the matrix, ME: 3-D Dimension sweep algorithm and 3-D Complexity Analysis; HW+AD: Improved mathematical notation and numerical validation experiments; HW+AD: theoretical analysis of the general N-dimensional Hessian matrix; All authors: General set-vectorization concepts, motivation, and revision/editing of mathematical formulation.

References

1. Beume, N., Naujoks, B., Emmerich, M.T.M.: SMS-EMOA: multiobjective selection based on dominated hypervolume. Eur. J. Oper. Res. **181**(3), 1653–1669 (2007). https://doi.org/10.1016/j.ejor.2006.08.008
2. Deutz, A.H., Emmerich, M., Wang, H.: The hypervolume indicator hessian matrix: analytical expression, computational time complexity, and sparsity. arXiv preprint arXiv:2211.04171 (2022)
3. Emmerich, M., Deutz, A., Beume, N.: Gradient-based/evolutionary relay hybrid for computing pareto front approximations maximizing the S-metric. In: Bartz-Beielstein, T., et al. (eds.) HM 2007. LNCS, vol. 4771, pp. 140–156. Springer, Heidelberg (2007). https://doi.org/10.1007/978-3-540-75514-2_11
4. Emmerich, M., Deutz, A.H.: Time complexity and zeros of the hypervolume indicator gradient field. In: Schuetze, O., et al. (eds.) EVOLVE - A Bridge between Probability, Set Oriented Numerics, and Evolutionary Computation III. SCI, vol. 500, pp. 169–193. Springer, Cham (2012). https://doi.org/10.1007/978-3-319-01460-9_8
5. Emmerich, M., Yang, K., Deutz, A., Wang, H., Fonseca, C.M.: A multicriteria generalization of Bayesian global optimization. In: Pardalos, P.M., Zhigljavsky, A., Žilinskas, J. (eds.) Advances in Stochastic and Deterministic Global Optimization. SOIA, vol. 107, pp. 229–242. Springer, Cham (2016). https://doi.org/10.1007/978-3-319-29975-4_12
6. Guerreiro, A.P., Fonseca, C.M., Emmerich, M.T., et al.: A fast dimension-sweep algorithm for the hypervolume indicator in four dimensions. In: CCCG, pp. 77–82 (2012)
7. Ishibuchi, H., Masuda, H., Nojima, Y.: A study on performance evaluation ability of a modified inverted generational distance indicator. Association for Computing Machinery, New York (2015). https://doi.org/10.1145/2739480.2754792
8. Kung, H.T., Luccio, F., Preparata, F.P.: On finding the maxima of a set of vectors. J. ACM **22**(4), 469–476 (1975). https://doi.org/10.1145/321906.321910
9. Paquete, L., Schulze, B., Stiglmayr, M., Lourenço, A.C.: Computing representations using hypervolume scalarizations. Comput. Oper. Res. **137**, 105349 (2022)
10. Schütze, O., et al.: A scalar optimization approach for averaged Hausdorff approximations of the Pareto front. Eng. Optim. **48**(9), 1593–1617 (2016)
11. Sosa Hernández, V.A., Schütze, O., Emmerich, M.: Hypervolume maximization via set based Newton's method. In: Tantar, A.-A., et al. (eds.) EVOLVE - A Bridge between Probability, Set Oriented Numerics, and Evolutionary Computation V. AISC, vol. 288, pp. 15–28. Springer, Cham (2014). https://doi.org/10.1007/978-3-319-07494-8_2
12. Sosa-Hernández, V.A., Schütze, O., Wang, H., Deutz, A.H., Emmerich, M.: The set-based hypervolume newton method for bi-objective optimization. IEEE Trans. Cybern. **50**(5), 2186–2196 (2020). https://doi.org/10.1109/TCYB.2018.2885974

13. Uribe, L., Bogoya, J.M., Vargas, A., Lara, A., Rudolph, G., Schütze, O.: A set based newton method for the averaged Hausdorff distance for multi-objective reference set problems. Mathematics **8**(10), 1822 (2020)
14. Wang, H., Deutz, A., Bäck, T., Emmerich, M.: Hypervolume indicator gradient ascent multi-objective optimization. In: Trautmann, H., et al. (eds.) EMO 2017. LNCS, vol. 10173, pp. 654–669. Springer, Cham (2017). https://doi.org/10.1007/978-3-319-54157-0_44
15. Wang, H., Emmerich, M., Deutz, A., Hernández, V.A., Schütze, O.: The Hypervolume Newton Method for Constrained Multi-objective Optimization Problems. Preprints (2022). https://doi.org/10.20944/preprints202211.0103.v1
16. Zitzler, E., Thiele, L.: Multiobjective optimization using evolutionary algorithms—a comparative case study. In: Eiben, A.E., Bäck, T., Schoenauer, M., Schwefel, H.-P. (eds.) PPSN 1998. LNCS, vol. 1498, pp. 292–301. Springer, Heidelberg (1998). https://doi.org/10.1007/BFb0056872
17. Zitzler, E., Thiele, L., Laumanns, M., Fonseca, C.M., da Fonseca, V.G.: Performance assessment of multiobjective optimizers: an analysis and review. IEEE Trans. Evol. Comput. **7**(2), 117–132 (2003). https://doi.org/10.1109/TEVC.2003.810758

On the Computational Complexity of Efficient Non-dominated Sort Using Binary Search

Ved Prakash[1], Sumit Mishra[1(✉)], and Carlos A. Coello Coello[2]

[1] Deptartment of CSE, Indian Institute of Information Technology Guwahati, Guwahati, India
{ved.prakash,sumit}@iiitg.ac.in
[2] Departamento de Computación, CINVESTAV-IPN, Mexico City, Mexico
ccoello@cs.cinvestav.mx

Abstract. Over the years, several approaches have been proposed to solve the problem of non-dominated sorting, which is one of the crucial steps in Pareto dominance-based multi-objective evolutionary algorithms (MOEAs). However, some of these approaches, even though they are correct, lack an in-depth analysis. In this paper, we focus on an approach known as Efficient Non-dominated Sort using Binary Search (ENS-BS) and show that the best case scenario presented in the paper for ENS-BS is not correct. We show this by providing a counter-example where the number of dominance comparisons is less than that reported in the original paper. This is done by obtaining a generic equation and getting the scenario inspired by this equation.

Keywords: Non-dominated sorting · Dominance comparisons · Time complexity

1 Introduction

Pareto dominance-based multi-objective evolutionary algorithms (MOEAs) rank the solutions in their population based on the Pareto-dominance relation. This ranking of solutions is done using non-dominated sorting. Non-dominated sorting is defined for points that belong to an M-dimensional space $\mathbb{R}^M$. In the field of evolutionary computation, each such M-dimensional point is an objective vector associated with some solution to an optimization problem. The set of such points is known as a MOEA's *population*.

The approaches for non-dominated sorting do not use any problem-related information such as the structure of an individual. Hence, we are not making any difference between the individuals and their objective vectors. Without loss of generality, let us assume that all objectives that correspond to the coordinates of the points need to be minimized. In such conditions, Pareto dominance is defined as follows: a point $p = \langle p_1, p_2, \ldots, p_M \rangle$ dominates a point $q = \langle q_1, q_2, \ldots, q_M \rangle$, denoted as $p \prec q$, if the two following conditions are satisfied:

M. Emmerich et al. (Eds.): EMO 2023, LNCS 13970, pp. 419–432, 2023.
https://doi.org/10.1007/978-3-031-27250-9_30

- $\forall m \in \{1, 2, \ldots, M\}$ $p_m \leq q_m$
- $\exists m \in \{1, 2, \ldots, M\}$ $p_m < q_m$.

When neither $p \prec q$ nor $q \prec p$, then points p and q are said to be non-dominated. Let $\mathbb{P} = \{\mathbb{P}_1, \mathbb{P}_2, \ldots, \mathbb{P}_N\}$ be a population of size N where each point in the population is represented using a vector of size M. In non-dominated sorting, the population $\mathbb{P}$ is divided into different fronts $\{F_1, F_2, \ldots, F_K\}$ where $1 \leq K \leq N$, such that:

- The union of all the fronts is equal to the population.
- All the fronts are disjoint.
- All the points in a particular front are non-dominated with each other.
- All the points of the first front are not dominated by any other point.
- For every point $p \in F_k$ where $2 \leq k \leq K$, $\exists q \in F_{k-1}$ such that $q \prec p$.

Non-dominated sorting provides a unique partition. A point is said to have rank k if it belongs to front F_k. Let the cardinality of a front F_k be n_k, so $N = \sum_{k=1}^{K} n_k$.

In the last 20 years, non-dominated sorting has attracted a lot of attention from the research community and the Fast non-dominated sorting algorithm [2] is one of the approaches which made it popular. The full complexity of the non-dominated sorting problem is not well understood. Unfortunately, the analysis of some of the approaches for non-dominated sorting is imperfect.

An approach based on the divide-and-conquer paradigm was proposed by Jensen [5] with time complexity $\mathcal{O}(N \log^{M-1} N)$. Jensen's approach is built on the approach proposed by Kung *et al.* [6] where it has been shown that the non-dominated points can be obtained in $\mathcal{O}(N \log N)$ time for $M = 2, 3$ and in $\mathcal{O}(N \log^{M-2} N)$ time for $M \geq 4$. This same time complexity has been discussed in [1]. Jensen's approach has a limitation that the points cannot share the same value for any objective. The author claimed that this limitation can be easily removed. However, Fortin et al. [3] showed that removing this limitation will lead to an increase in the worst-case time complexity. In recent years, there has been some work on the analysis of some of these approaches. In [9], it has been proved that the worst-case time complexity of Deductive Sort [7] is $\Theta(MN^3)$ contrary to the claimed worst-case time complexity of $\Theta(MN^2)$. It has also been shown that, after shuffling the input, the worst-case expected running time comes down to $\mathcal{O}(MN^2)$. Similarly, in [12], the best case time complexity of Deductive Sort [7] is shown to be $\mathcal{O}(MN\sqrt{N} + N^2)$ as opposed to the claimed $\mathcal{O}(MN\sqrt{N})$ time complexity. The worst-case time complexity of the Dominance Degree Approach for Non-dominated Sorting (DDA-NS) [20] is $\Theta(MN^2)$ in its original paper. However, the worst-case time complexity of DDA-NS has been proved to be $\Theta(MN^2 + N^3)$ in [10]. Recently, Filter Sort [17] has been proposed for non-dominated sorting. However, the authors of this paper have not performed its complexity analysis. In [8], it has been proved that Filter Sort has a worst-case complexity of $\Omega(N^3)$. In particular, a scenario has been presented which requires Filter Sort to perform $\Theta(N^3)$ dominance comparisons.

Recently, Nigam *et al.* [16] presented a scenario where the number of dominance comparisons performed by ENS-SS (sequential search version of Efficient Non-dominated Sort) is less than that reported in the original paper [19] in the best case scenario. However, the authors have not provided the best-case scenario. In this paper, we show the number of dominance comparisons between the points in the best case of ENS-BS (binary search version of Efficient Non-dominated Sort) [19], is not correct. For this purpose, we have obtained a generic equation that shows the number of dominance comparisons in the best case. Based on this equation, we have identified a scenario where ENS-BS performs a lower number of dominance comparisons than that reported in [19]. It is important to clarify that, in this paper, we are not claiming that the scenario which we have identified represents the best-case scenario.

2 Approach

The ENS approach works in two phases: (i) Presorting Phase and (ii) Ranking Phase. ENS is summarized in Algorithm 1. In the first phase, the points are sorted lexicographically based on the first objective. After presorting, the point which comes later in the sorted list will never dominate the former points. Thus, when two points p and q are compared, such that when q comes later than p in the sorted list, then there are only two possibilities: (i) p dominates q and (ii) p and q are non-dominated. In the second phase, the points are assigned a rank. ENS takes the points from the sorted list one by one and ranks them.

Algorithm 1. ENS FRAMEWORK

Input: $\mathbb{P} = \{\mathbb{P}_1, \mathbb{P}_2, \ldots, \mathbb{P}_N\}$: A population of size N in M-dimensional space
Output: $\{F_1, F_2, \ldots\}$: Set of fronts

1: Sort $\mathbb{P}$ in lexicographic order based on the first objective
2: $\texttt{NF} \leftarrow 0$ ▷ Number of fronts; Initially it is 0
3: $\mathbb{F} \leftarrow \emptyset$ ▷ Set of fronts; Initially it is empty as no front has been obtained
4: **for each** point $p \in \mathbb{P}$ **do**
5: $\quad \mathbb{F} \leftarrow$ INSERT-BS$(\mathbb{F}, p)$
6: **return** $\mathbb{F} = \{F_1, F_2, \ldots\}$ ▷ Set of non-dominated fronts

In this approach, if a point $p \in F_k$, then p is dominated by at least one point in the preceding front. As discussed in the *Presorting Phase*, a point can never be dominated by any succeeding point in the sorted population. Hence, a point needs to be only compared with those points which have already been assigned to some front.

The first point is assigned to the first front without any comparison. Any subsequent point is assigned to a front F_k if it is dominated by at least one point of all its previous fronts $F_1, F_2, \ldots, F_{k-1}$ and is not dominated by all the previously assigned points of front F_k. Here, one important thing to note is that it

Algorithm 2. INSERT-BS($\mathbb{F}, p$)

Input: $\mathbb{F}\{F_1, F_2, \ldots, F_{\texttt{NF}}\}$: Set of non-dominated fronts, p: Point for insertion into set of fronts
Output: Updated set of fronts after insertion of p

```
1: min ← 1
2: max ← NF
3: mid ← ⌊(1+NF)/2⌋
4: while TRUE do                                          ▷ p has not been ranked yet
5:     isDominated ← FALSE
6:     for u ← n_mid down to 1 do        ▷ Check for each point in F_mid sequentially
       starting from the last point
7:         if F_mid(u) ≺ p then
8:             isDominated ← TRUE                     ▷ p is dominated by F_mid(u)
9:             Break                                  ▷ p cannot be added to F_mid
10:    if isDominated = FALSE then ▷ p is not dominated by any of the point in F_mid
11:        if mid = min then                               ▷ Front at leaf is explored
12:            F_mid ← F_mid ∪ {p}                                 ▷ Add p to F_mid
13:            Break                                           ▷ p has been ranked
14:        else
15:            max ← mid
16:            mid ← ⌊(min+max)/2⌋                         ▷ Explore left sub-tree
17:    else
18:        if min = NF then                     ▷ Front at rightmost leaf is explored
19:            NF ← NF + 1                       ▷ Increment the number of fronts
20:            F_NF ← ∅                                      ▷ Create a new front
21:            F_NF ← F_NF ∪ {p}                ▷ Add p to the newly created front
22:            𝔽 ← 𝔽 ∪ {F_NF}                  ▷ Add new front to the set of fronts
23:            Break                                           ▷ p has been ranked
24:        else if mid = min then                          ▷ Front at leaf is explored
25:            F_mid+1 ← F_mid+1 ∪ {p}                           ▷ Add p to F_mid+1
26:            Break                                           ▷ p has been ranked
27:        else
28:            min ← mid + 1
29:            mid ← ⌊(min+max)/2⌋                        ▷ Explore right sub-tree
30: return 𝔽                         ▷ Updated set of fronts after insertion of p
```

is not necessary to compare a point with the points of all the previously obtained fronts. Thus, unnecessary comparisons can be avoided using the binary search based technique. This is because of the transitivity nature of the dominance relationship, *i.e.*, if point $p \prec q$ and $q \prec r$ then $p \prec r$. In this manner, a point can be assigned to front F_k by comparing it with the points of the already obtained $\log k$ fronts. The manner in which points are ranked using the binary search based strategy is summarized in Algorithm 2.

3 Best Case Analysis of ENS-BS

In the best case of ENS-BS, if a point $p \in F_k$, then p is dominated by the first point (to which it is compared) of all the $k-1$ previous fronts. As a binary search based strategy is used, the point p is compared with $\lceil \log k \rceil$ previous fronts. Since, there are n_k points in F_k, so the number of dominance comparisons *between points of different fronts* to obtain front F_k is $\lceil \log k \rceil n_k$. There are a total of K fronts, and therefore, the number of dominance comparisons *between points of different fronts*, in the best case is given by Eq. (1).

$$\mathtt{dC_{diff}} = \sum_{k=1}^{K} \lceil \log k \rceil n_k \tag{1}$$

Each of the n_k points of a front F_k should be compared with the already assigned points of the same front F_k. This means that the first point of F_k is not compared with any of the points of F_k. The second point of F_k is compared with the first point of F_k. The third point of F_k is compared with the (already assigned) two points of F_k and so on. So, the number of dominance comparisons *between points of the same front*, in the best case, to produce front F_k is obtained using Eq. (2).

$$\mathtt{dC}_k = \sum_{i=1}^{n_k} (i-1) = \frac{1}{2} n_k (n_k - 1) \tag{2}$$

As there are a total of K fronts, therefore, the number of dominance comparisons *between points of the same front* is given by Eq. (3).

$$\mathtt{dC_{same}} = \sum_{k=1}^{K} \frac{1}{2} n_k (n_k - 1) = \frac{1}{2} \sum_{k=1}^{K} n_k^2 - \frac{1}{2} \sum_{k=1}^{K} n_k = \frac{1}{2} \left[\sum_{k=1}^{K} n_k^2 \right] - \frac{1}{2} N \tag{3}$$

The number of dominance comparisons in the best case of ENS-BS is the sum of the dominance comparisons *between points of different fronts* and the dominance comparisons *between points of the same front.* The number of dominance comparisons in the best case of ENS-BS is given by Eq. (4).

$$\begin{aligned} \mathtt{dC} &= \mathtt{dC_{same}} + \mathtt{dC_{diff}} \\ &= \frac{1}{2} \left[\sum_{k=1}^{K} n_k^2 \right] - \frac{1}{2} N + \sum_{k=1}^{K} \lceil \log k \rceil n_k \end{aligned} \tag{4}$$

For a minimum number of dominance comparisons, the value of Eq. (4) should be minimum. For this, we need to know the number of fronts and the number of points per front. In [19], the authors claim that the minimum number of dominance comparisons occurs when all the points are in different fronts. This means that the number of fronts $K = N$ and the cardinality of each front is 1. When there is only one point in each front, then the number of dominance comparisons *between points of the same front* is 0. The total number of dominance comparisons is equal to the dominance comparisons *between points of different fronts* and is given by Eq. (5) [19]. In this analysis, we are assuming that $N = 2^x - 1$ where $x \geq 1$. The reason is that the scenario which we obtain is based on $N = 2^x - 1$. However, the scenario can be generalized for any positive value of N.

$$\begin{aligned}
\texttt{dC}_{\texttt{orig}} &= \sum_{i=1}^{N} \lceil \log i \rceil \\
&= 0 + 1 + \underbrace{[2+2]}_{\text{2 times}} + \underbrace{[3+3+3+3]}_{\text{4 times}} + \underbrace{[4+4+\cdots+4]}_{\text{8 times}} + \underbrace{[5+5+\cdots+5]}_{\text{16 times}} + \cdots + \\
&\quad \underbrace{[\{\log(N+1)-1\} + \{\log(N+1)-1\} + \cdots + \{\log(N+1)-1\}]}_{2^{\log(N+1)-2}\text{ times}} + \\
&\quad \underbrace{[\{\log(N+1)\} + \{\log(N+1)\} + \cdots + \{\log(N+1)\}]}_{2^{\log(N+1)-1}-1\text{ times}} \\
&= 0 + 1\cdot 2^0 + 2\cdot 2^1 + 3\cdot 2^2 + 4\cdot 2^3 + 5\cdot 2^4 + \cdots + \\
&\quad [\log(N+1)-1]\cdot\left[2^{\log(N+1)-2}\right] + \log(N+1)\cdot\left[2^{\log(N+1)-1} - 1\right] \\
&= 0 + 1\cdot 2^0 + 2\cdot 2^1 + 3\cdot 2^2 + 4\cdot 2^3 + 5\cdot 2^4 + \cdots + \\
&\quad [\log(N+1)-1]\cdot\left[2^{\log(N+1)-2}\right] + \log(N+1)\cdot\left[2^{\log(N+1)-1}\right] - \log(N+1) \\
&= \left[\sum_{i=1}^{\log(N+1)} i\cdot 2^{i-1}\right] - \log(N+1) \\
&= \left[\sum_{i=1}^{x} i\cdot 2^{i-1}\right] - x \qquad \text{As } N = 2^x - 1 \implies x = \log(N+1) \\
&= [(x-1)2^x + 1] - x \quad \text{(From Eq. (12))} \\
&= x2^x - 2^x + 1 - x \\
&= x(2^x - 1) - (2^x - 1) \\
&= (x-1)(2^x - 1) \\
&= [\log(N+1) - 1]\,N \\
&= N\log(N+1) - N
\end{aligned} \tag{5}$$

4 Identified Scenario

There are three terms in Eq. (4). The second term is fixed for a particular value of N. So, the first and third term play a role in obtaining the minimum number of dominance comparisons and thus determining the best-case scenario. Given a set of cardinalities of fronts, the first term will give the same value for any of the permutations of these cardinalities. However, the value of the third term will vary depending on the permutation. From the third term, it is clear that the cardinality of later fronts should be smaller than the cardinality of former fronts because a larger value is being multiplied with the cardinality of later fronts than with the former fronts.

The value of $\lceil \log k \rceil$ in Eq. (4) for the first front is $0 (= \lceil \log 1 \rceil)$. The same value for the second front is $1 (= \lceil \log 2 \rceil)$. The value of $\lceil \log k \rceil$ for the third and the fourth front is $2 (= \lceil \log 3 \rceil = \lceil \log 4 \rceil)$. Similarly, the value of $\lceil \log k \rceil$ for F_5, F_6, F_7, F_8 is 3. The value of $\lceil \log k \rceil$ for $F_9, F_{10}, \ldots, F_{16}$ is 4 and so on. This means that the same value of $\lceil \log k \rceil$ is multiplied with the cardinality of multiple fronts.

The value 0 is multiplied with the cardinality of one front (*i.e.*, F_1). The value 1 is also multiplied with the cardinality of one front (*i.e.*, F_2). The value 2 is multiplied with the cardinality of two fronts (*i.e.*, F_3, F_4). The value 3 is multiplied with the cardinality of four fronts (*i.e.*, F_5, F_6, F_7, F_8). The value 4 is also multiplied with the cardinality of eight fronts (*i.e.*, $F_9, F_{10}, \ldots, F_{16}$) and so on. In general, the value $x \geq 1$ is multiplied with the cardinality of 2^{x-1} fronts. So, we will keep the cardinality of multiple fronts the same. All those fronts whose cardinality is multiplied with the same value of $\lceil \log k \rceil$, will have the same cardinality.

Based on this aforementioned discussion, in our scenario, the number of fronts is $K = \frac{N+1}{2}$. For simplicity of the analysis, we assume that $N = 2^x - 1$ where $x \geq 1$. So $x = \log(N+1)$. In this scenario, the cardinality of the fronts is as follows.

- The cardinality of F_1 is x
- The cardinality of F_2 is $x - 1$
- The cardinality of F_3, F_4 is $x - 2$
- The cardinality of F_5, F_6, F_7, F_8 is $x - 3$
- The cardinality of $F_9, F_{10}, \ldots, F_{16}$ is $x - 4$
- The cardinality of $F_{17}, F_{18}, \ldots, F_{32}$ is $x - 5$

$\vdots$

- The cardinality of $F_{\frac{K}{2}}, F_{\frac{K}{2}+1}, \ldots, F_K$ is 1

In our scenario also, each point in a front is dominated by all the points in its preceding front which means that a point $p \in F_k$ will be only compared with one point in its previous fronts. The cardinality of all the fronts for the population size $\{2^3 - 1, 2^4 - 1, 2^5 - 1, 2^6 - 1\}$ is shown in Table 1. The process to obtain the cardinality of the fronts is summarized in Algorithm 3. In this Algorithm, `card[k]` stores the cardinality of front F_k.

Table 1. Cardinality of the population for ENS-BS

N	n_1	n_2	n_3	n_4	n_5	n_6	n_7	n_8	n_9	n_{10}	n_{11}	n_{12}	n_{13}	n_{14}	n_{15}	n_{16}	$n_{17}, n_{18}, \ldots, n_{32}$
7	3	2	1	1													
15	4	3	2	2	1	1	1	1									
31	5	4	3	3	2	2	2	2	1	1	1	1	1	1	1	1	
63	6	5	4	4	3	3	3	3	2	2	2	2	2	2	2	2	$1, 1, \ldots, 1$

Algorithm 3. CARDINALITY OF FRONTS

Require: N: Number of points in the population
Ensure: card[ ]: Cardinality of each front in the population

```
1: K ← (N+1)/2                                   ▷ Number of fronts
2: card[1...K] ← ∅          ▷ Array to store the cardinality of the fronts
3: x ← log(N + 1)
4: index ← 1
5: card[index] ← log(N + 1)                      ▷ Cardinality of first front
6: index ← index + 1
7: card[index] ← log(N + 1) − 1                  ▷ Cardinality of second front
   ▷ Obtain the cardinality of remaining fronts
8: for i ← 1 to x − 2 do
9:     for j ← 1 to 2^i do
10:        card[index] ← log(N + 1) − 1 − i
11:        index ← index + 1
12: Return card[ ]   ▷ Return the cardinality of each front in the population
```

Now, we obtain the number of dominance comparisons in our identified scenario. The number of dominance comparison consists of two factors: (1) the number of dominance comparisons *between points of the same front* and (2) the number of dominance comparisons *between points of different fronts.*

4.1 Dominance Comparisons Between Points of the Same Front

There is one front with cardinality x. Similarly, there is one front with cardinality $x - 1$. There are two fronts with cardinality $x - 2$. There are four fronts with cardinality $x - 3$. There are eight fronts with cardinality $x - 4$. There are sixteen fronts with cardinality $x - 5$ and so on. At the end, there are 2^{x-2} fronts with cardinality 1. For a front having cardinality z, the number of dominance comparisons among the points of such front is $\frac{1}{2}z(z-1)$ as all the points are compared with each other. Thus, the number of dominance comparisons *between points of the same front*, is given by Eq. (6).

$$\begin{aligned}
\text{dC}_{\text{same}} &= \frac{1}{2}n_1(n_1-1) + \frac{1}{2}n_2(n_2-2) + \cdots + \frac{1}{2}n_K(n_K-1) \\
&= \frac{1}{2}x(x-1) + \frac{1}{2}(x-1)(x-2) + 2\left[\frac{1}{2}(x-2)(x-3)\right] + \\
&\quad 2^2\left[\frac{1}{2}(x-3)(x-4)\right] + 2^3\left[\frac{1}{2}(x-4)(x-5)\right] + 2^4\left[\frac{1}{2}(x-5)(x-6)\right] + \\
&\quad \cdots + 2^{x-2}\left[\frac{1}{2}(x-(x-1))(x-x)\right] \\
&= \frac{1}{2}x(x-1) + \sum_{i=1}^{x-1} 2^{i-1}\left\{\frac{1}{2}(x-i)(x-i-1)\right\}
\end{aligned}$$

$$
\begin{aligned}
&= \frac{1}{2}x(x-1) + \frac{1}{4}\sum_{i=1}^{x-1} 2^i (x-i)(x-i-1) \\
&= \frac{1}{2}x(x-1) + \frac{1}{4}\sum_{i=1}^{x-1} 2^i (x^2 - x + i^2 - 2xi + i) \\
&= \frac{1}{2}x(x-1) + \frac{1}{4}(x^2 - x)\underbrace{\left[\sum_{i=1}^{x-1} 2^i\right]}_{A} + \frac{1}{4}\underbrace{\left[\sum_{i=1}^{x-1} i^2 2^i\right]}_{B} - \frac{1}{4}(2x-1)\underbrace{\left[\sum_{i=1}^{x-1} i2^i\right]}_{C} \\
&= \frac{1}{2}x(x-1) + \frac{1}{4}(x^2-x)(2^x-2) + \frac{1}{4}(x^2 2^x + 3\cdot 2^{x+1} - x2^{x+2} - 6) - \\
&\quad \frac{1}{4}(2x-1)(x2^x - 2^{x+1} + 2) \quad \text{From Eq. (13), (16) and (19)} \\
&= \frac{1}{2}x(x-1) + \frac{1}{4}(x^2 2^x - 2\cdot x^2 - x2^x + 2x) + \frac{1}{4}(x^2 2^x + 3\cdot 2^{x+1} - x2^{x+2} - 6) - \\
&\quad \frac{1}{4}(x^2 2^{x+1} - x2^{x+2} + 4x - x2^x + 2^{x+1} - 2) \\
&= \frac{1}{2}x(x-1) + \frac{1}{4}\left[(x^2 2^x + x^2 2^x - x^2 2^{x+1}) - (x2^x + x2^{x+2} - x2^{x+2} - x2^x) - \right. \\
&\quad \left. 2x^2 + 2x + 3\cdot 2^{x+1} - 6 - 4x - 2^{x+1} + 2\right] \\
&= \frac{1}{2}x(x-1) + \frac{1}{2}\left(2^{x+1} - x^2 - x - 2\right) \\
&= \frac{1}{2}\left(x^2 - x + 2^{x+1} - x^2 - x - 2\right) = \frac{1}{2}\left[2^{x+1} - 2x - 2\right] \\
&= 2^x - x - 1 = (N+1) - \log(N+1) - 1 = N - \log(N+1) \qquad (6)
\end{aligned}
$$

4.2 Dominance Comparisons Between Points of Different Fronts

The number of dominance comparisons *between points of different fronts* is given by Eq. (7).

$$
\begin{aligned}
\mathtt{dC_{diff}} &= \sum_{k=1}^{K} \lceil \log k \rceil n_k \\
&= 0\cdot n_1 + 1\cdot n_2 + 2(n_3 + n_4) + 3(n_5 + n_6 + n_7 + n_8) + 4(n_9 + n_{10} + \cdots + n_{16}) + \cdots + \\
&\quad \log K(n_{\frac{K}{2}} + n_{\frac{K}{2}+1} + \cdots + n_K) \\
&= 0\cdot x + 1\cdot (x-1) + 2\underbrace{[(x-2)+(x-2)]}_{\text{2 times}} + 3\underbrace{[(x-3)+(x-3)+(x-3)+(x-3)]}_{\text{4 times}} + \\
&\quad 4\underbrace{[(x-4)+(x-4)+\cdots+(x-4)]}_{\text{8 times}} + \cdots + \\
&\quad \log K \underbrace{[1+1+\cdots+1]}_{\frac{K}{2}\text{ times}} \\
&= (x-1) + 2\cdot 2\cdot (x-2) + 3\cdot 2^2\cdot (x-3) + 4\cdot 2^3\cdot (x-4) + \cdots + \\
&\quad (x-1)\cdot 2^{x-2}\cdot 1 \\
&= \sum_{i=1}^{x-1} i\cdot 2^{i-1}\cdot (x-i)
\end{aligned}
$$

$$
\begin{aligned}
&= \frac{1}{2}\sum_{i=1}^{x-1} x \cdot i \cdot 2^i - \frac{1}{2}\sum_{i=1}^{x-1} i^2 2^i \\
&= \frac{1}{2}x \underbrace{\left[\sum_{i=1}^{x-1} i2^i\right]}_{C} - \frac{1}{2}\underbrace{\left[\sum_{i=1}^{x-1} i^2 2^i\right]}_{B} \\
&= \frac{1}{2}x\left[x2^x - 2^{x+1} + 2\right] - \frac{1}{2}\left[x^2 2^x + 3\cdot 2^{x+1} - x2^{x+2} - 6\right] \quad \text{From Eq. (16) and (19)} \\
&= \frac{1}{2}\left[x^2 2^x - x2^{x+1} + 2x - x^2 2^x - 3\cdot 2^{x+1} + x2^{x+2} + 6\right] \\
&= \frac{1}{2}\left[-x2^{x+1} + 2x - 3\cdot 2^{x+1} + x2^{x+2} + 6\right] \\
&= \frac{1}{2}\left[x2^{x+1} - 3\cdot 2^{x+1} + 2x + 6\right] \\
&= \frac{1}{2}\left[2(N+1)\log(N+1) - 6(N+1) + 2\log(N+1) + 6\right] \\
&= (N+1)\log(N+1) - 3N + \log(N+1) \qquad (7)
\end{aligned}
$$

Thus, the number of dominance comparisons in our identified scenario by ENS-BS is given by Eq. (8).

$$
\begin{aligned}
\texttt{dC}_{\texttt{ident}} &= \texttt{dC}_{\texttt{same}} + \texttt{dC}_{\texttt{diff}} \\
&= [N - \log(N+1)] + [(N+1)\log(N+1) - 3N + \log(N+1)] \\
&= (N+1)\log(N+1) - 2N \qquad (8)
\end{aligned}
$$

We have obtained the number of dominance comparisons in the identified scenario for population sizes $\{2^3-1, 2^4-1, 2^5-1, \ldots, 2^{20}-1\}$ using Eq. (8). We have also obtained the number of dominance comparisons for the same population size using Eq. (5). These two dominance comparisons are shown in Fig. 1. From this figure, it is evident that the number of dominance comparisons in the identified scenario is less than that reported in [19]. The difference between the number of dominance comparisons in our identified scenario and the scenario reported in [19] is given by Eq. (9).

$$
\begin{aligned}
\texttt{diff} &= \texttt{dC}_{\texttt{orig}} - \texttt{dC}_{\texttt{ident}} \\
&= [N\log(N+1) - N] - [(N+1)\log(N+1) - 2N] \\
&= N - \log(N+1) \qquad (9)
\end{aligned}
$$

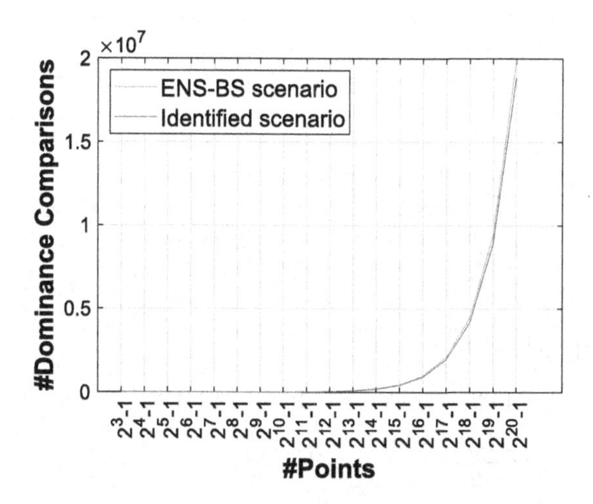

Fig. 1. Number of dominance comparisons performed by ENS-BS in the identified scenario and in the scenario reported in [19].

5 Conclusion and Future Work

In this paper, we have shown a scenario where the number of dominance comparisons performed by ENS-BS is less than that reported in the original paper [19] for the best case. Here, we are not claiming that the scenario which we have identified represents the best-case scenario. We have obtained an equation whose value provides the minimum number of dominance comparisons once we obtain the best-case scenario. Based on this equation, the scenario has been identified. Although the best-case time complexity does not matter much in most cases, still for the sake of completeness it is better to analyze such case properly as well. There have been some approaches like T-ENS [18] ENS-NDT [4], DCNS-BS [13,14], DCNSRC-BS [15], and Generalized Best Order Sort [11] which also follow the binary search based strategy to assign a point to its particular front. So, we will investigate whether the time complexity analysis of these approaches has some issues or not. Also, identifying the best-case scenario can be another research path that can be carried out in the future.

Acknowledgements. Carlos A. Coello Coello gratefully acknowledges support from CONACyT grant no. 2016-01-1920 (Investigación en Fronteras de la Ciencia 2016).

Appendix

$$
\begin{aligned}
S &= \sum_{i=1}^{x} i \cdot 2^{i-1} \\
&= 1 \cdot 2^0 + 2 \cdot 2^1 + \cdots + (x-1) \cdot 2^{x-2} + x \cdot 2^{x-1} \quad (10)\\
2S &= 1 \cdot 2^1 + 2 \cdot 2^2 + \cdots + (x-1) \cdot 2^{x-1} + x \cdot 2^x \quad (11)
\end{aligned}
$$

Subtract Eq. (11) from Eq. (10),

$$
\begin{aligned}
-S &= 1 \cdot 2^0 + 1 \cdot 2^1 + 1 \cdot 2^2 + \cdots + 1 \cdot 2^{x-1} - x \cdot 2^x \\
&= \frac{2^0 \left(2^x - 1\right)}{2-1} - x \cdot 2^x \\
&= 2^x - 1 - x \cdot 2^x \\
S &= x \cdot 2^x - 2^x + 1 \\
&= (x-1)2^x + 1 \quad (12)
\end{aligned}
$$

$$
A = \sum_{i=1}^{x-1} 2^i = 2\left(2^{x-1} - 1\right) = 2^x - 2 \quad (13)
$$

$$
\begin{aligned}
B &= \sum_{i=1}^{x-1} i^2 2^i \\
&= 1 \cdot 2 + 2^2 \cdot 2^2 + 3^2 \cdot 2^3 + \cdots + (x-1)^2 \cdot 2^{x-1} \quad (14)\\
2B &= 1 \cdot 2^2 + 2^2 \cdot 2^3 + 3^2 \cdot 2^4 + \cdots + (x-2)^2 \cdot 2^{x-1} + (x-1)^2 \cdot 2^x \quad (15)
\end{aligned}
$$

Subtract Eq. (15) from (14),

$$
\begin{aligned}
-B &= 1 \cdot 2 + 3 \cdot 2^2 + 5 \cdot 2^3 + \cdots + (2x-3) \cdot 2^{x-1} - (x-1)^2 \cdot 2^x \\
&= 2\left[1 + 3 \cdot 2 + 5 \cdot 2^2 + \cdots + (2x-3) \cdot 2^{x-2}\right] - (x-1)^2 \cdot 2^x \\
&= 2\left[x2^x - 2^x - 3 \cdot 2^{x-1} + 3\right] - (x^2 - 2x + 1) \cdot 2^x \\
&= x2^{x+1} - 2^{x+1} - 3 \cdot 2^x + 6 - x^2 2^x + x2^{x+1} - 2^x \\
&= x2^{x+2} - 2^{x+1} - 2^{x+2} - x^2 2^x + 6 \\
B &= x^2 2^x + 2^{x+2} + 2^{x+1} - x2^{x+2} - 6 \\
&= x^2 2^x + 3 \cdot 2^{x+1} - x2^{x+2} - 6 \quad (16)
\end{aligned}
$$

$$
\begin{aligned}
C &= \sum_{i=1}^{x-1} i2^i \\
&= 1 \cdot 2 + 2 \cdot 2^2 + 3 \cdot 2^3 + \cdots + (x-1)2^{x-1} \quad (17)\\
2C &= 1 \cdot 2^2 + 2 \cdot 2^3 + 3 \cdot 2^4 + \cdots + (x-2)2^{x-1} + (x-1)2^x \quad (18)
\end{aligned}
$$

Subtract Eq. (18) from (17),

$$\begin{aligned} -C &= 2 + 2^2 + 2^3 + \cdots + 2^{x-1} - (x-1)2^x \\ &= 2(2^{x-1} - 1) - (x-1)2^x \\ &= 2^x - 2 - x2^x + 2^x \\ &= 2^{x+1} - 2 - x2^x \\ C &= x2^x - 2^{x+1} + 2 \end{aligned} \tag{19}$$

References

1. Deb, K.: Multi-objective optimization using evolutionary algorithms: an introduction. In: Wang, L., Ng, A., Deb, K. (eds.) Multi-objective Evolutionary Optimisation for Product Design and Manufacturing, pp. 3–34. Springer, London (2011). https://doi.org/10.1007/978-0-85729-652-8_1
2. Deb, K., Pratap, A., Agarwal, S., Meyarivan, T.: A fast and elitist multiobjective genetic algorithm: NSGA-II. IEEE Trans. Evol. Comput. **6**(2), 182–197 (2002)
3. Fortin, F.A., Greiner, S., Parizeau, M.: Generalizing the improved run-time complexity algorithm for non-dominated sorting. In: Proceedings of Genetic and Evolutionary Computation Conference (GECCO'2013), pp. 615–622. ACM Press, New York, USA (2013), ISBN: 978-1-4503-1963-8
4. Gustavsson, P., Syberfeldt, A.: A new algorithm using the non-dominated tree to improve non-dominated sorting. Evol. Comput. **26**(1), 89–116 (2018)
5. Jensen, M.T.: Reducing the run-time complexity of multiobjective EAs: the NSGA-II and other algorithms. IEEE Trans. Evol. Comput. **7**(5), 503–515 (2003)
6. Kung, H.T., Luccio, F., Preparata, F.P.: On finding the maxima of a set of vectors. J. ACM **22**(4), 469–476 (1975)
7. McClymont, K., Keedwell, E.: Deductive sort and climbing sort: new methods for non-dominated sorting. Evol. Comput. **20**(1), 1–26 (2012)
8. Mishra, S., Buzdalov, M.: Filter sort is Ω (N^3) in the worst case. In: Bäck, T., et al. (eds.) PPSN 2020. LNCS, vol. 12270, pp. 675–685. Springer, Cham (2020). https://doi.org/10.1007/978-3-030-58115-2_47
9. Mishra, S., Buzdalov, M.: If unsure, shuffle: deductive sort is $\Theta(MN^3)$, but $\mathcal{O}(MN^2)$ in expectation over input permutations. In: Proceedings of Genetic and Evolutionary Computation Conference (GECCO 2020), pp. 516–523 (2020)
10. Mishra, S., Buzdalov, M., Senwar, R.: Time complexity analysis of the dominance degree approach for non-dominated sorting. In: Proceedings of the Genetic and Evolutionary Computation Conference Companion (GECCO'2020), pp. 169–170 (2020)
11. Mishra, S., Mondal, S., Saha, S., Coello Coello, C.A.: GBOS: generalized best order sort algorithm for non-dominated sorting. Swarm Evol. Comput. **43**, 244–264 (2018)
12. Mishra, S., Prakash, V.: Time complexity analysis of the deductive sort in the best case. In: Proceedings of the Genetic and Evolutionary Computation Conference Companion (GECCO'2021), pp. 337–338 (2021)
13. Mishra, S., Saha, S., Mondal, S.: Divide and conquer based non-dominated sorting for parallel environment. In: IEEE Congress on Evolutionary Computation (CEC 2016), pp. 4297–4304. IEEE Press, Vancouver, Canada (2016). ISBN: 978-1-5090-0623-6

14. Mishra, S., Saha, S., Mondal, S., Coello Coello, C.A.: A divide-and-conquer based efficient non-dominated sorting approach. Swarm Evol. Comput. **44**, 748–773 (2019)
15. Mishra, S., Saha, S., Mondal, S., Coello Coello, C.A.: Divide-and-conquer Based Non-dominated Sorting with Reduced Comparisons. Swarm Evol. Comput. **51**, 100580 (2019). article Number: UNSP 100580
16. Nigam, P., Mishra, S.: Counterexample to the best-case running time of efficient non-dominated sorting algorithm. In: Proceedings of the Genetic and Evolutionary Computation Conference Companion (GECCO'2022), pp. 798–800 (2022)
17. Wang, J., Li, C., Diao, Y., Zeng, S., Wang, H.: An efficient nondominated sorting algorithm. In: Proceedings of Genetic and Evolutionary Computation Conference (GECCO'2018), pp. 203–204. ACM (2018)
18. Zhang, X., Tian, Y., Cheng, R., Jin, Y.: A decision variable clustering-based evolutionary algorithm for large-scale many-objective optimization. IEEE Trans. Evol. Comput. **22**(1), 97–112 (2018)
19. Zhang, X., Tian, Y., Cheng, R., Yaochu, J.: An efficient approach to nondominated sorting for evolutionary multiobjective optimization. IEEE Trans. Evol. Comput. **19**(2), 201–213 (2015)
20. Zhou, Y., Chen, Z., Zhang, J.: Ranking vectors by means of the dominance degree matrix. IEEE Trans. Evol. Comput. **21**(1), 34–51 (2017)

Applications in Real World Domains

Evolutionary Algorithms with Machine Learning Models for Multiobjective Optimization in Epidemics Control

Krzysztof Michalak(✉)

Department of Information Technologies, Wrocław University of Economics and Business, Wrocław, Poland
krzysztof.michalak@ue.wroc.pl
http://krzysztof-michalak.pl

Abstract. This paper studies the use of machine learning models for multiobjective optimization of vaccinations used to control an epidemic spreading in a graph representing contacts between individuals. Graph nodes are parameterized by attributes which are known to affect the susceptibility of people to influenza and the disease transmission probability depends on the attributes of the node which can get infected. Instead of directly optimizing the assignment of vaccine doses to graph nodes, in the proposed approach an evolutionary algorithm is used to train a neural network, which is subsequently used to make decisions about vaccinating the nodes of the graph. In the paper, both a classifier and a regression model are used to select graph nodes for vaccination. The results obtained using the machine learning models improve over the results obtained by optimizing the assignment of vaccine doses to graph nodes. Importantly, the models trained on a certain problem instance can be used for selecting graph nodes for vaccination when other problem instances are solved.

Keywords: Vaccination optimization · DPEC · Neural networks · Graph-based optimization

1 Introduction

This paper studies the application of evolutionary algorithms to multiobjective optimization of vaccinations with the goal of limiting the spread of an epidemic which is simulated on a graph. The graph-based formalism can be used for representing threats spreading in various systems, such as epidemics, financial crises, wildfires and cascading failures in the infrastructure. Attempts to control spreading threats give rise to a number of optimization problems ranging from high-level abstractions to real-life applications. For example, in the Firefighter Problem (FFP) [9] the spreading of fire is simulated in discrete time steps on an undirected graph $G = \langle V, E \rangle$ with N_v nodes. The nodes can be in one of the states 'B' - burning, 'D' - defended and 'U' - untouched (neither burning nor

M. Emmerich et al. (Eds.): EMO 2023, LNCS 13970, pp. 435–448, 2023.
https://doi.org/10.1007/978-3-031-27250-9_31

defended). In each time step, a limited number of nodes N_f can be defended (set to the 'D' state), which makes them resistant to fire until the end of the simulation. The spreading of fire is, typically, deterministic: a node catches on fire if it is not defended and is adjacent to an already burning node. In most publications, solutions to the FFP are represented as permutations of N_v graph nodes, but other representations have also been studied in the literature [10,15]. In order to evaluate a solution $\pi \in \Pi_{N_v}$, the spreading of fire is simulated on the graph G and in each time step the first N_f yet untouched nodes are selected from the permutation π and set to the 'D' state. In the original version of the problem, the objective function is the number of nodes protected from fire at the end of the simulation, but FFP with a non-uniform node cost [14] as well as the multiobjective [13] version were also studied.

Another phenomenon observed in real-life networks which gives rise to optimization problems is the financial contagion [21]. Financial contagion is the spread of disturbances in the market caused by the fact that failing companies are unable to meet their obligations towards other companies leading to more and more companies failing following some initial shock. In order to simulate the spreading of bankruptcies on a graph, a threshold failure mechanism proposed by Watts [20] and used, among others, by Burkholz et al. [2] can be employed. Graph nodes can be protected by adjusting thresholds $\Theta_v \in [0,1]$ (for $v \in V$), which can be interpreted as the fractions of the assets set aside by companies as reserves for difficult times. Limiting the impact of the financial contagion can be studied as a multiobjective optimization problem in which the level of reserves has to be minimized along with the number of companies which fail as a result of the spreading wave of bankruptcies.

Last, but not least, the graph-based formalism can be used for studying epidemics. When an epidemic is simulated on a graph, the nodes can be set to states such as 'S' - susceptible, 'I' - infected, or 'R' - recovered. The outbreak of an epidemic is simulated by setting some nodes to the 'I' state and then the disease spreads via contacts represented as graph edges. Typically, the spreading of the disease is non-deterministic. For example, a node in the 'S' state can catch the disease from each infected ('I') node to which it is connected by an edge with the probability γ per a simulation time step. Based on such epidemic spreading model, the effectiveness of various epidemic control measures can be studied. For example, the possibility of vaccinating the nodes can be added to the model, by allowing the nodes to be set to the 'V' - vaccinated state [8,17]. Using the Susceptible-Vaccinated-Infected-Recovered (SVIR) model [18], vaccination optimization can be studied as a multiobjective optimization problem in which both the number of vaccine doses and the number of infected individuals are to be minimized. Studying such optimization problem is interesting from a practical perspective, because the obtained solutions can tell us how many disease cases we can expect when a given number of vaccine doses are distributed. Conversely, by studying the Pareto front produced by the optimization algorithm, we can determine how many vaccine doses we need to distribute, at least, to prevent the number of infected individuals from exceeding a certain limit. This, on the other hand is important, in order not to allow the epidemic to grow to such size that exceeds the capacity of the healthcare system. Also, Pareto fronts obtained when solving the vaccination optimization problem often show the effects of herd immunity - an effect which causes more nodes to be protected than just the vaccinated ones, because they are cut off from the infection source. Vaccination optimization using individual-based simulations is competitive to optimization using compartmental models, as some studies show [3]. Another advantage of this approach is that it allows assigning multiple attributes to graph nodes, such as the age, gender, or occupation, etc. in the case of people, and species in the case of animals.

In this paper, the multiobjective vaccination optimization problem is studied on graphs in which attributes known to affect the probability of getting influenza [7] are assigned to the nodes. This optimization problem is solved using evolutionary algorithms augmented by machine learning models. The paper is organized as follows. Section 2 defines the optimization problem along with the description of the attributes assigned to graph nodes. Section 3 describes the optimization algorithms studied in the paper. Section 4 presents the experimental setup and discusses the results. Section 5 concludes the paper.

2 Optimization Problem

The optimization problem studied in this paper is the bi-objective minimization of the number of vaccine doses and the number of individuals infected in a simulated epidemic. Because vaccinations are administered before the epidemic outbreak, the optimization problem represents a preemptive, rather than reactive, vaccination campaign. In such case, some well-known counter-epidemic strategies, such as ring vaccinations and acquaintance vaccinations, cannot be used, because at the time the vaccine is administered it is not known who will become infected when the epidemic starts. The Pareto front obtained by solving this optimization problems represents the best known trade-offs between the size of the vaccination campaign and the size of an epidemic outbreak. Naturally, it would be best to minimize both, because the vaccinations cost money and require organizational effort and the illnesses cause multiple problems, but, unfortunately, these objectives cannot be optimized at the same time. Therefore, a bi-objective optimization problem is defined, using simulations of epidemic outbreaks on a graph $G = \langle V, E \rangle$ with N_v nodes to determine the number of infected individuals. Solutions to this optimization problem are binary vectors of length N_v in which each element corresponds to one graph node and determines if this node should be vaccinated (elements equal 1) or not (elements equal 0). Formally, the problem is defined as:

$$\begin{aligned} &\text{minimize } F(x) = (f_1(x), f_2(x)) \\ &\text{subject to } x \in \Omega, \end{aligned} \tag{1}$$

where $\Omega = \{0,1\}^{N_v}$ is the solution space of this optimization problem.

The first objective is the number of vaccinated nodes: $f_1 = \sum_{i=1}^{N_v} x_i$. The second objective is the number of nodes infected in a simulated epidemic outbreak.

2.1 Simulation-Based Calculation of the Objective f_2

In order to determine the value of the objective $f_2(x)$, $N_{sim} = 10$ independent simulations are performed, each starting with a fraction $\alpha_{inf} = 0.01$ of initially infected nodes. In each run, the infected nodes are randomly selected with the uniform probability from all the nodes in V, so the solutions have to be general enough to stop the epidemic outbreak regardless of where it starts. The epidemic is simulated according to the Susceptible-Vaccinated-Infected-Recovered (SVIR) model [18]. At the beginning of the simulation run, nodes for which the elements of x equal 1 are set to the vaccinated ('V') state. Subsequently, $\alpha_{inf} \cdot N_v$ nodes are randomly selected with the uniform probability from all the nodes in V and if they were not vaccinated they are set to the infected ('I') state. All the other nodes remain in the susceptible ('S') state. Then, the epidemic is simulated in discrete time steps. The state of the graph at a time step t is represented as the vector of node states $s_t \in \{\text{'S'}, \text{'V'}, \text{'I'}, \text{'R'}\}^{N_v}$ and the state of a node $v \in V$ at a time step t is the element of that vector $s_t[v] \in \{\text{'S'}, \text{'V'}, \text{'I'}, \text{'R'}\}$. In each time step, the disease spreads from infected ('I') nodes to susceptible nodes ('S'). The disease transmission probability is calculated for each node as described in Sect. 2.3 using the real-life attributes described in Sect. 2.2. During the simulation, infected nodes recover from the disease with the probability $\beta = 0.1$ per a time step. Recovered nodes are set to the 'R' state and they are immune to the disease until the end of the simulation. Therefore, the epidemic is guaranteed to stop within a finite time period, which makes this model convenient for evaluating solutions to the optimization problem. After the simulation run number $s \in \{1, \ldots, N_{sim}\}$ stops, the value $f_2^{(s)}(x)$ is calculated as the number of nodes that got infected in that simulation run (technically, it suffices to count the nodes in the 'R' state at the end of the simulation). The value of the f_2 objective for the solution x is calculated by averaging the results of N_{sim} simulation runs: $f_2(x) = \frac{1}{N_{sim}} \sum_{s=1}^{N_{sim}} f_2^{(s)}(x)$.

2.2 Real-Life Attributes Affecting the Susceptibility to the Disease

Graph nodes are parameterized by attributes which are known to affect the susceptibility of people to influenza and the disease transmission probability depends on the attributes of the node which can get infected. The attributes used in this paper are based on the findings presented in the work of Guerrisi et al. [7], which discusses the results of a cohort study from 2012/13 to 2017/18. Consider an attribute, for example the 'Age'. For each value a_i of this attribute, the number of people for whom this value was recorded $N(a_i)$ is given in the aforementioned paper [7, Table 3], as well as the number of influenza cases $n(a_i)$ reported among people with the attribute value a_i. From the $N(a_i)$ and $n(a_i)$ values the probability of observing the attribute value a_i can be calculated as $P(a_i) = \frac{N(a_i)}{\sum_i N(a_i)}$ and the conditional probability of observing the attribute value a_i if the person is in the infected group can be calculated as $P(a_i|\text{inf}) = \frac{n(a_i)}{\sum_i n(a_i)}$. In Table 1, the attributes used in this paper are presented, along with the probabilities $P(a_i)$, $P(a_i|\text{inf})$ and the ratio $\frac{P(a_i|\text{inf})}{P(a_i)}$ which is used for determining the infection probability in Sect. 2.3.

Table 1. Real-life attributes affecting the susceptibility to the disease. Probabilities calculated using frequencies presented in [7, Table 3].

Attribute name/value	$P(a_i)$	$P(a_i\|\text{inf})$	$\frac{P(a_i\|\text{inf})}{P(a_i)}$
Gender			
Male	0.3952	0.3459	0.8754
Female	0.6048	0.6541	1.0814
Age			
[0 – 5)	0.0124	0.0237	1.9054
[5 – 15)	0.0460	0.0595	1.2935
[15 – 45)	0.2415	0.2792	1.1560
[45 – 65)	0.4036	0.4178	1.0352
[65 – 75)	0.2437	0.1887	0.7744
≥ 75	0.0528	0.0311	0.5899
Household			
Alone	0.1593	0.1562	0.9805
Child	0.3137	0.3663	1.1676
Adults	0.5269	0.4775	0.9061
Occupation			
Working	0.4778	0.5284	1.1057
Student	0.0892	0.1180	1.3231
Unemployed	0.0237	0.0300	1.2643
Retired	0.3683	0.2822	0.7661
Sick leave	0.0410	0.0415	1.0133
Residency			
Rural	0.1963	0.1931	0.9835
Urban	0.8037	0.8069	1.0040
Public transport			
No	0.8455	0.8216	0.9718
Yes	0.1545	0.1784	1.1543
Pets			
No	0.5482	0.5070	0.9249
Yes	0.4518	0.4930	1.0912
Contacts: patients			
No	0.8990	0.8853	0.9848
Yes	0.1010	0.1147	1.1353

Attribute name/value	$P(a_i)$	$P(a_i\|\text{inf})$	$\frac{P(a_i\|\text{inf})}{P(a_i)}$
Contacts: elderly			
No	0.8987	0.9024	1.0041
Yes	0.1013	0.0976	0.9640
Contacts: group			
No	0.6813	0.6494	0.9532
Yes	0.3187	0.3506	1.0999
Contacts: children			
No	0.7594	0.7123	0.9379
Yes	0.2406	0.2877	1.1960
Vaccination (last season)			
No	0.6201	0.6619	1.0673
Yes	0.3799	0.3381	0.8901
Smoking			
No	0.8932	0.8904	0.9969
Yes	0.1068	0.1096	1.0261
Comorbidity			
None	0.7550	0.7395	0.9795
Asthma	0.0563	0.0778	1.3823
Diabetes	0.0366	0.0316	0.8647
Heart	0.0961	0.0851	0.8858
Kidney	0.0059	0.0049	0.8236
Immunosup.	0.0255	0.0284	1.1148
Pulmonary	0.0247	0.0326	1.3225
Respiratory allergy			
No	0.6652	0.6153	0.9249
Yes	0.3348	0.3847	1.1492
BMI			
Underweight	0.0434	0.0438	1.0084
Normal	0.5772	0.5583	0.9671
Overweight	0.2709	0.2703	0.9977
Obese	0.1084	0.1276	1.1775

In order to assign the attributes to the graph nodes, N_v attribute vectors are randomly generated. Each value of an attribute is selected with the probability $P(a_i)$ given in Table 1, so, for example, the rural area of residency is assigned with the probability 0.1963, and urban area with the probability 0.8037. Because some of the attributes represent contacts, for each attribute vector the contacts intensity $N_{contacts}$ is calculated. The value of $N_{contacts}$ is set to the sum of values of attributes representing contacts (with patients, elderly, group, and children), so each of these attributes set to 1 increases $N_{contacts}$ by 1. The value of the 'Household' attribute increases $N_{contacts}$ by 1 if the value is 'Adults' and by 2 if the value is 'Child'. Attributes are assigned to graph nodes by sorting the attribute vectors in the ascending order with respect to the $N_{contacts}$ value, and sorting the graph nodes in the ascending order with respect to the node degree (the number of incident edges). Sorted attribute vectors are matched with sorted graph nodes, thereby ensuring that if the attributes indicate a low intensity of contacts the attribute vector is assigned to a low-degree graph node and if the attributes indicate a high intensity of contacts the attribute vector is assigned to a high-degree graph node.

2.3 Modelling Disease Transmission Probability Using Graph Node Attributes

When an epidemic is simulated, the disease spreads from infected ('I') nodes to susceptible nodes ('S') with a certain probability per a time step. The average transmission probability is $\gamma = 0.1$ per a time step and per each contact (graph edge connecting a susceptible node to an infected neighbour). However, the probability of contracting the disease by a node v depends on the attributes assigned to this node. Denote the vector of attributes assigned to the node v as $A(v) = [a_{i_1}^{(1)}, \ldots, a_{i_k}^{(k)}, \ldots, a_{i_{16}}^{(16)}]$, where $a_{i_k}^{(k)}$ is the i_k-th value of the k-th attribute. The probability $\gamma(v)$ of transmitting the disease to the node v from each of its infected neighbours (per a time step) can be calculated using the Bayes' theorem as:

$$\gamma(v) = \frac{\gamma \cdot P(A(v)|\mathrm{inf})}{P(A(v))}, \tag{2}$$

where $P(A(v)|\mathrm{inf})$ is the probability of observing the attribute vector $A(v)$ among infected individuals and $P(A(v))$ is the probability of observing the attribute vector $A(v)$ in the general population.

Because in the paper [7] only univariate frequencies are given, the probabilities $P(A(v)|\mathrm{inf})$ and $P(A(v))$ are approximated using the Naïve Bayes approach (assuming independence of the attributes), which is commonly used in machine learning for simplifying probabilistic modelling:

$$\gamma(v) = \frac{\gamma \cdot \prod_{k=1}^{16} P(a_{i_k}^{(k)}|\mathrm{inf})}{\prod_{k=1}^{16} P(a_{i_k}^{(k)})} = \gamma \cdot \prod_{k=1}^{16} \frac{P(a_{i_k}^{(k)}|\mathrm{inf})}{P(a_{i_k}^{(k)})}, \tag{3}$$

where $P(a_{i_k}^{(k)}|\mathrm{inf})$ is the probability of observing the i_k-th value of the k-th attribute among infected individuals and $P(a_{i_k}^{(k)})$ is the probability of observing the i_k-th value of the k-th attribute in the general population. Using Eq. (3), the probability $\gamma(v)$ of transmitting the disease to the node v from each of its neighbours (per a time step) can be calculated from the ratios $\frac{P(a_i|\mathrm{inf})}{P(a_i)}$ given in the last column of Table 1 and the average transmission probability γ. Denoting by $\mathcal{N}(v)$ the set of neighbours of the node v in the graph G, that is, $\mathcal{N}(v) = \{u \in V : \langle u, v \rangle \in E\}$, the overall probability $\overline{\gamma}(v)$ of the node v being infected at a given time step t can be calculated depending on the number of its infected neighbours:

$$\overline{\gamma}(v) = 1 - (1 - \gamma(v))^{|\{u \in \mathcal{N}(v) : s_t[u] = \text{'I'}\}|} . \quad (4)$$

In simulations, infection probabilities calculated in this section are used for determining if a node should transition from the 'S' (susceptible) to the 'I' (infected) state.

3 Optimization Algorithms

Optimization algorithms studied in this paper solve the optimization problem described in Sect. 2 simulating epidemics on graphs in which nodes have real-life attributes described in Sect. 2.2. The attributes of the nodes affect the probability of transmitting the disease, as described in Sect. 2.3. Some combinations of attribute values make the disease transmission more likely than the average transmission probability γ and some make it less likely. Also, the 'Contact' and the 'Household' attributes are correlated with the node degree in the graph, because attribute vectors with higher values of the contacts intensity $N_{contacts}$ are assigned to higher-degree nodes, as described in Sect. 2.2. Therefore, it can be expected, that the attributes assigned to graph nodes can be used to determine the best nodes to vaccinate (clearly, we prefer to vaccinate those nodes that are the most likely to contract the disease).

The solution space for the optimization problem studied in this paper is $\Omega = \{0, 1\}^{N_v}$. Each element in a solution $x \in \Omega$ corresponds to one node in the graph G on which the epidemic is simulated and determines if the node should be vaccinated ($x[v] = 1$) or not ($x[v] = 0$). In this paper, three optimization algorithms are compared, two of which use machine learning models (a classifier and a regression model) to decide which nodes to vaccinate.

3.1 EA-C

The EA-C algorithm uses a neural network as a classifier to determine which nodes to vaccinate. The network has $N_{in} = 18$ input neurons corresponding to 16 attributes listed in Table 1 and two elements of a weight vector λ used to direct the search towards different parts of the Pareto front in a similar manner as in the MOEA/D multiobjective evolutionary algorithm [11,22]. Because the weight vector λ is provided, the neural network can produce different results for solutions generated for different weights assigned to the objectives. The network has $N_{hid} = 10$ hidden neurons and produces one output which is used to classify the nodes. A solution for a given weight vector λ is obtained by feeding the attributes of each node $v \in V$ to the neural network along with the weight vector λ. If the output value is larger than 0.5 the node v is vaccinated and if the output value is smaller or equal 0.5 the node v is not vaccinated. In order to obtain the Pareto front, the weight vector is varied from $[0, 1]$ to $[1, 0]$ in 101 steps and for each λ one solution is generated. The algorithm generating solutions for the Pareto front in the EA-C algorithm is presented in Algorithm 1. The EA-C algorithm is a single objective evolutionary algorithm based on the Simple Genetic Algorithm (SGA) [19], that optimizes the weights of the neural network used for classifying the graph nodes. The number of weights (and biases) in this network is $N_{in} \cdot N_{hid} + N_{hid} + N_{hid} \cdot N_{out} + N_{out} = 201$, so the search space the EA-C algorithm operates on is $\mathbb{R}^{201}$. The EA-C uses four crossover operators: the single-point, the two-point and the uniform crossover [6, p. 52–53] along with the Simulated Binary Crossover (SBX) dedicated to real-valued genotypes [4]. For mutation, five operators reordering the elements of the genotype are used: displacement, insertion, inversion, scramble and transpose [12] along with two mutation operators dedicated to real-valued genotypes: the uniform mutation and the polynomial mutation [5]. The genetic operators are selected using an autoadaptation mechanism based on operator success rates [13]. The evaluation of each solution $x \in \mathbb{R}^{201}$ is obtained by using the elements of the vector x as weights for a neural network, generating a Pareto front using Algorithm 1, and evaluating this Pareto front using the hypervolume (HV) indicator [23].

Algorithm 1: Generating solutions for the Pareto front in EA-C.

```
Inputs:
    x ∈ R^201  - weights of the neural network
Output:
    PF         - the Pareto front

// A neural network with weights taken from x
M := MLP.Init(x)
// Generate solutions for the PF using different weight vectors
PF := ∅
for h := 0, ..., 100 do
    λ := [h/100, 1 − h/100]
    // Initially, each solution is a vector of N_v zeros
    s := [0, ..., 0]
    // Classify each node
    for v ∈ V do
        // Forward the attributes of the node v and the weight vector λ
        // through the network
        c := M([A[v], λ])
        if c > 0.5 then
            s[v] := 1
    PF := PF ∪ {s}

PF := RemoveDominated(PF)
return PF
```

3.2 EA-R

The EA-R algorithm uses a neural network as a regression model to rank the nodes with respect to the estimated preference for vaccination. Contrary to EA-C, a weight vector is not needed in this approach, so the network has $N_{in} = 16$ input neurons corresponding to 16 attributes listed in Table 1. The network has $N_{hid} = 10$ hidden neurons and produces one output which is used to rank the nodes. In order to obtain the Pareto front, the nodes are ranked by feeding the attributes corresponding to the nodes to the network. Then, $0\%, 1\%, \ldots, 100\%$ of the nodes are vaccinated in the order determined by the ranking. The algorithm generating solutions for the Pareto front in the EA-R algorithm is presented in Algorithm 2. The EA-R algorithm is a single objective evolutionary algorithm based on the Simple Genetic Algorithm (SGA) [19], that optimizes the weights of the neural network used for ranking the graph nodes. The number of weights (and biases) in this network is $N_{in} \cdot N_{hid} + N_{hid} + N_{hid} \cdot N_{out} + N_{out} = 181$, so the search space the EA-R algorithm operates on is $\mathbb{R}^{181}$. The EA-R uses the same genetic operators and autoadaptation mechanism as the EA-C algorithm. The evaluation of each solution $x \in \mathbb{R}^{181}$ is obtained by using the elements of the vector x as weights for a neural network, generating a Pareto front using Algorithm 2, and evaluating this Pareto front using the hypervolume indicator.

3.3 MOEA/D

The algorithm used as a baseline for comparison is the MOEA/D multiobjective evolutionary algorithm [11,22] performing the search directly on the solution space $\Omega = \{0, 1\}^{N_v}$. This algorithm uses three crossover operators: the single-point, the two-point and the uniform crossover [6, p. 52–53]. For mutation, five operators reordering the elements of the genotype are used: displacement, insertion, inversion, scramble and transpose [12] along with the classical bit-flip mutation [6, p. 40] mutating each position in the genotype independently of the others with the probability inversely proportional to the length of the genotype. As in the case of other algorithms studied in this paper, the genetic operators are selected using an autoadaptation mechanism [13].

4 Experiments and Results

In the experiments, REDS graphs [1] were used with real-life attributes described in Sect. 2.2 assigned to the nodes. The REDS graphs resemble naturally forming networks of contacts, because they are generated taking into account spatial relationships and the synergy effect. Nodes are uniformly placed on the unit square $[0, 1] \times [0, 1]$ with the R parameter controlling the radius within which the edges can be formed. Each node has the initial amount E of "social energy", and forming an edge incurs a cost proportional to its length D. The cost is discounted by the synergy factor S multiplied by the number of neighbours the two connected nodes have in common. In this paper, REDS graphs with $N_v =$

Algorithm 2: Generating solutions for the Pareto front in EA-R.

```
Inputs:
    x ∈ ℝ^181   - weights of the neural network
    V           - graph nodes
Output:
    PF          - the Pareto front

// A neural network with weights taken from x
M := MLP.Init(x)
// Rank graph nodes using the neural network
// The matrix A contains attributes of all nodes (one per row)
R := M(A)
V' := Sort(V, R);
// Generate solutions for the PF by vaccinating a different number of nodes
PF := ∅
for h := 0, ..., 100 do
    n := ⌈h/100 · N_v⌉
    // Vaccinate n nodes in the order in which they are sorted in V'
    s := [0, ..., 0]
    for i := 1, ... n do
        s[V'[i]] := 1
    PF := PF ∪ {s}

PF := RemoveDominated(PF)
return PF
```

$1000, \ldots, 10000$ were used with the remaining parameters set to $R = \frac{0.1}{\sqrt{N_v/1000}}$, $E = 0.5$, $R = \frac{1.0}{\sqrt{N_v/1000}}$. These parameter settings keep the average degree $\overline{k}$ of the obtained graphs at more or less the same level, which allows obtaining a similar epidemic dynamics on graphs of a different size (in this paper $\overline{k}$ varies from 28.9 for $N_v = 1000$ to 30.6 for $N_v = 10000$). The experiments presented here consisted of three phases:

1. Parameter tuning aimed at obtaining the best values of the parameters for each of the tested algorithms. The tuning was performed on a set of 10 problem instances with $N_v = 1000$ using the grid search approach with candidate values of the parameters: population size $N_{pop} \in \{50, 100, 200, 500\}$, crossover probability $P_{cross} \in \{0.5, 0.6, 0.7, 0.8, 0.9, 1.0\}$, and mutation probability $P^{(p)}_{mut} \in \{\frac{0.2}{N_v}, \frac{0.5}{N_v}, \frac{1.0}{N_v}, \frac{2.0}{N_v}, \frac{5.0}{N_v}\}$ for mutation operators applied to individual positions in the genotype (e.g. the bit-flip mutation), and $P^{(g)}_{mut} \in \{0.02, 0.04, 0.06, 0.08, 0.10\}$ for mutation operators applied to the whole genotype (e.g. the scramble mutation). For each set of parameter values each algorithm was run 10 times and the median hypervolume (HV) was calculated from these runs. Table 2 presents the values of parameters for which each of the algorithms attained the best results.

2. Comparison of the three tested algorithms, with the parameters set to the best values obtained in the previous phase, on 30 different problem instances with $N_v = 1000, \ldots, 10000$. For each graph size N_v each algorithm was run 30 times and the median HV was calculated from these runs. Table 3 presents the HV and the results of the Wilcoxon statistical test [16] with the null hypothesis stating the equality of the medians between EA-C and MOEA/D (and, respectively, EA-R and MOEA/D). The best result for each number of nodes N_v is underlined and the results of statistical comparison to MOEA/D as the reference method are shown. The (+) signs indicate results of the statistical test which rejected the null hypothesis stating the equality of the medians at the confidence level $\alpha = 0.05$ and thereby confirmed that EA-C (or EA-R) produced results superior to MOEA/D. The (–) signs indicate statistically significant results which were worse for EA-C (or EA-R) than for MOEA/D. The (=) signs indicate results for which the statistical test was not able to reject the null hypothesis at the confidence level $\alpha = 0.05$. Clearly, the EA-R was superior to MOEA/D in all the tests, while EA-C was better for larger graphs and for the smallest ones was worse than MOEA/D.
3. Testing the generalization capability of the neural networks produced by the EA-C and EA-R algorithms, which is desirable, because a good generalization capability means, that the models can be optimized once, and reused for many problem instances at no additional computational cost. For each of the 30 problem instances $\mathcal{I}_i$, $i \in \{1, \ldots, 30\}$ with a given graph size N_v, 29 Pareto fronts were generated using each of the neural networks obtained during the run of EA-C (and, respectively, EA-R) for the other 29 problem instances with the same graph size N_v (i.e. $\mathcal{I}_j$, $j \in \{1, \ldots, 30\} \backslash \{i\}$). Each of these 29 fronts was evaluated using the HV and the average of these 29 values was recorded as the result for the problem instance $\mathcal{I}_i$. This way, classifiers (and, respectively, regression models) pre-trained on each of the problem instances $\mathcal{I}_j$, $j \neq i$ were tested on the problem instance $\mathcal{I}_i$. Table 4 presents the median of the results obtained for problem instances $\mathcal{I}_i$, $i \in \{1, \ldots, 30\}$ for each graph size N_v compared to the results produced by MOEA/D. Results of the Wilcoxon statistical test are marked in this table in the same way as in Table 3. The results show, that pre-trained models are better than MOEA/D for $N_v \geq 2000$ (classification) and $N_v \geq 1500$ (regression).

Table 2. The values of the parameters obtained using the grid search approach.

Parameter name	EA-C	EA-R	MOEA/D
Population size N_{pop}	50	50	500
Crossover probability P_{cross}	0.8	0.6	0.6
Mutation probability per position $P_{mut}^{(p)}$	$\frac{2.0}{N_v}$	$\frac{2.0}{N_v}$	$\frac{0.2}{N_v}$
Mutation probability per genotype $P_{mut}^{(g)}$	0.08	0.08	0.02

Table 3. Comparison of the evolutionary algorithms on graphs with $N_v = 1000, \ldots, 10000$ using the hypervolume indicator.

N_v	MOEA/D	EA-C	EA-R	N_v	MOEA/D	EA-C	EA-R
1000	543330	533953 (−)	546081 (+)	2250	2633062	2673574 (+)	2716803 (+)
1250	836386	834414 (=)	849148 (+)	2500	3238913	3300603 (+)	3345401 (+)
1500	1183341	1196044 (+)	1216436 (+)	5000	12815375	13128431 (+)	13293698 (+)
1750	1600337	1621174 (+)	1648210 (+)	7500	28793532	29450989 (+)	29793720 (+)
2000	2078613	2117360 (+)	2151108 (+)	10000	50975073	52287150 (+)	52863685 (+)

Table 4. Comparison of the results obtained using pre-trained machine learning models to the results obtained by MOEA/D.

N_v	MOEA/D	Classifier	Regression	N_v	MOEA/D	Classifier	Regression
1000	543330	526838 (−)	536572 (−)	2250	2633062	2646497 (+)	2687114 (+)
1250	836386	821170 (−)	835245 (=)	2500	3238913	3272524 (+)	3314139 (+)
1500	1183341	1181147 (−)	1199050 (+)	5000	12815375	13074891 (+)	13210911 (+)
1750	1600337	1602026 (=)	1626966 (+)	7500	28793532	29370464 (+)	29665459 (+)
2000	2078613	2097411 (+)	2124237 (+)	10000	50975073	52151266 (+)	52695523 (+)

In all three phases of the experiments, the stopping condition was set to $max_{FE} = 10000$ solution evaluations. Even though the algorithms worked on different search spaces (weights of a neural network vs. binary vectors representing vaccine assignments to graph nodes), setting the same budget for all of them was motivated by the fact, that evaluating solutions using simulations is expensive and thus the number of solution evaluations is a good measure for representing the computational cost in this study.

5 Conclusion

In this paper, algorithms using machine learning models for the bi-objective optimization of vaccination assignments were studied. Instead of determining the assignment of vaccine doses to graph nodes, they optimize weights of a neural network which is subsequently used for selecting nodes to vaccinate. Proposed algorithms, especially EA-R which uses a regression model, outperformed the MOEA/D used as a reference method. Also, the optimized models show good generalization capability: on larger graphs they produce better results than MOEA/D, when applied to different problem instances than the ones for which they were optimized. Further work may include studying more machine learning models, different real-life attributes for diseases other than influenza, as well as comparing different metaheuristic optimization methods.

Acknowledgment. This work was supported by the Polish National Science Centre under grant no. 2015/19/D/HS4/02574. Calculations have been carried out using resources provided by Wroclaw Centre for Networking and Supercomputing (http://wcss.pl), grant No. 407.

References

1. Antonioni, A., Bullock, S., Tomassini, M.: REDS: an energy-constrained spatial social network model. In: Lipson, H., Sayama, H., Rieffel, J., Risi, S., Doursat, R. (eds.) ALIFE 14: The Fourteenth International Conference on the Synthesis and Simulation of Living Systems, pp. 368–375. MIT Press (2014)
2. Burkholz, R., Leduc, M., Garas, A., Schweitzer, F.: Systemic risk in multiplex networks with asymmetric coupling and threshold feedback. Phys. D **323–324**, 64–72 (2016)
3. Dalgıç, Ö.O., Özaltın, O.Y., Ciccotelli, W.A., Erenay, F.S.: Deriving effective vaccine allocation strategies for pandemic influenza: comparison of an agent-based simulation and a compartmental model. PLoS One **12**(2), 1–19 (2017)
4. Deb, K., Agarwal, R.: Simulated binary crossover for continuous search space. Complex Syst. **9**(2), 115–148 (1995)
5. Deb, K., Goyal, M.: A combined genetic adaptive search (GeneAS) for engineering design. Comput. Sci. Inform. **26**, 30–45 (1996)
6. Eiben, A.E., Smith, J.E.: Introduction to Evolutionary Computing, 2nd edn. Springer, Heidelberg (2015). https://doi.org/10.1007/978-3-662-44874-8
7. Guerrisi, C., et al.: Factors associated with influenza-like-illness: a crowdsourced cohort study from 2012/13 to 2017/18. BMC Public Health **19**, 879 (2019)
8. Harizi, I., Berkane, S., Tayebi, A.: Modeling the effect of population-wide vaccination on the evolution of COVID-19 epidemic in Canada. medRxiv (2021)
9. Hartnell, B.: Firefighter! An application of domination. In: 20th Conference on Numerical Mathematics and Computing (1995)
10. Hu, B., Windbichler, A., Raidl, G.R.: A new solution representation for the firefighter problem. In: Ochoa, G., Chicano, F. (eds.) EvoCOP 2015. LNCS, vol. 9026, pp. 25–35. Springer, Cham (2015). https://doi.org/10.1007/978-3-319-16468-7_3
11. Li, H., Zhang, Q.: Multiobjective optimization problems with complicated Pareto sets, MOEA/D and NSGA-II. IEEE Trans. Evol. Comput. **13**(2), 284–302 (2009)
12. Liu, C., Kroll, A.: Performance impact of mutation operators of a subpopulation-based genetic algorithm for multi-robot task allocation problems. Springerplus **5**(1), 1–29 (2016). https://doi.org/10.1186/s40064-016-3027-2
13. Michalak, K.: The Sim-EA algorithm with operator autoadaptation for the multiobjective firefighter problem. In: Ochoa, G., Chicano, F. (eds.) EvoCOP 2015. LNCS, vol. 9026, pp. 184–196. Springer, Cham (2015). https://doi.org/10.1007/978-3-319-16468-7_16
14. Michalak, K.: Estimation of distribution algorithms for the firefighter problem. In: Hu, B., López-Ibáñez, M. (eds.) EvoCOP 2017. LNCS, vol. 10197, pp. 108–123. Springer, Cham (2017). https://doi.org/10.1007/978-3-319-55453-2_8
15. Michalak, K.: Solving the parameterless firefighter problem using multiobjective evolutionary algorithms. In: Proceedings of the Genetic and Evolutionary Computation Conference Companion, GECCO 2019, pp. 1321–1328. ACM, New York (2019)
16. Rey, D., Neuhäuser, M.: Wilcoxon-signed-rank test. In: Lovric, M. (ed.) International Encyclopedia of Statistical Science, pp. 1658–1659. Springer, Heidelberg (2011). https://doi.org/10.1007/978-3-642-04898-2_616
17. Schlickeiser, R., Kröger, M.: Analytical modeling of the temporal evolution of epidemics outbreaks accounting for vaccinations. Physics **3**(2), 386–426 (2021)
18. Tornatore, E., Vetro, P., Buccellato, S.M.: SIVR epidemic model with stochastic perturbation. Neural Comput. Appl. **24**(2), 309–315 (2014)

19. Vose, M.D.: The Simple Genetic Algorithm: Foundations and Theory. MIT Press, Cambridge (1998)
20. Watts, D.J.: A simple model of global cascades on random networks. Proc. Natl. Acad. Sci. **99**(9), 5766–5771 (2002). https://doi.org/10.1073/pnas.082090499
21. Wu, F., Zhang, D., Ji, Q.: Systemic risk and financial contagion across top global energy companies. Energy Economics **97**, 105221 (2021)
22. Zhang, Q., Li, H.: MOEA/D: a multiobjective evolutionary algorithm based on decomposition. IEEE Trans. Evol. Comput. **11**(6), 712–731 (2007)
23. Zitzler, E., Thiele, L., Laumanns, M., Fonseca, C.M., da Fonseca, V.G.: Performance assessment of multiobjective optimizers: an analysis and review. IEEE Trans. Evol. Comput. **7**, 117–132 (2002)

Joint Price Optimization Across a Portfolio of Fashion E-Commerce Products

Sagnik Sarkar[1(✉)], Siddhartha Devapujula[1], Hrishikesh Vidyadhar Ganu[1], Chaithanya Bandi[2], Ravindra Babu Tallamraju[1], Chilamakurthi Vamsikrishna Satya[1], and Siddhant Doshi[1]

[1] Myntra Designs Pvt. Ltd., Bengaluru, India
{sagnik.sarkar,siddhartha.devapujula,hrishikesh.ganu, vamsi.chilamakurthi,siddhant.doshi}@myntra.com, bizchaba@nus.edu.sg, ravindrababu.t@gmail.com

[2] National University of Singapore, Singapore, Singapore

Abstract. Pricing is a key lever used by e-commerce companies to achieve "growth with profitability". Given the huge catalog size in e-commerce, most products have very close substitutes and complements. These complementary/substitute products result in influencing the demand for one another. Moreover, in the context of fashion, the utility of a product is mostly subjective. In categories like electronics, it's relatively easy to define the utility of a product based on its attributes, but the same is not directly applicable to fashion. Products with similar attributes can have different utilities for the customer and therefore can be priced differently. Taking these things into consideration we base our pricing strategy on the following 3-stage decision-making process: 1) identifying the items which influence each other 2) building demand models that include effects of demand transference 3) joint optimization of the prices to achieve revenue or profit margin targets. We discuss our contributions to building a real-world system that implements these 3 stages in the specific context of fashion e-commerce. Fashion e-commerce has its nuances when it comes to pricing compared to general e-commerce and we explain how we dealt with these difficulties. Moreover, in addition to the formulations, we also describe challenges faced in building working systems that scale to millions of products and hundreds of categories. In addition, we describe a unique approach to quantifying the dollar benefit under scenarios where true A/B testing is not possible for legal reasons. Lastly, we explain how this work has resulted in significant incremental revenue for a large fashion e-commerce company.

Keywords: E-commerce · Pricing · Optimization

1 Introduction

In the fashion e-commerce setting, millions of products are live on the platform every day. The discounts provided on these products are flexible and can be

M. Emmerich et al. (Eds.): EMO 2023, LNCS 13970, pp. 449–461, 2023.
https://doi.org/10.1007/978-3-031-27250-9_32

changed periodically, under certain business restrictions to meet revenue and gross-margin targets. Manual pricing strategies mostly rely on business heuristics that are derived from experience through trial and error. This requires a lot of manual effort that still leads to a loss of revenue due to sub-optimal pricing decisions. In this work, we propose a data-driven automatic pricing system that optimizes the revenue while meeting the business constraints for a large set of products across various product categories related to fashion. Products in the fashion category lack a distinctive name or any unique identification. Users typically describe their intent in a search query using product attributes, like, for purchasing a *Tshirt* user could search for {*tshirt*} or {*<brand >t-shirts*} or {*<color >t-shirts*}. Each of these queries would typically result in thousands of relevant products from the catalogue as opposed to queries in hard goods categories with well-defined products like "iPhone X 64 Gb" which would result in few relevant products. Since the intent isn't very specific, the user has a long list of products to choose from and hence price becomes a strong factor in influencing the purchase pattern of a user. For example, having a lower price can increase the demand for an item. But, this extra demand can be due to demand transference from other items which may lead to an overall reduction in the revenue. Therefore, we would want to consider the cross-product price sensitivities as well and perform a joint optimization. Our approach consists of 3 main stages namely, clustering, demand forecasting, and price optimization. In the clustering phase, we identify sets of products that affect each other's demand. Since fashion is an abstract domain, it becomes challenging to know the dependencies apriori and therefore we rely on customer interaction data to derive such groups of influencing products. Next, using the past sales data we train a demand forecasting model that can predict the sales of a product at any given price configuration. Finally, we use the parameters of the demand forecasting model and formulate the revenue maximization (subject to business constraints) problem as a Binary Quadratic Program (BQP). Although BQP problems are NP-hard, we use a standard mixed-integer linear programming (MILP) relaxation which is solved using an open-source MILP solver for hundreds of products within a practical computational time. In addition to the revenue and gross margin requirements, a major challenge lies in considering the age and current inventory of a product while pricing for it. Ideally, a product with low inventory can be priced less aggressively, i.e., with less discount, as compared to a product with high inventory, which would help to maintain a healthy inventory balance. Similarly, older products should be discounted higher than fresher products. To account for the age and inventory while generating the prices, we formulate a new objective for minimizing stockpiling at the end of the season. Since this objective is non-linear, we solve a multi-objective optimization problem using a genetic algorithm based approach and compare its efficacy with the MILP solutions.

Online experimentation is another challenge in our work. The standard A/B tests are not feasible in this scenario, as offering the same products at different prices to different customers would lead to legal issues. In the paper [11], authors propose a framework where a set of similar products are randomly divided into two buckets and each bucket gets a different treatment. After the experiment, the performance of both buckets is evaluated. In our case, this framework is not applicable because we consider products influencing each others demand. Therefore, the test and the control environments can influence each other. To solve this, we propose a novel framework that nullifies the interaction between the test and control environments thus approximating an A/B test. The main contributions of this work are summarized as follows:

- Novel pipeline for price optimization in the fashion e-commerce domain.
- Method to identify clusters of products that influence each other's demand.
- Demand forecasting models that fit in the proposed MILP optimization framework and compare their performances with each other.
- Multi-objective genetic algorithm based price optimization for inventory and age-aware pricing.
- Novel test framework for conducting on-field large-scale experiments and empirically demonstrating the effectiveness of the proposed method.

2 Related Work

Pricing optimization in general has been a heavily researched topic for the past few decades. The book by [14] provides a comprehensive survey of pricing methods used in various industries. Our work is closely related to works in static multi-product pricing for online retail. These frameworks typically include a demand forecasting step followed by revenue or profit maximization. Some recent works using this framework include [15] and [4]. Our optimization formulation is inspired by [8], but we apply it to the domain of fashion e-commerce with several differences in the overall framework. In the context of pricing decision support tools for retailers, there are works such as that of [2] for inventory clearance, [12] for dynamic retail pricing, and promotion planning. Our work is also closely related to myopic pricing policies which combine greedy price optimization with sequential estimation. In this framework, before making a price allocation in each time period, the seller first estimates the parameters of a demand model and then sets the prices that maximize the revenue or profit according to the most recent estimates of the parameters of the demand model. This framework has been theoretically shown [6,10] to suffer from the problem of "incomplete learning" because when this myopic pricing policy is repeated over several time periods, there is a threat that the demand model parameters do not converge to their true estimates with time due to lack of enough price dispersion. In our case, this problem is mitigated to a large extent because of parameter sharing between items and price experimentation in some parts of the inventory for which the pricing is controlled by external vendors.

3 Methodology

Demand Forecast is the most crucial element while pricing. The demand for a product can be influenced by its own price and the prices of other products on the platform. Apart from price, there would be several other factors influencing the demand. We model all these factors using the following generalized demand equation.

$$s_m(\mathbf{p}) = \alpha_m + \sum_{m'=1}^{M} \sum_{d=1}^{D} \beta_{mm'd} I_d(p_{m'}) \tag{1}$$

where $s_m(\mathbf{p})$ denotes the forecasted sales for the mth item, $\mathbf{p}$ denotes the price allocation vector for the given set of influencing products, $p_{m'}$ denotes the price of product m' and M denotes the number of products in the given set. α_m, $\beta_{mm'd}$, are the demand model parameters that are estimated from the past data. α_m captures the effect of non-price features like past sales, attributes of the product and any other external factors which do not depend on price and thus are constant as a function of $\mathbf{p}$. $\beta_{mm'd}$ denotes the price coefficients of the model, i.e., both the self price coefficients (for $m = m'$) and influencer price coefficients (for $m \neq m'$). The I_d functions denote the type of dependence of sales on price, for example, linear, quadratic, and logarithmic, to list a few. However, in this work, we have restricted our experiments to a linear dependence function for better interpretability.

According to the generalized form of demand as described above, the revenue forecast for the mth product can be given as:

$$rev_m(\mathbf{p}) = p_m s_m = \sum_{m'=1}^{M} p_m [\frac{\alpha_m}{M} + \sum_{d=1}^{D} \beta_{mm'd} I_d(p_{m'})] \tag{2}$$

Now, the total revenue for the set of products would be:

$$\begin{aligned} rev(\mathbf{p}) &= \sum_{m=1}^{M} \sum_{m'=1}^{M} p_m [\frac{\alpha_m}{M} + \sum_{d=1}^{D} \beta_{mm'd} I_d(p_{m'})] \\ &= \sum_{m=1}^{M} \sum_{m'=1}^{M} \eta_{mm'}(p_m, p_{m'}) \end{aligned} \tag{3}$$

Prices on the platform cannot be varied continuously. They should be modified in the steps of either the discounts or a fixed value. Hence, the basic optimization problem to maximize expected revenue becomes:

$$\begin{aligned} &\text{Maximize} \quad rev(\mathbf{p}) \\ &\text{subject to} \quad p_m \in \{P_{m1}, \ldots, P_{mK}\} \\ &\text{where} \quad m = 1, \ldots, M \end{aligned} \tag{4}$$

In the above formulation, $P_{m1}, \ldots, P_{mK}$, refers to the K possible price points for the product m. In our pipeline, we receive a set of base (initial) discounts from the business for each item and then we form the price points using the discount window [base−10, base−5, base, base+5, base+10]. The initial price configuration when the items are at base discounts is called the base price allocation.

3.1 Optimization

Once we have the demand for every product in the form described in Eq. 1, we need to optimize the revenue as discussed in the Eq. 4. The pricing strategy also needs to satisfy certain business requirements which can be modelled as described in the following sections. An exhaustive or a brute-force search would require $\Theta(K^M)$-time computation, and hence would be computationally intractable even if M is of the order of a few hundred. Hence we reformulate the problem to make it computationally tractable.

Binary Quadratic Program Formulation. Consider the binary variables $z_{m1}, \ldots, z_{mK} \in \{0, 1\}$ satisfying $\sum_{k=1}^{K} z_{mk} = 1$. Here, $z_{mk} = 1$ and $z_{mk} = 0$ refers to $p_m = P_{mk}$ and $p_m \neq P_{mk}$, respectively. Now, $\eta_{mm'}(p_m, p_{m'})$ in Eq. 3 can be rewritten as follows:

$$\eta_{mm'}(p_m, p_{m'}) = \mathbf{z}_m^\top Q_{mm'} \mathbf{z}_{m'}, \tag{5}$$

where $\mathbf{z}_m = [z_{m1}, \ldots, z_{mK}]^\top \in \mathbb{R}^{K \times 1}$, and $Q_{mm'} \in \mathbb{R}^{K \times K}$ is defined as:

$$Q_{mm'} = \begin{bmatrix} \eta_{mm'}(P_{m1}, P_{m'1}) & \cdots & \eta_{mm'}(P_{m1}, P_{m'K}) \\ \eta_{mm'}(P_{m2}, P_{m'1}) & \cdots & \eta_{mm'}(P_{m2}, P_{m'K}) \\ \vdots & \ddots & \vdots \\ \eta_{mm'}(P_{mK}, P_{m'1}) & \cdots & \eta_{mm'}(P_{mK}, P_{m'K}) \end{bmatrix}. \tag{6}$$

Therefore, the overall optimization problem, given in Eq. 4, can be rewritten as follows:

$$\begin{aligned} \text{Maximize} \quad & f(\mathbf{z}) := \mathbf{z}^\top Q \mathbf{z} \\ & \text{subject to} \quad \mathbf{z} \in \{0,1\}^{MK}, \\ & \sum_{k=1}^{K} z_{mk} = 1 \quad (m = 1, \ldots, M), \end{aligned} \tag{7}$$

where $\mathbf{z} = [z_{11}, \ldots, z_{1K}, z_{21}, \ldots, z_{2K}, \ldots, z_{M1}, \ldots, z_{MK}]$ and $Q \in \mathbb{R}^{MK \times MK}$ is given by

$$Q = \begin{bmatrix} Q_{11} & Q_{12} & \cdots & Q_{1n} \\ Q_{21} & Q_{22} & \cdots & Q_{2n} \\ \vdots & \vdots & \ddots & \vdots \\ Q_{n1} & Q_{n2} & \cdots & Q_{nn} \end{bmatrix}. \tag{8}$$

MILP Relaxation Method. For formulating a Mixed Integer Linear Programming (MILP) relaxation to the above-stated BQP, we introduce auxiliary variables $\bar{z}_{ij} \forall$ $(1 \leq i < j \leq KM)$ corresponding to $\bar{z}_{ij} = z_i z_j$. Using these new auxiliary variables, Eq. 7 can be re-formulated as:

$$\text{Maximize} \sum_{i=1}^{KM} q_{ii} z_i + \sum_{i=1}^{KM} \sum_{j=i+1}^{KM} (q_{ij} + q_{ji}) \bar{z}_{ij} \tag{9}$$

$$\text{subject to} \sum_{i=mK+1}^{mK+K} z_i = 1, (0 \leq m \leq M-1),$$

$$\bar{z}_{ij} = 0, (mK+1 \leq i < j \leq mK+K)$$

$$\sum_{i=mK+1}^{mK+K} \bar{z}_{ij} = z_j (mK+1 \leq i \leq mK+K < j)$$

$$z_i \in \{0,1\}, \bar{z}_{ij} \in \{0,1\}$$

where q_{ij} are the entries of the matrix Q. We use the open-source CBC solver [9] to solve the above MILP using the cutting-plane methods.

Business Requirements. While the solver can explore any combination of prices, the business has certain constraints such as guardrails around the prices, gross-margin percentage, etc. All such constraints can be expressed in a linear form and therefore can be handled in the MILP formulation. While the clustering and demand forecasting modules are agnostic of the business constraints, the optimization module has to be re-run on an ad-hoc basis whenever there is any change in business goals. Hence, the optimizer needs to have limited latency. In our case, a one-hour duration is the agreed upper limit for price generation from a business point of view.

Inventory and Age Aware Pricing. Typically in fashion e-commerce, items are seasonal. The perceived value of an item decreases with time as the fashion season progresses. For example, winter wears such as sweatshirts would be less often required and sold in the summer season. Therefore, generally, a discount gradient is followed such that the older items are sold at higher discounts than the new-coming fresh items. This avoids stockpiling at the end of the season. Other than the age of an item, the current inventory of the product must also be considered while pricing the item such that an item with a higher stock of inventory should be discounted more than an item with a lower stock of inventory, but of the same age. These business heuristics ensure that the current season products are more or less sold off before the end of the season. Taking all this into account we define a new objective called "aggregate estimated days spillover" denoted by *agg_eds*. Before defining this quantity we define another quantity called *doh* which stands for "days on hand" for a particular item's stock. It is defined as follows:

$$doh_i(\mathbf{p}) = \frac{current\ number\ of\ units\ of\ item\ available}{estimated\ rate\ of\ sales\ at\ price\ allocation\ \mathbf{p}} = \frac{I_i}{s_i(\mathbf{p})} \tag{10}$$

where i denotes the index of the ith item and $\mathbf{p}$ is the vector of price allocations. Having defined the *doh*, *agg_eds* is defined as follows:

$$agg\_eds(\mathbf{p}) = \sum_{i=1}^{M} max(0, age_i + doh_i(\mathbf{p}) - 180) \tag{11}$$

where age_i is the current age of item i and 180 (days) is considered the typical season length. Intuitively, $agg\_eds(\mathbf{p})$ approximately calculates the sum of the number of days by which an item will spill over into the next season across all the items in a cluster of size M, which ideally should be minimized. Since this objective is not linear with respect to the optimization variables, we cannot optimize for it in the MILP framework discussed previously. Therefore, we resort to a genetic algorithm (GA) based multi-objective optimization approach where we maximize revenue and gross margin and minimize *agg_eds*. We use the DEAP library [5] for implementing this solution. Specifically, each price allocation vector is an individual in the population of several solutions. We use a custom mutation function to ensure that each product has a unique price allocation. For crossover, we use the default two-point crossover function as provided in DEAP. The selection algorithm used is NSGA-II [3]. From the set of Pareto optimal solutions, we filter out those which don't satisfy the gross margin business constraints. We also filter out the solutions where the *agg_eds* objective is more than that of the base price allocation. Finally, from the set of filtered solutions, we select the one with the highest expected revenue.

3.2 Product Clusters

The problem formulation allows us the flexibility to assume that the demand for any product is dependent on the prices of all the products on the platform. However, in practice, products influence each other's demand within subsets. Clustering aims to identify such groups of products where there is demand transference amongst the products within a group but the demand transference across the groups is minimal. This reduces the number of optimization variables thus decreasing the compute time. Since the clusters can be optimized in parallel, it becomes possible to horizontally scale the optimization for pricing a very large number of products. Forming the product clusters consists of the following steps:

- Forming an item-item co-browse graph from user click-stream data where nodes represent items and edge weights are their co-browse frequencies.
- Learning item embeddings by applying the Deepwalk algorithm [13] to this item-item co-browse graph.
- Applying a constrained KMeans algorithm [1] on the item embeddings, generated in the previous step, and obtaining the final product clusters.

3.3 Demand Model

The next step after clustering involves predicting the future sales for a product, referred to as demand forecasting. Equation 1 represents the generalized form for the demand model that performs demand forecasting. We train the demand model using the historic sales data as the target variable and a relevant set of selected features. We consider broadly 2 types of features to predict sales, namely, the non-price features such as past sales of the item, the brand of the product, and list-views, to name a few, and price features such as the price of the product and the respective prices of its influencers. Ideally, we can assume that the demand for every item in a cluster is influenced by all the other items in that cluster. However, that significantly increases the number of parameters which might lead to overfitting. Hence, we choose the n nearest influencer products, in a way that ensures every item in the cluster has a direct or indirect influence on the demand for every other item in the cluster using the connected component analysis. We find $n = 5$ to be the most optimal value that satisfies the criterion through our offline analysis.

Regression Models. In this paper, we restrict the demand model to a linear form, i.e., the linear relation of the price features with the target variable is preserved. However, as an upper baseline, we also demonstrate the results of a fully connected neural network (FCN), which doesn't guarantee to preserve the linear relation. We train the following different types of demand forecasting models: 1) Linear Regression (LR), 2) FCN, 3) Linear Regression L1 Regularization (LRL1), 4) Linear Regression L2 Regularization (LRL2), and 5) Price-Linear Neural Network (PLN). As FCN cannot ensure the linear relation of the price features to the sales (target variable), so we customize the FCN to build PLN. In PLN, all the non-price features go as an input in the first layer and the price features are passed only in the penultimate layer, thereby ensuring a linear relation of the price features with the output.

3.4 Overall Price Recommendation Pipeline

In this section, we describe the flow of the overall price recommendation pipeline that is illustrated in Fig. 1. The pipeline starts with the ingestion of the user-browsing logs and the creation of an item-item co-occurrence graph using PySpark. The next step is to create the item embeddings within each ATG (Article-Type-Gender; example, Men's T-shirts) using the Deepwalk algorithm on which constrained K-Means is applied to obtain clusters within each ATG. We then identify the top n nearest neighbors for each item within its cluster. In the next step, we train a demand forecasting model to predict future sales. Using the learned demand model parameters, we derive the α_ms and $\beta_{mm'}$s which eventually are fed in the optimization. The optimization module takes these parameters for each product (as seen in Eq. 1) along with several business constraints. Since the clusters can be optimized independently of each other, we parallelize the optimization across spark clusters using PySpark. Finally, the output prices of the optimizer module are aggregated into a CSV file and consumed by the business team.

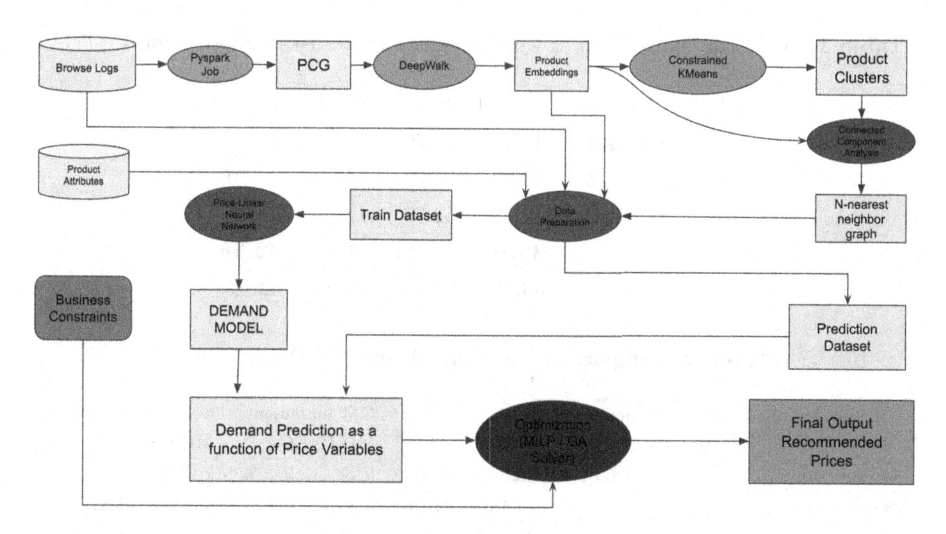

Fig. 1. Flow diagram

4 Offline Demand Model Evaluation

4.1 Data Preparation

The experiments were conducted on data with a volume of 2.96 million records for 4 ATGs. Each row in the data set represents a product, wherein the total sales of the product for the next T days serves as the target variable. The data spanned a total of 72k items across the 4 ATGs. The train and the test data sets are divided such that test data has only the date range that falls after a complete date range (gap of T days) in train data ensuring no data leakage. Here, the data was split into three sets: training set with 9 weeks of data, validation data of 1 week, and test set with 1 week of data. Here, we choose $T = 7$ for the experiments.

Performance Analysis. We use the Weighted Mean Absolute Percentage Error (WMAPE) as a metric to evaluate the performance of the demand models. WMAPE is one of the widely used metrics to measure the accuracy of demand predictions, which is calculated as the average of the absolute forecast errors for all products weighted by their actual sales. Table 1 shows the performance of various demand models in terms of their respective WMAPEs. Smaller WMAPE values indicate better forecast values. We see from Table 1 that PLN is highly competitive with FCN, which is an unrestricted model whereas PLN is bound to preserve the linear relation of price features with the target variable. There is just a minimal drop in the WMAPE going from FCN to PLN.

Table 1. WMAPE comparison of various demand forecasting models for 4 ATGs

Model	ATG-1	ATG-2	ATG-3	ATG-4
Linear regression	41.93	52.79	42.42	48.88
Lasso regression	43.09	52.79	43.90	49.70
Ridge regression	42.18	52.70	42.24	48.78
FCN	**33.10**	**35.23**	**35.98**	**42.56**
PLN	33.12	35.25	36.08	42.62

Table 2. Comparison between GA and MILP solutions.

Cluster AT name	Revenue increment		RGM increment		GM increment		*agg_eds* reduction	
	MILP	GA	MILP	GA	MILP	GA	MILP	GA
Jackets	11.59%	11.06%	50.65%	49.46%	9.2%	9.09%	7.95%	10.34%
Sweaters	14.56%	13.95%	23.15%	22.19%	3.35%	3.23%	−6.78%	3.81%
Dresses	42.38%	41.65%	42.53%	41.71%	0.03%	0.01%	1.9%	3.49%
Sweatshirts	19.9%	19.35%	63.45%	61.43%	10.46%	10.16%	1.94%	4.33%
Tops	16.28%	16.17%	52.59%	52.24%	9.24%	9.18%	0.69%	3.84%

5 Comparing GA Solutions to MILP Solutions

In Table 2, we broadcast the GA solutions, which are obtained considering multiple objectives, i.e., maximizing revenue and profits and minimizing the *agg_eds*, and compare them with the MILP solutions obtained using the MILP formulation for 5 random clusters. We report the percentage change of estimated (as per the demand model) revenue, gross margin and *agg_eds* metrics as compared to the base price allocation. We observe that the GA solution is quite competitive with the MILP solution in terms of revenue and gross margin. On the other hand, the GA solution consistently beats the MILP solution in terms of the *agg_eds* metric and thus comes out as a better approach for inventory and age-aware pricing.

6 Online Evaluation

A/B test is the usual way to measure the on-field performance of any new algorithm in the e-commerce domain. The users coming to the platform get uniformly allotted to test and control buckets and the performance of the buckets is compared. However, we are not allowed to show different prices for the same product to different customers at the same time. Hence, the possibility of doing an A/B test in our scenario gets eliminated. This persists as a huge obstacle in measuring the on-field impact of the algorithm.

6.1 Test Framework

We propose a framework where we would be evaluating performance across time. We need to consider the change in the environment over time and eliminate its impact on the experiment. As a first step, we randomly divide the clusters into two sets - A and B. We consider a time duration of $2t$ to perform one cycle of the experiment. The time duration $2t$ is divided into two equal segments t_1 and t_2 sequentially. During t_1, products in SetA are priced using the algorithm described in this paper whereas products in SetB are priced using the existing control algorithm. Later during t_2, we swap the treatments, i.e., SetB products get priced through the test algorithm, and SetA products are given the control treatment. The time period should be carefully chosen to reduce the impact of external factors (ex: day of the week, day of the month, etc) and also give enough time for the price effects to reflect in the demand. After consulting with the business teams, we arrived at the optimal time period of $t = 7$ days, which eliminates the "day of the week" effects on the demand. So, one cycle of testing takes two weeks to complete and such cycles are repeated several times to eliminate the effect of any external factors. We performed 4 such cycles of testing and evaluated the metrics.

6.2 Results

Revenue is the primary evaluation metric and the objective of the optimization module is set to be the same. But as we are testing across time periods, there can be external factors that may impact the test. Therefore we normalized the revenue by the number of impressions (list-views) since external factors that are not controlled by our pricing module can impact the impressions. Hence, normalizing by impressions will give a more accurate picture of our method's performance. The final metric used to evaluate the online performance is Revenue per impression (RPI), which is defined as:

$$RPI_{test} = (R_A t_1 + R_B t_2)/(I_A t_1 + I_B t_2) \tag{12}$$

$$RPI_{Control} = (R_B t_1 + R_A t_2)/(I_B t_1 + I_A t_2) \tag{13}$$

$$I_{C_i} = ((RPI_{test}/RPI_{Control}) - 1) * 100 \tag{14}$$

where $R_x t_k$ is the revenue generated from products of set x during time period t_k. $I_x t_k$ represents the impressions of the products of set x during time period t_k and I_{C_i} is the percentage improvement in RPI in the test cycle i.

Table 3 gives an overview of the performance improvement in all the test cycles. Final improvement, I_C is computed as the average of all I_{C_i} of all test cycles. We can see improvement in all the cycles, with an average improvement of 3.58% overall.

Table 3. Percentage of RPI Improvement for test cycles

Test cycle	I_{C_i}
1	4.16%
2	2.30%
3	4.86%
4	3.00%
I_C	**3.58%**

Statistical Validation. Since every product has been part of both the test and control algorithms at different time periods, we have the test and control RPI for each. We conduct two popular tests for comparing the significance of matched samples - paired t-test [7] and Wilcoxon signed-rank test [16]. While paired t-test assumes the difference of the samples follows a normal distribution, Wilcoxon test doesn't have such assumptions. Each cycle has three columns, namely, product, test RPI, and control RPI, wherein the p-value is computed for each cycle and averaged across all the test cycles. The p-values average across the 4 test cycles are 3.91e−112 and 6.04e−169 for the paired t-test and the Wilcoxon signed-rank test respectively, therefore indicating 100% significance in both the statistical tests.

7 Conclusion and Future Work

In this work, we propose a practical framework for revenue maximization through price optimization in the real-world setting of a fashion e-commerce company. Our framework is easily scalable to millions of products and can also handle practical business constraints. We discuss how non-linear regression methods can be applied in the MILP optimization framework and also report significant forecasting error reduction through such methods. We also propose a GA-based multi-objective inventory and age-aware pricing technique and compare its efficacy to the MILP solutions. We discuss challenges in evaluating the overall system due to legal issues with A/B testing and therefore propose an alternate method for evaluating such a system. The proposed framework has achieved significant revenue gains in several test cycles and is currently deployed as a price recommendation system on the platform for several months.

The current work applies to usual business (non-sale) days. Forecasting demand on sale days is challenging due to the multiple factors such as promotions, external events, holidays, etc. shaping the demand. In future, we would want to extend the model to sale days by adopting techniques from the dynamic pricing literature. Since demand forecasting accuracy is one of the crucial components of the framework, we would want to adopt more advanced deep learning-based demand forecasting techniques to our current optimization framework and thereby improve performance and robustness.

References

1. Bradley, P.S., Bennett, K.P., Demiriz, A.: Constrained k-means clustering. Microsoft Res. Redmond **20** (2000)
2. Caro, F., Gallien, J.: Clearance pricing optimization for a fast-fashion retailer. Ann. Oper. Res. **60**, 1404–1422 (2010). https://doi.org/10.2139/ssrn.1731402
3. Deb, K., Pratap, A., Agarwal, S., Meyarivan, T.: A fast and elitist multiobjective genetic algorithm: NSGA-II. IEEE Trans. Evol. Comput. **6**(2), 182–197 (2002). https://doi.org/10.1109/4235.996017
4. Ferreira, K.J., Lee, B.H.A., Simchi-Levi, D.: Analytics for an online retailer: demand forecasting and price optimization. Manuf. Serv. Oper. Manag. **18**(1), 69–88 (2016)
5. Fortin, F.A., Rainville, F.M.D., Gardner, M.A., Parizeau, M., Gagné, C.: DEAP: evolutionary algorithms made easy. J. Mach. Learn. Res. **13**(70), 2171–2175 (2012). http://jmlr.org/papers/v13/fortin12a.html
6. Harrison, J.M., Keskin, N.B., Zeevi, A.: Bayesian dynamic pricing policies: learning and earning under a binary prior distribution. Manage. Sci. **58**(3), 570–586 (2012)
7. Hsu, H., Lachenbruch, P.A.: Paired t test. Wiley StatsRef: statistics reference online (2014)
8. Ito, S., Fujimaki, R.: Optimization beyond prediction: prescriptive price optimization. In: Proceedings of the 23rd ACM SIGKDD International Conference on Knowledge Discovery and Data Mining, pp. 1833–1841 (2017)
9. Forrest, J.J., et al.: Coin-or/CBC: Version 2.10.5 (2020). https://doi.org/10.5281/zenodo.3700700
10. Keskin, N.B., Zeevi, A.: Dynamic pricing with an unknown demand model: asymptotically optimal semi-myopic policies. Oper. Res. **62**(5), 1142–1167 (2014)
11. Liu, J., Zhang, Y., Wang, X., Deng, Y., Wu, X.: Dynamic pricing on E-commerce platform with deep reinforcement learning. CoRR abs/1912.02572 (2019). http://arxiv.org/abs/1912.02572
12. Natter, M., Reutterer, T., Mild, A., Taudes, A.: Practice prize report-an assortmentwide decision-support system for dynamic pricing and promotion planning in diy retailing. Mark. Sci. **26**(4), 576–583 (2007)
13. Perozzi, B., Al-Rfou, R., Skiena, S.: DeepWalk: online learning of social representations. In: Proceedings of the 20th ACM SIGKDD International Conference on Knowledge Discovery and Data Mining, pp. 701–710 (2014)
14. Talluri, K.T., Van Ryzin, G., Van Ryzin, G.: The Theory and Practice of Revenue Management, vol. 1. Springer, Heidelberg (2004). https://doi.org/10.1007/b139000
15. Vakhutinsky, A., Mihic, K., Wu, S.M.: A prescriptive analytics approach to markdown pricing for an E-commerce retailer (2018). https://doi.org/10.13140/RG.2.2.35292.69767
16. Woolson, R.F.: Wilcoxon signed-rank test. Wiley encyclopedia of clinical trials, pp. 1–3 (2007)

Improving MOEA/D with Knowledge Discovery. Application to a Bi-objective Routing Problem

Clément Legrand[1(✉)], Diego Cattaruzza[2], Laetitia Jourdan[1], and Marie-Eléonore Kessaci[1]

[1] Univ. Lille, CNRS, Centrale Lille, UMR 9189 CRIStAL, 59000 Lille, France
{clement.legrand4.etu,laetitia.jourdan, marie-eleonore.kessaci}@univ-lille.fr

[2] Univ. Lille, CNRS, Inria, Centrale Lille, UMR 9189 CRIStAL, 59000 Lille, France
diego.cattaruzza@centralelille.fr

Abstract. Knowledge Discovery (KD) mechanisms (e.g. data mining, neural networks) receive more and more interest over the years. A KD mechanism uses an extraction procedure, namely K_{ext}, to discover knowledge, and an injection procedure, namely K_{inj}, to exploit knowledge. However such mechanisms are not often applied to multi-objective combinatorial problems, due to the optimization of many objectives, which can lead to learning conflicting knowledge. The key is to know how the components of the KD mechanism should coexist and interact with the knowledge. In this article, we work with the MOEA/D algorithm, and existing K_{inj} and K_{ext} components. We propose different interactions between the components of the KD mechanism, by using different numbers of knowledge groups (dedicated to the storage of the knowledge) and different strategies for the injection component. The variants are evaluated through the bi-objective Vehicle Routing Problem with Time Windows (bVRPTW). Our results show, that using five knowledge groups and an intensification strategy for the injection procedure leads to better results.

Keywords: Multi-objective optimisation · MOEA/D · Knowledge discovery · Routing problem

1 Introduction

When solving a discrete optimization problem, large parts of the search space are explored during the execution of the algorithm. However, most of the solutions encountered are simply ignored, while they can bring interesting knowledge about the search space. Indeed, this knowledge can guide the algorithm towards more interesting solutions [1]. On the other hand, it can also help the algorithm to avoid getting stuck in local optima, as explicitly defined in Tabu Search methods [10].

M. Emmerich et al. (Eds.): EMO 2023, LNCS 13970, pp. 462–475, 2023.
https://doi.org/10.1007/978-3-031-27250-9_33

In multi-objective problems, at least two conflicting functions are simultaneously optimized and the objective is to find the Pareto front of solutions. Over the years, many metaheuristics based on local search techniques and using evolutionary algorithms [6] have been designed to tackle these problems. Among the most popular algorithms, there are MOEA/D [28], NSGA-II [8], and their variants.

Using knowledge from explored solutions is helpful to reduce the search space or to focus on interesting parts of the space, and can improve the performances of the algorithms mentioned above. However extracting knowledge from solutions and then using it to guide the search is a complex task, which has not been highly explored in the literature. Considering the papers on that subject leads to the following terminology for *Knowledge Discovery* (KD) processes. A KD process is built upon two main procedures called *Knowledge Extraction* (K_{ext}) and *Knowledge Injection* (K_{inj}). The K_{ext} procedure aims to extract problem-related knowledge from one or several solutions. Then the extracted knowledge can be used by the K_{inj} procedure to build new solutions taking into account past iterations. However, given extraction and injection procedures for a specific problem, there exist a plethora of ways to integrate them within a metaheuristic.

In this article, we investigate how a KD mechanism can be integrated into MOEA/D. To that purpose, we consider a bi-objective Vehicle Routing Problem with Time Windows (bVRPTW). In this problem, we minimize both the total traveling time and the total waiting time of drivers. With these two objectives, we obtain more diverse and bigger fronts (in terms of cardinality) than those obtained when minimizing the number of vehicles and the total traveling time, which are the original optimized objectives. Moreover, considering the waiting time can lead to different applications (e.g. food delivery, medical transportation). We propose a large number of hybridization variants that are evaluated, showing that one, in particular, is statistically better than the others.

The remaining of the paper is organized as follows: Sect. 2 focuses on KD mechanisms and their link with combinatorial optimization. We present new strategies for the components of a KD process in Sect. 3. The KD mechanism is integrated into MOEA/D in Sect. 4. The bVRPTW is described in Sect. 5. Our experimental setup is presented in Sect. 6, and our protocol in Sect. 7. We show and discuss our results in Sect. 8. Finally, we conclude in Sect. 9.

2 Scientific Context

2.1 Knowledge Discovery in Metaheuristics

Hybridizing machine learning methods and metaheuristics has become quite common to solve combinatorial problems. The survey of Talbi [21] reviews a large panel of hybridizations that are frequently used in the literature. These hybridizations are divided into three categories depending on where the integration is performed: at a problem level, at a low level, or at a high level. A problem-level integration takes into account the characteristics of the problem itself (e.g. data relative to the instance considered) to guide the algorithm. A low-level integration focuses on solutions produced by algorithms. A relevant mechanism is able to analyze the structure of the solutions, learn from them, and then use this knowledge to improve the next steps. A high-level integration is interesting when several operators are available to solve a problem. A possible interest is to design automatically a problem-specific heuristic by selecting the most relevant operators to apply. In the following, we focus on low-level integration and learning from solutions, also called knowledge discovery (KD). KD can be realized either *online* or *offline* [7]. It is called online when it uses resources generated during the execution. Otherwise, it is called offline. Both of them have pros and cons. Online KD are often more adaptive and based on unsupervised methods, which may lead to a slow convergence rate. While offline methods are often supervised, and thus require huge amounts of data to be efficient.

Most KD processes are composed of an *extraction* mechanism (K_{ext}), where something is learned, and an *injection* mechanism (K_{inj}), which uses the extracted knowledge to find new promising solutions. A study of existing works in KD and its hybridization with metaheuristics [21] leads to four main questions: *What*/*Where*/*When*/*How* is the knowledge extracted/injected?

The question *What* is problem-dependent, since each problem may have specific relevant knowledge. In the context of this article, this question is answered in Sect. 5, where the problem is presented. Questions *Where* and *When* are algorithm-dependent since the extraction and injection steps have to be integrated into the process of the algorithm. Both of these questions are not the subject of this article, and thus not discussed here at length. However, these questions are answered in Sect. 4 for the specific case of our study. The question *How* deals with overall strategies used during the KD mechanism (e.g. intensification or diversification). The answer to this question should be adapted according to the category of the problem studied (multi-objective in our case). Our contribution focuses on this question and is detailed in Sects. 3 and 4.

2.2 Knowledge Integration in Multi-objective Optimization

In the literature, KD processes have received various interests mainly in single-objective optimization contexts. Especially in routing problems [1–3,15]. However, using KD processes in multi-objective combinatorial optimization is quite new and has not been widely investigated. Among the first works on this subject, we cite the paper of Wattanapornprom et al. [26]. In order to solve a bi-objective TSP they learn probabilities of arcs belonging to good solutions by using a reward and punishment system based on the solutions visited during the execution. The authors show that their learning procedure improves the performance of NSGA-II. The survey of Bandaru [4] regroups different data mining methods that can be used in multi-objective optimization. Recently, Moradi et al. [16] and Legrand et al. [12] proposed algorithms enhanced with learning mechanisms to solve routing problems. The former presented the MODLEM algorithm which uses decision trees updated during the execution to guide the algorithm through the search space. The latter designed a MOEA/D using a KD mechanism, that extracts sequences of customers from generated solutions and injects the most frequent ones in solutions to improve them.

3 Knowledge Discovery for Multi-objective Optimization

In this section, we propose an answer to the question *How* presented in Sect. 2.1. This question focuses on the interaction of the extraction and injection components with the knowledge itself. First, we present how knowledge groups are defined in Sect. 3.1. These groups allow the storage and the use of knowledge by the extraction and injection mechanisms. Section 3.2 is more focused on the possible strategies followed by both extraction and injection when interacting with the knowledge groups.

3.1 Definition of Knowledge Groups

One issue of KD mechanisms concerns the structure used to store the extracted knowledge to be injected. In multi-objective optimization, the fitness space is not in 1-Dimension and generally, the best solutions for one objective are not the same as for the other ones. We make the assumption that solutions sharing some similarities are more likely to be in the same region of the fitness space.

We propose to divide the fitness space into $k_{\mathcal{G}}$ regions each representing a *knowledge group*. The set of knowledge groups is denoted as $\mathcal{G}$. Therefore, a knowledge group is defined by a delimited region of the fitness space. The region can be either explicit (represented by equations) or implicit (represented by sets). If a solution belongs to the region of a knowledge group, then its associated knowledge is added to that group.

The number of knowledge groups and their construction within MOEA/D is discussed in Sect. 4.4.

3.2 Intensification and Diversification Strategies

Evolutionary algorithms use intensification and diversification mechanisms to explore the search space more in-depth or more largely. We propose to transpose these mechanisms of intensification and diversification to the KD for the extraction and injection mechanisms. On the one hand, we propose an intensification strategy, where the procedure has access to a small number of groups. The objective of the procedure is to focus on the same region of the fitness space, by exploring close regions. In that case, the knowledge is not widely shared between the groups. On the other hand, with a diversification strategy, the procedure has access to a large number of groups. The objective of the procedure is to explore different regions of the fitness space, by bringing diversity to the solutions. In that case, the knowledge can travel through the groups. The definition of these strategies for the integration of the KD into MOEA/D is discussed in Sect. 4.4.

4 MOEA/D Enhanced with Knowledge Discovery

4.1 MOEA/D

MOEA/D [28] is a genetic algorithm that approximates the Pareto front by decomposing the multi-objective problem into M several scalar objective subproblems. The scalarization is obtained by weighting each of the n objectives f_k

with a weight $w_k \in [0,1]$. Thus the fitness of a solution x for the subproblem i is the following quantity: $f(x|w^i) = \sum_{k=1}^{n} w_k^i \cdot f_k(x)$.

During an iteration, MOEA/D minimizes the i-th subproblem by using the solutions of its closest neighbors. The *neighborhood*, of size m, of a weight vector w^i is defined as the set of its m closest (for the euclidean distance) weight vectors among $\{w^1,\ldots,w^M\}$. Then the neighborhood $\mathcal{N}_m(i)$ of the i-th subproblem consists of the m subproblems defined with a weight vector belonging to the neighborhood of w^i. Note that each subproblem is associated with its best solution found during the execution.

At the start of MOEA/D, M weight vectors are given, then it works as follows. Initially, a random population (of size M) is generated and evaluated. The neighborhood (of size m) of each subproblem is also computed. When optimizing subproblem i, a random pair of solutions is selected from its neighborhood. The Partially Mapped crossover (PMX) is applied with probability p_{cro}, and only one solution is randomly kept. Then a Local Search (LS), described in Sect. 5, is applied with probability p_{mut}. Indeed the mutation is frequently replaced by an LS [11] in genetic algorithms. Finally, the resulting solution is added to the set S of solutions generated during the iteration, and a few neighbors of the subproblem i are updated. When all subproblems have been seen, S is merged with the archive A. If the termination criterion is reached, the nondominated solutions of A are returned, otherwise, a new iteration is started.

4.2 Construction of the Knowledge Groups and Strategies

As explained in Sect. 3.1, we use knowledge groups to store the extracted knowledge. Since we work with MOEA/D, we use the underlying subproblems to delimit the $k_{\mathcal{G}}$ groups. Note that regions are defined implicitly. We propose to characterize each knowledge group $\mathcal{G}_k \in \mathcal{G}$ by a vector $g^k = (g_1^k,\ldots,g_n^k) \in [0,1]^n$ satisfying $g_1^k + \ldots + g_n^k = 1$. Since the weight vectors of subproblems are chosen uniformly in that hyperplane, we also choose $k_{\mathcal{G}}$ uniformly distributed vectors in the same hyperplane, so that groups are balanced. In the following, we assume that we work in a bi-dimensional case. If $k_{\mathcal{G}} = 1$, then the group is associated with all the subproblems, thus the vector characterizing the group does not matter, and we set $g^1 = (0.5, 0.5)$. In the general case, when $k_{\mathcal{G}} \geq 2$, for $k \in \{1,\ldots,k_{\mathcal{G}}\}$, we characterize $\mathcal{G}_k$ with $g^k = (\frac{k-1}{k_{\mathcal{G}}-1}, 1 - \frac{k-1}{k_{\mathcal{G}}-1})$.

The definition of the regions of the groups is linked to the strategy followed by the extraction. We consider the M subproblems and their associated weight vectors defined in MOEA/D. Given a subproblem i of weight vector w^i, we can compute the set $\mathcal{N}_{\mathcal{G}}(i) = \{d(w^i, g^k) | 1 \leq k \leq k_{\mathcal{G}}\}$, where $d(w^i, g^k)$ represents the Euclidean distance between the i-th subproblem and the group $\mathcal{G}_k$. With this set, we can know how far each group is from the i-th subproblem. We propose to associate each subproblem with its $m_{\mathcal{G}}^{ext}$ closest groups. Therefore, the region of a group is the set of subproblems that are associated with that group. The smaller the value of $m_{\mathcal{G}}^{ext}$, the more intensive the extraction. We decide to keep only the most intensive strategy ($m_{\mathcal{G}}^{ext} = 1$) for the extraction. More precisely, if

x is a solution obtained while optimizing the subproblem i, then only the closest group to i (regarding $\mathcal{N}_{\mathcal{G}}(i)$) receives the knowledge extracted from x.

Concerning the injection, we introduce similarly a parameter $m_{\mathcal{G}}^{inj}$. It represents the number of groups that can provide the knowledge to be injected. The diversity increases along with the value of $m_{\mathcal{G}}^{inj}$. For the study, we keep only the two extreme values being 1 (for intensification) and M (for diversification). More precisely, when $m_{\mathcal{G}}^{inj} = 1$, only the closest group to the subproblem can provide the knowledge, and when $m_{\mathcal{G}}^{inj} = M$, it can be any group (chosen at random).

4.3 MOEA/D with Knowledge Discovery

In this section, we combine the elements described in the former section to obtain the framework shown in Algorithm 1. If lines 4, 12, and 17 are removed, then the algorithm becomes the variant of MOEA/D described in Sect. 4.1. At line 4, the procedure `createGroups` is called to create the vector of each group, as explained in Sect. 4.2. At line 12, the injection procedure $\mathrm{K_{inj}}$ is applied to the current solution x, using either an intensification ($m_{\mathcal{G}}^{inj} = 1$) or a diversification ($m_{\mathcal{G}}^{inj} = k_{\mathcal{G}}$) strategy as explained in Sect. 4.2. At line 17, the extraction procedure $\mathrm{K_{ext}}$ is used to extract the knowledge from the set of solutions generated during the iteration. Then it updates the closest group ($m_{\mathcal{G}}^{ext} = 1$) of the subproblem being optimized as explained in Sect. 4.2. In the following section, we instantiate the Algorithm 1 with different values of $k_{\mathcal{G}}$.

4.4 Experimental Variants

In this section, we present and discuss the different values of $k_{\mathcal{G}}$ retained for the study. Since the extraction is performed in an intensive manner, only the strategies for the injection are considered.

First of all, we consider the simplest case, where there is only one group ($k_{\mathcal{G}} = 1$). In that case, the intensification is equivalent to the diversification, leading to only one variant, the so-called *Base* algorithm.

It is known that solutions in the middle of the front (i.e. solutions that have an equivalent trade-off between the objectives) are the most difficult to obtain. Therefore we need to create at least $k_{\mathcal{G}} = 3$ to obtain a relevant decomposition. In this article, we limit the investigation to the case where the groups are uniformly spread along the front. Thus, two groups are focused on a specific objective, and an *intermediate* group gathers trade-off solutions. Hence there are two variants using three groups: A_{int}^{3} (resp. A_{div}^{3}), which uses an intensification (resp. diversification) strategy for the injection. Then we can refine the process to obtain $k_{\mathcal{G}} = 5$ (uniformly spread) groups in the decomposition, leading to two other variants: A_{int}^{5} and A_{div}^{5}. Moreover, we keep the extreme case where $k_{\mathcal{G}} = M$, creating as many groups as subproblems since it has been studied in [12]. In this case, each group is dedicated to one specific aggregation. More precisely, for $k \in \{1, \dots, k_{\mathcal{G}}\}$, $g^k = w^k$. However, it may lead to a waste of resources since a lot of redundant knowledge between groups may exist. The last two variants are: A_{int}^{M} and A_{div}^{M}.

Algorithm 1: Knowledge Discovery MOEA/D Framework.

```
Input: M weight vectors w^1, ..., w^M. The number k_G of knowledge groups and
       the strategy m_G^inj (resp. m_G^ext) for K_inj (resp. K_ext).
Output: The external archive A
/* Initialisation                                                        */
1  A ← ∅; S ← ∅
2  P ← random initial population (x^i for the i-th subproblem)
3  G ← createGroups(k_G)
4  for i ∈ {1, ..., M} do
5  |  N(i) ← indexes of the m closest weight vectors to w^i
6  |_ Obj^i ← {f_j(x^i) | 1 ≤ j ≤ n}
/* Core of the algorithm                                                 */
7  while not stopping criterion satisfied do
8  |  for i ∈ {1, ..., M} do
9  |  |  (i_1, i_2) ← Select(N(i))
10 |  |  x ← PMX(x^{i_1}, x^{i_2})
11 |  |  x ← K_inj(x, G, i, m_G^inj)
12 |  |  x ← LS(x)
13 |  |  S ← S ∪ {x}
14 |  |_ updateNeighbors(P, N(i), x)
15 |  A ← updateArchive(A, S)
16 |  G ← K_ext(G, S, m_G^ext)
17 |_ S ← ∅
18 return A
```

5 Bi-Objective Vehicle Routing Problem with Time Windows (bVRPTW)

5.1 Problem Description

The bVRPTW [23] is defined on a graph $G = (V, E)$, where $V = \{0, 1, \dots, N\}$ is the set of vertices and $E = \{(i, j) \mid i, j \in V\}$ is the set of arcs. It is possible to travel from i to j, incurring a travel cost c_{ij} and a travel time t_{ij}. Vertex 0 represents the depot where a fleet of K identical vehicles with limited capacity $\mathcal{Q}$ is based. Vertices $1, \dots, N$ represent the customers to be served, each one having a demand q_i and a time window $[a_i, b_i]$ during which service must occur. Vehicles may arrive before a_i. In that case, the driver has to wait until a_i to accomplish service incurring a waiting time. Arriving later than b_i is not allowed. It is assumed that all inputs are nonnegative integers. The bVRPTW calls for the determination of at most K routes such that the traveling cost and waiting time are simultaneously minimized and the following conditions are satisfied: (a) each route starts and ends at the depot, (b) each customer is visited by exactly one route, (c) the sum of the demands of the customers in any route does not exceed $\mathcal{Q}$, (d) time windows are respected.

5.2 Related Works

The original VRPTW aims to minimize the number of vehicles and the total traveling cost. In the literature, we find many lexicographic approaches that minimize the number of vehicles first and then the traveling cost. Nowadays, all Solomon's instances [20] can be optimally solved using an exact algorithm [17], however, the computational cost grows exponentially with the size of the instances. In practice, meta-heuristic algorithms can obtain a "good enough" solution in a short time and have the capacity to solve large-scale complex problems, which is more suitable for applications. Schneider et al. [19] proposed different granular neighborhoods to improve the local search performed. More recently Zhang et al. [27] designed a new Evolutionary Scatter Search with Particle Swarm Optimization, the so-called ESS-PSO, able to reach very good results on Solomon's instances in a small amount of time. Considering the multi-objective approaches, the literature is more sparse. Qi et al. [18] proposed a memetic algorithm based on MOEA/D to solve a bi-objective VRPTW. More recently, Moradi [16] integrated a learnable evolutionary model into a Pareto evolutionary algorithm.

5.3 Local Search and Knowledge Operators

The LS performed in Algorithm 1 is the same as described in [13]. Briefly, three neighborhood operators are used: swap, relocate, and 2-opt*. Initially, we shuffle the list of operators, so that they are not always applied in the same order. Then, for a given operator, we try to insert each customer to its best location, considering the possible moves allowed by the operator. If a better location is found for the customer, the process is repeated with another customer. When no more improving moves are found for all customers, the search stops, and the next operator is picked up.

Now we define the $\mathrm{K_{inj}}$ and $\mathrm{K_{ext}}$ mechanisms related to the bVRPTW. Both mechanisms are based on the work of Arnold et al. [1]. They introduced PILS, an optimization strategy that uses frequent patterns from high-quality solutions, to explore high-order local-search neighborhoods. PILS has been hybridized with the Hybrid Genetic Search (HGS) of Vidal et al. [25] and the Guided Local Search (GLS) of Arnold and Sörensen [2] to solve the Capacitated Vehicle Routing Problem (CVRP) with good results. Given a solution x of the problem, $\mathrm{K_{ext}}$ extracts all patterns of x with a size between 2 and $size_p$, a user-defined parameter. The depot is not considered inside patterns. Patterns are sequences of consecutive customers in a route. For instance, a route $r = (0, v_1, \ldots, v_{|r|}, 0)$, contains $max(|r| - k + 1, 0)$ patterns of size k. Then for each extracted pattern, its frequency inside the groups updated is incremented. $\mathrm{K_{inj}}$ tentatively injects N_{Inj} patterns in the current solution x. Only improving patterns are kept in the solution, leading to a kind of elitism selection for patterns. To select a pattern we proceed as follows. First, the size of the pattern is randomly chosen among $\{2, \ldots, size_p\}$. It allows to not bias the selection towards smaller, more numerous, patterns. Then the pattern is randomly chosen among the $N_{Frequent}$ most frequent patterns of the same size. Here $N_{Frequent}$ is also a parameter of the algorithm. When all the N_{Inj} patterns have been selected, they are injected one by one according to the following steps. Firstly arcs incident to a node of the pattern are removed and the nodes of the pattern are connected. This step creates several pieces of routes, that are reconnected to form a feasible solution. The reconnection is optimal, in the sense that all possibilities are tested. Because of time windows, we do not consider reversed patterns in our mechanism.

6 Experimental Setup

6.1 Solomon's Benchmark

We use Solomon's instances [20], of size 100, to evaluate the performance of all the seven variants presented in Sect. 4. This set is frequently used in the literature to evaluate the performance of multi-objective algorithms [9,16,18]. The set contains 56 instances divided into three categories according to the type of generation used, either R (random), C (clustered), or RC (random-clustered). The generation R (23 instances) randomly places customers in the grid, while the generation C (17 instances) tends to create clusters of customers. The generation RC (16 instances) mixes both generations. Each category is itself divided into two classes, either $1XX$ or $2XX$, according to the width of time windows. Instances of class $1XX$ have wider time windows than instances of class $2XX$, meaning that instances $2XX$ are more constrained.

6.2 Termination Criterion and Performance Assessment

The termination criterion of all the variants is set to 720 s. It allows us to obtain accurate and robust results. The quality of the fronts is evaluated with the unary hypervolume [29] (uHV), which measures the volume of the area dominated by the solutions of the front. Indeed, true Pareto fronts of the problem are not known, thus we can not use metrics that rely on them. For each instance, the two extreme points used to normalize the objectives of the solutions, are obtained through our experiments and are automatically updated when a new point is found. To compute the uHV we use the point (1.001, 1.001) as a reference. The experiments are run on two computers "Intel(R) Xeon(R) CPU E5-2687W v4 @ 3.00 GHz", with 24 cores each. The variants have been implemented using the jMetalPy framework [5]. The code is available at https://github.com/Clegrandlixon/kdmoopy.

6.3 Tuning

Each algorithm is tuned with irace [14] to find a good setting of the parameters. To perform the tuning, we generated 96 new instances of size 100, by using the method described by Uchoa et al. [24] to mimic Solomon's instances. Each variant uses the following parameters: M, the number of subproblems considered, and m the size of the neighborhood of each subproblem. The probabilities associated with each mechanism are p_{cro} for the crossover, p_{inj} for the injection, and p_{mut} for the LS. The granularity parameter δ [22] is used to prune the neighborhood during LS. The maximal size $size_p$ of the patterns extracted, and the number N_{Inj} of patterns injected, chosen among the $N_{Frequent}$ most frequent patterns. According to a preliminary study and existing works, we set $m = 1/4 \times M$ and $N_{Frequent} = 100$. We do not consider the number of groups $k_{\mathcal{G}}$ in the tuning, because we want to highlight its influence on the algorithm. We propose a different range of values for the seven remaining parameters (cf. Table 1), to define the configuration space in irace. We granted a budget of 2000 configurations over 8 iterations to irace. Each configuration is evaluated with the uHV metric. The best configurations are presented in Table 2.

We can remark that the number of subproblems is always below 60, which makes sense since small populations are often preferred in genetic algorithms. The granularity is almost always set to 25, which is coherent with existing studies in the literature on routing problems. The maximal size of patterns alternates between 5 and 7, which is close to the value recommended in [1]. Moreover, the probability of applying the LS seems low, but the LS is the most time-consuming step of the algorithm, mainly in the beginning when solutions are not optimized. With $p_{mut} = 0.10$ the LS represents already 50% of the running time. However, it represents only 60% when $p_{mut} = 0.25$. The second most time-consuming step is the injection mechanism. When $p_{inj} = 1.00$, it represents around 25% of the total running time, but this mechanism requires a constant cost during the execution contrarily to the LS.

Table 1. Parameter's space given to irace. The space contains 77175 configurations.

Name	Range
Population size: M	(20, 40, 60, 80, 100)
Granularity: δ	(10, 25, 50, 75, 100)
Probability of crossover: p_{cro}	(0.00, 0.10, 0.25, 0.50, 0.75, 0.90, 1.00)
Probability of mutation: p_{mut}	(0.00, 0.10, 0.25, 0.50, 0.75, 0.90, 1.00)
Maximum size of pattern: $size_p$	(5, 7, 10)
Number of patterns injected: N_{Inj}	(20, 40, 60)
Probability of injection: p_{inj}	(0.00, 0.10, 0.25, 0.50, 0.75, 0.90, 1.00)

7 Experimental Protocol

In our experiments, we investigate how the number of groups and the strategy followed by the injection impact the quality of the solutions returned.

Table 2. Best elite configurations returned by irace for each variant.

Params	Base	A^3_{int}	A^3_{div}	A^5_{int}	A^5_{div}	A^M_{int}	A^M_{div}
M	60	60	40	40	20	40	20
m	15	15	10	10	5	10	5
δ	50	25	25	25	25	25	25
p_{cro}	0.50	0.50	0.90	0.50	0.90	0.50	0.75
p_{mut}	0.10	0.10	0.10	0.25	0.10	0.10	0.25
$size_p$	5	5	7	7	5	5	5
N_{Inj}	60	20	40	60	60	40	40
p_{inj}	0.75	0.75	1.00	1.00	0.90	1.00	0.90

To that aim, each variant is executed 30 times on the 56 instances of size 100 of Solomon's benchmark. For each algorithm, the k-th run of an instance is executed with the seed $10(k-1)$, to compare the algorithms with the same seeds. We recall that the termination criterion is set to 720 s for all variants.

For each category of instance (either R, RC, or C), we compute the average uHV obtained over the 30 runs. Then we rank each variant on each instance and we compute the average rank on all the categories. We perform a Friedman test on the average uHV, to know if all algorithms are equivalent, and if it is not the case, we apply a pairwise Wilcoxon test with the *Bonferroni* correction to know which algorithms are statistically better. Finally, we define a fourth category *All*, containing all the instances, and we compute similarly the average ranks of each variant in that case.

8 Experimental Results and Discussion

In previous studies [12,13], we compared different instantiations of the framework with the original MOEA/D (i.e. without using the knowledge groups). It shows that using the knowledge discovery framework is beneficial. Table 3 (resp. Table 4) shows the average rank (resp. uHV) of each variant on each category of instance. The variant A^5_{int} always leads to the best average rank (1.46) and average uHV (0.828). Moreover, this variant returns statistically better results than the other variants. Hence it is interesting to use more than one group in a multi-objective context.

Using the diversification strategy with five groups worsen a lot the returned results. Indeed, A^5_{div} ranks 5.73 on average, which is the second highest rank. Only A^3_{int} has a higher rank. The other variant A^3_{div} has also a high rank, meaning that using three groups is clearly a wrong choice in that context.

The variants A^M_{int} and A^M_{div} provide average uHV that are close in value. It is 0.767 for A^M_{int} and 0.770 for A^M_{div}. The conclusion is similar if we look at each category separately. Hence, when many groups are used, there is not a significant difference between intensification and diversification strategies for the injection.

Surprisingly, the Base variant returns good results, except on clustered instances. Hence it is not interesting to use a too-large or a too-small number of groups. The goal is to provide a "good" intermediate value. Here, the best results are obtained with five groups, but further studies should investigate the behavior of the procedure with different numbers of groups or that consider the possibility to adapt the number of groups during execution.

Table 3. Average ranks of the variants according to their average uHV over the different categories of instance. Bold results are statistically significant.

Category	Base	A^3_{int}	A^3_{div}	A^5_{int}	A^5_{div}	A^M_{int}	A^M_{div}
R	2.52	6.65	4.09	**1.26**	5.59	4.61	3.28
RC	2.16	6.94	4.72	**1.56**	5.53	3.53	3.56
C	4.09	5.29	5.50	**1.62**	6.12	2.21	3.18
All	2.89	6.32	4.70	**1.46**	5.73	3.57	3.33

Table 4. Average uHV of the variants according to their average uHV over the different categories of instance. Bold results are statistically significant.

Category	Base	A^3_{int}	A^3_{div}	A^5_{int}	A^5_{div}	A^M_{int}	A^M_{div}
R	0.730	0.627	0.703	**0.764**	0.667	0.682	0.706
RC	0.738	0.590	0.695	**0.781**	0.665	0.713	0.705
C	0.889	0.848	0.848	**0.959**	0.831	0.934	0.919
All	0.780	0.684	0.745	**0.828**	0.716	0.767	0.770

9 Conclusion

Integrating a knowledge discovery mechanism into a metaheuristic requires taking into account a lot of design aspects, summarized by the questions: *What*, *Where*, *When*, and *How* should the knowledge be extracted and injected. In this article, we mainly focused on the question *How*, while we considered existing works to answer the other questions. In particular, to answer the *How* question we have to consider how should interact the extraction and injection components of the KD mechanism, to be as efficient as possible.

As a contribution, we defined the notion of knowledge groups, studied in the literature, by giving a construction for any number of groups in a bi-objective context. Moreover, we formalized the strategies that extraction and injection can follow, and we instantiated them to obtain an intensification and a diversification strategy. We integrated our propositions into a MOEA/D framework, and we tested them on a bVRPTW. The results showed that the variant using five knowledge groups with an intensification strategy for both the injection and extraction was statistically better than the others.

In the near future, our framework will be compared to different state-of-the-art algorithms (e.g. NSGA-II, MODLEM), and different problems will also be investigated (e.g. bTSP). The tuning phase performed by irace provided similar configurations for each of the variants. Hence it will be interesting to investigate, whether with the same parameter configuration for all variants similar conclusions can be reached. Moreover, the number of groups will be considered as a parameter to be tuned in future works, to see if irace achieves similar conclusions. We would also like to investigate more deeply the impact of the strategies presented for injection and extraction. Finally, we aim to create an adaptive algorithm, which automatically adapts the number of groups and the strategies followed by the operators.

References

1. Arnold, F., Santana, Í., Sörensen, K., Vidal, T.: PILS: exploring high-order neighborhoods by pattern mining and injection. Pattern Recognit. **116**, 107957 (2021)
2. Arnold, F., Sörensen, K.: Knowledge-guided local search for the vehicle routing problem. Comput. Oper. Res. **105**, 32–46 (2019)
3. Arnold, F., Sörensen, K.: What makes a VRP solution good? The generation of problem-specific knowledge for heuristics. Comput. Oper. Res. **106**, 280–288 (2019)
4. Bandaru, S., Ng, A.H., Deb, K.: Data mining methods for knowledge discovery in multi-objective optimization: part a-survey. Expert Syst. Appl. **70**, 139–159 (2017)
5. Benitez-Hidalgo, A., Nebro, A.J., Garcia-Nieto, J., Oregi, I., Del Ser, J.: jMetalPy: a python framework for multi-objective optimization with metaheuristics. Swarm Evolut. Comput. **51**, 100598 (2019)
6. Blot, A., Kessaci, M.É., Jourdan, L.: Survey and unification of local search techniques in metaheuristics for multi-objective combinatorial optimisation. J. Heuristics **24**(6), 853–877 (2018). https://doi.org/10.1007/s10732-018-9381-1
7. Corne, D., Dhaenens, C., Jourdan, L.: Synergies between operations research and data mining: the emerging use of multi-objective approaches. Eur. J. Oper. Res. **221**(3), 469–479 (2012)
8. Deb, K., Pratap, A., Agarwal, S., Meyarivan, T.: A fast and elitist multiobjective genetic algorithm: NSGA-ii. IEEE Trans. Evolut. Comput. **6**(2), 182–197 (2002)
9. Ghoseiri, K., Ghannadpour, S.F.: Multi-objective vehicle routing problem with time windows using goal programming and genetic algorithm. Appl. Soft Comput. **10**(4), 1096–1107 (2010)
10. Glover, F., Laguna, M.: Tabu Search. In: Du, DZ., Pardalos, P.M. (eds.) Handbook of Combinatorial Optimization, pp. 2093–2229. Springer, Boston, MA (1998). https://doi.org/10.1007/978-1-4613-0303-9_33
11. Knowles, J.D.: Local-search and hybrid evolutionary algorithms for Pareto optimization. PhD thesis, University of Reading Reading (2002)
12. Legrand, C., Cattaruzza, D., Jourdan, L., Kessaci, M.-E.: Enhancing moea/d with learning: application to routing problems with time windows. In: Proceedings of the GECCO Companion (2022)
13. Legrand, C., Cattaruzza, D., Jourdan, L., Kessaci, M.-E.: New neighborhood strategies for the bi-objective vehicle routing problem with time windows. In: Proceedings of MIC 2022 (2022)

14. López-Ibáñez, M., Dubois-Lacoste, J., Cáceres, L.P., Birattari, M., Stützle, T.: The irace package: iterated racing for automatic algorithm configuration. Oper. Res. Perspect. **3**, 43–58 (2016)
15. Lucas, F., Billot, R., Sevaux, M., Sörensen, K.: Reducing space search in combinatorial optimization using machine learning tools. In: Kotsireas, I.S., Pardalos, P.M. (eds.) LION 2020. LNCS, vol. 12096, pp. 143–150. Springer, Cham (2020). https://doi.org/10.1007/978-3-030-53552-0_15
16. Moradi, B.: The new optimization algorithm for the vehicle routing problem with time windows using multi-objective discrete learnable evolution model. Soft Comput. **24**(9), 6741–6769 (2020)
17. Pecin, D., Contardo, C., Desaulniers, G., Uchoa, E.: New enhancements for the exact solution of the vehicle routing problem with time windows. INFORMS J. Comput. **29**(3), 489–502 (2017)
18. Qi, Y., Hou, Z., Li, H., Huang, J., Li, X.: A decomposition based memetic algorithm for multi-objective vehicle routing problem with time windows. Comput. Oper. Res. **62**, 61–77 (2015)
19. Schneider, M., Schwahn, F., Vigo, D.: Designing granular solution methods for routing problems with time windows. Eur. J. Oper. Res. **263**(2), 493–509 (2017)
20. Solomon, M.M.: Algorithms for the vehicle routing and scheduling problems with time window constraints. Oper. Res. **35**(2), 254–265 (1987)
21. Talbi, E.-G.: Machine learning into metaheuristics: a survey and taxonomy. ACM Comput. Surv. (CSUR) **54**(6), 1–32 (2021)
22. Toth, P., Vigo, D.: The granular tabu search and its application to the vehicle-routing problem. Inf. J. Comput. **15**(4), 333–346 (2003)
23. Toth, P., Vigo, D.: Vehicle Routing: Problems, Methods, and Applications. SIAM, New Delhi (2014)
24. Uchoa, E., Pecin, D., Pessoa, A., Poggi, M., Vidal, T., Subramanian, A.: New benchmark instances for the capacitated vehicle routing problem. Eur. J. Oper. Res. **257**(3), 845–858 (2017)
25. Vidal, T., Crainic, T.G., Gendreau, M., Prins, C.: A unified solution framework for multi-attribute vehicle routing problems. Eur. J. Oper. Res. **234**(3), 658–673 (2014)
26. Wattanapornprom, W., Olanviwitchai, P., Chutima, P., Chongstitvatana, P.: Multi-objective combinatorial optimisation with coincidence algorithm. In: 2009 IEEE Congress on Evolutionary Computation, pp. 1675–1682. IEEE (2009)
27. Zhang, J., Yang, F., Weng, X.: An evolutionary scatter search particle swarm optimization algorithm for the vehicle routing problem with time windows. IEEE Access **6**, 63468–63485 (2018)
28. Zhang, Q., Li, H.: MOEA/D: a multiobjective evolutionary algorithm based on decomposition. IEEE Trans. Evolut. Comput. **11**, 6 (2007)
29. Zitzler, E., Thiele, L., Laumanns, M., Fonseca, C.M., Da Fonseca, V.G.: Performance assessment of multiobjective optimizers: an analysis and review. IEEE Trans. Evolut. Comput. **7**(2), 117–132 (2003)

The Prism-Net Search Space Representation for Multi-objective Building Spatial Design

Ksenia Pereverdieva[1], Michael Emmerich[1(✉)], André Deutz[1], Tessa Ezendam[2], Thomas Bäck[1], and Hèrm Hofmeyer[2]

[1] Leiden University, Leiden, The Netherlands
m.t.m.emmerich@liacs.leidenuniv.nl
[2] Eindhoven University of Technology, Eindhoven, The Netherlands

Abstract. A building spatial design (BSD) determines external and internal walls and ceilings of a building. The design space has a hierarchical structure, in which decisions on the existence or non-existence of spatial components determine the existence of variables related to these spaces, such as sizing and angles. In the optimization of BSDs it is envisioned to optimize various performance indicators from multiple disciplines in concert, such as structural, functional, thermal, and daylight performance. Existing representations of design spaces suffer from severe limitations, such as only representing orthogonal designs or representing the structures in parametric superstructure, allowing only for limited design variations. This paper proposes *prism nets* - a new way of representing the search space of BSDs based on triangulations defining space filling collections of triangular prisms that can be combined via coloring parameters to spaces. Prism nets can accommodate for non-orthogonal designs and are flexible in terms of topological variations. We follow the guidelines for representation and operator design proposed in the framework of metric-based evolutionary algorithms. The main contribution of the paper is a detailed discussion of the search space representation and corresponding mutation operators. Moreover, a proof of concept example demonstrates the integration into multi-objective evolutionary algorithms and provides first results on a simple, but reproducible, benchmark problem.

Keywords: Building spatial design · Mutation operators · Geometry optimization · Non-standard representations · Multi-disciplinary design

1 Introduction

One of the advantages of evolutionary algorithms, compared to most classical optimization algorithms, is that they can accommodate complex search spaces

The authors gratefully acknowledge financial support by NWO, The Netherlands, TOP Grant 18036: Excellent Buildings: Realistic geometries for optimal structural, thermal, and lighting performance.

M. Emmerich et al. (Eds.): EMO 2023, LNCS 13970, pp. 476–489, 2023.
https://doi.org/10.1007/978-3-031-27250-9_34

with variable dimension, such as the space of mathematical expressions [19], the chemical space consisting of molecules represented by chemical graphs [20], or various types of structures in engineering design [14] or neural architectures [15]. In the following we propose a non-standard representation of a search space for multi-objective BSD optimization.

The domain of building spatial design (BSD) is concerned with finding optimal layouts for buildings, including internal and external walls, floors and ceilings. The building spatial design crucially governs the performance of a building in terms of various performance indicators (objectives), such as energy performance (which is related to the outer surface area), structural performance (strength, stiffness, and stability), and daylight performance (related to the size and positioning of windows). In a previous project, an open source building spatial design optimization toolbox (BSO toolbox [4,7]) has been developed by researchers of the Eindhoven University of Technology, The Netherlands, and of Leiden University, The Netherlands. The toolbox supports the human designer in the task of multi-criteria and multi-disciplinary building spatial design. So far it is restricted to BSDs based on orthogonal space partitioning and it features building physics (energy performance) and structural engineering disciplines (structural performance) [6]. The BSO toolbox uses a collection of quad-hexahedrons to represent a Building Spatial Design (BSD). It then includes adaptive grammars that provide a discipline related design to the BSD (e.g. a structure system with among others flat shells, loads, and boundary conditions) including the properties for discipline specific analysis. The grammars can also function via evolutionary algorithms as described in Boonstra et al. [5]. Finally, the toolbox includes a Finite Element Method (FEM) simulation-based evaluation of the structural performance of BSDs, a Resistor Capacitor (RC) network based evaluation of thermal performance, and various design modification and constraint handling techniques. Another example of approach to BSD is generating floorplan designs [13]. Main features of this approach is simplicity of usage in practice and a new model of human-computer interaction. However, our approach allows more automation, non-orthogonal shapes, and optimization based on energy and structural performance of a building. The multiobjective optimization is accomplished by Pareto optimization using state-of-the-art optimization algorithms. The main optimization algorithm is a hybrid memetic multi-objective optimization algorithm [3] that is used to optimize layout choices, discrete variables as well as continuous variables (using local hypervolume gradient-based search [3]). Optimization is further explored by hybrid approaches that combine the algorithm with design process simulations [6]. The data generated during the optimization process can be interpreted by an explainability engine, which relates regions on the Pareto front to features of the building spacial design that are expressed in terms of the decision variables. A major downside of this system was that it was limited to orthogonal spatial designs, however, progress is made in allowing non-orthogonal BSD constrained to a collection of horizontal floor and vertical walls quad-hexahedrons [11]. Our vision in this new paper is to also represent more complex geometries of buildings in the BSO toolbox, namely,

BSDs with vertical walls but non-orthogonal floorplans or angles between walls. See Fig. 1 for examples of orthogonal designs, a realization of a design by the Dutch company 'De Twee Snoeken', and a non-orthogonal BSD.

To accomplish search spaces that comprise BSDs with more complex geometries we are going to propose the new *prism-net* representation. The *prism-net* search space accommodates all multi-floor building spatial designs with vertical and straight walls and horizontal floors and ceilings, and it can be integrated into evolutionary algorithms by augmenting it with mutation operators that will also be described in this paper. Importantly, the angles of corners of spaces are not restricted to right angles, allowing for more architectural freedom in the design and potentially a further improvement of the various design objectives. Together with the new *prism-net* representation we present a hierarchical mutation operator that encodes a scalable random modification of the building and is guided by the principles of mutation operator design as stated in Rudolph [18] for integer spaces and later refined in Droste and Wiesmann for metric spaces [10]. In brief, the principles are accessibility (every point should be accessible by a finite number of mutations from any other point), symmetry (reversibility), unbiasedness (maximum entropy), and scalability (of the mutation strength). They proved to lead to excellent results in evolutionary optimization when applied to non-standard search spaces such as integer vectors [18], binary decision diagrams [10] and (variable-dimensional) mixed-integer search spaces [16]. Our representation (the prism networks) consists of three levels - topological (triangulation of levels), categorical integer (assignment of prisms to spaces), and continuous (placement of corners or nodes of the triangulation). Constraints are introduced to express the concept of space in a building spatial design, and to accommodate practical needs, such as the avoidance of sharp corners, other geometrical preferences, and the connectivity of the building to the ground. Note that *this paper focuses on building representation for optimization, and not on benchmarking*, since there is no system with similar functionality to compare the results with.

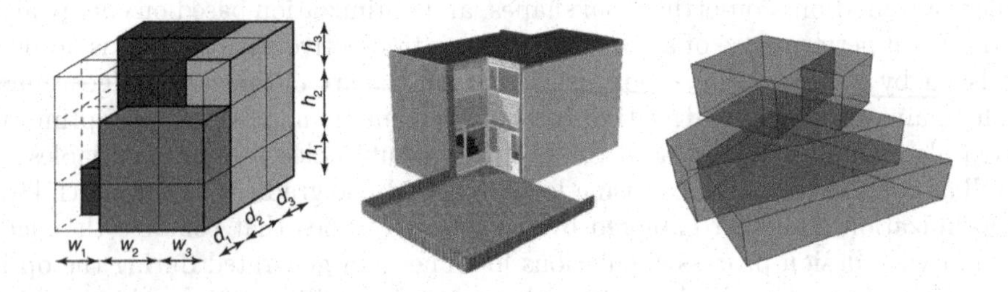

Fig. 1. Orthogonal BSDs (left) and a model of a building (middle), and BSDs based on quadrilateral floorplans with non-orthogonal elements (right).

The paper is structured as follows: In Sect. 2 we introduce the new search space representation for BSDs) and analyse its theoretical properties. In Sect. 3 we propose an hierarchical and scalable mutation operator for the BSDs that generates neighboring solutions in the search space. In Sect. 4 we show how the new representation can be integrated into existing evolutionary multi-objective optimization algorithms. Also first, reproducible, Pareto optimization results with simple to encode performance indicators are presented. We conclude this work with an outline to future research steps needed to fully integrate the BSD representation into real-world computer aided design optimization environments, such as the *BSO toolbox* (Sect. 5).

2 Search Space Representation

Firstly, let us introduce the three-dimensional Cartesian coordinate system $\mathbb{R}^3$. Each point can be described through three coordinates $\mathbf{p} = (x, y, z)$ with origin $O = (0, 0, 0)$. The plane $(x, y, 0)$, $x \in \mathbb{R}$, $y \in \mathbb{R}$ will be denoted by xOy.

The *first assumption* we make is that all ceilings and floors are parallel to the plane xOy and all walls are parallel to the z-axis. Hence we can express the building layout through a set of two-dimensional projections onto the plane xOy for each level. The number of levels L and the heights of levels are given by the variables: $(h_0, h_1, \ldots, h_L)$. Representation implies that the building can be devided into levels, but at the same time a space can be located on several levels. Note, that these are building spatial designs with flat roofs and ceilings.

The *second assumption* is that we are given a 3-D cuboid (more specifically, an axis aligned 3-D orthogonal polyhedron) V in which the building is positioned. For clarity of presentation, we might for now consider that the cuboid has sides parallel to the axes and one of the vertices coincides with the origin: $V = \{(x, y, z) : x \in [0, x_V], y \in [0, y_V], z \in [0, z_V]\}$, where x_V, y_V and z_V are the predefined maximal width, depth and height of a building correspondingly. Subsequently, the entire specified volume will be partitioned into prism shaped cells, of which some will be selected (active cells) and define the building, whereas non-selected cells partition the space not part of the building. Cells in the interior of the building can be combined to *spaces*, i.e., compartments of the BSDs the points of which are not separated by walls. Each cell is a triangular prism fully located on one of the levels. We will denote the number of cells as N_{cells} and the set of cells as $C = \{c_i\}, i \in \{1, 2, \ldots, N_{cells}\}$. Because of the first assumption we made and the fact that we require the triangular prisms to be confined to two adjacent levels, it is possible to describe each cell c_i in the following way: $c_i = [(x_{1i}, y_{1i}), (x_{2i}, y_{2i}), (x_{3i}, y_{3i}), l_i, s_i]$, where (x_{ki}, y_{ki}), – coordinates of the vertices of a triangular prism, $k \in \{1, 2, 3\}$, $l_i \in \{1, 2, \ldots, L\}$ – level on which the cell c_i is located (L denotes total number of levels), and $s_i \in \{0, 1, \ldots, N_{spaces}\}$ – integer categorical variables referred to as 'colors', defined further in the text (N_{spaces} is the maximum number of spaces to be represented). The set of prisms should form a partition of V and the sets of prisms for a given level should form a partition of that level, and all prisms should have non-zero volume, meaning that the vectors (x_{1i}, y_{1i}), (x_{2i}, y_{2i}), and (x_{3i}, y_{3i}) are not allowed to be co-linear. It is desirable to be able to create building designs having an external shape other than the outer volume V (e.g, a box), and spaces to have other shapes than triangular prisms. We introduce coloring scheme of triangular prisms to combine them into polygon-shaped spaces and hence building. Each cell c_i is associated with a non-negative integer variable $s_i \in \{0, \ldots, N_{colors}\}$. We will call the values of these integer variables *'colors'*, and they denote the space (an interior compartment that is not separated by walls) to which a cell belongs. The user-defined maximum for the number of spaces in the building is denoted by N_{colors}. If $s_i = 0$, then cell c_i is inactive, meaning that it does not belong to the building. If $s_i \in \{1, \ldots, N_{colors}\}$, then it is part of the building. If $s_i = s_j = n \neq 0$, $i \neq j$, $n \in \mathbb{N}$ then both cells c_i and c_j are parts of the same space s_n. See the example of one level in Fig. 2 (left). Here we see that only cells c_2, c_3, c_4, c_5, c_6, c_7, c_8, and c_{10} represent actual parts of the level. And cells c_2 and c_6 are combined into the space with color equal to 4 (red), cells c_5 and c_8 are combined into the space with color equal to 2 (yellow), cells c_7 and c_{10} are combined into the space with color equal to 1 (green), and cells c_3 and c_4 are combined into the space with color equal to 3 (purple). Other cells are technically present in the representation of this floor, but do not represent any part of the building.

Next we will define prism-nets as a data-structure and search space representation that is based on collections of triangular prisms of the aforementioned type and satisfy certain elementary constraints, given below. Here and further, "triangle" means the projection of triangular prisms (cells) on the xOy plane.

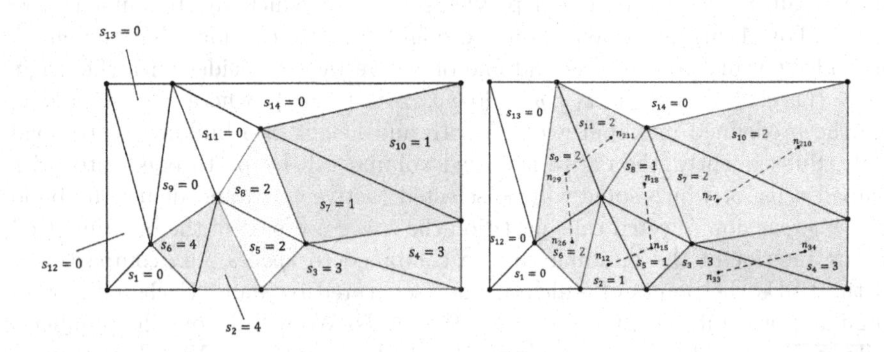

Fig. 2. Projection of one particular level to the plane xOy (left). Connectedness for the projection of one particular level to the plane xOy (right). (Color figure online)

Constraint 1: Non Overlap. Two cells on the same level should not overlap each other. It means that two triangles can intersect each other only in two cases: their intersection is a common vertex, or it is a common side of both triangles. Recall, that previously, we said that the space is partitioned by the prisms. However, this is not exactly true, because we allow the overlap to be of zero measure (that is overlap at the boundary). This way we can view prisms as closed sets.

Constraint 2: Complete Coverage. Every point in the volume V is covered by at least one cell.

Constraint 3: Connectivity of Spaces. As described earlier the color of a cell is a non-negative integer parameter which represents the space to which the cell belongs (positive values), or is zero for cells outside the building. It is necessary to prevent cases where cells with the same positive color are not connected to each other. Next, we consider separately two situations: when cells are on the same level and when they are not. To check if cells located on the same level with the same color are connected, we introduce a test based on the idea of a dual graph (see Fig. 2 (right)). The dual graph is constructed in the following way: nodes of the graph represent cells, and two nodes are connected if corresponding cells are connected (their intersection is either a face, a vertical edge, a vertex or a horizontal edge). We suggest to check if the dual graph is connected. The second condition is that the space has the same projection on every level, i.e. if we consider parts of the space belonging to the same level, or "layers" of the space, then all layers should have the same shape and location in 2-dimensional view. And moreover, a space should be located on adjacent levels.

Remark: Constraints 1–3 are intrinsic constraints, which means that they define constraints of the prism net representation. Constraint 1 and 2 guarantee that the projection of the prism net to the xOy plane is a triangulation for some set of nodes, which partitions the region V. Constraint 3 is intrinsic to the definition of spaces, making sure that spaces (regions of cells with the same color) are not separated by means of walls, floors or ceilings internally.

Definition 1. *(Properly colored) prism net: A* ***prism net*** *is a list of colored triangular prism cells* $c_i = [(x_{1i}, y_{1i}), (x_{2i}, y_{2i}), (x_{3i}, y_{3i}), l_i, s_i]$, $i = 1, \ldots, N_{cells}$ *positioned on level planes that are parallel to the ground-floor* ($z = 0$) *of a given outer cuboid* V, $l_i \in \{1, \ldots, N_{levels}\}$, *with colors* $s_i \in \{0, \ldots, N_{colors}\}$ *and contained in the cuboid* V. *The height of each triangular prism is the height difference of two consecutive level planes, of which* l_i *is the index of the lower of the two consecutive planes. In addition, in a prism-net also the 'partitioning' constraints* 1 *and* 2 *must be satisfied. – A* ***properly colored prism net (PCPN)*** *also satisfies constraint 3 (connectivity of spaces).*

Next we will define some further constraints on prism nets that turn out to be useful when implementing constraint checking or that are motivated by practical constraints for real buildings.

Constraint 4: Convex Polygon. Projections of spaces should form convex polygons. This constraint was added to keep the overall BSDs simple and to avoid costly constraint checking procedures. From a building engineering perspective, however, it is also possible to realize non-convex spaces, such as L-shaped spaces. In the current toolbox, only spaces can be handled with 4 corners. An L-shaped building can be exactly (or approximately) partitioned by triangular cells.

Constraint 5: Spaces should be Connected to the Ground. The next considered constraint is connectivity to the ground. Figure 3 illustrates different types of connection of the spaces. For our example problem we allow all types of connection except for the connection between the red space and all other spaces. To formulate this constraint more strictly we need to introduce a dual graph $G = (N, E)$. The set of nodes N corresponds to the set of spaces. Two nodes n_i, $n_j \in N$ are connected by an edge $e_{ij} \in E$ if the intersection of two corresponding spaces is not an empty set. Black lines in Fig. 3 (left) represent edges of the graph G. Constraint 5 is considered violated if there is no path from any space to at least one space on the ground level.

Remark: Note that here we allow the connection of spaces via a single point. Although this solution is feasible it is expected to perform poorly for a structural objective function.

Constraint 6: No Cavities. In this representation we would like to exclude the possibility of cavities. An example of a cavity mentioned above is illustrated in Fig. 3 (right) (the cell between "blue", "pink" and "green" spaces is the cavity). To perform such a check, it is necessary to determine the cells with a color value of zero which have a side on the border of the building, and make sure that all other cells with zero color value are connected to them. To do it we need to introduce a dual graph $G_0 = (N_0, E_0)$. The set of nodes N_0 corresponds to the set of cells with zero color value. Two nodes n_i, $n_j \in N$ are connected by an edge $e_{ij} \in E_0$ if the intersection of two corresponding cells has two vertices, i.e. the whole side (See Fig. 6). Constraint 6 is met if there is a path from every cell with zero color value to at least one cell with zero color value located on the boundary of the building. **Remark:** If constraint 4 and constraint 6 are met then on each level all cells belonging to the same space are connected and there are no cavities present in the building structure, and therefore external boundaries on each level of each space is one of a closed chain of sides.

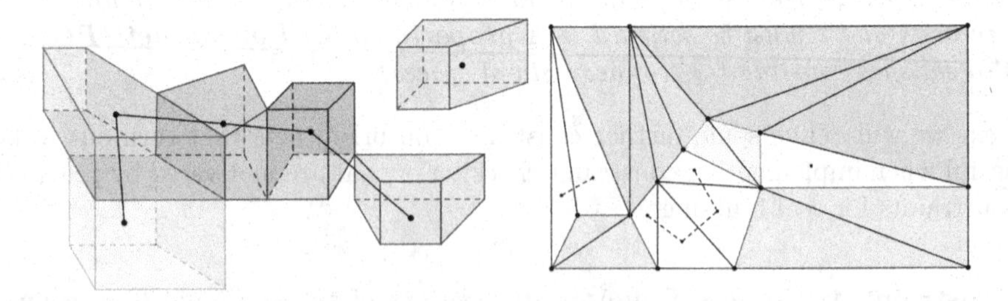

Fig. 3. Types of space connections (left). Allowed and not allowed location of cells with zero color value (right).

Constraint 7: Non Sharp Angles (optional). In addition we introduce a constraint that can be switched off by a user. This constraint avoids spaces with angles less than some threshold. A user selects the minimal allowable angle and after that we calculate each angle in each space and check it against the threshold.

3 Mutation Operators

In order to formulate search algorithms, it is essential to define operators that generate neighbors of a given design. Next we introduce *hierarchical mutation operators* that can be used in (multi-objective) evolutionary optimization to create variations of a given design in the prism-network representation. It consists of the three following main operations. (1) Topological mutation of the cell partitioning, (2) Changing the discrete s-values ('colors'), (3) Changing the continuous coordinates of the vertices.

By **topological mutation** we will refer to a building layout transformation that changes the triangulation of a convex quadrilateral space without changing the boundaries of spaces. We need to define possible operations of mutation in such a way that no constraint becomes violated. There are three topological mutations suggested: diagonal 'flip' (change of diagonal), adding a vertex, and deleting a vertex. The probability of applying each of them is determined by mutation rate. Next, we will focus on each of them in more detail. *Diagonal flip* chooses randomly one of the convex quadrilaterals formed by two triangle cells belonging to the same space and the same level. The common edge of two cells is the diagonal of mentioned quadrilateral. The mutation is a changing of this diagonal to the other diagonal of the quadrilateral as shown in Fig. 4a).

Adding a vertex splits into two cases: adding a vertex to an edge and adding a vertex to the interior of a cell. When adding a vertex occurs, first type of adding appears with probability 0.9 and the second one with probability 0.1. This ratio was picked empirically. For adding a vertex to an edge we randomly choose a side of a triangle. If two triangles have coinciding sides, we count them as one. Then we uniformly choose a point belonging to this side and add it as a vertex. And finally we split adjacent cells to avoid constraints violation. If the chosen side was on the boundary of two spaces (Fig. 4, c)) or if it was on the side inside a space (Fig. 4, e)), then we need to add two sides coming out of the selected vertex and corresponding cells. If the side was on the outer contour of the building structure (Fig. 4, d)), then we need to add only one side and corresponding cells. The second possible case is adding a vertex to the interior of a cell. We randomly choose a cell and uniformly select a point inside of it which becomes a vertex. And finally we add three sides and corresponding cells (see Fig. 4 b)).

Deleting a Vertex. Since we include the operation of adding a vertex then we also need a possibility of deleting a vertex so that the mutation operator is symmetrical. Firstly we determine the type of each vertex of the building structure. If the vertex is on the corner of the outer contour, then we cannot delete it. If the vertex is on the outer contour, but not on the corner, (Fig. 4, d) we allow

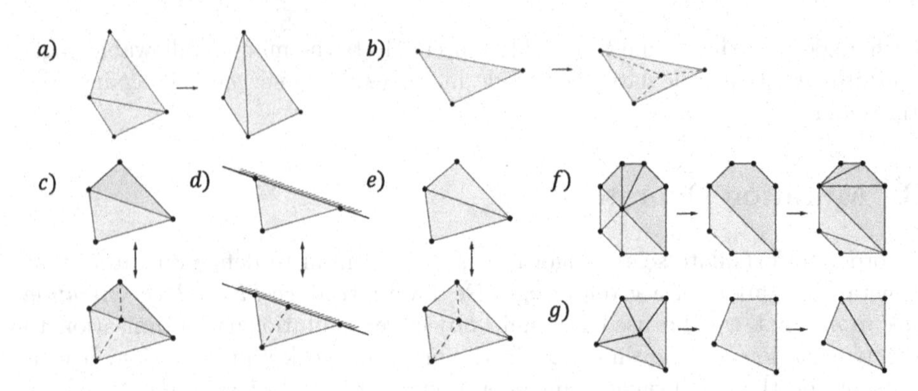

Fig. 4. Topological mutation: a) diagonal flip, b) adding a vertex to the interior of a cell, c)–g) adding a vertex to an edge/deleting a vertex.

to delete it only if it belongs exactly to one space (zero or non-zero space). If a vertex is in the building interior, then we allow to delete it only if it belongs to one (Fig. 4, e) or two (Fig. 4, c)) different spaces (zero or non-zero spaces). These conditions are justified by necessity of leaving only the unfolded angles after deleting a vertex. Secondly, we randomly select one of the vertices which are allowed to be deleted, we delete the vertex and combine the cells in a way that no constraint is violated as it is shown in Fig. 4 (c, d and e) for allowed cases. But sometimes additional triangulation might be needed. In Fig. 4 there are two cases when it is needed: f) with involvement of two spaces, g) in the interior of one space. We use standard Delaunay triangulation [8].

Next, we describe the **Discrete variable mutation** for the s-values (i.e. the 'colors'): Firstly we randomly choose the integer number from the set $\{0, \ldots, N_{spaces}\}$. If the chosen number is 0, then we randomly choose a cell with non-zero s-value and change it to 0. If the chosen number is not 0, we consider the space with s-value equal to the chosen number. From the set of cells belonging to other spaces and connected to the chosen space we pick a random number of cells and "color" them into the chosen color (s-value). After performing one of these two colorings, we check if all constraints are met. If one of them is violated, we skip this mutation. Finally, let us describe the **continuous parameter mutation**: The vertices of the triangulation can be moved to introduce topological changes on a particular level if the vertices are not on the corner of the building projection. If a randomly chosen vertex does not belong to the surface of polyhedron V, it is restricted to the space that is defined by the polygon formed by the triangles that are adjacent to the vertex. To move the vertex we randomly choose the angle from the half-open interval [0, 360), calculate the distance along the selected angle between the vertex and the polygon side, and use a truncated normal distribution with standard deviation equal to the obtained distance divided by 3 to generate the updated location for the vertex. If a randomly chosen vertex belongs to the surface of V, we identify the segment of the border on the projection to which the vertex belongs, randomly choose one of two directions, and similarly move the vertex according to the truncated normal distribution.

4 Integration to NSGA-II and SMS-EMOA

In order to test the new representation, we integrated it into two state-of-the-art evolutionary multi-objective optimization algorithms[1], the NSGA-II [9] and SMS-EMOA [1] algorithms, and performed multi-objective optimizations with two easy-to-reproduce example objective functions (see Github repository [17] for Python codes): (1) Minimize the external surface area, excluding the floor area (f_1). $f_1 := S_v + S_h \longrightarrow \min$, where S_v - surface area of all vertical external sides of the building, $S_h = \sum_{i=2}^{L}(s_i - s_{i-1})$, s_i – surface area of level i, L – number of levels. (2) Minimize the sum of deviations of space volumes from target predefined volumes. Here we specify the sizes of spaces and seek to minimize the absolute deviation from the prescribed sizes (f_2).$f_2 = \sum_{j=1}^{N_{spaces}} |V_j^a - V_j^d| \longrightarrow \min$, where V_j^a – actual volume of space j, V_j^d – predefined volume of space j.

These objective functions are motivated by resource efficient light-weight constructions. However, the ambition of the overall project is to state objectives that also include energy performance, which can for instance be measured using resistor networks, and structural performance, which can be computed using FEM simulations [6]. However, we would like to abstain in this paper from the details of simulation and are more interested in a problem that is reproducible and easy to understand for non-domain experts.

Three experiments were carried out with the described objectives with different values of the mutation rate. Each of the experiments contained 30 repeated runs. The NSGA-II and SMS-EMOA algorithm were tested. For all runs the same initial population (size: 10) was used. The initial population was set manually since randomized generation of building designs is to be done in future work. In the proposed experiment there are several invariants: the number of levels of the building is 2, the height of the building is 2 (1 for each floor), and the number of spaces is 3. The box V inside which the building is contained has dimensions $x_V = 5$, $y_V = 3$, and $z_V = 2$. Throughout the experiments, variables were limited by these values. The values of 100, 5 and 30 were chosen as the required space volumes for calculation of the second objective as V_1^d, V_2^d, and V_3^d correspondingly, and the value of 50 degrees was chosen as the minimum allowable angle of a space.

Since the values of the objectives differ significantly from each other, normalization is needed. The value of the surface area of no more than 75 was obtained experimentally, so it was decided to divide the absolute value of this objective

[1] Both algorithms feature parameterless selection in the bi-objective case. SMS-EMOA is highly competitive across a wide range of bi-objective problems [2]. NSGA-II is considered to be a commonly used algorithm for bi-objective optimization, whereas, for problems with more objectives, we would rather consider NSGA-III.

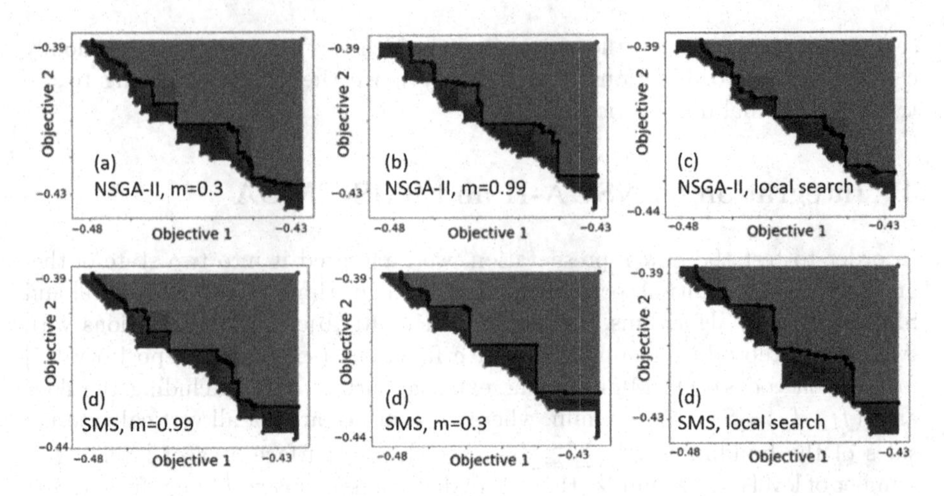

Fig. 5. Summary Pareto fronts (Empirical Attainment Levels [12]). Full red points are attained by all runs, full blue points by just one run, and for points with sliding shades of the color between blue and red are attained by 2 or more runs and less than 30 runs. The black curve marks the median attainment curve. (Color figure online)

by 150, and the value of the deviation from the specified volumes was divided by twice the sum of the required volumes. Thus, for the value of both objectives to lie in the interval $(0, 1)$ was achieved.

Each run of the NSGA-II algorithm was given a budget of 60 generations, within each the proposed mutation operator was used. The recombination operator was not used. Mutation rate in this case determined the probability of applying each of the mutations in the following sequence: topological mutation, discrete parameter mutation, continuous parameter mutation. If any of the mutations did not occur, the algorithm moved on to the next mutations in the list.

In the first of the experiments, the probability of using each of the mutations is 0.99, in the second with probability 0.3, and the third experiment can be considered as local search, in which the probability of topological and discrete mutations was 0.1, and continuous – 0.8. Firstly, points in objective space of all runs from 60 generations were combined and then sorted by means of Pareto dominance. Figure 5 (a)–(c) depicts the obtained Pareto non-dominated. We use *attainment curves* [12] in the plot (Attained by all runs (best), attained by half runs (median), attained by one run (worst)). The hyperparameter optimization is to be done in future work.

There are three examples of building designs presented in Fig. 6. Mutation rate 0.99 was set in order to obtain these designs. The knee point was chosen as the solution for which the objective 1 and objective 2 are closer to each other than for any other solutions among all 30 runs of the algorithm. Figure 6 also shows two extreme values: a design with a minimum value of objective 1 among all solutions and a design with a minimum value of objective 2. As you can see

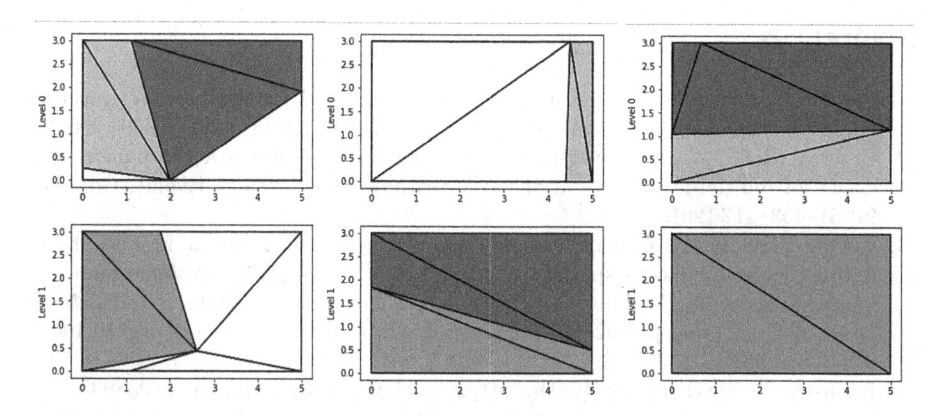

Fig. 6. Building design examples for NSGA-II with mutation rate 0.99: knee point and optimal solution in f_1 and in f_2.

from Fig. 6, optimization allows you to get extreme solutions corresponding to the objectives. Thus, for instance, to minimize the deviation of space volumes from the values of 100, 5 and 30 (for pink, blue and green spaces correspondingly), and for the maximum surface area, the algorithm outputs a design with a fully occupied volume V (Fig. 6, c)). And for minimal surface area the algorithm obtained a building design with a very small red space on level 0. None of the spaces disappeared, as the number of spaces was required to be constant. Three experiments were carried out with the described objectives with different values of the mutation rate. Each of the experiments contained 30 runs of the SMS-EMOA algorithm. Inputs used were completely the same as for NSGA-II algorithm. Obtained Pareto fronts are illustrated in Fig. 5 (d)–(f). Both algorithms produced almost linear Pareto fronts, however, Fig. 5 shows that the SMS-EMOA algorithm takes into account Objective 2 more than Objective 1 during optimization, unlike NSGA-II, where the solutions on Pareto fronts are distributed more evenly. Besides, for both algorithms Pareto fronts vary for different mutation rates. So, for the NSGA-II Pareto algorithm, the local search front is more sparse than for the mutation rate of 0.99 and 0.3, and for the SMS-EMOA algorithm, on the contrary, it is less sparse. The reasons for this behavior have yet to be understood. A promising idea is also to adapt mutation rates for the different mutation types by reinforcement learning [15].

5 Summary and Outlook

A new, non-orthogonal BSD representation was developed, equipped with a domain-specific hierarchical mutation operator, and integrated into MOEAs. The result forms an important step towards extending the design support systems (e.g. the BSO toolbox [4,7]). All data and a detailed description of algorithms are available in the GitHub repository [17].

References

1. Beume, N., Naujoks, B., Emmerich, M.: SMS-EMOA: multiobjective selection based on dominated hypervolume. Eur. J. Oper. Res. **181**(3), 1653–1669 (2007)
2. Bezerra, , L.C., Manuel López-Ibáñez, M., Stützle, T.: Automatic component-wise design of multiobjective evolutionary algorithms. IEEE Trans. Evolut. Comput. **20**(3), 403–417 (2016)
3. van der Blom, K., Boonstra, S., Wang, H., Hofmeyer, H., Emmerich, M.T.M.: Evaluating memetic building spatial design optimisation using hypervolume indicator gradient ascent. In: Trujillo, L., Schütze, O., Maldonado, Y., Valle, P. (eds.) NEO 2017. SCI, vol. 785, pp. 62–86. Springer, Cham (2019). https://doi.org/10.1007/978-3-319-96104-0_3
4. Boonstra, S.: BSO toolbox (2018). https://doi.org/10.5281/zenodo.3823893
5. Boonstra, S., van der Blom, K., Hofmeyer, H., Emmerich, M.T.: Conceptual structural system layouts via design response grammars and evolutionary algorithms. Autom. Constr. **116**, 103009 (2020)
6. Boonstra, S., van der Blom, K., Hofmeyer, H., Emmerich, M.T.: Hybridization of an evolutionary algorithm and simulations of co-evolutionary design processes for early-stage building spatial design optimization. Autom. Constr. **124**, 103522 (2021)
7. Boonstra, S., van der Blom, K., Hofmeyer, H., Emmerich, M.T., van Schijndel, J., de Wilde, P.: Toolbox for super-structured and super-structure free multi-disciplinary building spatial design optimisation. Adv. Eng. Inform. **36**, 86–100 (2018)
8. de Berg, M., Cheong, O., van Kreveld, M., Overmars, M.: Delaunay triangulations: height interpolation. In: Computational Geometry, pp. 191–218. Springer, Berlin, Heidelberg (2008). https://doi.org/10.1007/978-3-540-77974-2_9
9. Deb, K., Pratap, A., Agarwal, S., Meyarivan, T.A.M.T.: A fast and elitist multiobjective genetic algorithm: NSGA-II. IEEE Trans. Evolut. Comput. **6**(2), 182–197 (2002)
10. Droste, S., Wiesmann, D.: Metric based evolutionary algorithms. In: Poli, R., Banzhaf, W., Langdon, W.B., Miller, J., Nordin, P., Fogarty, T.C. (eds.) Genetic Programming. EuroGP 2000. LNCS, vol. 1802, pp. 29–43. Springer, Berlin, Heidelberg (2000). https://doi.org/10.1007/978-3-540-46239-2_3
11. Ezendam, T., Ezendam, T.T., Hofmeyer, H.H., Boonstra, S.S., Pauwels, P.P.: Two geometry conformal methods for the use in a multi-disciplinary non-orthogonal building spatial design optimisation framework. M.Sc.-thesis Eindhoven University of Technology, Department of the Built Environment, Structural Design Group (2021)
12. Fonseca, C.M., Guerreiro, A.P., López-Ibáñez, M., Paquete, L.: On the computation of the empirical attainment function. In: Takahashi, R.H.C., Deb, K., Wanner, E.F., Greco, S. (eds.) EMO 2011. LNCS, vol. 6576, pp. 106–120. Springer, Heidelberg (2011). https://doi.org/10.1007/978-3-642-19893-9_8
13. Grzesiak-Kopeć, K., Strug, B., Ślusarczyk, G.: Specification-driven evolution of floor plan design. In: Rudolph, G., Kononova, A.V., Aguirre, H., Kerschke, P., Ochoa, G., Tusar, T. (eds.) Parallel Problem Solving from Nature–PPSN XVII. PPSN 2022. LNCS, vol. 13399, pp. 368–381. Springer, Cham (2022). https://doi.org/10.1007/978-3-031-14721-0_26
14. Hajela, P., Lee, E., Lin, C.Y.: Genetic Algorithms in Structural Topology Optimization. In: Bendsoe, M.P., Soares, C.A.M. (eds.) Topology Design of Structures.

NATO ASI Series, vol. 227, pp. 117–133. Springer, Dordrecht (1993). https://doi.org/10.1007/978-94-011-1804-0_10
15. Igel, C., Kreutz, M.: Operator adaptation in evolutionary computation and its application to structure optimization of neural networks. Neurocomputing **55**(1–2), 347–361 (2003)
16. Li, R., et al.: Mixed integer evolution strategies for parameter optimization. Evolut. Comput. **21**(1), 29–64 (2013)
17. Pereverdieva, K.: Prism-Net Implementation (2022). https://doi.org/10.5281/zenodo.7430052
18. Rudolph, G.: An evolutionary algorithm for integer programming. In: Davidor, Y., Schwefel, H.-P., Männer, R. (eds.) PPSN 1994. LNCS, vol. 866, pp. 139–148. Springer, Heidelberg (1994). https://doi.org/10.1007/3-540-58484-6_258
19. Schmidt, M., Lipson, H.: Distilling free-form natural laws from experimental data. Science **324**(5923), 81–85 (2009)
20. van der Eelke, H., Kruisselbrink, J., Aleman, A., Emmerich, M.T., Bender, A., IJzerman, A.P.: Evolutionary design of selective adenosine receptor ligands. J. Cheminform. **2**(1), 1 (2010)

Selection Strategies for a Balanced Multi- or Many-Objective Molecular Optimization and Genetic Diversity: A Comparative Study

Susanne Rosenthal(✉)

Department of Computer Science, RFH - University of Applied Sciences, Cologne, Germany
susanne.rosenthal2@rfh-koeln.de

Abstract. The first step in drug design is the identification and optimization of lead molecules for therapeutic and diagnostic interventions. The analysis of molecular properties requires high laboratory evaluation costs. Computer-aided drug design provides effective approaches to optimize molecules with two general aims: firstly, the identification of candidate targets with several optimized physiochemical properties. Secondly, lead libraries have to be build with a broad range of compounds revealing a high genetic diversity among themselves with an at most similar behavior in bioactivity. MOEAs are nowadays established *in vitro* processes for molecular optimization problems with a continuous complexity increase. Therefore, MOEAs solving multi- and many-objective optimization problems with a suitable balance of convergence and genetic dissimilarity are challenging. For this purpose, a MOEA especially evolved for molecular optimization is enhanced by optionally two balancing survival selection strategies: a Pareto-based strategy is applied on a two-dimensional indicator problem consisting of a convergence and genetic diversity measure. The second strategy uses truncation selection based on a ranking measure referring to the convergence and genetic diversity measure. These configurations are compared to the recently proposed ad-MOEA with a specific environmental survival selection for multi- and many-objective optimization on four molecular optimization problems from 3 up to 6 objectives.

Keywords: Evolutionary algorithm · Balancing selection · Molecular optimization · Genetic diversity

1 Introduction

The aim of drug design in its first stage is the identification and optimization of lead molecules interacting with the disease-related target. Potential therapeutic leads have to provide a promising toxicity profile, have to bind - and only to bind

M. Emmerich et al. (Eds.): EMO 2023, LNCS 13970, pp. 490–503, 2023.
https://doi.org/10.1007/978-3-031-27250-9_35

- on the disease target without any adverse effects and possess a suitable 3D-structure to access 'drugability' [1]. For this purpose, computer-aided drug design is used to predict physiochemical properties of molecules. These methods are fast and cost-effective processes to identify and optimize promising therapeutic leads [2].

A single-objective Evolutionary algorithm (EA) for molecular optimization in a combined *in vitro* and *in silico* process has been evolved to reduce laboratory resources [3]. This algorithm provides exponential fitness improvement within the first 10 iterations and the convergence is slowed down to a linear improvement afterwards. This convergence behavior is further termed early convergence. Molecular optimization often require a simultaneous optimization of several physiochemical properties. For this purpose, a sophisticated version of this single-objective approach, termed COmponent-Specific Evolutionary Algorithm for Molecular Optimization (COSEA-MO), has been published and benchmarked on a 3- and 4-dimensional molecular optimization problem [4]. COSEA-MO also reveals exponential fitness improvement within the first 10 iterations decreasing afterwards. This is a mandatory feature for the application in a combined *in vitro* and *in silico* process since maximally 10 cycles of this combined process are usually performed in the laboratory. Furthermore, COSEA-MO has been enhanced for the application on many-objective molecular optimization problems by a winning-score based ranking method as survival selection providing again exponential fitness improvement within 10 iterations [5]. Further advantages of the this approach are parameter-free components. Thus, no parameter settings have to be done by the user which have a high impact on the performance. This is another important feature for the application in the laboratory.

The general aim of multi- and many-objective optimization is a suitable balance of convergence and diversity. This trade-off mainly depends on the survival selection strategy [7]. Diversity as the second main aim in an evolutionary process has to be interpreted differently in the context of molecular optimization [6]. Diversity in this application field refers to the variety of the genetic material among the candidate molecules which is further termed as Genetic Dissimilarity (GD). Therefore, the design of an enhanced version of COSEA-MO by optionally two balancing survival selection strategies referring to the aspect of early convergence and GD for multi- and many-objective molecular optimization is the main contribution of this work.

COSEA-MO optionally uses two different types of survival selection based on a convergence and GD indicator. The first strategy applies the Pareto dominance principle on a two-dimensional indicator problem instead of the common application on objective functions of the Multi- or Many-objective Optimization Problem (MOP or MaOP). This strategy directly optimizes the two generic aspects of molecular optimization problems, namely convergence and GD. A Winning Score Values (WSV) is used as convergence indicator [15] and a genetic diversity measure based on the matrix of Sneath [21] is applied to calculate the dissimilarity of the genetic material compared to a predefined reference peptid set. The second selection strategy of COSEA-MO uses a truncation strategy as directional selection concept based on a weighted summation of WSV and GD values. The COSEA-MO configurations with these two selection strategies are compared to ad-MOEA [7] with an environmental selection strategy on four molecular optimization problems. The evolution of this strategy is motivated by a trade-off between convergence and diversity for muti- and many-objective optimization and therefore has the same general aim as COSEA-MO. The environmental selection applies the Pareto dominance principle in combination with the convergence measure Sum of Normalized oBjectives (SoNB) and Crowding Distance (CD) promoting diversity.

The outline of this work is as follows: Sect. 2 gives an overview of the related and preliminary work. The proposed algorithm COSEA-MO with the two optional selection strategies as well as the recently published ad-MOEA are introduced in Sect. 3. Section 4 presents the simulation onsets and experiments. Section 5 concludes this work and gives an outlook on future work.

2 Preliminary Work

The performance of most Pareto-based MOEAs is significantly reduced with an increase of the objective number above three, commonly known as MaOP [8]. This fact is caused by an increase of non-dominated solutions. Obviously, these solutions become more and more incomparable regarding the Pareto dominance principle and the search behavior of those MOEA adapt a random search [9]. Furthermore, diversity as second aim in multi-objective optimization is less straightforward to define for MaOP. Numerous work has been done to improve MOEA for solving MaOP. These approaches can be classified as follows: algorithms with relaxing or alternative dominance relation [10,11], algorithms combining Pareto dominance and decomposition-based approaches [12] and algorithms combining the Pareto dominance principle with additional metrics, as in [7,13,14]. In the COmpressed-Objective Genetic Algorithm (COGA) [15], the increasing number of non-dominated solutions are further classified by a rank assignment based on the metric winning-score value. This value reflects the difference of the number of superior and inferior objectives between two solutions. The survival selection of COGA is applied on preference objectives consisting of the winning score values and a vicinity index which is used as density measure and reflects the level of solution clustering around a search location. The model of winning score value is used as convergence indicator in the proposed approach.

Less work has done to focus diversity in MaOPs. Diversity plays a central role in MOP and MaOP to prevent premature convergence to suboptimal solutions. In [20], a measure inspired by biodiversity has been introduced to accumulate dissimilarity in a population, especially for MaOPs. In [6], diversity has been re-interpreted in the application field of molecular optimization. Instead of the general diversity aspect, diversity of the genetic material among the candidate solutions on genotype level is focused. This works discusses different strategies to control and promote GD on various stages of an application-specific EA.

3 Proposed Approach and ad-MOEA

This section introduces the general framework of COSEA-MO and ad-MOEA. Both algorithms have the same general framework which is presented in Algorithm 1. Furthermore, COSEA-MO and ad-MOEA use the same variation operators and only differ in the components mating and survival selection.

Algorithm 1: General framework of COSEA-MO and ad-MOEA

Input: Population P_t, population size N, number of optimal solutions m, total number of generations T
Output: Next generation P_{t+1}
1: Random initialization of P_0;
2: **while** $t < T$ **do**
  $Q_t \leftarrow \mathit{MatingAndVariation}(P_t)$;
  $U_t \leftarrow P_t \cup Q_t$;
  $P_{t+1} \leftarrow \mathit{SelectionStrategy}(U_t)$;
  $t \leftarrow t + 1$;
end

The start population P_0 of size N is randomly initialized in both algorithms. The individuals represent peptides encoded as character strings symbolizing the 20 canonical amino acids. The offspring generation Q_t of size N is determined by selecting two parents of P_t for variation. COSEA-MO selects the parent individuals randomly, whereas ad-MOEA uses an adaptive mating selection strategy that is based on an assigned Weighted Rank (WK). The calculation of WK requires two sorting steps according to the Pareto rank and SoNB as well as the Pareto rank and CD. WK is the weighted summation of the two rank positions using a self-adaptive weight w. This weight is updated in each generation using the current weight, the number of non-dominated solutions, population size and objective number. Thereby, the mating strategy passes an adequate probability to appropriate solutions to participate in the offspring population. The variation operators are motivated by a suitable balance of global and local search. Deterministic dynamic variation operators are suitable operators to achieve this purpose. These operators are applied in the framework af COSEA-MO as well as ad-MOEA for a better performance comparison. A linear dynamic recombination operator and an adapted version of the deterministic dynamic mutation operator of Bäck and Schütz [22] is used to generate offspring (*MatingAndVariation*). The variation rates are adapted dynamically by predefined decreasing functions with the iteration progress: the recombination operator varies the number of recombination points by a linearly decreasing function

$$x_R(t) = \frac{l}{4} - \frac{l/4}{T} \cdot t,$$

where l is the peptide length, T the total number of the generations and t the index of the current generation. The adapted mutation operator determines the mutation probabilities via

$$p_{BS} = (a + \frac{l-2}{T-1}t)^{-1}$$

with $a = 5$. The mutation rates of the traditional operator are reduced by a higher value for a. After that, P_t and Q_t are combined to a population U_t of size $2N$. The succeeding generation P_{t+1} of size N in COSEA-MO is optionally determined by an Aspect-based Selection (AS) or Weighted-sum-based Selection (WS). ad-MOEA uses an environmental selection strategy (*SelectionStrategy*). These strategies are described in the following.

Aspect-Based Selection. The procedure of AS is described in Algorithm 2. The Pareto ranking principle of NSGA-II [16] is applied on a two-dimensional indicator-based maximization problem (line 3). Firstly, an average WSV is calculated and assigned to each individual in the population (line 1). This indicator reflects the individuals' quality. A WSV of the i-th solution is determined by summarizing the number of objectives m of the solution i that are superior to the corresponding objectives in a solution j minus the number of objectives in i that are inferior to j. The average WSV of a solution i is the sum of WSVs to each individual j in the population. This assignment ensures that solutions with high WSVs are close to the true Pareto front.

The average genetic dissimilarity (GD) of each solution to a predefined reference peptide set of size k is used as the second indicator to ensure genetic diversity in the population. GD of a peptide is calculated by averaging the dissimilarity values $D(i_j, r_j)$ of a peptide i to each peptide r in the reference set according to the dissmilarity matrix of Sneath (line 2). Here, GD of a peptide i to a peptide j is the sum of the dissimilarity values of all amino acid positions j. Since amino acids at each position of both peptides are compared, they have to be of the same length l. The values of WSV and GD are scaled to a range of 0 to 1 ensuring an equal impact on the fitness value.

The N-best individuals are selected in the succeeding generation based on the Pareto rank (line 4) and the volume dominance principle via binary tournament selection (line 5). The configuration of COSEA-MO with aspect-based selection is further termed COSEA-MO-AS.

Weighted-Sum-Based Selection. The selection procedure starts with assigning a fitness value to each individual of U_t by the following linear combination:

$$FV(i) = a \cdot WSV(i) + (1-a) \cdot GD(i) \quad (1)$$

with weight a reflecting the ratio of selection by convergence or GD indicator. The terms WSV and GD have to be maximized: peptides with an average high number of superior objectives relative to other members of the population and a high genetic diversity of the material to a predefined reference peptide set at the same time are preferred. The default value is $a = 0.5$ ensuring an equal impact of WSV and GD on the selection measure and providing a suitable balance of convergence and genetic diversity. Afterwards, the peptides in the population

Algorithm 2: Pseudo code of Aspect-based selection strategy

Input: Current population P_t with $|P_t| = 2N$, $P_{t+1} = \{\}$
Calculation of the two indicator values for each solution i:
1: $WSV(i) = \sum_{j=1}^{N} w_{ij}$ with $w_{ij} = sup_{ij} - inf_{ij}$;
2: $GD(i) = \sum_{r=1}^{k}(\frac{1}{l}\sum_{j=1}^{l} D(i_j, r_j))$;
Selection process:
3: Ranking of P_t according to (WSV, GD) into fronts F_i;
4: **while** $|P_{t+1}| + |F_i| < N$ **do**
  $P_{t+1} = P_{t+1} \cup F_i$; i++;
end
5: binary tournament selection: **while** $|P_{t+1}| < N$ **do**
  select $p_1, p_2 \in P_t \setminus \{P_{t+1}\}$:
  if $WSV(p_1) \cdot GD(p_1) < WSV(p_2) \cdot GD(p_2)$ add p_2 to P_{t+1} ;
  else add p_1 to P_{t+1};
end

are ordered according to $FV(i)$ and the best N candidates are selected for the succeeding generation. The configuration of COSEA-MO with WSV is further termed COSEA-MO-WS.

Environmental Selection. ad-MOEA uses an environmental selection strategy based on the concept of SoNB to promote converging capabilities. The selection process is mainly similar to NSGA-II [16] and performs as follows: Firstly, the Pareto dominance principle is applied on the combined parent and child population. SoNB and CD are calculated for each peptide. The first fronts F_i are included in the succeeding population as long as the population size N is not exceeded. The remaining peptides are selected from F_{i+1} based on the self-adaptive parameter w. Therefore, peptides are sorted in ascending order according to SoNB and are selected with a percentage of $w \cdot 100\% \cdot |F_{i+1}|$ based on SoNB. The remaining peptides are selected with a percentage of $(1 - w \cdot 100\%) \cdot |F_{i+1}|$ based on CD.

In the experiments, we analyze a further configuration of ad-MOEA. In this configuration we substitute CD to GD to promote genetic dissimilarity instead of the general evolutionary aspect of diversity. This configuration is further termed ad-MOEA-GD.

Table 1 gives an overview of the algorithm configurations with their abbreviations used in the experiments.

4 Experimental Setup

The performance of the configurations COSEA-MO-AS and COSEA-MO-WS are compared to ad-MOEA and the adaptive configuration ad-MOEA-GD on four differently dimensional molecular optimization problems according to the convergence behavior and average GD to a reference peptide. All experiments

Table 1. Algorithm configurations an their abbreviations

Algorithm	Selection strategy	Abbr.	Parameter
COSEA-MO	Aspect-based	COSEA-MO-AS	–
	Weighted-sum-based	COSEA-MO-WS(a)	Weight a in Eq. (1)
ad-MOEA	Environmental	ad-MOEA	–
	Environmental with GD	ad-MOEA-GD	–

are implemented in the open source jMetal library 4.5. [23] and uses the open source BioJava framework 4.2.0 [17]. Each experiment is run 30 times on each molecular optimization problem with 10 iterations and a population size of 100. The individuals are 20-mer peptides composed of the 20 canonical amino acids. Short peptides of length 20 are of specific interest because of their favorable properties as drugs.

In ad-MOEA, the initial self-adaptive weight value is chosen as 0.8. This has an impact on the WR value and more individuals are selected according to SoNB. Therefore, the aspect of convergence is stronger promoted than the aspect of diversity, represented by CD or GD respectively.

4.1 Molecular Optimization Problems

Four molecular optimization problems with 3 to 6 objective functions predicting physiochemical properties are used as experimental studies. Table 2 presents the composed physiochemical optimization problems with the used abbreviations: Needleman Wunsch Algorithm (NMW), Molecular Weight (MW), Average Hydrophilicity (Hydro), Instability Index (InstInd), Isoelectric Point (pI) and Aliphatic Index (aI). These molecular functions are provided by the BioJava library [24]. The physiochemical functions are shortly described in the following:

Table 2. Physiochemical functions of the different optimization problems

Dim.	Abbr.	Objective functions
3D	3D-MOP	NMW, MW, Hydro
4D	4D-MaOP	NMW, MW, Hydro, InstInd
5D	5D-MaOP	NMW, MW, Hydro, InstInd, pI
6D	6D-MaOP	NMW, MW, Hydro, InstInd, pI, aI

NMW is a method for the global sequence alignment of a solution to a predefined reference individual. It is used based on the common hypothesis that a high similarity between molecules refers to similar molecular properties.

MW is an important peptide property to ensure a good cell permeability. MW of a peptide sequence a of length l is calculated summarizing the mass of each amino acid (a_i) plus a water molecule:

$$MW(a) = \sum_{i=1}^{l} mass(a_i) + 17.0073(OH) + 1.0079(H),$$

where O (oxygen) and H (hydrogen) are the elements of the periodic system.

A common challenge of peptides in drug design is the solubility in aqueous solutions, especially peptides with stretches of hydrophobic amino acids. Therefore, Hydro is calculated by the hydrophilicity scale of Hopp and Woods [18] with a window size equal to the peptide length l. An average hydrophilicity value is assigned to each candidate peptide a using the scales for each amino acid a_i:

$$Hydro(a) = \frac{1}{l} \cdot (\sum_{i=1}^{l} hydro(a_i)).$$

The use of peptides as therapeutic agents is restricted by their instability. The InstInd is an indicator for this property and determined by the Dipeptide Instability Weight Values (DIWV) of each two consecutive amino acids in a peptide sequence. DIWV are provided by the GRP-Matrix [19]. These values are summarized and the final sum is normalized by the peptide length l:

$$InstInd(a) = \frac{10}{l} \sum_{i=1}^{l-1} DIWV(a_i, a_{i+1}).$$

pI of a peptide is characterized as the pH-value at which a peptide has a net charge of zero. A peptide has its lowest solubility in aqueous solutions at its pI. The pI value is calculated as follows: Firstly, the net charge for $pH = 7.0$ is determined. If this charge is positive, the pH at $7 + 3.5$ is calculated; otherwise the pH at $7 - 3.5$ is determined. This process is repeated until the modules of the charge is less or equal 0.0001.

aI of a peptide is characterized as the relative volume occupied by aliphatic side chains consisting of the amino acids alanine (Ala), valine (Val), isoleucine (Ile) and leucine (Leu). aI is regarded as a positive factor for the increase of thermostability. aI is calculated according to the formula:
$aI = X(Ala) + d \cdot X(Val) + e \cdot (X(Ile) + X(Leu))$, where $X(Ala)$, $X(Val)$, $X(Ile)$ und $X(Leu)$ are mole percent of the amino acids. The coefficients d and e are the relative volume at the valine side chain ($d = 2.9$) and Lei, Ile side chains ($e = 3.9$) to the side chain Ala.

These six objective functions comparatively act to reflect the similarity of a particular peptide and a pre-defined reference peptide:

$$f(\text{CandidatePept.}) := |f(\text{CandidatePept.}) - f(\text{ReferencePept.})|.$$

Therefore, the objective functions have to be minimized and the optimization problems are minimization problems. Furthermore, the objective values are normalized by the theoretical maximal value of each objective: $\bar{f}_k(x_i) = \frac{f_k(x_i)}{Max_k}$ for the k-objectives.

4.2 Performance Metrics

Two statistical metrics are chosen to evaluate the convergence performance and the average GD to a reference peptide set. In the experiments, this set comprises only one predefined individual. These metrics are applied on 20% approximately optimal individuals in each iteration for all algorithms. In the case of COSEA-MO-AS, the optimal individuals are selected by WSV in each generation. In COSEA-MO-WS, the optimal individuals are chosen by FV. The optimal individuals in the configurations of ad-MOEA are determined according to WR.

The Average Cuboid Volume (ACV) is used to measure the convergence behavior [25]. ACV calculates the averaged spanned space of each solution to an ideal reference point, which is usually known in real-world applications. The ACV indicator is given by

$$ACV = \frac{1}{n}\sum_{i=1}^{n}(\prod_{j=1}^{k}(x_{ij} - r_j)), \tag{2}$$

where n is the number of individuals that are evaluated, k the number of objectives and r_j the ideal point. The lower the ACV values, the better the convergence behavior since the molecular optimization problems have to be minimized. ACV as a simple statistical measure is preferred over traditional convergence metrics since it is independent of Pareto optimal solution sets which are usually unknown in real-world applications, of low computation cost, independent of the problem dimension and relative to the number of solutions allowing a comparison of differently sized candidate solution sets.

The genetic diversity performance is measured by averaging the GD values of each approximate candidate individual to the predefined reference peptide. Here, the averaged GD-values as genetic diversity indicator are not normalized.

4.3 Experimental Results

The performance results of COSEA-MO and ad-MOEA configurations are depicted in Fig. 1, 2. In the case of COSEA-MO-WS(0.7), the weight of FV is chosen as $a = 0.7$, which provided optimal performance results reffering to a suitable convergence and GD trade-off in previous experiments. Those experiments reveal that a higher value for weight a improves the convergence performance in MOP, whereas the default value is advisable for MaOPs. COSEA-MO-AS reveals the highest and therefore worst ACV-values with the highest average GD values at the same time. The other configurations are very similar in their convergence behavior, whereas COSEA-MO-WS(0.7) reveals slightly better results

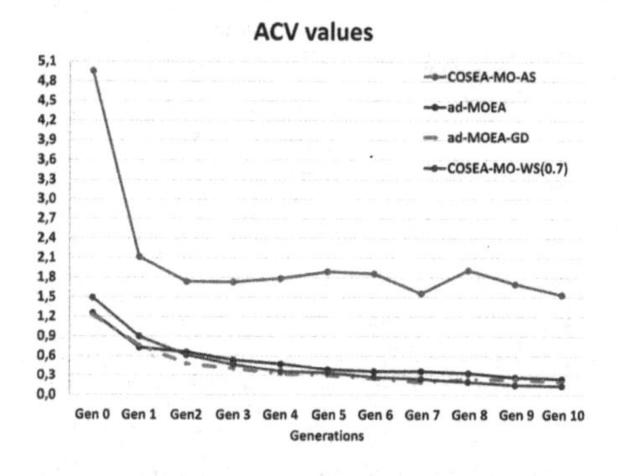

Fig. 1. 3D-MOP: ACV results

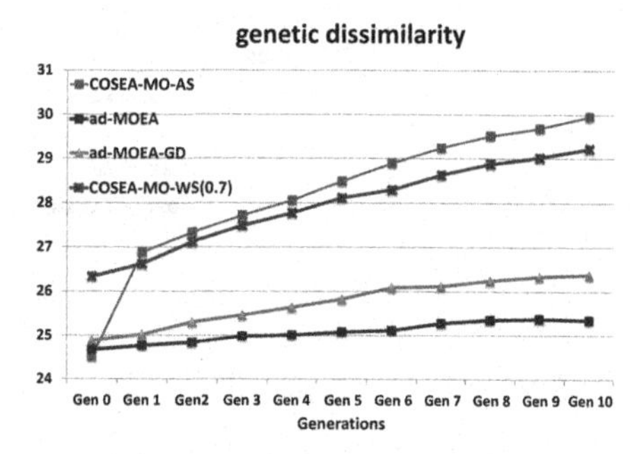

Fig. 2. 3D-MOP: average GD results

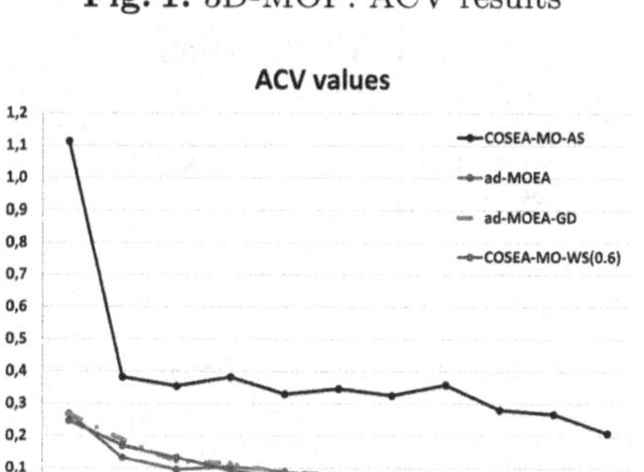

Fig. 3. 4D-MaOP: ACV results

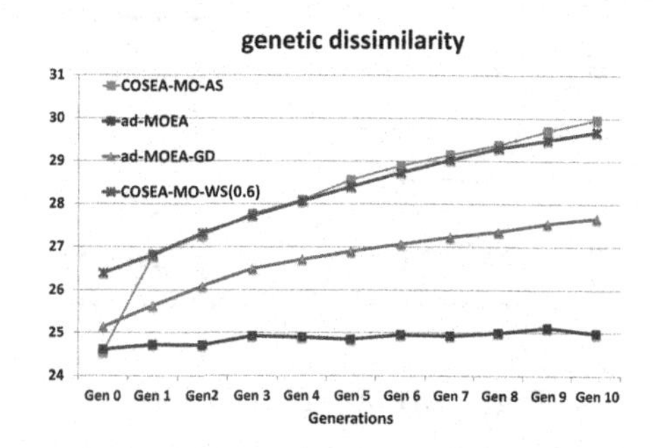

Fig. 4. 4D-MaOP: average GD results

with the second highest GD values. ad-MOEA has the worst GD values which are marginally improved for ad-MOEA-GD as a direct consequence of the substitution of CD with GD. A further performance improvement of ad-MOEA with a decrease of the self-adaptive weight w was impossible. This is a consequence of the survival selection strategy, where individuals are mainly Pareto front-based selected.

The 4D-MaOP performance results are generally comparable to 3D-MOP results (Fig. 3, 4). The optimal weight of FV is reduced to $a = 0.6$ in the case of COSEA-MO-WS(0.6). An adjusting of the GD values in the case of the COSEA-MO configurations is observable. Moreover, ad-MOEA-GD has noticeable higher GD values than ad-MOEA. In the case of ad-MOEA, it is observable that the identified Pareo front number in each generation is halved compared to the 3D-performance results with a higher individual number in each front. As a direct consequence, a higher individual number is selected by SoNB and CD or GD respectively compared to 3D-MOP.

The convergence results of the configruations in the case of the 5D-MaOP (Fig. 5, 6) as well as 6D-MaOP (Fig. 7, 8) are once again comparable to the previous ones: COSEA-MO-AS provides the worst ACV-values. The remaining

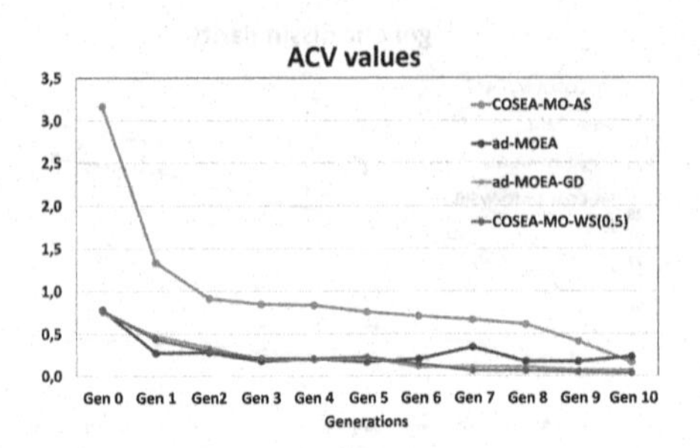

Fig. 5. 5D-MaOP: ACV results

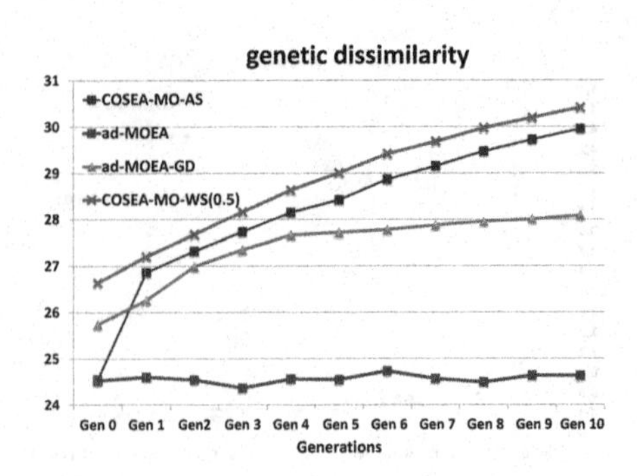

Fig. 6. 5D-MaOP: average GD results

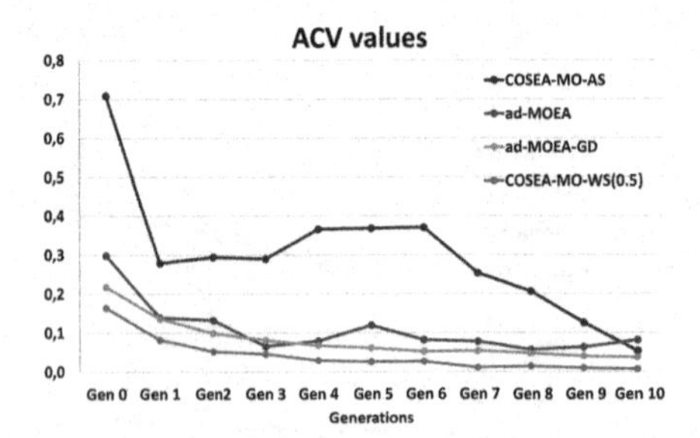

Fig. 7. 6D-MaOP: ACV results

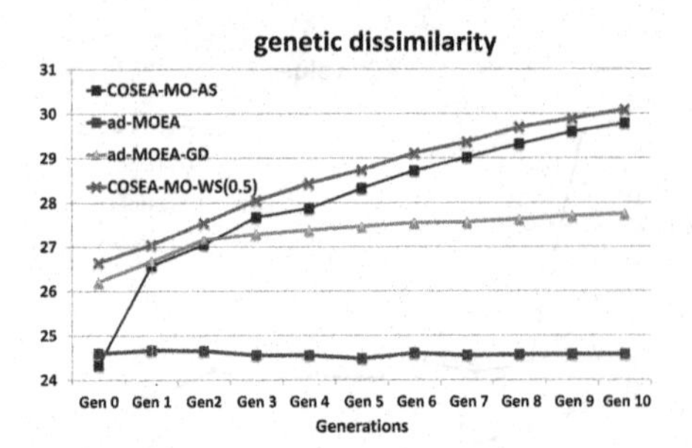

Fig. 8. 6D-MaOP: average GD results

configurations are comparable, whereas COSEA-MO-WS(0.5) provides merely better convergence results. The optimal weight of FV is default value $a = 0.5$ in COSEA-MO-WS(0.6) for both MaOPs. The ACV-values of ad-MOEA are more and more oscillating. In the case of the average GD values, COSEA-MO-WS(0.5) reveals the best results followed by COSEA-MO-AS and ad-MOEA-GD. The GD performance of ad-MOEA-GD is remarkably improved compared to the 3D-MOP and 4D-MaOP. It is recognizable that the number of Pareto fronts in the case of ad-MOEA is further decreased with an increase of the dimension number and the number of individuals in the first front. Especially in 6D-MaOP, the number of individuals in the first front exceeds the population number and therefore the individuals are mainly selected by SoNB and CD or GD respectively. Summerizing, the configuration COSEA-MO-WS reveals the best overall performance, but this optimal performance depends on the weights in FV. Promising results are also observable for ad-MOEA-GD in the case of 5D- and 6D-MaOP. This is a consequence of the fact, that the selection process is based on SoNB and GD, where the probability of selecting by SoNB and GD is given by the self-adaptive weight w. Generally, w is reduced from 0.8 to approximately 0.7 with an increase of the generation number. The configuration COSEA-MO-AS generally reveals the lowest convergence performance with the highest average GD values. Obviously, selecting individuals according to Pareto ranking based on the two indicator values WSV and GD does not result in an optimal convergence and genetic diversity balance. Thus, the weighted sum-based selection strategy or a probabilistic selection of individuals according to indicator values are more suitable strategies. However, ad-MOEA is generally of higher computation costs than COSEA-MO due to the two sortings for calculating WR.

5 Conclusion

This work presents a comparative study of three selection strategy types in EAs for molecular multi- and many-objective optimization: an aspect-based selection strategy which applies the Pareto-dominance principle on a two-dimension indicator problem reflecting the two aims convergence and diversity, a directional strategy as weighted summation of these two indicators and an environmental strategy combining Pareto-dominance principle with convergence and diversity indicators. Diversity as the second general aim in MOP and MaOP has been reinterpreted as GD in this specific application field of molecular optimization since GD among the candidate molecules is an important feature in drug design. The aspect-based and directional selection of COSEA-MO is compared to the original as well as adapted version of environmental selection in ad-MOEA. Generally, both algorithms only differ in their mating and survival selection strategies. The different configurations have been benchmarked on four multi- and many-objective molecular optimization problems. The experiments reveal an optimal balance of convergence and GD in the case of COSEA-MO with directional selection for all test cases. The configuration of ad-MOEA with GD instead of CD achieves second best performance results. ad-MOEA with CD as diversity indicator achieved the lowest GD results. These results highlight the difference of diversity in solution space in contrast to the required diversity in genotype space. COSEA-MO with aspect-based selection generally reveals lowest convergence performance, consequently a further indicator is required in this selection process for a better difference of the solutions quality. Though the results of ad-MOEA are very promising, the selection strategy is of higher computation cost due to two sorting steps of the combined children and parent population to assign rank values. Summarizing, the directional selection strategy outperforms the Pareto-based selection approaches.

For future work, the potential of improvement and generalization of the approaches aspect-based and directional selection in terms of higher dimensional MaOP as well as further application fields is focused, based on a deeper algorithm analysis of all approaches. Furthermore, these approaches will be compared to further evolutionary concepts promoting diversity, e.g. niching-based techniques. Since the optimal performance of the directional approach depends on a pre-defined weight, an improvement of this approach with a self-adaptive strategy is intended. A more theoretical aspect will be a deeper understanding of genotype diversity and amino acid dissimilarity in molecular optimization and its impact on molecular landscapes. These analytical results allow the control and improvement of the search behavior in further evolutionary strategies.

References

1. Lansdowne, L.E.: Target identification & validation in drug discovery. Technol. Netw. (2018). https://www.technologynetworks.com/drug-discovery/articles/target-identification-validation-in-drug-discovery-312290
2. Arya, H., Coumar, M.S.: Lead identification and optimization. Design and Development of Novel Drugs and Vaccines, pp. 31–63 (2021)
3. Röckendorf, N., Borschbach, M.: Molecular evolution of peptide ligands with custom-tailored characteristics. PLOS Comput. Biol. **8**(12) (2012). https://doi.org/10.1371/journal.pcbi.1002800
4. Rosenthal, S., Borschbach, M.: Design perspectives of an evolutionary process for multi-objective molecular optimization. In: Trautmann, H., et al. (eds.) EMO 2017. LNCS, vol. 10173, pp. 529–544. Springer, Cham (2017). https://doi.org/10.1007/978-3-319-54157-0_36
5. Rosenthal, S., Borschbach, M.: A winning score-based evolutionary process for multi-and many-objective peptide optimization. In: Proceedings of the 11th International Joint Conference on Computational Intelligence (IJCCI), pp. 49–58 (2019)
6. Rosenthal, S.: Diversity promoting strategies in a multi- and many-objective evolutionary algorithm for molecular optimization. In: Filipič, B., Minisci, E., Vasile, M. (eds.) BIOMA 2020. LNCS, vol. 12438, pp. 294–307. Springer, Cham (2020). https://doi.org/10.1007/978-3-030-63710-1_23
7. Palakonda, V., Mallipeddi, R.: An evolutionary algorithm for multi and many-objective optimization with adaptive mating and environmental selection. IEEE Access **8**, 82781–82796 (2020)
8. Li, H., Li, J., Tang, K., Yao, X.: Many-objective evolutionary algorithms: a survey. ACM Comput. Surv. **48**(1), 13 (2015)
9. Palakonda, V., Mallipeddi, R.: Pareto dominance-based algorithm with ranking methods for many-objective optimization. IEEE Access **5**, 11043–11053 (2017)
10. Batista, L., Campelo, F., Guimaraes, F. and Ramirez, J.: A comparison of dominance criteria in many-objective optimization problems. In: IEEE Congress of Evolutionary Computation (CEC), pp. 2359–2366 (2011)
11. Tian, Y., Cheng, R., Zhang, X., Su, Y., Jin, Y.: A strengthened dominance relation considering convergence and diversity for evolutionary many-objectvie optimization. IEEE Trans. Evol. Comput. **23**(2), 331–345 (2019)
12. Ishibuchi, H., Setoguchi, Y., Masuda, H., Nojima, Y.: Performance of decomposition-based many-objective algorithms strongly depends on Pareto front shapes. IEEE Trans. Neural Netw. **21**(2), 169–190 (2016). https://doi.org/10.1109/TEVC.2016.2587749
13. Li, M., Yang, S., Liu, X.: Shift-based density estimation for Pareto-based algorithms in many-objective optimization. IEEE Trans. Evol. Comput. **18**(3), 248–365 (2014)
14. Xiang, Y., Zhou, Y., Liu, M., Chen, Z.: A vector angle-based evolutionary algorithm for unconstrained many-objective optimization. IEEE Trans. Evol. Comput. **21**(1), 131–152 (2017)
15. Maneeratana, K., Boonlong, K., Chaiyaratana, N.: Compressed-objective genetic algorithm. In: Runarsson, T.P., Beyer, H.-G., Burke, E., Merelo-Guervós, J.J., Whitley, L.D., Yao, X. (eds.) PPSN 2006. LNCS, vol. 4193, pp. 473–482. Springer, Heidelberg (2006). https://doi.org/10.1007/11844297_48
16. Deb, K., Pratap, A., Agarwal, S., Meyarivan, T.: A fast and elitist multiobjective genetic algorithm: NSGA-II. IEEE Trans. Evol. Comput. **6**(2), 182–197 (2002)

17. Prlic, A., Yates, A., Spencer, E., et al.: BioJava: an open-source framework for bioinformatics (2018)
18. Hopp, T., Woods, K.: A computer program for predicting protein antigenic determinants. Mol. Immunol. **20**(4), 483–489 (1983)
19. Guruprasad, K., Reddy, B., Pandit, M.: Correlation between stability of a protein and its dipeptidecomposition: a novel approach for predicting in vivo stability of a protein from its primary structure. Protein Eng. **4**(2), 155–161 (1990)
20. Wang, H., Jin, Y., Yao, X.: Diversity assessment in many-objective optimization. IEEE Trans. Cybern. **47**, 1510–1522 (2017)
21. Sneath, P.: Relations between chemical structure and biological activity in peptides. J. Theor. Biol. **12**(2), 157–195 (1966)
22. Bäck, T., Schütz, M.: Intelligent mutation rate control in canonical genetic algorithms. In: Raś, Z.W., Michalewicz, M. (eds.) ISMIS 1996. LNCS, vol. 1079, pp. 158–167. Springer, Heidelberg (1996). https://doi.org/10.1007/3-540-61286-6_141
23. Nebro, A., Durillo, J.: jMetal: Metaheuristic Algorithms in Java (2019). http://jmetal.sourceforge.net/
24. BioJava: CookBook4.0. https://biojava.org/wiki/BioJava%3ACookBook4.0/
25. Rosenthal, S., Borschbach, M.: Average cuboid volume as a convergence indicator and selection criterion for multi-objective biochemical optimization. In: Emmerich, M., Deutz, A., Schütze, O., Legrand, P., Tantar, E., Tantar, A.-A. (eds.) EVOLVE – A Bridge between Probability, Set Oriented Numerics and Evolutionary Computation VII. SCI, vol. 662, pp. 185–210. Springer, Cham (2017). https://doi.org/10.1007/978-3-319-49325-1_9

A Multi-objective Evolutionary Framework for Identifying Dengue Stage-Specific Differentially Co-expressed and Functionally Enriched Gene Modules

Paramita Biswas[1] and Anirban Mukhopadhyay[2](✉)

[1] Department of Computer Science, Acharya Prafulla Chandra College, New Barrackpore, West Bengal, India
paramita@apccollege.ac.in

[2] Department of Computer Science and Engineering, University of Kalyani, Kalyani, West Bengal, India
anirban@klyuniv.ac.in

Abstract. Dengue infection threatens a significant proportion of the population of the world. The Dengue virus is an arthropod-borne single-stranded RNA virus. There are two main types of Dengue infection: Dengue Fever (DF) and Dengue Hemorrhagic Fever (DHF). Since only supportive clinical treatment is available for this endemic disease, the mortality rate due to Dengue hemorrhagic fever is relatively high. In the present work, a disease stage-specific differentially co-expressed gene module identification algorithm is developed to explore the relationship between the virus genomic particle and the host genomic substance for different stages of Dengue infection. The proposed algorithm is applied to a real-life Dengue patient expression dataset. Our algorithm uses a multi-objective framework and simultaneously optimizes topological dissimilarity and biological similarity among the genes for detecting disease stage-specific gene modules. Subsequently, we apply a disease stage-specific marker identification technique and found 16 and 13 potential markers for DF and DHF, respectively. The biological significance of the identified gene modules and stage-specific markers is also established.

Keywords: Multi-objective evolutionary optimization · Pareto optimality · Dengue virus · Differential gene co-expression · Stage-specific gene marker · Gene ontology

1 Introduction

Dengue has become a significant threat to humankind in tropical and subtropical regions. According to the World Health Organization (WHO), yearly, more than 390 million people are infected by this mosquito-borne Dengue virus (DENV) [1]. Each year, almost 500,000 cases of Dengue Hemorrhagic Fever (DHF)

M. Emmerich et al. (Eds.): EMO 2023, LNCS 13970, pp. 504–517, 2023.
https://doi.org/10.1007/978-3-031-27250-9_36

are reported worldwide with a 5% mortality rate (https://www.worldmosquitoprogram.org/en/learn/mosquito-borne-diseases/Dengue). Initially, all Dengue patients are febrile, and only a few progress to lethal infection. Early detection of DHF infection is difficult because the symptoms of DF and DHF are similar. The acute febrile disease can only be treated with supportive medical treatment, according to WHO guidelines. Dengue viruses contain one strand of RNA [6]. This RNA is referred to as positive-sense RNA because it can be translated directly into proteins. Dengue virus seizes the host's cellular mechanism for this RNA replication process. Therefore normal cellular activities of the host cell remain suspended. Patients' bodies further develop disease symptoms as a result. Antiviral drug development is necessary for healing the symptoms. This requires understanding how the virus genomic particles interact with the host cell's human genetic molecules.

A host cell's topological characteristics and functional cooperation among genes are crucial to gaining novel insights into viral pathogenesis. Therefore, biological network reconstruction for disease-related information retrieval can be posed as a multi-objective optimization problem [11]. We propose a method for detecting differentially expressed modules across disease stages. For this experiment, we used a real-life Dengue dataset with three different samples: normal human samples, DF patient samples, and DHF patient samples. We developed two differential co-expression networks using normal and DF samples. Subsequently, a multi-objective evolutionary module detection algorithm has been developed to identify genes with the highest degree of dissonance across normal and diseased states. We have customized the popular multi-objective genetic algorithm NSGA-II (Deb2002fast) to meet this need. Our proposed algorithm utilizes two important properties of a gene co-expression network as its objective functions. Identifying extreme topological dissimilarity within gene subsets across different co-expression networks is the first objective of our algorithm. This also aids in detecting how gene-gene interaction changes across different disease stages. We can also find which genes cooperate during disease progression this way. The functional similarity is another metric for measuring gene cooperation or dependencies. Hence, our second objective is to find functionally similar genes to ensure that the modules have similar functionalities. Having functionally coherent genes in a module shows that they can cooperate in disease progression. Detecting the changes in gene-gene interactions across different disease stages, and finding the functional responsibilities of interacting genes are equally important objective functions. Our algorithm optimizes both objectives simultaneously to identify disease-specific gene modules.

In the literature, there are several methods for detecting differentially expressed modules. Most of them optimize topological dissimilarity for differentially co-expressed module identification. Functional coherence between genes in modules may not be captured by topological dissimilarity alone. The proposed method identifies gene modules with high topological dissimilarity across different co-expression networks and significant functional cooperation among them. The identified modules and genetic markers were finally analyzed for biological relevance.

2 Methods

The proposed algorithm is illustrated in Fig. 1. In the figure, Control, DF, and DHF represent three phenotypic expression data. Co-expression networks related to control and DF and control and DHF are represented as network 1 and network 2, respectively. The NSGA-II-based module detection approach is applied separately to both networks. Modules are biologically validated with GO enrichment analysis and literature curation. We describe below how disease stage-specific modules are identified.

2.1 Computation of Co-expression Similarity of Genes

The co-expression analysis of genes indicates the dependency between genes for phenotypic traits. For each phenotype, a gene co-expression network was prepared. Nodes represent genes in gene co-expression networks. Each edge represents an association between two genes. Each edge is weighted according to its degree of association. To measure the strength of the association between gene pairs, Pearson correlation is used. It gives a completely connected network for each phenotype. The network is represented by an adjacency matrix. This is a similarity matrix $Sim_{n \times n}$ and each cell g_{ij} of this matrix represents the association strength between each gene pair i and j.

2.2 Computation of Differential Similarity of Genes for Different Disease States

We aim to identify disease stage-specific groups of genes in this article. Therefore, we need to build and study differential gene co-expression networks. Here we have studied how the neighbourhood associations of genes are changed in two different phenotypes. Correlation analysis's similarity matrix is hard thresholded,

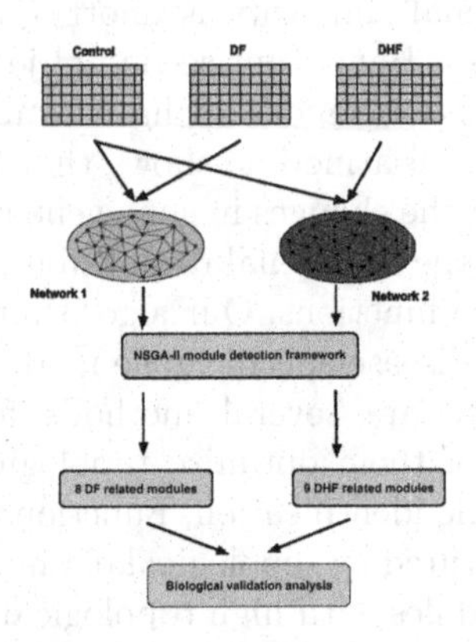

Fig. 1. Schematic representation of our Dengue stage-specific module detection and validation approach.

and gene pair co-expression values are further mapped into binary values using Eq. 1. From Fig. 2 it is clearly visible that the correlation value is equal to 0.5 giving a reasonable amount of highly correlated gene pairs. Hence we choose 0.5 as the hard threshold to binarize the data.

$$Sim(g_i, g_j) = \begin{cases} 1 & \text{if } Sim(g_i, g_j) \geq 0.5, \\ 0 & \text{if } Sim(g_i, g_j) < 0.5. \end{cases} \quad (1)$$

Jaccard Index is used here to measure the topological changes of a gene in different phenotypes. Here topological change indicates how the neighbourhood association strength of a gene changes across different phenotypic traits. The computation procedure is given in Eq. 2:

$$Diffsim_i = \frac{\left| Sim_i^{p1} \bigcap Sim_i^{p2} \right|}{\left| Sim_i^{p1} \bigcup Sim_i^{p2} \right|}, \quad (2)$$

where Sim_i^{p1} and Sim_i^{p2} represents the set of strongly associated neighbours of gene i in phenotype condition $p1$ and $p2$ respectively. If the neighbourhood association of a particular gene does not change in the mentioned phenotype conditions, it suggests that it has not actively taken part in disease progression. On the other hand, an extremely low neighbourhood association score of a gene suggests that the gene significantly contributes to the disease progression.

2.3 Computation of Semantic Similarity of Genes

The GO database provides a large set of biological and biochemical terms that describe the gene functionalities in a cell [17]. Here we have used a corpus-based semantic similarity computation approach proposed by Jiang and Conrath. Their proposed technique uses lexical taxonomies for calculating the semantic distance between the words and concepts [8]. Initially, a Directed Acyclic Graph (DAG) is created from the interrelated GO terms [15]. In the DAG, GO terms in different levels are linked by the relation "is a" or "part of" which indicate

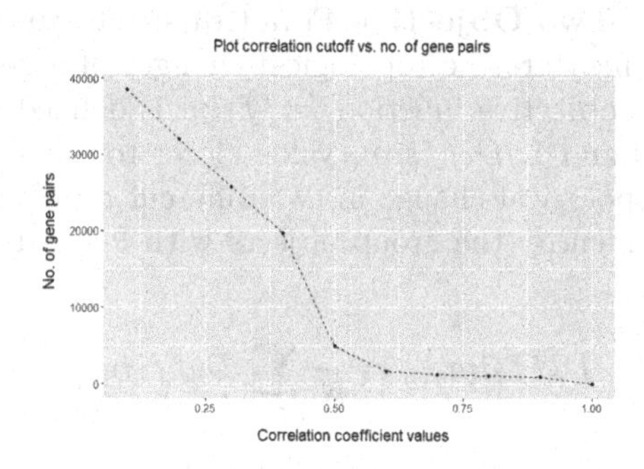

Fig. 2. Number of the gene pairs with respect to their correlation scores

a child GO term as a component or subclass of a parent GO term. The value of the information content (IC) of a GO term is obtained by estimating the probability of occurrence of this term in a large text corpus. The (IC) of a given term c is given by $IC(c) = -\ln p(c)$, where $p(c)$ is the probability of encountering the term c in the protein annotation dataset under consideration [9]. Jiang and Conrath proposed the simplified distance function to compute the dissimilarity between two genes (g_1, g_2) as given in Eq. 3.

$$Sdist(g_1, g_2) = IC(c_1) + IC(c_2) - 2 \times IC(LS(c_1, c_2)). \tag{3}$$

Genes' functional similarity depends on their GO terms' semantic distance. GO terms are retrieved from GO annotation database and organized as a Directed Acyclic Graph (DAG). In the DAG representation, p and c denote parent and child nodes respectively. GO_1 and GO_2 are the set of GO terms associated with genes $g1$ and $g2$, respectively. c_1 and c_2 are any arbitrary GO terms related to g_1 (gene1) and g_2 (gene2), respectively. $path(c_1, c_2)$ is the set of all the nodes in the shortest path from c_1 to c_2 in the DAG. $LS(c_1, c_2)$ represents the lowest subsumer of c_1 and c_2. In a taxonomy, the lowest subsumer of two concepts is the most specific common ancestor. Equation 3 shows the combined effect of the information contents only and ignores the factors related to local density, node depth, and link types. The $Sdist$ value will be low when two genes are well annotated and share high information content (related by precise GO term) [15]. This means that two genes are more functionally similar. Conversely, the larger value of $Sdist$ signifies functional dissimilarity between the genes.

2.4 Differentially Co-expressed Module Identification

The stage-specific differentially co-expressed module identification algorithm is discussed in this section. This algorithm is developed in three parts. In the first part, two objective functions are designed for finding the similarity between the genes. Next, a multi-objective optimization-based clustering technique is used to identify the optimized group of objects. Subsequently, the module hub is detected in the last part of the algorithm.

Representation of Two Objective Functions: The **first objective function** is designed to measure the topological changes of a gene in two different conditions. The first objective function $Diffsim$ is defined in Eq. 2. $Diffsim$ values lie between 0 and 1. $Diffsim_i$ value closer to 0 suggests that the gene i shows a greater topological change in two different conditions. Therefore, our first objective is to identify the group objects with lower differential similarity across two conditions.

$$Diffsim_{avg} = \frac{1}{N} \sum_{g \in M} Diffsim_g. \tag{4}$$

Equation 4 calculates the average differential similarity or topological dissimilarity of gene g within a module M. N denotes the number of genes in module M.

The **second objective function** is designed to measure the semantic similarity between the genes based on their corpus statistical information. The second objective function *Sdist* is defined in Eq. 3, and its values range between 0 and 1. Our objective is to identify genes with a higher semantic similarity score or lower *Sdist* score.

$$Sdist_{avg} = \frac{1}{N} \sum_{g_1, g_2 \in M} Sdist_{g_1, g_2}. \tag{5}$$

The above equation calculates the average semantic dissimilarity between genes g_1 and g_2 within a module M. N denotes the number of genes in module M.

Multi-objective Evolutionary Optimization: A Multi-objective optimization problem [5] aims to obtain a vector of decision variables X which satisfies the m inequality constraints $g_i(X) \geqslant 0, i = 1, 2, \ldots, m$, and p equality constraints $h_i(X) = 0, i = 1, 2, \ldots, p$, and optimizes k objective functions $f(X) = [f_1(X), f_2(X), \ldots, f_k(X)]^T$ [10]. A decision vector X^* is called a feasible solution if it satisfies all the constraints. The feasible solution space is denoted by $\mathcal{F}$. A multi-objective optimization technique is usually unable to provide a single feasible solution that optimizes all the objective functions simultaneously. Therefore Pareto-optimality and domination concepts are very useful for multi-objective optimization problems. The term domination for minimization problem can be stated as follows: let x_1 and x_2 be two solutions in $\mathcal{F}$. Then x_1 dominates x_2 if and only if $\forall i \in 1, 2, \ldots, k, f_i(x_1) \leqslant f_i(x_2)$, and $\exists i \in 1, 2, \ldots, k, f_i(x_1) < f_i(x_2)$. A solution $x^* \in \mathcal{F}$ is called Pareto-optimal if there exists no solution $x \in \mathcal{F}$ that dominates x^* [10]. It is impossible to improve a Pareto-optimal solution for one objective without affecting the other objectives [10]. The set of Pareto-optimal solutions is a non-dominated set.

NSGA-II-Based Module Identification: This work aims to identify dengue stage-specific gene modules. Therefore the construction of the objective functions is the most important part of this article. Here one objective function detects the topological differences of the gene modules across two different disease stages. Another objective function finds the functional similarities among the gene pairs within the modules. The second objective function inspects the functional coordination among genes during disease progression. Here we use Non-dominated Sorting Genetic Algorithm-II (NSGA-II) to set up a multi-objective framework to optimize the objective functions simultaneously. In our approach, each solution (chromosome) of the population is encoded as a binary string of length equal to the number of genes in the dataset. In a chromosome, a bit '1' indicates that the corresponding gene is selected, whereas a bit '0' indicates that the corresponding gene is not selected. Thus each chromosome represents a set of selected genes. The initial population set is populated with random binary strings. A set of non-dominated solutions is produced after a set of fitness computations, selections, crossovers, and mutations are carried out for a given number of generations. Each detected module's average differential dissimilarity score and average semantic similarity score are listed in Table 1.

Table 1. The average differential dissimilarity (Diff_Dissim) score and the average semantic similarity (Sem_Sim) score of each identified module in two datasets are listed below.

Module No. for Dataset 1	Average Diff_Dissim Score	Average Sem_Sim Score
1	0.933	0.866
2	0.938	0.854
3	0.939	0.855
4	0.949	0.877
5	0.946	0.847
6	0.964	0.862
7	0.945	0.851
8	0.936	0.849

Module No. for Dataset 2	Average Diff_Dissim Score	Average Sem_Sim Score
1	0.907	0.812
2	0.902	0.808
3	0.907	0.808
4	0.899	0.875
5	0.904	0.886
6	0.904	0.886
7	0.910	0.886
8	0.900	0.889
9	0.907	0.861

2.5 Most Promising Gene Identification

We obtained 8 and 9 stage-specific modules in dataset1 and dataset2, respectively. After a careful inspection of each module, we have found that 16 and 19 genes are present in all of the stage-specific modules of dataset1 and dataset2, respectively. These 16 and 19 genes share 6 genes. On inspection, it appears that the selected promising genes show high differential dissimilarity between normal and disease samples. The 16 genes from dataset1 have shown high differential dissimilarity scores in normal versus DF samples, but their differential dissimilarity scores are relatively lower in normal versus DHF samples. Similarly, 13 genes from dataset2 have shown higher differential dissimilarity scores in normal versus DHF samples than in normal vs DF samples. These genes appear to play a significant role in stage-specific disease progression. The differential dissimilarity scores of DF and DHF markers are listed in Table 2.

2.6 Statistical Significance of the Identified Modules

We have performed statistical tests to establish that the identified disease stage-specific gene modules and the disease stage related gene markers are not random; they significantly contribute to Dengue fever progression. We prepared simulated modules containing the same number of genes randomly selected from dataset1 and dataset2. We repeat the same process 100 times. We observe that randomly selected modules either showed high differential dissimilarity or high semantic similarity but failed to achieve both at a time. The simulated modules with their differential dissimilarity and semantic similarity scores are given in a supplementary file (https://sites.google.com/apccollege.ac.in/paramita/dengue-modules). A t-test was performed between the modules detected by our algorithm and randomly constructed modules. The t-test outcomes regarding both datasets

Table 2. DF and DHF related markers and their differential dissimilarity sore according to our experiment.

Sl. No.	DF Related Markers	Differential Dissimilarity Score
1.	**USP9X**	0.978
2.	**TSG101**	0.981
3.	**PKD1**	0.979
4.	**OXSR1**	0.972
5.	**GOLGA2**	0.970
6.	**FGA**	0.957
7.	**RNF125**	0.982
8.	**HECW1**	1.000
9.	**FN1**	0.981
10	**TCF7L2**	0.932
11.	**ZBTB17**	0.950
12.	**ZNF135**	1.000
13.	**ZNF365**	0.986
14.	**EVI1**	0.923
15.	**C19orf21**	0.933
16.	**NKAPL**	0.982

Sl. No.	DHF Related Markers	Differential Dissimilarity Score
1.	**TOX4**	0.925
2.	**TAF15**	0.872
3.	**TRAF4**	0.858
4.	**KAT5**	0.903
5.	**BAAT**	1.000
6.	**PAIP1**	0.905
7.	**UBE2I**	0.912
8.	**CHD3**	0.857
9.	**CAMK2B**	0.968
10.	**SMU1**	0.960
11.	ZBTB17	0.885
12.	ZNF135	0.927
13.	ZNF365	0.920
14.	EVI1	0.896
15.	C19orf21	0.900
16.	NKAPL	0.901
17.	**ARID2**	0.927
18.	**OXNAD1**	0.834
19.	**UBXD4**	0.973

are summarized in Table 3. According to the very small p-values generated by t-tests, it is clear that our algorithm detects statistically significant gene modules. We have used the Jennrich test to investigate whether our identified disease stage-specific gene markers are truly differentially co-expressed or not. This test compared two correlation matrices of DF-related markers corresponding to normal and DF stages. A similar process is followed for DHF-related markers. We have observed that the Jennrich test returned p-values 1.408×10^{-40} and 1.001×10^{-17} for DF-related markers and DHF-related markers, respectively. These low p-values are statistical evidence that identified disease stage-specific gene markers are truly differentially co-expressed.

Table 3. The t-test result for each differentially co-expressed module detected from dataset1 and dataset2 corresponds to their objective function values.

Module No. for dataset1	*p*-value for Objective1	*p*-value for Objective2
1	3.27e−13	1.03e−7
2	1.77e−14	9.55e−6
3	1.12e−16	7.77e−7
4	6.57e−14	3.35e−5
5	1.34e−14	4.02e−6
6	1.91e−13	2.62e−4
7	5.27e−13	3.61e−6
8	1.01e−15	9.97e−7

Module No. for dataset2	*p*-value for Objective1	*p*-value for Objective2
1	9.76e−12	5.86e−5
2	4.20e−10	2.81e−5
3	7.10e−10	6.97e−4
4	4.84e−10	3.29e−8
5	1.39e−8	1.69e−6
6	3.55e−9	9.88e−7
7	1.55e−9	6.09e−7
8	4.32e−8	1.48e−6
9	2.72e−11	4.25e−6

2.7 Comparison with Existing Methods

This method is compared with other popular differentially co-expressed module identification methods (CoXpress, DiffCoEx, DiffCoMO). In the study by Watson et al. [18] co-expression patterns from different phenotypic traits are detected first with an unsupervised network biology approach. CoXpress method applied a re-sampling approach to detect differentially co-expressed modules in one sample group but not in another. In DiffCoEx [16] topological dissimilarity between the co-expression networks is examined to obtain the differentially co-expressed modules. To develop a co-expressed network for each disease condition, DiffCoEx utilizes WGCNA (Weighted Gene Coexpression Network Analysis) framework. DiffCoMO identifies gene modules with higher module-wise distances between two co-expression networks and greater intra-module gene membership values over two infection stages [13]. These algorithms mostly focused on identifying the group of genes based on their differential signature across two disease conditions. In contrast, our algorithm not only detects the differentially co-expressed modules but also detects the modules having extreme functional similarities. For a fair comparison, we have chosen the two most important parameters: topological dissimilarity and semantic similarity of the gene modules.

In Fig. 3(a) and 3(b), the X-axis represents the topological dissimilarity, and the Y-axis represents the distribution of the topological dissimilarity within the detected modules. We can see from the figures above that our method can successfully detect the most dissimilar modules for parameter 1. Higher semantic dissimilarity within the gene modules indicates extreme biological similarity between the genes within the modules. Therefore, the second parameter is critical for detecting differentially co-expressed gene modules. In Fig. 3(c) and 3(d), the X-axis represents the semantic similarity, and the Y-axis represents the distribution of the semantic similarity within the detected modules. From Fig. 3(c), it appears that DiffCoMo and our method both successfully detect semantically similar gene modules from dataset1. However, in Fig. 3(d), our method more effectively identifies semantically similar gene modules from dataset2. It is obvious from these four figures that our proposed method detects differentially co-expressed gene modules across both disease conditions and performs better for both parameters than existing algorithms.

3 Biological Significance Study of the Identified Modules

Differentially co-expressed gene modules that the proposed algorithm has detected are studied further using DAVID database to reveal their biological significance. The significant GO terms related to differentially co-expressed DF and DHF gene modules are identified from dataset1 and dataset2 respectively, are listed in Table 4. We can observe that the DF gene modules 1, 2, 3, 4, 5, 6 & 7 and the DHF gene modules 4, 5, 6, 8 & 9 in Table 4 are related to several

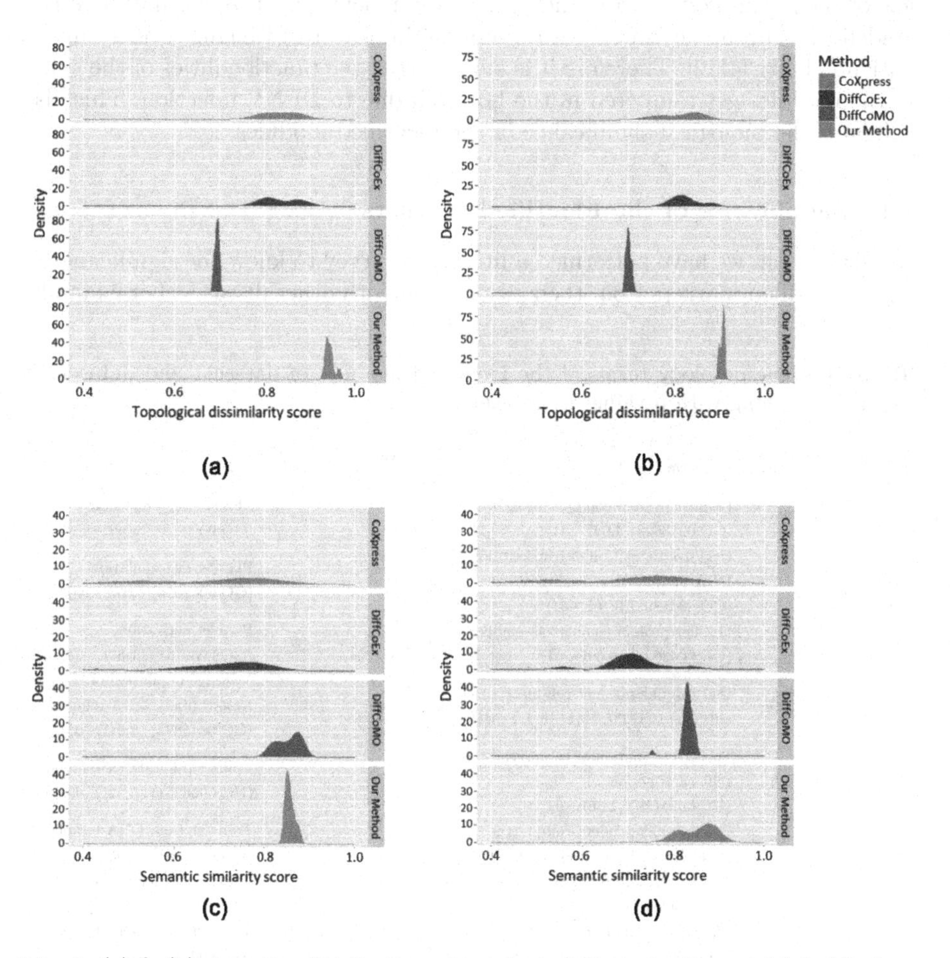

Fig. 3. (a) & (b) show the distribution of topological dissimilarity and (c) & (d) show the distribution of semantic similarity for CoXpress, DiffCoEx, DiffCoMO and our proposed algorithm for dataset1 and dataset2 respectively.

biological processes like Cellular protein complex assembly, Transcription, Regulation of transcription and Regulation of cell-cell adhesion. This observation suggests that the identified modules are actively taking part in the RNA synthesis process. Researchers have already found that DENV enters the host cell and replicates itself by using its cellular components. DENV tries to synthesize its viral RNA in the host cell. Therefore DENV targets the host factors related to the RNA synthesis process. This discussion biologically validates that identified gene modules are the potential targets of DENV in the host cell. In Table 4 the DF gene module 8 and the DHF gene modules 1, 2, 3 & 7 are related by the GO terms like Platelet activation and Platelet degranulation. Platelets or Thrombocytes are tiny blood cells. If blood vessels get damaged due to some injury, thrombocytes initiate the hemostasis process, preventing bleeding and fixing the damage. In a human body, if platelets are destroyed, or bone marrow makes very few of them, this condition is known as Thrombocytopenia. According to WHO guidelines, Thrombocytopenia is a potential indicator of Dengue severity in the patient's body [4,12]. Therefore it is evident that the functionalities of the genes in the modules get tampered in the host cell due to DENV infection. This also validates the biological significance of the identified modules.

3.1 Significance of the Identified Markers

In this section, we have performed a literature survey to know the significance of these identified markers from their association with other diseases. For example,

Table 4. Gene ontology terms of the identified modules of dataset1 and dataset2 is listed below along with p-values in brackets.

DF Module No.	# Genes	GO terms (BP)
1	29	Cellular protein complex assembly (GO:0043623)(4.04E−04)
2	37	Negative regulation of transcription, DNA-templated (GO:0045892)(6.10E−04)
3	37	Cellular protein complex assembly (GO:0043623) (6.69E−04)
4	19	Calcium-independent cell-matrix adhesion (GO:0007161) (8.91E−04)
5	28	Positive regulation of heterotypic cell-cell adhesion (GO:0034116) (1.36E−04)
6	18	Calcium-independent cell-matrix adhesion (GO:0007161) (4.14E−03)
7	34	Cellular protein complex assembly (GO:0043623)(5.62E−04)
8	36	Platelet degranulation (GO:0002576) (2.21E−06)

DHF Module No.	# Genes	GO terms (BP)
1	43	Platelet degranulation (GO:0002576)(8.39E−05)
2	39	Platelet degranulation (GO:0002576)(5.37E−05)
3	39	Platelet activation (GO:0030168) (1.60E−03)
4	30	Negative regulation of transcription, DNA-templated (GO:0045892) (8.91E−04)
5	23	Transcription, DNA-templated (GO:0006351) (5.62E−03)
6	23	Transcription, DNA-templated (GO:0006351) (4.05E−03)
7	31	Platelet degranulation (GO:0002576)(6.57E−04)
8	23	Transcription, DNA-templated (GO:0006351) (1.95E−04)
9	33	Transcription, DNA-templated (GO:0006351) (4.39E−03)

Table 5. Different types of disease associated with DF and DHF markers.

DF Markers	Viral Disease Association	Other Disease Association
USP9X	DENV, ZIKV, HPV, HIV-I	Breast Cancer, Prostate Cancer
TSG101	DENV, JEV, HIV-1	Breast Cancer, Uterine CC
PKD1	DENV, HCV , INFV	ADPKD
OXSR1	DENV, ANDV	OSCC, PD
ZBTB17	DENV, HPV-16 , ALKV	HCC, Hepatoblastoma
GOLGA2	DENV, HCV, INFV, HIV-I	AD, Prostate Cancer
FGA	DENV, HIV, INFV, HCV , ZIKV	GC, CRC
ZNF135	DENV, ALKV, HPV16	HNC, CC
ZNF365	DENV, EBV, HPV16 , HSV	Breast Cancer
RNF125	DENV, JEV, HIV	GBC, RCC
EVI1	DENV, HBV, HCV, Malaria	HCC, Cholangiocarcinoma, Ovarian and Lung cancer
HECW1	DENV, HIV-I	AD
FN1	DENV, HIV-I, HCV	RCC
TCF7L2	DENV, HIV-I, HCV, HBV, SARS-CoV-2	Type-2 Diabetes, CRC
C19orf21	DENV, ALKV, ZIKV	
NKAPL	DENV, CHB	HCC, Schizophrenia

DHF Markers	Virus Disease Association	Other Disease Association
TOX4	DENV, HIV-1	Breast Cancer
TAF15	DENV, ALKV, HIV-1, HCV	HCC, ALS, Frontotemporal dementia
TRAF4	DENV, ZIKV, CHIKV	Breast Cancer, SCLC, Ovarian Cancer, CRC, Prostate Cancer.
KAT5	DENV, HIV-1, HBV, HCV, CHIKV	HCC
BAAT	DENV, WNV, HCV	HCC
PAIP1	DENV, INFV	Breast Cancer, CC, Prostate Cancer
UBE2I	DENV, HIV, INFV, HCV	AD, Ovarian Cancer, Prostate Cancer
CHD3	DENV, HIV-1, INFV	
CAMK2B	DENV, INFV	Breast Cancer
SMU1	DENV, INFV, HIV-1	Adenocarcinoma
ARID2	DENV, ALKV, HBV, HCV	HCC, Type-2 Diabetes
OXNAD1	DENV, HPV16	OSCC, Skin Cancer
UBXN2A	DENV, ZIKV, WNV	

gene USP9X has shown more than 97% differential dissimilarity in its expression across normal vs DF samples. This gene is also present in each disease stage-specific module of dataset1. Therefore we have studied other works of literature that already have established the fact that the proteins related to these genes are significantly affected by the proteins of several viruses like ZIKV, HPV, and HIV-I [2,3,19]. Similarly, gene TAF15 has shown 87% differential dissimilarity across normal vs DHF samples in our experiment, and it is present in all the 9 disease stage-specific modules of dataset2. During the literature survey we have observed that protein products of TAF15 not only interact with DENV protein NS5 [6], but also take a crucial role in the progression of other viral diseases (e.g., ALKV, HIV-1, HCV) in the human body [7,14]. Literature survey findings of different types of disease associations with our identified disease stage-specific (DF and DHF) markers are summarized in Table 5. The evidential literature regarding each gene-disease association is given in a supplementary file (https://sites.google.com/apccollege.ac.in/paramita/dengue-modules). This significance study has shown that the most promising predicted marker genes are relevant to DENV infected disease progression and also related to other viral diseases. This confirms the importance and disease associations of the identified marker genes.

4 Conclusion

In this article, a multi-objective optimization-based disease stage-specific module detection algorithm has been proposed. This algorithm has been studied on a Dengue dataset with three samples (normal, DF and DHF). Our algorithm utilizes an NSGA-II-based multi-objective framework to detect topologically dissimilar but functionally similar genes modules.

The performance of our module detection algorithm has been compared with that of three popular module detection algorithms such as CoXpress, DiffCoEx and DiffCoMO. Our algorithm outperforms other methods in identifying biologically relevant modules and gene markers. The identified modules and the gene markers have been validated through gene ontology enrichment analysis. Moreover, we have performed a literature survey to reveal the disease associations of several identified markers. There is a lack of unified indication to measure the trade-off between different objective functions at the same time. In future, we try to design a unified indicator to measure the performance of our algorithm with other existing MOO algorithms. Moreover, our application to Dengue data demonstrates that it is capable of unlocking novel insights into other biological problems. Therefore, this may further help identify novel signalling pathways, biomarkers, and drug targets in other complex biological systems and diseases.

Acknowledgment. The work is supported by the MATRICS project grant (MTR/2020/000326) of SERB, DST, Govt. of India and research project grant (0083/RND/ET/KU10 /Jan-2021/1/1) from DST&BT, Govt. of West Bengal, India.

References

1. Bhatt, S., Gething, P.W., Brady, O.J., et al.: The global distribution and burden of dengue. Nature **496**(7446), 504–507 (2013)
2. Blais, D.R., Nasheri, N., McKay, C.S., et al.: Activity-based protein profiling of host-virus interactions. Trends Biotechnol. **30**(2), 89–99 (2012)
3. Carter, C.: Extensive viral mimicry of human proteins in AIDS, multiple sclerosis and other autoimmune disorders, late-onset and familial Alzheimer's disease and other genetic diseases. Nat. Precedings, 1 (2010)
4. da Costa Barros, T.A., de Oliveira-Pinto, L.M.: A view of platelets in dengue. In: Thrombocytopenia. IntechOpen (2018)
5. Deb, K., Pratap, A., Agarwal, S., et al.: A fast and elitist multiobjective genetic algorithm: NSGA-II. IEEE Trans. Evol. Comput. **6**(2), 182–197 (2002)
6. Dey, L., Mukhopadhyay, A.: DenvInt: a database of protein-protein interactions between dengue virus and its hosts. PLoS Neglected Trop. Dis. **11**(10), e0005879 (2017)

7. Geffin, R., Martinez, R., de las Pozas, A., et al.: Fingolimod induces neuronal-specific gene expression with potential neuroprotective outcomes in maturing neuronal progenitor cells exposed to HIV. J. Neurovirol. **23**(6), 808–824 (2017). https://doi.org/10.1007/s13365-017-0571-7
8. Jiang, J.J., Conrath, D.W.: Semantic similarity based on corpus statistics and lexical taxonomy. arXiv preprint cmp-lg/9709008 (1997)
9. Mazandu, G.K., Mulder, N.J.: Information content-based gene ontology functional similarity measures: which one to use for a given biological data type? PLoS ONE **9**(12), e113859 (2014)
10. Mukhopadhyay, A., Maulik, U., Bandyopadhyay, S., et al.: A survey of multiobjective evolutionary algorithms for data mining: part I. IEEE Trans. Evol. Comput. **18**(1), 4–19 (2013)
11. Naef, A., Abdullah, R., et al.: Multiobjective optimization to reconstruct biological networks. Biosystems **174**, 22–36 (2018)
12. Ojha, A., Nandi, D., Batra, H., et al.: Platelet activation determines the severity of thrombocytopenia in dengue infection. Sci. Rep. **7**(1), 1–10 (2017)
13. Ray, S., Maulik, U.: Identifying differentially coexpressed module during HIV disease progression: a multiobjective approach. Sci. Rep. **7**(1) (2017). Article number: 86. https://doi.org/10.1038/s41598-017-00090-2
14. Redwan, E.M., AlJaddawi, A.A., Uversky, V.N.: Structural disorder in the proteome and interactome of Alkhurma virus (ALKV). Cell. Mol. Life Sci. **76**(3), 577–608 (2019). https://doi.org/10.1007/s00018-018-2968-8
15. Sevilla, J.L., Segura, V., Podhorski, A., et al.: Correlation between gene expression and GO semantic similarity. IEEE/ACM Trans. Comput. Biol. Bioinf. **2**(4), 330–338 (2005)
16. Tesson, B.M., Breitling, R., Jansen, R.C.: DiffCoEx: a simple and sensitive method to find differentially coexpressed gene modules. BMC Bioinform. **11**(1), 1–9 (2010). Article number: 497
17. Wang, J.Z., Du, Z., Payattakool, R., et al.: A new method to measure the semantic similarity of GO terms. Bioinformatics **23**(10), 1274–1281 (2007)
18. Watson, M.: CoXpress: differential co-expression in gene expression data. BMC Bioinform. **7**(1) (2006). Article number: 509. https://doi.org/10.1186/1471-2105-7-509
19. Wen, F., Armstrong, N., Hou, W., et al.: Zika virus increases mind bomb 1 levels, causing degradation of pericentriolar material 1 (PCM1) and dispersion of PCM1-containing granules from the centrosome. J. Biol. Chem. **294**(49), 18742–18755 (2019)

Real-World Airline Crew Pairing Optimization: Customized Genetic Algorithm Versus Column Generation Method

Divyam Aggarwal[1(✉)], Dhish Kumar Saxena[2], Thomas Bäck[3], and Michael Emmerich[3]

[1] Optym, Whitefield, Bengaluru 560048, Karnataka, India
divyam.aggarwal@optym.com

[2] Indian Institute of Technology (IIT) Roorkee, Roorkee 247667, Uttarakhand, India
dhish.saxena@me.iitr.ac.in

[3] Leiden University, 2333 CA Leiden, The Netherlands
{t.h.w.baeck,m.t.m.emmerich}@liacs.leidenuniv.nl

Abstract. *Airline crew pairing optimization problem* (CPOP) aims to find a set of flight sequences (*crew pairings*) that cover all flights in an airline's highly constrained flight schedule at *minimum* cost. Since crew cost is second only to the fuel cost, CPOP solutioning is critically important for an airline. However, CPOP is NP-hard, and tackling it is quite challenging. The literature suggests, that when the CPOP's scale and complexity is reasonably limited, and an enumeration of all crew pairings is possible, then Metaheuristics are used, predominantly *Genetic Algorithms* (GAs). Else, *Column Generation* (CG) based Mixed Integer Programming techniques are used. Notably, as per the literature, a maximum of 45,000 crew pairings have been tackled by GAs. In a significant departure, this paper considers over 800 flights of a US-based large airline (with a monthly network of over 33,000 flights), and tests the efficacy of GAs by enumerating all 400,000+ crew pairings, apriori. Towards it, this paper proposes a domain-knowledge-driven customized-GA. The utility of incorporating domain-knowledge in GA operations, particularly *initialization* and *crossover*, is highlighted through suitable experiments. Finally, the proposed GA's performance is compared with a CG-based approach (developed in-house by the authors). Though the latter is found to perform better in terms of solution's cost-quality and run time, it is hoped that this paper will help in better understanding the strengths and limitations of domain-knowledge-driven customizations in GAs, for solving combinatorial optimization problems, including CPOPs.

Keywords: Airline crew pairing optimization · Combinatorial optimization · Genetic algorithms · Mixed integer programming · Column generation

1 Introduction

In Airline Scheduling Process, Airline Crew Scheduling (CS) is considered as one of the most important planning activities, since crew operating cost is the second largest after the fuel cost and even its marginal improvements may translate to millions of dollars annually. Over the past three decades, the Operations Research (OR) Society has

M. Emmerich et al. (Eds.): EMO 2023, LNCS 13970, pp. 518–531, 2023.
https://doi.org/10.1007/978-3-031-27250-9_37

given unprecedented attention to airline CS and proposed numerous optimization-based solution approaches. To meet the exponentially increasing demand over these years, the expansion of airline operations has lead to a tremendous increase in the number of flights, aircraft, and crew members to be scheduled, leaving the state-of-the-practices obsolete. Given this, it has become imperative to improve existing practices by leveraging recent technological advancements and enhanced computational resources.

Airline crew scheduling is a combination of challenging (*NP-hard* [13]) combinatorial optimization problems, namely, *crew pairing optimization* and *crew assignment* problems, which are tackled sequentially. The former problem aims to generate a set of flight sequences (each called a *crew pairing*) to cover all given flights at *minimum* cost, while satisfying several *legality* constraints linked to the federations' rules, airline-specific regulations, labor laws, etc. The latter problem aims to assign crew members to these optimally-generated pairings while satisfying the pairing and crew requirements. The scope of this research is limited to Airline Crew Pairing Optimization Problem (CPOP). Interested readers are referred to Aggarwal et al. [1] for a comprehensive review of the integration of other components of the airline scheduling process.

In CPOP, crew pairings have to satisfy multiple constraints to be classified as *legal*, and it is imperative to generate legal pairings in a time-efficient manner to assist the subsequent optimization search. Several legal pairing generation approaches, either based on a flight-network or a duty-network, have been proposed in the literature [2]. Based upon the scale of the CPOP being tackled, the pairing generation module can be invoked using two possible architectures– one wherein all pairings are enumerated a priori CPOP-solutioning, and the other wherein pairings are enumerated as and when required during the CPOP-solutioning. Regarding solution-methodologies, *mathematical programming techniques* and *metaheuristics*, are commonly employed. In the former category, *Column Generation* (CG) [20,21] is the most widely adopted technique, which is proven for efficiently solving large-scale CPOPs. It is an efficient search-space exploration technique, that iteratively generates only the pairings having a high potential of bringing in the associated cost benefits. In that, the original CPOP is relaxed into a Linear Programming Problem (LP/LPP); which is then solved iteratively by invoking an LP solver and generating new pairings by solving the corresponding pricing sub-problem(s) [16,20]. Finally, the resulting LPP solution is integerized using an integer programming (IP/IPP) solver or connection-fixing heuristics [29,32]. For more details, interested readers are referred to [3,10,31,32].

Among meta-heuristics, the most successful and widely adopted technique is *Genetic Algorithms* (GAs), which are population-based probabilistic-search heuristics, inspired by the theory of natural evolution [15]. GAs with customized operators are known to be successful in solving a variety of combinatorial optimization problems [7,8,22,23]. Several GA-based CPOP solution approaches, proposed in the literature, are broadly reviewed in Table 1. [31] is the first instance to customize a GA (using guided GA-operators) for solving a general class of SCPs. In that, the authors validated their proposed approach on small-scale synthetic test cases (with over 1,000 rows and just 10,000 columns). Notably, the literature review in the table could be summarised in two-fold. First, GAs presented by some instances– [7,11,18,27,33], have been validated using the flight networks of smaller airlines, operating in low-demand regions such as Greece, Turkey, etc. (leading to only a handful of all possible legal pair-

Table 1. An overview of the GA-based CPOP solution approaches from the literature

Literature Instances	Formulation+	Airline timetable	Flight data*		Airlines
			# Flights	# Pairings	
[7]	SCP	Did not solve CPOP	1,000	10,000	–
[18]	SPP	–	823	43,749	–
[27]	SCP	Daily	380	21,308	Multiple airlines
[17]	SCP	Monthly	2,100	11,981	Olympic airways
[33]	SCP	Monthly	710	3,308	Turkish airlines
[11]	–	–	506	11,116	Turkish airlines
[12]	SCP	–	714	43,091	Turkish airlines

+ SCP stands for Set-Covering Problem formulation and SPP stands for Set-Partitioning Problem formulation. * The provided values are the maximum among all the test-cases being used for validation.

ings, up to 45,000 pairings). These GAs become obsolete when scaled to even small flight networks of bigger airlines, operating in large geographical regions such as the USA, etc. Second, the results presented in some of these instances– [12,17], have been obtained by solving CPOPs formulated using only a subset of the original search-space (up to 12,000 pairings), i.e., all possible legal pairings are not used. In addition, [12] demonstrated that despite customizations, GAs failed to solve large-scale CPOPs with the same search-efficiency as small-scale CPOPs. Hence, it is imperative to develop GAs that can efficiently tackle CPOPs with bigger pairing-space, say up to a million.

In a significant departure from the existing GA-based approaches, this paper proposes a domain-knowledge-driven customized GA to efficiently tackle a CPOP with over 800 flights of a US-based large airline (operating over 33,000 monthly flights), by enumerating all possible crew pairings (over 400,000 pairings) a priori. In that, the GA operations, particularly initialization and crossover, are enhanced using domain-knowledge. Through suitable experiments, it is demonstrated that the proposed-GA is able to generate crew pairing solutions with varying characteristics such as low number of deadhead flights, crew-hotel-nights, etc., which are important KPIs used by airlines along with the crew pairing cost to evaluate the performance of their schedules. Another contribution of this paper is the insights shared on how well the proposed GA performs in comparison to a mathematical programming-based CPOP solution approach, on which the literature is mostly silent upon. For this comparison, a CG-based large-scale airline crew pairing optimizer (*CG-Optimizer*), developed in-house by the authors and validated by the research consortium's industrial sponsor– GE Aviation has been utilized. Though the CG-Optimizer is found to perform well in terms of the solution's cost quality and runtime, it is hoped that this paper will help better understand the strengths and limitations of domain-knowledge-driven customizations in GAs, for solving challenging combinatorial problems like CPOPs.

2 Airline Crew Pairing Optimization Problem

In CPOP, the input data includes an airline flight schedule with a finite number of flights, the pairings' costing criterion, and legality rules & regulations. As introduced before, a *crew pairing* is a sequence of flights to be flown by a crew member, beginning and ending at the same crew base. Other associated terminologies of CPOP are explained with the help of an example of a crew pairing, shown in Fig. 1. At times, a crew is required to be transported to an airport to fly their next flight. In such situations, the crew is transported as passengers in another flight, flown by another crew. Such a flight is called a *deadhead* or a *deadhead flight* for the transported crew. The presence of deadhead flights affects an airline's profit in two ways. First, the airline has to bear the loss of revenue on the passenger seats being occupied by the deadhead-ing crew, and the other is it has to pay the hourly wages to the deadhead-ing crew even when they are not servicing the flight. To maximize profits, airlines desire to minimize these deadheads as much as possible (ideally zero).

As mentioned in Sect. 1, it is imperative to develop a *legal crew pairing generation* approach to facilitate legal pairings to the optimization phase. In small- and medium-scale CPOPs, all legal pairings are generated explicitly before the optimization phase. The same approach is adopted in this work, and a duty-network-based parallel legal pairing generation algorithm [2] is used for generating all legal pairings explicitly. Interested readers are referred to [2] for an extensive review of the pairing generation literature too.

The goal of the optimization phase is to find a pairing subset from the generated set of all legal pairings to cover the given flights with the minimum cost possible. In literature, the CPOP is modeled either as a set-partitioning problem (SPP; each flight leg is allowed to be covered only once) or as a set-covering problem (SCP; over-coverage of flight legs i.e. deadheads are allowed). In this paper, the SCP formulation is adapted and modified to define the optimization problem for the proposed GA. Its mathematical model is presented in Sect. 3.2.

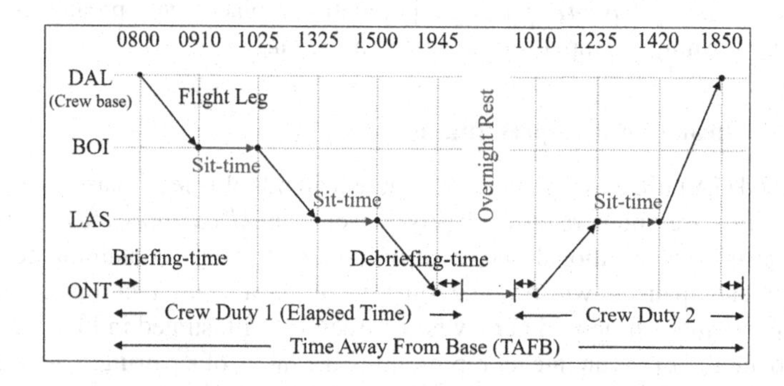

Fig. 1. A crew pairing beginning from *Dallas* (*DAL*) crew base

3 Genetic Algorithm

A customized-GA is proposed in this work to solve CPOPs for which enumeration and handling of the entire pairing set is computationally-tractable. Before starting the GA-search, the entire pairing set, denoted by *AllPairings*, is enumerated *a priori*. After pairing enumeration, the proposed GA tackles CPOP by formulating a fitness function based on its SCP formulation (given in Sect. 3.2). The GA-search starts by initializing the population using efficient pairing sets (each set $\subset$ *AllPairings*) by employing a novel initialization heuristic. Iteratively, the population is improved upon, each iteration referred to as a *generation*, by bringing-in new pairings from the remaining pairings' space using enhanced genetic operators.

The working of the proposed GA is explained in conjunction with the enhancement of its genetic operators, as described in the upcoming subsections. Notably, these genetic operators have either been enhanced or adopted from the GA-variants proposed in [7,17,28]. Now, the high-level pseudocode of the proposed GA, formalized in lines 1–13 of Algorithm 1, is explained below. In line 2, a set of chromosomes, notated as *InitialPop*, is generated by applying the novel initialization heuristic on *AllPairings*. In line 3, the fitness function value of chromosomes $\in$ *InitialPop* is computed. Lines 4–12 constitute GA-generations, which terminate as soon as the user-specific termination criterion is satisfied. This is followed by the selection of best-fit chromosomes $\in$ *InitialPop* (for generation = 1) or $\in$ *BestPop* (for generations > 1) that constitutes the parent population, notated as *ParentPop* (line 5). Subsequently, in line 6, the parent chromosomes reproduce to generate child chromosomes (set notated as *ChildPop*) via crossover operation. In line 7, child chromosomes $\in$ *ChildPop* are mutated to promote diversity in the solutions' pool. Notably, here, two different crossover and mutation operators are interchangeably used according to different settings of the proposed GA, as mentioned in Sect. 4. Being a combinatorial optimization problem, the generated child chromosomes may be infeasible with respect to flight coverage constraints. To re-install their feasibility, a *feasibility-repair heuristic* along with a *redundant-pairing removal heuristic* is applied to *ChildPop* (lines 8–9). Next, the fitness of chromosomes $\in$ *ChildPop* is computed (line 10). Finally, the chromosomes $\in$ *ChildPop* are combined with chromosomes $\in$ *ParentPop* using a population replacement operator, resulting in *BestPop* and forming the input for the next generation.

3.1 Novel Chromosome Representation

The proposed GA utilizes an architecture wherein all possible legal pairings, set denoted by *AllPairings*, are enumerated explicitly. Given that *AllPairings* may contain thousands of pairings, conventional binary-chromosome structure, containing genes corresponding to each pairing, will become impractical. As a result, a chromosome with 2-bits gene-encoding is proposed here, whose structure is illustrated in Fig. 2. In that, the first bit, notated as b_{1i}, is an integer representing the index of a pairing $p_i \in AllPairings$, which constitutes the chromosome. And the second bit, notated as b_{2i}, is a binary number representing the participation of the corresponding pairing for fitness evaluation ($= 1$, or $= 0$ otherwise). Moreover, to maintain diversity in the chromosome and prevent premature convergence, the chromosome structure, used in [28], is adapted here. As a result, a chromosome contains two parts, namely *expressed* and *unexpressed* parts, notated as $B^e : (b^e_1, b^e_2)$ and $B^u : (b^u_1, b^u_2)$, respectively. The former part involves pairings that participate in the fitness evaluation of the chromosome, whereas the latter part involves pairings that are not considered. However, these pairings ($\in$ unexpressed part) are utilized to preserve diversity with respect to the pairings in expressed part, so that diverse pairings can participate in the reproduction of child chromosomes. Moreover,

Algorithm 1: Pseudocode of the proposed GA and its constituting operators

```
1  begin
2      Proposed GA:
3      InitialPop ← Call Minimal-deadhead Initialization Heuristic(AllPairings)
4      Evaluate Fitness of InitialPop
5      while Termination criterion is not met do
6          ParentPop ← Selection operator(InitialPop/BestPop)
7          (Child1, Child2) ← Crossover(Parent1, Parent2)
               /* Crossover1 or Crossover2                    */
8          Mutation(ChildPop)             /* Mutation1 or Mutation2 */
9          Feasibility-repair Heuristic(ChildPop)
10             Redundant-pairing Removal Heuristic(ChildPop)
11         Fitness Evaluation(ChildPop)
12         BestPop ← Population Replacement(ParentPop ∪ ChildPop)
13     end
14     Minimal-deadhead Initialization Heuristic:
15     foreach chromosome ∈ InitialPop do
16         for expressed part do
17             Randomly select a zero-deadhead solution from AllPairings
18             if —all flights are not covered— then
19                 Select pairings from AllPairings w.r.t. the number of deadheads they are
                     bringing into the solution
20         end
21         for unexpressed part do
22             Randomly select pairings from AllPairings without replacement
23         end
24     end
25     Crossover2:
26     CombinedPairings ← Combined list of pairings in Parent1 and Parent2
27     foreach child chromosome do
28         for expressed part do
29             Randomly select a zero-deadhead solution from CombinedPairings
30         end
31         for unexpressed part do
32             Select pairings from CombinedPairings based on their dissimilarity with
                 expressed part (number of non-intersecting flights)
33         end
34     end
35     Procedure Redundant-pairing Removal Heuristic (ChildPop):
36     foreach pairing corresponding to index b^e_1i of each chromosome ∈ ChildPop do
37         if —b^e_2i = 1— then
38             set b^e_2i = 0
39             if —all flights are not covered in chromosome then
40                 set b^e_2i = 1
41     end
42 end
```

contrary to the chromosome structure used in [28], the length of the chromosome is kept fixed, and the length of expressed and unexpressed parts is allowed to vary during generations, making this a novel adaptation.

expressed part, e						unexpressed part, u					
b_{11}^{e}	b_{12}^{e}	b_{13}^{e}	.	.	b_{1x}^{e}	b_{11}^{u}	b_{12}^{u}	b_{13}^{u}	.	.	b_{1y}^{u}
b_{21}^{e}	b_{22}^{e}	b_{23}^{e}	.	.	b_{2x}^{e}	b_{21}^{u}	b_{22}^{u}	b_{23}^{u}	.	.	b_{2y}^{u}

Fig. 2. Chromosome structure

3.2 Fitness Evaluation

CPOP aims to minimize total crew pairing cost while covering all flights by at least one pairing. To evaluate the fitness of a chromosome, the fitness function is constructed using the set-covering problem formulation (SCP) of the airline CPOP [3,4], which is given as follows:

$$min\left\{\sum_{b^e_{1j}\in b^e_1} c_j.b^e_{2j} + \psi_D.\sum_{i\in S_F}\left(\sum_{b^e_{1j}\in b^e_1} a_{ij}.b^e_{2j} - 1\right)\right\},\ s.t. \sum_{b^e_{1j}\in b^e_1} a_{ij}.b^e_{2j} \geq 1\ \forall i \in S_F \quad (1)$$

In that, Eq. 1 represents the objective function and feasibility constraint of the airline CPOP. In that, S_F is the flight set to be covered; c_j is the cost of pairing p_j given by b^e_{1j}; ψ_D is the deadhead penalty-cost set by airlines; a_{ij} is a binary constant representing the coverage of flight f_i by pairing p_j ($= 1$, or $= 0$ otherwise); and b^e_{2j} is the binary decision variable (given by the binary bit of the expressed part of the chromosome), representing the selection of pairing p_j in the solution ($= 1$, or $= 0$ otherwise). Notably, the pairing and deadhead costs (objective function components in Eq. 1) are two objectives to be minimized. Here, a weighted scalarization approach is used for combining them, wherein deadhead flights are penalized using the penalty cost assigned by the industrial sponsors.

3.3 Minimal-Deadhead Initialization Heuristic

Generally, chromosomes in the initial population are generated using randomly selected genes to allow for exploratory search upfront. However, in single-objective problems like CPOP, it is imperative to constitute the initial population with diverse and reasonably good-quality chromosomes, supporting the initial-exploration stage while expediting the convergence. In this work, an effective initialization heuristic, referred to as *Minimal-deadhead Initialization Heuristic* (lines 14–24 of Algorithm 1), is proposed, which induces pairings that bring a lesser number of deadhead-flights into the solution. To generate the expressed part, first, a zero-deadhead pairing set ($\subset$ *AllPairings*) is chosen randomly (lines 16–20). If all flights are not covered in the expressed part, then more pairings (from *AllPairings*) are introduced in the increasing order of the number of deadhead flights they bring in. The unexpressed part is initialized using randomly selected pairings from *AllPairings* until the finite length of the chromosome is exhausted (lines 21–23).

3.4 Selection

This operator selects parent chromosomes from the input population according to their fitness-function values. Here, a binary tournament selection operator [14] is utilized. It creates N sets of two randomly-selected chromosomes (N being the population size) and selects the fittest chromosome to constitute the resulting parent population– *ParentPop*.

3.5 Crossover

In crossover, new child chromosomes are reproduced by transforming genetic information from parent chromosomes using different strategies, such as one-point crossover, two-point crossover, uniform crossover, fusion crossover [7], etc. Here, two specific crossovers are studied and compared, namely *Crossover1* and *Crossover2*. The former is the fusion crossover [7] that has been widely adopted in CPOP's literature. In that, a fitness-based probability is used to decide the gene of which parent chromosome will pass on to the child chromosome. The latter is an adaption of a domain-knowledge-driven greedy crossover [28], which was originally proposed to improve the convergence of the GA-search. Its pseudocode is given in lines 26–35 of Algorithm 1. In that, the child chromosome's expressed part is constructed using a zero-deadhead pairing set, which is selected randomly from the combined pool of pairings in the parent chromosomes. And, its unexpressed part is formed using the remaining pairings on the basis of their dissimilarity with the expressed part, i.e., the flights they are covering differently compared to the expressed part.

3.6 Mutation

In mutation, certain genes of the child chromosomes (from crossover) are altered to prevent premature convergence. Here, two widely-adopted mutation strategies are studied and compared. The first is a bit-flip mutation, referred to as *Mutation1*, and the other is the mutation proposed in [17], referred as *Mutation2*, which utilizes density of the fittest solution in the population. In *Mutation1*, if an ith gene gets selected for mutation, then b_{2i} bit is flipped from 0 to 1 or vice-versa. Whereas in *Mutation2*, if an ith gene is selected for mutation, then b_{2i} is mutated from 0 to 1 or vice-versa, based on a probability equivalent to the percentage of 1s in the fittest individual.

3.7 Feasibility-Repair Heuristic

It is well-known that in combinatorial optimization problems such as CPOP, crossover and mutation operations may render the child chromosomes infeasible, leading to the requirement of a feasibility-repair step. A heuristic, proposed in [7], is adapted in this work by adding a redundant-pairing removal step. In that, for each uncovered flight in the infeasible chromosome, a pairing ($\in$ *AllPairings*) with minimum value of a quality index (defined as *Cost of pairing/Number of uncovered flights the pairing covers*) is selected. After this, a redundant-pairing removal heuristic is proposed (lines 37–43 of Algorithm 1), which finds and removes the pairings with zero contribution in the overall flight coverage of the chromosome.

3.8 Population Replacement

The last step is the population replacement step wherein *ParentPop* & *ChildPop* are combined to select the *ParentPop* for next generation, notated as *BestPop*. There exists two main strategies, namely generational and steady-state. Here, generational strategy is adopted in which selects best N chromosomes out of N parent and N child chromosomes to constitute *BestPop*.

4 Computational Experiments

All the computational experiments in this research work are performed with a real-world test-case, which includes 839 flights and crew based on a single home base–Dallas, USA (*DAL*). This test-case has been extracted from the networks of US-based big airlines (operating upto 33,000 monthly flights with upto 15 crew bases), provided by the research consortium's industrial sponsors– GE Aviation. It is found that 430,873 legal crew pairings are possible for this test-case, which is enormously huge in comparison to the amount of pairings dealt in the existing GA-based approaches (Sect. 1). In this research work, all the algorithms are implemented using *Python* and executed using just-in time (JIT) compiler– *PyPy*, improving the computational speeds by a great extent. All computations are performed on a HP Z640 workstation (2 X Intel® Xeon® Processor E5-2630v3 @2.40 GHz and 8-Cores/16-Threads, enabled with parallelization capabilities).

The parameter settings of the proposed GA, used for the experiments in this research, are given in Table 2. It is observed that on increasing the GA's population size, the number of GA-generations may decrease as it may bring more diversity in the population's solution-quality at each generation. However, each generation's time may increase proportionately. Overall, this may not drastically degrade the final runtime-performance. Hence, the population size here is selected accordingly. For the termination of the proposed GA, its overall runtime is selected as the termination criterion

Table 2. GA parameter settings

Parameters	Value
Population size	24
Termination	5000 s
Chromosome length	$100 + MaxLen(InitialPop)$
Crossover rate	0.9
Mutation rate	$3 \cdot (1/ChromosomeLen)$

Table 3. GA configurations

Operators	GA1	GA2	GA3	GA4
Initialization		✓	✓	✓
Mutation1	✓	✓		
Mutation2			✓	✓
Crossover1	✓	✓	✓	
Crossover2				✓

instead of the number of generations, given different strategies being used in different settings of GAs being compared here. The chromosome length has been selected in accordance to the best practice solutions. In [6], $(1/ChromosomeLen)$ is proposed as the lower bound for the optimal mutation rate. However, during experiments, it is observed that this lower bound shall be inflated by some factor (here, 3) in order to prevent premature convergence. This is also in alignment with the observations of the authors in [31] with variable mutation rate.

In this research work, variants of GA-operators are proposed which are either developed by the authors or adapted from the variants present in the literature. To solve the above-mentioned airline test-case and similar problems, it is imperative to find the most effective combination of these operators. Towards this, four configurations of the GA are implemented and tested in this work, the structure of whom are shown in Table 3. For each of these GA-configurations, ten runs, initialized with different random seeds (uniformly distributed between 0 and 1), are performed. The experimental results of these runs are summarized in Table 4 and the comparative plots are shown in Fig. 3. First, the merits of using the proposed minimal-deadhead initialization heuristic are assessed. For this, the best solution among the initial populations generated in GA1-runs (using random initialization) and GA2-runs (using the proposed initialized heuristic), are compared, as recorded in first two rows of Table 4. It is observed that the characteristics of the best initial solution from the GA2-runs (number of deadheads and total cost) are reasonably very good compared to those of GA1-runs. Notably, the initialization runtime for these GA-configurations are similar, as the additional runtime consumed by the proposed heuristic is compensated by the runtime required to repair the infeasible solutions obtained using random initialization in GA1-runs. Moreover, GA2-runs lead to a better-cost crew pairing solution (best solution across all seeds) compared to the GA1-runs. These observations endorse the effectiveness of using the proposed initialization heuristic.

Second, the merits of the proposed mutation strategies are assessed. For this, the GA-configurations– GA2 (using *Mutation1*) and GA3 (using *Mutation2*) are compared. From the results recorded in Table 4, it is observed that GA3-runs lead to a better crew pairing solution (in terms of both cost and number of deadheads) compared to the GA2-runs. However, the difference between them is marginal, equalizing the effects of both mutation strategies. Consecutively, *Mutation2* is considered for the subsequent experiments.

Table 4. Experimental results of the GA-runs

Runtime (sec)	GAs	Crew pairing cost (USD)			# Deadheads		
		$\bar{x} \pm \sigma$	Best	Worst	$\bar{x} \pm \sigma$	Best	Worst
70	GA1	2,649,823 $\pm$ 57,559	2,494,649	2,710,084	1,095 $\pm$ 45	977	1,151
	GA2	1,417,223 $\pm$ 9,380	1,398,427	1,430,115	156 $\pm$ 06	149	164
5000	GA1	980,226 $\pm$ 23,091	964,857	1,037,504	40 $\pm$ 04	35	49
	GA2	1,195,229 $\pm$ 225,555	957,832	1,430,115	98 $\pm$ 61	35	164
	GA3	1,192,104 $\pm$ 228,745	949,591	1,430,115	98 $\pm$ 61	30	164
	GA4	993,209 $\pm$ 5,337	987,638	1,001,487	09 $\pm$ 04	06	21

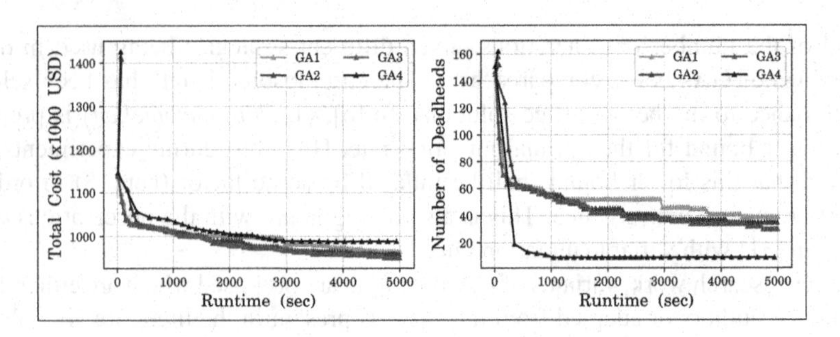

Fig. 3. Characteristic plots of the GA-runs

Third, the merits of the proposed crossover strategies are assessed. For this, the GA-configurations– GA3 (using *Crossover1*) and GA4 (using *Crossover2*), are compared. From the tabulated results and plot of GA4 in Fig. 3, it is quite evident that the proposed *Crossover2* strategy is highly effective in reducing the number of deadheads that too in a very less runtime. However, the cost of the final crew pairing solution from GA4-runs is marginally poorer than those of the GA3-runs. On further analyzing the crew pairings of best solution from GA4-runs, it is observed that the majority of pairings contain very less number of flight legs, each referred to as a *short-pairing*. Hence, with such short-pairings, the solution contains a large number of pairings to cover all 839 flights. Moreover, during the GA search, the current best solution, dominated by large number of short-pairings, becomes too rigid to allow any large-pairing (covering a large number of flights, contrary to a short-pairing) to enter the solution, hence, stopping the search at local optima.

As mentioned before, a large-scale column generation based airline crew pairing optimizer (*CG-Optimizer*) is used in this research to assess the performance of the proposed GA-configurations, and to share the insights on how well a highly-customized GA performs in comparison to advanced mathematical programming techniques. *CG-Optimizer* is developed in-house by the authors as part of the overall research project, and has been tested and validated on real-world, large-scale and complex flight networks provided by GE Aviation. The exhaustive details of *CG-Optimizer* are presented in the technical report– [3]. The final crew pairing solution of *CG-Optimizer* is compared with the best solutions of the proposed GA configurations, and the results are recorded in Table 5. From the tabulated results, it is quite evident that the crew pairing

Table 5. Crew pairing solutions of CG-Optimizer and proposed GA configurations

Algorithms	Total cost (USD)	# Deadheads	# Pairings	%age cost gap
CG-Optimizer	850,303	02	142	0
GA1	964,858	39	169	13.47
GA2	957,833	35	172	12.65
GA3	949,592	30	171	11.68
GA4	987,639	09	242	16.15

solution offered by *CG-Optimizer* is of superior quality than any of the solutions offered by the proposed GA configurations, with minimum percentage cost difference being 11.7%. Moreover, the number of deadheads as well as the number of pairings involved are minimal in the solution offered by *CG-Optimizer*. This endorses the fact that CG-based (mathematical programming) CPOP solution approaches are highly effective in solving CPOPs with moderately sized flight networks, compared to a GA-based solution approach despite several domain-knowledge-driven enhancements.

5 Conclusion

This paper proposes a domain-knowledge-driven customized GA, with enhanced genetic operations, particularly initialization and crossover, to efficiently tackle a CPOP with over 800 flights of a US-based large airline (operating over 33,000 monthly flights), by enumerating all 400,000+ crew pairings a priori. The proposed minimal-deadhead initialization heuristic is effective in achieving a better-initial solution compared to a random initialization strategy (with 78% better cost and 555% lesser deadheads) in approximately similar runtime. On assessing the performance of two widely-adopted mutation operators, it is found that both perform similarly with *Mutation2* performing marginally better than *Mutation1*. A deadhead-minimizing crossover operator, *Crossov-er2*, is also proposed which is found to be effective in reducing the number of deadheads significantly within a short runtime.

In addition to the above, this paper shares insights on the comparison of customized GAs with CG-based CPOP solution approaches. For this, the performance of the proposed GA is compared viz-a-viz CG-based large-scale optimizer (*CG-Optimizer*, developed by the authors) to solve large-scale CPOPs with over billion-plus legal pairings, 4,000 flights, and 15 crew bases. Though it is found that the crew pairing solution offered by *CG-Optimizer* is of superior quality than any of the solutions offered by the proposed GA (with minimum percentage cost difference being 11.7%), it is believed that this paper will serve as a template to better understand the strengths and limitations of domain-knowledge-driven customizations in GAs (other metaheuristics) for solving combinatorial optimization problems, including CPOPs.

Notably, *Crossover2* favors deadhead minimization, leading to the selection of short-pairings and driving the GA-search towards local optima. Towards it, search-space expansion heuristics [9], and variable mutation rates [7] could be adapted. This work paves the way for a detailed multi-objective study of the airline CPOP under realistic assumptions. Moreover, an important future direction is to investigate the trade-off between crew operating cost and robustness against delays (by adding slack time to the duration of given flights). This slack time can be based on the likelihood of delays (obtained using a machine learning model) and/or based on the systemic importance of flight connections. Lastly, the emergent trend of utilizing machine learning capabilities to assist combinatorial optimization using metaheuristics or mathematical programming may also hold promise to improve the current propositions [5, 19, 24–26, 30].

Acknowledgement. This research work is supported by MEITY, India [grant 13(4)/2015-CC&BT]; NWO, the Netherlands; and GE Aviation, India. Thanks to the industrial sponsor's (GE Aviation) team members: Saaju Paulose, Arioli Arumugam and Rajesh Alla for their invaluable support in successfully completing this research. Notably, during this research, the first author (Divyam Aggarwal) was a Ph.D. Candidate at IIT Roorkee, India.

References

1. Aggarwal, D., Saxena, D.K., Emmerich, M.: Interdependence and integration among components of the airline scheduling process. In: Paper presented at the 21st World Conference of the Air Transport Research Society (ATRS 2017), Antwerp, Belgium, 5–8 July 2017 (2017). http://www.optimization-online.org/DB_FILE/2020/05/7774.pdf
2. Aggarwal, D., Saxena, D.K., Emmerich, M., Paulose, S.: On large-scale airline crew pairing generation. In: 2018 IEEE Symposium Series on Computational Intelligence (SSCI), pp. 593–600. IEEE (2018)
3. Aggarwal, D., Saxena, D.K., Paulose, S., Bäck, T., Emmerich, M.: Airline crew pairing optimization framework for large networks with multiple crew bases and hub-and-spoke subnetworks. arXiv preprint arXiv:2003.03994 (2020)
4. Aggarwal, D., Saxena, D.K., Paulose, S., Bäck, T., Emmerich, M.: A novel column generation heuristic for airline crew pairing optimization with large-scale complex flight networks. arXiv preprint arXiv:2005.08636 (2020)
5. Aggarwal, D., Singh, Y.K., Saxena, D.K.: On learning combinatorial patterns to assist large-scale airline crew pairing optimization. arXiv preprint arXiv:2004.13714 (2020)
6. Bäck, T.: Optimal mutation rates in genetic search. In: Proceedings of the Fifth International Conference on Genetic Algorithms. Morgan Kaufmann, San Mateo (1993)
7. Beasley, J.E., Chu, P.C.: A genetic algorithm for the set covering problem. Eur. J. Oper. Res. **94**(2), 392–404 (1996)
8. Deb, K., Myburgh, C.: Breaking the billion-variable barrier in real-world optimization using a customized evolutionary algorithm. In: Proceedings of the Genetic and Evolutionary Computation Conference 2016, pp. 653–660 (2016)
9. Demirel, N.Ç., Deveci, M.: Novel search space updating heuristics-based genetic algorithm for optimizing medium-scale airline crew pairing problems. Int. J. Comput. Intell. Syst. **10**(1), 1082–1101 (2017)
10. Desaulniers, G., Lessard, F., Saddoune, M., Soumis, F.: Dynamic constraint aggregation for solving very large-scale airline crew pairing problems. SN Oper. Res. Forum **1**(3), 1–23 (2020). https://doi.org/10.1007/s43069-020-00016-1
11. Deveci, M., Demirel, N.C.: A hybrid genetic algorithm for airline crew pairing optimization. In: Economic and Social Development: Book of Proceedings, p. 118 (2016)
12. Deveci, M., Demirel, N.Ç.: Evolutionary algorithms for solving the airline crew pairing problem. Comput. Ind. Eng. **115**, 389–406 (2018)
13. Garey, M.R., Johnson, D.S.: Computers and Intractibility: A Guide to the Theory of NP-Completeness, vol. 44. W. H. Freeman & Company, New York (1979)
14. Goldberg, D.E., Deb, K.: A comparative analysis of selection schemes used in genetic algorithms. In: Foundations of Genetic Algorithms, vol. 1, pp. 69–93. Elsevier (1991)
15. Holland, J.H.: Adaptation in Natural and Artificial Systems: An Introductory Analysis with Applications to Biology, Control, and Artificial Intelligence. MIT Press, Cambridge (1992)
16. Irnich, S., Desaulniers, G.: Shortest path problems with resource constraints. In: Desaulniers, G., Desrosiers, J., Solomon, M.M. (eds.) Column Generation, pp. 33–65. Springer, Boston (2005). https://doi.org/10.1007/0-387-25486-2_2

17. Kornilakis, H., Stamatopoulos, P.: Crew pairing optimization with genetic algorithms. In: Vlahavas, I.P., Spyropoulos, C.D. (eds.) SETN 2002. LNCS (LNAI), vol. 2308, pp. 109–120. Springer, Heidelberg (2002). https://doi.org/10.1007/3-540-46014-4_11
18. Levine, D.: Application of a hybrid genetic algorithm to airline crew scheduling. Comput. Oper. Res. **23**(6 SPEC. ISS.), 547–558 (1996)
19. Liu, S., Zhang, Y., Tang, K., Yao, X.: How good is neural combinatorial optimization? arXiv preprint arXiv:2209.10913 (2022)
20. Lübbecke, M.E.: Column generation. Wiley Encyclopedia of Operations Research and Management Science (2010)
21. Lübbecke, M.E., Desrosiers, J.: Selected topics in column generation. Oper. Res. **53**(6), 1007–1023 (2005)
22. Maskooki, A., Deb, K., Kallio, M.: A customized genetic algorithm for bi-objective routing in a dynamic network. Eur. J. Oper. Res. **297**(2), 615–629 (2022)
23. Mittal, S., Aggarwal, D., Saxena, D.K.: Innovative design of hydraulic actuation system for operator fatigue reduction and its optimization. In: Salagame, R.R., Ramu, P., Narayanaswamy, I., Saxena, D.K. (eds.) Advances in Multidisciplinary Analysis and Optimization. LNME, pp. 225–233. Springer, Singapore (2020). https://doi.org/10.1007/978-981-15-5432-2_19
24. Mittal, S., Saxena, D.K., Deb, K., Goodman, E.D.: A learning-based innovized progress operator for faster convergence in evolutionary multi-objective optimization. ACM Trans. Evol. Learn. Optim. (TELO) **2**(1), 1–29 (2021)
25. Morabit, M., Desaulniers, G., Lodi, A.: Machine-learning-based column selection for column generation. Transp. Sci. **55**(4), 815–831 (2021)
26. Morabit, M., Desaulniers, G., Lodi, A.: Machine-learning-based arc selection for constrained shortest path problems in column generation. arXiv preprint arXiv:2201.02535 (2022)
27. Ozdemir, H.T., Mohan, C.K.: Flight graph based genetic algorithm for crew scheduling in airlines. Proc. Joint Conf. Inf. Sci. **5**(3–4), 1003–1006 (2000)
28. Park, T., Ryu, K.R.: Crew pairing optimization by a genetic algorithm with unexpressed genes. J. Intell. Manuf. **17**(4), 375–383 (2006)
29. Parmentier, A., Meunier, F.: Aircraft routing and crew pairing: updated algorithms at air france. Omega **93**, 102073 (2020)
30. Shen, Y., Sun, Y., Li, X., Eberhard, A., Ernst, A.: Enhancing column generation by a machine-learning-based pricing heuristic for graph coloring. In: Proceedings of the AAAI Conference on Artificial Intelligence, vol. 36, pp. 9926–9934 (2022)
31. Vance, P., et al.: A heuristic branch-and-price approach for the airline crew pairing problem. Technical report lec-97-06, Georgia Institute of Technology, Atlanta (1997)
32. Zeren, B.I., Özkol, İ: A novel column generation strategy for large scale airline crew pairing problems. Expert Syst. Appl. **55**, 133–144 (2016)
33. Zeren, B., Özkol, İ: An improved genetic algorithm for crew pairing optimization. J. Intell. Learn. Syst. Appl. **04**(01), 70–80 (2012)

Multiobjective Optimization of Evolutionary Neural Networks for Animal Trade Movements Prediction

Krzysztof Michalak[1(✉)] and Mario Giacobini[2]

[1] Department of Information Technologies, Wrocław University of Economics and Business, Wrocław, Poland
krzysztof.michalak@ue.wroc.pl
[2] Department of Veterinary Sciences, University of Torino, Turin, Italy
mario.giacobini@unito.it
http://krzysztof-michalak.pl

Abstract. The analysis of animal trade movements plays a crucial role in understanding the spreading of zoonotic diseases in livestock. This article addresses the problem of predicting sending or receiving animal transports by farms and other premises. Two recurrent neural network models are used for this task: classical Recurrent Neural Networks (RNNs) and Long Short-Term Memory (LSTM) networks. Optimization of neural network weights is performed using the MOEA/D algorithm with the goal of obtaining good trade-offs between the false positive (FP) and true positive (TP) rates. The results show, that neural classifiers optimized on historical data (in this article taken from the years 2017–2019) can be used for making predictions on future data (in this article taken from the year 2020) without a serious degradation of the classification quality. In the experiments, the overall performance of the RNN model was better than that of the LSTM model, however, the LSTM performed slightly better than the RNN in the range of lower FP rates. The results of this study motivate further research on using predictive models for optimizing counter-epidemic measures, for example vaccination campaigns.

Keywords: Dynamic networks · RNN · LSTM · Zoonotic diseases

1 Introduction

This article presents an application of evolutionary neural networks to prediction of animal trade movements. Transporting animals is important from business perspective, but it also constitutes a way for animal diseases to spread [4]. Therefore, analyzing the patterns of the movements is of interest for researchers studying zoonotic diseases [1]. Analyzing the movements can help understanding the disease spread [4] and restricting the movements can be used as an epidemic control measure [11,12]. Even though some authors attempt to analyze the network of movements using static models [5], important aspects of the transportation network are discovered using dynamic network modelling [23].

M. Emmerich et al. (Eds.): EMO 2023, LNCS 13970, pp. 532–545, 2023.
https://doi.org/10.1007/978-3-031-27250-9_38

To date, various tools have been applied to the prediction of animal trade movements, such as Random Forests [16] and linear regression models [24]. In this article, recurrent neural networks (RNNs) [20] are used for predicting when a transport of animals is sent or received by a premise. Recurrent neural networks are well-suited for handling temporal data [19]. Applications of RNNs include language modelling [7], prediction of power consumption in buildings [21], load forecasting in power grids [25], prediction of delays in computer networks [2], modelling human mobility [6] and stress detection [22]. Importantly, RNNs have been shown to be able to predict rare events, such as earthquakes [3]. This capability of RNNs is important for the application shown in this article, because sending or receiving of animal transports by any given premise can be separated by long intervals of inactivity.

This article is structured as follows. Section 2 describes the problem of predicting transports sent and received by farms and other premises and formulates this problem as a classification task. Section 3 presents neural networks used for performing the classification along with the description of attribute encoding. Section 4 describes the optimization problem studied in this article. Section 5 describes the experiments and presents the results. Section 6 concludes the article.

2 Prediction of Animal Transports

The problem studied in this article is a problem of predicting sending and receiving of animal transports by premises (farms, pastures and lairage centers). The study is based on a real life dataset of animal trade movements in Italy in years 2017–2020. The dynamic network of animal transports described by this dataset forms a graph $G = \langle V, E \rangle$ in which nodes (premises) are described by attributes listed in Table 1 and edges (transports) are described by attributes listed in Table 2. Because the dates of the movements are not always precisely recorded (there can be a delay between the real transport and the recorded date), the movements are aggregated in 7-day periods, which is a commonly followed practice [1]. It is important to note, that the edges in graph G only exist for a short period of time, which makes this network different from other dynamic networks, such as social networks, in which contacts can appear and disappear, but usually last for a longer period of time. This characteristic of the animal transport network is of importance when studying the spreading of infectious diseases, because the disease can spread only when a transport of animals occurs between two given premises.

For each time instant t and node $v \in V$ incoming and outgoing transports can occur. In this article, such events are predicted independently, which gives rise to a two-label classification problem:

$$\langle \mathcal{A}(v), \{\mathcal{M}^{(t')} : t' = 1, \ldots, t-1\} \rangle \mapsto \langle C_{in}, C_{out} \rangle, \qquad (1)$$

where the labels $C_{in}, C_{out} \in \{0, 1\}$ represent the occurrence of incoming and outgoing transports, respectively, in the time instant t. The attributes of the node v (constant in time) are denoted $\mathcal{A}(v)$ and the sequence of movements that occurred at the node v before the time instant t is denoted $\{\mathcal{M}^{(t')} : t' = 1, \ldots, t-1\}$. The $\langle , \rangle$ symbol represents elements treated jointly as inputs (attributes of the node v and the sequence of movements that occurred at that node) and outputs (C_{in}, C_{out} labels) of the classification model.

Table 1. Attributes describing nodes (premises).

Name	Type	Values	Num. nodes
Premise type	Enum	Farm	238666
		Pasture	14924
		Lairage	670
Production profile	Enum	Meat	111828
		Milk	37926
		Mixed	43487
		Wool	196
		Rearing	276
		RearingAndSlaughter	203
		Slaughter	125
		SelfConsumption	11888
		Zoo	61
		Unknown	48270
Species	Enum	Buffalo	2543
		Cattle	141038
		Goat	13767
		Sheep	81988
		Unknown	14924
Reproductors	Bool	No	122453
		Yes	131807
Lat	Real	[2.594194, 18.50745]	
Lon	Real	[32.20722, 47.5927]	

3 Neural Networks

In this article, two neural models are used for performing the classification described in Sect. 2: the classical Recurrent Neural Network (RNN) [15] and the Long Short-Term Memory (LSTM) neural network [27]. In both networks, the hidden state $h^{(t)}$ is kept and the LSTM additionally keeps the cell state $c^{(t)}$. Both networks take a sequence of inputs $x^{(t)}$, $t = 1, \ldots$ and produce a sequence of outputs $y^{(t)}$. The input vectors $x^{(t)}$ contain attributes $\mathcal{A}(v)$ of the node $v \in V$ for which the predictions are made (constant in time) and the attributes of

Table 2. Attributes describing edges (movements).

Name	Type	Values
Movement date	Date	
Number of animals	Int	
Transported animal species	Enum	Buffalo
		Cattle
		Goat
		Sheep
		Unknown
The other premise*) type	Enum	Farm
		Pasture
		Lairage

*) Source premise for incoming transports, destination premise for outgoing transports.

incoming and outgoing movements at the timestep t (see Tables 1 and 2). For movements, the date attribute is not used, because it is represented by the time step t, and the number of movements that occurred in the timestep t at the node v and number of transported animals are added to the input vector. The nominal (enum) attributes are encoded using the one-hot encoding, that is, for each of those attributes a binary vector is formed with a value of one set at the position corresponding to the value of the attribute (e.g. 'Buffalo' species is encoded as 10000 and 'Goat' species as 00100). The number of inputs is $N_{in} = 74$ for both networks. Because the problem studied in this article is a multilabel (two-label) classification problem, $N_{out} = 2$ outputs are produced using sigmoid (logistic) activation functions in the output layer of the RNN, and by adding a layer with sigmoid activation functions on top of the LSTM layer. One of the outputs corresponds to an incoming transport predicted at the time step $t+1$, and the other output corresponds to an outgoing transport predicted at the time step $t+1$. The calculations for the classical Recurrent Neural Network (RNN) are performed according to the equations:

$$a_1^{(t)} = W_1 x^{(t)} + U h^{(t-1)} + b_1 \tag{2}$$

$$h^{(t)} = \tanh\left(a_1^{(t)}\right) \tag{3}$$

$$a_2^{(t)} = W_2 h^{(t)} + b_2 \tag{4}$$

$$y^{(t)} = \text{sigmoid}\left(a_2^{(t)}\right) \tag{5}$$

where:

$x^{(t)}$ - the network input,
$a_1^{(t)}$, $a_2^{(t)}$ - activations of the hidden and output layer, respectively,
$h^{(t)}$ - the hidden state,

$y^{(t)}$ - the network output,
W_1 - an $N_{hid} \times N_{in}$ matrix of input-to-hidden layer connections weights,
W_2 - an $N_{out} \times N_{hid}$ matrix of hidden-to-output layer connections weights,
U - an $N_{hid} \times N_{hid}$ matrix of recurrent connections weights,
b_1 - a vector with N_{hid} elements, containing the biases for the hidden layer,
b_2 - a vector with N_{out} elements, containing the biases for the output layer.

In the RNN network, the weights are W_1, W_2, U, b_1, and b_2, so the total number of adjustable parameters is $k = N_{hid} \cdot (N_{in} + N_{hid} + N_{out} + 1) + N_{out}$. Substituting the number of inputs $N_{in} = 74$ and outputs $N_{out} = 2$ we get $k = (N_{hid})^2 + 77N_{hid} + 2$.

The calculations for the Long Short-Term Memory (LSTM) neural network are performed according to the equations:

$$f^{(t)} = \text{sigmoid}\left(W_f x^{(t)} + U_f h^{(t-1)} + b_f\right) \tag{6}$$

$$i^{(t)} = \text{sigmoid}\left(W_i x^{(t)} + U_i h^{(t-1)} + b_i\right) \tag{7}$$

$$o^{(t)} = \text{sigmoid}\left(W_o x^{(t)} + U_o h^{(t-1)} + b_o\right) \tag{8}$$

$$c^{(t)} = f^{(t)} \circ c^{(t-1)} + i^{(t)} \circ \tanh\left(W_c x^{(t)} + U_c h^{(t-1)} + b_c\right) \tag{9}$$

$$h^{(t)} = o^{(t)} \circ \tanh\left(c^{(t)}\right) \tag{10}$$

$$y^{(t)} = \text{sigmoid}\left(W_y h^{(t)} + b_y\right) \tag{11}$$

where:
$x^{(t)}$ - the network input,
$f^{(t)}$ - the forget gate state,
$i^{(t)}$ - the input gate state,
$o^{(t)}$ - the output gate state,
$c^{(t)}$ - the cell state,
$h^{(t)}$ - the hidden state,
$y^{(t)}$ - the network output,
W_f, W_i, W_o, W_c - $N_{hid} \times N_{in}$ weight matrices,
U_f, U_i, U_o, U_c - $N_{hid} \times N_{hid}$ weight matrices,
b_f, b_i, b_o, b_c - bias vectors with N_{hid} elements,
W_y - an $N_{out} \times N_{hid}$ weight matrix,
b_y - a bias vector with N_{out} elements,

and the $\circ$ symbol denotes the element-wise multiplication of vectors. The LSTM has $k = 4 \cdot N_{hid} \cdot (N_{in} + N_{hid} + 1) + N_{out} \cdot (N_{hid} + 1)$ adjustable parameters. Substituting the number of inputs $N_{in} = 74$ and outputs $N_{out} = 2$ we get $k = 4(N_{hid})^2 + 302N_{hid} + 2$. Clearly, the LSTM model has roughly four times more parameters than the RNN model.

4 Optimization Problem

In the experiments, multiobjective optimization of weights of neural networks was performed with the goals of minimizing f_1 - the false positive (FP) rate of classification performed by the optimized neural network and maximizing f_2 - the true positive (TP) rate. The search space for this optimization problem is $\Omega = \mathbb{R}^k$, where $k = (N_{hid})^2 + 77N_{hid} + 2$ for the classical Recurrent Neural Network (RNN) with N_{hid} hidden neurons and $k = 4(N_{hid})^2 + 302N_{hid} + 2$ for the Long Short-Term Memory (LSTM) neural network with N_{hid} hidden neurons. In order to evaluate a solution $x \in \Omega$, the elements of the real vector x are used as weights for a neural network (an RNN or an LSTM) and the network is used to perform classification for each premise (graph node $v \in V$) and time instant $t = 1, \ldots$ in the time interval used for optimization (years 2017–2019 in this article). For each time instant t, the network gets the attributes $\mathcal{A}$ of the node v (constant in time) and incoming and outgoing movements at the node v in the time instant t as inputs. It produces two outputs $y^{(t)}_{incoming}, y^{(t)}_{outgoing} \in [0, 1]$, which are used for predicting the occurrence of incoming and outgoing movements at the node v in the time instant $t + 1$. If the value of $y^{(t)}_{incoming}$ (respectively, $y^{(t)}_{outgoing}$) is smaller than 0.5 it is predicted that incoming (respectively, outgoing) movements will not occur at the node v in the time instant $t+1$. If the value of $y^{(t)}_{incoming}$ (respectively, $y^{(t)}_{outgoing}$) is larger or equal 0.5 it is predicted that incoming (respectively, outgoing) movements will occur at the node v in the time instant $t + 1$. The predictions are compared to the actual movements, that is, the movements that indeed occurred at the node v in the time instant $t + 1$. By calculating the fraction of classifications (nodes and time instants) which correctly or incorrectly predicted the incoming or outgoing movements, the two objectives are calculated:

f_1 - the false positive (FP) rate: the fraction of situations when no movement has occurred for which the classifier predicted a movement,
f_2 - the true positive (TP) rate: the fraction of situations when a movement has occurred for which the classifier predicted a movement.

Obviously, the f_1 objective has to be minimized, because it represents situations when the classifier mistakenly predicts a movement and in reality there is none. The f_2 objective has to be maximized, because it represents situations when a movement is correctly predicted.

5 Experiments and Results

In the experiments, multiobjective optimization described in Sect. 4 was performed using the MOEA/D multiobjective evolutionary algorithm [13,28] with solutions represented as real vectors in $\Omega = \mathbb{R}^k$. The evolutionary algorithm used four crossover operators: the single-point crossover, the two-point crossover, the uniform crossover [10, p. 52–53], and the Simulated Binary Crossover (SBX) [8].

Seven mutation operators were used: five operators reordering the elements of the genotype: displacement, insertion, inversion, scramble and transpose [14] and two operators dedicated to real-valued genotypes: the uniform mutation and the polynomial mutation [9]. When a genetic operator had to be selected, an autoadaptation mechanism based on operator success rates [17] was used. Because the discussed optimization problem is a multiobjective one, the optimization algorithm produced Pareto fronts containing solutions non-dominated with respect to two objectives: f_1 - the false positive rate (minimized) and f_2 - the true positive rate (maximized). The quality of the Pareto fronts was measured using the hypervolume (HV) quality indicator [29].

The experiments consisted of two phases:

1. Tuning of the number of hidden neurons N_{hid} in the neural networks, and the parameters of the evolutionary algorithm.
2. Testing the generalization capability of the neural models optimized for the time interval 2017–2019 on data from the year 2020.

5.1 Parameter Tuning

In the first phase of the experiments, the best number of hidden neurons N_{hid} in the neural networks was selected, and the parameters of the evolutionary algorithm were tuned. The number of hidden neurons was selected from four possible values $N_{hid} \in \{2, 5, 10, 20\}$. For each neural model (RNN and LSTM), and the number of hidden neurons N_{hid}, the parameters of the evolutionary algorithm were tuned using the grid search approach with the following candidate values:

- Population size $N_{pop} \in \{50, 100, 200, 500\}$.
- Crossover probability $P_{cross} \in \{0.5, 0.6, 0.7, 0.8, 0.9, 1.0\}$.
- Mutation probability $P^{(p)}_{mut} \in \{\frac{0.2}{k}, \frac{0.5}{k}, \frac{1.0}{k}, \frac{2.0}{k}, \frac{5.0}{k}\}$ (where k is the genotype length) for mutation operators applied to individual positions in the genotype (e.g. the uniform mutation), and $P^{(g)}_{mut} \in \{0.02, 0.04, 0.06, 0.08, 0.10\}$ for mutation operators applied to the whole genotype (e.g. the scramble mutation).

For each set of values of these parameters, 10 runs of the optimization algorithm were performed with the stopping condition of $max_{FE} = 10000$ solution evaluations, and the set of parameter values for which the best (highest) median hypervolume was obtained was selected. In this phase of the experiments no additional validation of the obtained results (such as the cross-validation) was performed. Instead, in the experiments described in Sect. 5.2, the neural models were optimized using data from years 2017–2019 and were subsequently used for making predictions on data from the year 2020 (unseen during training). The values of parameters for which the best results were attained for each neural model are presented in Table 3. Figure 1 shows the median hypervolume values attained in the parameter tuning phase. Each plot shows the results for varying values of one of the parameters with the remaining parameters set to the best values obtained for a given neural model (Table 3). Clearly, the smallest number of hidden neurons ($N_{hid} = 2$) was suitable for both the LSTM and the RNN. As for the population size, both models produced the best results when optimized using an evolutionary algorithm running with the population of $N_{pop} = 100$ solutions. For crossover and mutation probabilities it is harder to draw a firm conclusion, as the quality of the results varies and different settings seem to be the most suitable for different models.

Table 3. The best values of the network size and evolutionary algorithm parameters obtained using the grid search approach for the two neural network models studied in this article.

Parameter name	Neural model	
	RNN	LSTM
Number of hidden neurons N_{hid}	2	2
Population size N_{pop}	100	100
Crossover probability P_{cross}	0.9	1.0
Mutation probability per position $P_{mut}^{(p)}$	$\frac{5.0}{k}$	$\frac{1.0}{k}$
Mutation probability per genotype $P_{mut}^{(g)}$	0.10	0.06
Obtained median hypervolume (HV)	0.8328	0.8282
Standard deviation	0.0090	0.0206

From this phase of experiments, it can be concluded, that small neural networks (with $N_{hid} = 2$ hidden neurons) perform best for the studied problem. Classical Recurrent Neural Networks (RNNs) performed somewhat better than Long Short-Term Memory (LSTM) networks as indicated by a slightly higher hypervolume (0.8328 vs. 0.8282). Also, the dispersion of the results (measured using the standard deviation) was smaller for the RNNs than for the LSTMs.

5.2 Testing the Generalization Capability of the Neural Models

In this phase of the experiments, the neural models were optimized using data from years 2017–2019 and were subsequently used for making predictions on data from the year 2020. This phase of the experiments was performed in order to test the generalization capability of the neural models, that is, the capability to perform well on data not seen during training (optimization). The optimization of neural networks was performed using the parameters obtained in Sect. 5.1 using movements from years 2017–2019 as a continuous sequence in time. After the weights of the neural networks had been optimized using the evolutionary algorithm, the networks made predictions on movements from the year 2020, which were fed to the neural networks as a new sequence, starting from $t = 1$. Each neural model (RNN and LSTM) was optimized 30 times with the stopping

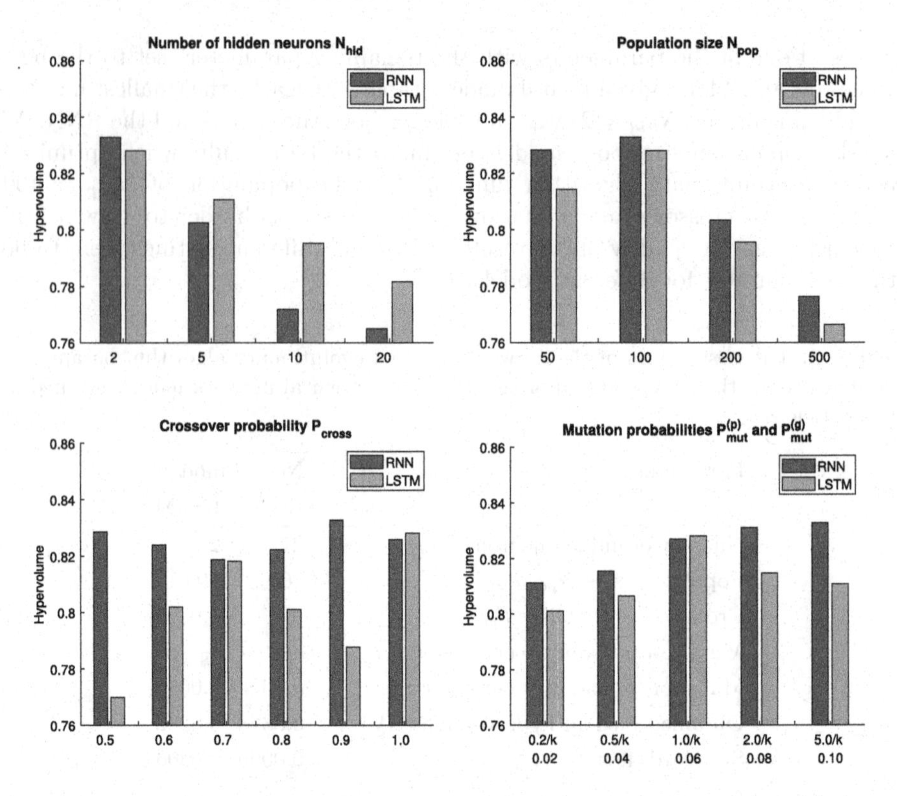

Fig. 1. The median hypervolume values attained in the parameter tuning phase. Each plot shows the results for varying values of one of the parameters with the remaining parameters set to the best values obtained for a given neural model (Table 3).

condition of $max_{FE} = 10000$ solution evaluations, and was subsequently used for making predictions. The evolutionary algorithm produces an entire Pareto front of solutions (neural networks) which differ with respect to the values of the false positive (FP) rate (objective f_1) and the true positive (TP) rate (objective f_2) attained on data from years 2017–2019. When these solutions are used for setting weights of neural networks which make predictions for the year 2020, each solution attains some other values of the FP rate (objective f_1) and the TP rate (objective f_2). It is worth noticing, that, because the objectives in the optimization problem studied in this article are the FP rate and the TP rate, the Pareto fronts can be interpreted as the Receiver Operating Characteristic (ROC) of the neural classifiers generated by the evolutionary algorithm. Similarly, the hypervolume indicator calculated for these fronts is the equivalent of the Area Under Curve (AUC) classification quality measure. The hypervolume (AUC) value calculated from non-dominated solutions generated for the year 2020 when using the RNNs was 0.6904 and when using LSTMs was 0.6777.

Figures 2 and 3 show the Pareto fronts (ROC curves) for both neural network models tested in this article obtained by taking non-dominated solutions from the union of results from 30 runs performed for each neural model. The FP and TP rates for solutions optimized on data from years 2017–2019 are marked by black dots and the FP and TP rates attained when making predictions for the year 2020 are marked by red dots. Naturally, the results for data on which the models were optimized (years 2017–2019) are better than the results obtained when making prediction on future data (year 2020), but the degradation of classification quality is not very large.

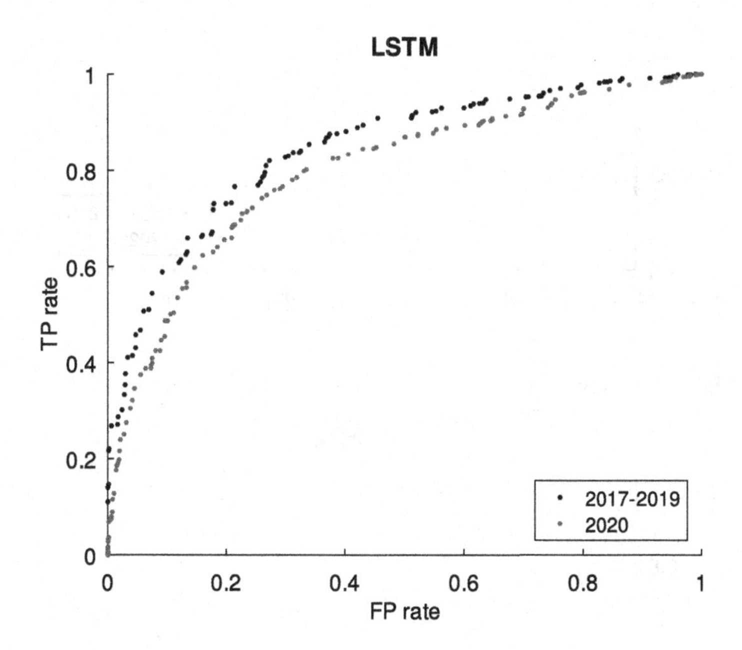

Fig. 2. The Pareto fronts (ROC curves) for the LSTM neural network.

Figure 4 shows a comparison of the results obtained when using the LSTM (orange) and RNN (blue) neural networks on future movements taken from the year 2020. Interestingly, the LSTM produced better results than the RNN for medium values of the FP rate (in the range [0.2, 0.4], approximately). As noted above, the hypervolume was better for the RNNs (0.6904 vs. 0.6777), which can be attributed to slightly better RNN performance for very large values of the FP rate (top-right corner in Fig. 4). However, from practical perspective, it is more important to obtain better TP rate values for possibly low FP rate values, because when FP rate gets close to 1.0 the classifier generates an enormous number of false alarms. Therefore, the LSTM model can be expected to be better-suited for practical applications than the RNN model.

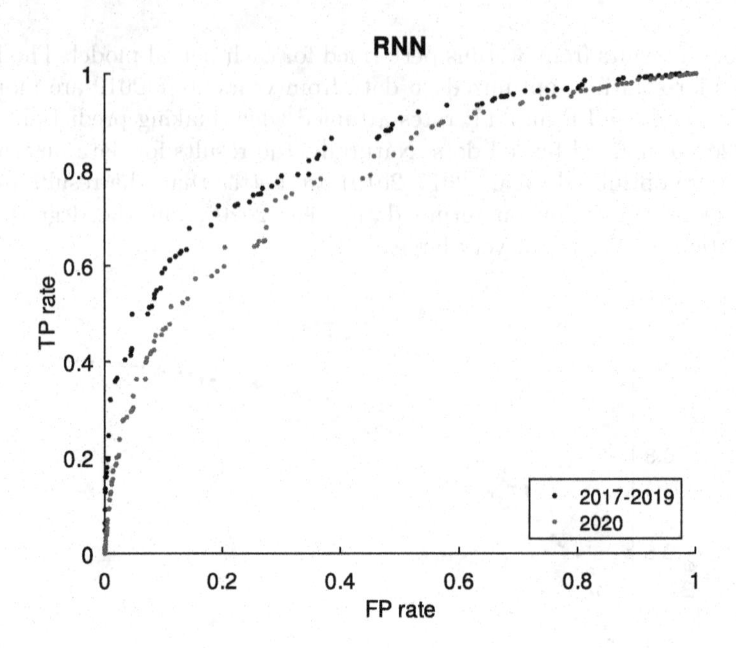

Fig. 3. The Pareto fronts (ROC curves) for the RNN neural network.

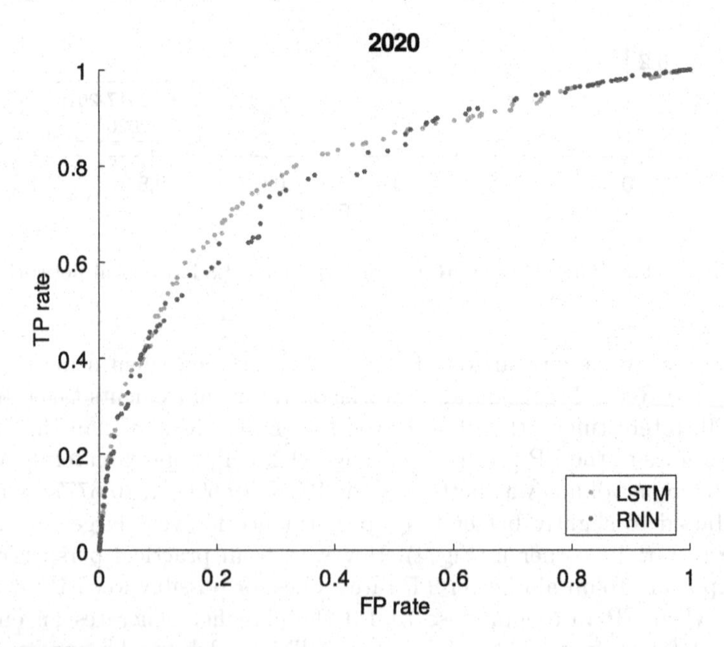

Fig. 4. The Pareto fronts (ROC curves) obtained using both neural models for the data from the year 2020.

6 Conclusion

In this article, multiobjective optimization of evolutionary neural networks was studied, with the goal of predicting animal trade movements for farms and other premises, which send and receive animals when running their businesses. Two neural models were optimized using the MOEA/D algorithm: the Long Short-Term Memory (LSTM) model and the classical Recurrent Neural Network (RNN) model. Both neural models used in the experiments shown a similar performance, with the LSTM performing slightly better on future data in the lower range of the false positive rate values, which probably makes it better-suited for practical applications.

Because animal movements allow infectious diseases to spread, predicting these movements is important from the epidemics control perspective. This article is a first stepping stone in the direction of using predictive models for planning future counter-epidemic actions based on historical records of animal movements. Future work may include using recurrent neural networks as predictive models for optimizing counter-epidemic measures, such as vaccination campaigns. Other directions for extended studies in the area presented in this article were suggested by one of the reviewers. These suggestions are interesting, but, because of limited space, they had to be left for future work. One of the possibilities is to compare the evolutionary training of neural networks with canonical algorithms such as the Scaled Conjugate Gradient (SCG) algorithm [18]. Another direction of work could be to use larger (deep) neural networks for classification. Evolutionary algorithms can work with large search spaces effectively, so evolutionary optimization of graph-to-sequence neural models such as Graph2Seq [26] can be considered. On the other hand, in this article, smaller neural models worked better for the studied problem. Thus, further studies are necessary to assess the applicability of deep neural networks to the animal movements prediction problem.

Acknowledgment. This work was supported by the Polish National Science Centre under grant no. 2015/19/D/HS4/02574. We acknowledge that the results of this research have been achieved using the DECI resource ARIS based in Greece at the National Infrastructures for Research and Technology S.A. (GRNET) with support from the PRACE aisbl under project ID DYNNETOPT. The authors would like to acknowledge Informatica Area Prevenzione of the ASL CN1 of Piedmont for the data and useful discussion.

References

1. Bajardi, P., Barrat, A., Natale, F., Savini, L., Colizza, V.: Dynamical patterns of cattle trade movements. PLOS One **6**(5), 1–19 (2011)
2. Belhaj, S., Tagina, M.: Modeling and prediction of the internet end-to-end delay using recurrent neural networks. J. Netw. **4**, 528–535 (2009)
3. Berhich, A., Belouadha, F.Z., Kabbaj, M.I.: A location-dependent earthquake prediction using recurrent neural network algorithms. Soil Dyn. Earthq. Eng. **161**, 107389 (2022)

4. Bigras-Poulin, M., Barfod, K., Mortensen, S., Greiner, M.: Relationship of trade patterns of the Danish swine industry animal movements network to potential disease spread. Prev. Vet. Med. **80**(2), 143–165 (2007)
5. Büttner, K., Krieter, J., Traulsen, A., Traulsen, I.: Static network analysis of a pork supply chain in Northern Germany - characterisation of the potential spread of infectious diseases via animal movements. Prev. Vet. Med. **110**(3), 418–428 (2013)
6. Capanema, C.G., de Oliveira, G.S., Silva, F.A., Silva, T.R., Loureiro, A.A.: Combining recurrent and graph neural networks to predict the next place's category. Ad Hoc Netw. **138**, 103016 (2023)
7. De Mulder, W., Bethard, S., Moens, M.F.: A survey on the application of recurrent neural networks to statistical language modeling. Comput. Speech Lang. **30**(1), 61–98 (2015)
8. Deb, K., Agarwal, R.: Simulated binary crossover for continuous search space. Complex Syst. **9**(2), 115–148 (1995)
9. Deb, K., Goyal, M.: A combined genetic adaptive search (GeneAS) for engineering design. Comput. Sci. Inform. **26**, 30–45 (1996)
10. Eiben, A.E., Smith, J.E.: Introduction to Evolutionary Computing, 2nd edn. Springer, Heidelberg (2015). https://doi.org/10.1007/978-3-662-44874-8
11. Ezanno, P., Arnoux, S., Joly, A., Vermesse, R.: Rewiring cattle trade movements helps to control bovine paratuberculosis at a regional scale. Prev. Vet. Med. **198**, 105529 (2022)
12. Hidano, A., Carpenter, T.E., Stevenson, M.A., Gates, M.C.: Evaluating the efficacy of regionalisation in limiting high-risk livestock trade movements. Prev. Vet. Med. **133**, 31–41 (2016)
13. Li, H., Zhang, Q.: Multiobjective optimization problems with complicated Pareto sets, MOEA/D and NSGA-II. IEEE Trans. Evol. Comput. **13**(2), 284–302 (2009)
14. Liu, C., Kroll, A.: Performance impact of mutation operators of a subpopulation-based genetic algorithm for multi-robot task allocation problems. Springerplus **5**(1), 1–29 (2016). https://doi.org/10.1186/s40064-016-3027-2
15. Marhon, S.A., Cameron, C.J.F., Kremer, S.C.: Recurrent neural networks. In: Bianchini, M., Maggini, M., Jain, L.C. (eds.) Handbook on Neural Information Processing. Intelligent Systems Reference Library, vol. 49, pp. 29–65. Springer, Berlin, Heidelberg (2013). https://doi.org/10.1007/978-3-642-36657-4_2
16. Marsot, M., Canini, L., Janicot, S., Lambert, J., Vergu, E., Durand, B.: Predicting veal-calf trading events in France. Prev. Vet. Med. **209**, 105782 (2022)
17. Michalak, K.: The Sim-EA algorithm with operator autoadaptation for the multiobjective firefighter problem. In: Ochoa, G., Chicano, F. (eds.) EvoCOP 2015. LNCS, vol. 9026, pp. 184–196. Springer, Cham (2015). https://doi.org/10.1007/978-3-319-16468-7_16
18. Moller, M.F.: A scaled conjugate gradient algorithm for fast supervised learning. Neural Netw. **6**, 525–533 (1993)
19. Salehinejad, H., Sankar, S., Barfett, J., Colak, E., Valaee, S.: Recent advances in recurrent neural networks (2018)
20. Salem, F.M.: Recurrent Neural Networks. Springer, Heidelberg (2022)
21. Sendra-Arranz, R., Gutiérrez, A.: A long short-term memory artificial neural network to predict daily HVAC consumption in buildings. Energy Build. **216**, 109952 (2020)
22. Sharma, S.D., Sharma, S., Singh, R., Gehlot, A., Priyadarshi, N., Twala, B.: Deep recurrent neural network assisted stress detection system for working professionals. Appl. Sci. **12**(17), 8678 (2022)

23. Vidondo, B., Voelkl, B.: Dynamic network measures reveal the impact of cattle markets and alpine summering on the risk of epidemic outbreaks in the Swiss cattle population. BMC Vet. Res. **14**, 88 (2018)
24. Vlad, I.M., Beciu, S., Ladaru, G.R.: Seasonality and forecasting in the Romanian trade with live animals. Agric. Agric. Sci. Proc. **6**, 712–719 (2015). Conference Agriculture for Life, Life for Agriculture
25. Wen, L., Zhou, K., Yang, S.: Load demand forecasting of residential buildings using a deep learning model. Electr. Power Syst. Res. **179**, 106073 (2020)
26. Xu, K., Wu, L., Wang, Z., Feng, Y., Sheinin, V.: Graph2Seq: graph to sequence learning with attention-based neural networks. CoRR abs/1804.00823 (2018)
27. Yu, Y., Si, X., Hu, C., Zhang, J.: A review of recurrent neural networks: LSTM Cells and network architectures. Neural Comput. **31**(7), 1235–1270 (2019)
28. Zhang, Q., Li, H.: MOEA/D: a multiobjective evolutionary algorithm based on decomposition. IEEE Trans. Evol. Comput. **11**(6), 712–731 (2007)
29. Zitzler, E., Thiele, L., Laumanns, M., Fonseca, C.M., da Fonseca, V.G.: Performance assessment of multiobjective optimizers: an analysis and review. IEEE Trans. Evol. Comput. **7**, 117–132 (2002)

Transfer of Multi-objectively Tuned CMA-ES Parameters to a Vehicle Dynamics Problem

André Thomaser[1,2(✉)], Marc-Eric Vogt[1], Anna V. Kononova[2], and Thomas Bäck[2]

[1] BMW AG, Munich, Germany
{andre.thomaser,marc-eric.vogt}@bmw.de
[2] LIACS, Leiden University, Leiden, The Netherlands
{a.m.thomaser,a.kononova,t.h.w.baeck}@liacs.leidenuniv.nl

Abstract. The conflict between *computational budget* and *quality of found solutions* is crucial when dealing with expensive black-box optimization problems from the industry. We show that through multi-objective parameter tuning of the Covariance Matrix Adaptation Evolution Strategy on benchmark functions different optimal algorithm configurations can be found for specific computational budgets and solution qualities. With the obtained Pareto front, tuned parameter sets are selected and transferred to a real-world optimization problem from vehicle dynamics, improving the solution quality and budget needed. The benchmark functions for tuning are selected based on their similarity to a real-world problem in terms of Exploratory Landscape Analysis features.

Keywords: Multi-objective · Parameter tuning · CMA-ES · Transfer learning · Vehicle dynamics · Exploratory landscape analysis

1 Introduction

The Covariance Matrix Adaptation Evolution Strategy (CMA-ES) [16,18] is a class of iterative heuristic algorithms for solving generally non-linear, non-convex, single objective, continuous optimization problems by finding a solution within the feasible set $\mathcal{X} \subset \mathbb{R}^n$ that minimizes the objective function f [27] (Sect. 2). CMA-ES has been successfully applied to many real-world optimization problems including topology optimization [13] and hyperparameter optimization of neural networks [30].

Apart from the landscape of the objective function, the performance of CMA-ES on a specific optimization problem is determined by several parameters, and also, by the variant of CMA-ES [5]. In order not to concern the user with the task of selecting the appropriate parameters of an algorithm from a wide range of different settings and variants, automatic parameter tuning as an optimization problem itself was proposed [4,14]. Hence, two optimization problems can be distinguished: solving the original problem and parameter tuning [12]. Components of the former optimization problems are the original problem and the

M. Emmerich et al. (Eds.): EMO 2023, LNCS 13970, pp. 546–560, 2023.
https://doi.org/10.1007/978-3-031-27250-9_39

algorithm to find an optimal solution for this problem; while the latter consists of the algorithm and a meta-algorithm to find optimal parameters for the algorithm to solve the original problem. The quality of solutions for the original problem is called *fitness* and the quality of the parameters of the algorithm is called *utility* [12].

When measuring the utility two main metrics can be formulated: finding the best possible solution within a given budget (fixed-budget) and finding a solution as quickly as possible with a given target quality (fixed-target) [6]. In this paper we investigate the *conflict* of these two objectives when tuning the parameters of an algorithm: solution quality and used budget. We perform multi-objective parameter tuning to obtain a Pareto front, also called "performance front" [11], consisting of non-dominated parameter sets that satisfy both the quality and budget objective (Sect. 5). Since the real-world black-box optimization problem from vehicle dynamics design (Sect. 3) is computationally expensive to evaluate, we conduct and investigate the parameter tuning on similar functions from the black-box optimization benchmark (BBOB) [17].

Relating the performance of an algorithm on synthetic benchmark functions to real-world optimization problems for transfer learning is a difficult task [41]. One way is to assess the similarity between the real-world problem and the benchmark functions. Therefore, we use the same approach as in previous work [45] by performing *Exploratory Landscape Analysis* (ELA) [32] (Sect. 4).

2 Covariance Matrix Adaptation Evolution Strategy

In every generation g of the CMA-ES [16], a population x consisting of λ offspring is sampled from a multivariate normal distribution with mean value $m^{(g)} \in \mathbb{R}^n$, covariance matrix $C^{(g)} \in \mathbb{R}^{n \times n}$ and standard deviation $\sigma^{(g)} \in \mathbb{R}_{>0}$:

$$x_k^{(g+1)} \sim m^{(g)} + \sigma^{(g)} \mathcal{N}(0, C^{(g)}) \quad \forall\, k = 1, ..., \lambda. \tag{1}$$

Moreover, the mean value $m^{(g)}$, the covariance matrix $C^{(g)}$ and the standard deviation $\sigma^{(g)}$ are adapted in each generation as described below. With the given weights w_i, the new mean value $m^{(g+1)}$ is calculated as the weighted average of μ selected parents from the population:

$$m^{(g+1)} = m^{(g)} + c_m \sum_{i=1}^{\mu} w_i (x_{i:\lambda}^{(g+1)} - m^{(g)}). \tag{2}$$

The covariance matrix $C^{(g)}$ is updated with the evolution path $p_c^{(g)} \in \mathbb{R}^n$:

$$\begin{aligned} C^{(g+1)} = {} & (1 - c_1 - c_\mu \sum_{i=1}^{\lambda} w_i) C^{(g)} \\ & + c_1 \underbrace{p_c^{(g+1)} p_c^{(g+1)^T}}_{\text{rank-one update}} + c_\mu \underbrace{\sum_{i=1}^{\lambda} w_i \, y_{i:\lambda}^{(g+1)} (y_{i:\lambda}^{(g+1)})^T}_{\text{rank-}\mu\text{ update}}, \end{aligned} \tag{3}$$

$$p_c^{(g+1)} = (1 - c_c)p_c^{(g)} + \sqrt{c_c(2 - c_c)\mu_{eff}}\frac{m^{(g+1)} - m^{(g)}}{\sigma^{(g)}}, \tag{4}$$

$$\mu_{eff} = (\sum_{i=1}^{\mu} w_i^2)^{-1}, \quad y_{i:\lambda}^{(g+1)} = \frac{x_{i:\lambda}^{(g+1)} - m^{(g)}}{\sigma^{(g)}}, \tag{5}$$

and the standard deviation $\sigma^{(g)}$ is updated with the conjugate evolution path $p_\sigma^{(g)} \in \mathbb{R}^n$:

$$\sigma^{(g+1)} = \sigma^{(g)} \exp\left(\frac{c_\sigma}{d_\sigma}\left(\frac{\left\|p_\sigma^{(g+1)}\right\|}{E\left\|\mathcal{N}(0, I)\right\|} - 1\right)\right), \tag{6}$$

$$p_\sigma^{(g+1)} = (1 - c_\sigma)p_\sigma^{(g)} + \sqrt{c_\sigma(2 - c_\sigma)\mu_{eff}}C^{(g)^{-\frac{1}{2}}}\frac{m^{(g+1)} - m^{(g)}}{\sigma^{(g)}}. \tag{7}$$

The strategy parameters λ, μ, c_1, c_c, c_{mu}, c_σ define the optimization behavior of CMA-ES and can themselves be optimized for specific functions or groups of functions [3,51].

Several CMA-ES variants have been developed so far. In this paper the following variants are considered [36,46]:

1. **Active Update** [23]: Extends the adaptation of the covariance matrix with the most successful individuals by also considering the least successful individuals with a negative factor and therefore actively reducing the probability of searching in unpromising directions.
2. **Elitism** [46]: In the standard (μ, λ)-CMA-ES the μ children replace the λ parents. In the $(\mu + \lambda)$-CMA-ES the children and parents together form the next population.
3. **Mirrored Sampling** [8]: Only half of the search points of a new population are sampled from a multivariate normal distribution, the other half is the mirror image of the first one. Mirrored sampling increases the uniform distribution of the search points.
4. **Orthogonal Sampling** [48]: Orthonormalizes the vectors of the search points with the Gram-Schmidt process [7].
5. **Threshold Convergence:** [39]: Prevents the algorithm from getting stuck in a local optimum by requiring the mutation vectors to reach a length threshold. The threshold decreases after each generation.
6. **Step-Size Adaptation:** The standard step-size control in CMA-ES is Cumulative Step-Size Adaptation (CSA) [16]. Two-Point Step-Size Adaptation (TPA) [15] uses two search points from the population and Median Success Rule (MSR) [1] uses the median fitness of the offspring to an individual from the previous iteration to adjust the step-size.
7. **Weighted Recombination** [19]: Recombination is accomplished in CMA-ES by adjusting the mean values m with a weight vector w_i.

The combination of the different variants with respect to the optimization problem can improve the performance of CMA-ES [46,47].

3 Real-World Problem Description

The anti-lock braking system (ABS) [26] and the variable damper control (VDC) [35] can significantly improve driving safety by reducing the braking distance. The ABS adjusts the brake pressure so that the brake slip remains within the optimal range, thus preventing the wheels from locking and increasing the brake performance. The VDC improves brake performance by adjusting the damper constants of the shock absorbers, which influence the wheel load and, therefore, the braking force.

The *emergency straight-line full-stop braking maneuver with ABS fully engaged* [21] is a standard maneuver in the automotive industry for assessing the braking performance of a vehicle. The braking distance y is defined as the integral of the vehicle longitudinal velocity over time from velocity $v_s = 100\,\text{km/h}$ at time t_s to $v_e = 0\,\text{km/h}$ at time t_e:

$$y = \int_{t_s}^{t_e} v(t)\, dt. \tag{8}$$

Mechanical vehicle and its control systems, the driver and the environment can be simulated via a two-track model implemented in Simulink [44]. To accurately model the steady-state and transient behavior of the tires under slip conditions the MF-Tyre/MF-Swift tire model [42], which is based on Pacejka's Magic Formula [38] is used. On a standard workstation[1], one full simulation run takes about 15 to 20 min.

The objective is to find a parameter setting x within the lower bound B_{lb} and upper bound B_{ub} that minimizes the braking distance $y(x)$. In total 28 ABS and two VDC parameters are considered.

4 Exploratory Landscape Analysis for Transfer Learning

Exploratory Landscape Analysis. (ELA) [32] quantifies high-level properties of the landscape of an optimization problem [33]. The *flacco* package provides a wide collection of ELA features [25]. The total of 68 single features, structured in six sets, are appropriate for the considered real-world problem (Sect. 3): classical ELA (distribution, level, meta) [32], information content [34], dispersion [31], linear model, nearest better clustering [24,40] and principal component.

We use the instance-generating mechanism of the BBOB function suite [17] and consider five instances of each function. To calculate the features, a Sobol' design [37,43] with 16384 samples in $[-5, 5]^{30}$ is used. The resulting set of computed features is filtered [29] to exclude features with zero standard deviation across all problems and feature pairs with Spearman's rank correlation [28] above 0.99. This leaves 39 features. The feature values are then min-max-scaled to $[0, 1]$ for equal weighting.

[1] HP Workstation Z4 G4 Intel Xeon W-2125 4.00 GHz/4.50 GHz 8.25MB 2666 4C 32 GB DDR4-2666 ECC SDRAM.

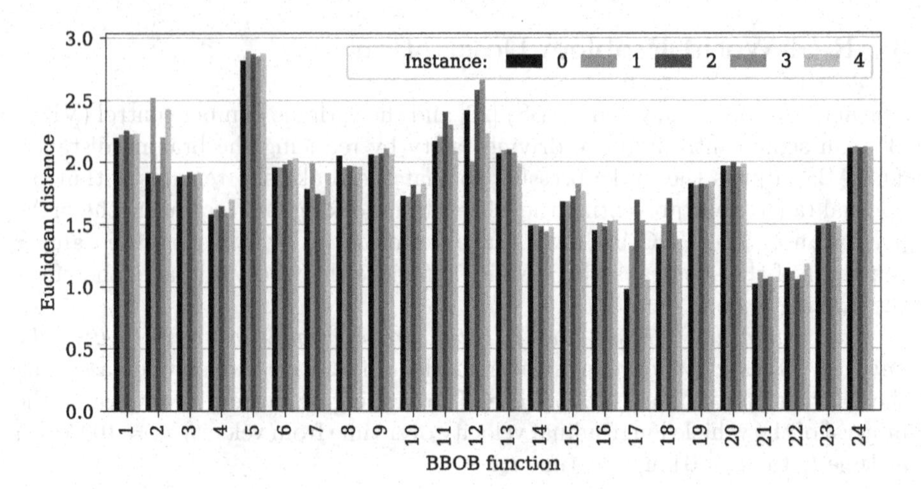

Fig. 1. Euclidean distances in the 39-dimensional ELA feature space between the considered real-world problem and 24 BBOB functions. The five instances {0, 1, 2, 3, 4} of each BBOB function are considered.

We define the similarity of two problems p_1 and p_2 as the Euclidean distance d between their feature vectors F_{p_1} and F_{p_2}:

$$d(p_1, p_2) = \|F_{p_1} - F_{p_2}\|_2 . \tag{9}$$

Using such distance, we investigate the similarity between the real-world problem and the five instances of each BBOB function based on the 39-dimensional feature vectors and conclude that such distances vary across functions and instances (Fig. 1). BBOB functions f_{17}, f_{21} are selected as *similar* to the considered real-world problem. Furthermore, to test the transferability of optimal parameters across functions, we augment the tuning reference with BBOB's sphere function f_1 as an example of a *dissimilar reference.*

5 Multi-objective Tuning of Algorithm's Parameters on Reference Functions

We define as a *meta-algorithm* an algorithm used to find the optimal set of parameters θ^* for an optimization algorithm A to solve the original real-world problem. Since the real-world problem is computationally expensive to evaluate, such meta-optimization can be performed on another (similar) problem or to increase generalisability on a set of problems – both such tasks exemplify transfer learning. In the following, we call such a function set a *tuning reference* Π. For each considered BBOB function f_j we use the five instances $i \in \{0, ..., 4\}$ as a tuning reference: $\Pi_{f_j} = \{f_{j_0}, ..., f_{j_4}\}$.

To compare the quality of a solution found by an algorithm across different problem instances and functions, we consider the distance $\Delta f = f - f^*$ to the known optimum f^* of a BBOB function f as *cost function* C:

$$C = \Delta f = f - f^*. \quad (10)$$

Considering the probabilistic nature of the algorithms, to obtain statistically meaningful results, we conduct $n_{opt} = 100$ optimization runs on each problem of the tuning reference Π. A *performance measure* over the obtained cost values $(c_1, ..., c_{n_{opt}})$ can then be calculated by a statistic h (e.g., mean, median). To focus more on average performance than peak performance and reduce the variance, we consider the median and the standard deviation across the n_{opt} optimization runs for each problem in the tuning reference:

$$h(c_1, ..., c_{n_{opt}}) = median(c_1, ..., c_{n_{opt}}) + std(c_1, ..., c_{n_{opt}}). \quad (11)$$

The quality of a parameter set θ for an algorithm A on the tuning reference Π, in the following referred to as *utility* $u(\theta, \Pi)$, is then assessed as the average performance measure h over the n_p problems in the tuning reference Π:

$$u(\theta, \Pi) = \frac{1}{n_p} \sum_{i=1}^{n_p} h(C(A(\theta), \Pi_i)). \quad (12)$$

In addition to the quality of the solution found across several optimization runs, the wall-clock time spent to find this solution is another crucial performance criterion of an optimization algorithm. On the real-world problem, the wall-clock time needed by the algorithm to generate the next population is negligible. The wall-clock time can be reduced mainly by running several simulations in parallel. Thus, the time for an optimization run correlates not directly with the evaluation budget n_{eval}, but with the number of serial iterations n_{iter}. In each iteration $n_{parallel}$ solution candidates can be evaluated simultaneously. The number of iterations depends on the population size λ, if $\lambda = n_{parallel}$, n_{iter} is equal to the number of generations $\frac{n_{eval}}{\lambda}$:

$$n_{iter} = \frac{n_{eval}}{\lambda} \left\lceil \frac{\lambda}{n_{parallel}} \right\rceil. \quad (13)$$

We assess a parameter set of CMA-ES based on two *conflicting objectives*: the utility u (Eq. 12) as a measurement of the quality of found solutions and the iteration budget n_{iter} (Eq. 13). Many algorithms exist for multi-objective hyperparameter optimization [20]. We employ the NSGA-II [10] implementation from the hyper-parameter optimization tool Optuna [2] as the *meta-algorithm*. To find a so-called performance front of Pareto optimal parameter sets for the CMA-ES algorithm, the meta-algorithm has a budget of 10,000 evaluations.

The set of algorithm's parameters being searched and assessed via multi-objective parameter tuning is specified in Table 1, based on the modular CMA-ES implementation [36,46]. Because of practical limitations in software licenses and computational resources required for the considered real-world problem, a maximum of 30 simulations can be executed in parallel. Therefore, only population sizes which are multiples of 30 are considered for CMA-ES. Infeasible

Table 1. Parameters and variants of CMA-ES with their value space for multi-objective parameter tuning with NSGA-II.

Hyperparameter	Description	Space
λ	Number of children derived from parents	{30, 60, 90}
μ_r	Ratio of parents selected from population	[0.2, 0.8]
σ_0	Initial standard deviation	]0, 1]
C_1	Learning rate rank-one update	]0, 1]
C_c	Learning rate covariance matrix adaption	]0, 1]
C_μ	Learning rate rank-μ update	]0, 1]
C_σ	Learning rate step size control	]0, 1[
Active update	Covariance matrix update variation	{on, off}
Elitism	Strategy of the evolutionary algorithm	$\{(\mu, \lambda), (\mu + \lambda)\}$
Mirrored sampling	Mutations are the mirror image of another	{on, off}
Orthogonal	Orthogonal sampling	{on, off}
Threshold	Length threshold for mutation vectors	{on, off}
Adaptation σ	How to adapt the step size σ	{CSA, TPA, MSR}
Weights	Weights for recombination	{default, equal, $\frac{1}{2}^{\lambda}$}

solutions generated during the run of CMA-ES are corrected using the "saturate" method [9].

We set the iteration budget to 4200 function evaluations to practically be able to use the same budget once optimized parameters are transferred to the real-world problem (where such budget translates to about two days of simulations for one optimization run). The meta-algorithm can select the following iteration budgets: $\{10, 20, 30, 40, 60, 80, 100, 140\}$.

5.1 Results

A set of parameters θ is tuned based on the quality of found solutions $u(\theta, \Pi)$ on the tuning reference Π and the iteration budget n_{iter}. We obtained k Pareto optimal parameter sets $\theta^*_{\Pi_{f_j},i}$, $i \in \{1, ..., k\}$ from each tuning reference Π_{f_j}, $j \in \{1, 17, 21\}$. For further investigations, we assess these parameter sets on the tuning references $\Pi_{f_{17}}$ and $\Pi_{f_{21}}$ (Fig. 2).

The resulting tuned parameter sets are shown in Table 2. It is important to mention that across all tuned parameter sets CSA is used, mirrored and orthogonal sampling are on and "elitist" is off.

Optimal parameter sets for 140 and 10 iterations on $\Pi_{f_{17}}$ and $\Pi_{f_{21}}$ are additionally assessed for all considered intermediate iteration budgets (marked with circles in Fig. 2). If the optimization run is stopped earlier or continued beyond the optimal iteration budget, another tuned parameter set would have performed better on average. It turns out that performance of the tuned parameter sets is very *sensitive* to the iteration budget – runs of the algorithm with parameters tuned for one budget result in significantly worse performance on other budgets.

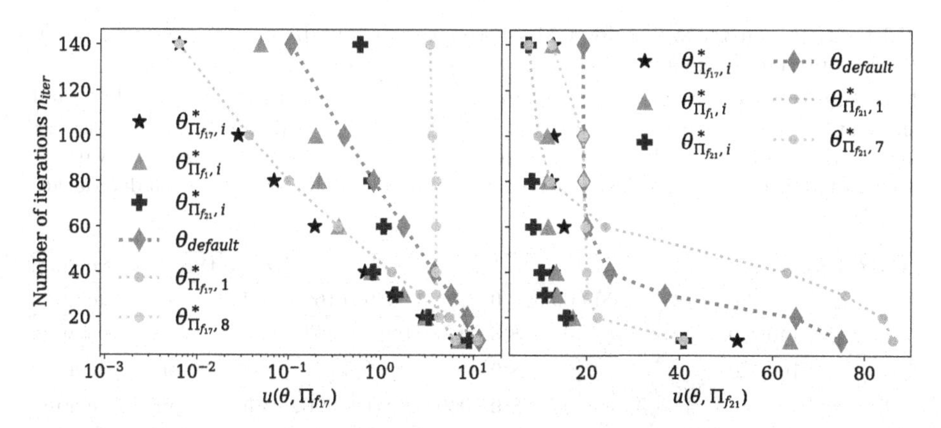

Fig. 2. Values of the quality of found solutions $u(\theta, \Pi)$ (Eq. 12) and the iteration budgets n_{iter} on the tuning references $\Pi_{f_{17}}$ (left) and $\Pi_{f_{21}}$ (right) for the obtained k Pareto optimal parameter sets $\theta^*_{\Pi_{f_j},i}$, $i \in \{1, ..., k\}$ from each tuning reference Π_{f_j}, $j \in \{1, 17, 21\}$ (Table 2). Each tuning reference consists of five instances of the corresponding BBOB function. Setting $\theta_{default}$ refers to a default parameter set (not tuned) recommended in the modular CMA-ES implementation with an adjusted population size of 30. The optimal parameter sets for 140 and 10 iterations on $\Pi_{f_{17}}$, $\Pi_{f_{21}}$ and $\theta_{default}$ are assessed for all considered iteration budgets. Symbols are connected with dotted lines when the same CMA-ES parameter set is used.

For example, $\theta^*_{\Pi_{f_{17}},8}$ (marked with yellow circles in Fig. 2, left) is tuned for 10 iterations and performs worse than the default parameters for iteration budgets grater than 30 (shown as rhombi). The other way around, $\theta^*_{\Pi_{f_{21}},1}$ (marked with silver circles in Fig. 2, right) is tuned for 140 iterations and is worse than the default parameters (shown as rhombi) for iteration budgets less than 80. Thus, a tuned parameter set performs better only for a *specific* iteration budget.

The obtained Pareto front from the multi-objective parameter tuning in Fig. 2 shows the conflict between the budget spent and the solution quality found on a set of optimization problems. With increasing the number of iterations the quality of the found solutions generally increases, exceptions can occur because of the variance across the n_{opt} optimization runs. However, the improvement decreases as the number of iterations increases. In particular, on $\Pi_{f_{21}}$, the quality of the solutions found does not increase significantly beyond 60 iterations, both for the tuned parameter sets and for the default parameter set. Thus, the obtained Pareto front indicates which number of iterations is suitable and at what point even a tuned parameter set for a larger iteration budget is *not recommended.*

Despite the dissimilarity of the landscapes of f_1, f_{17} and f_{21}, *a transfer of CMA-ES parameters* tuned on reference Π_{f_1} improves the performance of CMA-ES compared to the default parameter set on $\Pi_{f_{17}}$ and $\Pi_{f_{21}}$ (marked with triangles and diamonds in Fig. 2). Also, transferring the parameter sets tuned on $\Pi_{f_{17}}$ improves the performance of CMA-ES on $\Pi_{f_{21}}$ and vice versa, except for $\theta^*_{\Pi_{f_{21}},1}$ on $\Pi_{f_{17}}$ (marked with crosses and stars in Fig. 2).

Table 2. Parameter values for CMA-ES of the default parameter set $\theta_{default}$ and Pareto optimal parameter sets $\theta^*_{\Pi_{f_j},i}$ on the tuning references Π_{f_j}, $j \in \{1, 17, 21\}$. Each tuning reference consists of five instances of the BBOB function f_j. The search space of the parameters is given in Table 1. CSA, mirrored, orthogonal sampling is used and elitist is not across all tuned parameter sets. In the default settings these variants are all off. For parameter values between zero and one, higher values are shaded darker.

CMA-ES	n_{iter}	λ	μ_r	σ_0	c_1	c_c	c_μ	c_σ	active	threshold	weights
$\theta_{default}$	-	30	0.5000	0.2000	0.0020	0.1208	0.0049	0.1600	off	off	default
$\theta^*_{\Pi_{f_{17}},1}$	140	30	0.4506	0.3151	0.0038	0.0060	0.0241	0.3038	off	on	default
$\theta^*_{\Pi_{f_{17}},2}$	100	30	0.5282	0.3277	0.0038	0.0060	0.0241	0.4072	off	on	default
$\theta^*_{\Pi_{f_{17}},3}$	80	30	0.4506	0.3249	0.0038	0.0060	0.0241	0.4502	off	on	default
$\theta^*_{\Pi_{f_{17}},4}$	60	30	0.4506	0.2819	0.0038	0.0060	0.0241	0.6545	off	on	default
$\theta^*_{\Pi_{f_{17}},5}$	40	30	0.4506	0.4409	0.0038	0.2904	0.0246	0.9613	on	off	default
$\theta^*_{\Pi_{f_{17}},6}$	30	30	0.4506	0.4409	0.0038	0.2904	0.0246	0.9613	on	off	default
$\theta^*_{\Pi_{f_{17}},7}$	20	30	0.7140	0.5780	0.0852	0.0060	0.0246	0.9712	on	off	default
$\theta^*_{\Pi_{f_{17}},8}$	10	30	0.7140	0.7145	0.2886	0.0060	0.0246	0.9327	on	off	default
$\theta^*_{\Pi_{f_{21}},1}$	140	60	0.7208	0.8618	0.0026	0.7898	0.1301	0.1266	on	on	default
$\theta^*_{\Pi_{f_{21}},2}$	80	60	0.7208	0.8618	0.0026	0.7133	0.1254	0.5105	off	on	default
$\theta^*_{\Pi_{f_{21}},3}$	60	60	0.7208	0.8618	0.0026	0.7898	0.1301	0.8456	on	on	default
$\theta^*_{\Pi_{f_{21}},4}$	40	30	0.7208	0.7145	0.0852	0.0135	0.0173	0.9817	off	on	default
$\theta^*_{\Pi_{f_{21}},5}$	30	30	0.7751	0.7145	0.0852	0.0135	0.0173	0.9817	off	on	default
$\theta^*_{\Pi_{f_{21}},6}$	20	30	0.7208	0.7145	0.0852	0.0135	0.0173	0.9448	off	on	default
$\theta^*_{\Pi_{f_{21}},7}$	10	30	0.7208	0.2824	0.2761	0.0060	0.0173	0.9145	off	on	$\frac{1}{2}^{\lambda}$
$\theta^*_{\Pi_{f_1},1}$	140	30	0.4651	0.6867	0.0096	0.5414	0.0056	0.7292	off	off	default
$\theta^*_{\Pi_{f_1},2}$	100	30	0.4651	0.4948	0.0096	0.0021	0.0056	0.9712	off	off	default
$\theta^*_{\Pi_{f_1},3}$	80	30	0.4651	0.4948	0.0096	0.0021	0.0056	0.9712	off	off	default
$\theta^*_{\Pi_{f_1},4}$	60	30	0.4651	0.4948	0.0330	0.0021	0.0056	0.9712	off	off	default
$\theta^*_{\Pi_{f_1},5}$	40	30	0.4651	0.4948	0.0429	0.0021	0.0056	0.9712	on	off	default
$\theta^*_{\Pi_{f_1},6}$	30	30	0.4651	0.5557	0.1044	0.0021	0.0056	0.9712	on	off	default
$\theta^*_{\Pi_{f_1},7}$	20	30	0.4651	0.5557	0.1044	0.0021	0.0056	0.9712	on	off	default
$\theta^*_{\Pi_{f_1},8}$	10	30	0.4651	0.5852	0.3947	0.0021	0.0056	0.9712	off	on	default

Looking at individual values in the tuned parameter sets, the used variants mirrored and orthogonal sampling across all tuned parameter sets indicate a general increase in utility of CMA-ES for different problems. The population size was increased from the default value of 14 to at least 30 because of the number of parallel evaluations. In accordance with [48, 49], mirrored and orthogonal sampling especially improves the exploration effects of a large population. The initial step size σ_0 is higher than the default step size for all tuned parameter sets using the larger population size for an initial higher exploration. The tuned parameters are also adjusted to features of the problem and the algorithm that are identical or similar on the other problems like dimension and population size. *Transfer learning* thus can improve the performance of the algorithm.

A major difference between $\Pi_{f_{17}}$ and $\Pi_{f_{21}}$ is the comparatively worse utility of the solutions found regardless of the parameter set. On $\Pi_{f_{21}}$ the algorithm often gets stuck in a local optimum and does not find a solution near the global optimum. Therefore, exploration is especially important on $\Pi_{f_{21}}$, resulting in significant differences in parameter values of $\theta^*_{\Pi_{f_{21}},i}$ to $\theta^*_{\Pi_{f_1},i}$ and $\theta^*_{\Pi_{f_{17}},i}$. An open question is whether, instead of one single long run, two or three shorter runs result in better peak performance.

It is of interest to note that the learning rates mostly decrease or increase constantly across the iterations for the tuned parameter sets, especially c_σ. Also active covariance matrix update variation is mostly on for lower number of iterations and length threshold for mutation vectors is set to on for higher numbers of iterations. This highlights the need to conduct more research in the direction of landscape-aware adaptive parameter tuning [22].

5.2 Transfer of Tuned Parameters to the Real-World Problem

We transfer parameter sets obtained from the tuning references to the real-world problem from vehicle dynamics design and analyze whether they also improve the search performance of CMA-ES compared to the default parameter set. One optimization run with 140 iterations on the real-world problem takes about two days. Therefore, only two optimization runs for the same parameter set are conducted and only 40 and 140 iterations are considered as budget for one optimization run. Thus, we transfer the optimal parameter sets on the considered tuning references Π_{f_1}, $\Pi_{f_{17}}$, $\Pi_{f_{21}}$ for 140 iterations $\theta^*_{\Pi_{f_1},1}$, $\theta^*_{\Pi_{f_{17}},1}$, $\theta^*_{\Pi_{f_{21}},1}$ and for 40 iterations $\theta^*_{\Pi_{f_1},5}$, $\theta^*_{\Pi_{f_{17}},5}$, $\theta^*_{\Pi_{f_{21}},4}$ to the real-world problem (Fig. 3). The initialization of the mean value $m^{(0)}$ is set to the default parametrization of the control parameters for all optimizations runs.

Overall, small differences between the curves can be observed. For an iteration budget of 140 (Fig. 3, left), the solutions found with $\theta^*_{\Pi_{f_{21}},1}$ tend to be worse compared to the other parameter sets, especially at the beginning of the optimization run. This is the price for the higher exploration of the search-space. Unexpectedly, by far the shortest braking distance is found by the first run of CMA-ES with the parameter set $\theta^*_{\Pi_{f_1},1}$ (tuned on a dissimilar function f_1) within only 41 iterations and the second run can compete with the others as well. This confirms the observations on the tuning references (Sect. 5.1), where a transfer of $\theta^*_{\Pi_{f_1},i}$ also improved the performance of CMA-ES compared to the default parameter set. Tuning the parameters of CMA-ES to general problem characteristics and algorithm settings like problem dimension and population size improves the search behavior.

With an iteration budget of only 40, the variance across the found braking distances increases compared to 140 iterations. Overall, the best parameter sets

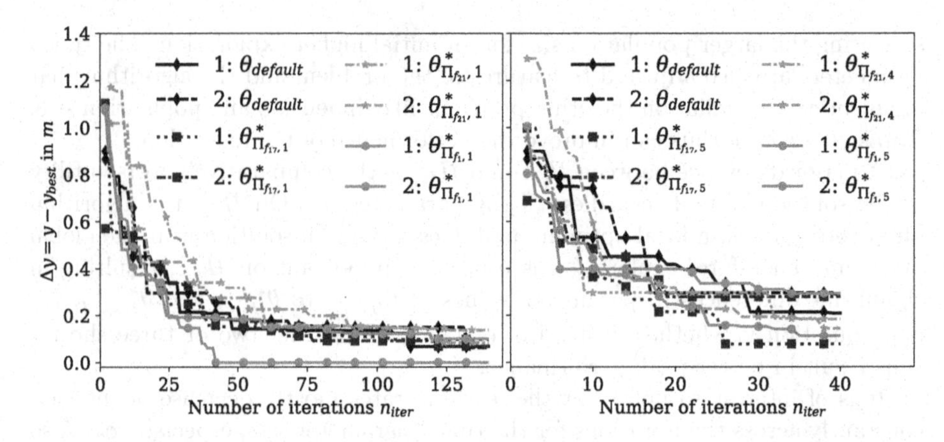

Fig. 3. Distance to the best known braking distance attainable on the considered real-world problem within intermediate computational budgets by CMA-ES configurations tuned for iteration budgets of 140 (left) and 40 (right). Line colours define the tuning reference Π_{f_j} (where index $j \in \{1, 17, 21\}$ defines the BBOB function, five instances were considered) on which CMA-ES (Table 2) has been tuned. Two independent runs are shown for each CMA-ES parameter set. Setting $\theta_{default}$ refers to a default parameter set recommended in the modular CMA-ES implementation with an adjusted population size of 30 (not tuned). (Color figure online)

for 40 iterations tend to find shorter braking distances faster than the default parameter set $\theta_{default}$.

In summary, similar phenomena in the performance of CMA-ES on benchmark functions can also be observed on the real-world problem, encouraging further investigation of transfer learning.

6 Conclusion

In this paper, we tuned different parameters and variants of CMA-ES with the two objectives of computational budget needed and quality of the solution found on functions taken from the BBOB benchmark test set. A tuned parameter set is only optimal for a given budget and problem or set of problems, so a Pareto front consisting of different parameter sets for each set of problems was derived. In order not to tune the parameters to a specific budget, the area under the empirical cumulative distribution function curve could be an alternative objective [50].

The use of certain variants of CMA-ES results in an improvement across all problems considered. One reason for this is the adaptation to general problem characteristics and algorithm settings. For example, orthogonal and mirrored sampling generally improve search performance for relatively large populations, while a higher initial step size and "threshold" improve exploration of the large search space. A simple solution besides tuning variants of CMA-ES would be to provide simple heuristics and recommendations (rules of thumb).

The values of the parameter sets tuned on different sets of problems differ significantly, but lead to similar results on the real-world problem from vehicle dynamics. A new best solution on the real-world problem was found by a tuned parameter set on the sphere function f_1. The improvement of this solution is 1.8 times better than the improvement of the best solution in a Sobol' design with 16384 samples compared to the default parameterization of the real-world problem.

The similarity of the considered benchmark functions to the real-world problem was quantified by the Euclidean distance of Exploratory Landscape Analysis feature values. The assumption of correlation between similarity of two problems quantified by ELA features and the difficulty for an algorithm configuration of solving them needs further investigations.

Acknowledgements. This paper was written as part of the project newAIDE under the consortium leadership of BMW AG with the partners Altair Engineering GmbH, divis intelligent solutions GmbH, MSC Software GmbH, Technical University of Munich, TWT GmbH. The project is supported by the Federal Ministry for Economic Affairs and Climate Action (BMWK) on the basis of a decision of the German Bundestag.

The authors would like to thank Jacob de Nobel and Diederick Vermetten for their support with the modular CMA-ES implementation.

References

1. Ait Elhara, O., Auger, A., Hansen, N.: A median success rule for non-elitist evolution strategies: study of feasibility. In: Proceedings of the 15th Annual Conference on Genetic and Evolutionary Computation, GECCO 2013, Association for Computing Machinery, New York, pp. 415–422 (2013). https://doi.org/10.1145/2463372.2463429
2. Akiba, T., Sano, S., Yanase, T., Ohta, T., Koyama, M.: Optuna: a next-generation hyperparameter optimization framework. In: Proceedings of the 25th ACM SIGKDD International Conference on Knowledge Discovery & Data Mining, KDD 2019, Association for Computing Machinery, New York, pp. 2623–2631 (2019). https://doi.org/10.1145/3292500.3330701
3. Andersson, M., Bandaru, S., Ng, A.H., Syberfeldt, A.: Parameter tuned CMA-ES on the CEC'15 expensive problems. In: 2015 IEEE Congress on Evolutionary Computation (CEC), pp. 1950–1957 (2015). https://doi.org/10.1109/CEC.2015.7257124
4. Bäck, T.: Evolutionary Algorithms in Theory and Practice: Evolution Strategies, Evolutionary Programming, Genetic Algorithms. Oxford University Press Inc, Oxford (1996). https://doi.org/10.1093/oso/9780195099713.001.0001
5. Bäck, T., Foussette, C., Krause, P.: Contemporary Evolution Strategies. Natural Computing Series, 1st edn. Springer, Berlin (2013)
6. Bartz-Beielstein, T., et al.: Benchmarking in Optimization: Best Practice and Open Issues. Technical report (2020). http://arxiv.org/2007.03488arxiv.org/pdf/2007.03488
7. Björck, Å.: Numerics of gram-Schmidt orthogonalization. Linear Algebra Appl. **197**, 297–316 (1994). https://doi.org/10.1016/0024-3795(94)90493-6

8. Brockhoff, D., Auger, A., Hansen, N., Arnold, D.V., Hohm, T.: Mirrored sampling and sequential selection for evolution strategies. In: Schaefer, R., Cotta, C., Kołodziej, J., Rudolph, G. (eds.) PPSN 2010. LNCS, vol. 6238, pp. 11–21. Springer, Heidelberg (2010). https://doi.org/10.1007/978-3-642-15844-5_2
9. Caraffini, F., Kononova, A.V., Corne, D.: Infeasibility and structural bias in differential evolution. Inf. Sci. **496**, 161–179 (2019)
10. Deb, K., Pratap, A., Agarwal, S., Meyarivan, T.: A fast and elitist multiobjective genetic algorithm: NSGA-II. IEEE Trans. Evol. Comput. **6**(2), 182–197 (2002). https://doi.org/10.1109/4235.996017
11. Dréo, J.: Using Performance Fronts for Parameter Setting of Stochastic Metaheuristics. In: Proceedings of the 11th Annual Conference Companion on Genetic and Evolutionary Computation Conference: Late Breaking Papers, pp. 2197–2200. ACM Conferences, Association for Computing Machinery, New York (2009). https://doi.org/10.1145/1570256.1570301
12. Eiben, A.E., Smit, S.K.: Parameter tuning for configuring and analyzing evolutionary algorithms. Swarm Evol. Comput. **1**(1), 19–31 (2011). https://doi.org/10.1016/j.swevo.2011.02.001
13. Fujii, G., Takahashi, M., Akimoto, Y.: CMA-ES-based structural topology optimization using a level set boundary expression-application to optical and carpet cloaks. Comput. Methods Appl. Mech. Eng. **332**, 624–643 (2018). https://doi.org/10.1016/j.cma.2018.01.008
14. Grefenstette, J.: Optimization of control parameters for genetic algorithms. IEEE Trans. Syst. Man Cybern. **16**(1), 122–128 (1986). https://doi.org/10.1109/TSMC.1986.289288
15. Hansen, N.: CMA-ES with Two-Point Step-Size Adaptation. Technical report RR-6527, INRIA (2008). https://www.hal.inserm.fr/INRIA/inria-00276854
16. Hansen, N.: The CMA Evolution Strategy: A Tutorial. Technical report (2016). https://arxiv.org/pdf/1604.00772
17. Hansen, N., Finck, S., Ros, R., Auger, A.: Real-Parameter Black-Box Optimization Benchmarking 2009: Noiseless Functions Definitions. Technical report RR-6829, INRIA (2009). https://hal.inria.fr/inria-00362633/
18. Hansen, N., Ostermeier, A.: Adapting arbitrary normal mutation distributions in evolution strategies: the covariance matrix adaptation. In: Proceedings of the IEEE International Conference on Evolutionary Computation, pp. 312–317 (1996). https://doi.org/10.1109/ICEC.1996.542381
19. Hansen, N., Ostermeier, A.: Completely derandomized self-adaptation in evolution strategies. Evol. Comput. **9**(2), 159–195 (2001). https://doi.org/10.1162/106365601750190398
20. Hernández, A.M., van Nieuwenhuyse, I., Rojas-Gonzalez, S.: A survey on multi-objective hyperparameter optimization algorithms for Machine Learning. ArXiv (2021). https://arxiv.org/pdf/2111.13755.pdf
21. International Organization for Standardization: ISO 21994:2007 - Passenger cars - Stopping distance at straight-line braking with ABS - Open-loop test method (2007)
22. Jankovic, A., Eftimov, T., Doerr, C.: Towards feature-based performance regression using trajectory data. In: Castillo, P.A., Jiménez Laredo, J.L. (eds.) EvoApplications 2021. LNCS, vol. 12694, pp. 601–617. Springer, Cham (2021). https://doi.org/10.1007/978-3-030-72699-7_38
23. Jastrebski, G.A., Arnold, D.V.: Improving evolution strategies through active covariance matrix adaptation. In: IEEE International Conference on Evolutionary Computation, pp. 2814–2821 (2006). https://doi.org/10.1109/CEC.2006.1688662

24. Kerschke, P., Preuss, M., Wessing, S., Trautmann, H.: Detecting funnel structures by means of exploratory landscape analysis. In: Proceedings of the 2015 Annual Conference on Genetic and Evolutionary Computation, pp. 265–272. ACM Digital Library, Association for Computing Machinery, New York (2015). https://doi.org/10.1145/2739480.2754642
25. Kerschke, P., Trautmann, H.: Comprehensive feature-based landscape analysis of continuous and constrained optimization problems using the r-package flacco. In: Bauer, N., Ickstadt, K., Lübke, K., Szepannek, G., Trautmann, H., Vichi, M. (eds.) Applications in Statistical Computing. SCDAKO, pp. 93–123. Springer, Cham (2019). https://doi.org/10.1007/978-3-030-25147-5_7
26. Koch-Dücker, H.-J., Papert, U.: Antilock braking system (ABS). In: Reif, K. (ed.) Brakes, Brake Control and Driver Assistance Systems. BPAI, pp. 74–93. Springer, Wiesbaden (2014). https://doi.org/10.1007/978-3-658-03978-3_6
27. Kochenderfer, M.J., Wheeler, T.A.: Algorithms for Optimization. The MIT Press, Cambridge and London (2019)
28. Kokoska, S., Zwillinger, D.: CRC Standard Probability and Statistics Tables and Formulae, CRC Press, Boca Raton (2000). https://doi.org/10.1201/b16923
29. Long, F.X., van Stein, B., Frenzel, M., Krause, P., Gitterle, M., Bäck, T.: Learning the characteristics of engineering optimization problems with applications in automotive crash. In: Proceedings of the Genetic and Evolutionary Computation Conference. GECCO 2022, Association for Computing Machinery, New York, (2022). https://doi.org/10.1145/3512290.3528712
30. Loshchilov, I., Hutter, F.: CMA-ES for Hyperparameter Optimization of Deep Neural Networks (2016). https://arxiv.org/abs/1604.07269
31. Lunacek, M., Whitley, D.: The dispersion metric and the CMA evolution strategy. In: Proceedings of the 8th Annual Conference on Genetic and Evolutionary Computation, p. 477. Association for Computing Machinery (2006). https://doi.org/10.1145/1143997.1144085
32. Mersmann, O., Bischl, B., Trautmann, H., Preuss, M., Weihs, C., Rudolph, G.: Exploratory landscape analysis. In: Lanzi, P.L. (ed.) Proceedings of the 13th Annual Conference on Genetic and Evolutionary Computation, ACM Conferences, ACM, New York, pp. 829–836 (2011). https://doi.org/10.1145/2001576.2001690
33. Mersmann, O., Preuss, M., Trautmann, H.: Benchmarking evolutionary algorithms: towards exploratory landscape analysis. In: Schaefer, R., Cotta, C., Kołodziej, J., Rudolph, G. (eds.) PPSN 2010. LNCS, vol. 6238, pp. 73–82. Springer, Heidelberg (2010). https://doi.org/10.1007/978-3-642-15844-5_8
34. Muñoz, M.A., Kirley, M., Halgamuge, S.K.: Exploratory landscape analysis of continuous space optimization problems using information content. IEEE Trans. Evol. Comput. **19**(1), 74–87 (2015). https://doi.org/10.1109/TEVC.2014.2302006
35. Niemz, T.: Reducing Braking Distance by Control of Semi-Active Suspension. Dissertation, Technische Universität Darmstadt (2007). https://tuprints.ulb.tu-darmstadt.de/912/
36. de Nobel, J., Vermetten, D., Wang, H., Doerr, C., Bäck, T.: Tuning as a Means of Assessing the Benefits of New Ideas in Interplay with Existing Algorithmic Modules. Technical report (2021). https://arxiv.org/pdf/2102.12905
37. Owen, A.B.: Scrambling sobol' and niederreiter-xing points. J. Complex. **14**(4), 466–489 (1998). https://doi.org/10.1006/jcom.1998.0487
38. Pacejka, H.B., Bakker, E.: The magic formula tyre model. Veh. Syst. Dyn. **21**(sup001), 1–18 (1992). https://doi.org/10.1080/00423119208969994

39. Piad-Morffis, A., Estévez-Velarde, S., Bolufé-Röhler, A., Montgomery, J., Chen, S.: Evolution strategies with thresheld convergence. In: 2015 IEEE Congress on Evolutionary Computation (CEC), pp. 2097–2104 (2015). https://doi.org/10.1109/CEC.2015.7257143
40. Preuss, M.: Improved topological niching for real-valued global optimization. In: Chio, C., et al. (eds.) EvoApplications 2012. LNCS, vol. 7248, pp. 386–395. Springer, Heidelberg (2012). https://doi.org/10.1007/978-3-642-29178-4_39
41. Sala, R., Müller, R.: Benchmarking for metaheuristic black-box optimization: perspectives and open challenges. In: 2020 IEEE Congress on Evolutionary Computation (CEC), pp. 1–8. IEEE (2020). https://doi.org/10.1109/CEC48606.2020.9185724
42. Siemens Digital Industries Software: Tire Simulation & Testing (2020). https://www.plm.automation.siemens.com/global/en/products/simulation-test/tire-simulation-testing.html
43. Sobol', I.M.: On the distribution of points in a cube and the approximate evaluation of integrals. Comput. Math. Math. Phys. **7**(4), 86–112 (1967). https://doi.org/10.1016/0041-5553(67)90144-9
44. The MathWorks Inc: Simulink (2015). https://www.mathworks.com/'
45. Thomaser, A., Kononova, A.V., Vogt, M.E., Bäck, T.: One-shot optimization for vehicle dynamics control systems: towards benchmarking and exploratory landscape analysis. In: Proceedings of the Genetic and Evolutionary Computation Conference Companion, pp. 2036–2045. Association for Computing Machinery, New York (2022). https://doi.org/10.1145/3520304.3533979
46. van Rijn, S., Wang, H., van Leeuwen, M., Bäck, T.: Evolving the structure of evolution strategies. In: 2016 IEEE Symposium Series on Computational Intelligence (SSCI), pp. 1–8 (2016). https://doi.org/10.1109/SSCI.2016.7850138
47. van Rijn, S., Wang, H., van Stein, B., Bäck, T.: Algorithm configuration data mining for cma evolution strategies. In: Proceedings of the Genetic and Evolutionary Computation Conference, GECCO 2017, Association for Computing Machinery, New York, pp. 737–744 (2017). https://doi.org/10.1145/3071178.3071205
48. Wang, H., Emmerich, M., Bäck, T.: Mirrored orthogonal sampling with pairwise selection in evolution strategies. In: Proceedings of the 29th Annual ACM Symposium on Applied Computing, SAC 2014, Association for Computing Machinery, New York, pp. 154–156 (2014). https://doi.org/10.1145/2554850.2555089
49. Wang, H., Emmerich, M., Bäck, T.: Mirrored orthogonal sampling for covariance matrix adaptation evolution strategies. Evol. Comput. **27**(4), 699–725 (2019). https://doi.org/10.1162/evco_a_00251
50. Ye, F., Doerr, C., Wang, H., Bäck, T.: Automated configuration of genetic algorithms by tuning for anytime performance. IEEE Trans. Evol. Comput. **26**(6), 1526–1538 (2022). https://doi.org/10.1109/TEVC.2022.3159087
51. Zhao, M., Li, J.: Tuning the hyper-parameters of CMA-ES with tree-structured Parzen estimators. In: 2018 Tenth International Conference on Advanced Computational Intelligence (ICACI), pp. 613–618 (2018). https://doi.org/10.1109/ICACI.2018.8377530

Multi-criteria Decision Making and Interactive Algorithms

Preference-Based Nonlinear Normalization for Multiobjective Optimization

Linjun He[1,2], Yang Nan[1], Hisao Ishibuchi[1(✉)], and Dipti Srinivasan[2]

[1] Department of Computer Science and Engineering, Southern University of Science and Technology (SUSTech), Shenzhen, China
nany@mail.sustech.edu.cn, hisao@sustech.edu.cn

[2] Department of Electrical and Computer Engineering, National University of Singapore, Singapore, Singapore
dipti@nus.edu.sg

Abstract. Normalization is commonly used in multiobjective evolutionary algorithms (MOEAs) in order to handle multiobjective optimization problems with differently-scaled objectives. The goal of normalization is to obtain uniformly-distributed solutions over the entire Pareto front. However, in practice, such a uniform solution set may not be a well-distributed solution set for decision making when the desired distribution of solutions is not uniform. To obtain a well-distributed solution set that meets the desired distribution, in this paper, we propose a preference-based nonlinear normalization method that transforms the objective space based on the probability integral transform theorem. As a result, the use of a standard MOEA to search for uniformly-distributed solutions in the transformed objective space leads to a desired well-distributed solution set. The proposed method is incorporated in three different MOEAs (i.e., a Pareto dominance-based MOEA, a decomposition-based MOEA, and an indicator-based MOEA). Experimental results demonstrate the flexibility and effectiveness of the proposed method. Our code is available at https://github.com/linjunhe/moea-pn.

Keywords: Evolutionary multiobjective optimization (EMO) · Preference incorporation · Decision making · Normalization

1 Introduction

Real-world multiobjective optimization problems (MOPs) usually have multiple conflicting and differently-scaled objectives [20,29]. To solve such problems, various multiobjective evolutionary algorithms (MOEAs) have been proposed in recent years [21]. In recently proposed MOEAs, normalization is usually used before environmental selection to handle badly-scaled MOPs [6,15,23,24,38]. Various studies have been conducted to examine and improve normalization methods (see Section II-B). In such studies on normalization, researchers usually implicitly assume that the desired distribution of solutions on each objective is

M. Emmerich et al. (Eds.): EMO 2023, LNCS 13970, pp. 563–577, 2023.
https://doi.org/10.1007/978-3-031-27250-9_40

uniform. As a result, the goal of normalization is to obtain uniformly-distributed solutions over the entire Pareto front in a normalized objective space.

However, for some real-world applications where the law of diminishing returns holds, a uniformly-distributed solution set may not be a well-distributed solution set for decision making [10]. For example, let us assume that we are looking for a car for our personal use based on the following two objectives: maximization of the maximum speed and minimization of the price. For the first objective, presentation of uniformly-distributed solutions to the decision maker may be acceptable when he/she does not articulate any preferences. However, for the second objective, the price distribution of available cars is generally not uniform but positively skewed [10,34] as illustrated in Fig. 1(a). This naturally raises a question: which is a better solution set between the following two sets of prices ($\times$1000$) of 10 candidate cars for decision making?

- Positively skewed distribution (see Fig. 1(b)): $A = \{40, 70, 90, 100, 120, 140, 170, 210, 270, 400\}$.
- Uniform distribution (see Fig. 1(c)): $B = \{40, 80, 120, 160, 200, 240, 280, 320, 360, 400\}$.

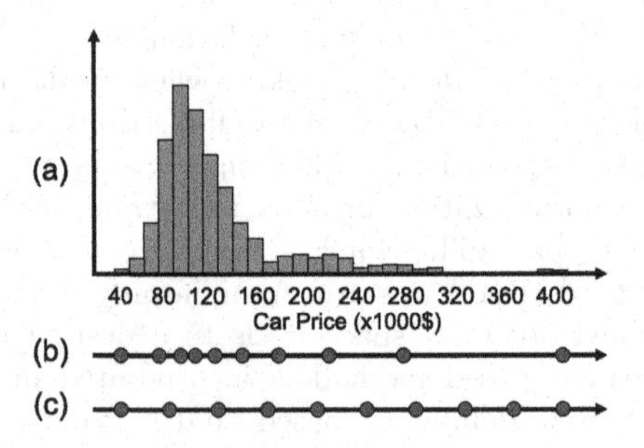

Fig. 1. Illustration of (a) histogram of car price, (b) positively skewed solutions (well-distributed for decision making), and (c) uniformly-distributed solutions.

As pointed out in [34], the presentation of candidate car set A with a biased distribution will be more useful for most people than candidate car set B with a uniform distribution. This is because the distribution of A is similar to the distribution of cars in the car market.

To obtain a well-distributed solution set like A for decision making, in this paper, a preference-based nonlinear normalization method is proposed. The contributions of this paper can be summarized as follows.

- We propose a preference-based nonlinear normalization method. Based on the preference (i.e., the desired distribution of solutions based on collected data), the objective space is transformed according to the probability integral transform theorem, such that the search of uniformly-distributed solutions in the transformed space results in well-distributed solutions in the original objective space for decision making.

- We discuss the relation between the proposed normalization method and the conventional linear normalization method. Experimental results show that the conventional linear normalization method is a special case of the proposed method.
- The proposed method can be incorporated in any existing MOEAs in a plug-in manner. This is different from existing preference incorporation methods (see Section II-C) that need a specific modification in the environmental selection mechanism of each MOEA. We incorporate the proposed method in different MOEAs to demonstrate its flexibility and effectiveness.

The rest of the paper is organized as follows. Preliminary knowledge on multiobjective optimization, linear normalization, and preference-based MOEAs are presented in Sect. 2. In Sect. 3, the proposed preference-based nonlinear normalization is presented, and its relation to linear normalization is discussed. In Sect. 4, comprehensive experiments are conducted to verify the discussed relation and to demonstrate the flexibility and effectiveness of the proposed method. In Sect. 5, we conclude the paper.

2 Preliminaries

2.1 Multiobjective Optimization Problem

A multiobjective optimization problem (MOP), which aims to minimize m conflicting objectives at the same time, can be written as follows.

$$\begin{aligned} &\text{Minimize} \quad \mathbf{f}(\mathbf{x}) = (f_1(\mathbf{x}), f_2(\mathbf{x}), \ldots, f_m(\mathbf{x}))^T, \\ &\text{subject to} \quad \mathbf{x} \in \Omega, \end{aligned} \tag{1}$$

where $f_i(\mathbf{x})$ is the i-th objective function and $\mathbf{x}$ is an n-dimensional decision vector in the feasible region $\Omega \subseteq \mathbb{R}^n$. Due to the conflicting nature of the objectives, the MOP has a set of Pareto optimal solutions, called the Pareto set. The image of the Pareto set in the objective space is called the Pareto front (PF).

2.2 Linear Normalization

To deal with MOPs with differently-scaled objectives, objective space normalization is usually performed before environmental selection of an MOEA. Each objective function in (1) is usually linearly transformed as follows.

$$\widetilde{f}_i(\mathbf{x}) = \frac{f_i(\mathbf{x}) - z_i^{\text{lb}}}{z_i^{\text{ub}} - z_i^{\text{lb}}}, i \in \{1, 2, \ldots, m\}, \tag{2}$$

where $\widetilde{f}_i(\mathbf{x})$ is the i-th normalized objective function, and z_i^{lb} and z_i^{ub} are the lower and upper bounds of the i-th objective function, respectively.

Investigation on normalization has attracted a lot of researchers' attention. As pointed out in [16,18], normalization methods can affect the performance of decomposition-based MOEAs in both positive and negative ways. Fukumoto and Oyama [12] and Liu et al. [25] investigated the impact of normalization methods for constrained decomposition-based MOEAs and multi-modal MOEAs, respectively. He et al. [17] analyzed the relation between normalization methods and weight vector scaling methods for decomposition-based MOEAs. A metric was proposed in [13] for investigating normalization methods. To make use of the advantages of normalization and reduce its negative effects, several new normalization methods were proposed. Blank et al. [3] proposed a normalization method characterized by extreme point preservation. Dynamic normalization methods were designed based on a sigmoid function in [14] or a step function in [28]. Wang et al. [36] proposed to use surrogate-based search to improve normalization bounds. Among these studies on normalization, researchers usually implicitly assume that the desired distribution of solutions on each objective is uniform. The proposed preference-based nonlinear normalization method in this paper does not rely on this assumption.

2.3 Preference Incorporation

Generally, decision makers are often interested in a small region of the PF instead of the entire PF, known as the region of interest (ROI). To search for the ROI, various approaches have been proposed to incorporate preference into MOEAs. These approaches can be roughly divided into the following four categories.

- *Objective Comparison-based Approaches.* Relative importance of each objective can be described by weights specified by the decision maker, by linguistic labels obtained from pairwise comparisons between objectives, or by pairwise trade-off information provided by the decision maker. This information is then used to modify the Pareto dominance [5,11], crowding operator [27], or quality indicator [40] to bias the population towards the ROI.
- *Solution Ranking-based Approaches.* Pairwise comparisons between solutions are made by the decision maker to learn a utility function. The learned utility function is then used to modify the dominance relation [7,19], crowding operator [1], or both of them [4] in order to identity the ROI.
- *Reference Point-based Approaches.* The decision maker's preference is articulated by a reference point or a set of reference points. Solutions close to the reference point(s) are then prioritized by modifying the crowding operator [8,26], dominance [39], or quality indicator [30] to guide the search towards the reference point(s).
- *Desirability Function-based Approaches.* For each objective, two thresholds (i.e., an absolutely satisfying objective value and a marginally infeasible objective value) are provided by the decision maker. These thresholds serve as parameters of desirability functions, by which the objective functions are transformed [35].

Most existing approaches directly incorporate preference information into the environmental selection mechanisms of MOEAs (e.g., modifying the dominance relation, the crowding operator, and the quality indicator). The proposed method focuses on the normalization part of MOEAs. The preference is incorporated by nonlinearly normalizing the objective space without any modifications on the environmental selection mechanisms of the original MOEAs. Note that the desirability function-based approach [35] also transforms the objective space. However, our method is different from [35] as follows.

1. The transformation in [35] is based on a desirability function and the decision maker is asked to provide an absolutely satisfying objective value and a marginally infeasible objective value. Our method transforms the objective space based on the probability integral transform theorem when the desired distribution of solutions (i.e., the distribution of collected data) is available.
2. The goal of [35] is to search for uniformly-distributed solutions in the ROI. Our method targets for a well-distributed solution set that meets the desired distribution of solutions for decision making.
3. The approach in [35] is designed for hypervolume-based MOEAs while our method can be integrated with any existing MOEAs.

3 Proposed Preference-Based Nonlinear Normalization

In this section, the proposed preference-based nonlinear normalization method is presented. The goal of the proposed method is to adjust the distribution of solutions for each objective. For each objective, the desired distribution of solutions can be either inferred from collected data or specified by the decision maker.

With the desired distribution, the objective is transformed by the corresponding cumulative distribution function (CDF). This transformation can be understood by the probability integral transform theorem [33]: Suppose that a random variable X has a continuous distribution for which the CDF is Φ. Then $\Phi(X)$ is a random variable having a standard uniform distribution. This theorem ensures that the desired distribution of solutions for the original objective function is converted into a uniform distribution after such transformation. As a result, we can use a standard MOEA to search for uniformly-distributed solutions in the transformed objective space. The obtained solutions are well-distributed in the original objective space. The details of the proposed nonlinear transformation are presented as follows.

Collected Data. The desired distribution of solutions can be modeled by collected data like Fig. 1(a). Since the original distribution of collected data is usually unknown, we cannot compute the exact CDF. Instead, we compute the empirical CDF. The empirical CDF is an estimate of the CDF that generates the points in the sample, and it converges with the probability of one to the original distribution according to the Glivenko-Cantelli theorem [32]. For a data set $\{x_1, x_2, ..., x_n\}$, the empirical CDF is calculated as follows:

$$\Phi(x) = \frac{1}{n} \left|\{x_i | x_i \leq x, i = 1, 2, \ldots, n\}\right|, \tag{3}$$

where $|\cdot|$ measures the cardinality of a set. In other words, the value of the empirical CDF at a given point x is the proportion of observations that are less than or equal to x.

Note that the empirical CDF is a step function that makes a discrete jump of size $1/n$ at each of the n data points. Due to its discreteness, the empirical CDF cannot be directly used as a continuous transformation function for each objective. To transform objective values at points other than the original data points, linear interpolation is performed by connecting each midpoint of adjacent two jumps (e.g., adjacent data points) in the empirical CDF to smooth the step function.

Preference Distribution. When data are unavailable, the distribution can be specified by the decision maker. We use the beta distribution to model the decision maker's preference due to its ability to take a great diversity of shapes using only two positive real number parameters α and β. By specifying the two parameters, the decision maker can express his/her preference for the desired distribution of solutions for each objective as shown in Fig. 2(a). For example, the distribution with $\alpha = 1$ and $\beta = 10$ means that the decision maker prefers to have more solutions with small objective values. As an extreme case, $\alpha = \beta = 1$ means that the decision maker has no preference about the distribution of desired solutions.

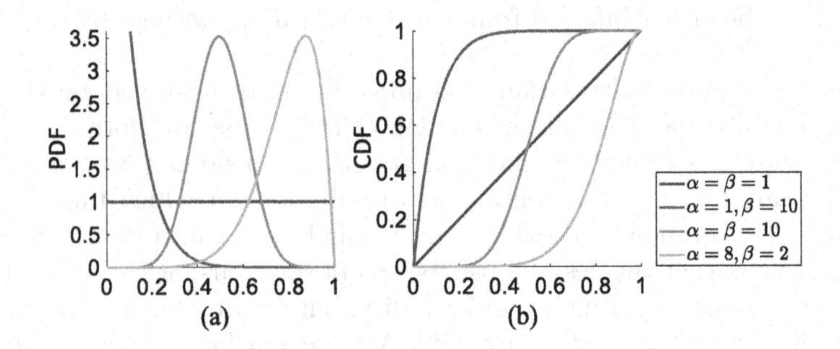

Fig. 2. Example of (a) probability density functions (PDFs) and (b) their corresponding cumulative distribution functions (CDFs) of the beta distribution with different values of α and β.

With the articulated preference distribution as a beta distribution, the objective function is transformed by the following transformation function:

$$\Phi(x \mid \alpha, \beta) = \frac{1}{B(\alpha, \beta)} \int_0^x t^{\alpha-1}(1-t)^{\beta-1} dt, \tag{4}$$

where $B(\cdot)$ is the beta function. In practice, Eq. (4) is the cumulative distribution function (CDF) of the beta distribution. Figure 2(b) shows the corresponding CDFs for the PDFs in Fig. 2(a). The slope in each CDF shows how quickly the objective value is changing after transformation. A steeper slope in the CDF means that more solutions are preferred while a gentler slope means that less solutions are preferred.

If the decision maker has no preference for an objective, the desired distribution is specified as a uniform distribution (i.e., $\alpha = \beta = 1$). For the uniform distribution $U(a, b)$, where a and b are the minimum and maximum values, its CDF is $\Phi(x) = (x - a)/(b - a)$ for $a \leq x \leq b$. By replacing x with $f_i(\mathbf{x})$, we have $\Phi(f_i(\mathbf{x})) = (f_i(\mathbf{x}) - a)/(b - a)$. When a and b are the lower and upper bounds as in (2), such transformation is exactly the same as the linear normalization. As shown in Fig. 2(b), the CDF of the uniform distribution (i.e., $\alpha = \beta = 1$) is linear. The proposed transformation performs a linear mapping from the original objective values to the range $[0, 1]$, which is exactly the same as the linear normalization. That is, when the uniform distribution is specified, the proposed normalization method is equivalent to the common linear normalization method.

Incorporation in MOEAs and Indicators. The proposed nonlinear normalization method can be easily incorporated into any MOEAs in a plug-in manner. This is because the proposed method focuses on the normalization part of MOEAs, which is an independent algorithmic component. In most existing preference-based MOEAs, the environmental selection mechanism of each algorithm is modified from its base MOEA. Such modification only works for that specific MOEA. On the contrary, the proposed method enables any MOEAs to search in a transformed objective space. In this paper, we incorporate the proposed normalization method into three MOEAs, one from each categories: SPEA2 [42] (a Pareto dominance-based MOEA), NSGA-III [6] (a decomposition-based MOEA), and SMS-EMOA [2] (an indicator-based MOEA). The resulting algorithms are denoted as SPEA2-PN, NSGA-III-PN, and SMS-EMOA-PN, respectively.

To evaluate the solutions obtained by preference-based MOEAs using reference points, Li et al. [22] transforms the obtained solutions using reference points, and the standard performance indicators are used. Inspired by [22], we use the proposed normalization method to transform the obtained solutions. After the transformation, the standard performance indicators can be used directly to evaluate the obtained solutions. In this paper, we use the hypervolume (HV) [43] and pure diversity (PD) [37] indicators. In the transformed objective space, these indicators are referred to as P-HV and P-PD.

4 Experimental Studies

In this section, we experimentally examine the proposed normalization method. First, the relation between the proposed method and the conventional linear normalization method is examined. Then, the proposed method is incorporated into different MOEAs and is examined on test problems under different preferences. We also visually examine the obtained solutions in the original and transformed objective spaces. Our experiments are conducted on PlatEMO [31]. In all the

Table 1. Average P-HV values over 51 runs obtained by the original SPEA2 and its two variants with different normalization methods.

Problem	SPEA2	SPEA2-N	SPEA2-PN
SZDT1	6.9959e-1 (2.39e-3) −	7.0343e-1 (1.04e-3) ≈	- **7.0383e-1 (3.06e-4)**
SZDT2	4.2694e-1 (8.65e-4) −	- **4.2935e-1 (2.65e-4)** ≈	4.2923e-1 (4.06e-4)
SZDT3	5.5570e-1 (4.68e-2) −	- **5.7666e-1 (2.25e-2)** ≈	5.7659e-1 (3.01e-2)
+/ − / ≈	0/3/0	0/0/3	

examined algorithms, the population size is set to 20 in order to clearly show the effect of preference incorporation. The evaluation of 50,000 solutions is used as the termination condition. Each algorithm is executed 51 times on each test problem. The Wilcoxon rank-sum test with a significance level of 0.05 is used to validate the statistical significance. The three symbols "+", "−", and "≈" mean that an algorithm is significantly better than, significantly worse than, or statistically similar to the baseline algorithm, respectively.

4.1 Relation to Linear Normalization

We have discussed the relation between the conventional linear normalization method and the proposed preference-based normalization method in Sect. 3. To experimentally demonstrate such relation, we compare the original SPEA2, SPEA2 with the linear normalization (denoted as SPEA2-N), and SPEA2 with the proposed preference-based normalization (SPEA2-PN). In SPEA2-PN, a uniform distribution (i.e., $\alpha = \beta = 1$) is applied to each objective. Since the proposed method does not modify the original SPEA2 and introduces no additional parameters, the algorithm parameters recommended in the original SPEA2 are used.

We choose ZDT1-3 [41] to examine the three algorithms. ZDT1 and ZDT2 have connected convex and connected concave PFs, respectively. ZDT3 has a disconnected PF with both concave and convex parts. Since our focus in this paper is badly-scaled MOPs, we modified the objectives of each test problem such that the first objective has the range $[0, 1000]$ and the second objective has the range $[0, 1]$. The modified MOPs with badly-scaled PFs are called SZDT1-3.

The average P-HV values and the standard deviation values are presented in Table 1. The original SPEA2 is significantly worse than SPEA2-PN (i.e., with the proposed normalization method) on the three badly-scaled test problems, while the results obtained by SPEA2-PN are statistically similar to these obtained by SPEA2-N (i.e., with the linear normalization method).

In Fig. 3, we show the the final population obtained by each algorithm on SZDT1 in a single run with the medium P-HV value. We can see that the original SPEA2 is not able to obtain uniformly distributed solutions on the PF of SZDT1, while SPEA2 with each normalization method obtains a uniform solution set. This is because the original SPEA2 maintains the diversity only relying the objective f_1 with a large scale since the value of the other objective f_2 is neglectable due to the lack of normalization. Theoretically, the same

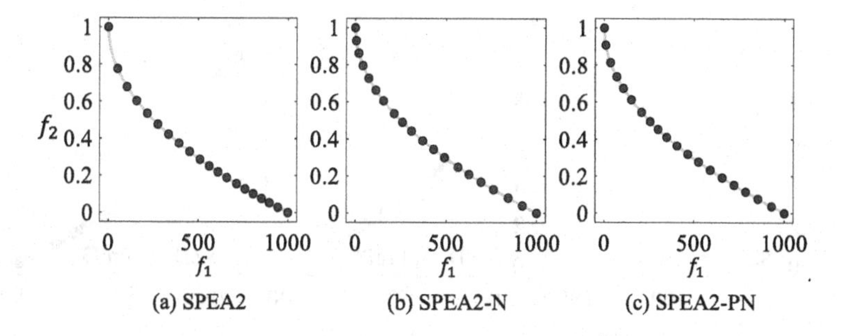

Fig. 3. Solutions obtained by the original SPEA2 and its two variants with different normalization methods on SZDT1.

results will be obtained from SPAE2-N and SPEA2-PN with $\alpha = \beta = 1$. Minor differences between Fig. 3(a) and (b) are due to randomness (e.g., different initial populations). These results clearly show that the proposed preference-based normalization method performs similarly as the conventional linear normalization method when the uniform distribution is applied for each objective in the proposed method.

4.2 Incorporation into Different MOEAs

In order to show the flexibility and effectiveness of the proposed normalization method, we incorporate it into SPEA2, NSGA-III, and SMS-EMOA. The resulting algorithms are denoted as SPEA2-PN, NSGA-III-PN, and SMS-EMOA-PN, respectively. We consider three specifications of preference (α, β) (see Fig. 2):

- Pref 1: $(1, 10)$ for f_1 and $(1, 1)$ for f_2,
- Pref 2: $(10, 10)$ for f_1 and $(1, 1)$ for f_2,
- Pref 3: $(10, 1)$ for f_1 and car price data [9] (see Fig. 1(a)) for f_2.

The MOEAs are examined by comparing their original version with its variant using the proposed normalization method. We use P-HV and P-PD to evaluate the ability of each algorithm to obtain solutions with the desired distribution. The results are presented in Tables 2 and 3. Compared with the baseline algorithms, we can see that MOEAs with the proposed method is able to find better solutions under different preferences in terms of both P-HV and P-PD. That is, the proposed normalization method is able to change the search behavior of different MOEAs and enables them to search for solutions with the desired distribution and good convergence.

The obtained solutions in a single run with the median P-HV values among 51 runs of SMS-EMOA-PN under each preference setting are shown in Fig. 4 for SZDT1-3. We can see that solutions with different distributions are found when different preference settings are used regardless of the PF shape. For example, with the first type of preference, the obtained solutions concentrate on the upper left corner of the PF as shown in Fig. 4.

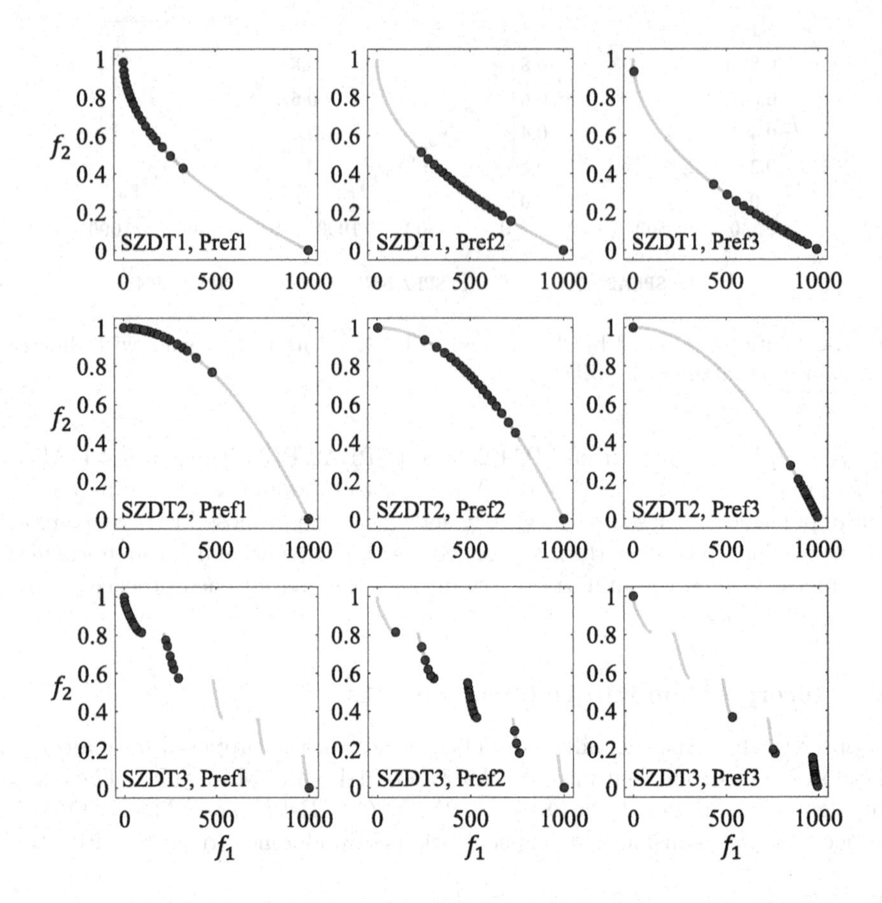

Fig. 4. Solutions obtained by SMS-EMOA-PN on SZDT1-3 with different preferences.

Table 2. Average P-HV values over 51 runs obtained by SPEA2, NSGA-III, SMS-EMOA and their variants incorporated with the proposed method.

	Problem	SPEA2		NSGA-III		SMS-EMOA	
		Original	Proposed	Original	Proposed	Original	Proposed
Pref 1	SZDT1	3.6668e−1 −	**3.8084e−1**	3.4133e−1 −	**3.7897e−1**	3.5272e−1 −	**3.8395e−1**
	SZDT2	1.8173e−1 −	**1.8256e−1**	1.8217e−1 −	**1.8255e−1**	1.8035e−1 −	**1.8438e−1**
	SZDT3	2.7447e−1 −	**2.7805e−1**	2.6290e−1 −	**2.7609e−1**	2.7270e−1 ≈	**2.7527e−1**
Pref 2	SZDT1	7.3480e−1 −	**7.4153e−1**	7.2809e−1 −	**7.3734e−1**	7.3669e−1 −	**7.4386e−1**
	SZDT2	3.6777e−1 −	**3.7531e−1**	3.6882e−1 −	**3.7541e−1**	3.7044e−1 −	**3.8079e−1**
	SZDT3	5.9604e−1 ≈	**5.9973e−1**	5.9299e−1 −	**5.9639e−1**	5.9323e−1 −	**6.0059e−1**
Pref 3	SZDT1	7.6441e−1 −	**8.2860e−1**	7.0662e−1 −	**8.1430e−1**	7.6629e−1 −	**8.2969e−1**
	SZDT2	3.8772e−1 −	**4.8247e−1**	4.1563e−1 −	**4.7781e−1**	3.8747e−1 −	**4.8571e−1**
	SZDT3	2.5070e−1 −	**3.5554e−1**	2.3818e−1 −	**3.2576e−1**	1.8924e−1 −	**3.1968e−1**
	+/ − / ≈	0/8/1		0/9/0		0/8/1	

Table 3. Average P-PD values over 51 runs obtained by SPEA2, NSGA-III, SMS-EMOA and their variants incorporated with the proposed method.

	Problem	SPEA2		NSGA-III		SMS-EMOA	
		Original	Proposed	Original	Proposed	Original	Proposed
Pref 1	SZDT1	1.0294e+3 −	**1.2592e+3**	0.8314e+3 −	**1.2403e+3**	0.7924e+3 −	**1.1645e+3**
	SZDT2	3.7400e+2 −	**6.2384e+2**	3.6494e+2 −	**6.0748e+2**	3.5841e+2 −	**4.5318e+2**
	SZDT3	7.1915e+2 −	**8.3424e+2**	7.2647e+2 −	**9.7171e+2**	6.8284e+2 −	**8.8513e+2**
Pref 2	SZDT1	1.0118e+3 −	**1.2125e+3**	0.9623e+3 −	**1.0573e+3**	0.9885e+3 −	**0.9909e+3**
	SZDT2	1.2875e+3 −	**1.2915e+3**	1.3405e+3 −	**1.4146e+3**	1.0512e+3 −	**1.1568e+3**
	SZDT3	6.6243e+2 ≈	**6.7921e+2**	6.6102e+2 −	**7.6149e+2**	6.2770e+2 −	**7.3463e+2**
Pref 3	SZDT1	1.0248e+3 −	**1.2914e+3**	1.0318e+3 −	**1.2765e+3**	0.8897e+3 −	**1.0341e+3**
	SZDT2	1.0297e+3 −	**1.5667e+3**	1.0449e+3 −	**1.3278e+3**	0.9596e+3 −	**1.2921e+3**
	SZDT3	3.7906e+2 −	**8.3445e+2**	4.6532e+2 −	**8.1875e+2**	3.1576e+2 −	**8.5598e+2**
	+/ − / ≈	0/8/1		0/9/0		0/9/0	

4.3 Analysis in the Transformed Objective Space

In the previous subsection, we demonstrated the effectiveness and flexibility of the proposed normalization method. Here, we analyze the search behavior in the transformed objective space. The solution set obtained by each of the three algorithms (SPEA2-PN, NSGA-III-PN and SMS-EMOA-PN) on SZDT1 with Pref 2 is shown in each figure in Fig. 5 in the original and transformed spaces.

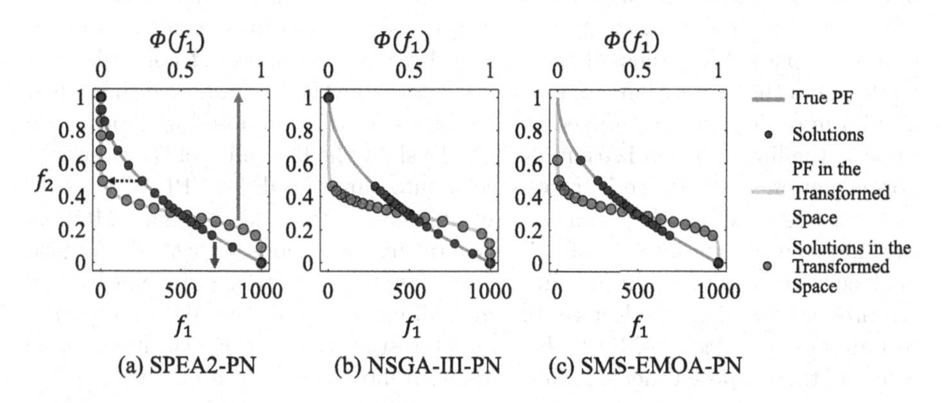

Fig. 5. Solutions obtained by (a) SPEA2-PN, (b) NSGA-III-PN, and (c) SMS-EMOA-PN on SZDT1 with Pref 2 in the original and transformed spaces.

To understand the search behavior of the proposed algorithms, we can take a look at the transformed objective space (with the top x-axis labeled as $\Phi(f_1)$) in Fig. 5. The true PF of SZDT1 (dark gray curve) is transformed to the light gray curve by the proposed nonlinear normalization method. Since the linear normalization is used for f_2, the f_2 value of each solution has no change whereas the location of each solution (i.e., f_1 value) is changed by the nonlinear transformation (see the dark dotted arrow in Fig. 5(a)). Uniformly-distributed blue solutions are obtained by SPEA2 in the transformed objective space. That is, by searching for the uniformly distributed blue solutions using SPEA2 with no modification in the transformed objective space, we can obtain the red solutions with desired distribution in the original space.

In addition, we can see that the distributions of the blue solutions by different algorithms are slightly different in Fig. 5. For SPEA2-PN, the obtained blue solutions are uniformly distributed due to the k-th nearest distance used in SPEA2. For NSGA-III-PN, the blue solutions close to each weight vector in NSGA-III are obtained. For SMS-EMOA-PN, the blue solutions that maximizes the HV value are obtained. This explains why different solution sets are obtained by the three algorithms in the original objective space even with the same preference.

5 Conclusion

In this paper, we proposed a preference-based nonlinear normalization method. Different from existing preference incorporation methods where the preference is incorporated by modifying the environmental selection mechanisms of existing MOEAs, we related preference with normalization. The preference is articulated in the form of a desired distribution of solutions, and then is incorporated into the proposed normalization method to transform the objective space according to the probability integral transform theorem. The proposed method enables any MOEAs to search for uniformly-distributed solutions in a transformed objective space and results in solutions with the desired distribution in the original objective space. We discussed the relation between the proposed normalization method and the conventional linear normalization method. We showed that when a uniform distribution is applied to each objective, the proposed method is the same as the linear normalization method. To show the flexibility of the proposed normalization method, we incorporated it into three MOEAs: SPEA2, NSGA-III, and SMS-EMOA. Experimental results showed that the standard MOEAs can find solutions of interest after incorporating the proposed method. We also analyzed the obtained solutions in the transformed space to clearly explain why the proposed method is effective. In this preliminary work, we only reported the results on two-objective MOPs. It is an interesting future research direction to examine the proposed method on MOPs with more than two objectives.

Acknowledgements. This work was supported in part by National Natural Science Foundation of China (Grant No. 61876075), Guangdong Provincial Key Laboratory (Grant No. 2020B121201001), the Program for Guangdong Introducing Innovative and Enterpreneurial Teams (Grant No. 2017ZT07X386), The Stable Support Plan Program of Shenzhen Natural Science Fund (Grant No. 20200925174447003), Shenzhen Science and Technology Program (Grant No. KQTD2016112514355531), and the National Research Foundation Singapore under its AI Singapore Programme (Award Number: [AISG-RP-2018-004]).

References

1. Battiti, R., Passerini, A.: Brain-computer evolutionary multiobjective optimization: a genetic algorithm adapting to the decision maker. IEEE Trans. Evol. Comput. **14**(5), 671–687 (2010)
2. Beume, N., Naujoks, B., Emmerich, M.: SMS-EMOA: multiobjective selection based on dominated hypervolume. Eur. J. Oper. Res. **181**(3), 1653–1669 (2007)
3. Blank, J., Deb, K., Roy, P.C.: Investigating the normalization procedure of NSGA-III. In: Proceedings of the International Conference on Evolutionary Multi-Criterion Optimization, vol. 11411, pp. 229–240. East Lansing, MI, USA (2019)
4. Branke, J., Greco, S., Słowiński, R., Zielniewicz, P.: Interactive evolutionary multiobjective optimization driven by robust ordinal regression. Bulletin of the Polish Academy of Sciences. Tech. Sci. **58**(3), 347–358 (2010)
5. Branke, J., Kaußler, T., Schmeck, H.: Guidance in evolutionary multi-objective optimization. Adv. Eng. Softw. **32**(6), 499–507 (2001)
6. Deb, K., Jain, H.: An evolutionary many-objective optimization algorithm using reference-point-based nondominated sorting approach, part I: solving problems with box constraints. IEEE Trans. Evol. Comput. **18**(4), 577–601 (2014)
7. Deb, K., Sinha, A., Korhonen, P.J., Wallenius, J.: An interactive evolutionary multiobjective optimization method based on progressively approximated value functions. IEEE Trans. Evol. Comput. **14**(5), 723–739 (2010)
8. Deb, K., Sundar, J.: Reference point based multi-objective optimization using evolutionary algorithms. In: Proceedings of the Genetic and Evolutionary Computation Conference, pp. 635–642. Seattle Washington, USA (2006)
9. Desai, A.: 100,000 UK used car data set, Version 3. https://www.kaggle.com/datasets/adityadesai13/used-car-dataset-ford-and-mercedes. Accessed 26 Sept 2022
10. Englmaier, F., Schmöller, A., Stowasser, T.: Price discontinuities in an online market for used cars. Manage. Sci. **64**(6), 2754–2766 (2018)
11. Fernandez, E., Lopez, E., Lopez, F., Coello, C.A.C.: Increasing selective pressure towards the best compromise in evolutionary multiobjective optimization: the extended NOSGA method. Inf. Sci. **181**(1), 44–56 (2011)
12. Fukumoto, H., Oyama, A.: Impact of estimation method of ideal/nadir points on practically-constrained multi-objective optimization problems for decomposition-based multi-objective evolutionary algorithm. In: IEEE Symposium Series on Computational Intelligence, pp. 2138–2145. Xiamen, China (2019)
13. He, L., Ishibuchi, H., Srinivasan, D.: Metric for evaluating normalization methods in multiobjective optimization. In: Proceedings of the Genetic and Evolutionary Computation Conference, pp. 403–411. Lille, France (2021)
14. He, L., Ishibuchi, H., Trivedi, A., Srinivasan, D.: Dynamic normalization in MOEA/D for multiobjective optimization. In: Proceedings of the IEEE Congress on Evolutionary Computation. Glasgow, Scotland, United Kingdom (2020)
15. He, L., Ishibuchi, H., Trivedi, A., Wang, H., Nan, Y., Srinivasan, D.: A survey of normalization methods in multiobjective evolutionary algorithms. IEEE Trans. Evol. Comput. **25**(6), 1028–1048 (2021)
16. He, L., Nan, Y., Shang, K., Ishibuchi, H.: A study of the naïve objective space normalization method in MOEA/D. In: IEEE Symposium Series on Computational Intelligence, pp. 1834–1840. Xiamen, China (2019)

17. He, L., Shang, K., Nan, Y., Ishibuchi, H., Srinivasan, D.: Relation between objective space normalization and weight vector scaling in decomposition-based multi-objective evolutionary algorithms. In: IEEE Transactions on Evolutionary Computation (2022) (Early Access)
18. Ishibuchi, H., Doi, K., Nojima, Y.: On the effect of normalization in MOEA/D for multi-objective and many-objective optimization. Complex Intell. Syst. **3**(4), 279–294 (2017)
19. Köksalan, M., Karahan, I.: An interactive territory defining evolutionary algorithm: iTDEA. IEEE Trans. Evol. Comput. **14**(5), 702–722 (2010)
20. Kumar, A.: A benchmark-suite of real-world constrained multi-objective optimization problems and some baseline results. Swarm Evol. Comput. **67**, 100961 (2021)
21. Li, B., Li, J., Tang, K., Yao, X.: Many-objective evolutionary algorithms: a survey. ACM Comput. Surv. **48**(1), 1–35 (2015)
22. Li, K., Deb, K., Yao, X.: R-metric: evaluating the performance of preference-based evolutionary multiobjective optimization using reference points. IEEE Trans. Evol. Comput. **22**(6), 821–835 (2018)
23. Liu, S., et al.: A self-guided reference vector strategy for many-objective optimization. IEEE Trans. Cybern. **52**(2), 1164–1178 (2022)
24. Liu, Y., Ishibuchi, H., Masuyama, N., Nojima, Y.: Adapting reference vectors and scalarizing functions by growing neural gas to handle irregular Pareto fronts. IEEE Trans. Evol. Comput. **24**(3), 439–453 (2020)
25. Liu, Y., Ishibuchi, H., Yen, G.G., Nojima, Y., Masuyama, N., Han, Y.: On the normalization in evolutionary multi-modal multi-objective optimization. In: Proceedings of the IEEE Congress on Evolutionary Computation, pp. 1–8. Glasgow, United Kingdom (2020)
26. Narukawa, K., Setoguchi, Y., Tanigaki, Y., Olhofer, M., Sendhoff, B., Ishibuchi, H.: Preference representation using Gaussian functions on a hyperplane in evolutionary multi-objective optimization. Soft. Comput. **20**(7), 2733–2757 (2016)
27. Rachmawati, L., Srinivasan, D.: Incorporating the notion of relative importance of objectives in evolutionary multiobjective optimization. IEEE Trans. Evol. Comput. **14**(4), 530–546 (2010)
28. Saxena, D.K., Kapoor, S.: On timing the nadir-point estimation and/or termination of reference-based multi- and many-objective evolutionary algorithms. In: Proceedings of the International Conference on Evolutionary Multi-Criterion Optimization, vol. 11411, pp. 191–202. East Lansing, MI, USA (2019)
29. Tanabe, R., Ishibuchi, H.: An easy-to-use real-world multi-objective optimization problem suite. Appl. Soft Comput. **89**, 106078 (2020)
30. Thiele, L., Miettinen, K., Korhonen, P.J., Molina, J.: A preference-based evolutionary algorithm for multi-objective optimization. Evol. Comput. **17**(3), 411–436 (2009)
31. Tian, Y., Cheng, R., Zhang, X., Jin, Y.: PlatEMO: a MATLAB platform for evolutionary multi-objective optimization. IEEE Comput. Intell. Mag. **12**(4), 73–87 (2017)
32. Tucker, H.G.: A generalization of the Glivenko-Cantelli theorem. Ann. Math. Stat. **30**(3), 828–830 (1959)
33. Rohatgi, V.K., Saleh, A.K.M.E.: An introduction to probability and statistics. John Wiley & Sons (2015)
34. Vosper, S., Mercure, J.F.: Assessing the effectiveness of South Africa's emissions-based purchase tax for private passenger vehicles: a consumer choice modelling approach. J. Energy South Afr. **27**(4), 25–37 (2016)

35. Wagner, T., Trautmann, H.: Integration of preferences in hypervolume-based multiobjective evolutionary algorithms by means of desirability functions. IEEE Trans. Evol. Comput. **14**(5), 688–701 (2010)
36. Wang, B., Singh, H.K., Ray, T.: Adjusting normalization bounds to improve hypervolume based search for expensive multi-objective optimization. Complex Intell. Syst, pp. 1–17 (2021)
37. Wang, H., Jin, Y., Yao, X.: Diversity assessment in many-objective optimization. IEEE Trans. Cybern. **47**(6), 1510–1522 (2017)
38. Xiang, Y., Zhou, Y., Yang, X., Huang, H.: A many-objective evolutionary algorithm with Pareto-adaptive reference points. IEEE Trans. Evol. Comput. **24**(1), 99–113 (2020)
39. Yi, J., Bai, J., He, H., Peng, J., Tang, D.: ar-MOEA: A novel preference-based dominance relation for evolutionary multiobjective optimization. IEEE Trans. Evol. Comput. **23**(5), 788–802 (2019)
40. Zitzler, E., Brockhoff, D., Thiele, L.: The hypervolume indicator revisited: on the design of Pareto-compliant indicators via weighted integration. In: Proceedings of the International Conference on Evolutionary Multi-Criterion Optimization, pp. 862–876. Matsushima, Japan (2007)
41. Zitzler, E., Deb, K., Thiele, L.: Comparison of multiobjective evolutionary algorithms: empirical results. Evol. Comput. **8**(2), 173–195 (2000)
42. Zitzler, E., Laumanns, M., Thiele, L.: SPEA2: improving the strength Pareto evolutionary algorithm. TIK-Report 103 (2001)
43. Zitzler, E., Thiele, L.: Multiobjective optimization using evolutionary algorithms - a comparative case study. In: Proceedings of the International Conference on Parallel Problem Solving from Nature, pp. 292–301. Amsterdam, The Netherlands (1998)

Incorporating Preference Information Interactively in NSGA-III by the Adaptation of Reference Vectors

Giomara Lárraga(✉), Bhupinder Singh Saini, and Kaisa Miettinen

University of Jyvaskyla, Faculty of Information Technology, P.O. Box 35 (Agora), FI-40014 University of Jyvaskyla, Finland
{giomara.g.larraga-maldonado,bhupinder.s.saini,kaisa.miettinen}@jyu.fi

Abstract. Real-world multiobjective optimization problems involve decision makers interested in a subset of solutions that meet their preferences. Decomposition-based multiobjective evolutionary algorithms (or MOEAs) have gained the research community's attention because of their good performance in problems with many objectives. Some efforts have been made to propose variants of these methods that incorporate the decision maker's preferences, directing the search toward regions of interest. Typically, such variants adapt the reference vectors according to the decision maker's preferences. However, most of them can consider a single type of preference, the most common being reference points. Interactive MOEAs aim to let decision-makers provide preference information progressively, allowing them to learn about the trade-offs between objectives in each iteration. In such methods, decision makers can provide preferences in multiple ways, and it is desirable to allow them to select the type of preference for each iteration according to their knowledge. This article compares three interactive versions of NSGA-III utilizing multiple types of preferences. The first version incorporates a mechanism that adapts the reference vectors differently according to the type of preferences. The other two versions convert the preferences from the type selected by the decision maker to reference points, which are then utilized in two different reference vector adaptation techniques that have been used in *a priori* MOEAs. According to the results, we identify the advantages and drawbacks of the compared methods.

Keywords: Multiobjective optimization · Interactive methods · Decision making · Multiobjective evolutionary algorithms · Decomposition-based MOEAs · NSGA-III

1 Introduction

Real-world problems typically involve multiple conflicting objective functions to be optimized simultaneously. Because of such conflict among the objectives, the so-called multiobjective optimization problems (MOPs) do not have a single

M. Emmerich et al. (Eds.): EMO 2023, LNCS 13970, pp. 578–592, 2023.
https://doi.org/10.1007/978-3-031-27250-9_41

solution, but a set of trade-off solutions called Pareto front. However, solving these problems aims to help a DM, which is a domain expert, to find their most preferred solution.

In the operations research field, multiobjective optimization methods have been classified according to the role of the DM during the solution process [25]. *No preference* methods do not involve the DM at all, being suitable when the DM does not have clear expectations of the final solution. *A priori* methods only ask for preference information from the DM at the beginning of the solution process. These methods are mainly utilized when the DM already knows the trade-offs between the objectives and their preferences are clear. *A posteriori* methods consider the DM's preferences after approximating the Pareto optimal set. Finally, *interactive* methods allow the DM to provide preference information iteratively during the solution process. In each iteration, the DM can learn about the trade-offs among the objectives and utilize this new insight for updating the preference information.

DMs can express their preferences in multiple ways, e.g., by providing desirable values for each objective function, specifying desirable ranges for each objective function, comparing pairs of solutions, etc. The type of preferences utilized in the solution process should depend on the DM, as according to their experience, they may feel more comfortable using a specific type of preference. In interactive methods, there should also be possible to allow the DM to change the type of preferences according to their needs [1].

MOEAs have been successfully applied to approximate the Pareto front of multiobjective optimization problems. There are mainly three types of MOEAs [35]: dominance-based, indicator-based, and decomposition-based. Decomposition-based MOEAs have received much attention from the research community in recent years for maintaining a good performance even when the number of objectives is increased, as opposed to domination-based methods [18]. These methods decompose an MOP into multiple single-objective optimization problems or multiple simpler MOPs, which are then optimized collaboratively. Such decomposition is performed using a set of reference vectors (RVs)[1] and a scalarization function. To find a representative set of near-Pareto optimal solutions, the RVs are generated uniformly in a simplex. The most popular decomposition-based MOEAs are MOEA/D [36], RVEA [5], and NSGA-III [10]. These methods typically do not consider the preferences of the DM during the solution process. However, multiple variants of such methods have been proposed that utilize preference information a priori or interactively (e.g., [14,15,19,33]).

Decomposition-based MOEAs with preference incorporation generally utilize the preferences to adapt the RVs, directing the search toward the region of interest. However, most of them work with a single type of preference, reference points being the most common. Interactive RVEA [15] is the only decomposition-based MOEA that can handle multiple types of preferences. This method allows DM to choose among four types of preferences: reference points, preferred ranges,

[1] RVs also known as weight vectors or reference points in different MOEAs. To avoid confusion, we will continue to use the term "RVs" only.

and preferred and non-preferred solutions. Then, the RVs are adapted according to the selected type of preference. As this approach does not need any other modification in the structure of the MOEA, the RV adaptation technique can be used in other decomposition-based MOEAs (e.g., MOEA/D [17]).

The performance of interactive RVEA using multiple types of preferences has been analyzed in various articles [2, 17]. From the results, it has been shown that the method's performance decreases when preferred ranges are utilized in most of the test instances. However, it is difficult to identify the reason behind this behavior, as the method has not been compared with other MOEAs using different types of preferences.

In this article, we compare three interactive versions of NSGA-III. All of them work by adapting the RVs using the preference information. However, the adaptation technique is performed differently in all versions. The first version utilizes the RV adaptation techniques employed in interactive RVEA. In this case, the RVs are adapted differently depending on the type of preference. The other two versions incorporate an intermediate step that converts the preferences from the type given by the DM to reference points. We do this, as reference points are the most common type of preference utilized by the RV adaptation techniques. This preference conversion allows us to adopt the existing RV adaptation techniques allowing the DM to select between multiple types of preferences. The main contributions of this article are the following:

- We incorporate the RVs adaptation technique from interactive RVEA into NSGA-III. Both methods are compared utilizing reference points and preferred ranges in different benchmark problems to identify their main advantages and disadvantages.
- We propose a simple conversion technique for transforming preferred ranges and preferred solutions into reference points. The resulting reference points are utilized as input for two different RV adaptation techniques: the one employed by R-NSGA-III [33] and NUMS [19]. In addition, we utilize these techniques in an interactive method for the first time.
- We compare three interactive versions of NSGA-III utilizing multiple types of preference information. These methods are tested with various benchmark problems considering 5, 7, and 9 objective functions. As involving real DMs will be very time consuming, the experimentation was conducted using an artificial DM [2], which enabled inexpensive comparison of interactive methods without real DMs.

The rest of the article is organized as follows. Section 2 discusses the background of multiobjective optimization and evolutionary algorithms. A brief overview of the a priori and interactive versions of NSGA-III available in the literature is presented in Sect. 3. Interactive RVEA, RNSGA-III, and NUMS are also described in the same section. The proposed interactive versions of NSGA-III and the mechanism to convert preferred ranges and preferred solutions into reference points are described in Sect. 4. Section 5 compares the three interactive versions of NSGA-III presented in this article using various benchmark problems. Finally, we conclude this article in Sect. 6.

2 Background

2.1 Multiobjective Optimization

A multiobjective optimization problem can be mathematically formulated as follows:

$$\begin{aligned} \text{minimize} \quad & \mathbf{F}(\mathbf{x}) = (f_1(\mathbf{x}), \ldots, f_k(\mathbf{x})) \\ \text{subject to} \quad & \mathbf{x} \in S, \end{aligned} \tag{1}$$

where f_i $(i = 1, \ldots, k)$ are the k conflicting objective functions (with $k \geq 2$). $S \subset \mathbb{R}^n$ is the feasible set of decision vectors $\mathbf{x} = (x_1, ..., x_n)^T$ with n decision variables. There is a corresponding objective vector $\mathbf{F}(\mathbf{x})$ for every feasible decision vector $\mathbf{x}$. The problem can involve equality and inequality constraints that must be satisfied by the decision vectors for them to be feasible. It is not possible for all of the objective functions in (1) to reach their optimal values simultaneously due to the conflicts between them. Given two solutions $\mathbf{x}^* \in S$ and $\mathbf{x}' \in S$, $\mathbf{x}^*$ dominates $\mathbf{x}'$ if and only if $f_i(\mathbf{x}^*) \leq f_i(\mathbf{x}')$ for all $i = 1, \ldots k$, and $f_j(\mathbf{x}^*) < f_j(\mathbf{x}')$ for at least one index $j = 1, \ldots, k$. If there is no solution $\mathbf{x} \in S$ that dominates solution $\mathbf{x}^* \in S$, then $\mathbf{x}^*$ is Pareto optimal. The set of all Pareto optimal solutions is known as Pareto optimal set, and the corresponding objective vectors compose the Pareto front.

The best and worst objective function values in the Pareto front are known as ideal $\mathbf{z}^*$ and a nadir $\mathbf{z}^{nad}$ points, respectively. The ideal point can be computed by minimizing each objective function separately. Usually, calculating the nadir point requires computing the whole Pareto set. However, it can be approximated using a pay-off table [25], or other means [11].

In real-world problems, a DM is usually involved in the solution process. A DM is typically interested in a part of the Pareto front close to their preferences, named region of interest. DM can express their preference information in multiple ways [3,23]. In this article, we are interested in the following types of preferences:

- Giving a reference point $\mathbf{r} = (r_1, ..., r_k)$, where each r_i is a desirable value (also known as aspiration level) for the objective function f_i $(i = 1, ..., k)$.
- Selecting t (with $t \geq 1$) most preferred solutions from a solution set. We denote them by $PS = [\mathbf{ps}_1, ..., \mathbf{ps}_t]$.
- Specifying preferred ranges for the objective functions. The preferred range for an objective f_i $(i = 1, ..., k)$ is denoted by $[f_i^l, f_i^u]$, being f_i^l and f_i^u the lower and upper bounds, respectively. Then, the preferred ranges for all the objectives is a k-dimensional hyper-box $PR = [f_1^l, f_1^u] \times, ..., \times [f_k^l, f_k^u]$.

In interactive methods, the DM provides preference information progressively. The intervals in which MOEAs ask for preference information from the DM are known as iterations. In most cases, iterations take place every N_{gen} generations, where N_{gen} is a parameter that is set before the method begins. Typically, two stages can be observed in an interactive solution process: the learning and decision phases [24]. During the learning phase, the DM explores different parts of the Pareto front until they find a region of interest. Then, in the decision phase, the DM tries to fine-tune the solutions in the region of interest until finding the most preferred solution.

2.2 Evolutionary Algorithms

MOEAs are capable of generating an approximation of the Pareto optimal front in a single run [9]. They can be classified [35] according to their structure into: *dominance-based MOEAs*, which compare solutions utilizing Pareto dominance-based mechanisms; *indicator-based MOEAs* that use quality indicators as selection criteria; and *decomposition-based MOEAs*, which decompose the MOP into multiple single objective optimization problems or a set of simpler MOPs, which are optimized collaboratively.

In this article, we are interested in decomposition-based MOEAs. These methods need two main components: a set of RVs (typically uniformly distributed in the objective space) and a scalarization function. RVs can be generated using a simplex lattice design [7]. The number of vectors generated by such a method is given by $\binom{q+k-1}{k-1}$, where q is a parameter to control the density of the solutions and k is the number of objective functions. Scalarizing functions map objective vectors to real-value scalars and are used to evaluate the solutions of a section of the objective space. Such sections evolve in the direction of the RV associated with them. Among the most well-known decomposition-based MOEAs are MOEA/D [36], RVEA [5], and NSGA-III [10].

3 Related Work

Multiple versions of NSGA-III with a priori and interactive preference incorporation have been proposed in the literature. Such methods typically modify the distribution of the RVs according to the preferences provided by the DM. As a result, the method does not provide an approximation of the complete Pareto front, but the obtained solutions focus on a region of interest. In this section, we describe the existing versions of NSGA-III that consider the preferences of the DM. We classify these methods according to the type of preference incorporation: a priori and interactive. We also describe some methods utilized in the rest of the article: interactive RVEA, R-NSGA-III, and NUMS.

A Priori Methods: In the article where NSGA-III was proposed [10], the authors suggested a mechanism to incorporate preferences a priori by filtering the RVs utilized by the method. A similar approach was used by Yan et al. [34] and da Silva et al. [8,30,31]. Cheng et al. [6] proposed an a priori version of NSGA-III that requires the DM to specify a central RV, which is utilized to adapt the complete set of RVs. P-NSGA-III [4] also modifies the RVs according to the preference information, but it asks the DM for information about the importance of objectives. Finally, R-NSGA-III [33] receives one or multiple reference points from the DM, which are used to adapt the RVs toward one or multiple regions of interest. The RVs are updated at each generation, as the process involves normalizing the reference points utilizing an approximation of the ideal and nadir points, which change depending on the obtained solutions.

Interactive Methods: Mnasri et al. [27] proposed PI-NSGAIII-VF, the first interactive version of NSGA-III. This method is a hybrid of NSGA-III and a strategy for incorporating DM's preferences on any MOEA: PI-EMO-VF [12]. PI-NSGAIII-VF asks the DM to classify a set of solutions during each iteration. Such information is utilized to approximate a value function progressively. Then, a stopping condition is set up according to this value function. T-NSGA-III [22] modifies the selection procedure of NSGA-III to incorporate the DM's preferences in the form of preferred ranges. Such ranges are utilized to transform the objective values into a new coordinate system defined by the upper and lower bounds for each objective function specified by the DM. I-NSGA-III-PLVF [20] asks the DM to score a set of solutions. The method learns a value function using the provided scores, which is then employed to model the DM's preferences. IOPIS-NSGA-III [29] asks the DM for a reference point, which is utilized to create a (typically lower-dimensional) preference-incorporated space (consisting of a set of scalarization functions) to reformulate the optimization problem. Finally, PI-NSGA-III-PC-INK [26] modifies the dominance relation using polyhedral cones constructed using the preference information. This method shows a set of solutions in each iteration, from which the DM can select the preferred one.

Interactive RVEA. Interactive RVEA [15] is the only decomposition-based MOEA that provides the DM with multiple options to give preferences. The method initializes a set of uniformly distributed RVs, which are adapted differently according to the type of preferences selected by the DM. For reference points and preferred solutions, the RVs are redirected toward the preference. For non-preferred solutions, the RVs closer to such solutions are removed, while the rest are kept. A Latin hyper-cube sampling is utilized for preferred ranges. Then, the obtained vectors are normalized into unit vectors, which replace the initial RVs. This method needs a parameter $v \in (0, 1)$ to control the spread of the RVs. A small value of v results in RVs close to the preference information, while a value close to 1 will produce more sparse RVs. In the rest of this article, we will refer to this RV adaptation technique as *IRA*.

R-NSGA-III. R-NSGA-III [33] is an a priori method that requires the DM to provide one or multiple reference points. As a result, the method will provide one set of solutions close to each of the provided reference points. R-NSGA-III adapts the RVs at each generation to incorporate the preferences of the DM. First, the reference points provided by the DM are normalized using the normalization procedure of NSGA-III. Then, a set of uniformly distributed RVs is generated and shrunk using a spread parameter that controls the size of the region of interest. The intercepts of the unit hyperplane and the vectors from the ideal point to each normalized reference point are computed. Finally, the obtained values are utilized to shift the shrunken RVs to the unit hyperplane. The same procedure is performed for each reference point provided by the DM. In addition, the extreme points of the hyperplane are also added to the set of RVs. In the rest of this article, we will refer to this technique as *RPA*.

NUMS. Li et al. [19] proposed a nonuniform mapping scheme (NUMS) to map a set of uniformly distributed RVs on a canonical simplex to new positions close to a reference point provided by the DM. In this case, the mapping function is nonlinear and is determined by an RV's position in relation to the pivot point. A pivot point represents the region of interest on a simplex and determines its position. The nonuniform mapping is utilized to bias the RVs toward the pivot point. The method gives the option of keeping the boundary of the simplex or not. In addition, a spread parameter is needed for this method, which is a value between 0 and 1, representing the relative ratio of the size of the region of interest with respect to the Pareto front. This method has been applied to MOEA/D both a priori and interactively.

4 Proposal

In this article, we propose three interactive versions of NSGA-III that can utilize multiple types of preference information: reference points, preferred ranges, and preferred solutions. The proposed methods do not change the structure of NSGA-III but only modify the distribution of the RVs to obtain solutions in a region of interest.

The first interactive NSGA-III proposed on this article incorporates the IRA technique to handle multiple types of preferences. We will refer to this method as *iNSGA-III-IRA*. It is worth noting that the possibility of having this type of method has been mentioned in [29]. However, the authors compared this method with interactive RVEA using only reference points. Although some experiments with interactive RVEA have shown a lower performance when using preferred ranges [2], it is difficult to identify if this behavior is related to the performance of RVEA or the IRA technique. In this article, we compare iNSGA-III-IRA with interactive RVEA utilizing reference points, preferred ranges, and preferred solutions to identify the potential and drawbacks of the IRA technique in problems with different features.

The second and third methods proposed in this article adopt a preference conversion layer that converts preferred ranges and preferred solutions into reference points. The preference conversion layer allows us to utilize reference-point-based RV adaptation techniques from the literature but with other types of preference. In a real-world scenario, the DM is unaware of the preferences conversion, as it is an intermediate layer between the user interface and the method. The conversion of the preference information is performed as follows:

- *Preferred ranges:* Let $PR = [[f_i^l, f_i^u], \ldots, [f_k^l, f_k^u]]$ be a preferred range given by the DM. We can obtain a reference point $\boldsymbol{r} = [f_1^l, f_2^l, \ldots, f_k^l]$, or alternatively $\boldsymbol{r} = [f_1^u, f_2^u, \ldots, f_k^u]$.
- *Preferred solutions:* If the DM selects a single solution ($\boldsymbol{ps}$) in an iteration, it can be directly utilized as a reference point. If more than one solution is selected (PS), a reference point can be computed as $\boldsymbol{r} = [max(PS_1), max(PS_2), \ldots, max(PS_k)]$.

It is worth noting that when selecting a set of preferred solutions, the DM can provide preferences with different ideologies in mind. The case we considered here is the simplest one, where the DM chooses solutions close to each other, indicating that they are interested in a specific region of interest. However, this is not always the case, as it is also possible that the DM selects solutions on different parts of the Pareto front. The interpretation in such a case can be divided into two. The first option is that the DM is interested in multiple regions of interest and wants to obtain more solutions for all of them. In such a case, the conversion would not lead to a single reference point but a set of them. This can easily be handled by the RPA technique, as it can distribute the RVs among multiple regions of interest. However, if a technique such as NUMS is utilized, it would be necessary to make separate runs for each reference point. The second cause is related to the learning process of the DM. Suppose the solutions are too sparse in the Pareto front. In that case, it can also mean that the DM still does not have a clear idea of their preferences and that the method should keep providing solutions in a region of interest that cover most of the Pareto front. Both types of interpretations together with different mechanisms to compute a suitable reference point are subject to further research.

In this article, we utilize two reference point-based RV adaptation techniques with NSGA-III: NUMS, and RPA (we refer to these algorithms as iNSGA-III-NUMS and iNSGA-III-RPA, respectively). To make these methods interactive, we run them multiple times changing the preference information. As the experiments only involve benchmark problems, no further changes are needed. However, it is worth noting that when utilizing interactive methods in real-world problems, some general properties need to be considered to reduce the cognitive load of the DM [1,16,32].

Figure 1 illustrates the structure of one iteration for each proposed method. The yellow line represents the stages involved in iNSGA-III-IRA. In this case, the preferences of the DM are received directly by the IRA method, as it can already handle multiple types of preference. Then, both iNSGA-III-NUMS (blue line) and iNSGA-III-RPA (red line) convert the preferences into reference points, which are utilized as input for the RV adaptation technique. For all the proposed methods, the adapted reference vectors are used by NSGA-III to direct the search toward a region of interest.

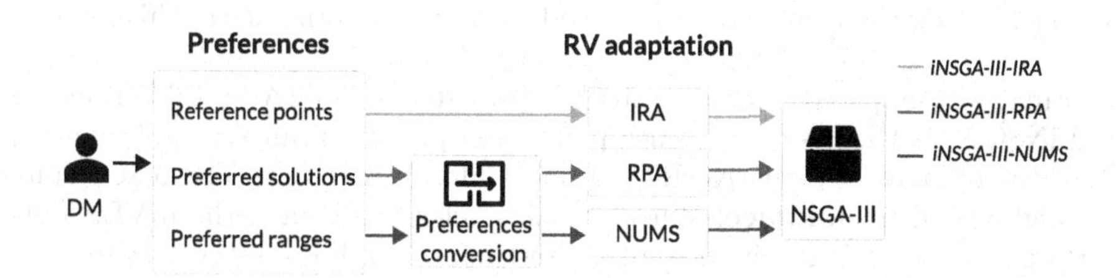

Fig. 1. Structure of an iteration of each of the methods proposed in this article.

Our proposal is the first attempt to unify preference information provided by the DM, which is an important research topic for both a priori and interactive methods [28]. In addition, although we only consider two RV adaptation techniques, a similar structure can be utilized with other methods.

5 Algorithmic Comparison

In this section, we compare the three interactive versions of NSGA-III presented in this article utilizing ADM-II [2], an artificial DM capable of comparing interactive methods using reference points, preferred ranges, and preferred and non-preferred solutions. The performance evaluation using ADM-II is divided into two stages: one for the learning phase and another for the decision phase. The main difference between the two stages is how ADM-II computes the preference information. For the learning phase, the preferences are distributed on the Pareto front, indicating that the DM is still exploring multiple regions of interest. For the decision phase, the preferences are closer to a specific region of interest, fine-tuning the solutions belonging to it. Each iteration, ADM-II generates a reference point that is utilized in the method and when applying some of the performance indicators (e.g., R-IGD). If the method requires preferred ranges, they are obtained by perturbing the reference point. This feature of ADM-II will allow us to evaluate if the solutions provided by the proposed methods are in the same region of interest after changing the type of preferences.

For the experimentation, we considered four iterations for the learning phase and three for the decision phase. After each iteration, ADM-II computes the performance indicators for the solutions obtained by the methods. Then, a cumulative indicator value is calculated for each phase as suggested in [2]. We run the methods 15 times for each problem with a given number of objectives. We considered two problems of the DTLZ benchmark [13] in the experiments: DTLZ1 and DTLZ3, with 5, 7, and 9 objectives. The number of variables is given by $10 + k - 1$. The maximum number of generations for each iteration was set to 200 for each method. The spread parameter was set to 0.2 for all the methods.

We utilized two performance indicators with ADM-II: R-IGD [21] and the composite front contribution (CFC) [2]. R-IGD measures the convergence and diversity of the obtained solutions. On the other hand, CFC measures the number of non-dominated solutions provided by each method when constructing a composite front. The more non-dominated solutions a method has contributed to constructing the composite front, the better the performance of this method is.

First, we compared interactive RVEA (referred as iRVEA in Table 1 and 2) and iNSGA-III-IRA using reference points and preferred ranges. Table 1 and 2 show the obtained cumulative R-IGD and CFC, respectively. The best results are highlighted in **boldface**. When utilizing reference points, the iRVEA outperforms iNSGA-III-IRA in most of the test instances, both in the learning and decision phase, according to the cumulative R-IGD and CFC. Such values mean that the solutions obtained by iRVEA are closer to the preference information and that this method produces more non-dominated solutions than iNSGA-III-IRA. When using preferred ranges, the R-IGD values indicate a better performance of iNSGA-III-IRA in DTLZ1, while iRVEA obtained better results for DTLZ3. However, according to the cumulative CFC, iNSGA-III-IRA obtained more non-dominated solutions in most test instances.

Table 1. Cumulative R-IGD for iRVEA and iNSGA-III using reference points and preferred ranges.

Problem	k	Phase	Reference points				Preferred ranges			
			iRVEA		iNSGA-III-IRA		iRVEA		iNSGA-III-IRA	
			Mean	Std	Mean	Std	Mean	Std	Mean	Std
DTLZ1	5	Learning	**2.5333**	0.1772	2.5910	0.2128	2.7213	0.3402	**2.6263**	0.3677
		Decision	1.9421	0.2115	**1.8894**	0.2673	2.1656	0.2291	**1.9068**	0.4622
	7	Learning	**2.7599**	0.1845	3.1748	0.4511	3.1315	0.6090	**3.0019**	0.8780
		Decision	**2.0264**	0.3609	3.0600	1.1414	3.0205	0.5322	**2.5845**	1.1048
	9	Learning	**2.6071**	0.1974	2.7505	0.2968	3.2852	0.5054	**2.9947**	0.5919
		Decision	**1.8139**	0.1492	2.5000	0.6562	3.1749	0.6887	**2.3387**	1.0481
DTLZ3	5	Learning	**0.2460**	0.2358	0.3289	0.3686	**0.5332**	0.1669	1.0590	0.7600
		Decision	**0.3913**	0.6630	0.4887	1.1339	**0.2312**	0.2518	0.9985	0.9803
	7	Learning	**0.6357**	0.5161	0.7592	0.2355	**1.9082**	0.4623	3.6905	1.4945
		Decision	**0.6928**	0.7482	0.7321	0.7137	**0.7134**	0.4282	3.3938	1.4959
	9	Learning	**0.6371**	0.3334	0.8224	0.4365	**0.2361**	0.7682	3.6308	1.3574
		Decision	**0.4576**	0.3578	0.8221	0.7066	**0.4497**	0.3196	3.4340	1.9351

Then we compared iNSGA-III-IRA, iNSGA-III-NUMS, and iNSGA-III-RPA using reference points, preferred ranges, and preferred solutions. It is worth noting that R-IGD cannot be utilized to compare methods using preferred solutions. For this reason, we only considered the cumulative CFC for that type of preference. The obtained results are shown in Table 3. The best results are highlighted in **boldface**. According to the cumulative R-IGD and CFC, iNSGA-III-IRA showed a better performance than the compared methods in most test instances when utilizing reference points. For preferred ranges, the solutions obtained by iNSGA-III-RPA are closer to the preference information and provide more non-dominated solutions than the compared methods. For preferred solutions, we selected five solutions as preferences for each iteration. According to the results, iNSGA-III-IRA obtained many more non-dominated solutions. However, this is due to how the IRA technique handles sets of preferred solutions. It computes a region of interest for each solution the DM selects without considering if they overlap. The other compared methods convert the preferred solutions into a reference point, making it easier to compare their performance. Among them, iNSGA-III-RPA obtained more non-dominated solutions in most test instances.

5.1 Discussion

When comparing iNSGA-III-IRA and interactive RVEA we could notice that the latter obtained better results when using reference points. Although iNSGA-III-IRA obtained better R-IGD values only on DTLZ1 utilizing preferred ranges, we can notice that it produces more non-dominated solutions than the compared method in both DTLZ1 and DTLZ3. A more extensive experimentation is needed to have a clearer idea of the features of the problems in which iNSGA-III-IRA

Table 2. Cumulative CFC for iRVEA and iNSGA-III using reference points and preferred ranges.

Problem	k	Phase	Reference points				Preferred ranges			
			iRVEA		iNSGA-III-IRA		iRVEA		iNSGA-III-IRA	
			Mean	Std	Mean	Std	Mean	Std	Mean	Std
DTLZ1	5	Learning	443.2	68.0835	**454.8**	53.7788	111.6	112.6634	**485.6**	31.7055
		Decision	**387.6**	2.2000	369.1	8.1664	283.5	206.3745	**357.9**	34.0307
	7	Learning	**343.5**	18.7150	138.5	57.1896	116.2	99.8637	**261.4**	60.8805
		Decision	**269.7**	0.6403	199.1	79.6485	168.0	183.7145	**235.4**	22.0191
	9	Learning	**671.1**	40.2006	257.8	149.4957	218.4	187.6205	**574.7**	60.7339
		Decision	**517.4**	1.6248	465.0	71.8053	249.4	293.6914	**443.5**	56.3777
DTLZ3	5	Learning	367.4	111.3240	**441.2**	85.5871	413.3	128.8682	**477.0**	43.0976
		Decision	309.8	148.9495	**334.6**	72.7010	202.3	89.5255	**358.0**	40.9072
	7	Learning	**292.7**	62.9842	169.6	95.9252	**346.1**	109.1040	221.0	91.3893
		Decision	**214.5**	98.5792	192.7	52.2323	131.1	172.7637	**176.4**	66.6381
	9	Learning	**611.2**	75.5299	415.0	130.3580	**698.8**	96.7831	532.0	102.3426
		Decision	**407.2**	180.1426	367.8	132.7764	301.2	222.4984	**409.1**	92.2339

outperforms interactive RVEA. Also, at the moment it is not possible to compare the performance of these methods using preferred and non-preferred solutions due to the lack of performance indicators utilizing these types of preferences.

After comparing iNSGA-III-NUMS and iNSGA-III-RPA with iNSGA-III-IRA we could notice that the preference conversion did not affect the quality of the solutions. According to the obtained results, the IRA method performs best using reference points. However, utilizing the preference conversion mechanism helps get better results with preferred ranges and preferred solutions. It is worth noting that iNSGA-III-RPA performed better than iNSGA-III-NUMS in most test instances. Considering other types of preference conversions and more RV adaptation techniques are interesting future research directions.

Using ADM-II enabled inexpensive comparison of interactive methods without real DMs. However, it needs performance indicators to evaluate the solutions provided by each method. There are only a few performance indicators for MOEAs with preference incorporation. In addition, most of them can only compare methods that utilize reference points. ADM-II can use R-IGD for preferred ranges due to the mechanism utilized to generate the preference information. When generating the preferred ranges, ADM-II employs an equivalent reference point, which is then used in the performance indicator. Comparing methods that utilize preferred solutions is still an open problem. In this case, although the cumulative CFC gives us an idea of the number of non-dominated solutions provided by each method, it is difficult to measure the quality of such solutions without an additional quality indicator.

Table 3. Cumulative R-IGD and CFC for iNSGA-III-IRA, iNSGA-III-NUMS, and iNSGA-III-RPA. Column k indicates the number of objectives, P is the phase of the solution process (L: learning, D: decision), and the method names are shorten as IRA, NUMS, and RPA.

Problem	k	P	R-IGD						CFC					
			IRA		NUMS		RPA		IRA		NUMS		RPA	
			Mean	Std	Mean	Std	Mean	Std	Mean	Std	Mean	Std	Mean	Std
Reference points														
DTLZ1	5	L	**2.5171**	0.1017	3.5837	0.6634	2.5746	0.1360	425.0	81.0086	**452.2**	65.0443	409.9	113.7580
		D	1.9301	0.1754	2.0761	0.2814	**1.9064**	0.1693	**387.1**	3.4482	360.2	7.3185	367.6	7.5657
	7	L	**2.5484**	0.1501	9.2395	1.0619	2.7300	0.3176	**329.4**	39.0620	108.8	66.5219	132.3	49.3053
		D	**1.8378**	0.1492	6.0133	2.2759	1.9784	0.3551	**269.0**	1.5492	47.8	38.6000	190.5	79.6382
	9	L	**2.5164**	0.1352	5.6678	1.1235	2.5593	0.1840	**655.8**	63.3574	1.9	3.2696	422.4	133.4820
		D	**1.8940**	0.3560	4.7159	1.2630	1.9412	0.4530	**517.6**	2.1541	2.2	4.9759	493.3	4.4508
DTLZ3	5	L	0.3032	0.1339	0.4785	0.0955	**0.2405**	0.0782	336.6	117.5884	**444.7**	66.8581	412.4	79.1052
		D	0.4480	0.6420	0.2555	0.1301	**0.1335**	0.0482	293.5	146.2315	363.4	3.2311	**370.2**	6.9397
	7	L	**0.6406**	0.2752	6.7688	2.5678	0.6795	0.6611	**257.4**	80.5409	67.7	27.7995	157.5	72.7451
		D	**0.4848**	0.4233	4.4562	3.2389	1.3165	1.9153	**199.3**	96.5578	13.5	10.1316	178.5	60.7260
	9	L	**0.5366**	0.2089	9.0865	1.2953	0.6724	0.3866	**629.7**	60.9230	94.6	46.7979	497.1	63.9022
		D	**0.3523**	0.2050	7.4008	0.6856	0.8317	1.1832	**514.5**	8.9917	24.2	16.7021	435.7	69.2272
Preferred ranges														
DTLZ1	5	L	2.9547	0.3119	2.9987	0.4754	**2.4959**	0.1699	91.5	118.4924	**482.6**	18.3314	443.9	66.3151
		D	1.9964	0.1886	2.1821	0.5939	**1.8669**	0.2768	101.0	135.5699	**363.1**	10.2806	362.0	20.7605
	7	L	3.0137	0.3383	8.2075	1.6523	**2.4159**	0.1714	49.3	51.7012	76.1	57.9473	**312.6**	48.8512
		D	2.7146	0.9704	4.4261	2.8460	**1.7150**	0.0142	35.6	68.3728	12.4	21.7127	**251.6**	1.3565
	9	L	3.2866	0.3085	6.1723	1.6570	**2.5955**	0.2913	294.8	141.4205	25.2	42.2583	**644.0**	46.4909
		D	2.1963	0.5379	4.4676	1.2529	**1.7571**	0.1213	202.8	276.1850	17.5	42.3019	**492.9**	4.7634
DTLZ3	5	L	0.3937	0.1353	1.6153	0.2997	**0.1137**	0.0433	215.7	135.9059	446.4	70.3210	**464.3**	57.3882
		D	0.3835	0.3890	1.0444	0.1871	**0.1062**	0.0416	122.7	95.7831	365.0	4.5826	**368.2**	13.3626
	7	L	1.7638	0.4380	8.2372	2.0041	**1.0111**	0.6701	**331.8**	122.8184	73.2	25.4982	284.8	32.3691
		D	0.6916	0.8202	5.9940	2.4421	**0.2821**	0.1333	113.1	124.3627	20.7	18.9634	**192.7**	51.9077
	9	L	1.7413	0.3917	9.8321	0.6294	**1.1918**	0.7182	**695.1**	104.4581	152.9	74.4613	595.2	50.4892
		D	0.5013	0.5152	6.8605	2.0710	**0.2970**	0.1104	261.0	217.8454	31.9	26.4290	**422.9**	76.1294
Preferred solutions														
DTLZ1	5	L	-	-	-	-	-	-	**1205.2**	407.4918	398.3	96.2310	440.6	80.3308
		D	-	-	-	-	-	-	**1866.3**	37.4514	349.6	36.9681	377.5	2.2472
	7	L	-	-	-	-	-	-	**1237.9**	307.3807	101.1	59.6899	300.3	76.0514
		D	-	-	-	-	-	-	**1267.5**	12.3713	48.4	39.4416	254.0	1.5492
	9	L	-	-	-	-	-	-	**4053.6**	753.0172	16.7	47.4406	560.6	103.1079
		D	-	-	-	-	-	-	**2492.2**	5.4000	11.9	34.3874	485.7	21.8406
DTLZ3	5	L	-	-	-	-	-	-	**1599.5**	345.5605	404.2	107.6362	463.5	54.9877
		D	-	-	-	-	-	-	**1895.4**	3.3226	348.7	21.6474	378.6	1.4967
	7	L	-	-	-	-	-	-	**1508.1**	409.6194	65.6	33.4610	194.3	91.0066
		D	-	-	-	-	-	-	**1263.8**	24.1031	28.8	20.4392	245.7	21.9456
	9	L	-	-	-	-	-	-	**4075.1**	704.5970	136.7	60.7833	583.7	96.0532
		D	-	-	-	-	-	-	**2472.6**	25.2198	51.2	42.5977	485.4	15.6729

6 Conclusions

In this article, we proposed three interactive versions of NSGA-III. These methods incorporate the preferences of the DM by adapting the set of RVs. iNSGA-III-IRA utilizes the RV adaptation technique used by interactive RVEA. However, when comparing both methods, we could identify that the performance of interactive RVEA is better when utilizing reference points. However, when using preferred ranges, iNSGA-III-IRA obtained better results for one of the test problems with 5, 7, and 9 objectives. Extended experimentation considering more realistic problems is needed to identify the types of problems in which each method performs better. We also proposed a mechanism to convert preferred ranges and preferred solutions into reference points. The obtained reference points were utilized as input for the NUMS and RPA techniques. The performance of iNSGA-III-NUMS and iNSGA-III-RPA compared with iNSGA-III-IRA suggest that the obtained results still reflect the preferences after the conversion. In addition, iNSGA-III-RPA obtained better results than the compared methods using preferred ranges. Converting between more types of preferences is subject to further research and also considering other types of RV adaptation techniques.

Acknowledgments. This research was supported by the Academy of Finland (grant number 322221). The research is related to the thematic research area DEMO (Decision Analytics utilizing Causal Models and Multiobjective Optimization, jyu.fi/demo) of the University of Jyväskylä.

References

1. Afsar, B., Miettinen, K., Ruiz, F.: Assessing the performance of interactive multiobjective optimization methods: a survey. ACM Comput. Surv. **54**(4), 1–27 (2021). https://doi.org/10.1145/3448301
2. Afsar, B., Ruiz, A.B., Miettinen, K.: Comparing interactive evolutionary multiobjective optimization methods with an artificial decision maker. Complex Intell. Syst. **2021**, 1–17 (2021). https://doi.org/10.1007/S40747-021-00586-5
3. Bechikh, S., Kessentini, M., Ben Said, L., Ghédira, K.: Chapter four - preference incorporation in evolutionary multiobjective optimization: a survey of the state-of-the-art. In: Hurson, A.R. (ed.) Advances in Computers, vol. 98, pp. 141–207. Elsevier (2015). https://doi.org/10.1016/bs.adcom.2015.03.001
4. Bi, X., Yu, D., Liu, J., Hu, Y.: A preference-based multi-objective algorithm for optimal service composition selection in cloud manufacturing. Int. J. Comput. Integr. Manuf. **33**(8), 751–768 (2020). https://doi.org/10.1080/0951192X.2020.1775298
5. Cheng, R., Jin, Y., Olhofer, M., Sendhoff, B.: A reference vector guided evolutionary algorithm for many-objective optimization. IEEE Trans. Evol. Comput. **20**(5), 773–791 (2016). https://doi.org/10.1109/TEVC.2016.2519378
6. Cheng, R., Rodemann, T., Fischer, M., Olhofer, M., Jin, Y.: Evolutionary many-objective optimization of hybrid electric vehicle control: from general optimization to preference articulation. IEEE Trans. Emerg. Topics Comput. Intell. **1**(2), 97–111 (2017). https://doi.org/10.1109/TETCI.2017.2669104
7. Cornell, J.A.: Experiments with Mixtures: Designs, Models, and the Analysis of Mixture Data. Wiley, Hoboken (2011)
8. Da Silva, I.R.S., De Alencar, J.E.A., De Andrade Lira Rabelo, R.: A preference-based multi-objective demand response mechanism. In: 2020 IEEE Congress on Evolutionary Computation, CEC 2020 - Conference Proceedings. Institute of Electrical and Electronics Engineers Inc. (2020). https://doi.org/10.1109/CEC48606.2020.9185875

9. Deb, K.: Multi-Objective Optimization using Evolutionary Algorithms. Wiley, Chichester (2001)
10. Deb, K., Jain, H.: An evolutionary many-objective optimization algorithm using reference-point-based nondominated sorting approach, part I: solving problems with box constraints. IEEE Trans. Evol. Comput. **18**(4), 577–601 (2014). https://doi.org/10.1109/TEVC.2013.2281535
11. Deb, K., Miettinen, K., Chaudhuri, S.: Towards an estimation of nadir objective vector using a hybrid of evolutionary and local search approaches. IEEE Trans. Evol. Comput. **14**(6), 821–841 (2010)
12. Deb, K., Sinha, A., Korhonen, P.J., Wallenius, J.: An interactive evolutionary multiobjective optimization method based on progressively approximated value functions. IEEE Trans. Evol. Comput. **14**(5), 723–739 (2010). https://doi.org/10.1109/TEVC.2010.2064323
13. Deb, K., Thiele, L., Laumanns, M., Zitzler, E.: Scalable Test Problems for Evolutionary Multiobjective Optimization, pp. 105–145. Springer, London (2005). https://doi.org/10.1007/1-84628-137-7_6
14. Gong, M., Liu, F., Zhang, W., Jiao, L., Zhang, Q.: Interactive MOEA/D for multi-objective decision making. In: Proceedings of the 13th Annual Conference on Genetic and Evolutionary computation - GECCO 2011. ACM, New York (2011)
15. Hakanen, J., Chugh, T., Sindhya, K., Jin, Y., Miettinen, K.: Connections of reference vectors and different types of preference information in interactive multiobjective evolutionary algorithms. In: 2016 IEEE Symposium Series on Computational Intelligence, SSCI 2016, Proceedings. IEEE (2017). https://doi.org/10.1109/SSCI.2016.7850220
16. Lárraga, G., Miettinen, K.: A general architecture for generating interactive decomposition-based MOEAs. In: Rudolph, G., Kononova, A.V., Aguirre, H., Kerschke, P., Ochoa, G., Tušar, T. (eds.) PPSN 2022. LNCS, pp. 81–95. Springer, Cham (2022). https://doi.org/10.1007/978-3-031-14721-0_6
17. Lárraga, G., Miettinen, K.: Interactive MOEA/d with multiple types of preference information. In: Proceedings of the Genetic and Evolutionary Computation Conference Companion, pp. 1826–1834. GECCO 2022, Association for Computing Machinery, New York, NY, USA (2022). https://doi.org/10.1145/3520304.3534013
18. Li, K.: Decomposition multi-objective evolutionary optimization: From state-of-the-art to future opportunities. CoRR abs/2108.09588 (2021). arxiv.org/abs/2108.09588
19. Li, K., Chen, R., Min, G., Yao, X.: Integration of preferences in decomposition multiobjective optimization. IEEE Trans. Cybern. **48**(12), 3359–3370 (2018). https://doi.org/10.1109/TCYB.2018.2859363
20. Li, K., Chen, R., Savic, D., Yao, X.: Interactive decomposition multiobjective optimization via progressively learned value functions. IEEE Trans. Fuzzy Syst. **27**(5), 849–860 (2019). https://doi.org/10.1109/TFUZZ.2018.2880700
21. Li, K., Deb, K., Yao, X.: R-metric: evaluating the performance of preference-based evolutionary multiobjective optimization using reference points. IEEE Trans. Evol. Comput. **22**(6), 821–835 (2018). https://doi.org/10.1109/TEVC.2017.2737781
22. Li, L., Chen, H., Li, J., Jing, N., Emmerich, M.: Preference-based evolutionary many-objective optimization for agile satellite mission planning. IEEE Access **6**, 40963–40978 (2018). https://doi.org/10.1109/ACCESS.2018.2859028
23. Miettinen, K., Hakanen, J., Podkopaev, D.: Interactive nonlinear multiobjective optimization methods. In: Greco, S., Ehrgott, M., Figueira, J.R. (eds.) Multiple Criteria Decision Analysis. ISORMS, vol. 233, pp. 927–976. Springer, New York (2016). https://doi.org/10.1007/978-1-4939-3094-4_22

24. Miettinen, K., Ruiz, F., Wierzbicki, A.P.: Introduction to multiobjective optimization: interactive approaches. In: Branke, J., Deb, K., Miettinen, K., Słowiński, R. (eds.) Multiobjective Optimization. LNCS, vol. 5252, pp. 27–57. Springer, Heidelberg (2008). https://doi.org/10.1007/978-3-540-88908-3_2
25. Miettinen, K.: Nonlinear Multiobjective Optimization. Kluwer Academic Publishers, Boston (1999)
26. Mnasri, S., Nasri, N., Alrashidi, M., van den Bossche, A., Val, T.: IoT networks 3D deployment using hybrid many-objective optimization algorithms. J. Heuristics **26**(5), 663–709 (2020). https://doi.org/10.1007/s10732-020-09445-x
27. Mnasri, S., Nasri, N., Van Den Bossche, A., Val, T.: 3D indoor redeployment in IoT collection networks: a real prototyping using a hybrid PI-NSGA-III-VF. In: 2018 14th International Wireless Communications and Mobile Computing Conference, IWCMC 2018, pp. 780–785. Institute of Electrical and Electronics Engineers Inc. (2018). https://doi.org/10.1109/IWCMC.2018.8450372
28. Purshouse, R.C., Deb, K., Mansor, M.M., Mostaghim, S., Wang, R.: A review of hybrid evolutionary multiple criteria decision making methods. In: 2014 IEEE Congress on Evolutionary Computation (CEC), pp. 1147–1154 (2014). https://doi.org/10.1109/CEC.2014.6900368
29. Saini, B.S., Hakanen, J., Miettinen, K.: A new paradigm in interactive evolutionary multiobjective optimization. In: Bäck, T., et al. (eds.) PPSN 2020. LNCS, vol. 12270, pp. 243–256. Springer, Cham (2020). https://doi.org/10.1007/978-3-030-58115-2_17
30. Santos Da Silva, I.R., De Andrade Lira Rabelo, R., Rodrigues, J.J., Carvalho, A.: A multi-objective approach for energy management in a microgrid scenario. In: 2020 5th International Conference on Smart and Sustainable Technologies, SpliTech 2020. Institute of Electrical and Electronics Engineers Inc. (2020). https://doi.org/10.23919/SpliTech49282.2020.9243847
31. da Silva, I.R., Rabêlo, R.d.A., Rodrigues, J.J., Solic, P., Carvalho, A.: A preference-based demand response mechanism for energy management in a microgrid. J. Cleaner Prod. **255**, 120034 (2020). https://doi.org/10.1016/j.jclepro.2020.120034
32. Thiele, L., Miettinen, K., Korhonen, P.J., Molina, J.: A preference-based evolutionary algorithm for multi-objective optimization. Evol. Comput. **17**(3), 411–436 (2009). https://doi.org/10.1162/evco.2009.17.3.411
33. Vesikar, Y., Deb, K., Blank, J.: Reference point based NSGA-III for preferred solutions. In: 2018 IEEE Symposium Series on Computational Intelligence (SSCI), pp. 1587–1594 (2018). https://doi.org/10.1109/SSCI.2018.8628819
34. Yan, J., Deng, H.: Generation of large-bandwidth x-ray free electron laser with evolutionary many-objective optimization algorithm. Phys. Rev. Accelerators Beams **22**(2), 020703 (2019). https://doi.org/10.1103/PhysRevAccelBeams.22.020703
35. Zhang, J., Xing, L.: A survey of multiobjective evolutionary algorithms. In: 2017 IEEE International Conference on Computational Science and Engineering (CSE) and IEEE International Conference on Embedded and Ubiquitous Computing (EUC), vol. 1, pp. 93–100 (2017). https://doi.org/10.1109/CSE-EUC.2017.27
36. Zhang, Q., Li, H.: MOEA/D: a multiobjective evolutionary algorithm based on decomposition. IEEE Trans. Evol. Comput. **11**(6), 712–731 (2007). https://doi.org/10.1109/TEVC.2007.892759

A Systematic Way of Structuring Real-World Multiobjective Optimization Problems

Bekir Afsar(✉), Johanna Silvennoinen, and Kaisa Miettinen

University of Jyvaskyla, Faculty of Information Technology, P.O. Box 35 (Agora), FI-40014 University of Jyvaskyla, Finland
{bekir.b.afsar,johanna.silvennoinen,kaisa.miettinen}@jyu.fi

Abstract. In recent decades, the benefits of applying multiobjective optimization (MOO) methods in real-world applications have rapidly increased. The MOO literature mostly focuses on problem-solving, typically assuming the problem has already been correctly formulated. The necessity of verifying the MOO problem and the potential impacts of having an incorrect problem formulation on the optimization results are not emphasized enough in the literature. However, verification is crucial since the optimization results will not be meaningful without an accurate problem formulation, not to mention the resources spent in the optimization process being wasted.

In this paper, we focus on the MOO problem structuring, which we believe deserves more attention. The novel contribution is the proposed systematic way of structuring MOO problems that leverages problem structuring approaches from the literature on multiple criteria decision analysis (MCDA). They are not directly applicable to the formulation of MOO problems since the objective functions in the MOO problem depend on decision variables and constraint functions, whereas MCDA problems have a given set of solution alternatives characterized by criterion values. Therefore, we propose to elicit expert knowledge to identify decision variables and constraint functions, in addition to the objective functions, to construct a MOO problem appropriately. Our approach also enables the verification and validation of the problem before the actual decision making process.

Keywords: Problem structuring · MOO problem formulation · Eliciting expert knowledge · Identifying objectives · Decision making · Stakeholder interviews

1 Introduction

The concept of optimization refers to making the best use of a given situation and resources. In particular, optimization is the process of determining the values of decision variables to optimize (minimize or maximize) the values of objective functions since, in real-world problems, we are typically faced with multiple conflicting objective functions, such as decreasing expenses and maximizing profit.

M. Emmerich et al. (Eds.): EMO 2023, LNCS 13970, pp. 593–605, 2023.
https://doi.org/10.1007/978-3-031-27250-9_42

Such problems are known as *multiobjective optimization (MOO) problems*, and they involve two or more conflicting objective functions that must be minimized or maximized simultaneously. Typically, several compromise solutions with varying tradeoffs exist, and the preference information of a decision maker (DM), who is an expert in the problem domain, is required to identify the most preferred solution [37].

A MOO problem typically consists of three main components [40]: *Objective functions* define the mathematical representations of aspects what characterize the goodness of the decision to be made. *Decision variables* represent the choices that must be made. *Constraint functions (constraints)* are, e.g., (in)equality constraints or lower and upper bounds that limit the values of the decision variables. Real-world problems may have many objective functions, decision variables, and constraints. Identifying all these components at once may be difficult for a DM or an analyst (who is an expert in modeling MOO problems and can apply optimization methods to solve MOO problems). Based on experiences (e.g., [13,23,43]), several iterations are often needed in which the optimization is performed based on some initial versions of the MOO problem, and some objective functions, decision variables, and (or) constraints are eliminated or modified, or new ones added based on the initial results. Furthermore, some objective functions may sometimes be converted to constraints or vice versa. In this paper, we consider deterministic MOO problems.

For real-world MOO problems, the problem formulation is essential, where the three components mentioned above are to be identified. However, the importance of problem formulation is often neglected in the literature [28] while the focus is on developing methods for solving MOO problems, assuming that the problem has already been formulated. Often, benchmark problems are used to demonstrate the use of the proposed methods. Accordingly, most studies lack information on how the problems have been formulated, i.e., without sufficient information on how the functions involved and decision variables have been identified.

Multiple criteria decision analysis (MCDA) is a sub-discipline of operation research that facilitates the systematic evaluation of alternatives in terms of multiple, conflicting criteria (e.g., [7,31]). The criteria in MCDA and the objective functions in MOO problems basically mean the same. In practice, however, the problems are very different since alternatives in MCDA are explicitly given, while in MOO, they are implicitly described with objective functions and decision variables. Thus, the approaches developed to structure MCDA problems are not directly applicable to formulating MOO problems.

Modeling an accurate and adequate MOO problem may need many brainstorming sessions between domain experts (whom we here refer to as DMs or stakeholders even though there may be many domain experts involved in the problem formulation) and the analyst. In some cases, it can take a lot of time to formulate an appropriate MOO problem. For example, it is mentioned in [13] that three iterations with different problem formulations were needed to meet the needs of a shape optimization problem in an air intake ventilation system of a tractor cabin. Among other changes, the number of objective functions varied. Besides, some objective functions may be first incorrectly formulated (e.g., [23,43]). Thus, having a verified and validated MOO problem is crucial, as it directly affects the decision. In other words, a poorly formulated MOO problem may result in erroneous or misleading solutions.

In this paper, we propose a generic systematic way of structuring deterministic real-world MOO problems. Our systematic approach consists of a four-step methodology, with each step showing the corresponding methods and tools for obtaining a verified and validated MOO problem prior to solving it (i.e., conducting the real decision making process). We begin by eliciting the DM's knowledge to identify the decisions that must be made and the objectives that must be met. Importantly, we list explicit questions to be asked in structuring the problem. We then verify and validate the MOO problem after constructing it based on the identified objective functions, decision variables, and constraints. This process is iterative, where the DM and the analyst are involved.

The remainder of this paper is organized as follows: In Sect. 2, we introduce the basic concepts and provide a brief literature review. Section 3 is devoted to the proposed systematic way of structuring real-world optimization problems. We discuss the possible ways of analyzing the qualitative data gathered through stakeholder interviews and mention the limitations of applying the proposed approach in different application domains in Sect. 4. Finally, we draw conclusions in Sect. 5.

2 Background

In this section, we present the basic concepts required to understand the proposed systematic way of structuring MOO problems. In addition, we provide a brief literature review to demonstrate the gap in the literature as well as the differences in problem structuring for the MOO and MCDA problems.

2.1 Basic Concepts

In real-world applications, we may have different ways to evaluate the identified objectives. An *analytical objective* function can be described analytically by a mathematical expression and is evaluated by solving the mathematical function formulation. However, this is not always possible in real-world applications, where the objective function may not be known or cannot be represented as a mathematical function [28]. In this case, the collected data (e.g., from physical experiments or from past performance) can be used to formulate objective functions, and we refer to them as *data-driven objective* function. On the other hand, some objectives are evaluated using computer simulations, called *simulation-based objective* functions. This means that the output of a simulator is needed for objective function evaluation. In what follows, we use the shorter term *objectives* instead of objective functions (and the same for constraints).

As previously stated, a MOO problem involves a set of conflicting objectives (to be minimized or maximized) as characteristics of a good decision that the DM wants to make. A set of feasible solutions can be generated by a set of decision variable values in the feasible region, subject to constraints. Constraints are the conditions that should be satisfied and can be represented by mathematical functions. Sometimes, the values of the decision variables are also limited by lower and upper values, known as boundary constraints. A solution is called feasible if it satisfies the constraints. Typically, no solution exists to optimize all objectives simultaneously. Instead, we have so-called *Pareto optimal* solutions [37]. A feasible solution is called Pareto optimal if it is not *dominated* by any other feasible solution, i.e., improving any objective value always implies sacrificing in at least one of the others. Thus, there are tradeoffs among the objectives. Pareto optimal solutions are mathematically incomparable, and to identify the *most preferred solution* as the final decision, we need preference information of a DM. By a solution (decision making) process, we mean finding the most preferred solution.

MOO methods can be divided into three classes based on when preference information is incorporated in the decision making process [26,37]. In *a priori* methods, the DM first provides preference information before the optimization, and solutions reflecting the DM's preferences are generated. On the other hand, in *a posteriori* methods, a representative set of Pareto optimal solutions is generated first, and then the preference information of the DM is used to select the most preferred one among them. Finally, in *interactive* methods, the DM provides preference information iteratively to direct the solution process and to obtain specific Pareto optimal solutions that reflect their preference information. Interactive methods are useful because they allow the DM to participate in the decision making process iteratively while also learning about tradeoffs among the objectives and available solutions, as well as the feasibility of the preferences [38].

2.2 Brief Literature Review

Structuring MCDA problems is a well-studied research field. Similar to MOO problem formulation, identifying criteria is an important phase, as different sets of criteria result in different decisions [11]. In problem structuring, analysts (often called facilitators in MCDA) ask DMs to list their criteria. On the other hand, this is often considered *more of an art than a science* [29]. While many MCDA studies have assumed that a well-structured problem is available [8], in the late 1990s, Keeney's work on value-focused thinking emphasized the need for effective problem structuring [30]. Accordingly, the general problem structuring methods presented for rational analysis [41] were integrated with MCDA (e.g., [3,6,17]).

Furthermore, experimental studies were conducted to structure MCDA problems in practice (see, e.g., [8,20], and the survey [36] and references therein). For meaningful analysis, identifying key and fundamental criteria instead of having a large set of criteria was emphasized in [35], and the characteristics of good criteria including, e.g., completeness, conciseness, non-redundancy, understandability, measurability and preferential independence, were listed. Moreover, some recent studies used online surveys to elicit criteria from a large number of individuals without the assistance of an analyst (or facilitator) (see, e.g., [2,22]).

The proposed methods (e.g., in [36] and references therein) first ask stakeholders to state criteria in the domain in brainstorming sessions. Then, a facilitator shows a master list (pre-generated set of criteria) and asks stakeholders to select and match some of the criteria in the master list with their stated criteria. Finally, they select the final set of criteria from the self-generated criteria in addition to the recognized ones from the master list. To apply this kind of method, we need at least a master list which may not be the case for a new problem domain. How to create a list of criteria from scratch is still an open research question.

A few real-world studies focusing on MOO problem formulation have been published. The authors of [47] proposed a holistic approach allowing the reformulation of the constructed initial optimization problem by adding or removing objectives and (or) decision variables. Their approach was applied to a complex real-world problem for the conceptual design of indoor sports buildings. They utilized simulators in building design, and the objectives were identified according to the output data of the simulators. Similarly, in [46], the objectives of the MOO problem were identified for chemical processes by utilizing the available data, and machine learning model(s) for some or all objectives are fitted whenever applicable.

In [24], several (semi)structured interviews with stakeholders were used to formulate a preliminary design problem of wood-based insulating materials. In [5], the same method was applied in formulating the MOO problem of microfiltration of skim milk. Unfortunately, the complete set of questions used in stakeholder interviews was not shared. Thus, published studies in the MOO literature lack information on how the MOO problem was formulated through stakeholder interviews. To advance research in structuring MOO problems, the procedures must be fully published and reported in detail in order to increase reliability and reproducibility so that others can utilize them in other application domains.

3 Systematic Way of Structuring MOO Problems

We propose a generic systematic way of structuring MOO problems, which is not application-specific. With the proposed approach, one can identify the components of MOO problems through structured interviews with stakeholders and then construct the MOO problem by utilizing the existing knowledge in the domain. Since ensuring the correctness of the MOO problem at once is not easy, especially for large-scale problems, i.e., with many objectives and decision variables, we propose an iterative process that is repeated until the MOO problem is verified and validated. In what follows, we first present the main steps in general and then provide further details for each step. Overall, the main steps of the proposed systematic approach are the following:

- **Step 1:** Identify components of the MOO problem (objectives, decision variables, and constraints) through stakeholder interviews.
- **Step 2:** Construct the MOO problem. (It should be noted that some constraints can also be constructed using the options indicated below.)
  - **Analytical objectives:** Formulate the mathematical functions representing the objectives.
  - **Data-driven objectives:** Construct surrogate models based on the data available to derive data-driven objectives.
  - **Simulation-based objectives:** Configure the simulator using identified components of the MOO problem. Basically, the input of the simulator is decision variables, and simulation results are used to evaluate objectives.
- **Step 3:** Verify the (overall) applicability of the MOO problem via a preliminary assessment by the analyst. (The analyst generates random points in the feasible region for the decision variables, evaluates objectives and constraints, and checks the generated results.) If verification fails, go to Step 2.
- **Step 4:** Validate the MOO problem until the DM is convinced of the correctness of the MOO problem. (The analyst generates nondominated solutions via some selected (a posteriori) MOO methods and presents them to the DM. The analyst then asks if the tradeoffs among the objectives are clear and meaningful.) If the validation fails, go to Step 2. Otherwise, terminate the process.

In what follows, we discuss each step of the proposed approach. For **Step 1**, we propose a list of interview questions in Table 1 to be used in the structured interviews with the domain experts (or stakeholders) to identify the objectives, decision variables, and constraints. We cannot ask for these details directly (it would be highly difficult to verbalize and define the problem to be solved straight as objectives, decision variables, and constraints), but we need to elicit the knowledge of domain experts and understand their needs and requirements. The key aim is to understand the decision to be made and the data availability. Typically, analysts ask casual (informal) questions to domain experts and try to formulate the MOO problem based on their understanding. However, this process is up to the analysts' capabilities and the way they ask questions. We, therefore, propose a systematic and comprehensive list of interview questions. We have selected some of the questions proposed in [30] and added some new ones to obtain information for decision variables and constraints (they are not considered in structuring MCDA problems). The interview questions first aim to get a general overview and understanding of the problem and then seek to formulate the problem technically. For example, the decision to be made is an abstract thought at a general level at the beginning; on the other hand, decision variables try to capture the decision to be made at a technical level.

Step 1 starts with a round of interviews. First, we propose to apply these interview questions to each stakeholder individually to ensure that each stakeholder's perspectives are considered in structuring the problem. The interview method utilized can vary if changes to the order of the questions are to be made to be able to elicit the required expert knowledge. If the question list is utilized

Table 1. Proposed list of interview questions to be used in stakeholder interviews. Note: Questions written in italics are from [30].

Aim	Questions
A wish list	*What do you want? What do you value? What should you want?*
Decisions	What is the nature of the decision you must make? What kind of actions are to be taken?
Available information	What information is needed for you to take the decision? How do you get that information? What kind of data is available?
Goals	What perspectives characterize the goodness of your decision? *What are your aspirations?*
Strategic objectives	*What are your ultimate objectives? What are your values that are fundamental?*
Generic objectives	*What environmental, social, economic or safety objectives are important?*
Structuring objectives	*Why is that objective important? How can you achieve it? What do you mean by this objective?*
Quantification of objectives	*How would you measure achievement of an objective?*
Decision variables	How can these objectives be accomplished? What kind of information do you need to evaluate your objectives? What factors/variables affect your objective values?
Structuring decision variables	Which decision variables influence which objectives?
Constraints	What are the limiting factors of your objectives and (or) decision variables? Are there any lower and (or) upper bounds for decision variables?

in the depicted order, a method of a structured interview can be used (e.g., [21]). The output lists problem characteristics from each stakeholder's point of view once the qualitative data (responses given to the interview questions) are analyzed. The interview data needs to be transcribed into the textual format and can be analyzed with various analysis methods depending on the context of the study, such as with conventional, directed, or summative qualitative content analysis [25] or with thematic analysis, e.g., [10]. We then propose to have a joint workshop, including all the stakeholders, to adjust and finalize the components of the MOO problem.

In **Step 2**, the analyst constructs the MOO problem utilizing the output of Step 1. According to the responses given to the question of available information, the analyst can understand which objectives and constraints have to be modeled based on 1) mathematical function formulations, 2) the available data, or

3) calling a simulator. For the first option, 1), the analyst writes a piece of programming code that represents the mathematical function. In the second option, 2), if we have data originating from the problem domain for some objectives and constraints, the analyst constructs surrogate models based on the data available. In this, the analyst first pre-processes the data since it can be incomplete, noisy, and heterogeneous [28] and then needs to select the best-suited surrogate model based on their previous experiences. Another option is to apply an automatic surrogate model selector (e.g., [42]) on the available data. In [42], the authors use the data from known optimization problems to train several surrogate models (e.g., polynomials [34], neural networks [4], radial basis functions [12], support vector machines [44] and stochastic models such as Kriging [33] or Bayesian modeling [45]) and extract features from the trained surrogate models. These features are then used to identify the best surrogate modeling technique for a given new data set. If some of the objectives and constraints are to be evaluated based on the output of a simulator, 3), the analyst may set up the simulator based on the identified components for the objective and constraint in question, which may involve pre- and post-processing. Pre-processing is needed to configure the input of the simulator using the decision variables, while post-processing is necessary to evaluate the corresponding objective and constraint values from the output of the simulator.

Step 3 is applied to technically verify the MOO problem before solving it. First, the analyst generates random decision variable values (e.g., uniform random distribution or normal distribution with some mean and standard deviation in the feasible region) and evaluates and checks the feasibility of objectives and generated results to verify the MOO problem. Here, a MOO problem is said to be verified if the MOO problem is correctly programmed, different types of objectives and constraints (e.g., analytical, data-based, or simulator-based) are appropriately constructed, and generated results are meaningful considering the given input values. If this step fails, the analyst goes to Step 2, checks the details of the MOO problem to see whether there is an error, and fixes errors (e.g., the code is corrected or the post-processing of the simulator is restored). This process terminates when the MOO problem appears to be constructed correctly to the best knowledge of the analyst.

Finally, in **Step 4**, the analyst generates a set of nondominated solutions using some optimization methods (e.g., a posteriori evolutionary MOO methods [14,28]) and presents these solutions to the DM. The DM can select some solutions to study the tradeoffs among the objectives. A MOO problem is said to be validated if the tradeoffs among the objectives are meaningful. If the DM is not happy with the generated solutions, the DM can decide to ask the analyst to revisit the constructed MOO problem. If any error is found, the analyst fixes the MOO problem formulation. Furthermore, if the DM so desires, the analyst converts some objectives into constraints (or vice versa). The analyst then generates new nondominated solutions using the revisited MOO problem. Besides validating the MOO problem, the analyst can also study the time spent in generating nondominated solutions. If the MOO problem is computationally demanding, computationally inexpensive surrogate models can be used to replace expensive (objective or constraint) functions. This process continues until the DM is happy with the generated solutions and sure about the correctness of the solutions to the best of their knowledge.

Once these steps in the proposed systematic approach are carried out successfully, we have a verified and validated MOO problem, which can be trusted and used in the actual decision making process. After this, the analyst and the DM can decide which type of method is applied to solve the problem. They can apply, e.g., any interactive multiobjective optimization method [38], including interactive evolutionary ones [27] if the DM wants to direct the solution process with one's preference information.

One can also apply interactive methods as a part of the verification process as proposed in [43]. There, a so-called augmented interactive MOO method is introduced, which incorporates verification and solution processes.

4 Discussion

The results of the qualitative content analysis based on the listed interview questions can be further considered to be analyzed with different method combinations to support the structuring of the problem. One possibility is to utilize cognitive mapping and causal maps. Next, we discuss the logic and main benefits of these kinds of approaches.

After obtaining results from the stakeholder interview transcripts with a chosen qualitative content analysis method, the results (commonly in the form of descriptive categories with sub-categories) can be further elaborated with causal maps for the group discussion (negotiation) purposes to enable understandability and fluency within the group in structuring the MOO problem. Causal maps are efficient in structuring problems as they enable rich representation of ideas modeled as exhaustive networks of argument chains [39]. A causal map can be constructed with the aid of an analyst (or a facilitator) directly from the negotiation process. However, for a detailed understanding of the problem and its reliable formulation, a more in-depth data collection and analysis method combination can be used [16]. As causal maps are often constructed in a group setting, they can precede the procedure of cognitive mapping and construction of individual cognitive maps [15]. A cognitive map is a representation of thoughts of a problem that is derived from the process of cognitive mapping (i.e., mapping an individual's thinking structure and contents about a problem [15]). Cognitive mapping is based on Kelly's personal construct theory [32] (for instructions on how to conduct cognitive mapping, see [1]). Usually, cognitive maps are based on interview data [15] and are presented as networks of nodes and arrows indicating relations represented by an individual.

Thus, the whole procedure of formulating MOO problems could benefit from utilizing a combination of research methods, for example, from interviews to different mappings. The starting point of the problem structuring procedure is beneficial to be based on eliciting expert knowledge of the problem domain through verbalizations to gain a reliable understanding of the phenomenon under investigation. Often this is done via expert interviews (see, e.g., [9]). In addition, thinking-aloud protocols are an efficient way of eliciting expert knowledge through verbalizations [18,19], from which cognitive maps can be created for further problem formulation. Also, from interview data, cognitive maps can be created via cognitive mapping from each interviewee separately and then via integration into an initial group map to be elaborated as a causal map in a group discussion, searching for a consensus for structuring the problem.

The applicability of the proposed approach in different application domains depends on the resources dedicated to the problem structuring. We assume that the stakeholders are convinced that having a verified and validated MOO problem to get reliable results is more important than merely solving the problems based on unstructured foundations that might not reflect the problem in real-life. However, since the main foundation of the proposed approach is getting the experts' knowledge to identify the components of the MOO problem, it is limited by the experts' reliability and willingness to participate in different steps. In addition, analysts lacking enough knowledge of qualitative data analysis (e.g., in compiling cognitive maps) may be another challenge to applying the proposed approach.

5 Conclusions

We have observed that structuring a MOO problem has not been studied enough in the literature. In this paper, we have proposed a systematic way of structuring verified and validated MOO problems. The proposed approach identifies which objectives to optimize and how decision variables and constraints affect each objective. We first identify the objectives, decision variables, and constraints of a MOO problem through structured stakeholder interviews. To support this interview in a concrete manner, we have proposed a list of interview questions. In the second step, the analyst constructs the MOO problem utilizing the results of the qualitative analysis of the interview responses. We have also proposed possible qualitative content analysis methods to be utilized as well as further analysis methods of cognitive mapping and causal maps to structure the problem based on the interview data. Third, the analyst verifies the MOO problem with randomly generated values for decision variables. Finally, the DM validates the MOO problem by studying some generated solutions with the help of the analyst. This is an iterative process. In this way, the DM gains enough confidence in the formulated MOO problem as it is verified and validated.

This paper is aimed at supporting the task of structuring MOO problems. Our approach covers all aspects of constructing MOO problems, from identifying components to modeling and testing before the actual solution process. We have presented our approach in a general manner, and it can be applied in any real application domain. Because of space limitations, we could not consider an example to apply the proposed approach. However, we plan future studies on applying it in different case studies to demonstrate its applicability.

Acknowledgements. We thank Professor Theodor Stewart and Professor Francisco Ruiz for their helpful comments. This research was partly funded by the Academy of Finland (grant number 322221) and is related to the thematic research area Decision Analytics utilizing Causal Models and Multiobjective Optimization (DEMO), jyu.fi/demo, at the University of Jyvaskyla.

References

1. Ackermann, F., Eden, C., Cropper, S.: Getting Started with Cognitive Mapping. Banxia Software, Kendal (1992)
2. Aubert, A.H., Schmid, S., Beutler, P., Lienert, J.: Innovative online survey about sustainable wastewater management: What young swiss citizens know and value. Environ. Sci. Policy **137**, 323–335 (2022)
3. e Costa, C.A.B., Ensslin, L., Émerson, C., Cornêa, E.C., Vansnick, J.C.: Decision support systems in action: Integrated application in a multicriteria decision aid process. Eur. J. Oper. Res. **113**(2), 315–335 (1999)
4. Beale, H.D., Demuth, H.B., Hagan, M.: Neural Network Design. PWS Publishing, Boston (1996)
5. Belna, M., Ndiaye, A., Taillandier, F., Agabriel, L., Marie, A.L., Gésan-Guiziou, G.: Formulating multiobjective optimization of 0.1 μm microfiltration of skim milk. Food Bioprod. Process. **124**, 244–257 (2020)
6. Belton, V., Ackermann, F., Shepherd, I.: Integrated support from problem structuring through to alternative evaluation using COPE and V · I · S · A. J. Multi-Criteria Decis. Anal. **6**(3), 115–130 (1997)
7. Belton, V., Stewart, T.: Multiple Criteria Decision Analysis: An Integrated Approach. Springer, New York (2002)
8. Belton, V., Stewart, T.: Problem structuring and multiple criteria decision analysis. In: Ehrgott, M., Figueira, J., Greco, S. (eds.) Trends in Multiple Criteria Decision Analysis. International Series in Operations Research & Management Science, vol. 142, pp. 209–239. Springer, Boston (2010). https://doi.org/10.1007/978-1-4419-5904-1_8
9. Bogner, A., Littig, B., Menz, W.: Interviewing Experts. Springer, Cham (2009)
10. Braun, V., Clarke, V.: Thematic Analysis. American Psychological Association, Washington (2012)
11. Brownlow, S., Watson, S.: Structuring multi-attribute value hierarchies. J. Oper. Res. Soc. **38**(4), 309–317 (1987)
12. Buhmann, M.D.: Radial Basis Functions: Theory and Implementations. Cambridge University Press, Cambridge (2003)
13. Chugh, T., Sindhya, K., Miettinen, K., Jin, Y., Kratky, T., Makkonen, P.: Surrogate-assisted evolutionary multiobjective shape optimization of an air intake ventilation system. In: Proceedings of the 2017 IEEE Congress on Evolutionary Computation, pp. 1541–1548. IEEE (2017)
14. Deb, K.: Multi-objective optimisation using evolutionary algorithms: An introduction. In: Wang, L., Ng, A.H.C., Deb, K. (eds.) Multi-objective Evolutionary Optimisation for Product Design and Manufacturing, pp. 3–34. Springer, London (2011). https://doi.org/10.1007/978-0-85729-652-8_1
15. Eden, C.: Analyzing cognitive maps to help structure issues or problems. Eur. J. Oper. Res. **159**(3), 673–686 (2004)
16. Eden, C., Ackermann, F.: Making Strategy: The Journey of Strategic Management. Sage (1998)

17. Ensslin, L., Dutra, A., Ensslin, S.R.: MCDA: A constructivist approach to the management of human resources at a governmental agency. Int. Trans. Oper. Res. **7**(1), 79–100 (2000)
18. Ericsson, K.A., Hoffman, R.R., Kozbelt, A., Williams, A.M.: The Cambridge Handbook of Expertise and Expert Performance. Cambridge University Press, Cambridge (2018)
19. Ericsson, K.A., Simon, H.A.: Verbal reports as data. Psychol. Rev. **87**(3), 215 (1980)
20. Franco, L.A., Montibeller, G.: Problem structuring for multicriteria decision analysis interventions. Wiley Encyclopedia Oper. Res. Manage. Sci. (2010)
21. Given, L.M.: The Sage Encyclopedia of Qualitative Research Methods. Sage, Newcastle upon Tyne (2008)
22. Haag, F., Zürcher, S., Lienert, J.: Enhancing the elicitation of diverse decision objectives for public planning. Eur. J. Oper. Res. **279**(3), 912–928 (2019)
23. Hämäläinen, J., Miettinen, K., Tarvainen, P., Toivanen, J.: Interactive solution approach to a multiobjective optimization problem in paper machine headbox design. J. Optim. Theory Appl. **116**(2), 265–281 (2003). https://doi.org/10.1023/A:1022453820000
24. Hobballah, M.H., Ndiaye, A., Michaud, F., Irle, M.: Formulating preliminary design optimization problems using expert knowledge: Application to wood-based insulating materials. Expert Syst. Appl. **92**, 95–105 (2018)
25. Hsieh, H.F., Shannon, S.E.: Three approaches to qualitative content analysis. Qual. Health Res. **15**(9), 1277–1288 (2005)
26. Hwang, C.L., Masud, A.: Multiple Objective Decision Making - Methods and Applications: A State-of-the-Art Survey. Springer, Cham (1979)
27. Jaszkiewicz, A., Branke, J.: Interactive multiobjective evolutionary algorithms. In: Branke, J., Deb, K., Miettinen, K., Słowiński, R. (eds.) Multiobjective Optimization. LNCS, vol. 5252, pp. 179–193. Springer, Heidelberg (2008). https://doi.org/10.1007/978-3-540-88908-3_7
28. Jin, Y., Wang, H., Sun, C.: Data-Driven Evolutionary Optimization. SCI, vol. 975. Springer, Cham (2021). https://doi.org/10.1007/978-3-030-74640-7
29. Keeney, R.L.: Structuring objectives for problems of public interest. Oper. Res. **36**(3), 396–405 (1988)
30. Keeney, R.L.: Value-focused Thinking: A Path to Creative Decisionmaking. Harvard University Press, Cambridge (1996)
31. Keeney, R.L., Raiffa, H., Meyer, R.F.: Decisions With Multiple Objectives: Preferences and Value Trade-offs. Cambridge University Press, Cambridge (1993)
32. Kelly, G.: Personal Construct Psychology. Norton, Nueva York (1955)
33. Kleijnen, J.P.: Kriging metamodeling in simulation: A review. Eur. J. Oper. Res. **192**(3), 707–716 (2009)
34. Madsen, J.I., Shyy, W., Haftka, R.T.: Response surface techniques for diffuser shape optimization. AIAA J. **38**(9), 1512–1518 (2000)
35. Marttunen, M., Haag, F., Belton, V., Mustajoki, J., Lienert, J.: Methods to inform the development of concise objectives hierarchies in multi-criteria decision analysis. Eur. J. Oper. Res. **277**(2), 604–620 (2019)
36. Marttunen, M., Lienert, J., Belton, V.: Structuring problems for multi-criteria decision analysis in practice: A literature review of method combinations. Eur. J. Oper. Res. **263**(1), 1–17 (2017)
37. Miettinen, K.: Nonlinear Multiobjective Optimization. Kluwer Academic Publishers, Boston (1999)

38. Miettinen, K., Ruiz, F., Wierzbicki, A.P.: Introduction to multiobjective optimization: Interactive approaches. In: Branke, J., Deb, K., Miettinen, K., Słowiński, R. (eds.) Multiobjective Optimization. LNCS, vol. 5252, pp. 27–57. Springer, Heidelberg (2008). https://doi.org/10.1007/978-3-540-88908-3_2
39. Montibeller, G., Belton, V.: Causal maps and the evaluation of decision options-a review. J. Oper. Res. Soc. **57**(7), 779–791 (2006)
40. Nocedal, J., Wright, S.J.: Numerical Optimization. Springer, Cham (1999). https://doi.org/10.1007/978-0-387-40065-5
41. Rosenhead, J.: What's the problem? An introduction to problem structuring methods. Interfaces **26**(6), 117–131 (1996)
42. Saini, B.S., Lopez-Ibanez, M., Miettinen, K.: Automatic surrogate modelling technique selection based on features of optimization problems. In: Proceedings of the Genetic and Evolutionary Computation Conference Companion, pp. 1765–1772. GECCO 2019, Association for Computing Machinery (2019)
43. Sindhya, K., Ojalehto, V., Savolainen, J., Niemistö, H., Hakanen, J., Miettinen, K.: Coupling dynamic simulation and interactive multiobjective optimization for complex problems: An APROS-NIMBUS case study. Expert Syst. Appl. **41**(5), 2546–2558 (2014)
44. Smola, A.J., Schölkopf, B.: A tutorial on support vector regression. Stat. Comput. **14**(3), 199–222 (2004). https://doi.org/10.1023/B:STCO.0000035301.49549.88
45. Snoek, J., Larochelle, H., Adams, R.P.: Practical Bayesian optimization of machine learning algorithms. In: Pereira, F., Burges, C.J.C., Bottou, L., Weinberger, K.Q. (eds.) Advances in Neural Information Processing Systems 25, pp. 2951–2959. Curran Associates. (2012)
46. Wang, Z., Li, J., Rangaiah, G.P., Wu, Z.: Machine learning aided multi-objective optimization and multi-criteria decision making: Framework and two applications in chemical engineering. Comput. Chem. Eng. **165**, 107945 (2022)
47. Yang, D., Ren, S., Turrin, M., Sariyildiz, S., Sun, Y.: Multi-disciplinary and multi-objective optimization problem re-formulation in computational design exploration: A case of conceptual sports building design. Autom. Constr. **92**, 242–269 (2018)

IK-EMOViz: An Interactive Knowledge-Based Evolutionary Multi-objective Optimization Framework

Abhiroop Ghosh(✉), Kalyanmoy Deb, Ronald Averill, and Erik Goodman

Computational Optimization and Innovation Laboratory, Michigan State University, East Lansing, MI 48824, USA
{ghoshab1,kdeb,averillr,goodman}@egr.msu.edu
https://www.coin-lab.org/

Abstract. The knowledge and intuition of experienced users for practical optimization problems are often underutilized in academic research. Such knowledge, formulated as inter-variable relationships, can assist an optimization algorithm in finding good solutions faster. User-provided information can be utilized at the beginning of the optimization, or during the optimization in an interactive fashion. In this paper, we propose IK-EMOViz, a software framework to allow discovery and use of knowledge from and to an EMO algorithm interactively. Key knowledge common to current non-dominated solutions are extracted using rule learning methods and shared with the decision-makers (DMs) through a easy-to-comprehend visualization tool. Learned knowledge are then filtered and vetted by DMs are communicated to the EMO algorithm using the same visualization tool. EMO algorithm then processes the filtered knowledge and integrates them with its search operators. Repeated such interactions have resulted in faster convergence to the final trade-off set on a large-scale engineering design problem. In addition, the effect of asynchronous and synchronous interactivity is also evaluated to make the proposed interactive optimization procedure more pragmatic.

Keywords: Interactive optimization · Knowledge extraction · Graphical user interface · Multi-objective optimization

1 Introduction

For practical multi-objective optimization problems (MOPs), keeping the user in the loop may be beneficial because of several reasons. First, user's knowledge accrued over many years of dedicated time spent on the problem can be suitably leveraged. An algorithm may find it time-consuming or simply difficult to gather an equivalent amount of knowledge from scratch. Second, consulting the user intermittently during the optimization process may generate a natural interest through ownership by the user on the final solution, thereby resulting in a willful acceptance of the final solution. Besides the practical importance of

M. Emmerich et al. (Eds.): EMO 2023, LNCS 13970, pp. 606–619, 2023.
https://doi.org/10.1007/978-3-031-27250-9_43

co-solving a problem with human users, computational researchers often dismiss such collaborative methodologies as subjective and of lesser quality research. However, there is merit for both types of studies. If problem-solving is the main goal, the use of algorithmic rigor and human expertise, if available, should both be embraced. This study is an attempt to involve human users to interact with evolutionary multi-objective optimization (EMO) algorithms in solving problems quickly by utilizing an appropriate combination of their acquired knowledge.

An obvious way to involve users is to develop an interactive optimization framework, where the user can provide guidance during the optimization run. Users can provide information in multiple ways, such as aspiration levels [5], relative importance of objective functions [14], pairwise solution comparison [12], etc. Decision support system software like NAUTILUS Navigator [17] and FACTS Analyzer [15,18] play an important role in ensuring a smooth user experience while interacting with the optimization algorithm.

Problem knowledge can also be learned automatically. *Innovization* studies [2,4] are good examples, where additional problem information were extracted from high-performing solutions in the form of simple mathematical rules such as power laws ($x_i x_j^b = c$). Such rules can also be found during the optimization run and used to speed up convergence [6–9,16].

The IK-EMO method [8] uses an automatic knowledge extraction method in an interactive optimization framework in order to achieve faster convergence and take user feedback into account. In order for IK-EMO to be effective in practice, a user-friendly visualization software is necessary. This paper introduces such a software framework (IK-EMOViz) and demonstrates its visualization features on 120 and 820-member truss design problems. In addition, this study evaluates a practicality providing the decision-makers time to analyze obtained knowledge, while the algorithm is allowed to continue to run in the background.

In the remainder of the paper, we provide a detailed description of the proposed IK-EMOViz in Sect. 2. Section 3 describes the scalable 3D truss design problem with two objectives and presents the results of the IK-EMOViz procedure. Conclusions are drawn in Sect. 4.

2 Proposed IK-EMO Visualizer (IK-EMOViz)

IK-EMO Visualizer, or IK-EMOViz is a software implementation of user interactivity in the IK-EMO [8] framework. In this section, we briefly cover the basic IK-EMO components shown in Fig. 1 and how they relate to IK-EMOViz.

2.1 User-Provided Knowledge Before the Optimization

An n-variable problem, we can have $\frac{n(n-1)}{2}$ pairwise variable interactions. The IK-EMO framework seeks to reduce the number of interactions that needs to be learned by defining variable clusters or groups [7–9]. As an example, consider a seven-variable problem with two non-interacting variable groups, $G_1 = \{1,4,6\}$, and $G_2 = \{3,5\}$. For G_1, all three pairwise combinations (x_1, x_4), (x_4, x_6), and (x_1, x_6) can be checked for existence of any possible relationships. A similar

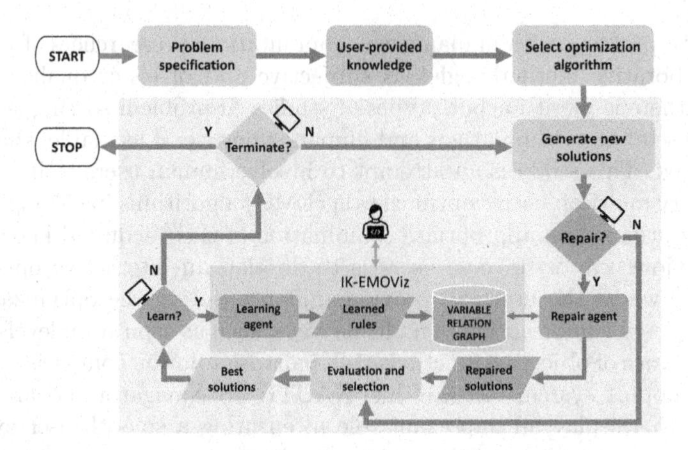

Fig. 1. Interactive knowledge-based EMO framework (IK-EMO), adapted from [8]. Blue blocks represent a normal EMO. Green blocks represent the automated knowledge extraction and application. Information from the learning agents and variable relation graphs is presented to the user using IK-EMOViz. The user, in turn, can provide feedback using the same interface. (Color figure online)

process is repeated for G_2. Since x_2 and x_7 are not part of any group, they are assumed to not exhibit any significant pattern, and can be ignored from knowledge extraction analysis.

2.2 Automatic Knowledge-Extraction During Optimization

Any automated knowledge-extraction method or a *learning agent*, used as a part of the IK-EMO framework, needs to analyze high-performing non-dominated (ND) solutions obtained by the EMO algorithm for knowledge extraction. A learning interval (T_L) is defined as the number of generations or solution evaluations (SEs) after which a new set of rules are learned. Knowledge in this case are mathematical relations or rules existing between one or more variable pairs among the ND solutions. The learned rules are stored by IK-EMO in a data structure known as a variable relation graph (VRG) [8].

In this paper, we restrict the rules to simple power laws [4], however the concept can be extended to extract other structures of rules [1,10] as well. For example, $x_i x_j^b = c$ represents a power law rule between two variables x_i and x_j, where b and c are constants. Power laws are able to represent a wide variety of rules, such as proportionate ($b = -1$) or inversely proportionate ($b = 1$) relationships among two variables. A special case of the power law where $b = 0$ is referred to as a constant rule, since it involves only one variable taking a constant value ($x_i = c$). This type of rule can occur if multiple high-performing solutions are expected to have in common a fixed value of a specific variable [11].

2.3 Knowledge Application Through Repair Operators

After a learning agent learns the constant rules and power laws, the next step is to apply these rules to newly-generated solutions using a repair agent. An important issue is to determine how much to adhere to each learned rule [8]. Three repair operators with different rule adherence schemes are used in this study: tight to loose adherence RA1 to RA3, described in the next paragraph. A repair interval (T_R) is defined as the number of generations or SEs between any two repair operations. The order of repair for each new solution is determined by a graph-traversal algorithm [8] which generates a variant of the VRG created by the learning agent.

A constant rule repair agent applies the rule $x_i = \kappa_i$ to a particular solution $\mathbf{x}^{(k)}$ by setting the variable $x_i^{(k)}$ to κ_i. Constant rules are always implemented with tight adherence. For a power law rule $x_i {x_j}^b = c$, either x_i or x_j is selected as the base (independent) variable based on which the other variable will be repaired in the VRG. If, for a particular solution $\mathbf{x}$, x_j is selected as the base variable, x_i is set as follows: $x_i = \frac{c}{x_j^b}$. Depending on the rule adherence scheme, the c-value used for repair may be modified to $x_i {x_j}^b = c_r$. RA1 uses $c_r = c$ (tight adherence); RA2 uses $c_r \in \mathcal{N}(c, \sigma_c)$ (medium adherence), and RA3 uses $c_r \in \mathcal{N}(c, 2\sigma_c)$ (loose adherence), where σ_c is the standard deviation of learned c-values from current ND solutions.

An ensemble approach (RA-E) combining the above three options (RA1, RA2, and RA3) is also included [8]. RA-E can switch adherence schemes based on the successful survival of solutions repaired using each of the three schemes, with a fourth option added where no repair to an offspring is performed.

2.4 User Feedback Through IK-EMOViz's Graphical User Interface

IK-EMOViz is implemented in Python using Plotly and Dash [13], which provides a browser-based graphical user interface (GUI) for the IK-EMO framework. IK-EMOViz allows the user to access real-time data about the optimization run such as convergence indicators, scatter plots, and parallel coordinate plots (PCP) through separate widgets. Figure 2 shows an example instance of IK-EMOViz where the user wants to analyze the results of a bi-objective optimization problem after 60 generations. The optimization progress in this case is represented by a hypervolume (HV) [19] evolution plot (left-most plot in Fig. 2) over the generations completed. The plot is dynamically updated as each generation is completed. Indicators other than HV can also be used, if desired. The scatter plot widget (middle plot) shows the entire population in the objective space with the ND solutions marked in orange and dominated solutions marked in blue. By using the generator slider shown below the plot, the user can also check the objective vectors of the population in any earlier generation. The software saves all earlier populations and can display any earlier population, if desired. The PCP plot (right-most plot) gives a visualization of the objective and decision variable values together. An additional widget is available whereby the user can visualize any solution from the Pareto front scatter plot, but for brevity, it is not shown here.

The 'IK-EMO Controls' widget has three buttons (not shown here). The first button can pause or resume the optimization run. This is used if the user wants the algorithm to wait till he/she analyzes the results and provides feedback, also

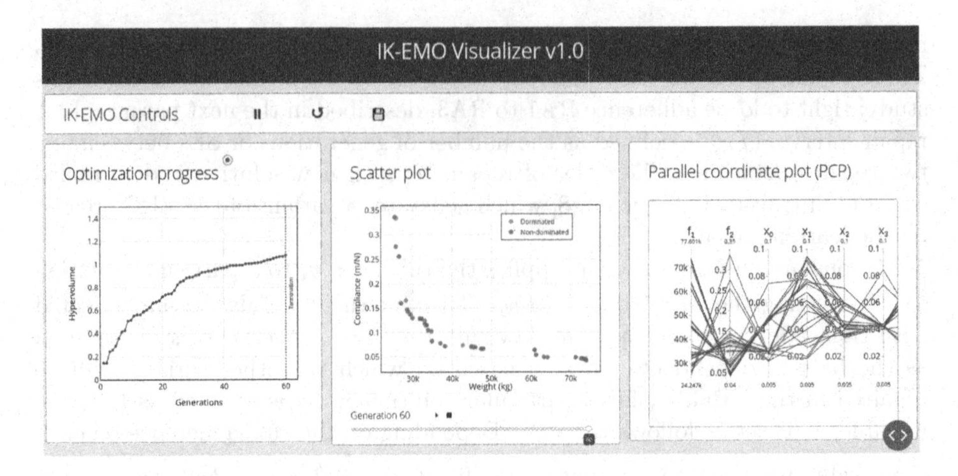

Fig. 2. IK-EMOViz graphical user interface showing the optimization progress, scatter plots, and parallel coordinate plot widgets describing the results of a bi-objective optimization problem which was terminated after 60 generations.

known as 'synchronous interaction'. The second button is used to refresh all the widgets except the one showing optimization progress to get the latest data. This functionality is necessary if the user did not pause the optimization run ('asynchronous interaction') and wants to see the updated results and their associated rules. The third button saves any user feedback which will be considered by IK-EMO in subsequent generations.

Another important widget displays the latest rules and the corresponding VRGs generated by the learning agent for all variable groups. A group selector menu allows the user to switch among multiple groups. Figure 3 shows the rule list and VRG for Group 2 variables of the example problem. Two variables found to possess a significant relation are connected by a gray edge whose thickness is proportional to the rule score. Nodes are colored on the basis of their degree, with reddish nodes indicating a high number of connected edges, and bluish nodes representing a low number of connected edges.

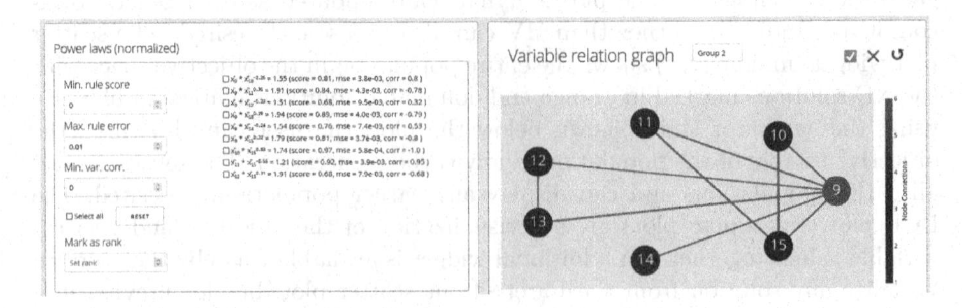

Fig. 3. IK-EMOViz graphical user interface showing the full VRG for variable group 2 for a specific optimization problem.

After analyzing the VRG and associated rules, the user provides feedback by modifying one or more nodes in the VRG. The following operations can be performed on the learned rules.

- Exclusion: The user may select to remove certain rules provided by the algorithm, based on their knowledge of the problem. The VRG will be updated by removing the corresponding edges.
- Selection: The user may wants to keep only certain rules. In that case, the corresponding VRG edges will be retained and the rest will be deleted.
- Filtering: If there are a large number of rules, the user can choose to filter them based on criteria like rule scores, variable correlations, etc.
- Ranking: The user may provide a ranking of rules (rank 1 is most preferred) provided by the algorithm. The algorithm will then try to implement the rules according to the ranks, as demonstrated in [8].

Figure 3 shows a portion of the IK-EMOViz GUI for the example problem considered previously. For Group 2, the list of power laws obtained is shown on the left and the corresponding VRG is shown on the right. Figure 4 shows the selected rules by the user achieved by clicking the corresponding check-boxes. On clicking the green tick button (top corner in right plot), the VRG is updated with only the selected rules. For example, Node 12 in Fig. 3 is now absent in Fig. 4, since the user did not select any rule involving variable x_{12}. This type of operation can be useful when the user only wants to select a few rules involving a few important variables.

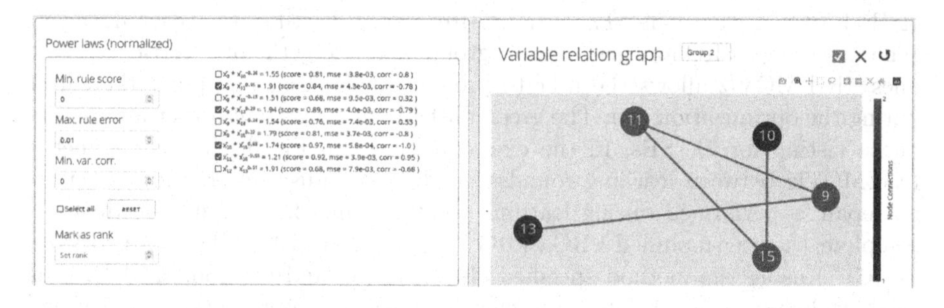

Fig. 4. IK-EMOViz graphical user interface showing a rule selection operation.

Figure 5 shows the resulting VRG when some rules are excluded from Fig. 3 by selecting them in the rule list and clicking the cross button on the top right of the VRG. This is useful when the user only wants to exclude specific rules provided by the algorithm.

2.5 Synchronous vs Asynchronous User Interaction

Pausing the optimization until the user finishes providing feedback ensures the user always has access to the latest data. This is also referred to as synchronous

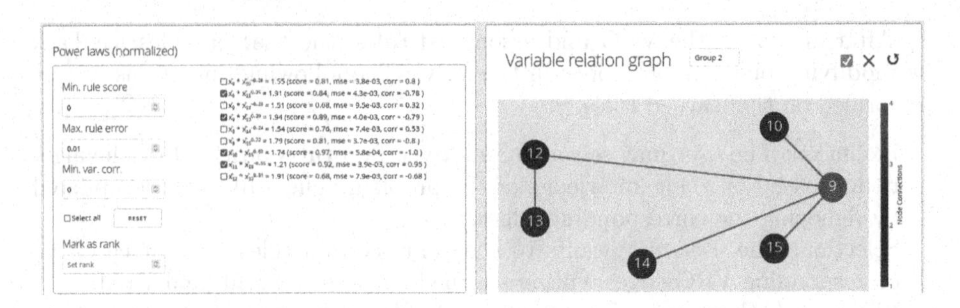

Fig. 5. IK-EMOViz graphical user interface showing a rule exclusion operation.

interaction. However, the user can take a longer time to analyze the results and provide his/her feedback. It can be more practical to continue running the optimization in the background while the user analyzes the intermediate solutions and their associated rules. This is known as asynchronous interaction. In the previous section, the pause optimization functionality of IK-EMOViz was introduced. It allows the user to switch between synchronous and asynchronous interaction modes.

Apart from the learning interval (T_L) and repair interval (T_R), we introduce another quantity, the user feedback time (T_U). This is defined as the number of generations or SEs that can be performed during the time the user is analyzing the results and preparing the feedback. T_U is undefined for synchronous user interaction. Figure 6 illustrates the asynchronous user interaction mechanism of IK-EMOViz, assuming $T_L = T_R$. After every T_L SEs, the learning agent generates a new VRG for each variable group, marked by the blue vertical dashed lines. IK-EMOViz allows the user to provide feedback at any point thereafter during the optimization run. The green dashed lines represent a user interaction phase lasting for T_U SEs. In the example shown in Fig. 6, the user launches IK-EMOViz between learning rounds 2 and 3. Since user interaction is active, no repair is performed during learning rounds 3 and 4. Once user feedback is complete, the user-modified VRG ($\text{VRG}^{(U)}$) is merged with the last learned VRG ($\text{VRG}^{(4)}$) using the method specified elsewhere [8], and the combined VRG is used by the repair agent. Thus, a larger T_U means the user is making a decision based on potentially outdated information and may not be as efficient as using the latest knowledge.

3 Truss Design Problem

For this study, a scalable truss optimization problem [8] is used. The aim is to minimize two objectives: (a) weight, and (b) compliance of a truss subject to a load on all the top nodes of the truss. Member radii ($\mathbf{r}$), and length of the vertical members ($\mathbf{L}_v$) are the two types of decision variables. Stress and displacement in each member is constrained to lie below threshold values.

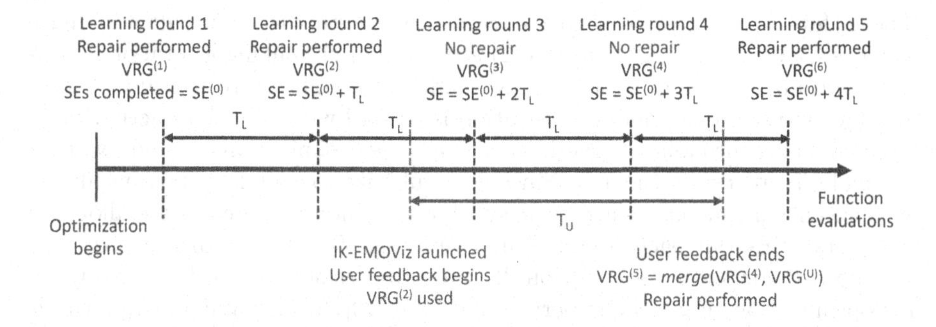

Fig. 6. Asynchronous user interaction example with $T_L = T_R$. Blue dashed lines represent the learning phases taking place at intervals of T_L SEs. Green dashed lines represent a user interaction phase. (Color figure online)

3.1 Experimental Settings

In this paper, we use two instances of the truss design problem, a truss with 120 members, 36 nodes, 129 decision variables and 156 non-linear constraints, and another larger truss with 820 members, 236 nodes, 879 decision variables, and 1056 non-linear constraints. Variable groups were created similar to [8] according to the physical orientation of the truss beams, shown in Table 1. The 120-member truss is used to illustrate how IK-EMOViz can allow the user to obtain insights about an optimization problem. NSGA-II [3] is chosen as the optimization algorithm of IK-EMO in this paper. Population size is set as 40 and the maximum number of generations is set as 100. After 100 generations, IK-EMOViz is launched. The results are presented in the next section.

Table 1. Variable groups for the 120 and 820-member truss design problems.

Group	Variable type	Variable indices	
		120-member truss	820-member truss
G_1	l_i of vertical members	$[120 - 128]$	$[820 - 878]$
G_2	r_i of top longitudinal members	$[8 - 15], [24 - 31]$	$[0 - 57], [116 - 173]$
G_3	r_i of bottom longitudinal members	$[0 - 7], [16 - 23]$	$[58 - 115], [174 - 231]$
G_4	r_i of vertical members	$[36 - 53]$	$[236 - 353]$

The 820-member truss design problem is used to illustrate the power of IK-EMO in finding good solutions with periodic user interactions. NSGA-II is run for a maximum number of generations of 12,500 (set by trial-and-error process and required to work with 879 variables), thus giving a total computational budget of 500,000. To ensure consistency, three artificial users – U1, U2, and U3 are created to select rules from the all the generated rules by our rule learning method with scores above 0.9, 0.7, and 0.5, respectively. Thus, user U1 chooses fewer rules compared to U2 and U3. For each user, four rule adherence schemes (RA1, RA2, RA3, and RA-E) are considered, making a total of 12 separate optimization runs. Moreover, two different scenarios are considered: synchronous interaction, where optimization is paused when user interaction takes place, and asynchronous interaction, in which the optimization does not wait for the analysis process by the user. Since we would like to complete the runs after a fixed execution time is achieved, the synchornous interaction cases are allocated less overall SEs (for each round of user analysis, T_U SEs are discounted). For each user-repair agent combination, 20 runs are performed, involving a total of 480 optimization runs. The hypervolume (HV) [19] metric value for each run is recorded.

3.2 Experimental Results and Discussion on the 120-Member Truss

For the 120-member truss, the Pareto front and HV evolution plots are shown in Fig. 7. The figures are extracted from IK-EMOViz. From the Pareto-optimal solutions found till now, power law rules are extracted. Group G_1 representing the length of the vertical members are shown in Fig. 8 highlighted in green. From

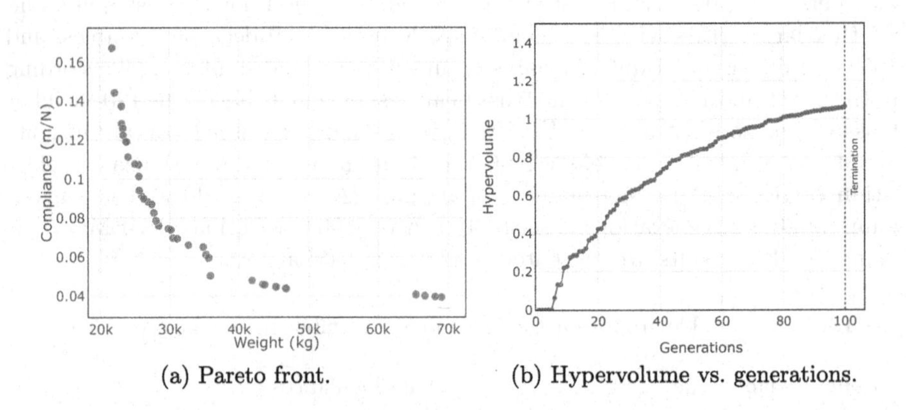

(a) Pareto front. (b) Hypervolume vs. generations.

Fig. 7. Pareto front and hypervolume plot for a 120-member truss optimization problem obtained from IK-EMOViz.

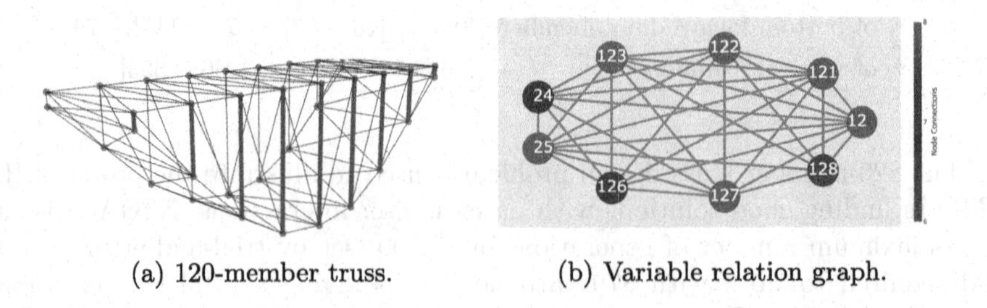

(a) 120-member truss. (b) Variable relation graph.

Fig. 8. A Pareto-optimal truss design with group G_1 highlighted in green and the corresponding variable relation graph obtained from IK-EMOViz. (Color figure online)

Fig. 8b, it can be seen that all the variables are related to each other through some significant power law. However, through the functionality of IK-EMOViz we can perform some simplifications. Variables 120–128 represent consecutive vertical members shown in Fig. 8a from right to left. So we perform rule selection and select rules of the form $x_i x_{i+1}^b = c$ where $i = 120, 121, \ldots, 127$. The reduced set of power law rules thus obtained for G_1 along with their rule compliance values are given in Table 2. Rule compliance is defined as the proportion of ND solutions that follow a power law. The corresponding VRG is shown in Fig. 9a.

Table 2. Power law rules found for group G_1.

Rule no	Power law	Rule compliance
1	$x_{120}x_{121}^{-0.57} = 0.92$	1.00
2	$x_{121}x_{122}^{-1.00} = 0.87$	1.00
3	$x_{122}x_{123}^{-0.92} = 0.96$	1.00
4	$x_{123}x_{124}^{-0.82} = 1.05$	1.00
5	$x_{124}x_{125}^{-1.13} = 0.98$	1.00
6	$x_{125}x_{126}^{-0.78} = 1.14$	1.00
7	$x_{126}x_{127}^{-1.06} = 1.10$	1.00
8	$x_{127}x_{128}^{-1.72} = 1.03$	1.00

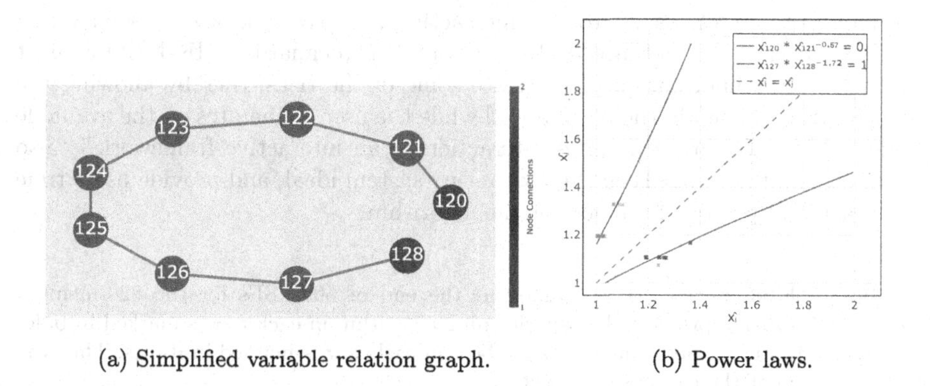

(a) Simplified variable relation graph. (b) Power laws.

Fig. 9. Simplified variable relation graph of group G_2 and power laws between pairs of variables (x_{120}, x_{121}) and (x_{127}, x_{128}) obtained from IK-EMOViz.

From Table 2, let us consider two power laws $\phi_1(\mathbf{x}) = x_{120}x_{121}^{-0.57} - 0.92 = 0$ and $\phi_2(\mathbf{x}) = x_{127}x_{128}^{-1.72} - 1.03 = 0$. x_{120} and x_{128} represent the length of the vertical members at each end, and x_{121} and x_{127} represent the adjacent vertical members, respectively. Figure 9b obtained from IK-EMOViz plots the two power laws over the normalized variable range of [1, 2] and plots another line $x_i = x_j$. The variables of the ND set from which the power laws are extracted are also shown. An interesting observation is that the $\phi_1(\mathbf{x})$ lies above and $\phi_2(\mathbf{x})$ lies below the $x_i = x_j$ line. This indicates that $x_{120} < x_{121}$ and $x_{127} > x_{128}$ among good solutions. Analyzing the rest of the power laws in Table 2 and the loading condition of the truss, we observe that $x_{120} < x_{121} < \ldots < x_{124}$ and $x_{124} > x_{125} > \ldots > x_{128}$. This indicates that the length of the vertical members increase from the end to the middle of the truss, which is an expected from an engineering intuition for the specific support and loading conditions [7]. IK-EMO is able to successfully extract these rules and IK-EMOViz provides useful functionality to the user to understand the rules. Such a knowledge derived from intermediate iterations of an optimization run will be reassuring to an inquisitive user.

3.3 Experimental Results and Discussion on the 820-Member Truss

The final median HV obtained at the end of 500k SEs for each user-repair agent combination are shown in Table 3. The row-wise best is marked in bold with the statistically similar performing algorithms marked in italics. The column-wise best is marked by a gray box. It can be seen that user U2 who selects a moderate amount of rules (supported by 70% or more ND solutions) is the best performer for all the repair agents. Moreover, for each user, RA2 with medium adherence scheme is the best algorithmic strategy in 4 out of 6 cases, and RA-E (ensemble scheme) has a the best performance in 3 out of 6 cases. This shows that an intermediate amount of rule usage combined with a medium level of rule adherence works the best, as was also observed in another study [8]. In all cases, asynchronous interaction results in a better performance due to the use of more SEs. While asynchronous interaction runs the risk of the user making a decision based on outdated rules, the built-in safeguards of IK-EMO against incorrect user information [8] mitigate some of the risks, thereby allowing the optimization to run in the background while the user deliberates on the available information. The asynchronous interaction in an interactive framework is also practical and does not keep the computing system ideal, and provide users time to grasp and prepare feedback for the algorithm.

Table 3. Final median HV obtained at the end of 500k SEs for the 820-member truss optimization problem. Best-performing algorithm in each row is marked in bold. Statistically similar performance to the best in each row is marked in italics. The gray boxes represent the column-wise best.

Repair agent	U1		U2		U3	
	Sync	Async	Sync	Async	Sync	Async
None (base)	0.79	0.79	0.79	0.79	0.79	0.79
RA1	0.81	0.98	0.88	**1.01**	0.82	0.95
RA2	0.84	*1.06*	0.91	**1.08**	0.77	0.99
RA3	0.81	0.96	0.86	**0.98**	0.79	0.90
RA-E	0.85	1.02	0.88	**1.08**	0.79	0.99

For user U2 and repair agent RA2, the PF and HV plots comparing synchronous and asynchronous interaction cases are shown in Fig. 10. Figure 10a shows how asynchronous interaction can provide better solutions compared to synchronous interaction. The HV plot in Fig. 10b shows a portion of the optimization run between 50 k and 200 k SEs with user interaction instances being marked in red. Asynchronous interaction results in a better median HV.

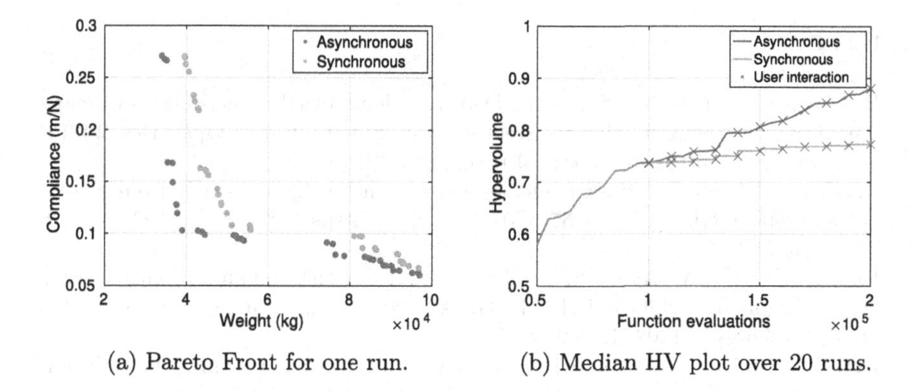

(a) Pareto Front for one run.

(b) Median HV plot over 20 runs.

Fig. 10. Pareto fronts and median hypervolume plots obtained by IK-EMO with U2, RA2, and asynchronous interaction for 820-member truss design problem.

4 Conclusions and Future Work

This study has introduced a knowledge-based interactive optimization tool IK-EMOViz for executing a better-informed optimization study. Through a 120-member truss design problem, we have shown how IK-EMOViz can help the user to obtain useful insights about an optimization problem in the form of simplistic relationships among variables and visualization of the interaction through relationship graphs. In addition, the user can also provide their own feedback by providing their preferences on obtained relationships using the tool. IK-EMOViz also allows the user to perform an asynchronous interaction by utilizing computing resources in the background while understanding and analyzing the obtained relationships. As demonstrated on an 820-member truss problem, the option of asynchronous interaction can result in a better performance. Moreover, moderately few relationships chosen by the user applied with moderate adherence within the EMO algorithm have found to produce better results.

Our future work will focus on enhancing the functionality of IK-EMOViz by allowing direct modification of rule parameters. The user should also be able to introduce new rules if they are not present in the rule set extracted by the learning agents. Other modifications can include introduction of new repair agents or disabling existing ones. The VRG widget can also be made interactive by turning it into a 3D model, allowing the user to directly modify the VRG. Nevertheless, this study provides evidence of the importance of a robust software implementation for any knowledge-based interactive optimization method that combines human knowledge and machine intelligence in executing an optimization task faster than either of them alone. IK-EMOViz opens a communication channel between the user and the optimization algorithm through scientific visualization and analytics. This will benefit users desiring greater involvement in the optimization process, especially for practical problems.

References

1. Bandaru, S., Aslam, T., Ng, A.H., Deb, K.: Generalized higher-level automated innovization with application to inventory management. Eur. J. Oper. Res. **243**(2), 480–496 (2015). https://doi.org/10.1016/j.ejor.2014.11.015
2. Bandaru, S., Deb, K.: Higher and lower-level knowledge discovery from Pareto-optimal sets. J. Glob. Optim. **57**, 281–298 (2013). https://doi.org/10.1007/s10898-012-0026-x
3. Deb, K., Pratap, A., Agarwal, S., Meyarivan, T.: A fast and elitist multiobjective genetic algorithm: NSGA-II. IEEE Trans. Evol. Comput. **6**(2), 182–197 (2002). https://doi.org/10.1109/4235.996017
4. Deb, K., Srinivasan, A.: Innovization: Innovating design principles through optimization. In: GECCO 2006 - Genetic and Evolutionary Computation Conference, vol. 2, pp. 1629–1636 (2006). https://doi.org/10.1145/1143997.1144266
5. Deb, K., Sundar, J.: Reference point based multi-objective optimization using evolutionary algorithms. In: GECCO 2006 - Genetic and Evolutionary Computation Conference, vol. 1, pp. 635–642. ACM Press, New York (2006). https://doi.org/10.1145/1143997.1144112
6. Gaur, A., Deb, K.: Adaptive use of innovization principles for a faster convergence of evolutionary multi-objective optimization algorithms. In: Proceedings of the 2016 Genetic and Evolutionary Computation Conference, New York, USA, pp. 75–76 (2016). https://doi.org/10.1145/2908961.2909019
7. Ghosh, A., Deb, K., Averill, R., Goodman, E.: Combining user knowledge and online *Innovization* for faster solution to multi-objective design optimization problems. In: Ishibuchi, H., et al. (eds.) EMO 2021. LNCS, vol. 12654, pp. 102–114. Springer, Cham (2021). https://doi.org/10.1007/978-3-030-72062-9_9
8. Ghosh, A., Deb, K., Goodman, E., Averill, R.: An interactive knowledge-based multi-objective evolutionary algorithm framework for practical optimization problems [manuscript submitted for publication] (2022). https://doi.org/10.48550/arXiv.2209.08604
9. Ghosh, A., Deb, K., Goodman, E., Averill, R.: A user-guided innovization-based evolutionary algorithm framework for practical multi-objective optimization problems. Eng. Optim., 1–13 (2022). https://doi.org/10.1080/0305215X.2022.2144275
10. Ghosh, A., et al.: Interpretable AI agent through nonlinear decision trees for lane change problem. In: 2021 IEEE Symposium Series on Computational Intelligence (SSCI), pp. 01–08 (2021). https://doi.org/10.1109/SSCI50451.2021.9659552
11. Ghosh, A., Goodman, E., Deb, K., Averill, R., Diaz, A.: A large-scale bi-objective optimization of solid rocket motors using innovization. In: 2020 IEEE Congress on Evolutionary Computation (CEC), pp. 1–8 (2020). https://doi.org/10.1109/CEC48606.2020.9185861

12. Greco, S., Matarazzo, B., Słowiński, R.: Interactive evolutionary multiobjective optimization using dominance-based rough set approach. In: IEEE Congress on Evolutionary Computation (CEC) 2010 (2010). https://doi.org/10.1109/CEC.2010.5585982
13. Hossain, S.: Visualization of bioinformatics data with dash bio. In: Proceedings of the 18th Python in Science Conference (2019). https://dash.plot.ly/dash-bio
14. Miettinen, K., Ruiz, F., Wierzbicki, A.P.: Introduction to multiobjective optimization: interactive approaches. In: Branke, J., Deb, K., Miettinen, K., Słowiński, R. (eds.) Multiobjective Optimization. LNCS, vol. 5252, pp. 27–57. Springer, Heidelberg (2008). https://doi.org/10.1007/978-3-540-88908-3_2
15. Ng, A.H., Bernedixen, J., Moris, M.U., Jägstam, M.: Factory flow design and analysis using internet-enabled simulation-based optimization and automatic model generation. In: Proceedings - Winter Simulation Conference, pp. 2176–2188 (2011). https://doi.org/10.1109/WSC.2011.6147930
16. Ng, A.H.C., Dudas, C., Boström, H., Deb, K.: Interleaving innovization with evolutionary multi-objective optimization in production system simulation for faster convergence. In: Nicosia, G., Pardalos, P. (eds.) LION 2013. LNCS, vol. 7997, pp. 1–18. Springer, Heidelberg (2013). https://doi.org/10.1007/978-3-642-44973-4_1
17. Ruiz, A.B., Ruiz, F., Miettinen, K., Delgado-Antequera, L., Ojalehto, V.: NAUTILUS navigator: free search interactive multiobjective optimization without trading-off. J. Glob. Optim. **74**(2), 213–231 (2019). https://doi.org/10.1007/s10898-019-00765-2
18. Smedberg, H., Bandaru, S.: Interactive knowledge discovery and knowledge visualization for decision support in multi-objective optimization. Eur. J. Oper. Res. **306**(3), 1311–1329 (2022). https://doi.org/10.1016/J.EJOR.2022.09.008
19. Zitzler, E., Thiele, L.: Multiobjective optimization using evolutionary algorithms - A comparative case study. In: Eiben, A.E., Back, T., Schoenauer, M., Schwefel, HP. (eds) Parallel Problem Solving from Nature—PPSN V. PPSN 1998.Lecture Notes in Computer Science, vol. 1498 LNCS, pp. 292–301. Springer Verlag, Cham (1998). https://doi.org/10.1007/bfb0056872

An Interactive Decision Tree-Based Evolutionary Multi-objective Algorithm

Seyed Mahdi Shavarani[1(✉)], Manuel López-Ibáñez[1], Richard Allmendinger[1], and Joshua Knowles[1,2]

[1] Alliance Manchester Business School, University of Manchester,Manchester, UK
{seyedmahdi.shavarani,manuel.lopez-ibanez}@manchester.ac.uk
[2] Schlumberger Cambridge Research, Cambridgeshire, UK

Abstract. Recent research using machine decision makers has revealed that some leading interactive evolutionary multi-objective optimization algorithms do not perform robustly with respect to interactions with preference models (and biases) posited to be representative of human Decision Makers (DMs). In order to model preferences better, we propose an explainable interactive method that uses decision trees to automate (fast) pairwise comparisons based on trade-offs of two given solutions. To cancel out possible biases and errors in estimations, we use the trained tree in holistic comparisons to determine solutions that survive each generation. We test our new method with respect to two different preference models (Tchebychef and Sigmoid) on problems from 2 to 10 objectives, and control both the number of interactions available and various biases. The results suggest the superiority of our method in learning the DM's preferences and in terms of the utility value of the final solution returned by the algorithm compared with some well-known interactive methods.

Keywords: Interactive evolutionary multi-objective optimization · Decision tree · Machine decision maker · Preference learning · Bias

1 Introduction

Many real-life optimization problems have several conflicting objectives to be optimized simultaneously [22]. Due to the conflicting nature of objective functions in such Multi-Objective Optimization Problems (MOOPs), it is generally not possible to have a single solution where all the objectives attain their optimal values. Instead, the general goal of solving MOOPs is to reach a small subset of Pareto-optimal solutions with interesting trade-offs or the most preferred one thereof. Evolutionary Multi-Objective Optimization Algorithms (EMOAs) naturally work with a population of solutions and can generate a representation of the Pareto Front (PF) in a single run. Hence they are well aligned to the requirements of MOOPs. However, with an increase in the number of objectives,

Supplementary Information The online version contains supplementary material available at https://doi.org/10.1007/978-3-031-27250-9_44.

M. Emmerich et al. (Eds.): EMO 2023, LNCS 13970, pp. 620–634, 2023.
https://doi.org/10.1007/978-3-031-27250-9_44

EMOAs lose their selection pressure, and their performance declines exponentially [7,16,17]. Interactive algorithms compensate for this issue by building a preference model of the Decision Maker (DM) and generating only those parts of the PF that are interesting to the DM [1,5]. Such *interactive* EMOAs (iEMOAs) alternate between the decision-making and optimization phase in order to reduce computational costs and support the DM in reaching a desirable solution, while incurring minimal cognitive effort.

Despite the potential of iEMOAs as described above, unfortunately scant evidence exists to date to attest to how well such algorithms perform under realistic conditions [1,19]. The difficulty is partly due to the challenge of testing with human DMs. Rather, recent studies, exploiting a Machine DM (MDM) to enable statistical testing over sufficient interactive runs [19,25], have indicated that iEMOAs may not perform as expected under realistic conditions. Specifically, their performance declines with a higher number of objectives or when the elicited preference information is affected by human-specific biases, such as inconsistent decisions [25] and fatigue [31]. Thus, even the algorithms that perform well under ideal conditions seem to lack much robustness to biases and other complications that are typical in real-life situations. The primary sources of the problem should be traced back to the core features of an interactive method, which include interaction style, preference model, and the way the preferences are exploited inside the method to direct the search towards the Most Preferred Solution (MPS) [28].

In re-designing these components, we suggest adopting Decision Trees (DTs). DTs have been widely used in learning user preferences [9,10,27,29], but not, to our knowledge, in interactive methods. As non-parametric supervised learning methods, DTs can be seen as a piece-wise constant approximation [6] that delivers classification by recursive partitioning of the solutions space. Utilizing DTs does not require any assumptions about the DM's preference model. DTs are competitive with other learning methods in terms of accuracy and speed, superior in terms of interpretability and comprehensiveness [9], and thus, are very popular among classification techniques [30]. DTs are easy to understand, interpret and visualize (compared with black-box models). This characteristic is particularly beneficial in helping human DMs understand and trust the process and final results [2]. Recently, there have been an emphasis on the explainability of the learning methods in interactive methods and particularly it has been encouraged to employ explainable learning methods, such as DTs for preference learning [20]. DTs do not require any pre-processing or scaling of the data, which can be a tricky task in MOOPs with an unknown PF, and they can work with different types of data, from categorical to continuous [2]. DTs naturally detect relevant features and simplify the learned model [27,29].

We use binary classification DTs to learn the desirable trade-offs in pairwise comparisons and to predict the preferred solution from the DM's perspective. DTs overcome problems of other learning-to-rank algorithms, which are susceptible to significant errors when confronted with small biases and non-idealities such as inconsistent decisions [9]. The learned preferences are local to the area of the PF that is being explored in that stage of the interaction, which is a desirable property. The experimental results indicate that the proposed algorithm, which we call DT-based EMOA (DTEMOA), achieves competitive performance when compared with other iEMOAs in terms of the desirability of the final solution returned and robustness to biased or inconsistent preferences.

In what follows, our proposed DTEMOA algorithm, and the way DTs are utilized in it, are explained in Sect. 2. Our experimental setup is laid out in Sect. 3, and the results are discussed in Sect. 4. Finally, conclusions and future research directions are provided in Sect. 5.

2 Methods

We focus on indirect preference elicitation in our method, where the DM is asked to provide a ranking of a subset of non-dominated solutions at each interaction. We do not attempt to learn a complete ordering of a set of solutions. Instead, similar to [23], we try to learn how the DM performs pairwise comparisons. As in [14], the preferences of the DM are modeled by simple rules. We use DTs to generate such rules to predict the preferred (winner) solution in pairwise comparisons.

DTs consist of nodes that partition data according to a rule. Our method is based on the DT known as CART [6] using a Gini impurity measure to identify good partitions of the data.

2.1 Decision Tree-Based EMOA (DTEMOA)

Preference Elicitation. In each interaction, a small sample of solutions is randomly selected from the population ($S \subseteq pop$, $|S| = N^{\text{sol}}$) and their objective vectors $Z = \{\mathbf{z} = \mathbf{f}(\mathbf{x}) |\ \mathbf{x} \in S\}$ are presented to the DM. The DM ranks the options based on their utility. There is no requirement for a complete ordering of the solutions; the algorithm can handle a partial ordering. However, we assume the DM provides a complete ordering for simplicity. Let $a \succ b$ denote that a is preferred over b. We represent the total order provided by the DM with the vector $\mathbf{r}$ such that the rank of $\mathbf{z}_i$ is r_i (lower rank values are better) and $\mathbf{z}_i \succ \mathbf{z}_j \iff r_i < r_j$.

Preference Learning. We aim to learn a DM's preferences from the ranking provided as well as predicting (simulating) her preference when evaluating the trade-offs of two objective vectors. DTEMOA infers pairwise orderings from the elicited preference information that can be in the form of complete or partial ranking of a subset of solutions. To construct the training set T, we use the well-known pairwise transformation [15], i.e., for each pair of objective vectors $\mathbf{z}_i, \mathbf{z}_j \in Z$, we create a training example $\mathbf{z}_i - \mathbf{z}_j$ labeled with:

$$\phi(\mathbf{z}_i - \mathbf{z}_j) = \text{sign}(r_j - r_i) = \begin{cases} +1 & \text{if } \mathbf{z}_i \succ \mathbf{z}_j \\ -1 & \text{otherwise;} \end{cases} \tag{1}$$

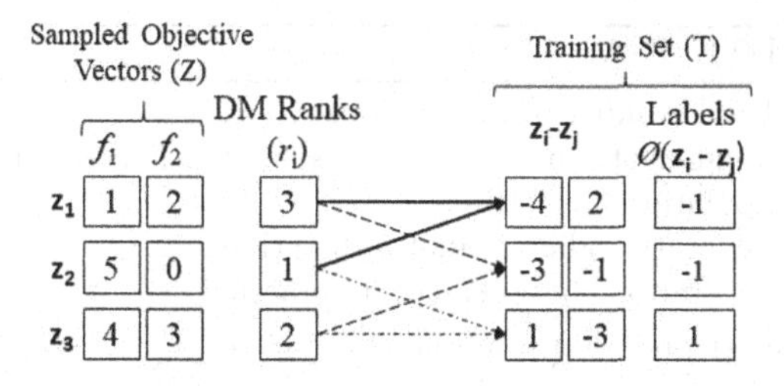

Fig. 1. Construction of the training set from the ranked solutions over a problem with 2 objectives. The number of solutions ranked by the DM is $|Z| = 3$. For each pair of ranked solution ($\mathbf{z}_i, \mathbf{z}_j \in Z$) an example, labeled by $\phi(\mathbf{z}_i - \mathbf{z}_j)$, is added to the training set. The size of the training set is $\binom{|Z|}{2}$.

i.e., an example is labeled +1 if the first solution is preferred over the second one, and −1, otherwise. An instance of such a process is given in Fig. 1, where $Z \subset \mathbb{R}^2$, $|Z| = 3$, and each possible pairwise comparison of the ranked set generates an example of the training set T.

In the next step, T is used to train the DT to predict the preferred solution in pairwise comparisons of unseen data as well as the probability of such a decision. Subsequent interactions add more training examples to T, increasing the accuracy of the model. To predict the preferred solution when comparing two objective vectors $\mathbf{z}$ and $\mathbf{z}'$, the trade-off vector $\mathbf{z} - \mathbf{z}'$ is given as the input to the trained DT, which predicts its class (label) $\phi(\mathbf{z} - \mathbf{z}') \in \{+1, -1\}$ indicating whether $\mathbf{z}$ is preferred over $\mathbf{z}'$. The probability of the sample being in class $c \in \{+1, -1\}$ at leaf node o is calculated by $\frac{N_o^c}{N_o}$, where N_o^c is the number of samples of class c in node o, and N_o is the total number of samples in o.

Determination of Solution Score for Sorting. Each solution in the population is given a score used to sort solutions. To calculate the score of a solution $\mathbf{x}$, it is compared with all other solutions in the population and the probability that it is preferred over the compared one is calculated using the trained DT. The sum of all such probabilities is the score of the solution $\mathbf{x}$:

$$score(\mathbf{x}) = \sum_{\substack{\mathbf{x}' \in pop \\ \mathbf{x} \neq \mathbf{x}'}} \Pr\{\mathbf{f}(\mathbf{x}) \succ \mathbf{f}(\mathbf{x}')\} = \sum_{\substack{\mathbf{x}' \in pop \\ \mathbf{x} \neq \mathbf{x}'}} \Pr\{\phi(\mathbf{f}(\mathbf{x}) - \mathbf{f}(\mathbf{x}')) = 1\} \qquad (2)$$

Integrating DTs into EMOAs. Given the above steps, let us explain how the preferences of the DM are considered in an EMOA to guide the search toward the most preferred parts of the PF. We use NSGA-II [11] as the underlying optimizer. After some generations of the NSGA-II, the algorithm stops to make the first interaction, elicits a ranking from a subset of solutions and consequently builds the training set T as explained above. A DT is trained using T to predict the preferred solution in pairwise comparisons. In the following generations, the

Algorithm 1: DTEMOA

Input: N^{int} : Total number of interactions
N^{sol} : Number of solutions evaluated by the DM per interaction
pop : Population of solutions
gen_1 : Generations before first interaction
gen_i : Generations between two interactions

Output: The most preferred solution

1 $T \leftarrow \emptyset$
2 $pop \leftarrow$ run NSGA-II for gen_1 generations
3 **for** 1 **to** N^{int} **do**
4 $Z \leftarrow$ select N^{sol} solutions
5 $\mathbf{r} \leftarrow$ ask the DM to rank the solutions in Z
6 $T \leftarrow T \cup \{(\mathbf{z}_i - \mathbf{z}_j, \phi(\mathbf{z}_i - \mathbf{z}_j)) \mid \forall \mathbf{z}_i, \mathbf{z}_j \in Z, \phi(\mathbf{z}_i - \mathbf{z}_j) = \text{sign}(r_j - r_i)\}$
7 Train DT using T
8 $pop \leftarrow$ run NSGA-II for gen_i generations
replacing crowding distance with *score* (Eq. 2)
9 **return** best $\mathbf{x} \in pop$ ranked first by non-dominated sorting and then *score*

DT is used to calculate the scores of the solutions (Eq. 2) to sort solutions with the same non-dominated sorting rank, i.e., the solution scores replace the crowding distance of NSGA-II to differentiate between non-dominated solutions. The solutions with a higher score have a better chance of survival and participation in mating and the generation of new offspring. The pseudo-code of the algorithm is illustrated in Algorithm 1.

When dealing with DMs, it is crucial to gain the DM's trust and confidence in the process and the results. When using DTs, learned trees can be conveniently visualized, summarizing all the rules elicited from the data set. Such a visualization is illustrated in Fig. 2, showing a DT that was built based on preference information elicited from the DM in an experiment on the DTLZ1 problem.

3 Experimental Design

This section outlines details pertaining to the Machine DM (MDM) we use to simulate a real DM, the benchmark problems, algorithms considered, performance metrics, and algorithm parameters settings as used in the subsequent experimental study.

3.1 Machine Decision Maker (MDM)

To simulate a real DM in the experiments, we have used the MDM introduced in [19] and improved in [25]. The MDM is composed of a utility function (UF) that simulates the true preferences of the DM and simulations of cognitive biases and other non-ideal decision-making behaviors that can happen in interactions with the DM. Shavarani et al. [25] proposed using a version of the Sigmoid

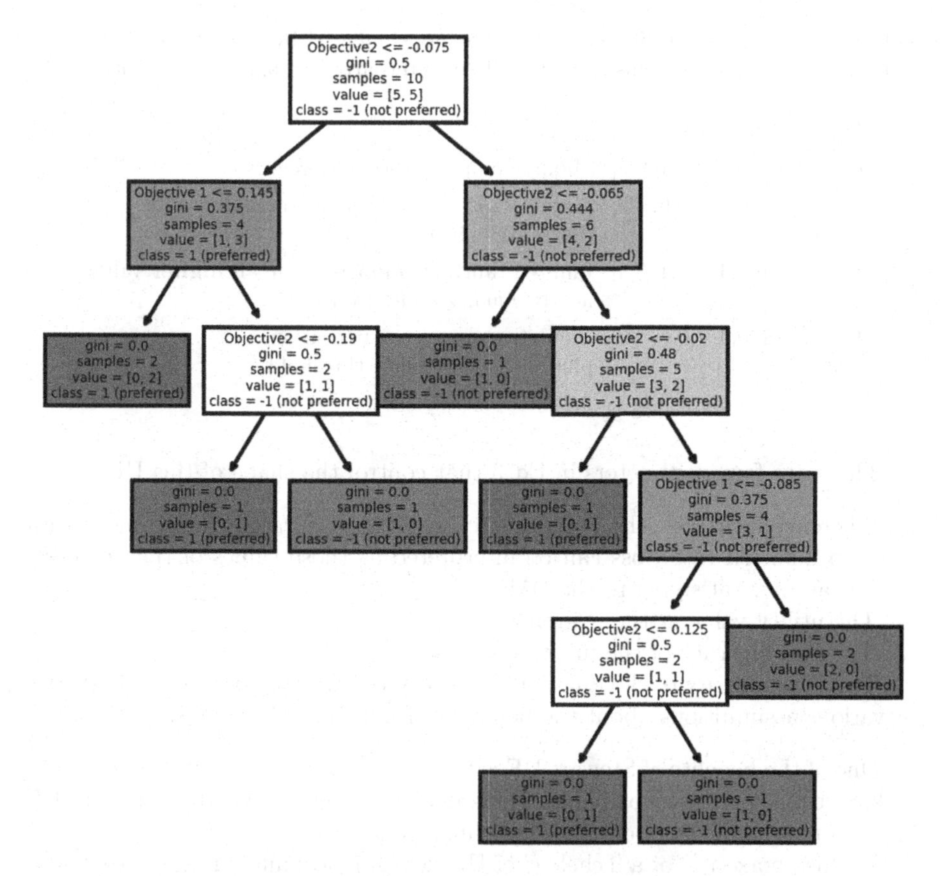

Fig. 2. A decision tree example built on the information elicited in one interaction on problem DTLZ1 with $m = 2$ objectives. Here objective l indicates the trade-off value for l^{th} objective, i.e., $z_{il} - z_{jl}$ when comparing solution i with solution j. The *samples* field indicates the number of samples that fall on that node. The cth element of *value* in each node o indicates N_o^c, the number of samples that belong to the cth class. Gini impurity is a measure (from 0 to 0.5) of the quality of the split.

UF introduced by Stewart [26] for experimenting on iEMOAs and modified it for minimization problems without violating any of its underlying assumptions. The modified UF used here (and hereafter called Stewart UF) is formulated as follows:

$$U'(\mathbf{z}) = \sum_{i=1}^{m} w_i u_i(z_i) \tag{3}$$

$$u_i(z_i) = \begin{cases} \lambda_i + \dfrac{(1-\lambda_i)(1-e^{-\beta_i(\tau_i - z_i)})}{1-e^{-\beta_i \tau_i}} & \text{if } 0 \leq z_i \leq \tau_i \\ \dfrac{\lambda_i \cdot (e^{\alpha_i(1-z_i)} - 1)}{e^{\alpha_i(1-\tau_i)} - 1} & \text{if } \tau_i < z_i \leq 1 \end{cases}$$

Table 1. Description of different DM behaviors simulated by combinations of τ_i and λ_i for minimization problems (based on its counterpart for maximization [26]).

Type	τ_i	λ_i	Description
1	[0.6, 0.9]	[0.1, 0.4]	Mainly compensatory preferences
2	[0.6, 0.9]	[0.6, 0.9]	Mainly compensatory preferences, but with sharp preference threshold
3	[0.1, 0.4]	[0.1, 0.4]	Limited range of compensation, all **higher** values nearly equally undesirable
4	[0.1, 0.4]	[0.6, 0.9]	Limited range of compensation, plus sharp preference threshold

There are four parameters in Eq. 3 that control the shape of this UF:

τ_i: Reference level, i.e., the point where losses are separated from gains with a steep inflation rate (loss can be interpreted as those values of the objective that are not satisfying to the DM).
λ_i: The utility value at the reference level.
α_i: The non-linearity of the function over losses.
β_i: The non-linearity of the function over the gains. Having $\alpha_i > \beta_i > 0$ satisfies various assumptions about the behavior of human DMs [25,26].

One of the benefits of Stewart UF is the ability to simulate different decision-making behaviors, as depicted in Table 1. It has been shown that Stewart UF imposes more difficulties on the algorithms [25].

We also make use of a Tchebychef UF, which is formulated as follows:

$$U(\mathbf{z} = \mathbf{f}(\mathbf{x})) = \max_{i=1,\ldots,m} w_i |z_i - z_i^*|, \tag{4}$$

where w_i is the weight of each objective function, and $\mathbf{z}^*$ is the ideal or Utopian point. Since the selected benchmark problems are to be minimized and preserve consistency, it is assumed that the DM prefers solutions with lower utility values.

Aside from the tests under ideal conditions, we use the MDM capability of simulation of inconsistencies in the decisions of the DM. The MDM adds a normally distributed random noise with mean 0 and variance σ. The variance σ controls the amount of noise. To investigate the robustness of the algorithms, we test with $\sigma \in \{0.005, 0.01, 0.1, 0.2\}$.

3.2 Benchmark Problems

We follow [3,18] and perform our experiments on well-known DTLZ benchmark problems [12]. DTLZ1, DTLZ2, DTLZ7 are selected from the DTLZ test suite. Each of these problems exposes different difficulties to the algorithm and allows one to investigate its performance from various aspects. DTLZ1 contains $11^k - 1$ local PFs, and each of them can attract an EMOA before reaching the global PF.

DTLZ2 investigates the performance of an EMOA in getting close to true PF. Finally, DTLZ7 has 2^{m-1} disconnected Pareto-optimal regions in the objective space. It is used to check the diversity of the solutions. The problem dimension n and the dimension of the objective space m are selected in a way that n is $m+4$ for DTLZ1, $m+9$ for DTLZ2 and $m+19$ for DTLZ7, as suggested in [12]; we consider $m \in \{2, 4, 10\}$. Also to make the problems more challenging, for DTLZ1 we follow [3] and limit x_i to $[0.25, 0.75]$, and for DTLZ2 we map x_i to $x_i/2 + 0.25$ as suggested by [8]. We consider all possible combinations of different problem sets, number of objective functions, different value functions and different decision-making behaviors. Full details about the set of 45 test configurations can be found in the supplementary materials [24].

3.3 Evaluating Performance and Competing Algorithms

Interactive methods are supposed to develop the best solution or a small subset of non-dominated solutions that maximize the DM's satisfaction. Thus, it makes sense to evaluate the performance of the interactive methods based on the utility value of the returned solution [1]. We compare the performance of DTEMOA against two state-of-the-art algorithms: BCEMOA [3] and iTDEA [18]. Preference learning in BCEMOA is similar to the way it is done by DTEMOA in that both learn to rank solutions, however, BCEMOA uses support vector machines (SVM) for preference learning. iTDEA uses a completely different preference learning scheme, where the preferences of the DM are reflected in the search process by prioritizing solutions in the proximity of the DM's selected solution. Both BCEMOA and DTEMOA use NSGA-II as their underlying optimizer. iTDEA still uses the same non-domination and mating methods, but in each generation only one solution is created and whether or not it is accepted to the population depends on the position of that solution in the objective space and its distance to its closest solution in the population. The iTDEA is sensitive to the threshold parameter that controls the acceptable distance for the new solution. A large threshold would prevent solutions to enter the population, while a small one would make the population grow uncontrolled and increase computational costs, specially in many-objective problems. These similarities and differences were the main motivation behind selecting these two algorithms.

3.4 Algorithm Parameter Settings

The parameters of the selected algorithms are illustrated in Table 2. iTDEA generates and evaluates only one solution per generation. Thus, the number of generations is set to $N^{\text{gen}} = 80\,000$ to have an equal number of objective evaluations in all compared algorithms. As suggested in [18], the first interaction of iTDEA happens after $\frac{N^{\text{gen}}}{3}$ generations, and the number of generations between each subsequent interaction is given by $gen_i = \frac{N^{\text{gen}}}{2(N^{\text{int}}-1)}$. Thus, with a larger number of interactions, gen_1 becomes smaller. The experimental study will investigate the impact of different numbers of interactions, N^{int}. The initial

Table 2. Parameter settings of the iEMOAs. In addition, total solution evaluations is 80 000, the population size $|pop| = 200$, the number of iterations is $N^{\text{int}} \in \{2, 3, 4\}$, and the DM ranks $N^{\text{sol}} = 5$ solutions per interaction. After N^{int} interactions, the algorithms continue running without further interaction until reaching N^{gen} generations.

Parameter	iTDEA	BCEMOA	DTEMOA
Total generations (N^{gen})	80 000	400	400
Generations before 1st interaction (gen_1)	$N^{\text{gen}}/3$	200	200
Generations between further interactions (gen_i)	$\frac{N^{\text{gen}}}{2(N^{\text{int}}-1)}$	20	20

and final territory parameters of iTDEA are respectively set to 0.1 and 0.00001 in problems with two objectives, which was one of the alternatives suggested in [18]. For problems with $m = 4$ and 10 objectives, these values change to 0.5 and 0.25, respectively. Any smaller values for these parameters would make the size of the archive population large and the computational costs unaffordable. Other parameters for BCEMOA and iTDEA are set as suggested in [3,18]. The hyper-parameters of the DT in DTEMOA and SVM in BCEMOA are tuned by grid search cross-validation at each interaction [24].

We run experiments with a different number of interactions $N^{\text{int}} \in \{2, 3, 4\}$ to test the effect of interactions on the results. Our initial experiments indicate that a higher number of interactions does not seem to increase the quality of the solutions substantially. Each experiment is repeated 40 times with different random seeds.

Implementations. BCEMOA, DTEMOA, iTDEA, MDM and utility functions are implemented in Python version 3.7.6. The NSGA-II and DTLZ benchmark implementations were acquired from the Pygmo library 2.16.0 [4], the SVM ranking model from the Preference Learning Toolbox [13] powered by scikit-learn 0.23.1 [21]. Scikit-learn is also used to implement the DT models.

4 Results and Discussion

This section is divided into two parts. In the first part, we focus on our proposed preference learning technique and provide some insight into its performance. In the second part, the performance of the proposed DTEMOA is compared with iTDEA and BCEMOA.

4.1 Assessing Ranking Performance

One of the main differentiating characteristics of iEMOAs is the way they adapt to the DM's preferences and the way they learn and model the preferences [28]. The methods compared in our study differ significantly in how they interact and adapt to the DM. Both BCEMOA and DTEMOA try to learn a model

Table 3. Mean accuracy (and standard deviation) of preference learning in DTEMOA and BCEMOA on the DTLZ7 problem for different UFs. The different number of interactions indicate how the algorithms exploit the accumulated preference data over interactions. The number of solutions presented to the DM at each interaction is $N^{\mathrm{sol}} = 5$. The best accuracy for each UF, m and N^{int} is highlighted in bold face.

UF	N^{int}	$m = 2$		$m = 4$		$m = 10$	
		BCEMOA	DTEMOA	BCEMOA	DTEMOA	BCEMOA	DTEMOA
Stewart	2	79.4(11.3)	84.8(12.1)	84.2(7.1)	83(7.2)	68.1(7.2)	62.2(6.5)
	3	82.9(9)	84.9(11.4)	82.2(7)	82.9(5.7)	69.5(5.8)	67(6)
	4	85.7(7.7)	86.5(5.5)	81.6(7.1)	83.5(2.9)	71.1(6.1)	69.3(6.3)
	5	**87.5(8.1)**	84(11.3)	81(6.7)	81.8(6.5)	70.3(4.5)	**71.3(7.1)**
Tchebychef	2	**100(0)**	99.9(0.2)	75.5(11.4)	95(10.1)	67.3(5.8)	96.4(8.5)
	3	**100(0)**	99.7(1.1)	74(7.8)	96.8(8.8)	67.1(6.5)	98.8(3.4)
	4	**100(0)**	**100(0)**	75.6(5.8)	97(9.6)	67.7(5.8)	97.2(9.9)
	5	**100(0)**	**100(0)**	76.7(5.1)	**98.2(9.9)**	68.3(6)	**100(0)**

to rank solutions but use different learning techniques. We can directly compare their accuracy independently of other algorithmic aspects. We evaluate the accuracy of the preference learning models of BCEMOA and DTEMOA on a randomly-generated population of 400 solutions for DTLZ7 problem with different number of objectives $m \in \{2, 4, 10\}$. It is also interesting to see how each algorithm exploits accumulated information elicited in different interactions. Thus, we simulate 5 interactions. Before each interaction, NSGA-II is used to evolve the population for 20 generations, and 5 solutions are selected randomly from the non-dominated front and ranked by a UF. The preference learning of both algorithms is applied to the elicited data. Then the accuracy of each model in ranking the non-dominated solutions in the population is measured by counting the proportion of correct pairwise rankings to all possible pairwise rankings:

$$\mathrm{Acc} = 100 \cdot \frac{\sum_{i=1}^{|pop|-1} \sum_{j=i+1}^{|pop|} \mathrm{I}\{r_i < r_j \wedge \hat{r}_i < \hat{r}_j\}}{\sum_{i=1}^{|pop|-1} \sum_{j=i+1}^{|pop|} \mathrm{I}\{r_i < r_j\}}, \tag{5}$$

where I is the indicator function which is equal to 1 if the given condition is true, otherwise 0; $r_i = U(\mathbf{f}(\mathbf{x}_i))$ is the true rank of the solution $\mathbf{x}_i \in pop$ according to the UF, and $\hat{r}_i$ is the rank predicted by the learning models for the same solution. The experiments are repeated 40 times for each problem.

Table 3 summarizes the results. In particular, the accuracy of both methods is similar with 2 objectives. However, the preference learning technique in DTEMOA performs better when the number of objectives increases.

Table 4. Performance of the algorithms over different number of interactions under ideal conditions. The utility values are averaged over all tests with all possible combinations of different problems, number of objective functions, different UFs and different decision-making behaviors. The results are rounded to 3 decimal places. The *p*-values obtained from the Wilcoxon test using Holm's method to adjust for multiple comparisons are also reported.

N^{int}	Mean utility value			Wilcoxon test's *p*-values		
	BCEMOA	DTEMOA	iTDEA	DTEMOA vs. BCEMOA	DTEMOA vs. iTDEA	BCEMOA vs. iTDEA
2	0.206	0.184	0.214	0.000	0.032	0.192
3	0.206	0.179	0.259	0.000	0.000	0.000
4	0.203	0.177	0.278	0.000	0.000	0.000

4.2 Comparison of the Performance with Other iEMOAs

The overall performance of the algorithms, measured in terms of the true utility of the returned solution (lower values are better), over all tests is illustrated in Tables 4 and 5. Table 4 summarizes the performance of the algorithms over different number of interactions. Table 5 show the results over different number of objective functions categorized into experiments under ideal conditions and those with simulation of inconsistent decisions. The results of this table are over experiments with 3 interactions, which is reasonable number of interactions based on the observations in Table 4. The results are aggregated over all tests with different UFs, problems and decision making behaviors. DTEMOA generally returns a solution with a better true utility than BCEMOA and iTDEA. The Wilcoxon test indicates that the differences in utility are significant (*p*-value ≈ 0). The second important observation, as illustrated in Table 5, is the robustness of the DTEMOA towards inconsistencies in the DM's decisions. Considering all the results over all tests, the performance of DTEMOA has declined by 1.6% when inconsistent decisions are simulated; this change is insignificant compared to the deterioration in the performance of BCEMOA (17%) and iTDEA (3.5%). Furthermore, the superiority of DTEMOA becomes more significant when the number of objectives increases.

Due to space limitations, we have not presented results for each test but reported only the aggregated results. The interested reader is referred to the supplementary materials [24] for the full set of results.

Table 5. Comparing the performance of algorithms over various number of objective functions (m). The utility values are averaged over all tests with all possible combinations of different problem sets, different UFs and different decision-making behaviors ($N^{\text{int}}= 3$). The results are rounded to 3 decimal places. Algorithms are compared pairwise, reporting the p-value for a Wilcoxon test using Holm's method to adjust for multiple comparisons.

	Under ideal conditions						With simulation of DM inconsistencies					
	mean utility value			p-value			mean utility value			p-value		
m	BCEMOA	DTEMOA	iTDEA	DTEMOA vs. BCEMOA	DTEMOA vs. iTDEA	BCEMOA vs. iTDEA	BCEMOA	DTEMOA	iTDEA	DTEMOA vs. BCEMOA	DTEMOA vs. iTDEA	BCEMOA vs. iTDEA
2	0.270	0.270	0.256	1.000	1.000	1.000	0.292	0.266	0.268	0.000	0.012	0.055
4	0.215	0.193	0.319	0.002	0.000	0.000	0.244	0.200	0.328	0.000	0.000	0.000
10	0.122	0.077	0.097	0.000	0.652	0.000	0.116	0.081	0.103	0.000	0.158	0.000

5 Conclusions

There have been many improvements in the field of interactive methods, and yet it seems there is room for more. Recent research on benchmarking of these methods has revealed that they may not perform as expected when confronted with non-ideal conditions that were not anticipated in their development [1,25]. At present, the field has little idea of how to explain or predict which components of iEMOA methods are most vulnerable to non-idealities or what conditions affect their performance the most. However, we believe the core of any interactive method is the preference model and the accuracy of these models in reflecting the DM's preferences is crucial in the optimization process. With these observations, this study proposed an innovative preference model and learning approach using decision trees to predict the result of pairwise comparisons from the DM's perspective to manage the convergence direction of the population. The method's performance is found to be stable in many-objective problems because more objectives translate to more attributes that can participate in classification. Further, when the number of objectives increases, DTs can naturally identify the most important ones. Our method is also found to be more robust to biases as we make holistic comparisons to cancel out any estimation errors in single predictions. In contrast, other learning methods such as SVM rely on kernel performance, and complications such as higher dimensions, non-linearity, and biases make kernel selection an intensive task and generally less accurate.

Another critical advantage of DTs in this context is their intuitive interpretation and ease of visualization, both being essential elements in gaining the

DM's trust in the process and the results. We used the suggested experimental design of [25] to evaluate the performance of the algorithm and examine the efficiency of the method. Further research may extend this study by investigating the application of weighted decision trees to emphasize the most recently elicited information. Another interesting research direction is to explore the use of decision tree regressors instead of classifiers in the preference model.

Reproducibility. We make source code and data for reproducing our results publicly available as supplementary materials [24] to motivate further research in this direction.

References

1. Afsar, B., Miettinen, K., Ruiz, F.: Assessing the performance of interactive multiobjective optimization methods: a survey. ACM Computing Surveys **54**(4) (2021). https://doi.org/10.1145/3448301
2. Allen, J., Moussa, A., Liu, X.: Human-in-the-loop learning of qualitative preference models. In: Barták, R., Brawner, K.W. (eds.) Proceedings of the Thirty-Second International Florida Artificial Intelligence Research Society Conference, pp. 108–111. AAAI Press (2019)
3. Battiti, R., Passerini, A.: Brain-computer evolutionary multiobjective optimization: a genetic algorithm adapting to the decision maker. IEEE Trans. Evol. Comput. **14**(5), 671–687 (2010). https://doi.org/10.1109/TEVC.2010.2058118
4. Biscani, F., Izzo, D., Yam, C.H.: A global optimisation toolbox for massively parallel engineering optimisation. Arxiv preprint arXiv:1004.3824 (2010). http://arxiv.org/abs/1004.3824
5. Branke, J., Deb, K., Miettinen, K., Słowiński, R. (eds.): Multiobjective Optimization: Interactive and Evolutionary Approaches. Lecture Notes in Computer Science, vol. 5252. Springer, Heidelberg (2008)
6. Breiman, L., Friedman, J., Stone, C.J., Olshen, R.A.: Classification and Regression Trees. CRC Press (1984)
7. Brockhoff, D., Zitzler, E.: Are all objectives necessary? On dimensionality reduction in evolutionary multiobjective optimization. In: Runarsson, T.P., Beyer, H.-G., Burke, E., Merelo-Guervós, J.J., Whitley, L.D., Yao, X. (eds.) PPSN 2006. LNCS, vol. 4193, pp. 533–542. Springer, Heidelberg (2006). https://doi.org/10.1007/11844297_54
8. Brockhoff, D., Zitzler, E.: Improving hypervolume-based multiobjective evolutionary algorithms by using objective reduction methods. In: Proceedings of the 2007 Congress on Evolutionary Computation (CEC 2007), pp. 2086–2093. IEEE Press, Piscataway (2007). https://doi.org/10.1109/CEC.2007.4424730
9. Cheng, W., Hühn, J., Hüllermeier, E.: Decision tree and instance-based learning for label ranking. In: Danyluk, A.P., Bottou, L., Littman, M.L. (eds.) Proceedings of the 26th International Conference on Machine Learning, ICML 2009, pp. 161–168. ACM Press, New York (2009). https://doi.org/10.1145/1553374.1553395
10. Dalip, D.H., Gonçalves, M.A., Cristo, M., Calado, P.: Exploiting user feedback to learn to rank answers in Q&A forums: a case study with stack overflow. In: Proceedings of the 36th International ACM SIGIR Conference on Research and Development in Information Retrieval, SIGIR 2013, pp. 543–552. Association for Computing Machinery, New York (2013). ISBN 9781450320344. https://doi.org/10.1145/2484028.2484072

11. Deb, K., Pratap, A., Agarwal, S., Meyarivan, T.: A fast and elitist multi-objective genetic algorithm: NSGA-II. IEEE Trans. Evol. Comput. **6**(2), 182–197 (2002). https://doi.org/10.1109/4235.996017
12. Deb, K., Thiele, L., Laumanns, M., Zitzler, E.: Scalable test problems for evolutionary multiobjective optimization. In: Abraham, A., Jain, L., Goldberg, R. (eds.) Evolutionary Multiobjective Optimization. Advanced Information and Knowledge Processing, pp. 105–145. Springer, London (2005). https://doi.org/10.1007/1-84628-137-7_6
13. Farrugia, V.E., Martínez, H.P., Yannakakis, G.N.: The preference learning toolbox (2015). arXiv:1506.01709
14. Greco, S., Matarazzo, B., Słowiński, R.: Interactive evolutionary multiobjective optimization using dominance-based rough set approach. In: Ishibuchi, H., et al. (eds.) Proceedings of the 2010 Congress on Evolutionary Computation (CEC 2010), pp. 1–8. IEEE Press, Piscataway (2010)
15. Herbrich, R., Graepel, T., Obermayer, K.: Support vector learning for ordinal regression. In: ICANN 1999: Proceedings of the 9th International Conference on Artificial Neural Networks, pp. 97–102 (1999). https://doi.org/10.1049/cp:19991091
16. Ishibuchi, H., Tsukamoto, N., Nojima, Y.: Evolutionary many-objective optimization: a short review. In: Proceedings of the 2008 Congress on Evolutionary Computation (CEC 2008), pp. 2419–2426. IEEE Press, Piscataway (2008). https://doi.org/10.1109/CEC.2008.4631121
17. Khare, V., Yao, X., Deb, K.: Performance scaling of multi-objective evolutionary algorithms. In: Fonseca, C.M., Fleming, P.J., Zitzler, E., Thiele, L., Deb, K. (eds.) EMO 2003. LNCS, vol. 2632, pp. 376–390. Springer, Heidelberg (2003). https://doi.org/10.1007/3-540-36970-8_27
18. Köksalan, M.: Karahan, İ: an interactive territory defining evolutionary algorithm: iTDEA. IEEE Trans. Evol. Comput. **14**(5), 702–722 (2010). https://doi.org/10.1109/TEVC.2010.2070070
19. López-Ibáñez, M., Knowles, J.: Machine decision makers as a laboratory for interactive EMO. In: Gaspar-Cunha, A., Henggeler Antunes, C., Coello, C.C. (eds.) EMO 2015. LNCS, vol. 9019, pp. 295–309. Springer, Cham (2015). https://doi.org/10.1007/978-3-319-15892-1_20
20. Misitano, G., Afsar, B., Larraga, G., Miettinen, K.: Towards explainable interactive multiobjective optimization: R-XIMO. Auton. Agents Multi-Agent Syst. **36**(42) (2022). https://doi.org/10.1007/s10458-022-09577-3
21. Pedregosa, F., et al.: Scikit-learn: machine learning in python. J. Mach. Learn. Res. **12**, 2825–2830 (2011)
22. Purshouse, R.C., Deb, K., Mansor, M.M., Mostaghim, S., Wang, R.: A review of hybrid evolutionary multiple criteria decision making methods. COIN Report 2014005, Computational Optimization and Innovation (COIN) Laboratory, University of Michigan, USA (2014)
23. Quan, G., Greenwood, G.W., Liu, D., Hu, S.: Searching for multiobjective preventive maintenance schedules: combining preferences with evolutionary algorithms. Eur. J. Oper. Res. **177**(3), 1969–1984 (2007). https://doi.org/10.1016/j.ejor.2005.12.015
24. Shavarani, S.M., López-Ibáñez, M., Allmendinger, R., Knowles, J.D.: An interactive decision tree-based evolutionary multi-objective algorithm: supplementary material (2022). https://doi.org/10.5281/zenodo.7429806

25. Shavarani, S.M., López-Ibáñez, M., Knowles, J.D.: Realistic utility functions prove difficult for state-of-the-art interactive multiobjective optimization algorithms. In: Chicano, F., Krawiec, K. (eds.) Proceedings of the Genetic and Evolutionary Computation Conference, GECCO 2021, pp. 457–465. ACM Press, New York (2021). https://doi.org/10.1145/3449639.3459373
26. Stewart, T.J.: Robustness of additive value function methods in MCDM. J. Multi-Criteria Decis. Anal. **5**(4), 301–309 (1996)
27. Xia, F., Zhang, W., Li, F., Yang, Y.: Ranking with decision tree. Knowl. Inf. Syst. **17**(3), 381–395 (2008). https://doi.org/10.1007/s10115-007-0118-y
28. Xin, B., Chen, L., Chen, J., Ishibuchi, H., Hirota, K., Liu, B.: Interactive multiobjective optimization: a review of the state-of-the-art. IEEE Access **6**, 41256–41279 (2018). https://doi.org/10.1109/ACCESS.2018.2856832
29. Yu, P.L.H., Wan, W.M., Lee, P.H.: Decision tree modeling for ranking data. In: Fürnkranz, J., Hüllermeier, E. (eds.) Preference Learning, pp. 83–106. Springer, Heidelberg (2011). https://doi.org/10.1007/978-3-642-14125-6_5. ISBN 978-3-642-14125-6
30. Zhao, H., Ram, S.: Constrained cascade generalization of decision trees. IEEE Trans. Knowl. Data Eng. **16**(6), 727–739 (2004). https://doi.org/10.1109/TKDE.2004.3
31. Zujevs, A., Eiduks, J.: Adaptive multi-objective optimization procedure model. In: Scientific Proceeding of Riga Technical University, vol. 5, pp. 46–54 (2008)

Author Index

M. Emmerich et al. (Eds.): EMO 2023, LNCS 13970, pp. 635–636, 2023.
https://doi.org/10.1007/978-3-031-27250-9

Printed in the United States
by Baker & Taylor Publisher Services